PRINCIPLES OF
ANATOMY AND PHYSIOLOGY

PRINCIPLES OF
ANATOMY AND PHYSIOLOGY

Gerard J. Tortora
Bergen Community College
Paramus, NJ

Nicholas Peter Anagnostakos
Bergen Community College
Paramus, NJ

Canfield Press San Francisco

A Department of Harper & Row, Publishers, Inc.

New York   Evanston   London

Illustrator: Russell Peterson
Designer: Rita Naughton
Cover: Suzanne Fridley

This book was set in 10 point Caledonia by Kingsport Press
Printed and bound by Kingsport Press

Harper & Row, Publishers, Inc.
10 East 53rd Street
New York, NY 10022

Library of Congress Cataloging in Publication Data

Tortora, Gerard J., 1940–
    Principles of anatomy and physiology.

    Includes index.
    1. Human physiology. 2. Anatomy, Human. I. Anagnostakos,
Nicholas Peter, 1924–   joint author.
II. Title. DNLM: 1. Anatomy. 2. Homeostasis
3. Physiology. QS4 T712p
QP34.5.T67   612   75–1214
ISBN 0-06-388770-3

75 76 77 78 79   10 9 8 7 6 5 4 3 2

PREFACE

AUDIENCE

Principles of Anatomy and Physiology is designed for use in introductory courses in human anatomy and physiology. It assumes no previous study of the body or of the physicochemical principles on which body structure and function are based. The text has been written especially for students in allied health-science programs, such as nursing education, medical assistant, physician's assistant, medical laboratory technology, radiologic technology, inhalation therapy, dental hygiene, physical therapy, mortuary science, medical records, and other paramedical-oriented programs. However, because of the scope of the text, we feel it may be used also by students in other programs, such as biological sciences, premedical, predental, science technology, liberal arts, and physical education.

OBJECTIVES

The subject matter of human anatomy and physiology is an exceedingly large and complex body of knowledge to present in an introductory course. Accordingly, we have emphasized unifying concepts and data considered critical to a basic understanding and working knowledge of the human body. We have selected data that support or explain these concepts or that are generally considered basic to an introduction to the subject matter, and we have rejected data unessential to this objective. Our second primary objective is to present the material at a reading level that can be handled by the average student enrolled in health-science programs leading to diploma as well as associate and baccalaureate degrees. Of course, we have not avoided technical vocabulary or vital, but difficult, concepts. Instead, we have attempted to develop step-by-step, easy-to-comprehend explanations of each concept and have avoided needlessly difficult nontechnical vocabulary and syntax. We feel that these goals are, in fact, essential to good textbook writing at any level.

THEMES

Two major themes dominate the text: homeostasis and pathology. Throughout the book, the student is shown how normal anatomy and physiology is maintained by dynamic counterbalancing forces. Pathology is viewed as a disruption in homeostasis. Accordingly, we have presented a large number of clinical topics and have contrasted them with specific normal processes.

ORGANIZATION

The book is organized into five principal areas of concentration: (1) Organization of the Human Body, (2) Principles of Support and Movement, (3) Control Systems, (4) Maintenance of the Human Body, and (5) Continuity.

Unit 1, Organization of the Human Body, is designed to provide an understanding of the structural and functional levels of the body from molecules to organ systems. The first chapter introduces the concept of homeostasis and defines negative and positive feedback systems. The second chapter in the unit contains all the physicochemical principles and data required for an understanding of the physiology presented in the rest of the text. The first half of the chapter (the atom, chemical reactions, and inorganic compounds) is intended for the student who has never taken a chemistry course. The second half (organic compounds) will probably be essential reading for all students.

Chapters 3 to 6 deal with cell, tissue, and organ system levels of organization. A generalized cell is used to demonstrate the cellular level, and a series of illustrations is provided for each cell part. Each illustration series includes an electron micrograph of the part, an adjacent labeled line drawing of the electron micrograph, and a small illustration of the whole cell, with the part under consideration in color. The discussion of the cell ends with a description and schematic drawing of protein synthesis. We have not listed specific codes for specific amino acids. But we have given a general description of how base sequence in hereditary material determines amino acid sequence in proteins, which, in turn, determines structural and functional properties of the body. This principle is applied elsewhere in the book. For example, errors in metabolism and the special type of hemoglobin associated with sickle cell anemia are covered in a later unit. Tissue organization is presented through descriptions of the structure, functions, and locations of the principal kinds of epithelium and connective tissue, excluding bone and blood. The histology of bone, blood, muscle, and nerve tissue is dealt with later under the relevant organ systems. A discussion of inflammation and tissue repair is also included. The skin and its accessory organs is utilized to acquaint the student with the organization of organs and systems.

The final chapter in the unit prepares the student for the study of gross anatomy by introducing him to the overall structural plan of the body. Included are short sections on body cavities, regions and planes of the body, and directional terms.

Unit 2, Principles of Support and Movement, analyzes the anatomy and physiology of the skeletal system, articulations, and the skeletal muscle system. Histology and physiology of cardiac and smooth muscle are also included. The gross anatomy of bones and muscles has been organized into a series of exhibits. Most of these exhibits contain a drawing or drawings of the bone or muscles under consideration, and all contain written descriptions in either tabular or highly condensed narrative form. Roentgenograms of bones are included to provide students with an opportunity to transfer knowledge to clinical situations. Many students find muscle identification an onerous chore. To help the student, we have provided the following learning aids: The illustrations of muscles are shown with duplicates of the drawings that are used for bone identification. In this way, the student is given consistent points of reference. We have also presented a brief section on the criteria for naming skeletal muscles, and each muscle exhibit contains a listing of appropriate prefixes, roots, and suffixes and their definitions.

Unit 3, Control Systems, emphasizes the importance of the nerve impulse in the immediate maintenance of homeostasis, the role of receptors in providing information about the internal and external environment, and the importance of hormones in maintaining long-range homeostasis. The material on gross anatomy of the brain is geared toward explaining functional interrelationships between the major parts of the brain and between the brain and the rest of the body. Clinical topics relate structural and/or physiological disruptions in specific areas of the nervous system to symptoms. Factors that affect conduction across synapses are discussed, and Parkinsonism is used as an example of disruption in impulse conduction. In the chapter on the endocrine system, emphasis is placed on the regulation of hormone secretion through feedback systems, and a flowchart is presented for each system. Discussion of the effects and regulation of the sex hormones is omitted here because we feel that these topics can be dealt with more logically in the chapters on reproduction and development. The unit ends with a relatively thorough explanation of how the general stress syndrome operates, how it differs from homeostatic responses, and why it is protective. We feel that this topic provides background for some of the current concepts of disease and also presents a good review of some of the material in the unit.

Unit 4, Maintenance of the Human Body, is designed to show the student how the body maintains itself on a day-to-day basis through such activities as respiration, digestion, cellular metabolism, urine production, and buffer systems. Antigen-antibody response is a prevailing theme in the chapters on the circulatory system, and topics such as the role of the thymus in immunity and the problem of transplant rejection are included. The presentation of blood pressure and flow has been carefully constructed, and we hope this material is clearer in *Principles of Anatomy and Physiology* than it is in most competing texts.

Digestion is treated regionally so that the anatomy and physiology are integrated. For example, the student learns the anatomy and physiology of the mouth and its accessory organs and the digestive processes that occur within the mouth before he continues to the next segment of the gastrointestinal tract. A subsequent chapter deals with the fate of absorbed foods. Here, emphasis is placed on the relationships of protein and fat metabolism to the glucose pathway. A chart showing integrated metabolism summarizes much of this material. The role of vitamins and minerals in enzyme-coenzyme systems is mentioned. Specific vitamins and minerals, their sources, known and suspected functions, and deficiency symptoms are listed in exhibits.

Restoration and maintenance of fluid balance and blood pH is an important area of knowledge for students considering careers in health fields.

Consequently, the last two chapters in the unit pay particular attention to these topics.

Unit 5, Continuity, emphasizes the relationship of the endocrine system to sexual development, regulation of menstrual and ovarian cycles, and maintenance of pregnancy. The unit concludes with a few simple and basic concepts of classical genetics so that the health student will have some background for understanding the inheritance of genetic disorders.

SPECIAL FEATURES

The book contains a number of special learning aids for students, including the following:

1. Student Objectives appear at the beginning of each chapter. Each objective describes a knowledge or skill the student should acquire as he studies the chapter. (See Note to Student for an explanation of how the objectives can be utilized.) End-of-chapter Review Questions and Problems are designed specifically to help meet the stated objectives. In addition, each chapter Summary in Outline provides a checklist of major topics the student should learn.
2. The health-science student is generally expected to learn a great deal about the anatomy of certain organ systems—specifically, bones, skeletal muscles, blood vessels, and nerves. In these high anatomy areas, we have pulled the anatomical details out of the narrative and placed them in Exhibits, most of which contain illustrations. This method organizes the data and deemphasizes rote learning of concepts presented in the narrative.
3. An unusually large number of clinical applications are described. The topics provide review of normal body processes and allow the student to see why the study of anatomy and physiology is so fundamental to a career in any of the health fields. Topics include such interest sparkers as acupuncture and pain, smoking and cancer, and the birth process. In addition, glossaries of selected medical terminology appear at the ends of most chapters.
4. Many roentgenograms are used.
5. A two-part glossary appears at the end of the book. The first part contains prefixes, suffixes, and combining words; the second part contains important terms used in the text. A pronunciation key is included.

SUPPLEMENTARY MATERIALS

Ancillary materials are also available for use with *Principles of Anatomy and Physiology.* These include

1. A complimentary Teachers' Guide. For each chapter of the textbook, the Guide contains a résumé, a listing of key instructional concepts, problem-solving essay questions, and selected audiovisual materials. A directory of the distributors of the audiovisual materials is also provided.
2. A complimentary Question Card File. The File is a booklet containing objective test questions for each chapter in the textbook. The questions have been carefully designed to evaluate student understanding of data, concepts, clinical situations, and their applications.
3. A slide set consisting of fifty slides selected from illustrations in the textbook. It is available through Harper & Row, Publishers, Inc.
4. Laboratory and Study Manual. The Manual is geared specifically to the textbook in terms of content and illustrations. It contains modules that guide the student in making microscopic examinations, recording data, and performing gross examinations, and physiological experiments. A unique feature of the Manual is the incorporation of selected modules that deal with clinical considerations and applications only. At the discretion of the instructor, some of the modules may be utilized as out-of-class assignments or as part of student evaluation. The Manual is in preparation.

Gerard J. Tortora
Nicholas P. Anagnostakos

ACKNOWLEDGEMENTS

Since the inception of this textbook, Canfield Press has provided us with the services of several individuals who are specialists in their respective scientific disciplines. These people gave us invaluable assistance through their reviews of the manuscript. Among those to whom we wish to express our deepest gratitude are the following: W. Henry Hollinshead, Professor of Anatomy at the University of North Carolina School of Medicine, reviewed Chapters 1 to 12; Lawrence M. Elson, Professor of Anatomy at City College of San Francisco, reviewed Chapters 13 to 25; June C. Abbey, Acting Chairman of the Department of Nursing in Biological Dysfunction at the University of California, San Francisco Medical Center, reviewed the entire manuscript. Michael Robinson, Professor of Biology at Tarrant County Junior College, Texas, and Carlo Vecchiarelli, Professor of Biology at Chabot College, Hayward, California, both took long and hard looks at the first draft of the manuscript. Antoinette M. Anastasia, Dean of the College of Liberal Arts and former Chairman of the Biology Department at Fairleigh Dickinson University, NJ, reviewed most of the manuscript during its early stages of development. G. Karl Ludwig, M.D., Kaiser Hospital, South San Francisco, reviewed the material dealing with disorders. All these individuals assisted us in our attempt to develop an accurate, logical, and pedagogical presentation of the material.

We are especially pleased with the outstanding quality of the line drawings. The bulk of the artwork was prepared by Russell Peterson, who provided us with numerous insights regarding the conceptualization of the art and the correlation of the art with the textual material. Nelva B. Richardson drew the anatomical and physiological illustrations in Chapters 20 to 25. Heather Kortebein and Barbara Hack prepared many biological and flowchart illustrations throughout the book. Selected pieces of art were also reviewed by Drs. Hollinshead and Elson.

Myra Schachne did the time-consuming photographic research. The many individuals, publishers, and companies who provided photographs, photomicrographs, and electron micrographs are acknowledged in appropriate captions.

The editorial assistance provided by Canfield Press for the development of the project has been outstanding. We wish to express appreciation to R. Wayne Oler, Editor-in-Chief and Associate Publisher, who personally supervised many phases of the project and offered us all the resources of Canfield Press to successfully complete the project; and Ann K. Ludwig, Editor, who undertook the burden of coordinating various phases of the project.

We are also indebted to Janet Wagner, Developmental Editor, who always provided us with the necessary combination of guidance, inspiration, and encouragement; and copy editors Linda Harris and Kathy LaMar.

Early drafts of the manuscript were typed by Geraldine Tortora and Christine Anagnostakos; the final manuscript was typed by Florence Campbell. We are also indebted to Margaret MacMillan for her secretarial assistance.

NOTE TO THE STUDENT

At the beginning of each chapter is a listing of Student Objectives. *Before* you read the chapter, please read the objectives carefully. Each objective is a statement of a skill or knowledge we would like you to acquire. To meet these objectives you will have to do several things. Obviously, you must read the chapter very carefully, and, if there are sections of the chapter you do not understand after one reading, reread the sections before continuing. In conjunction with your reading, pay particular attention to the figures and Exhibits; they have been carefully coordinated to what you are reading. At the end of each chapter are two other guides you may find useful. The first of these guides, Chapter Summary in Outline, is a summary of important topics discussed in the chapter. This section is designed to consolidate the essential points covered in the chapter so you may recall and relate them to each other. The second guide, Review Questions and Problems, is a series of questions designed specifically to help you to meet your objectives. After you have answered the questions, you should return to the beginning of the chapter and reread the objectives to see whether or not you have met your goal.

CONTENTS

This textbook is dedicated to

Teachers who have inspired us,
Students we hope to motivate, and
Our families, who have encouraged us

ORGANIZATION OF THE HUMAN BODY

CHAPTER 1

AN INTRODUCTION TO THE HUMAN BODY

STUDENT OBJECTIVES

After you have read this chapter, you should be able to:

1. Define physiology and anatomy with its subdivisions

2. Determine the relationship between structure and function

3. Compare the levels of structural organization within the human body

4. Define a cell, tissue, organ, system, and organism

5. Explain why homeostasis is a state that results in normal body activities and why the inability to achieve homeostasis is a condition which leads to malfunction and related disorders

6. Identify the effects of stress on homeostasis

7. Describe the interrelationships of body systems in maintaining homeostasis

8. Compare the role of the endocrine and nervous systems in maintaining homeostasis

9. Define a feedback system and explain its role in homeostasis

10. Contrast the homeostasis of blood pressure through nervous control and blood sugar level through hormonal control

You are about to begin a study of the human body in order to learn how your body is organized and how it functions. The study of the human body involves many branches of science, each of which contributes to a more comprehensive understanding of the parts of your body and how they work. Once you learn how your body normally works, you can understand what happens to your body when it is injured, diseased, or placed under stress.

Two branches of science that will help you understand your body parts and functions are anatomy and physiology. **Anatomy** refers to the study of *structure* and the relationships among structures. Anatomy is a very broad science, and the study of structure becomes more meaningful when specific aspects of the science are considered. For example, *gross anatomy* deals with structures that can be studied without using a microscope. Another kind of anatomy, *systemic anatomy,* covers particular systems of the body, such as the system of nerves, spinal cord, and brain, or the system of heart, blood vessels, and blood. *Regional anatomy* is a division of anatomy dealing with a specific region of the body, such as the head, neck, chest, or abdomen. *Developmental anatomy* is the study of development from the fertilized egg to the adult form. This branch of anatomy, also called *embryology,* is generally restricted to the study of development from the fertilized egg to the period just before birth. Other branches of anatomy are *pathological anatomy,* the study of structural changes caused by disease, and *histology,* the microscopic study of the structure of tissues and cells.

Whereas anatomy and its branches deal with structures of the body, **physiology** deals with the *functions* of the body parts. In other words, physiology is concerned with how a part of the body actually works. As you will see in later chapters, physiology cannot be completely separated from anatomy. Each structure of the body is custom-modeled to carry out a particular set of functions. For instance, the interior of the nose is lined with hairs that allow the nose to perform the function of filtering dust from inhaled air. Bones are able to function as rigid supports for the body because they are constructed of hard minerals. In a sense, then, the structure of a part determines what functions it will perform. In turn, body functions often influence the size, shape, and health of the structures. For example, glands perform the function of manufacturing chemicals. Some of these chemicals stimulate bones to build up minerals so they become hard and strong. Other chemicals cause the bones to give up some of their minerals so they do not become too thick or too heavy. Anatomy cannot really be understood without a knowledge of physiology, and vice versa. For this reason, you will learn about the human body by studying its structures and functions together.

HOW ARE YOU PUT TOGETHER?

The human body is an organism consisting of levels of structural organization that are associated with each other in several ways. An **organism** is a total living form. Some organisms, like the amoeba, consist of only one cell. Most organisms, including humans, consist of millions of cells. The lowest level of structural organization, the *chemical level,* includes all chemical substances essential for maintaining life. All these chemicals are made up of atoms joined together in various ways (Figure 1–1). The chemicals, in turn, are put together to form the next higher level of organization, the *cellular level.* **Cells,** as you probably know, are the basic structural and functional units of the organism. Among the many kinds of cells found in your body are muscle cells, nerve cells, and blood cells. Figure 1–1 shows several isolated cells from the lining of the stomach. Each of these cells has a different structure, and each performs a different job.

From the cellular level, the next higher level of structural organization is the *tissue level.* **Tissues** are groups of cells that are joined together to perform specific functions. For example, when the isolated cells shown in Figure 1–1 are joined

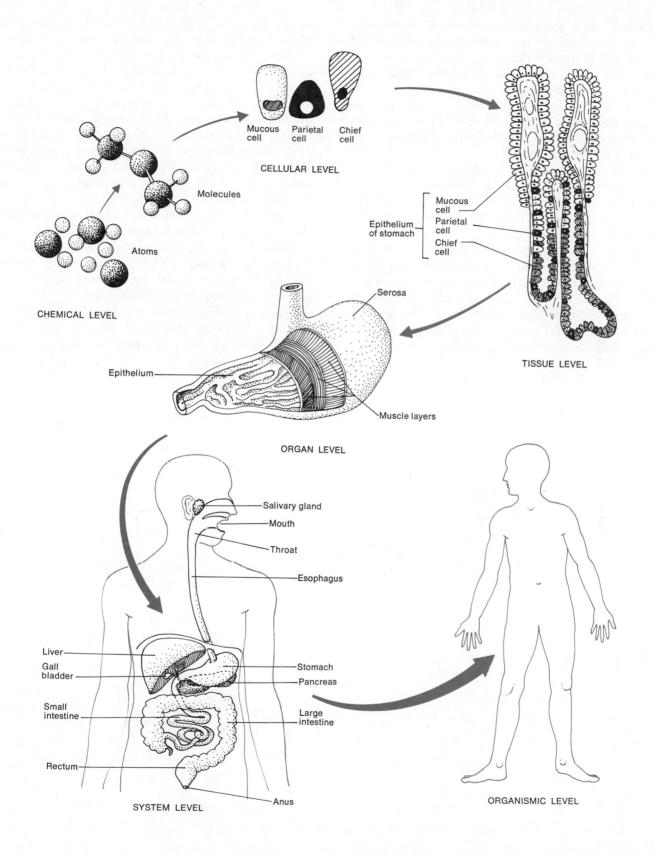

Figure 1–1. Levels of structural organization.

together, they form a tissue called *epithelium*, which lines the stomach. Each kind of cell in the tissue has a specific function. Mucous cells produce mucus, a slime that lubricates food as it passes through the stomach. Parietal cells produce acid in the stomach, and chief cells produce enzymes needed to digest proteins. Other examples of tissues in your body are muscle tissue, bone tissue, and nervous tissue.

In many places in the body, different kinds of tissues are joined together to form an even higher level of organization, the *organ level*. **Organs** are groups of two or more tissues that perform a particular function. Examples of organs are the heart, liver, lungs, brain, and stomach. Referring to Figure 1–1, you will see that the stomach is an organ since it consists of two or more tissues. Three of the tissues that make up the stomach are shown here. The serous tissue layer (also called the serosa) protects the stomach and reduces friction when the stomach moves and rubs against other organs. The muscle tissue layers of the stomach contract to mix food and pass the food on to the next digestive organ. The epithelial tissue layer produces mucus, acid, and enzymes.

The next higher level of structural organization in the body is the *system level*. A **system** consists of an association of organs that have a common function. The digestive system, which functions in the breakdown of food, is comprised of the mouth, saliva-producing glands, throat, esophagus, stomach, small intestine, large intestine, rectum, liver, gallbladder, and pancreas. All the parts of the body functioning with each other constitute the total organism.

In the chapters that follow, you will examine the structure and function of the following body systems: integumentary (pertaining to the skin), skeletal, muscular, nervous, endocrine (pertaining to hormones), circulatory, respiratory, digestive, urinary, and reproductive. Exhibit 1–1 contains a listing of these systems, their representative organs, and their general functions.

The systems are presented in the Exhibit in the order in which they will be studied in later chapters.

Exhibit 1–1. PRINCIPAL SYSTEMS OF THE HUMAN BODY, REPRESENTATIVE ORGANS, AND FUNCTIONS

SYSTEM AND DIAGRAM OF REPRESENTATIVE ORGANS · SYSTEM AND DIAGRAM OF REPRESENTATIVE ORGANS

1. INTEGUMENTARY

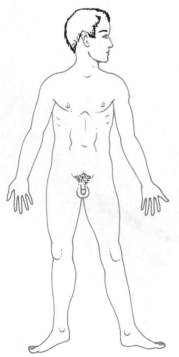

The skin and its associated structures such as hair, nails, and glands

FUNCTION: Protects the body, regulates body temperature, and eliminates wastes

2. SKELETAL

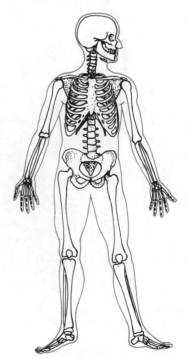

All the bones of the body

FUNCTION: Supports and protects the body, gives leverage, produces blood cells, and stores minerals

SYSTEM AND DIAGRAM OF REPRESENTATIVE ORGANS

3. MUSCULAR

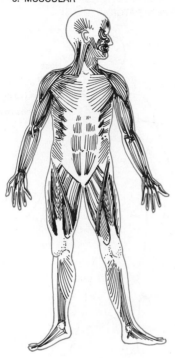

All the muscle tissue of the body

FUNCTION: Allows movement, maintains posture, and produces heat

5. ENDOCRINE

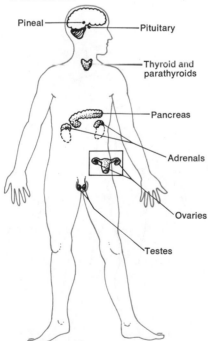

All glands that produce hormones

FUNCTION: Regulates body activities through hormones

SYSTEM AND DIAGRAM OF REPRESENTATIVE ORGANS

4. NERVOUS

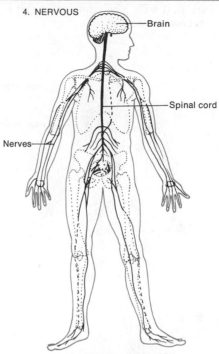

Brain, spinal cord, nerves, sense organs, e.g., eye and ear

FUNCTION: Regulates body activities through nerve impulses

6. BLOOD VASCULAR

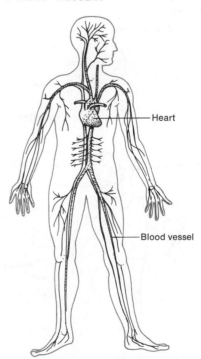

Blood, heart, and blood vessels

FUNCTION: Distributes nutrients to cells, eliminates wastes from cells, carries oxygen and carbon dioxide to and from cells, maintains the acid-base balance of the body, and protects against disease

Exhibit 1–1. PRINCIPAL SYSTEMS OF THE HUMAN BODY, REPRESENTATIVE ORGANS, AND FUNCTIONS *(cont'd.)*

SYSTEM AND DIAGRAM OF REPRESENTATIVE ORGANS	SYSTEM AND DIAGRAM OF REPRESENTATIVE ORGANS

7. LYMPH VASCULAR

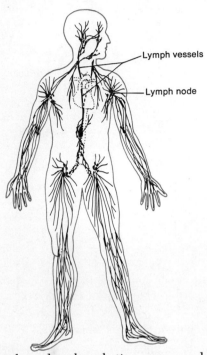

Lymph vessels

Lymph node

Lymph, lymph nodes, lymphatic organs, and lymph vessels

FUNCTION: Filters the blood and protects the body against disease

8. RESPIRATORY

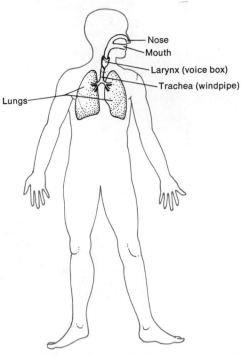

Nose
Mouth
Larynx (voice box)
Trachea (windpipe)
Lungs

A series of tubes leading into and out of the lungs

FUNCTION: Obtains oxygen, eliminates carbon dioxide, regulates the acid-base balance of the body

9. DIGESTIVE

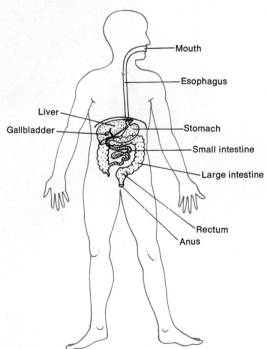

Mouth
Esophagus
Liver
Gallbladder
Stomach
Small intestine
Large intestine
Rectum
Anus

A long tube and associated organs such as the liver and gallbladder

FUNCTION: Performs the physical and chemical breakdown of food for use by the body, eliminates solid wastes

10. URINARY

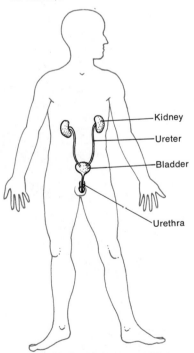

Kidney
Ureter
Bladder
Urethra

Organs that produce and eliminate urine

FUNCTION: Eliminates wastes, regulates fluid balance, maintains the acid-base balance of the body

SYSTEM AND DIAGRAM OF REPRESENTATIVE ORGANS

11. REPRODUCTIVE

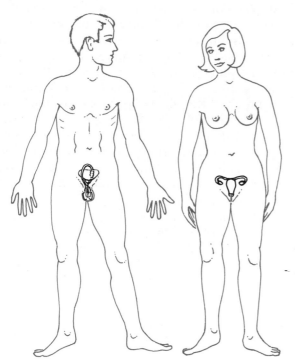

Organs that produce reproductive cells (sperms and ova)

FUNCTION: Reproduces the species

HOMEOSTASIS: PHYSIOLOGICAL BALANCE

The concept of homeostasis is perhaps the central theme of the entire field of human physiology. For this reason, homeostasis is also a central theme of this text and deserves special attention here. **Homeostasis** is the condition in which the internal environment of the body remains relatively constant. The term *homeo* means same, whereas *stasis* pertains to standing still. In order for the cells of the body to survive, the composition of the surrounding body fluids must be controlled precisely at all times. The body fluid that exists outside the body cells is called **extracellular fluid** and is found in two principal places. The fluid filling the microscopic spaces between the cells of tissues is called *intercellular fluid*, or *interstitial fluid* (*inter* = between). The extracellular fluid found in blood vessels is referred to as *plasma* (Figure 1–2). Among the substances in extracellular fluid are gases, nutrients, and electrically charged chemical particles called ions—all needed for the maintenance of life. Extracellular fluid circulates through the blood and lymph vessels and from there moves into the spaces between the tissue cells. Thus, it is in constant motion throughout the body. Essentially, all body cells are surrounded by the same fluid environment, and, for this reason, extracellular fluid is often designated as the internal environment of the body.

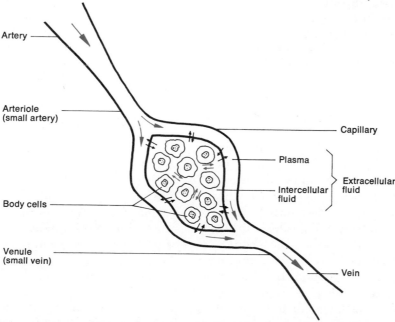

Figure 1–2. The internal environment of the body. Extracellular fluid circulates through arteries and arterioles and then into minute blood vessels called capillaries. From there it moves into the spaces around the body cells. Next, the fluid returns to the capillaries, passes through the venules, and is emptied into the veins.

An organism is said to be in homeostasis when its internal environment (1) contains exactly the right concentrations of gases, nutrients, and ions, (2) has an optimal temperature, and (3) has an optimal pressure for the health of the cells. When homeostasis is disturbed, ill health results. If the body fluids are not eventually brought back into balance, death occurs.

Homeostasis in all organisms is continually disturbed by **stress**, which is any stimulus that creates an imbalance in the internal environment. The stress may come from the external environment in the form of heat, cold, loud noises, or lack of oxygen. Or the stress may originate within the body as, for example, high blood pressure, pain, tumors, or unpleasant thoughts. Most stresses are mild and routine. For instance, muscular exercise creates temporary chemical and temperature imbalances in the internal environment. Occasionally, though, the body is subjected to extreme stresses such as poisoning, overexposure, severe infection, and surgical operations.

Fortunately, the body has many regulating (homeostatic) devices that oppose the forces of stress and bring the internal environment back into balance. In fact, high resistance to stress is a striking feature of all organisms. Consider the following examples. Some people live in deserts where the daytime temperatures easily reach 120°F. Others work outside all day in subzero weather. Yet everyone's internal body temperature remains at 98.6°F. Mountain climbers exercise strenuously at high altitudes, where the oxygen content of the air is low. But once they adjust to the new altitude, they do not suffer from oxygen shortage. The extremes in temperature and in oxygen content of the air are stresses from the external environment, yet the body is able to compensate and remain in homeostasis. The muscular exercise performed by the mountain climber is an example of an internal stress. Walter Cannon, the American physiologist who coined the term homeostasis, noted, for example, that the heat produced by the muscles during strenuous exercise would curdle the body's proteins if the body did not have some way of dissipating heat quickly by sweating. In addition to heat, muscles that are being exercised also produce a great deal of lactic acid, the acid of sour milk. If the body did not have a homeostatic mechanism for reducing the amount of the acid, the extracellular fluid would become acidic and destroy the cells.

Every body structure, from the cellular to the system level, contributes in some way to keeping the internal environment within normal limits. The homeostatic function of the circulatory system, for example, is to keep the fluids of all parts of the body constantly mixed. When we are at rest, fresh blood is circulated throughout the entire body about once every minute. But when we are active and our muscles need nutrients more rapidly, the heart quickens its pace and pumps fresh blood to the organs five times a minute. In this way, the circulatory system helps compensate for the stress of increased activity.

The respiratory system (the lungs and air passageways) offers another example of a homeostatic mechanism in the body. The respiratory system obtains oxygen and removes carbon dioxide, a waste product that could be harmful to body cells if its concentration were too high. The cells of the body need more oxygen and produce more carbon dioxide when they are very active. Therefore, during periods of activity, the respiratory system must work faster to keep the oxygen in the extracellular fluid from falling below normal limits and to keep excessive amounts of carbon dioxide from accumulating.

The digestive system and related organs help to maintain the homeostasis involved in providing nutrients and removing wastes. As circulating blood passes through the organs of digestion, the products of digestion are transported to the body fluids so they can be used as nutrients by the cells. The liver, kidneys, endocrine glands, and other organs help to alter or store the products of digestion in various ways. The kidneys help to remove the wastes produced by cells after they have utilized the nutrients.

Controlling homeostasis

The homeostatic mechanisms of the body, such as those performed by the circulatory, respiratory, and digestive systems, are themselves subject to control by the nervous system and the endocrine system. The nervous system regulates homeostasis by detecting when the body is headed out of its balanced state and by sending messages to the proper organs to counteract the stress. For instance, when the muscle cells are active, they take a great deal of oxygen from the blood. They also give off carbon dioxide, which is picked up by the blood. Certain nerve cells detect the chemical changes occurring in the blood and send a message to the brain. The brain then sends a message to the heart to pump blood more quickly to the lungs so that the blood can give up its excess carbon dioxide and take on more oxygen. Simultaneously, the brain sends a message to the lungs to breathe faster so that the carbon dioxide can be exhaled and more oxygen can be inhaled.

Homeostasis is also controlled by the endocrine system, which consists of a series of glands that secrete chemical regulators, called hormones, into the blood. Whereas nerve impulses coordinate homeostasis very rapidly, hormones coordinate homeostasis much more slowly. Both means of control are directed toward the same end. Two specific examples of homeostasis that can be considered here are the nervous control of blood pressure and the hormonal control of blood sugar level.

Homeostasis of blood pressure: Nervous control

Blood pressure is the force exerted by blood against the walls of the blood vessels, especially the arteries. It is determined primarily by three factors: the rate and strength of the heartbeat, the amount of blood, and the resistance offered by the arteries as blood passes through them. The resistance of the arteries results from the chemical properties of the blood and the size of the arteries at any given time.

If some stress, either internal or external, causes the heartbeat to speed up, the following sequence occurs (Figure 1–3): As the heart pumps faster, it pushes more blood into the arteries per minute—a situation that raises pressure in the arteries. The higher pressure is detected by pressure-sensitive cells in the walls of certain

arteries, which send nerve impulses to the brain. The brain interprets the message and sends impulses to the heart to slow the heart rate.

The nervous control of a relatively constant blood pressure is an example of a feedback system. A **feedback system** is any circular situation in which information is fed back into the system. As will be seen, a feedback system may be either negative or positive. In the case of regulating blood pressure, the body itself may be considered the system. The *input* is the information picked up by the pressure-sensitive cells, and the *output* is the return toward normal blood pressure. Looking at Figure 1–3, you will see that the system runs in a circle. The pressure-sensitive cells are able to detect when the blood pressure begins to normalize, and this output is fed back into the system, even after the return to homeostasis has begun. In other words, the cells report back to the brain on the changed blood pressure, and, if the pressure is still too high, the brain continues to send out impulses to slow the heartbeat. Since this feedback system reverses the direction of the initial condition from a rising to a falling blood pressure, the system is called a *negative feedback system*. If, instead, the brain had signaled the heart to beat even faster and the blood pressure had kept on rising, the system would have been a *positive feedback system*. Most of the feedback systems of the body are, however, negative feedback systems.

Referring again to Figure 1–3, you will see that a second feedback control is also involved in maintaining a normal blood pressure. Small arteries, called arterioles, have muscular walls

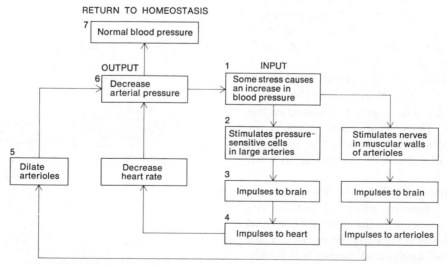

Figure 1–3. Homeostasis of blood pressure. Note that the output is fed back into the system, and the system continues to lower blood pressure until there is a return to normal conditions.

that normally "squeeze" the hollow tubes which the blood flows through. When the blood pressure increases, sensors in the arterioles send messages to the brain, and the brain signals the muscular walls to relax so that the diameter of the small tubes increases. As a consequence, the blood flowing through the arterioles is offered less resistance, and blood pressure drops back to normal. In sum, then, a suddenly increased blood pressure is brought back to normal by mechanisms that slow the heartbeat and widen the diameter of the arterioles. Conversely, a fall in blood pressure would be counteracted by increased heart activity and a decrease in the diameter of the arterioles. The net result, in either case, is a change in blood pressure back to normal.

Homeostasis of blood sugar level: Hormonal control

An example of homeostatic regulation by hormones is the maintenance of blood sugar level. This sugar, called glucose, is found in blood at a certain level and is one of the body's most important sources of energy. Under normal circumstances, the concentration of sugar in the blood averages about 90 milligrams/100 milliliters of blood.[1] This level is maintained primarily by two hormones secreted by the pancreas: insulin and glucagon. Let us assume that you have just eaten some candy. The sugar in the candy is broken down by the organs of digestion and then moves from the digestive tract into the blood. The sugar then becomes a stress because it raises the blood sugar level above normal. In response to this stress, the cells of the pancreas are stimulated to secrete insulin (Figure 1–4a). Once insulin gets into the blood, it has two principal effects. First, it increases sugar consumption by cells. That is, it accelerates the rate at which body cells take in sugar. Blood sugar level is lowered by this action since the sugar moves from the blood into body cells. Second, insulin accelerates the process by which sugar is stored in the liver and muscles. Thus, even more sugar is removed from the blood. In essence, the action of insulin decreases blood sugar concentration until it returns to normal.

The other hormone produced by the pancreas, glucagon, has the opposite effect of insulin. Assume that you have not eaten for several hours and your blood sugar level is steadily decreasing. Lack of sugar is now the stress, and, under this condition, the cells of the pancreas are instead stimu-

[1] 1 ounce = 28.35 grams; 1 gram = 1,000 milligrams (mg.); 1 quart = 946.4 milliliters (ml.).

lated to secrete glucagon (Figure 1–4b). This hormone accelerates the process by which sugar stored in the liver is sent back into the blood stream. The blood sugar level is thus increased until it returns to normal. In summary, then, the blood sugar level is regulated by two hormones. Insulin lowers the level of blood sugar if the concentration increases above normal, and glucagon raises the level of blood sugar if it has fallen below normal.

In the chapters that follow, each system of the body can be considered a structural and functional complex which is attempting to maintain homeostasis. When a particular system achieves homeostasis, it functions normally. Significant deviations from homeostasis result in abnormal conditions or *pathological* (disease) *conditions*. In order to establish this concept clearly, many of the chapters cover the normal anatomy and physiology of a system first and then examine various disorders related to each system.

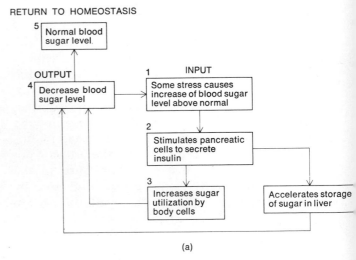

(a)

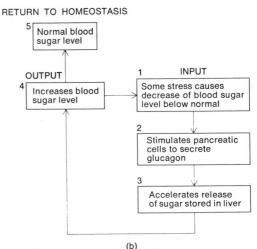

(b)

Figure 1–4. Homeostasis of blood sugar level.

Chapter summary in outline

ANATOMY AND PHYSIOLOGY

1. Anatomy is the study of structure and how structures are related to each other.

2. Subdivisions of anatomy include gross anatomy (macroscopic), systemic anatomy (systems), regional anatomy (regions), embryology (development prior to birth), pathological anatomy (disease), and histology (microscopic study of tissues and cells).

3. Physiology is the study of how structures function.

HOW ARE YOU PUT TOGETHER?

1. The human body consists of levels of structural organization from the chemical level to the system level.

2. The chemical level is represented by all the atoms and molecules in the body. The cellular level consists of cells. The tissue level is represented by tissues. The organ level consists of body organs, and the system level is represented by organs that work together to perform a more general function.

3. The human body is a collection of structurally and functionally integrated systems.

HOMEOSTASIS

1. Homeostasis is a condition in which the internal environment of the body (extracellular fluid) remains relatively constant.

2. Stress is any stimulus that creates an imbalance in the internal environment.

3. If a stress acts on the body, homeostatic mechanisms attempt to counteract the effects of the stress and bring the system back to normal.

4. All body systems attempt to maintain homeostasis.

Controlling homeostasis

1. Homeostasis is controlled by the nervous and endocrine systems.

2. Examples include the nervous control of blood pressure and the hormonal control of blood sugar level.

3. Blood pressure is determined by the rate and force of the heartbeat, the amount of blood, and by arterial resistance. A rising blood pressure falls back toward normal due to a negative feedback system of blood pressure regulation.

4. A normal blood sugar level is maintained by the actions of insulin and glucagon. Insulin causes a rising sugar level to fall back to normal. Glucagon brings a low level up to normal.

Review questions and problems

1. Define anatomy. How does each subdivision of anatomy help you understand the structure of the human body? Define physiology.

2. Construct a diagram to illustrate the levels of structural organization that characterize the body. Be sure to define each level.

3. Outline the function of each system of the body, and list several organs that comprise each system.

4. Define homeostasis. What is extracellular fluid? Why is it called the internal environment of the body?

5. How is stress related to homeostasis? What is a stress? Give several examples.

6. How is homeostasis related to normal and abnormal conditions in the body?

7. Substantiate this statement: "Homeostasis is a cooperative effort of all body parts."

8. What systems of the body control homeostasis? Explain.

9. Discuss briefly how the regulation of blood pressure and blood glucose level are examples of homeostasis.

CHAPTER 2

CHEMICAL LEVEL OF ORGANIZATION

STUDENT OBJECTIVES

After you have read this chapter, you should be able to:

1. Identify by name and symbol the principal chemical elements of the human body

2. Explain, by diagramming, the structure of an atom

3. Define a chemical reaction as a function of electrons in incomplete outer energy levels

4. Describe ionic bond formation in a molecule of NaCl

5. Discuss covalent bond formation as the sharing of outer energy level electrons

6. Identify and compare each kind of chemical reaction

7. Define the type of energy involved when chemical bonds are formed and broken

8. Define and distinguish between inorganic and organic compounds

9. Discuss the functions of water as a solvent, suspending medium, chemical reactant, heat absorber, and lubricant

10. List and compare the properties of acids, bases, and salts

11. Define pH as the degree of acidity or alkalinity of a solution

12. Describe the role of a buffer system as a homeostatic mechanism that maintains the pH of a body fluid

13. Describe and compare the structure and functions of carbohydrates, lipids, and proteins

14. Differentiate between dehydration synthesis and hydrolysis of organic molecules

15. Define the homeostatic role of enzymes as catalysts

16. Contrast the structure of DNA and RNA

17. Define the roles of DNA in heredity and protein synthesis

18. Describe the functions of RNA in protein synthesis

Many of the common substances that you eat and drink, such as water, sugar, ordinary table salt, and cooking oil, play vital roles in keeping you alive. In this chapter, you will learn something about how the molecules of these substances function in your body. Fundamental to this study is a knowledge of basic chemistry and chemical processes. To understand the nature of the matter you are made from and the various changes that this matter goes through in your body, you will need to know which chemical elements are present in the human organism and how they interact. The chapter is organized into two main parts. Part I reviews a few basic principles of chemistry. Those of you who already understand atomic structure, bonding, and chemical reactions may skip Part I and proceed to the second part of the chapter. Part II provides important information on the chemical life processes going on in your body. The principles discussed here will be needed for an understanding of later chapters in the text.

PART I: BASIC CHEMISTRY FOR LIVING SYSTEMS

All living and nonliving things consist of **matter,** which is anything that occupies space and possesses mass. Matter may exist in a solid, liquid, or gaseous state. All forms of matter are made up of a limited number of building units called *chemical elements.* At present, scientists can recognize 105 different elements. Elements are designated by letter abbreviations, usually derived from the first or second letter of the Latin or English name for the element. Such letter abbreviations are called *chemical symbols.* Examples of chemical symbols are H (hydrogen), C (carbon), O (oxygen), N (nitrogen), Na (sodium), K (potassium), Fe (iron), and Ca (calcium).

Approximately 23 elements are found in the human organism (Figure 2–1). Carbon, hydrogen, oxygen, and nitrogen comprise about 96 percent of the body's weight. These four elements together with phosphorus and calcium constitute approximately 99 percent of the total body weight. Eighteen other chemical elements compose the remaining 1 percent of the matter in the human organism.

Atomic structure of elements

Each element is made up of units of matter called **atoms.** An element is simply a quantity of matter composed of atoms that are all of the same type. For instance, a handful of the element carbon, such as pure coal, contains only carbon atoms. A tank of oxygen contains only oxygen atoms. Measurements indicate that the smallest atoms are less than 250,000,000 of an inch in diameter, and the largest atoms are 50,000,000 of an inch in diameter. In other words, if 50,000,000 of the largest atoms were placed end to end, they would measure approximately 1 inch in length.

An atom consists of two basic parts: the nucleus and the electrons (Figure 2–2). The centrally located *nucleus* contains positively charged particles called *protons* (p^+) and uncharged particles called *neutrons* (n). Because each of the protons has one positive charge, the nucleus itself is positively charged. The second basic part of an atom contains the *electrons* (e^-). These are negatively charged particles that spin around the nucleus in approximately oval-shaped orbits. The number of electrons in an atom always equals the number of protons. Since each electron carries one negative charge, the negatively charged electrons and the positively charged protons balance each other, and the atom is electrically neutral.

What makes the atoms of one element different from the atoms of another element? The answer lies in the number of protons. Looking at Figure 2–3, you will notice that the hydrogen atom contains one proton. The helium atom contains two. The carbon has six, and so on. Each different kind of atom has a different number of protons in its nucleus. The number of protons in an atom is called the atom's *atomic number.* Therefore, we can say that each kind of atom, or element, has a different atomic number.

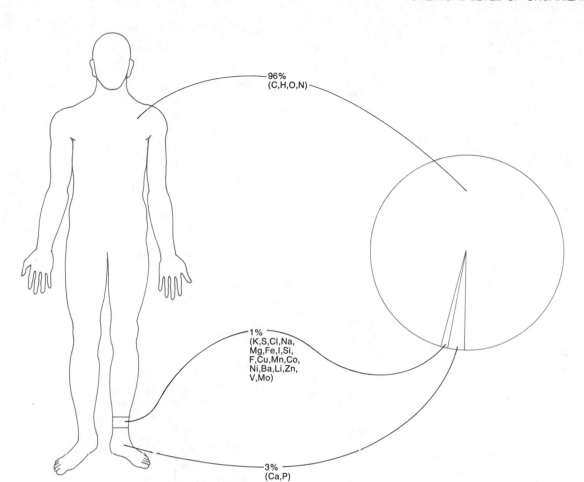

Figure 2–1.   Percentages of various chemical elements found in the human organism. Key to chemical symbols: Ba = barium, C = carbon, Ca = calcium, Cl = chlorine, Co = cobalt, Cu = copper, F = fluorine, Fe = iron, H = hydrogen, I = iodine, K = potassium, Li = lithium, Mo = molybdenum, Mg = magnesium, Mn = manganese, N = nitrogen, Na = sodium, Ni = nickel, O = oxygen, P = phosphorus, S = sulfur, Si = silicon, V = vanadium, and Zn = zinc.

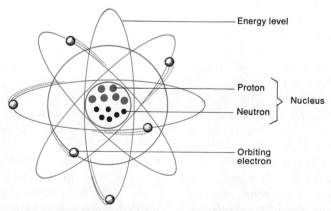

Figure 2–2.   The structure of an atom. In this highly simplified version of a carbon atom, note the centrally located nucleus, which contains neutrons and protons. The electrons orbit about the nucleus at varying distances from its center.

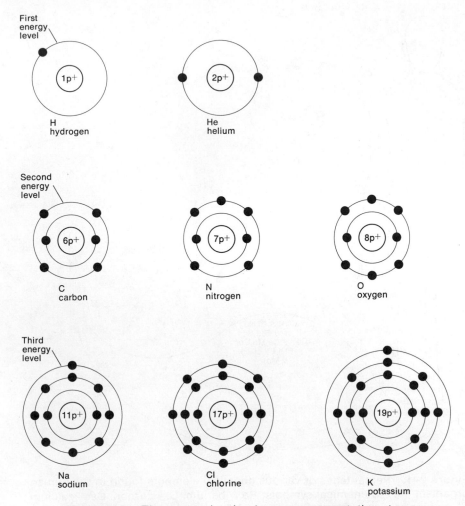

Figure 2–3. The energy levels of some representative atoms.

How atoms combine

When atoms combine with other atoms, or when they later break apart from other atoms, the process is called a *chemical reaction*. Because chemical reactions are the foundation of all life processes, it is important to study their nature in some detail.

The electrons are the part of the atom that actively participate in chemical reactions. As we have mentioned, the electrons spin around the nucleus in orbits. In Figure 2–3, the orbits are represented as concentric circles lying at different distances from the nucleus. We call these orbits *energy levels*. Notice that each orbit has a maximum number of electrons which it can hold. For instance, the energy level, or orbit, nearest the nucleus never holds more than two electrons, no matter what element we are talking about. This orbit can be referred to as the first energy level. The second energy level holds a maximum

of eight electrons. The third level also can hold a maximum of eight electrons.

An atom always attempts to fill its outermost orbit with the maximum number of electrons that it can hold. In order to do this, the atom may give up an electron, take on an electron, or share an electron with another atom—whichever is easiest. For instance, take a look at the chlorine atom. Its outermost energy level, which happens to be the third level, has seven electrons. However, the third level of any atom can hold a maximum of eight electrons. Chlorine can thus be visualized as having a shortage of one electron. In fact, chlorine usually does try to pick up an extra electron. Sodium, by contrast, has only one electron in its outer level. This again happens to be the third energy level. It is much easier for sodium to get rid of the one electron in its third energy level than it is to fill the third level by taking on seven more electrons. A few atoms, like helium, have completely filled outer energy levels and do not need to gain or lose electrons.

Atoms with incompletely filled outer energy levels, like sodium and chlorine, tend to combine with other atoms in a chemical reaction. During the reaction, the atoms can trade off or share electrons and, thereby, fill their outer energy levels. Atoms that already have filled outer levels generally do not participate in chemical reactions for the simple reason that they do not need to gain or lose electrons. When two or more atoms combine in a chemical reaction, the resulting combination is called a **molecule.** A molecule may contain two atoms of the same kind, as in the hydrogen molecule: H_2. The subscript 2 indicates that there are two hydrogen atoms in the molecules. Molecules may also be formed by the reaction of two or more different kinds of atoms, as in the hydrochloric acid molecule: HCl. Here, an atom of hydrogen is attached to an atom of chlorine. A molecule that contains at least two different kinds of atoms is called a **compound.** Hydrochloric acid, which is present in the digestive juices of the stomach, is an example of a compound. A molecule of hydrogen is not a compound.

The atoms in a molecule are held together by

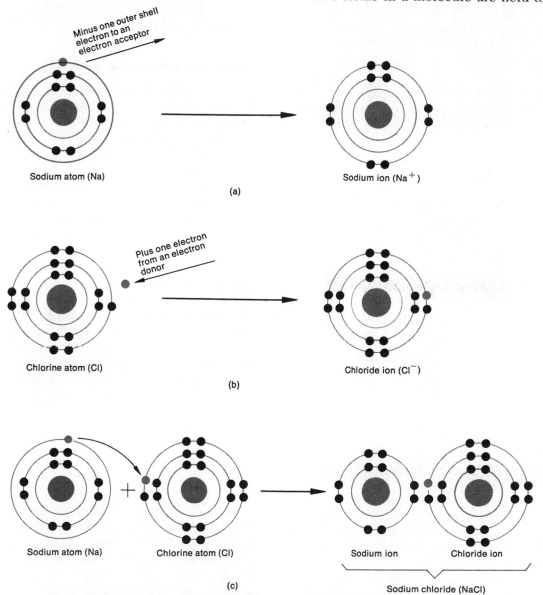

Figure 2–4. The formation of an ionic bond. (a) An atom of sodium attains stability by passing a single electron to an electron acceptor. The loss of this single electron results in the formation of a sodium ion (Na^+). (b) An atom of chlorine attains stability by accepting a single electron from an electron donor. The gain of this single electron results in the formation of a chloride ion (Cl^-). (c) When the Na^+ and the Cl^- ions are combined, they are held together by an attraction called an ionic bond, and a molecule of NaCl is formed.

forces of attraction called *chemical bonds*. Here, we shall consider the two basic types of chemical bonding: ionic bonds and covalent bonds.

Ionic bonds: The transfer of electrons

As we have mentioned, atoms are electrically neutral because the number of positively charged protons equals the number of negatively charged electrons. But when an atom gains or loses electrons, this balance is destroyed. If the atom gains electrons, it acquires an overall negative charge. Such a negatively or positively charged atom or group of atoms is called an *ion*. *If the atom loses electrons, it acquires an overall positive charge.*

As an example, consider the sodium ion (Figure 2–4a). The sodium atom (Na) has 11 protons and 11 electrons, with one electron in its outer energy level. When sodium gives up the single electron in its outer level, it is left with 11 protons and only 10 electrons. The atom now has an overall positive charge of one (+1). This positively charged sodium atom is called a sodium ion, and it is written as Na^+.

Another example is the formation of the chloride ion (Figure 2–4b). Chlorine has a total of 17 electrons, with 7 of them in the outer energy level. Since this energy level can hold 8 electrons, chlorine tends to pick up an electron that has been lost by another atom. By accepting an electron, chlorine acquires a total of 18 electrons. However, it still has only 17 protons in its nucleus. The chloride ion (Cl^-) therefore has one negative charge.

The positively charged sodium ion (Na^+) and the negatively charged chloride ion (Cl^-) attract each other. This is explained by the principle that unlike charges attract one another. The attraction, called an *ionic bond*, holds the two atoms together, and a molecule is formed (Figure 2–4c). The formation of this molecule, sodium chloride (NaCl) or table salt, is one of the most common examples of ionic bonding. We may now define an ionic bond as an attraction between atoms in which one atom loses electrons and another atom gains electrons. Generally, atoms whose outer energy level is less than half-filled lose electrons and form positively charged ions. Examples of such ions are potassium ion (K^+), calcium ion (Ca^{2+}), iron ion (Fe^{2+}), and sodium ion (Na^+). By contrast, atoms whose outer energy level is more than half-filled tend to gain electrons and form negatively charged ions. Examples of these ions include iodine ion (I^-), chloride ion (Cl^-), and sulfur ion (S^{2-}).

Notice that an ion is always symbolized by writing the chemical abbreviation followed by the number of positive (+) or negative (−) charges which the ion acquires.

Hydrogen is an example of an atom whose outer level is exactly half-filled. The first energy level can hold two electrons, but in hydrogen atoms it contains only one (see Figure 2–3). Hydrogen may lose its electron and become a positive ion (H^+). This is precisely what happens when hydrogen combines with chlorine to form hydrochloric acid (H^+Cl^-). However, hydrogen is equally capable of forming another kind of bond altogether, called a covalent bond.

Covalent bonds: The sharing of electrons

The second kind of chemical bond to be considered here is the *covalent bond*. This kind of bond is far more common in organisms than is the ionic bond. When a covalent bond is formed, neither of the combining atoms loses or gains an electron. Instead, the two atoms share a pair of electrons. Let us look again at the hydrogen atom. One way that a hydrogen atom can fill its outer energy level is to combine with another hydrogen atom to form the molecule H_2 (Figure 2–5a). In the H_2 molecule, the two atoms share a pair of electrons. Each hydrogen atom has its own electron plus one electron from the other atom. The shared pair actually circles the nuclei of both atoms. Therefore, the outer energy levels of both atoms are filled. When only one pair of electrons is shared between atoms, as in the H_2 molecule, a *single covalent bond* is formed. For convenience, a single covalent bond is expressed as a single line between the atoms (H—H). When two pairs of electrons are shared between two atoms, a *double covalent bond* is formed, which is expressed as two parallel lines ($=$) (Figure 2–5b). A *triple covalent bond*, expressed by three parallel lines ($\equiv$), occurs when three pairs of electrons are shared (Figure 2–5c).

Methane (CH_4) is an example of covalent bonding between atoms of different elements (Figure 2–5d). Commonly called swamp gas, methane is responsible for the odor of decaying vegetation. The same principles that apply to covalent bonding between atoms of the same element also apply here. The outer energy level of the carbon atom can hold eight electrons but has only four. Each hydrogen atom can hold two electrons but has only one. Consequently, in the methane molecule, the carbon atom shares four pairs of electrons. One pair is shared with each hydrogen atom. Each of the four carbon electrons orbits around both the carbon nucleus and a hydrogen nucleus. Each hydrogen electron circles around its own nucleus and the carbon nucleus.

Elements whose outer energy levels are half-filled, such as hydrogen and carbon, form covalent bonds quite easily. In fact, carbon always forms covalent bonds. It never becomes an ion. However, many atoms whose outer energy levels are more than half-filled also form covalent bonds. An example is oxygen, which is shown in Figure

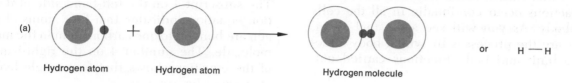

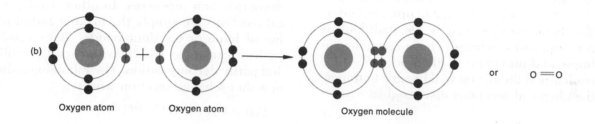

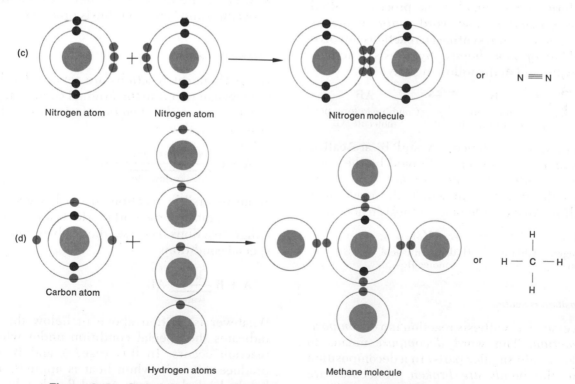

Figure 2–5. Shown is the formation of a covalent bond between atoms of the same element and between atoms of different elements. (a) A single covalent bond between two hydrogen atoms. (b) A double covalent bond between two oxygen atoms. (c) A triple covalent bond between two nitrogen atoms. (d) Single covalent bonds between a carbon atom and four hydrogen atoms. The representations on the far right are another way of showing the covalent bonds. Each straight line between atoms is a single covalent bond.

2–5b. We shall not go into the reasons why some atoms tend to form covalent bonds rather than ionic bonds. You should, however, understand the basic principles involved in bond formation. Chemical reactions are nothing more than the making or breaking of bonds between atoms. And these reactions occur continually in all the cells of your body. As you will see again and again, reactions are the processes by which body structures are built and body functions carried out.

Chemical reactions

Students are often frightened by the idea of studying reactions. But you may be reassured that you do not have to become a chemist in order to understand this book. In this section, we shall look at four basic types of reactions. These reactions are very simple and important to life processes. Once you have learned them, you will be able to understand the chemical reactions discussed later.

Synthesis reactions

When two or more atoms, ions, or molecules combine to form new molecules, the process is called a *synthesis reaction*. The word *synthesis* means to put together, and synthesis reactions involve the *making of new bonds*. Synthesis reactions can be expressed in the following way:

$$A \quad + \quad B \quad \rightarrow \quad AB$$

Atom, ion, or molecule A   Atom, ion, or molecule B   Combine to form new molecule AB

The combining substances, A and B, are called the *reactants;* the substance formed by the combination is called the *end product*. The arrow indicates the direction in which the reaction is proceeding. An example of a synthesis reaction is:

$$H^+ \quad + \quad Cl^- \quad \rightarrow \quad HCl$$

Hydrogen ion   Chloride ion   Hydrochloric acid molecule

Decomposition reactions

The reverse of a synthesis reaction is a *decomposition reaction*. The word *decompose* means to break down into smaller parts. In a decomposition reaction, the *bonds are broken*. Molecules are broken down into simpler molecules, ions, or atoms. A decomposition reaction occurs in this way:

$$AB \quad \rightarrow \quad A \quad + \quad B$$

Molecule AB breaks down into   Atom, ion, or molecule A   Atom, ion, or molecule B

Under the proper conditions, methane can decompose into carbon and hydrogen, as follows:

$$CH_4 \quad \rightarrow \quad C \quad + \quad 4H$$

Methane molecule   One carbon atom   Four hydrogen atoms

The subscript 4 on the left-hand side of the reaction equation indicates that four atoms of hydrogen are bonded to one carbon atom in the methane molecule. The number 4 on the right-hand side of the equation shows that four single hydrogen atoms have been set free.

Exchange reactions

All chemical reactions are based on synthesis or decomposition processes. In other words, chemical reactions are simply the making and/or breaking of ionic or covalent bonds. Many reactions, such as *exchange reactions*, are partly synthesis and partly decomposition. The following indicates how an exchange reaction works:

$$AB + CD \rightarrow AD + BC$$

Here the bonds between A and B and between C and D are broken in a decomposition process. New bonds are then formed between A and D and between B and C in a synthesis process.

Reversible reactions

When chemical reactions are reversible, the end product can revert to the original combining molecules. A *reversible reaction* is indicated by two arrows, as follows:

$$A + B \xrightleftharpoons[\text{breaks down to}]{\text{combines with}} AB$$

Some reversible reactions occur because neither the reactants nor the end products are very stable. Other reactions reverse themselves only under special conditions:

$$A + B \xrightleftharpoons[\text{water}]{\text{heat}} AB$$

Whatever is written above or below the arrows indicates the special condition under which the reaction occurs. In this case, A and B react to produce AB only when heat is applied, and AB breaks back down into A and B only when water is added. Figure 2–6 summarizes the basic kinds of chemical reactions that can occur.

The energy of chemical reactions

Some form of energy is involved whenever bonds between atoms in molecules are formed or broken

GENERAL NATURE   SPECIFIC EXAMPLE

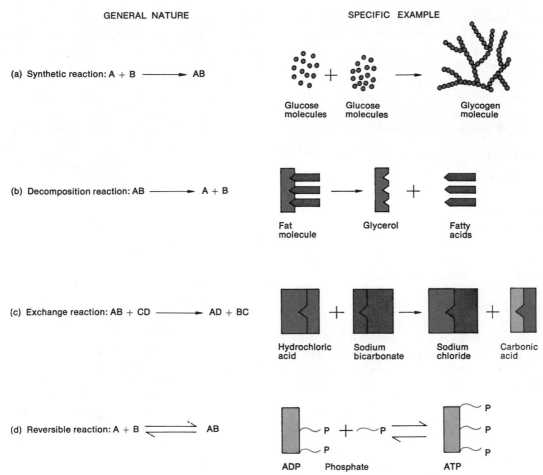

(a) Synthetic reaction: A + B ⟶ AB

Glucose molecules   Glucose molecules   Glycogen molecule

(b) Decomposition reaction: AB ⟶ A + B

Fat molecule   Glycerol   Fatty acids

(c) Exchange reaction: AB + CD ⟶ AD + BC

Hydrochloric acid   Sodium bicarbonate   Sodium chloride   Carbonic acid

(d) Reversible reaction: A + B ⇌ AB

ADP   Phosphate   ATP

Figure 2–6. Kinds of chemical reactions. (a) Synthetic reaction. When linked together as shown, molecules of glucose form a molecule of glycogen. Glucose is a sugar that is your primary source of energy. Glycogen is a storage form of that sugar found in your liver and skeletal muscles. (b) Decomposition reaction. Shown in the example is a molecule of fat breaking down into glycerol and fatty acids. This particular reaction occurs whenever you eat and digest a food containing fat. (c) Exchange reaction. In this exchange, atoms of different molecules are exchanging with each other. Shown is a buffer reaction in which your body eliminates strong acids to help you to maintain homeostatis. (d) Reversible reaction. The molecule of ATP (adenosine triphosphate) is an important source of stored energy. When you need the energy, the ATP breaks down into ADP (adenosine diphosphate) and PO_4 (phosphate group), and energy is released. The cells of your body reconstruct ATP by using the energy of the foods you eat to attach ADP to PO_4. All the examples shown are discussed in more detail in subsequent chapters.

during the chemical reactions taking place in the body. When a chemical bond is formed, energy is required. Conversely, when a bond is broken, energy is released. This means that synthesis reactions need energy in order to occur, whereas decomposition reactions give off energy. The building processes of the body, such as the construction of bones, the growth of hair and nails, and the replacement of injured cells, occur basically through synthesis reactions. The breakdown of foods, on the other hand, occurs through decomposition reactions. When foods are decom-

posed, they release energy that can be used by the body for its building processes. Released energy can also be used to warm the body by taking the form of *heat energy*. Foods can be partially broken down into compounds that can be stored in the body. Later, when added energy is needed, the body finishes breaking down these reserve compounds. This stored energy is called *potential energy*.

In Chapter 3, we shall be interested in another form of energy, *kinetic energy*, which is also very important to life processes. Unlike potential, or stored, energy, kinetic energy is the energy of motion. Gas molecules have a great deal of kinetic energy, which means that they move large distances quickly. The molecules of solids, by contrast, have little kinetic energy and remain close together. If heat is applied to any matter, some of the heat energy is converted to kinetic energy, and the molecules move more quickly.

Metabolism—an example of chemical reactions in the body

The word **metabolism** stands for all the synthesis and decomposition reactions that are occurring in the body. A metabolic reaction is simply a chemical reaction that goes on inside you. When we say that a person has a high metabolism, we mean that the chemical reactions in his body are proceeding at an unusually fast rate. The decomposition reactions are occurring so quickly that foods tend to be broken down completely before the body has a chance to store them. Consequently, the person with a high metabolism can usually eat a great deal without gaining weight. Because of the tremendous amounts of energy given off by rapid decomposition reactions, the person also has a great deal of energy for activities as well as a good deal of heat energy. For this reason, people with high metabolisms appear to have a lot of "nervous" energy and often complain that they are too hot.

Chemical reactions in the bodies of people with low metabolisms are carried out abnormally slowly. Food is broken down slowly. Much of it is only partially broken down and then stored. These people tend to gain weight easily, have little energy, and are always cold. Because their synthesis reactions are also slowed down, their bodies build up new structures very slowly. Wounds, for instance, often take a long time to heal.

PART II: CHEMICAL LIFE PROCESSES

With the basics of chemistry under your belt, you can apply your knowledge of chemicals, bonding, and energy to the life processes of the body. You now need to know the major chemicals of the body and learn why they are important to life itself. As you read the following material, you will begin to find answers to questions such as: Why do we need to drink water? What foods are vital to the body's growth and health? What is a gene?

Most of the chemicals in the body exist in the form of compounds. Biologists and chemists divide these compounds into two principal classes: inorganic compounds, which lack carbon, and organic compounds, which always contain carbon. **Inorganic compounds** are usually small, ionically bonded molecules that are vital to body functions. They include water, many salts, such as NaCl, and many acids, such as HCl. **Organic compounds** are held together mostly or entirely by covalent bonds. They tend to be very large molecules and are therefore good building blocks for body structures. Organic compounds present in the body include carbohydrates, lipids, proteins, nucleic acids, and ATP.

Inorganic substances

The importance of water

One of the most important, as well as the most abundant, inorganic substances in the human organism is water. In fact, with few exceptions, such as the enamel of teeth and bone tissue, water is by far the most abundant material in all cells. For example, about 60 percent of red blood cells, 75 percent of muscle tissue, and 92 percent of blood plasma consist of water. Although there is no specific amount of water that must be present in living matter, the average water content is 65 to 75 percent. The following functions of water explain why it is such an important compound in living systems:

1. Water is an excellent solvent and suspending medium. A *solvent* is a liquid or gas in which some other material (solid, liquid, or gas), called a *solute*, has been dissolved. The combination of solvent plus solute is called a **solution,** and one common example of a solution is salt water. A solute, such as salt in water, never settles out of its solution. The solute can be retrieved only through a chemical reaction or, in some cases, by boiling off the solvent. In a *suspension*, by contrast, the suspended material mixes with the liquid or suspending medium, but it will eventually settle out of the mixture. An example of a suspension is cornstarch and water. If the two materials are shaken together, a milky mixture will form. After the mixture sits for a while, however, the water clears at the top, and the cornstarch settles to the bottom.

The solvating property of water is very important to your health and survival. For example, water in the blood forms a solution with some of the oxygen you breathe in, allowing the oxygen to be carried to your body cells. Water in the blood also dissolves a large part of the carbon dioxide that is carried from the cells to the lungs to be breathed out. Furthermore, if the surfaces of the air sacs in your lungs are not moist, oxygen cannot dissolve and, therefore, cannot move into your blood to be distributed throughout your body. As another example, both smell and taste depend upon water as a solvent. Substances smelled or tasted must be partially in solution. Otherwise, nerve endings cannot react to their presence and the sensations of smell and taste are not registered. Your survival could depend on your ability to smell or taste a potentially dangerous substance. Water is also the solvent that carries nutrients into and wastes out of your body cells.

As a suspending medium, water is also vital to your survival. Many large organic molecules are suspended in the water of your body cells. These molecules are consequently able to come in contact with other chemicals, allowing various chemical reactions to occur. Certainly, you could not remain healthy if important chemical reactions did not take place in the proper manner or at the proper speed.

2. Water can participate in chemical reactions. During digestion, for example, water can be added to large nutrient molecules in order to break them down into smaller molecules. This kind of breakdown is necessary if the body is to utilize the energy in nutrients. Water molecules are also used in synthesis reactions, that is, reactions in which smaller molecules are built into larger ones. Such synthesis reactions occur in the production of hormones and enzymes.

3. Water absorbs and releases heat very slowly. In comparison to other substances, water requires a large amount of heat to increase its temperature and a great loss of heat to decrease its temperature. Because of this property, water maintains a more constant body temperature than other solvents, despite fluctuations in environmental temperatures. Water thus helps to maintain a homeostatic body temperature.

4. Water serves as a lubricant in various regions of the body. It is a major part of mucus and other lubricating fluids. Lubrication is especially necessary in the chest and abdomen, where internal organs touch and slide over each other. It is also necessary at joints, where bones, ligaments, and tendons rub against each other. In the digestive tract, water moistens foods to ensure their smooth passage.

Acids, bases, and salts

When molecules of inorganic acids, bases, or salts are dissolved in water in the body cells, they undergo *ionization* or *dissociation*. That is, they break apart into ions. An **inorganic acid** may be defined as a substance that dissociates into one or more *hydrogen ions* (H^+) and one or more negative ions, called *anions* (Figure 2–7a). An **inorganic base,** by contrast, dissociates into one or more *hydroxyl ions* (OH^-) and one or more positive ions, called *cations* (Figure 2–7b). A **salt,** when dissolved in water, dissociates into one or more positive ions and one or more negative ions, neither of which is H^+ or OH^- (Figure 2–7c).

Many inorganic salts are found in the body. Some occur in cells, whereas others occur in the body fluids, such as lymph, blood, and the extracellular fluid of tissues. These salts are dissociated

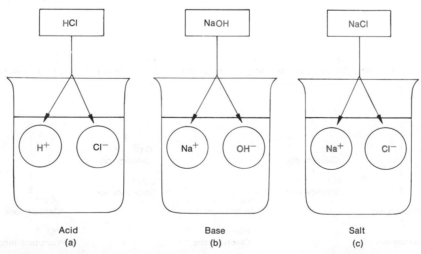

Figure 2–7. The ionization of inorganic acids, bases, and salts. (a) When placed in water, hydrochloric acid, HCl, dissociates into H^+ ions. (b) When the base sodium hydroxide, NaOH, is placed in water, it dissociates into OH^- ions. (c) When table salt, NaCl, is placed in water, it dissociates into positive and negative ions, neither of which is H^+ nor OH^-.

into both cations and anions. The ions of salts are the sources of many essential chemical elements. (See Exhibit 2–1.) Examples of how salts dissociate into ions that provide essential chemical elements are shown in Figure 2–8. Chemical analyses reveal that sodium and chloride ions are present in higher concentrations than other ions in extracellular body fluids. Inside the cells, phosphate and potassium ions are more abundant than other ions. Chemical elements such as sodium, phosphate, potassium, or iodine are present in the body only in chemical combination with other elements or as ions. Their presence as free, un-ionized atoms could be instantly fatal. A list of representative elements found in the body is shown in Exhibit 2–1.

Your acid-base balance

The fluids of your body must maintain a fairly constant balance of acids and bases. In solutions such as those found in body cells or in extracellular fluids, acids dissociate into hydrogen ions

Exhibit 2–1. REPRESENTATIVE CHEMICAL ELEMENTS

CHEMICAL ELEMENT	COMMENT	CHEMICAL ELEMENT	COMMENT
Oxygen (O)	Constituent of water and organic molecules; functions in cellular respiration	Chlorine (Cl)	Anion of NaCl, an important salt; important in water movement between cells
Carbon (C)	Found in every organic molecule	Sulfur (S)	Component of many proteins, especially the contractile protein of muscle
Hydrogen (H)	Constituent of water, all foods, and most organic molecules	Potassium (K)	Required for growth and important in nerve conduction and muscle contraction
Nitrogen (N)	Component of all protein molecules and nucleic acid molecules	Sodium (Na)	Cation of NaCl; structural component of bone; essential in blood to maintain water balance; needed for nerve conduction
Calcium (Ca)	Constituent of bone and teeth; required for blood clotting, hormone synthesis, membrane integrity, and contraction of muscle	Magnesium (Mg)	Component of many enzymes
		Iodine (I)	Vital to functioning of thyroid gland
Phosphorus (P)	Component of many proteins, nucleic acids, and ATP; required for normal bone and tooth structure; found in nervous tissue	Iron (Fe)	Essential component of hemoglobin and respiratory enzymes

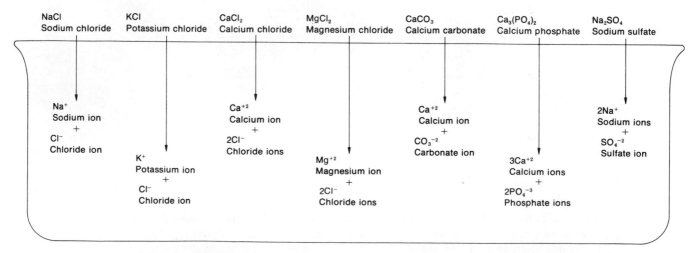

Figure 2–8. Dissociation of representative salts into ions that provide essential chemical elements for the body.

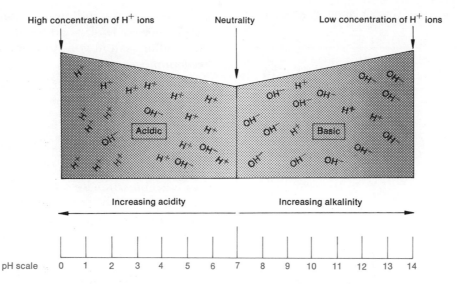

Figure 2–9. The pH scale. At pH 7 (neutrality), the concentration of H⁺ and OH⁻ ions is equal. A pH value below 7 indicates an acid solution; that is, there are more H⁺ ions than OH⁻ ions. The lower the numerical value of pH, the more acid the solution is because the H⁺ ion concentration becomes progressively greater. A pH value above 7 indicates a basic solution—there are more OH⁻ ions than H⁺ ions. The higher the numerical value of pH, the more alkaline the solution is because the OH⁻ ion concentration becomes progressively greater. A change in one whole number on the pH scale represents a tenfold change from the previous concentration.

(H⁺) and anions. Bases, on the other hand, break down into hydroxyl ions (OH⁻) and cations. The more hydrogen ions that exist in a solution, the more acid the solution is. Conversely, the more hydroxyl ions that exist in a solution, the more basic, or alkaline, the solution is. The term **pH** is used to describe the degree of *acidity* or *alkalinity* of a given solution. Biochemical reactions, or reactions that occur in living systems, are extremely sensitive to even small changes in the acidity or alkalinity of the environments in which they occur. In fact, H⁺ and OH⁻ ions are involved in all biochemical processes, and the functions of cells are modified greatly by any departure from narrow limits of normal H⁺ and OH⁻ concentrations. For this reason, the acids and bases that are constantly formed in the body must be kept in balance.

The degree of acidity or alkalinity of a solution is expressed by a *pH scale*. Numerically, the pH scale runs from 0 to 14 (Figure 2–9). A solution that is 0 on the scale has very many H⁺ ions and no OH⁻ ions. A solution that rates 14 has very many OH⁻ ions and no H⁺ ions. The midpoint in the scale is 7, where the concentration of H⁺ and OH⁻ ions is equal. Any substance that has a pH of 7, such as pure water, is neutral. Any solution that has more H⁺ ions than OH⁻ ions is an *acid solution* and has a pH below 7. If a solution

has more OH⁻ ions than H⁺ ions, it is an *alkaline solution* and has a pH above 7. A change of one whole number on the pH scale represents a tenfold change from the previous concentration. That is, a pH of 2 indicates 10 times fewer H⁺ ions than a pH of 1. A pH of 3 indicates 10 times fewer H⁺ ions than a pH of 2 and 100 times fewer H⁺ ions than a pH of 1.

The buffer system

Although the pH of body fluids may differ, the normal limits for the various fluids are generally quite specific and narrow. Exhibit 2–2 shows the pH values for some body fluids such as blood, urine, saliva, and semen. These are compared with some common substances that you have probably tasted or touched. Even though strong acids and bases are continually taken into the body, the pH levels of these various body fluids remain relatively constant. The mechanisms that maintain these homeostatic pH values in the body are called the **buffer systems**.

The essential function of a buffer system is to react with strong acids or bases that are taken into or produced by the body and replace them with weak acids or bases which can change the normal pH values only slightly. Strong acids (or bases) ionize relatively easily and contribute many H⁺

Exhibit 2–2. NORMAL pH VALUES OF REPRESENTATIVE
SUBSTANCES

SUBSTANCE	pH VALUE
Gastric juice (digestive juice of the stomach)	1.2–3.0
Lemon juice	2.2–2.4
Grapefruit	3.0
Cider	2.8–3.3
Pineapple juice	3.5
Tomato juice	4.2
Clam chowder	5.7
Urine	5.5–7.5
Saliva	6.5–7.5
Milk	6.6
Pure water	7.0
Blood	7.35–7.45
Semen (fluid containing sperm)	7.35–7.50
Cerebrospinal fluid (fluid associated with nervous system)	7.4
Pancreatic juice (digestive juice of the pancreas)	7.1–8.2
Eggs	7.6–8.0
Bile (liver secretion that aids in fat digestion)	7.6–8.6
Milk of magnesia	10.0–11.0
Lime water	12.3

(or OH⁻) ions to a solution. They therefore change the pH drastically. Weak acids (or bases) do not ionize as easily. They contribute fewer H^+ (or OH^-) ions and have relatively little effect on the pH. The chemicals that change strong acids or bases into weak ones are called buffers and are found in the body's fluids. Of the several buffer systems, we shall examine only the *carbonic acid–bicarbonate buffer system*, which is the most important one in extracellular fluid.

Like most of the buffer systems of the body, the carbonic acid-bicarbonate buffer system consists of a *pair* of compounds. One of these compounds is a *weak acid*, and the other is a *salt of that acid*. The weak acid of the buffer pair is *carbonic acid* (H_2CO_3), and the salt of the acid is *sodium bicarbonate* ($NaHCO_3$). In solution, the members of this buffer pair dissociate as follows:

Acid
component: $H_2CO_3 \rightleftharpoons H^+ + HCO_3^-$
Carbonic acid   Hydrogen ion   Bicarbonate ion

Salt
component: $NaHCO_3 \rightleftharpoons Na^+ + HCO_3^-$
Sodium bicarbonate   Sodium ion   Bicarbonate ion

Consider the following situation. If a strong acid, such as HCl, is added to extracellular fluid, the buffer system goes to work, and the following reaction occurs:

$$HCl + NaHCO_3 \rightleftharpoons NaCl + H_2CO_3$$

Hydrochloric acid (strong acid)   Sodium bicarbonate (salt of the bicarbonate buffer system)   Sodium chloride (salt)   Carbonic acid (weak acid)

The chloride ion of HCl and the sodium ion of sodium bicarbonate combine to form NaCl, a substance that has no effect on pH. The hydrogen ion of the HCl could greatly lower pH by making the solution more acid, but this H^+ ion combines with the bicarbonate ion (HCO_3^-) of sodium bicarbonate to form carbonic acid, a weak acid that lowers pH only slightly. In other words, due to the action of the buffer system, the strong acid, HCl, has been replaced by a weak acid, and the pH remains relatively constant.

Now suppose a strong base, such as sodium hydroxide (NaOH), is added to the extracellular fluid. The following reaction takes place:

$$NaOH + H_2CO_3 \rightleftharpoons H_2O + NaHCO_3$$

Sodium hydroxide (strong base)   Carbonic acid (weak acid of the bicarbonate buffer system)   Water   Sodium bicarbonate (salt)

In this reaction, the OH^- ion of sodium hydroxide could greatly raise the pH of the solution by making it more alkaline. However, the OH^- ion combines with an H^+ ion of carbonic acid and forms water, a substance that has no effect on pH. In addition, the Na^+ ion of sodium hydroxide combines with the bicarbonate ion (HCO_3^-) to form sodium bicarbonate, a salt that also has no effect on pH. Thus, due to the action of the buffer system, the strong base is replaced by water and salt, and the pH remains relatively constant.

Organic substances

Organic substances are chemical compounds that contain carbon and usually hydrogen and oxygen as well. Carbon has several properties that make it particularly useful to living organisms. For one thing, it can react with one to several hundred other carbon atoms to form large molecules of many different shapes. This means that the body can build many kinds of compounds out of carbon, hydrogen, and oxygen. Each compound can be especially suited for a particular structure or function. The large size of carbon molecules and the fact that they do not dissolve easily in water make them very useful materials for building body structures. Carbon compounds are mostly or entirely held together by covalent bonds, and they tend to decompose easily. This means that organic compounds are also a good source of energy. Ionic compounds are not good energy sources because they form new ionic bonds as soon as the old ones are broken.

Quick-energy carbohydrates

A fairly large and diverse group of organic compounds found in the body is the **carbohydrates**, also known as sugars and starches. The carbohydrates assume a number of important functions in living systems. A few carbohydrates form structural units. For instance, one type of sugar is a building block of genes and is instrumental in determining heredity. Some carbohydrates are converted to proteins and to fats or fatlike substances, which are used to build structures and provide an emergency source of energy. Other carbohydrates function as food reserves. One example is glycogen, which is stored in the liver and skeletal muscles. The principal function of carbohydrates, however, is to provide the most readily available source of energy to sustain life.

Carbon, hydrogen, and oxygen are the elements found in carbohydrates. The ratio of hydrogen to oxygen atoms is always two to one. This can be seen in the formulas for carbohydrates such as ribose ($C_5H_{10}O_5$), glucose ($C_6H_{12}O_6$), and sucrose ($C_{12}H_{22}O_{11}$). Although there are some exceptions, the general formula for carbohydrates is $(CH_2O)_n$, where n symbolizes three or more CH_2O units. Carbohydrates can be divided into three major groups: monosaccharides, disaccharides, and polysaccharides.

MONOSACCHARIDES. Simple sugars are called *monosaccharides*. These compounds contain from three to seven carbon atoms and cannot be broken down into simpler sugar molecules. Simple sugars with three carbons in the molecule are called trioses. The number of carbon atoms in the molecule is indicated by the prefix *tri*. There are also tetroses, or four-carbon sugars, pentoses (five-carbon sugars), hexoses (six-carbon sugars), and heptoses (seven-carbon sugars). Pentoses and hexoses are exceedingly important to the human organism. For example, the pentose, deoxyribose, is a component of genes. The hexose, glucose, is the main energy-supplying molecule of the body.

DISACCHARIDES. A second group of carbohydrates, called the *disaccharides*, consists of two monosaccharides joined chemically. In the process of disaccharide formation, two monosaccharides combine to form a disaccharide molecule, and a molecule of water is lost. This reaction is referred to as *dehydration synthesis*. The word *dehydration* means the loss of water. The following reaction shows disaccharide formation. Molecules of the monosaccharides glucose and fructose combine to form a molecule of the disaccharide sucrose, or cane sugar:

$$C_6H_{12}O_6 \ + \ C_6H_{12}O_6 \ \rightarrow C_{12}H_{22}O_{11} + \ H_2O$$

Glucose   Fructose   Sucrose   Water
(monosaccharide) (monosaccharide) (disaccharide)

You may be puzzled to see that glucose and fructose have the same chemical formulas. Actually, they are different monosaccharides since the relative positions of the oxygens and carbons are different in the two different molecules. (See Figure 2–10.) The formula for sucrose is $C_{12}H_{22}O_{11}$ and not $C_{12}H_{24}O_{12}$ since a molecule of H_2O is lost in the process of disaccharide formation. In any dehydration synthesis, there will be a loss of a molecule of water. Along with this water loss, there will be the synthesis of two small molecules, such as glucose and fructose, into one large, more complex molecule, such as sucrose (Figure 2–10). In a similar manner, the dehydration synthesis of two monosaccharides such as glucose and galactose will form the disaccharide lactose, or milk sugar.

Disaccharides can also be broken down into smaller, simpler molecules by adding water. This reverse chemical reaction is called *digestion* or *hydrolysis*, which means to cut by water. A molecule of sucrose, for example, may be digested into its components of glucose and fructose by the addition of water. The mechanism of this reaction also is represented in Figure 2–10.

POLYSACCHARIDES. The third major group of carbohydrates, called the *polysaccharides*, consists of eight or more monosaccharides joined together through dehydration synthesis. Polysaccharides have the formula $(C_6H_{10}O_5)_n$. Like disaccharides, polysaccharides can be broken into their constituent sugars through hydrolysis reactions. Unlike monosaccharides or disaccharides, however, they usually lack the characteristic sweetness of sugars like fructose or sucrose and are usually not soluble in water. One of the important polysaccharides is glycogen.

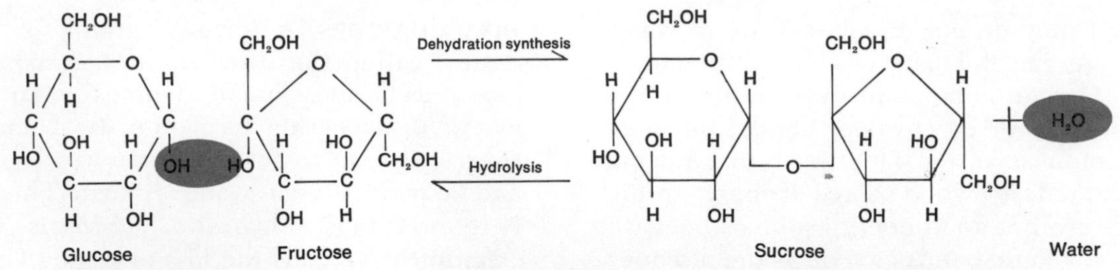

Figure 2–10. Dehydration synthesis and hydrolysis of a molecule of sucrose. In the dehydration synthesis (read from left to right), the two smaller molecules, glucose and fructose, are joined to form a larger molecule of sucrose. Note the loss of a water molecule. In hydrolysis (read from right to left), the larger sucrose molecule is broken down into the two smaller molecules, glucose and fructose. Here, a molecule of water is added to sucrose for the reaction to occur.

The versatile lipids

A second group of organic compounds that is important to the human organism is the **lipids.** Like carbohydrates, lipids are composed of carbon, hydrogen, and oxygen, but they do not have a two-to-one ratio of hydrogen to oxygen. Most lipids are insoluble in water, but they readily dissolve in solvents such as alcohol, chloroform, and ether. Since lipids are a very large and diverse group of compounds, only one kind, the *fats,* will be discussed here. Pertinent information regarding other lipids is provided in Exhibit 2–3. Some information in the Exhibit will mean more to you as you read later chapters.

A molecule of fat consists of two basic components. The first component is called *glycerol,* and the second is a group of compounds called *fatty acids* (Figure 2–11). A single molecule of fat is formed when a molecule of glycerol combines with three molecules of fatty acids. This reaction, like the one described for disaccharide formation, is also a dehydration synthesis reaction. During hydrolysis, a single molecule of fat is broken down into its constituent fatty acids and glycerol.

Fats represent the human organism's most highly concentrated source of energy. In fact, they provide twice as many Calories (a form of energy) per weight as either carbohydrates or proteins. In general, however, fats are about 10 to 12 percent less efficient as body fuels than are carbohydrates. A great amount of the fat Calorie is wasted and thus not available for the body to use.

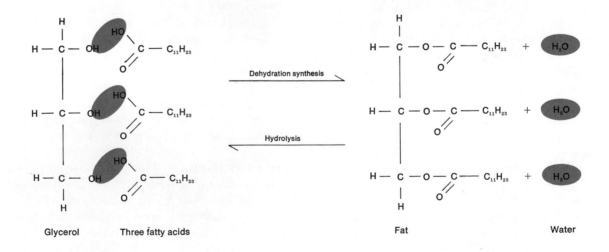

Figure 2–11. In the dehydration synthesis of a fat (read from left to right), one molecule of glycerol combines with three fatty acid molecules, and there is a loss of three molecules of water. In the hydrolysis of a fat (read from right to left), a molecule of fat is broken down into a single molecule of glycerol and three fatty acid molecules upon the addition of three molecules of water.

Exhibit 2–3. RELATIONSHIPS OF REPRESENTATIVE
LIPIDS TO THE HUMAN ORGANISM

LIPIDS	RELATIONSHIP
I. Fats	Protection, insulation, source of energy
II. Phospholipids	
A. Lecithin	Major lipid component of cell membranes; constituent of plasma
B. Cephalin and sphingomyelin	Found in high concentrations in nerves and brain tissue
III. Steroids	
A. Cholesterol	Constituent of all animal cells, blood, and nervous tissue; suspected relationship to heart disease and "hardening of the arteries"
B. Bile salts	Substances that emulsify or suspend fats prior to their digestion
C. Vitamin D	Produced in skin upon exposure to ultraviolet radiation; necessary for bone growth and development
D. Estrogens	Sex hormones produced in large quantities by females
E. Androgens	Sex hormones produced in large quantities by males
IV. Porphyrins (lipid portions of organic molecules)	
A. Hemoglobin	Oxygen-transporting pigment in red blood cells
B. Bile pigments	Bilirubin, a reddish pigment, and biliverdin, a greenish pigment, are both formed from hemoglobin and are responsible for the brown color of feces
C. Cytochromes	Coenzymes that are involved in the respiration of all cells
V. Other lipoid substances	
A. Carotenes	Pigment in egg yolk, carrots, and tomatoes; vitamin A is formed from carotenes; retinene, also formed from vitamin A, is a photoreceptor in the retina of the eye
B. Vitamin E	"Antisterility" vitamin
C. Vitamin K	Vitamin that promotes blood clotting and prevents excessive bleeding

Building and regulating proteins

A third principal group of organic compounds is **proteins**. These compounds are much more complex in structure than the carbohydrates or lipids. They are also responsible for much of the structure of body cells and are related to many physiological activities. For example, proteins in the form of enzymes speed up many essential biochemical reactions. Other proteins assume a necessary part in muscular contraction. Antibodies are proteins that provide the human organism with defenses against invading microbes. And some hormones that regulate body functions are also proteins. A classification of proteins on the basis of their functions is shown in Exhibit 2–4.

Chemically, proteins always contain carbon, hydrogen, oxygen, and nitrogen. Many proteins also contain sulfur and phosphorus. Just as monosaccharides are the building units of sugars and fatty acids and glycerol are the building units of fats, *amino acids* are the building blocks of proteins. In protein formation, amino acids combine

Exhibit 2–4. CLASSIFICATION OF PROTEINS ON THE BASIS
OF FUNCTION

NATURE OF PROTEIN	DESCRIPTION
I. Structural proteins	Proteins that form the structural framework of various parts of the body. Examples: keratin in the skin, hair, and fingernails and collagen in bone and connective tissue.
II. Regulatory proteins	Proteins that function as hormones and regulate various physiological processes. Examples: insulin, which regulates blood sugar, and adrenalin, which regulates the diameter of blood vessels.
III. Contractile proteins	Proteins that serve as contractile elements in muscle tissue. Examples: myosin and actin.
IV. Immune proteins	Proteins that serve as antibodies to protect the body against invading microbes. Example: gamma globulin.
V. Transport proteins	Proteins that transport vital substances throughout the body. Example: hemoglobin, which transports oxygen.
VI. Catalytic proteins	Proteins that act as enzymes and function by controlling the kinds and rates of biochemical reactions. Examples: salivary amylase, pepsin, and lactase.

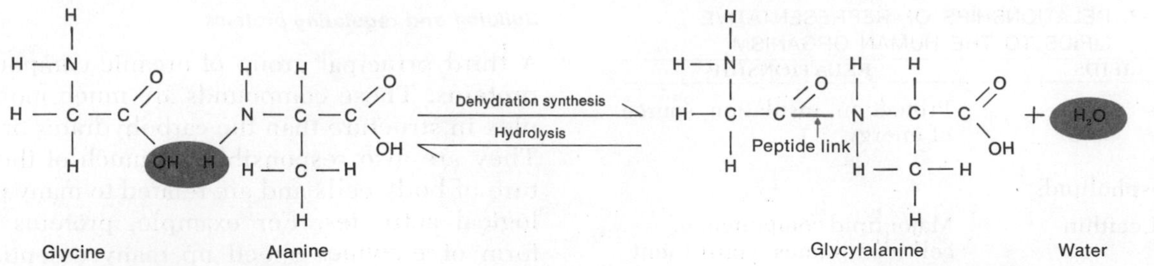

Figure 2–12. Protein formation. When two or more amino acids are chemically united, the resulting bond between them is called a peptide link. In the example shown, glycine and alanine, the two amino acids, are joined to form the dipeptide, glycylalanine. At the point where water is lost, the peptide link is formed.

to form more complex molecules, while water molecules are lost. The process is a dehydration synthesis reaction, and the bonds formed between amino acids are called *peptide links* (Figure 2–12).

When two amino acids combine, a *dipeptide* results. Adding another amino acid to a dipeptide produces a *tripeptide*. Further additions of amino acids result in the formation of *polypeptides*, which are large protein molecules. At least 20 different amino acids are found in proteins. An extremely large variety of proteins is possible because each variation in the number or sequence of amino acids can produce a different protein. The situation is similar to using an alphabet of 20 letters to form words. Each letter could be compared to a different amino acid, and each word would be a different protein.

ENZYMES, THE CATALYSTS. Proteins that are produced by living cells to catalyze or speed up many of the reactions in the body are **enzymes.** Protein in the form of an enzyme acts as a *catalyst*, which is any chemical substance that speeds up a reaction without being permanently altered by the reaction.

For any chemical or biochemical reaction to occur, a certain amount of energy is required. The amount of energy needed is called the *activation energy*. As stated earlier, energy can be transformed from one state to another. When heat energy is added to molecules, some of the heat is transformed to kinetic energy, and thus the energy of the molecules is increased. One way to cause most of the molecules to obtain an activation-level energy is to heat them. Unfortunately, heat happens to destroy the body's proteins. The role of an enzyme, then, is to decrease the amount of energy needed to start the reaction.

Exactly how enzymes lower activation energies is not fully understood. However, we do know that an enzyme attaches itself to one of the reacting molecules, called a *substrate*, and the two form

a temporary *enzyme-substrate complex.* Many kinds of enzymes exist, but each kind can attach to only one kind of substrate. Apparently, the enzyme molecule must fit perfectly with the substrate molecule like pieces of a jigsaw puzzle (Figure 2–13). If the enzyme and substrate do not fit properly, no reaction occurs.

When an enzyme-substrate complex is formed, the substrate molecule can react with other molecules in a decomposition or synthesis reaction. After the reaction is completed, the products of the reaction move away from the enzyme, and the enzyme is free to attach to another substrate molecule. The whole process of attachment, reaction, and detachment takes place very quickly, and most enzymes are capable of interacting with 1,000 substrate molecules per minute.

A word should be said about naming enzymes so that you will be able to recognize them easily. Many enzymes are named by adding the suffix *ase* to the name of their substrates. For example, the enzyme that is involved in breaking down sucrose is called *sucrase*. Enzymes that hydrolyze or break down proteins are classified as *proteases*. Other enzymes are named by their action. *Dehydrogenase* enzymes, for example, remove hydrogen atoms from a substrate. The important point to remember here is that the suffix *ase* denotes an enzyme.

You may have rightly assumed by now that enzymes are very important to your body's overall

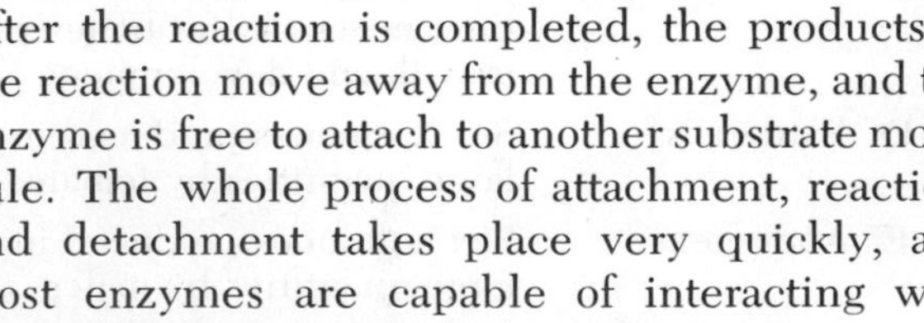

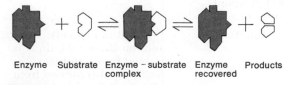

Enzyme   Substrate   Enzyme – substrate   Enzyme   Products
complex   recovered

Figure 2–13. Enzyme action in a decomposition reaction. The enzyme and substrate molecules combine to form an enzyme-substrate complex. During combination, the substrate is changed into products. Once the products are formed, the enzyme is recovered (moves away from the products) and may be used again to catalyze a similar reaction.

homeostasis. They quickly catalyze chemical reactions, and they also determine the rate and kind of reactions that occur.

Regulating nucleic acids

Exceedingly large organic molecules containing carbon, hydrogen, oxygen, nitrogen, and phosphorus are **nucleic acids.** The term nucleic acid derives from the fact that these compounds were first discovered in the nuclei of cells. Whereas the basic structural units of proteins are amino acids, the basic units of nucleic acids are called *nucleotides,* which are described below. Nucleic acids are divided into two principal kinds: **DNA,** or **deoxyribonucleic acid,** and **RNA,** or **ribonucleic acid.**

A molecule of DNA is a chain composed of repeating nucleotide units. Each nucleotide of DNA consists of three basic parts (Figure 2–14a): (1) It contains one of four possible *nitrogen bases,* which are ring-shaped structures containing atoms of C, H, O, and N. The nitrogen bases found in DNA are named adenine, thymine,

cytosine, and guanine. (2) It contains a pentose sugar called *deoxyribose.* (3) It has a phosphoric acid called the *phosphate group.* The nucleotides are named according to the nitrogen base that is present. Thus, a nucleotide containing thymine is called a *thymine nucleotide.* One containing adenine is called an *adenine nucleotide,* and so on.

The chemical composition of the DNA molecule was known before 1900, but it was not until 1953 that a model of the organization of the chemicals was constructed. This model was proposed by J. D. Watson and F. H. C. Crick on the basis of data from many investigations. Figure 2–14b shows the following structural characteristics of the DNA molecule: (1) The molecule consists of two strands with crossbars. The strands twist about each other in the form of a *double helix* so that the shape resembles a twisted ladder. (2) The uprights of the DNA "ladder" consist of alternating phosphate groups and the deoxyribose portions of the nucleotides. (3) The rungs of the ladder contain paired nitrogen bases. As shown, adenine always pairs off with thymine, and cytosine always pairs off with guanine.

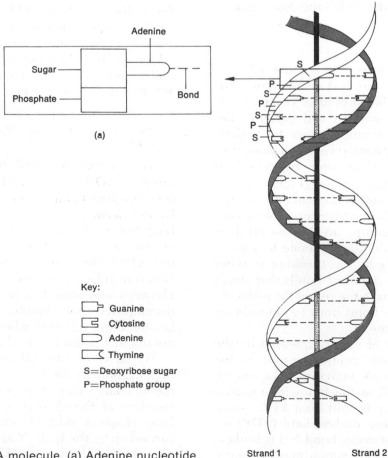

Key:

⬡⊐ Guanine
⬡⊏ Cytosine
⬡ Adenine
⬡⊏ Thymine

S = Deoxyribose sugar
P = Phosphate group

Strand 1   Strand 2

Figure 2–14. The DNA molecule. (a) Adenine nucleotide. (b) Portion of an assembled DNA molecule.

(a)

(b)

As you probably know, cells contain hereditary material called genes. However, you may not have known that a gene is simply a segment of a DNA molecule. Our genes determine which traits we inherit, and they control all the activities that take place in our cells throughout a lifetime. When a cell divides, its hereditary information is passed on to the next generation of cells. The passing of information is possible because of the unique structure of DNA. This will be discussed further in Chapter 3. For now, we can say that the two basic functions of DNA are to control the activities of body cells and to transmit hereditary information to future generations.

RNA, the second principal kind of nucleic acid, differs from DNA in several aspects. In some cases, RNA is double stranded like DNA, but, in others, it is only single stranded. At times, RNA appears as a combination of both double and single strands. The sugar in the RNA nucleotide is the pentose ribose. And RNA does not contain the nitrogen base thymine. Instead of thymine, RNA has a nitrogen base called uracil. At least three different kinds of RNA have been identified in cells. Each type of RNA has a specific role in reacting with DNA to help regulate protein synthesis reactions.

ATP and stored energy

A high-energy nucleotide that is extremely important to the life of the cell is **adenosine triphosphate,** or **ATP.** This substance is found universally in all living systems and performs the essential function of storing energy for various cellular activities. ATP consists of three phosphate groups and an adenosine unit composed of adenine and the sugar ribose (Figure 2–15). The two phosphate groups at the end opposite the adenosine unit are called high-energy phosphate groups. When these groups are broken off from the rest of the molecule, they liberate four to six times as much energy as the breaking of other chemical bonds. The chemical bonds that attach these phosphate groups to the rest of the molecule are indicated by wavy bond lines (~) to indicate their high-energy content.

The large amounts of energy released by the splitting off of phosphate groups are used by the cell to carry on its basic activities. The energy is supplied to the cell in the following manner. If one phosphate group is split from ATP, a compound called **adenosine diphosphate (ADP)** results. Because a high-energy bond (~) is broken, energy is released. This reaction may be represented as follows:

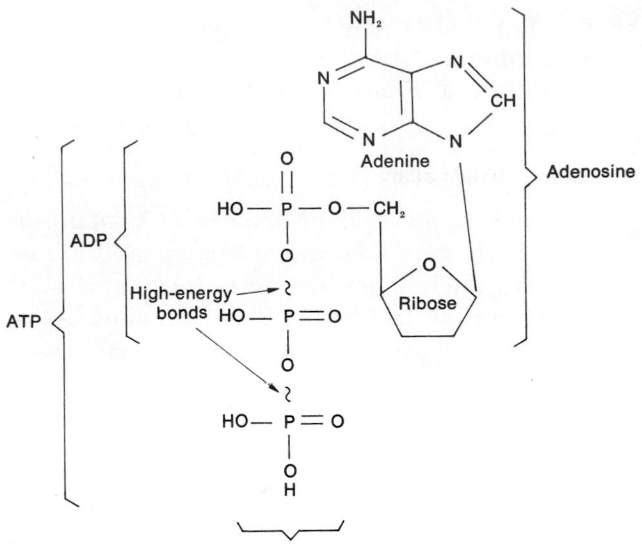

Figure 2–15. The structure of ATP.

$$\text{ATP} \;\rightleftharpoons\; \text{ADP} \;+\; \text{P} \;+\; \text{E}$$

Adenosine triphosphate   Adenosine diphosphate   Phosphate   Energy

The energy supplied by the conversion of ATP into ADP is constantly being used by the cell. Since the supply of ATP at any given time is limited, a mechanism exists to replenish it. By way of this mechanism, a phosphate group is added to ADP to manufacture more ATP. Logically, energy is required to manufacture ATP. The reaction may be represented as follows:

$$\text{ADP} \;+\; \text{P} \;+\; \text{E} \;\rightleftharpoons\; \text{ATP}$$

Adenosine diphosphate   Phosphate   Energy   Adenosine triphosphate

The energy required to attach a phosphate group to ADP is supplied by various decomposition reactions taking place in the cell, particularly by the decomposition of glucose. Why does the body not use the energy from the decomposition of glucose and avoid going to the trouble of making ATP? The answer is that glucose cannot be stored in cells. Some glucose can be converted to glycogen and stored in the liver, but the rest is decomposed immediately. ATP, by contrast, can be stored in every cell, where it provides potential energy that is not released until needed.

We have tried to outline the chemical level of organization in the body and introduce some of the molecules that are important to life. An understanding of these areas is a necessary basis for later chapters dealing with chemical reactions carried on by the body. You thus may wish to refer often to this chapter to aid your understanding of subsequent discussions.

Chapter summary in outline

PART I: BASIC CHEMISTRY FOR LIVING SYSTEMS

1. Matter is anything that occupies space and has mass. It is made of building units called chemical elements.

2. Carbon, hydrogen, oxygen, and nitrogen comprise 96 percent of body weight. These elements together with phosphorus and calcium comprise 99 percent of body weight.

Atomic structure of elements

1. Units of matter of all chemical elements are called atoms.

2. Atoms consist of a nucleus, which contains protons and neutrons, and orbiting electrons moving in energy levels.

3. The total number of electrons of an atom is its atomic number.

How atoms combine

1. The electrons are the part of an atom that actively participate in chemical reactions. An atom attempts to fill its outer energy level through bonding.

2. A molecule is the smallest unit of two or more combined atoms. A molecule containing two or more different kinds of atoms is a compound.

3. In an ionic bond, outer energy level electrons are transferred from one atom to another. The transfer forms ions, whose unlike charges attract each other and form ionic bonds.

4. In a covalent bond, there is a sharing of pairs of outer energy level electrons.

Chemical reactions

1. Synthetic reactions involve the combination of reactants to produce a new molecule. Bonds are formed.

2. In decomposition reactions, a substance breaks down into other substances. Bonds are broken.

3. Exchange reactions involve the replacement of one atom or atoms by another atom or atoms.

4. In reversible reactions, end products can revert to the original combining molecules.

5. When chemical bonds are formed, energy is needed. When bonds are broken, energy is released. This is known as chemical bond energy.

PART II: CHEMICAL LIFE PROCESSES

1. Inorganic substances usually lack carbon, contain ionic bonds, resist decomposition, and dissolve readily in water.

2. Organic substances always contain carbon and usually have hydrogen. Most organic substances contain covalent bonds and are insoluble in water.

Inorganic substances

1. Water is the most abundant substance in the body. It is an excellent solvent and suspending medium, participates in chemical reactions, absorbs and releases heat slowly, and lubricates.

2. Acids, bases, and salts ionize into ions in water. An acid ionizes into H^+ ions, and a base ionizes into OH^- ions. A salt ionizes into neither H^+ nor OH^- ions. Cations are positively charged ions, and anions are negatively charged ions.

3. The pH of different parts of the body must remain fairly constant for the body to remain healthy. On the pH scale, 7 represents neutrality. Values below 7 indicate acid solutions, and values above 7 indicate alkaline solutions.

4. The pH values of different parts of the body are maintained by buffer systems, which usually consist of a weak acid and the salt of that acid. Buffer systems eliminate excess H^+ ions and excess OH^- ions in order to maintain pH homeostasis.

Organic substances

1. Carbohydrates are either sugars or starches that provide most of the energy needed for life. They may be monosaccharides, disaccharides, or polysaccharides. Carbohydrates, and other organic molecules, are joined together to form larger molecules with the loss of water by a process called dehydration synthesis. In the reverse process, called hydrolysis or digestion, larger molecules are broken down into smaller ones upon the addition of water.

2. Lipids are a diverse group of compounds that includes fats, steroids, pigments, and vitamins. Fats protect, insulate, provide energy, and are stored.

3. Proteins are constructed from amino acids. They give structure to the body, regulate processes, provide protection, cause muscles to contract, and transport substances.

4. Enzymes are proteins produced by the body. They catalyze, or speed up, chemical reactions. Enzymes act upon substrates by lowering the activation energy needed for reaction.

5. DNA and RNA are nucleic acids consisting of nitrogen bases, sugar, and phosphate groups. DNA is a double helix and is the primary chemical in genes. RNA differs slightly in structure and chemical composition from DNA and is concerned largely with protein synthesis reactions.

6. The principal energy-storing molecule in the body is ATP. When its energy is liberated, it is decomposed to ADP. ATP is manufactured from ADP using the energy supplied by the food you eat.

Review questions and problems

1. What is the relationship of matter to the body?

2. Define a chemical element. List the chemical symbols for ten different chemical elements. Which chemical elements comprise the bulk of the human organism?

3. What is an atom? Can you diagram the positions of the nucleus, protons, neutrons, and electrons in an atom of oxygen and an atom of nitrogen? What is an atomic number?

4. What is meant by an energy level?

5. How are chemical bonds formed? Distinguish between an ionic bond and a covalent bond. Give at least one example of each.

6. Can you determine how a molecule of $MgCl_2$ is ionically bonded? Magnesium has two electrons in its outer shell. Construct a diagram to verify your answer.

7. Refer to Figure 2–5b, c, and see if you can determine why there is a double covalent bond between atoms in an oxygen molecule (O_2) and a triple covalent bond between atoms in a nitrogen molecule (N_2).

8. What are the four principal kinds of chemical reactions? How is energy related to chemical reactions?

9. Can you identify what kind of reaction each of the following represents?

   (a) $H_2 + Cl_2 \rightarrow 2HCl$

   (b) $3NaOH + H_3PO_4 \rightarrow Na_3PO_4 + 3H_2O$

   (c) $CaCO_3 + CO_2 + H_2O \rightarrow Ca(HCO_3)_2$

   (d) $HNO_3 \rightarrow H^+ + NO_3^-$

   (e) $NH_3 + H_2O \rightleftharpoons NH_4^+ + OH^-$

10. How do inorganic compounds differ from organic compounds? List and define the principal inorganic and organic compounds that are important to the human body.

11. What are the essential functions of water in the body? Distinguish between a solution and a suspension.

12. Define an acid, a base, and a salt. How does the body acquire some of these substances? List some functions of the chemical elements furnished by ions of salts.

13. What is pH? Why is it important to maintain a relatively constant pH? What is the pH scale? List the normal pH values of some common fluids, biological solutions, and foods.

14. Refer to Exhibit 2–2, and select the two substances whose pH values are closest to neutrality. Is the pH of milk or the pH of cerebrospinal fluid closer to 7? Is the pH of bile or the pH of urine farther from neutrality? If there are 100 OH^- ions at a pH of 8.5, how many OH^- ions are there at a pH of 9.5?

15. What are the components of a buffer system? What is the function of a buffer pair? Diagram and explain how the carbonic acid-bicarbonate buffer system of extracellular fluid maintains a constant pH even in the presence of a strong acid or a strong base. How is this an example of homeostasis?

16. Why are the reactions of buffer pairs more important with strong acids and bases than with weak acids and bases?

17. Define a carbohydrate. Why are carbohydrates important to the body? How are carbohydrates classified?

18. Compare dehydration synthesis and hydrolysis. Why are they important?

19. How do lipids differ from carbohydrates? What are some relationships of lipids to the body?

20. Define a protein. What is a peptide linkage? Discuss the classification of proteins on the basis of function.

21. Distinguish between an enzyme and a substrate. List some of the principal characteristics of enzymes. Relate the concept of activation energy to enzyme action. How do enzymes maintain homeostasis?

22. What is a nucleic acid? How do DNA and RNA differ with regard to chemical composition, structure, and function?

23. What is ATP? What is the essential function of ATP in the human body? How is this function accomplished?

CHAPTER 3

CELLULAR LEVEL OF ORGANIZATION

STUDENT OBJECTIVES

After you have read this chapter, you should be able to:

1. Define and list a cell's generalized parts

2. Describe the molecular organization of the cell membrane

3. List the factors related to semipermeability of the cell membrane

4. Define diffusion, osmosis, filtration, dialysis, active transport, phagocytosis, and pinocytosis

5. Describe the chemical composition and list the functions of cytoplasm

6. Describe two general functions of a cell nucleus

7. Distinguish between agranular and granular endoplasmic reticulum

8. Define the function of ribosomes

9. Describe the role of the Golgi complex in the synthesis, storage, and secretion of glycoproteins

10. Describe the function of mitochondria as "powerhouses of the cell"

11. Explain why a lysosome in a cell is called a "suicide packet"

12. Describe the structure and function of centrioles in cellular reproduction

13. Distinguish between the structural and functional differences of cilia and flagella

14. Define and list several examples of a cell inclusion

15. Define and list several examples of an extracellular material

16. Describe the sequence of events involved in cell division

17. Describe the significance of cell division

18. Define a gene

19. Describe the sequence of events involved in protein synthesis

After studying some of the chemicals that make up the body, you can now find out how these chemicals are organized into parts of cells and how the chemicals function in cells. A study of the body at the cellular level of organization is important to a total understanding of the structure and function of the body because many activities essential to life occur in cells and many disease processes originate in cells. A cell may be defined as the basic, living, structural and functional unit of the body and, in fact, of all organisms. It is the smallest structure capable of performing all the activities vital to life. **Cytology** is the specialized branch of science concerned with the study of cells. In this chapter, we shall concentrate on the structure of cells, the functions of cells, and the reproduction of cells. A series of illustrations accompanies each cell structure that you study. The first illustration shows the location of the structure within the cell. The second is an *electron micrograph,* which is a photograph taken with a high-powered electron microscope.[1] The third

[1] The electron microscope can magnify an object up to 200,000 times its size. In comparison, the light microscope that you probably use in your laboratory magnifies objects 1,000 to 2,000 times their size (1,000–2,000×).

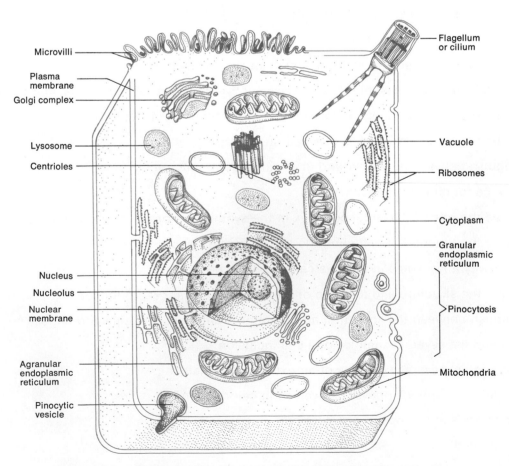

Figure 3–1. Generalized cell based on electron microscope studies.

illustration that accompanies each structure is a drawing of the electron micrograph. The drawing will clarify some of the small details by exaggerating their outlines.

THE GENERALIZED CELL

In the discussions that follow, we shall speak about a **generalized cell,** which is a composite of many different kinds of cells found in the body. The characteristics of specific cells will be discussed in later chapters as parts of the systems to which they belong. Bone cells, for example, will be discussed in the chapter on the skeleton. Examine the generalized cell illustrated in Figure 3–1, and keep in mind that no such single cell actually exists.

For convenience, we shall divide the generalized cell into four principal parts:

1. The *cell surface,* or the outer, limiting membrane separating the internal parts of the cell from the extracellular fluid and external environment

2. *Cytoplasm,* or the ground substance of the cell in which organelles are embedded

3. *Organelles,* the cellular components that are highly specialized for particular cellular activities

4. *Inclusions,* or the secretions and storage areas of cells

Extracellular materials, which are substances external to the cell surface, will also be examined in connection with cells.

THE CELL SURFACE

The exceedingly thin structure that separates one cell from other cells and from the external environment is called the **plasma membrane,** or **cell membrane** (Figure 3–2). Recent electron microscopy studies have shown that the plasma membrane

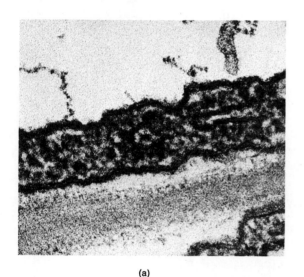
(a)

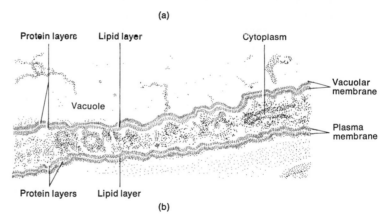

(a)

Protein layers   Lipid layer   Cytoplasm

Vacuole

Vacuolar membrane

Plasma membrane

Protein layers   Lipid layer

(b)

(a)

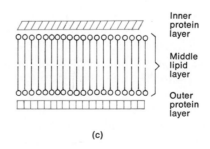

Inner protein layer

Middle lipid layer

Outer protein layer

(c)

Figure 3–2. The plasma membrane. The generalized cell indicates the position of the plasma membrane. (a) Electron micrograph of the detailed structure of the membrane magnified 225,000X. (b) Labeled diagram of the electron micrograph. (c) Presumed position of chemicals that comprise the membrane. [*(a) courtesy of Dr. Myron C. Ledbetter, Brookhaven National Laboratory.*]

ranges from 65 to 100 angstroms in thickness, a dimension below the limits of the light microscope.[2] In electron micrographs (Figure 3–2a, b), the membrane appears as a three-layered structure. The structure consists of two dark lines, each about 30 Å thick, which are separated by an intermediate light line of about the same thickness. According to the most commonly accepted theory of membrane structure, the dark lines are composed of protein. The intermediate light area is a double layer of lipid molecules (Figure 3–2c). Other features of the membrane structure are areas that look like breaks along the membrane surface. These breaks appear at intervals and range in size from 7 to 10 Å in diameter. Researchers suspect they may be pores.

The basic functions of the plasma membrane are to enclose the components of the cell and to serve as a boundary through which substances must pass to enter or exit the cell. One important characteristic of the plasma membrane is that it permits certain ions and molecules to enter or exit the cell but restricts the passage of others. For this reason, cell membranes are described as **semipermeable**. In general, plasma membranes are freely permeable to water. In other words, they let water into and out of the cell. However, they act as barriers to the movement of almost all other substances. The ease with which a substance passes through a membrane is called the membrane's *permeability* to that substance. The permeability of plasma membrane appears to be a function of several factors. We can list four of these factors:

1. *Weight of the entering molecules.* Large-sized molecules cannot pass through the cell membrane. Water, glucose, and amino acids are relatively small molecules and can enter and exit the cell easily. However, most proteins, which consist of many amino acids linked together, seem to be too large to pass through the membrane. Many scientists believe that the giant-sized molecules do not enter the cell because they are larger than the diameters of the suspected membrane pores.
2. *Solubility in lipids.* Substances that dissolve easily in lipids pass through the membrane more readily than other substances since a major part of the plasma membrane consists of lipid molecules.

3. *Charge on ions.* The charge of an ion attempting to cross the plasma membrane can determine how easily the ion can enter or leave the cell. This is true because the protein portion of the membrane is capable of ionization. If an ion has a charge opposite that of the membrane, it is attracted to the membrane and passes through more readily. Conversely, if the ion attempting to cross the membrane has the same charge as the membrane, it is repulsed by the membrane, and its passage is restricted. This phenomenon conforms to the rule of physics that opposite charges attract, whereas like charges repel each other.
4. *Presence of carrier molecules.* Plasma membranes contain special molecules called carriers that are capable of attracting and transporting substances across the membrane regardless of size, ability to dissolve in lipids, or membrane charge. This mechanism will be considered later in the chapter.

These four aspects of the plasma membrane work together to determine whether the membrane will be permeable to the various substances attempting to enter or leave the cell. Before discussing further aspects of the membrane, we need to consider some of the processes involved when substances move from one region to another.

Moving materials across plasma membranes

The mechanisms whereby substances move across the plasma membrane are important to the life of the cell. Certain substances, for example, must move into the cell to support life, whereas waste materials or substances that may be harmful must be moved out of the cell. The processes involved in these movements may be divided into two broad categories, depending upon whether the cell participates in the process by expending energy. Accordingly, the process may be classed as either passive or active. In *passive processes*, substances move across plasma membranes without any help from the cell. The substances move, on their own, from an area where their concentration is greater to an area where their concentration is less. The substances could also be pushed through the cell membrane by pressure from an area where the pressure is greater to an area where it is less. These phenomena will be explained shortly. In *active processes*, by contrast, the cell contributes energy and assumes a role in moving the substance across the membrane. Both active and passive processes should be understood in the study of cell physiology.

[2] One **angstrom**, usually written as 1 Å, = $1/250,000,000$ inch. Another microscopic unit of measurement is the **micron** (μ), which is equal to $1/25,000$ inch.

Passive processes

DIFFUSION. A passive process called **diffusion** occurs when there is a *net* or greater movement of molecules or ions from a region of high concentration to a region of low concentration. The movement from high to low concentration continues until the molecules are evenly distributed. This point of even distribution is called *equilibrium*. The difference between the higher and lower concentrations is called the *concentration gradient*. Molecules moving from the high-concentration area to the low-concentration area are said to move *down* or *with* the concentration gradient. Consider the following example. If a dye pellet is placed in a beaker filled with water, the color of the dye is seen immediately around the pellet. At increasing distances from the pellet, the color becomes lighter (Figure 3–3). If the beaker is observed some time later, however, the water solution will be a uniform color. This happens because the dye molecules possess kinetic energy, which causes them to move about at random, dispersing them throughout the entire area. The dye molecules move down the concentration gradient from an area of high concentration to an area of low concentration. The water molecules also move from a high-concentration to a low-concentration area. When dye molecules and water molecules are evenly distributed among themselves, equilibrium is reached and diffusion ceases, even though molecular movements continue. As another example of diffusion, consider what would happen if you opened a bottle of perfume in a room. The perfume molecules would diffuse until an equilibrium is reached between the perfume molecules and the air molecules in the room.

In the examples cited, no membranes were involved. Diffusion may occur, however, through semipermeable membranes in the body. One of the best examples of this kind of diffusion in the human body is the movement of oxygen from the blood into the cells and the movement of carbon dioxide from the cells back into the blood. This will be discussed further in Chapter 19.

OSMOSIS. Another passive process by which materials move across membranes is **osmosis.** Unlike diffusion, this process specifically refers to the movement of water molecules through a semipermeable membrane from an area of high water concentration to an area of lower water concentration. Once again, a simple apparatus may be used to demonstrate the process. The apparatus shown in Figure 3–4 consists of a tube, which is constructed from cellophane, a semi-

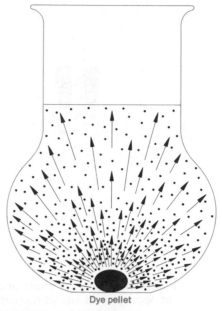

Dye pellet

Figure 3–3. The principle of diffusion. Molecules of the dye pellet move from the high-concentration region to the low-concentration region. At the same time, water molecules move from areas of high concentration to the areas of low concentration. The movement of both types of molecules will continue until dye molecules and water molecules are evenly dispersed in the beaker.

permeable membrane. The cellophane tube is filled with a colored 20 percent sugar solution. The upper portion of the cellophane tube is plugged with a rubber stopper through which a glass tubing is fitted. The cellophane tube is placed into a beaker containing pure water. Initially, the relative concentrations of water on either side of the semipermeable membrane are different. There is a lower concentration of water inside the cellophane tube than there is outside of it. As a result of this difference, water moves from the beaker into the cellophane tube. There is no movement of sugar from the cellophane tube into the beaker, however, since the cellophane is impermeable to molecules of sugar. This is because sugar molecules are too large to go through the pores of the membrane. As water movement into the cellophane tube continues, the sugar solution becomes increasingly diluted and begins to move up the glass tubing. After a period of time, the water that has accumulated in the cellophane tube and the glass tubing exerts a downward pressure that forces water molecules back out of the cellophane tube and into the beaker. When water molecules leave the cellophane tube and enter the tube at the same rate, equilibrium is reached.

Osmosis may also be understood by considering the effects of different water concentrations on

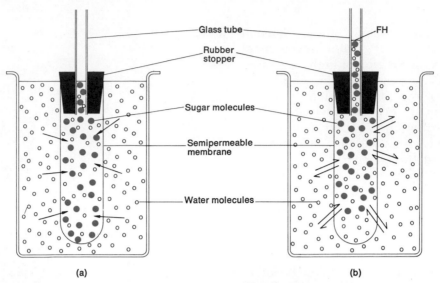

Glass tube
Rubber stopper
Sugar molecules
Semipermeable membrane
Water molecules
FH
(a)
(b)

Figure 3–4. The principle of osmosis. (a) Apparatus at the start of the experiment. (b) Apparatus at equilibrium. In (a), the cellophane tube contains a 20 percent sugar solution and is immersed in a beaker of distilled water. The arrows indicate that water molecules can pass freely into the cellophane tube but that sugar molecules are held back by the semipermeable membrane. As water moves into the cellophane tube by osmosis, the sugar solution is diluted, and the volume of the solution in the cellophane tube increases. This increased volume is shown in (b), with the sugar solution moving up the glass tubing. The final height reached (FH) occurs at equilibrium. At this point, the number of water molecules leaving the cellophane tube is equal to the number of water molecules entering the tube.

red blood cells. If the normal shape of a red blood cell is to be maintained, the cell must be placed in an **isotonic solution** (Figure 3–5a). This is a solution in which the concentrations of water molecules and solute molecules are the same on both sides of the semipermeable membrane. In the case of red blood cells, the concentrations of water and solute in the extracellular fluid outside the red blood cell must be the same as the concentration of the extracellular fluid. Under ordinary circumstances, a 0.85 percent NaCl solution is isotonic for red blood cells. In this condition, water molecules enter and exit the cell at the same rate,

allowing the cell to maintain its normal shape. A different situation results if red blood cells are placed in a solution that has a lower concentration of solutes and, therefore, a higher concentration of water. This is called a **hypotonic solution.** In this condition, water molecules enter the cells faster than they can leave. This causes the red blood cells to swell and eventually burst (Figure 3–5b). A good hypotonic solution is distilled water, which has no solute molecules at all. On the other hand, a **hypertonic solution** has a higher concentration of solutes and a lower concentration of water than the red blood cells. One example of a hypertonic solution is a 10 percent NaCl solution. In such a solution, water molecules move out of the cells faster than they can enter. This causes the cells to shrink (Figure 3–5c). Quite obviously, red blood cells may be greatly impaired or destroyed if placed in solutions that deviate significantly from the isotonic state.

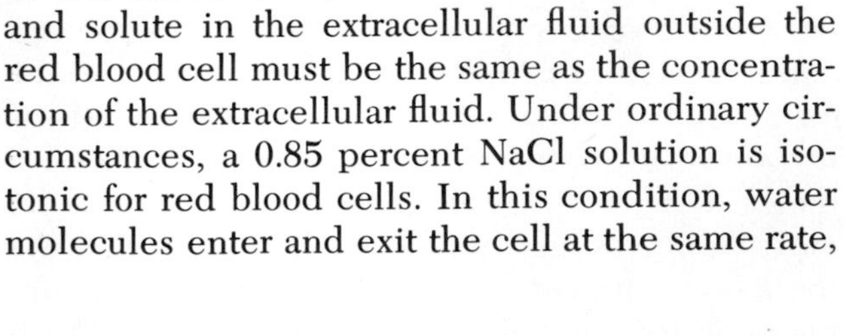

FILTRATION. A third passive process involved in moving materials in and out of cells is **filtration.** This process involves the movement of solvents such as water and dissolved substances such as sugar across a semipermeable membrane by mechanical pressure. Such a movement is always from an area of higher pressure to an area of lower

Isotonic
Hypotonic
Hypertonic
Water
Water
Water
(a)
(b)
(c)

Figure 3–5. The principle of osmosis applied to red blood cells. Shown here are the effects on red blood cells when placed in (a) an isotonic solution, (b) a hypotonic solution, and (c) a hypertonic solution.

pressure and continues as long as a pressure difference exists. Most small- to medium-sized molecules can be forced through a cell membrane by pressure. An example of filtration occurs in the kidneys, where the blood pressure supplied by the heart forces water and urea through thin cell membranes of tiny blood vessels and into the kidney cells. In this basic process, protein molecules are retained by the body since they are too large to be forced through the cell membranes of the kidney cells. Harmful substances, such as urea, however, are small enough to be forced through and eliminated.

DIALYSIS. The final passive process to be considered is **dialysis,** the process by which the artificial kidney works. Dialysis involves the separation of small molecules from large molecules by diffusion and osmosis through a semipermeable membrane. For example, assume that a solution containing molecules of various sizes is placed in a tube that is permeable only to the smaller molecules. The tube is then placed in a beaker of distilled water. Eventually, the smaller molecules will move from the tube into the water in the beaker, and the larger molecules will be left behind. This principle of dialysis is employed in artificial kidneys. In the operation of an artificial kidney, the blood of the patient is passed into a dialysis tube outside the patient's body. The dialysis tube takes the place of the patient's kidneys. As the blood moves through the tube, waste products pass from the blood into a solution surrounding the dialysis tube. At the same time, certain nutrients are passed from the solution into the blood. The blood is then returned to the body.

Active processes

We shall now turn our attention to processes in which cells actively participate in moving substances across membranes. In these processes, the cell must expend energy. By participating in the transport of substances, the cell can even move them against a concentration gradient. Among the active processes we shall consider are active transport, phagocytosis, and pinocytosis.

ACTIVE TRANSPORT. The process by which substances, usually ions, are transported across cell membranes from an area of lower concentration to an area of higher concentration is called **active transport** (Figure 3–6a). Although the exact mechanism is not known, the following sequence is believed to occur:

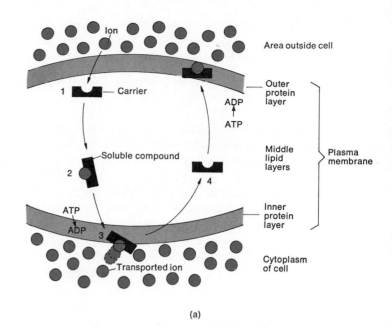

(a)

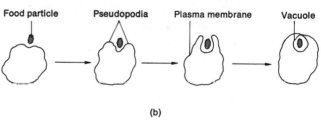

(b)

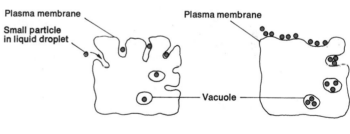

(c)

Figure 3–6. Active processes. (a) Mechanism of active transport. (b) Phagocytosis. (c) Two variations of pinocytosis. In the variation on the left, the ingested substance enters a channel formed by the plasma membrane and becomes enclosed in a vacuole at the base of the channel. In the variation on the right, the ingested substance becomes enclosed in a vacuole that forms and detaches at the surface of the cell.

1. An ion outside the cell membrane is attached to an enzymelike carrier molecule located in or on the cell membrane.
2. The ion-carrier complex forms a compound that is soluble in the lipid portion of the membrane.
3. The compound moves toward the interior portion of the membrane where it is split by enzymes.

4. The ion is then transported into the cell, and the carrier returns to the surface of the membrane to pick up another ion.

The energy for the attachment and release of the carrier molecule is supplied by ATP.

PHAGOCYTOSIS. Another active process by which cells take in substances across the plasma membrane is called **phagocytosis,** or "cell eating" (Figure 3–6b). In this process, projections of cytoplasm, called *pseudopodia,* engulf solid particles exterior to the cell. Once the particle is surrounded, the membrane folds inwardly, forming a membrane sac around the particle. This newly formed sac, called a *digestive vacuole,* breaks off from the outer cell membrane, and the solid material inside the vacuole is digested. Indigestible particles are removed from the cell by a reverse phagocytosis. This process is important because molecules that would normally be restricted from crossing the cell membrane can enter the cell. The phagocytic white blood cells of the body make up an important defense mechanism. Through phagocytosis, the white blood cells destroy bacteria and other foreign substances. More will be said about this process in Chapter 16.

PINOCYTOSIS. The final active process to be discussed is called **pinocytosis,** or "cell drinking." In this process, the engulfed material consists of a liquid rather than a solid (Figure 3–6c). Moreover, no cytoplasmic projections are formed. Instead, the liquid is attracted to the surface of the membrane. The membrane folds inwardly, surrounds the liquid, and detaches from the rest of the intact membrane. Whereas relatively few cells are capable of phagocytosis, many cells may carry on pinocytosis. Examples include cells in the kidneys and urinary bladder.

Unusual plasma membranes

Electron microscope studies have revealed that plasma membranes of certain cells contain a number of modifications. That is, they have different structures for very specific purposes. For example, the membranes of some of the cells lining the small intestines have small, cylindrical projections called **microvilli** (Figure 3–7a). These fingerlike projections enormously increase the absorbing area of the cell surface. A single cell may have as many as 3,000 microvilli, and a square millimeter of intestine may contain as many as 200 million microvilli.

Another membrane modification is found in the rod and cone cells of the eye. These cells serve as photoreceptors, or light-receiving cells. The upper portion of each rod cell contains two-layered, disc-shaped membranes called **sacs** that contain the pigments involved in vision (Figure 3–7b). A final example of a membrane modification is the **myelin sheath** that surrounds portions of nerve cells (Figure 3–7c). It is believed that the myelin sheath protects the nerve and is related to the nutrition of the nerve.

CYTOPLASM

The living matter inside the cell's plasma membrane and external to the nucleus is called **cytoplasm** (Figure 3–8a, b). It is the matrix or ground substance of the cell in which a variety of organelles and inclusions are found. Physically, cytoplasm may be described as a thick, semitransparent, elastic fluid containing suspended particles. Chemically, cytoplasm is 75 to 90 percent water plus solid components. Proteins, carbohydrates, lipids, and inorganic substances comprise the bulk of the solid components. The inorganic substances and most carbohydrates are soluble in water and are present as a true solution. The majority of organic compounds, however, are found as colloids, or particles that remain suspended in the surrounding ground substance. Since the particles of a colloid bear electrical charges that repel each other, they remain suspended and separated from each other.

Functionally, cytoplasm is the substance in which chemical reactions occur. The cytoplasm receives raw materials from the external environment and converts the raw materials into usable energy by decomposition reactions. Cytoplasm is also the site where new substances are synthesized for cellular use. It packages chemicals for transport to other parts of the cell or other cells of the body and facilitates the excretion of waste materials.

ORGANELLES

Despite the myriad chemical activities occurring simultaneously in the cell, there is little interference of one reaction with another. This is so because the cell has a system of compartmentalization that is provided by structures collectively called **organelles.** These structures are specialized portions of the cell that assume various roles in growth, maintenance, repair, and control. An understanding of the structure and function of

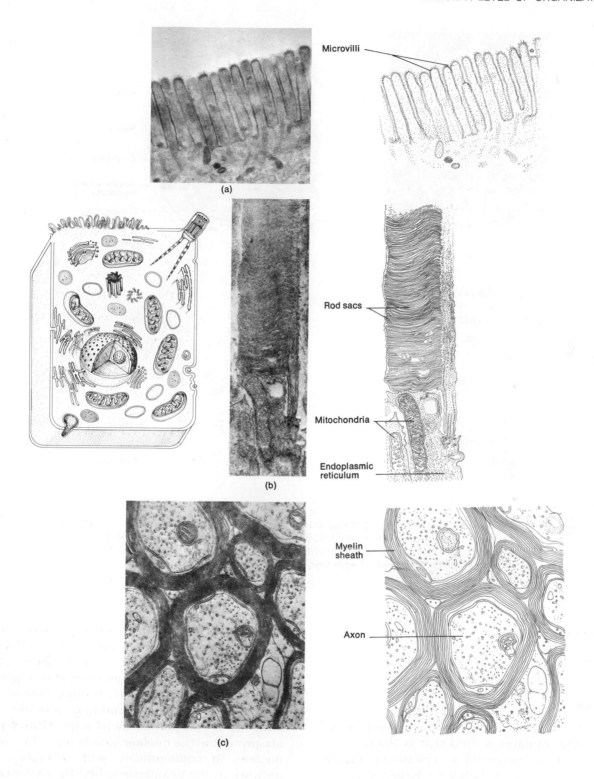

(a)

(b)

(c)

Microvilli

Rod sacs

Mitochondria

Endoplasmic
reticulum

Myelin
sheath

Axon

Figure 3–7. Modified plasma membranes. (a) Microvilli. Appearing on the left is an electron micrograph of a portion of small intestine magnified 20,000X. The labeled diagram appears at the right. (b) Rod cell sacs. On the left is an electron micrograph of a portion of a rod cell of the eye magnified 2,000X. The labeled diagram appears at the right. (c) Myelin sheath. On the left is an electron micrograph of a myelin sheath seen in cross section at a magnification of 20,000X. The labeled diagram is on the right. *(Electron micrographs courtesy of E. B. Sandborn, M.D., Université de Montréal.)*

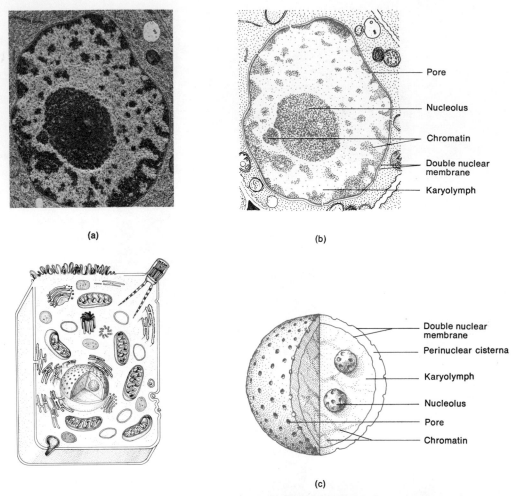

(a)

(b)

Pore

Nucleolus

Chromatin

Double nuclear membrane

Karyolymph

Double nuclear membrane

Perinuclear cisterna

Karyolymph

Nucleolus

Pore

Chromatin

(c)

Figure 3–8. Cytoplasm and the nucleus. (a) Electron micrograph of cytoplasm and the nucleus at a magnification of 31,600X. (b) Labeled diagram of the electron micrograph. (c) Diagrammatic representation of a nucleus with two nucleoli. [*(a) courtesy of Dr. Myron C. Ledbetter, Brookhaven National Laboratory.*]

representative organelles will aid you in understanding subsequent discussions of systems in the body.

The nucleus

Generally a spherical- or oval-shaped organelle, the **nucleus** contains a fluid that is thicker than the fluid of the surrounding cytoplasm (Figure 3–8a, b). In addition to being the largest structure in the cell, the nucleus controls cellular structure, directs many cellular activities, and contains the hereditary factors of the cell, called genes. Certain cells, such as mature red blood cells and cells in the center of the lens of the eye, do not have nuclei. These cells carry on only limited chemical activity and are not capable of growth or reproduction.

Structurally, the nucleus is separated from the cytoplasm by a double membrane called the *nuclear membrane* (Figure 3–8b, c). Between the two layers of the nuclear membrane is a space referred to as the *perinuclear cisterna*. This arrangement of the nuclear membrane resembles the structure of the plasma membrane. Minute pores are present in the nuclear membrane, allowing the nucleus to communicate with a membranous network in the cytoplasm called the endoplasmic reticulum. This network will be described shortly. Substances entering and exiting the nucleus are believed to pass through the tiny pores. Inside the nucleus, three prominent structures are visible. The first of these is a gel-like substance called *karyolymph*. One or more spherical bodies called the *nucleoli* are also present. These structures are composed primarily of RNA and assume

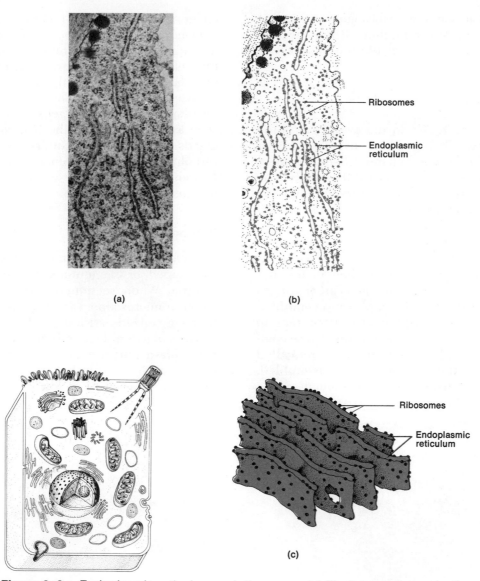

(a)   (b)

Ribosomes

Endoplasmic
reticulum

Ribosomes

Endoplasmic
reticulum

(c)

Figure 3–9. Endoplasmic reticulum and ribosomes. (a) Electron micrograph of the endoplasmic reticulum and ribosomes at a magnification of 76,000X. (b) Labeled diagram of the electron micrograph. (c) Diagrammatic representation of the endoplasmic reticulum and ribosomes. [*(a) courtesy of Dr. Myron C. Ledbetter, Brookhaven National Laboratory.*]

a role in directing protein synthesis. Finally, there is the *genetic material* consisting principally of DNA. When the cell is not reproducing, the genetic material appears as a threadlike mass and is called *chromatin.* Prior to cellular reproduction the chromatin shortens and thickens into rod-shaped bodies called *chromosomes.* These bodies also will be discussed subsequently.

Endoplasmic reticulum and ribosomes

Within the cytoplasm, there is a system consisting of pairs of parallel membranes enclosing narrow cavities of varying shapes. This system is known as the **endoplasmic reticulum,** or **ER** (Figure 3–9). The ER, in other words, is a network of canals running through the entire cytoplasm. These canals are continuous with both the plasma membrane and nuclear membrane. It is believed that the ER provides a surface area for chemical reactions, a pathway for transporting molecules within the cell, and a storage area for synthesized molecules. Attached to the outer surfaces of the ER are exceedingly small, dense, spherical bodies called **ribosomes.** In these areas, the ER is referred to as *granular,* or rough, reticulum. Portions of the ER that have no ribosomes are called *agranular*, or smooth, reticulum. Ribosomes are thought to serve as the sites of protein synthesis in the cell.

Golgi complex

Another structure found in the cytoplasm is the **Golgi complex.** This structure usually consists of four to eight flattened channels, stacked upon each other with expanded areas at their ends. The stacked elements are called *cisternae,* and the expanded, terminal areas are referred to as *Golgi vacuoles* (Figure 3–10). Generally, the Golgi complex is located near the nucleus and is directly connected, in parts, to the ER. The Golgi complex functions in the synthesis of carbohydrates. Recent evidence indicates that carbohydrates synthesized by the Golgi complex are combined with proteins synthesized by the ribosomes to form carbohydrate-protein complexes. These complexes of carbohydrate and protein are called glycoproteins. As the glycoproteins are assembled, they accumulate in the flattened channels of the Golgi complex. The channels expand and form

Golgi vacuoles. After a certain critical size is reached, the vacuoles pinch off from the channel, migrate through the cytoplasm, and pass out of the cell through the plasma membrane. Outside the plasma membrane, the vacuoles rupture and release their contents. The Golgi complex is well developed and highly active in secretory cells such as those found in the pancreas and salivary glands. Essentially, the Golgi complex synthesizes carbohydrates and combines them with proteins. It then packages the resulting glycoprotein and secretes it from the cell.

Mitochondria

Throughout the cytoplasm appear small, spherical, rod-shaped, or filamentous structures called **mitochondria.** When sectioned and viewed under an electron microscope, each reveals an elaborate internal organization (Figure 3–11). A mitochondrion consists of a double membrane similar in structure to the plasma membrane. The outer mitochondrial membrane is smooth, but the inner membrane is thrown into a series of folds called *cristae.* The center of the mitochondrion is referred to as the

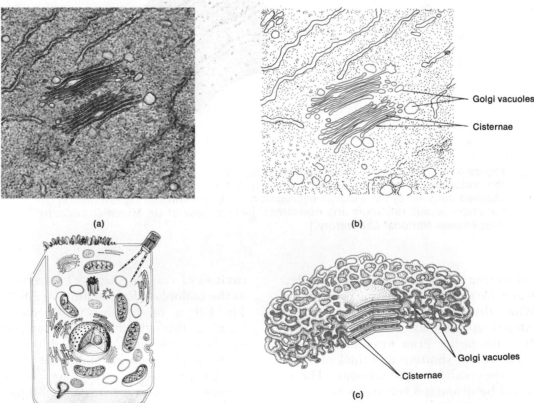

(a)

(b)

Golgi vacuoles

Cisternae

Golgi vacuoles

Cisternae

(c)

Figure 3–10. The Golgi complex. (a) Electron micrograph of two Golgi complexes at a magnification of 78,000X. (b) Labeled diagram of the electron micrograph. (c) Diagrammatic representation of the Golgi complex. Can you find the agranular endoplasmic reticulum in the electron micrograph? [*(a) courtesy of Dr. Myron C. Ledbetter, Brookhaven National Laboratory.*]

matrix. Because of the nature and arrangement of the cristae, the inner membrane provides an enormous surface area for chemical reactions. It is believed that enzymes involved in energy-releasing reactions which form ATP are arranged on the cristae. Mitochondria are frequently called the "powerhouses of the cell" because of the central role they play in the production of ATP.

Lysosomes

When viewed under the electron microscope, **lysosomes** appear as membrane-enclosed spheres somewhat smaller than mitochondria (Figure 3–12). Unlike mitochondria, however, lysosomes have only a single membrane and lack detailed structure. Moreover, they contain powerful digestive enzymes capable of breaking down many kinds of molecules. These enzymes are also capable of digesting bacteria that may enter the cell. White blood cells, which ingest bacteria by phagocytosis, contain large numbers of lysosomes. Scientists have wondered why these powerful enzymes do not also destroy their own

cells. The suspected reason is that the lysosome membrane in a healthy cell is impermeable to enzymes so they cannot move out into the cytoplasm. When a cell is injured, though, the lysosomes release their enzymes. The enzymes then promote reactions that break the cell down into its chemical constituents. The chemical remains are either reused by the body or excreted. Because of this function, lysosomes have been called "suicide packets."

Centrosome

A rather dense area of cytoplasm, generally spherical in shape and located near the nucleus, is called the **centrosome** or **centrosphere**. Within the centrosome is a pair of cylinder-shaped structures, the **centrioles** (Figure 3–13). Each centriole is composed of a ring of nine evenly spaced bundles. Each bundle, in turn, consists of three hollow tubules. The two centrioles are situated so that the long axis of one is at right angles to the long axis of the other. Centrioles assume a role in cell reproduction—a process

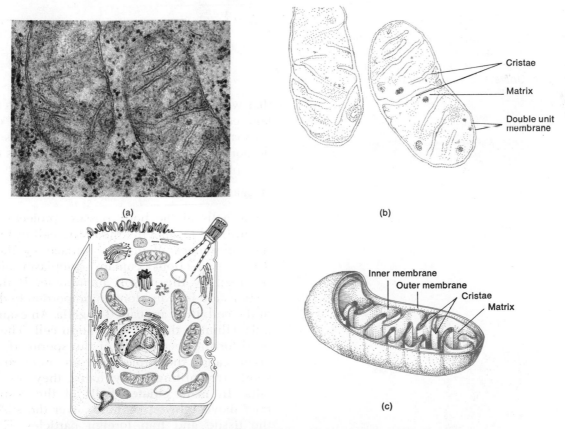

(a)

(b)

(c)

Figure 3–11. Mitochondria. (a) Electron micrograph of two mitochondria at a magnification of 20,000X. (b) Labeled diagram of the electron micrograph. (c) Diagrammatic representation of a mitochondrion. [*(a) courtesy of Dr. Myron C. Ledbetter, Brookhaven National Laboratory.*]

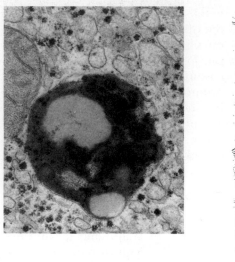

(a)

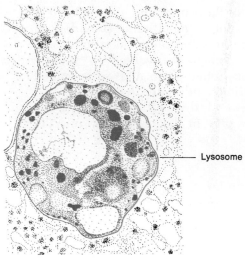

Lysosome

(b)

Figure 3–12. The lysosome. (a) An electron micrograph of a lysosome magnified 55,000X. (b) Labeled diagram of the electron micrograph. [*(a) courtesy of Dr. F. Van Hoof, Universite Catholique de Louvain.*]

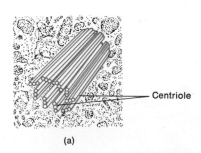

Centriole

(a)

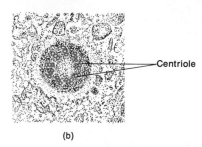

Centriole

(b)

Figure 3–13. Centrioles. Diagrams of centrioles seen in (a) longitudinal and (b) cross sections.

that will be described later in this chapter. Certain cells, such as mature nerve cells, do not possess a centrosome. As a result, these cells do not reproduce. This is why nerve cells cannot be replaced if they are destroyed.

Flagella and cilia

Some cells of the body possess projections that are utilized for moving the entire cell or for moving substances along the surface of the cell. These projections contain cytoplasm and are bounded by the plasma membrane. If the projections are few and long in proportion to the size of the cell, they are called **flagella.** An example of a flagellum is the tail of a sperm cell. The tail is used for the locomotion of the sperm. If, on the other hand, the projections are numerous and short, resembling many hairs, they are called **cilia.** In man, ciliated cells of the respiratory tract move lubricating fluids over the surface of the tissue and trap foreign particles. Electron microscopy has revealed that there is no fundamental structural difference between cilia and flagella (Figure 3–14).

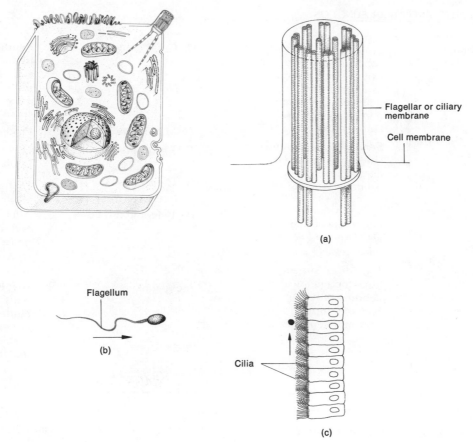

Figure 3–14. Flagella and cilia. (a) Diagrammatic representation of a flagellum or cilium. (b) Flagellum of a sperm cell. (c) Cilia of the respiratory tract moving a particle upward toward the mouth.

Organelles, whether they are nuclei or flagella, work continuously to maintain the life of the cell. As part of their activities, some of the organelles manufacture products that are stored in the cells or secreted. The stored products are called inclusions, whereas the secreted products become part of the extracellular materials that lie outside the cells.

CELL INCLUSIONS

The **cell inclusions** are a rather large and diverse group of chemical substances. These products are principally organic in nature and may appear or disappear at various times in the life of the cell. Some inclusions, such as hemoglobin crystals and melanin, are kept within the cell. *Hemoglobin crystals* lie inside red blood cells. They perform the function of attaching to oxygen molecules and carrying the oxygen to other cells. *Melanin* is a pigment stored in the cells of the skin, hair, and eyes. It protects the body by screening out harmful ultraviolet rays from the sun. Other inclusions are temporarily stored in the cell.

When the body needs them, they are released. One example is *glycogen*, a polysaccharide that is stored in liver and skeletal muscle cells. When the body requires quick energy, the cell organelles break down the glycogen into glucose and release the glucose. *Lipids*, which are stored in fat cells, may be decomposed when the body runs out of carbohydrates for producing energy. Still other inclusions are secreted fairly continuously by the cell. An example is mucus, which is produced by cells that line organs. Its function is to provide lubrication. The major parts of the cell and their functions are summarized in Exhibit 3–1.

EXTRACELLULAR MATERIALS

The substances that lie outside cells are collectively called **extracellular materials.** They include the body fluids, which provide a medium for dissolving, mixing, and transporting substances. They include secreted inclusions like mucus. And they also include some special substances that form the matrix, or mold, in which some cells are embedded.

Exhibit 3–1. CELL PARTS AND THEIR FUNCTIONS

PART	FUNCTIONS	PART	FUNCTIONS
I. Plasma membrane	Protects and allows substances to enter or exit the cell through osmosis, filtration, dialysis, active transport, phagocytosis, and pinocytosis.	D. Golgi complex	Synthesizes carbohydrates, combines carbohydrates with proteins, and packages materials for secretion.
II. Cytoplasm	Serves as the ground substance in which chemical reactions occur.	E. Mitochondria	Produce ATP.
		F. Lysosomes	Digest chemicals and foreign microbes.
III. Organelles		G. Centrioles	Form spindles during cell division.
A. Nucleus	Controls cellular activities and contains genes.	H. Flagella and cilia	Afford movement of cell or movement of particles along surface of cell.
B. Endoplasmic reticulum	Provides a surface area for chemical reactions; provides a pathway for transporting chemicals; serves as a storage area.	IV. Inclusions	Involved in overall body functions. Include materials retained in cell (hemoglobin), reserve materials (glycogen, fats), and secretions (mucus).
C. Ribosomes	Act as sites of protein synthesis.		

The matrix materials are produced by certain cells and are deposited outside their plasma membranes. The matrix supports the cells, binds them together, and gives strength and elasticity to the tissue. Some of the matrix materials are *amorphous*, which means they have no specific shape. These include hyaluronic acid and chondroiten sulfate. *Hyaluronic acid* is a viscous, fluidlike substance that binds cells together, lubricates joints, and maintains the shape of the eyeballs. *Chondroiten sulfate* is a jellylike substance that provides support and adhesiveness in cartilage, bone, heart valves, the cornea of the eye, and the umbilical cord. Other matrix materials are *fibrous*, or threadlike. Fibrous materials provide strength and support for tissues. Among these are **collagen,** or *collagenous fibers*. Collagen is found in all kinds of connective tissue, especially in bones, tendons, and ligaments. **Reticulin,** also called *reticular fibers*, is a matrix material that forms a network around fat cells, nerve fibers, muscle cells, and blood vessels. **Elastin,** found in *elastic fibers*, is a substance that gives elasticity to the skin and to the tissues which form the walls of blood vessels.

CELL DIVISION

Most of the cell activities mentioned thus far maintain the life of the cell on a day-to-day basis. However, cells become damaged, diseased, or wear out and die. Moreover, new cells must be produced for growth. The vital process by which cells are replaced is called cell division.

Cell division is the process by which cells reproduce themselves. For our purposes, assume that cell division or, more appropriately, nuclear division, may be one of two kinds. The first kind of division is the process by which a single parent cell duplicates itself. This process is known as mitosis and cytokinesis. It is the process by which body cells replace themselves. The second kind of division is a mechanism by which sperm and egg cells are produced. This process is called meiosis, which is the mechanism that enables the reproduction of an entirely new organism. Meiosis will be discussed in the chapter on developmental processes and inheritance (Chapter 25).

Mitosis

The function of mitosis and cytokinesis is to replace cells in the body. The process ensures that each new daughter cell has the same *number* and *kind* of chromosomes as the original parent cell. After the process is complete, the two daughter cells have the same hereditary material and genetic potential as the parent cell. This kind of cell division results in an increase in the number of body cells. Mitosis and cytokinesis are, therefore, the means by which dead or injured cells are replaced and also the means by which cells are added for body growth. In a 24-hour period, the average human adult loses about 500 million cells from different parts of the body. Obviously, these cells must be replaced. Cells that have a short life span, such as the cells of the outer layer of skin, the cornea of the eye, and the digestive tract, are continually being replaced. The succession of events that takes place during mitosis and cytokinesis are plainly visible under a microscope after the cells have been stained in the laboratory.

When a cell reproduces itself, it must duplicate its chromosomes so that its heredity may be passed on to succeeding generations of cells. As you may recall from Chapter 2, a **chromosome** is a highly coiled DNA molecule made up of pairs of nucleotides. A chromosome, in turn, consists of **genes,** which are groups of nucleotides on the chromosome. Each human chromosome consists of about 20,000 genes.

The process called **mitosis** is the duplication of chromosomes and the distribution of the two sets of chromosomes into two separate and equal nuclei. For convenience, biologists break down the process into four stages: prophase, metaphase, anaphase, and telophase. These are arbitrary classifications, and mitosis is actually a continuous process, one stage merging imperceptibly into the next. Interphase is the stage that occurs between consecutive cell divisions.

When a cell is carrying on every life process except for division, it is said to be in **interphase** (Figure 3–15a). One of the principal events of interphase is the duplication of DNA. When DNA duplicates, its helical structure partially uncoils (Figure 3–16). Those portions of DNA that remain coiled stain darker than the uncoiled portions. This unequal distribution of stain causes the DNA to appear as a granular mass called **chromatin.** (See Figure 3–15a.) During uncoiling, DNA separates at the points where the nitrogen bases are connected. Each exposed nitrogen base then picks up a complementary nitrogen base (with associated sugar and phosphate group) from the cytoplasm of the cell. This uncoiling and complementary base pairing continues until each of the two original DNA strands is matched and joined with two newly formed DNA strands. The net effect is that the original DNA molecule has become two DNA molecules.

During interphase the cell is also synthesizing most of its RNA and proteins. It is producing chemicals so that all cellular components can be doubled during division. When you look at an interphase cell under a microscope, you will notice that the nucleus has a clearly defined membrane, nucleoli, karyolymph, and chromatin. As interphase progresses, you will also see two pairs of centrioles. The centrioles divide and the resulting two pairs of centrioles separate. Once a cell completes its activities during interphase, mitosis begins.

During **prophase** (Figure 3–15b), the centrioles move apart and project a series of radiating fibers called *asters.* The centrioles move to opposite poles of the cell and become connected by another system of fibers called *spindle fibers.* Simultaneously, the chromatin has been shortening and thickening into chromosomes. The nucleoli have become less distinct, and the nuclear membrane has disappeared. Each prophase "chromosome" is actually composed of a pair of separate structures called *chromatids.* Each chromatid is a complete chromosome, made of a double-stranded DNA molecule. Each chromatid is attached to its chromatid pair by a small spherical body called the *centromere.* During prophase, the chromatid pairs move toward the nuclear region of the cell.

During **metaphase** (Figure 3–15c), the second stage of mitosis, the chromatid pairs line up on the equatorial plane of the spindle fibers. The centromere of each chromatid pair attaches itself to a spindle fiber. The lengthwise separation of the chromatids now takes place. Each centromere divides, and the independent chromatids begin moving to opposite poles of the cell.

Anaphase (Figure 3–15d), the third stage of mitosis, is characterized by the continued movement of complete sets of chromatids, or chromosomes, to opposite poles of the cell. During this movement, the centromeres that are attached to the spindle fibers seem to drag the trailing parts of the chromosomes toward opposite poles.

Telophase (Figure 3–15e), the final stage of mitosis, consists of a series of events that are approximately opposite those of prophase. By now, two identical sets of chromosomes have reached opposite poles. New nuclear membranes begin to enclose them. The chromosomes start to assume their chromatin form. Finally, nucleoli reappear, and the spindle fibers disappear. The formation of two nuclei identical to those in cells of interphase terminates telophase. A mitotic cycle has been completed (Figure 3–15f).

The time required for a complete mitotic cycle varies with the kind of cell, its location, and the influence of certain factors such as temperature. Furthermore, the different stages of mitosis are

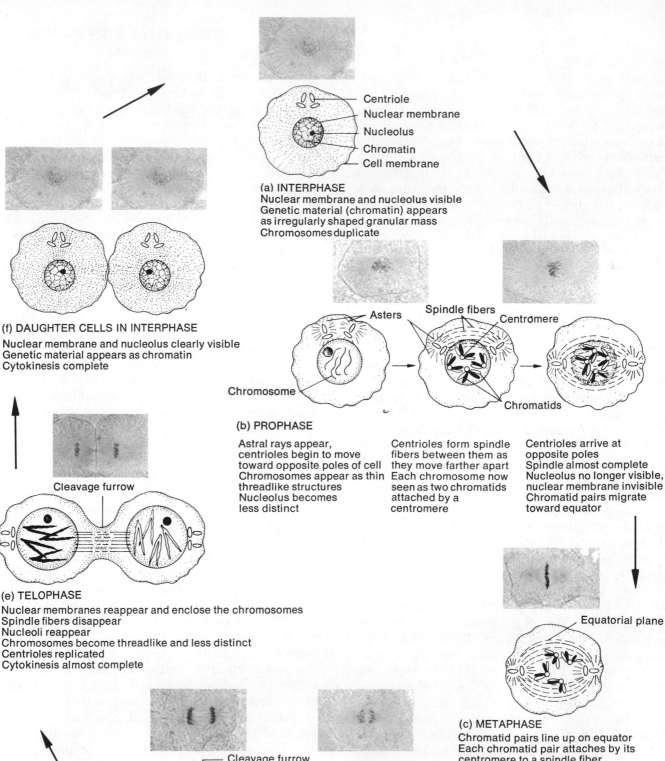

Centriole
Nuclear membrane
Nucleolus
Chromatin
Cell membrane

(a) INTERPHASE
Nuclear membrane and nucleolus visible
Genetic material (chromatin) appears
as irregularly shaped granular mass
Chromosomes duplicate

(f) DAUGHTER CELLS IN INTERPHASE

Nuclear membrane and nucleolus clearly visible
Genetic material appears as chromatin
Cytokinesis complete

Asters   Spindle fibers   Centromere

Chromosome

Chromatids

(b) PROPHASE

Astral rays appear,
centrioles begin to move
toward opposite poles of cell
Chromosomes appear as thin
threadlike structures
Nucleolus becomes
less distinct

Centrioles form spindle
fibers between them as
they move farther apart
Each chromosome now
seen as two chromatids
attached by a
centromere

Centrioles arrive at
opposite poles
Spindle almost complete
Nucleolus no longer visible,
nuclear membrane invisible
Chromatid pairs migrate
toward equator

Cleavage furrow

(e) TELOPHASE

Nuclear membranes reappear and enclose the chromosomes
Spindle fibers disappear
Nucleoli reappear
Chromosomes become threadlike and less distinct
Centrioles replicated
Cytokinesis almost complete

Equatorial plane

(c) METAPHASE

Chromatid pairs line up on equator
Each chromatid pair attaches by its
centromere to a spindle fiber
at equator of spindle
Lengthwise separation of chromatid pairs
occurs and centromeres divide

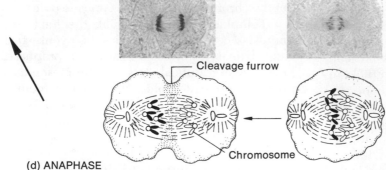

Cleavage furrow

Chromosome

(d) ANAPHASE

Two complete sets of single-stranded
chromosomes migrate toward opposite
poles of the cell
Cytokinesis may start

Chromosomes move toward opposite poles of cell

Figure 3–15. Cell division: mitosis and cytokinesis. Shown here are photomicrographs (photos taken through a light microscope) and diagrammatic representations of the various stages of division in whitefish eggs. Read the sequence
starting at (a), and move clockwise until you complete the cycle. *(Photomicrographs courtesy of Carolina Biological Supply Company.)*

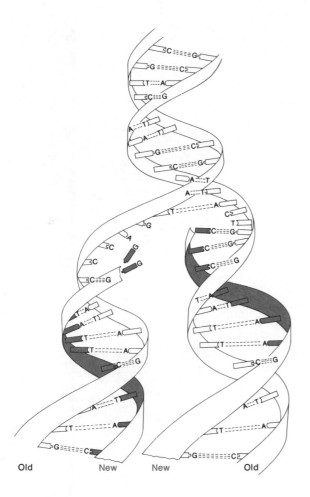

Figure 3–16. Duplication of DNA. Each strand of the double helix separates by breaking the bonds between nucleotides. New nucleotides attach at the proper sites, and two new strands of DNA are paired off with the two old strands. After duplication, the two DNA molecules, each consisting of a new and an old strand, return to their helical structure.

not equal in duration. Prophase is usually the longest stage, lasting from one to several hours. Metaphase is considerably shorter, ranging from 5 to 15 minutes. Anaphase is the shortest stage, lasting from 2 to 10 minutes. Telophase lasts from 10 to 30 minutes. These lengths of time are only relative, however, and should not be taken as exact limits.

Cytokinesis

The division of the cytoplasm, called **cytokinesis,** often begins in late anaphase and terminates in telophase. Cytokinesis begins with the formation of a cleavage furrow that runs around the cell at its equator. The furrow progresses inward, re-

sembling a constricting ring, and cuts completely through the cell to form two separate portions of cytoplasm (Figure 3–15d to f).

GENE ACTION

Mitosis and cytokinesis are vital processes because they ensure that the daughter cells have identical sets of chromosomes and, therefore, identical genes. Let us take a look at genes and see why they are so important to life. A **gene** is a group of nucleotides on a DNA molecule that serves as the master mold for manufacturing a specific protein. Each gene consists of about 1,000 pairs of nucleotides, which appear in a *specific* sequence on the DNA molecule. No two genes have exactly the same sequence of nucleotides. This fact is the key to heredity. Gene action is thought to occur in the following way.

Each gene directs the manufacture of a particular kind of RNA. The gene segments of DNA manufacture RNA in much the same fashion as the entire DNA molecule duplicates itself during mitosis (Figure 3–17a). The major differences between DNA and RNA synthesis are the following:

1. In RNA synthesis, the sugar ribose is used instead of deoxyribose, and the nitrogen base uracil is used instead of thymine.
2. Each RNA molecule is molded by only a segment of the DNA molecule. The RNA molecule, therefore, does not have nearly the number of nucleotides as the DNA molecule has.

Several different types of RNA are manufactured by genes. We shall look at two of them, messenger RNA and transfer RNA. After a molecule of *messenger RNA (mRNA)* is produced, it leaves the nucleus and travels to a ribosome. Meanwhile, a number of *transfer RNA (tRNA)* molecules also leave the nucleus. Each *t*RNA molecule migrates into the cytoplasm and attaches itself to an amino acid. It should be noted that each kind of *t*RNA attaches to only one kind of amino acid (Figure 3–17b). The *t*RNA molecules then carry their amino acids to the *m*RNA that is lying near the ribosome (Figure 3–17c). The *m*RNA serves as a mold for the manufacture of a protein. This resembles somewhat the process by which DNA serves as a mold for the production of RNA. Each group of nitrogen bases on the *m*RNA attracts one particular kind of *t*RNA with its amino acid and no other. The sequence and kind of bases in the *m*RNA group determine

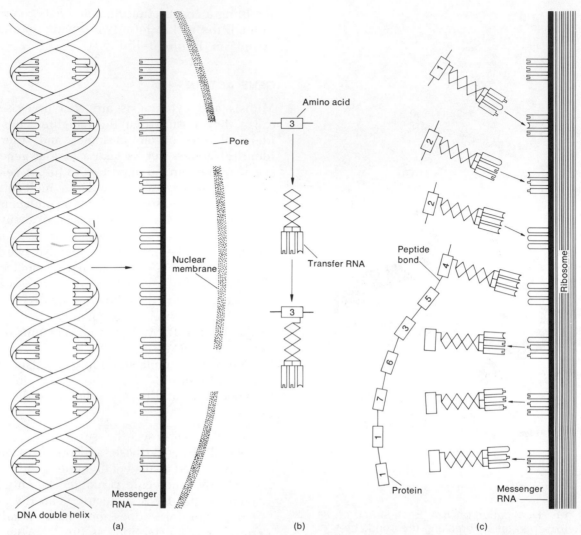

Figure 3–17. Protein synthesis. (a) In the nucleus, the DNA helix separates, and one strand of the helix serves as a mold for the formation of messenger RNA. The messenger RNA leaves the nucleus and moves to a ribosome. (b) In the cytoplasm, transfer RNA molecules attach to amino acids. Each specific kind of transfer RNA attaches to a specific amino acid. (c) Transfer RNAs bring their amino acids to the messenger RNA on the ribosome. The transfer RNAs temporarily bond to the messenger RNA according to a specific pairing of nitrogen bases. In the process, the amino acids are brought into close contact and peptide bonds form between them. The chain of bonded amino acids is the synthesized protein.

which *t*RNA molecules will be attracted. Thus, the *m*RNA determines the order in which the amino acids line up. As the *t*RNAs pair off with their respective *m*RNA groups, a line of amino acids forms alongside the *m*RNA molecule. Peptide bonds are then formed between the amino acids, and a protein is made.

The important point to remember is that the base sequence of the gene determines the sequence of the bases in the *m*RNA. The sequence of the bases in the *m*RNA then determines the order and kind of amino acids that will go into the protein. Thus, each gene is responsible for making a particular type of protein.

You may ask: "What do proteins have to do with heredity?" The answer is "Everything." Enzymes, as you remember, are proteins. Very few chemical reactions can take place in the body without the proper enzyme to act as a catalyst. The chemical reactions involved in manufacturing cilia, for instance, require special types of enzymes. Other enzymes are required for forming cell membranes, microvilli, and so on. Thus, the enzymes manufactured in our cells largely determine the structure and function of our cells. In turn, the structure and functions of our individual cells determine the appearance of our bodies and the activities that we can perform.

Chapter summary in outline

THE GENERALIZED CELL

1. A cell is the basic, living, structural and functional unit of the body.
2. A generalized cell is a composite that represents various cells of the body.
3. Cytology is the science concerned with the study of cells.
4. The principal parts of a cell are the plasma membrane, cytoplasm, organelles, and inclusions. Extracellular materials are manufactured by the cell and deposited outside the plasma membrane.

THE CELL SURFACE

1. The plasma membrane, or cell membrane, surrounds the cell and separates it from other cells and the external environment.
2. The plasma membrane is composed of two layers of proteins surrounding one layer of lipids. It is believed that the membrane contains pores.
3. The semipermeable nature of the membrane restricts the passage of certain substances. Substances can pass through the membrane depending on their molecular weight, lipid solubility, charges, and carriers.

Moving materials across plasma membranes

1. Passive processes involve the kinetic energy of individual molecules.
2. Diffusion is the net movement of molecules or ions from an area of higher concentration to an area of lower concentration until an equilibrium is reached.
3. Osmosis is the movement of water through a semipermeable membrane from an area of higher water concentration to an area of lower water concentration.
4. Filtration is the movement of water and dissolved substances across a semipermeable membrane by pressure.
5. Dialysis is the separation of small molecules from large molecules by diffusion and osmosis through a semipermeable membrane.
6. Active processes involve the use of ATP by the cell.
7. Active transport is the movement of ions across a cell membrane from lower to higher concentration. This process relies upon the participation of carriers.
8. Phagocytosis is the ingestion of solid particles by pseudopodia. It is an important process used by white blood cells to destroy bacteria that enter the body.
9. Pinocytosis is the ingestion of a liquid by the plasma membrane. In this process, the liquid becomes surrounded by a vacuole.

Unusual plasma membranes

1. The membranes of certain cells are structured for specific functions.
2. Microvilli are fingerlike projections of the digestive system that increase the surface area for absorption of food.
3. Rod cells of the eye contain sacs that pick up light for vision.
4. The myelin sheath of nerve cells protects, aids impulse conduction, and provides nutrition.

CYTOPLASM

1. The cytoplasm is the living matter inside the cell that contains organelles and inclusions.
2. It is composed mostly of water plus proteins, carbohydrates, lipids, and inorganic substances. The chemicals in cytoplasm are either in solution or in a colloid, or suspended, form.
3. Functionally, cytoplasm is the medium in which chemical reactions occur.

ORGANELLES

1. These are specialized structures that carry on specific activities.

The nucleus

1. Usually the largest organelle, the nucleus controls cellular activities.
2. Cells without nuclei, such as mature red blood cells, do not grow or reproduce.
3. The parts of the nucleus include the nuclear membrane, nucleoli, and genetic material (DNA).

Endoplasmic reticulum and ribosomes

1. The ER is a network of parallel membranes, continuous with the plasma membrane and nuclear membrane.
2. It functions in chemical reactions, transportation, and storage.
3. Granular ER has ribosomes attached to it. Agranular ER does not contain ribosomes. Ribosomes are small spherical bodies that serve as sites of protein synthesis.

Golgi complex

1. This structure consists of four to eight flattened channels stacked on each other.
2. In conjunction with the ER, the Golgi complex synthesizes glycoproteins.
3. It is particularly prominent in secreting cells such as those in the pancreas or salivary glands.

Mitochondria

1. These structures consist of a smooth outer membrane and a folded inner membrane. The inner folds are called cristae.
2. The mitochondria are called "powerhouses of the cell" because ATP is produced within them.

Lysosomes

1. Lysosomes are spherical structures containing digestive enzymes.
2. They are found in large numbers in white blood cells, which carry on phagocytosis.
3. If the cell is injured, lysosomes release enzymes and digest the cell. For this reason, they are called "suicide packets."

Centrosome

1. The dense area of cytoplasm containing the centrioles is called a centrosome.
2. Centrioles are paired cylinders arranged at right angles to one another. They assume an important role in cell reproduction.

Flagella and cilia

1. These cell projections have the same basic structure and are used in movement.
2. If projections are few and long, they are called flagella. If they are numerous and hairlike, they are called cilia.
3. The flagellum on a sperm cell serves to move the entire cell. The cilia on cells of the respiratory tract move foreign matter along the cell surfaces toward the throat for elimination.

CELL INCLUSIONS

1. These chemical substances are produced by cells. They may be stored, may participate in chemical reactions, and may have recognizable shapes.

2. Examples of cell inclusions are glycogen, hemoglobin crystals, mucus, and melanin.

EXTRACELLULAR MATERIALS

1. These are all the substances that lie outside the cell membrane.

2. They provide support and a medium for the diffusion of nutrients and wastes.

3. Some, like hyaluronic acid, are amorphous, or have no shape. Others, like collagen, are fibrous, or threadlike.

CELL DIVISION

1. The kind of cell division that results in the formation of new cells is called mitosis and cytokinesis. Cell division that results in the production of sperm and egg cells is termed meiosis.

2. Mitosis and cytokinesis replace and add body cells. Prior to mitosis and cytokinesis, the DNA molecules, or chromosomes, replicate themselves so that the same chromosomal complement can be passed on to future generations of cells.

3. A cell carrying on every life process except division is said to be in interphase.

4. Mitosis, division of the nucleus, consists of prophase, metaphase, anaphase, and telophase.

5. Cytokinesis, division of the cytoplasm, occurs in late anaphase and terminates in telophase.

GENE ACTION

1. Genes consist of about 1,000 pairs of nucleotides on a DNA molecule.

2. DNA directs the manufacture of proteins through RNA as follows:

   (a) DNA unwinds and the unpaired nucleotides serve as a mold for the synthesis of RNA. This RNA is called messenger RNA.

   (b) Messenger RNA travels to a ribosome. Other kinds of RNA, called transfer RNA, attach to amino acids and move to the ribosome.

   (c) At the ribosome, transfer RNA attaches the amino acids to the messenger RNA according to the sequence determined by the DNA in the gene.

   (d) Proteins thus synthesized may serve as enzymes.

Review questions and problems

1. Define a cell. What are the five principal portions of a cell? What is meant by a generalized cell?

2. Discuss the structure of the plasma membrane. What factors determine the permeability of the plasma membrane? How are plasma membranes modified for various functions?

3. What are the major differences between active processes and passive processes in moving substances across plasma membranes?

4. Define and give an example of each of the following: diffusion, osmosis, filtration, active transport, phagocytosis, and pinocytosis.

5. Compare the effect on red blood cells of an isotonic, hypertonic, and hypotonic solution.

6. Discuss the chemical composition and physical nature of cytoplasm. What is its function?

7. What is an organelle? By means of a diagram, indicate the structure and describe the function of the following organelles: nucleus, endoplasmic reticulum, ribosomes, Golgi complex, mitochondria, lysosomes, centrioles, cilia, and flagella.

8. Define a cell inclusion. Provide examples and indicate their functions.

9. What is an extracellular material? Give examples and the functions of each.

10. How does DNA duplicate itself?

11. Discuss mitosis and cytokinesis with regard to stages. What are the characteristics of each stage, the relative duration, and the importance?

12. Summarize the steps involving gene action in protein synthesis.

13. To increase your knowledge of the parts of a cell, label the cell shown below. In addition, summarize the function of each part. When you have finished, refer to Figure 3–1 to see how well you labeled the cell.

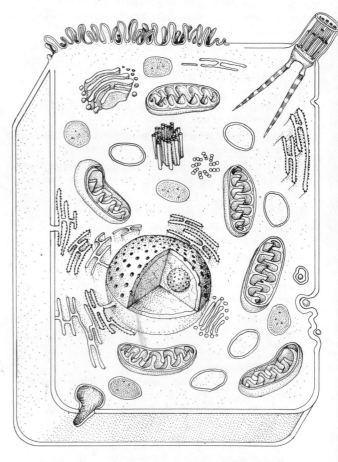

CHAPTER 4

TISSUE LEVEL OF ORGANIZATION

STUDENT OBJECTIVES

After you have read this chapter, you should be able to:

1. Define a tissue

2. Classify the tissues of the body into four major types

3. Describe the distinguishing characteristics of epithelial tissue

4. Contrast the structural and functional differences of covering, lining, and glandular epithelium

5. Compare the shape of cells and the layering arrangements of covering and lining epithelium

6. List the structure, function, and location of simple, stratified, and pseudostratified epithelium

7. Define a gland

8. Distinguish between exocrine and endocrine glands

9. Describe the distinguishing characteristics of connective tissue

10. Contrast the structural and functional differences between embryonal and adult connective tissues

11. Describe the ground substance, fibers, and cells that comprise connective tissue

12. List the structure, function, and location of loose connective tissue, adipose tissue, collagenous, elastic, and reticular connective tissue

13. List the structure, function, and location of the three types of cartilage

14. Define a membrane

15. List the location and function of mucous, serous, synovial, and cutaneous membranes

16. List and describe the symptoms of tissue inflammation

17. Outline the stages of the inflammatory response

18. Describe the conditions necessary for tissue repair

19. Contrast the regenerative capacities of various tissues of the body

20. Describe the importance of nutrition, adequate circulation, and age to tissue repair

The cells discussed in the preceding chapter are highly organized units, but they do not function as isolated units. Instead, they work together in a group of somewhat similarly constructed cells, called a tissue. We shall now examine how the body operates at the tissue level of organization.

A **tissue** is an aggregation of cells working together to perform a specialized activity. For example, some tissues of the body function in moving body parts. Others move food through body organs. Some tissues protect and support the body. And still others function to produce chemicals such as enzymes and hormones. Depending on their functions and structure, the various tissues of the body are classified into four principal types: (1) epithelial tissue, which covers body surfaces, lines body cavities, and forms glands; (2) connective tissue, which protects and supports the body and its organs and binds organs together; (3) muscular tissue, which is responsible for movement; and (4) nervous tissue, which initiates and transmits nerve impulses that coordinate body activities. Some of these tissues will be discussed in later chapters as parts of a particular system. Others, such as epithelial tissue and most connective tissues, will be treated in detail in this chapter.

EPITHELIAL TISSUE

The tissues falling into this main category carry out many activities in the body, ranging from protection to secretion. **Epithelial tissue,** or more simply **epithelium,** may be divided into two subtypes: (1) *covering and lining epithelium* and (2) *glandular epithelium.* (See Exhibit 4–1.) Covering and lining epithelium forms the outer covering of external body surfaces and the outer covering of some internal organs. It lines the body cavities and the interiors of the respiratory and digestive tracts, blood vessels, and ducts. Along with nervous tissue, it comprises the parts of the sense organs that are sensitive to stimuli such as light and sound. Glandular epithelium constitutes the secreting portion of glands.

Both types of epithelium consist largely or entirely of closely packed cells that contain very little intercellular material. In other words, there is little fluid or few fibers between cells. In addition, the epithelial cells are arranged in continuous sheets that may be either single or multilayered. Nerves run through these sheets, but blood vessels do not. The vessels that supply nutrients and remove wastes are located in underlying connective tissue. Epithelium overlies and adheres firmly to connective tissue, which holds the epithelium in position and prevents it from being torn. The surface of attachment between epithelium and connective tissue is a thin layer of modified connective tissue called the *basement membrane.* Since all epithelium is subjected to a certain degree of wear, tear, and injury, its cells can divide and produce new cells to replace those that are destroyed. These general characteristics are found in both types of epithelial tissue. We can now see how the two types differ, by first looking at covering and lining epithelium.

Covering and lining epithelium

Covering and lining epithelium is arranged in several different ways, and the arrangement is related to its location and function. If the epithelium is specialized for absorption or filtration and is located in an area that has minimal wear and tear, the cells of the tissue are arranged in a single layer. Such an arrangement is called *simple epithelium.* If the epithelium is not specialized for absorption or filtration and is found in an area with a high degree of wear and tear, then the cells are stacked in several layers. This tissue is referred to as *stratified epithelium.* A third, less common arrangement of epithelium is called *pseudostratified.* Like simple epithelium, pseudostratified epithelium has only one layer of cells. However, some of the cells do not reach the surface—an arrangement that gives the tissue a multilayered, or stratified, appearance. (See Exhibit 4–1.) The pseudostratified cells that do reach the surface either secrete mucus or contain cilia that move mucus and foreign particles for eventual elimination from the body.

In addition to classifying covering epithelium according to the number of its layers, it may also be categorized by cell shape. The cells may be flat, cubelike, columnar or may resemble a cross between shapes. *Squamous* cells are flattened, scalelike, and fitted together to form a mosaic. *Cuboidal* cells are cube-shaped when viewed in cross section. *Columnar* cells are long and cylindrical, appearing as rectangles set on their ends. *Transitional* cells look like a combination of shapes and are found where there is a great degree of distention or expansion in the body. Transitional cells on the bottom layer of an epithelial tissue may range in shape from cuboidal to columnar. In the intermediate layer, they may be cuboidal or polyhedral. Transitional cells in the superficial layer may range from cuboidal to squamous, depending on how much they are pulled out of shape during certain body functions.

Considering layers and cell type in combination, we may classify covering and lining epithelium as follows:

1. Simple
   (*a*) Squamous
   (*b*) Cuboidal
   (*c*) Columnar
2. Stratified
   (*a*) Squamous
   (*b*) Cuboidal
   (*c*) Columnar
   (*d*) Transitional
3. Pseudostratified

Exhibit 4–1. SUMMARY OF EPITHELIAL TISSUES

NAME AND DESCRIPTION	DIAGRAM	MICROSCOPIC APPEARANCE	LOCATION AND FUNCTION
I. Covering and lining epithelium			In areas where filtration and absorption occur; these areas generally receive minimal wear and tear. Filtration, secretion, and absorption.
A. Simple Cells are arranged into a single layer.			
1. Squamous Single layer of flat, scalelike cells; large, centrally located nucleus.			Lines air sacs of lungs, glomerular capsule of kidneys, crystalline lens of eyes, and eardrum. Called endothelium when it lines heart, blood, and lymph vessels and forms capillaries. Called mesothelium when it lines body visceral organs. Filtration and absorption.
2. Cuboidal Single layer of cube-shaped cells; centrally located nucleus.			Covers surface of ovary; lines inner surface of cornea and lens of eye, kidney tubules, and smaller ducts of many glands. Secretion and absorption.
3. Columnar Single layer of rectangular cells; nuclei at bases of cells.			Nonciliated variety contains goblet cells. This type lines stomach, small and large intestines, digestive glands, and gallbladder. Secretion and absorption.

Key to labels: (1) nucleus, (2) basement membrane, (3) connective tissue layer, (4) goblet cell, (5) ciliated cell.

NAME AND DESCRIPTION	DIAGRAM	MICROSCOPIC APPEARANCE	LOCATION AND FUNCTION
3. Columnar *(cont'd.)*			Ciliated variety also contains goblet cells. It lines upper respiratory tract, fallopian tubes, and ducts of the testes. Moves mucus by ciliary action.
B. Stratified Cells are arranged in several layers.			In areas where no filtration and absorption occur; generally receive maximal wear and tear. Protection, secretion, distention, waterproofing.
1. Squamous Several layers of cells; deeper layers are cuboidal to columnar; superficial layers are flat and scalelike; basal cells replace surface cells as they are shed.			Nonkeratinizing variety lines wet surfaces such as mouth, esophagus, part of epiglottis, and vagina. Protection. Keratinizing variety forms outer layer of skin. Protection.
2. Cuboidal Two or more layers of cube-shaped cells.			Ducts of adult sweat glands. Protection.
3. Columnar Several layers of polyhedral cells; only superficial layer is columnar.			Lines part of male urethra and some larger excretory ducts. Protection and secretion.
4. Transitional Resembles stratified squamous nonkeratinizing tissue, except superficial cells are larger and more rounded.			Lines urinary tract. Permits distention.

Exhibit 4–1. SUMMARY OF EPITHELIAL TISSUES *(cont'd.)*

NAME AND DESCRIPTION	DIAGRAM	MICROSCOPIC APPEARANCE	LOCATION AND FUNCTION
C. Pseudostratified Not a true stratified tissue; nuclei of cells are present at different levels; some cells do not reach surface.	(5) (1) (4) (2) (3)		Lines larger excretory ducts of many large glands and male urethra; ciliated variety with goblet cells lines larger respiratory passages and some ducts of male reproductive system. Secretion and movement of mucus by ciliary action.
II. Glandular epithelium **A. Exocrine glands** Secrete products into ducts.	Duct Secreting cells		Sweat, oil, and wax glands of the skin; digestive glands such as salivary glands, which secrete into mouth cavity. Produce perspiration, oil, and wax; produce digestive enzymes.
B. Endocrine glands Secrete products (hormones) directly into blood.	Blood capillaries Secreting cells		Pituitary gland at base of brain; thyroid gland near voice box; adrenal glands above kidneys. Produce hormones that regulate various body activities.

Key to labels: (1) nucleus, (2) basement membrane, (3) connective tissue layer, (4) goblet cell, (5) ciliated cell.
Photomicrograph sources: Simple squamous (×450), stratified squamous (×500), stratified transitional (×100), pseudostratified (×450), exocrine and endocrine (×400) courtesy of Dr. Donald I. Patt, from *Comparative Vertebrate Histology*, Donald I. Patt and Gail R. Patt, Harper & Row, Publishers, Inc., New York, 1969. Simple cuboidal and columnar courtesy of Carolina Biological Supply Company. Stratified cuboidal and columnar courtesy of Dr. Edward J. Reith, from *Atlas of Descriptive Histology*, Edward J. Reith and Michael H. Ross, Harper & Row, Publishers, Inc., New York, 1970.

Simple epithelium

SIMPLE SQUAMOUS EPITHELIUM. This type of simple epithelium consists of a single layer of flat, scalelike cells. When viewed from the surface, this epithelium resembles a tiled floor (Exhibit 4–1). The nucleus of each cell is centrally located and is oval or spherical in shape. Since simple squamous epithelium has only one layer of cells, it is highly adapted to the functions of diffusion and filtration. Thus, we find simple squamous epithelium lining the air sacs of the lungs, where oxygen is exchanged with carbon dioxide. It is present in the part of the kidney that filters the blood. It is also found in very delicate structures such as the crystalline lens of the eye and the lining of the eardrum. Simple squamous epithelium is found in parts of the body that have little wear or tear. A tissue that is very similar to simple squamous epithelium is called endothelium. *Endothelium* lines the heart and the blood and lymph vessels and forms the walls of capillaries. The term *mesothelium* is applied to another simple squamous epithelium-like tissue that lines the ventral body cavity and covers the viscera. Strictly speaking, endothelium and mesothelium are not epithelial tissue because their origins in the embryo are different. But for all practical purposes, you may consider them to be simple squamous epithelium.

SIMPLE CUBOIDAL EPITHELIUM. When viewed from the top, the cells of simple cuboidal epithelium appear as polygons fitted closely together. The cuboidal nature of the cells is obvious only when the tissue is sectioned at right angles (Exhibit 4–1). Like simple squamous epithelium, these cells possess a centrally located nucleus. Simple cuboidal epithelium is found covering the surfaces of the ovaries and lining the inner surfaces of the cornea. It lines the kidney tubules, where water is reabsorbed into the blood, and lines the smaller ducts and secreting units of many glands, such as the thyroid. This tissue performs the functions of secretion and absorption.

SIMPLE COLUMNAR EPITHELIUM. The surface view of simple columnar epithelium is similar to that of simple cuboidal tissue. When sectioned at right angles to the surface, however, the cells appear somewhat rectangular. The nuclei are located near the bases of the cells (Exhibit 4–1). Simple columnar epithelium is modified in several ways, depending on its location and function in the body. Simple columnar epithelium lines the stomach, the small and large intestines, the digestive glands, and the gallbladder. In such sites, the cells protect the underlying tissues. Many of them are also modified so that they can aid in food-related activities. In the small intestine especially, the plasma membranes of the cells are folded into many fingerlike projections called *microvilli*. (See Figure 3–7a.) The microvilli arrangement increases the surface area of the plasma membrane and thereby allows digested nutrients and fluids to diffuse into the body at a faster rate. Interspersed among the typical columnar cells of the intestine are other modified columnar cells called *goblet cells*. These cells, which secrete mucus, are so named because the mucus accumulates in the upper half of the cell, causing the area to bulge out. The whole cell resembles a goblet or wine glass. Mucus in this tissue serves as a lubricant that prevents friction between the food and the walls of the digestive tract. A third modification of columnar epithelium is found in cells with hairlike processes called *cilia*. In portions of the upper respiratory tract, ciliated columnar cells are interspersed with goblet cells. Mucus secreted by the goblet cells forms a film over the respiratory surface. This film traps foreign particles that are breathed in. The cilia, which wave in unison, move the mucus and foreign particles toward the throat, where it can be swallowed or spit out. In this way, only air is allowed to enter the lungs. Ciliated columnar epithelium, combined with goblet cells, is also found in the uterus and uterine tubes of the female reproductive system and in certain ducts of the testes of the male.

Stratified epithelium

In contrast to simple epithelium, stratified epithelium consists of at least two layers of cells. This means that stratified epithelium is relatively durable and can protect underlying tissues from the outside environment and from wear and tear. Some stratified epithelium cells are also involved in secretion.

STRATIFIED SQUAMOUS EPITHELIUM. In the layers of this type of epithelium, the more superficial cells are flat, whereas the deeper cells vary in shape from cuboidal to columnar (Exhibit 4–1). The basal, or bottom, cells are continually multiplying by cell division. As the newly produced cells grow in size, they compress the cells on the surface and push them outward. According to this growth pattern, basal cells continually shift upward and outward. As they move farther away from the deep layer and their source of blood, they become dehydrated, shrink, and grow harder. Once at the surface, the cells are rubbed off. New cells continually emerge, are sloughed off, and replaced.

One form of stratified squamous is called *stratified squamous nonkeratinizing epithelium*. This type of tissue is found on wet surfaces that are subjected to considerable wear and tear and that do not perform the function of absorption. Such surfaces include the insides of the mouth, the gullet, and the vagina. Another form of stratified squamous is called *stratified squamous keratinizing epithelium*. The surface cells of this type of epithelium are modified into a tough, resistant layer of material containing keratin. *Keratin* is a protein that is waterproof, resistant to friction, and relatively impervious to bacterial invasion. The outer layer of skin consists of this tissue.

STRATIFIED CUBOIDAL EPITHELIUM. This relatively rare type of epithelium is found primarily in the ducts of sweat glands of adults (Exhibit 4–1). It sometimes consists of more than two layers of cells. Its function is mainly protective.

STRATIFIED COLUMNAR EPITHELIUM. Like stratified cuboidal, this type of tissue is also relatively infrequent in the body. Usually the basal layer or layers consist of shortened, irregularly polyhedral cells. Only the superficial cells are columnar in form (Exhibit 4–1). This kind of epithelium lines part of the male urethra and some larger excretory ducts. It functions in protection and secretion.

TRANSITIONAL EPITHELIUM. This kind of epithelium is very much like stratified squamous nonkeratinizing epithelium. The distinction is that the outer layer of cells in transitional epithelium tend to be large and rounded rather than flat (Exhibit 4–1). This feature allows the tissue to be stretched without the danger of the outer cells breaking apart from one another. When stretched, they are drawn out into squamouslike cells. Because of this arrangement, transitional epithelium lines hollow structures that are subjected to expansion from within, such as the urinary bladder. Its obvious function is to prevent a rupture of the epithelium.

Pseudostratified epithelium

The third category of covering and lining epithelium is called *pseudostratified epithelium.* The nuclei of the cells in this kind of tissue are at varying positions, some toward the surface and some in the basal region. Even though all the cells are attached to the basement membrane in a single layer, some of the cells do not reach the surface (Exhibit 4–1). This gives the impression of a multi-layered tissue, which is the reason for the designation *pseudo*stratified epithelium. It lines the larger excretory ducts of many glands and parts of the male urethra. In addition, pseudostratified epithelium may be ciliated and found with goblet cells. In this form, it lines the larger respiratory passages and some of the ducts of the male reproductive system.

The various kinds of epithelium that we have discussed so far are arranged in sheets which cover or line a body surface. Even though some of the covering and lining cells release products such as mucus, secretion is just a secondary function of this kind of epithelium. We shall now discuss a type of epithelium whose primary function is secretion.

Glandular epithelium

The function of glandular epithelium is secretion, which is accomplished by glandular cells that lie in clusters below the covering and lining epithelium. A **gland** can be one cell or a group of highly specialized epithelial cells that secrete substances either into ducts or into the blood. The production of such substances always requires active work by the glandular cells and results in an expenditure of energy. On the basis of this distinction, all glands of the body are classified as exocrine or endocrine glands (Exhibit 4–1). *Exocrine glands* secrete their products into ducts or tubes that empty out at the surface of the covering and lining epithelium. The product of an exocrine gland may be released at the surface of the skin or released into a hollow organ. The secretions of exocrine glands include enzymes, oil, and sweat. But these glands never secrete hormones. Examples of exocrine glands are sweat glands, which eliminate water, some ions, and nitrogenous wastes, and salivary glands, which secrete a digestive enzyme. *Endocrine glands,* by contrast, are ductless and, consequently, must secrete their products directly into the blood. The secretions of endocrine glands are always hormones. Examples of endocrine glands include the pituitary, thyroid, and adrenal glands.

Epithelial tissues cover body surfaces and line body cavities and organs. In these locations, epithelium functions in protection, filtration, absorption, and distention. Epithelium, as a component of glands, also produces various secretions such as hormones, digestive enzymes, sweat, and oil. Connective tissues, in comparison, not only protect but also provide shape for body organs, support body organs, and attach organs to each other. We shall now examine some of the structural characteristics, locations, and functions of connective tissue.

CONNECTIVE TISSUE

The most abundant tissue in the body is **connective tissue.** This binding and supporting tissue usually has a rich blood supply. The cells are widely scattered, rather than closely packed. And the tissue has a great deal of intercellular material, called the *matrix.* In contrast to epithelium, connective tissues do not occur on free surfaces, such as the surfaces of a body cavity or the external surface of the body. The general functions of connective tissues are protection, support, and the binding together of various organs.

The intercellular substance in a given connective tissue largely determines the qualities of the tissue. These substances are nonliving and may consist of fluid, semifluid, or mucoid (mucus-like) material. In cartilage, the intercellular material is firm. In bone, the substance is quite rigid. The living parts of connective tissue are the cells, which produce the intercellular substances. The cells may also store fats, produce new blood cells, ingest bacteria and cell debris, form anticoagulants, or give rise to antibodies, which protect against diseases.

The various kinds of connective tissue may be classified in several ways depending upon the criteria employed. We shall classify them as follows:

Embryonal connective tissue

Connective tissue that is present only in the embryo or fetus is called *embryonal connective tissue* (Exhibit 4–2). Whereas the term *embryo* refers to a developing human from fertilization through the first 2 months of pregnancy, a *fetus* is regarded as a developing human from the third month of pregnancy to birth. One example of connective tissue found only in the embryo is *mesenchyme.* This is the tissue from which all other connective tissues eventually arise. Mesenchyme may be observed under the skin and along the developing bones of the embryo. Another kind of embryonal connective tissue is *mucous connective tissue,* which is found only in the fetus. This tissue, also called *Wharton's jelly,* is located in the umbilical cord of the fetus where it provides support for the wall of the cord.

Adult connective tissue

Adult connective tissue is connective tissue that exists in the newborn and that does not change after birth. On the basis of kinds of cells and the nature of intercellular substance present, adult connective tissue is subdivided into several kinds.

Connective tissue proper

Connective tissue that has a more or less fluid intercellular material and a fibroblast (described below) as the typical cell is termed *connective tissue proper.* Five examples of such tissues may be distinguished.

LOOSE CONNECTIVE, OR AREOLAR. This type of tissue is one of the most widely distributed connective tissues in the body. Structurally, it consists of fibers and several kinds of cells embedded in a semifluid ground substance (Exhibit 4–2). This ground substance consists of a viscous material called hyaluronic acid. Normally, the thick consistency of this material impedes the movement of substances through the tissue. However, if an enzyme called hyaluronidase is injected into the tissue, hyaluronic acid changes to a watery consistency. This is of great clinical importance because the reduced viscosity hastens the absorption and diffusion of injected drugs and fluids through the tissue and thus can lessen tension and pain.

The three types of fibers embedded between the cells of loose connective tissue are collagenous fibers, elastic fibers, and reticular fibers. *Collagenous,* or *white, fibers* are very tough and resistant to a pulling force, yet are somewhat flexible. These fibers often occur in bundles. They are composed of many minute, wavy fibers called fibrils lying parallel to one another. The bundle arrangement affords a great deal of strength. Chemically, collagenous fibers consist of the protein collagen. *Elastic,* or *yellow, fibers,* by contrast, are smaller than collagenous fibers and freely branch and rejoin one another. Elastic fibers consist of a protein called elastin. These fibers also provide some strength and have a high degree of elasticity. *Reticular fibers* are protein-polysaccharide complexes. They are very fine fibers that branch extensively. Some authorities believe that reticular fibers are immature collagenous fibers. Like collagenous fibers, reticular fibers provide support and strength.

The cells in loose connective tissue are numerous and varied. The majority of the cells are *fibroblasts,* which are large, flat cells with branching processes. If the tissue is injured, the fibroblasts are believed to form collagenous fibers. Some evidence suggests that fibroblasts also form elastic fibers and the viscous ground substance. Other cells found in loose connective tissue are called *macrophages,* or *histiocytes.* They are irregular in form with short branching processes. These cells are capable of engulfing bacteria and cellular debris by the process of phagocytosis. Thus, they provide an important defense for the body. Macrophages are found in the liver, lymph organs, and bone marrow. A third kind of cell in loose connective tissue is the *plasma cell.* These cells are small and either round or irregularly shaped. They give rise to antibodies and, accordingly, provide a defensive mechanism through immunity. Plasma cells are found in many places of the body, but the majority are found in connective tissue, especially in that of the digestive

Exhibit 4–2. SUMMARY OF CONNECTIVE TISSUES

NAME AND DESCRIPTION	DIAGRAM	MICROSCOPIC APPEARANCE	LOCATION AND FUNCTION
I. Embryonal Connective tissue present only in embryo or fetus.			
A. Mesenchyme Consists of highly branched mesenchymal cells embedded in a fluid substance with scattered fibers.		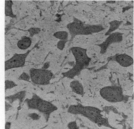	Under skin and along developing bones of the embryo. Forms all other kinds of connective tissue.
B. Mucous Consists of flattened or spindle-shaped cells embedded in a mucus-like substance containing collagenous fibers.		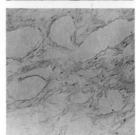	Umbilical cord of fetus. Support.
II. Adult Connective tissue present without change after birth. **A. Connective tissue proper** Consists of a fluid intercellular material with a fibroblast as the typical cell.			
1. Loose or areolar Consists of fibers (collagenous and elastic) and several kinds of cells (fibroblasts, macrophages, plasma cells, and mast cells) embedded in a semifluid.	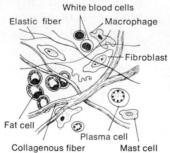	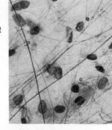	Subcutaneous layer of skin, mucous membranes, blood vessels, nerves, body organs. Strength, elasticity, support, phagocytosis, produces antibodies, and produces an anticoagulant.
2. Adipose Contains fibroblasts specialized for fat storage; cells have a "signet-ring" shape.	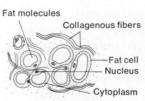		Subcutaneous layer of skin, around heart and kidneys, marrow of long bones, padding around joints. Reduces heat loss through skin, serves as food reserve, supports, and protects.
3. Collagenous Collagenous, or white, fibers predominate and are arranged in bundles; fibroblasts are in rows between bundles.	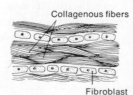		Forms tendons, ligaments, aponeuroses, membranes around various organs, and fasciae. Provides strong attachment between various structures.

Diagram labels (Loose or areolar): White blood cells, Elastic fiber, Macrophage, Fibroblast, Fat cell, Plasma cell, Collagenous fiber, Mast cell

Diagram labels (Adipose): Fat molecules, Collagenous fibers, Fat cell, Nucleus, Cytoplasm

Diagram labels (Collagenous): Collagenous fibers, Fibroblast

NAME AND DESCRIPTION	DIAGRAM	MICROSCOPIC APPEARANCE	LOCATION AND FUNCTION
4. Elastic Elastic, or yellow, fibers predominate and branch freely; fibroblasts present in spaces between fibers.	 Elastic fibers	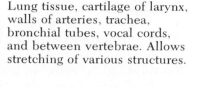	Lung tissue, cartilage of larynx, walls of arteries, trachea, bronchial tubes, vocal cords, and between vertebrae. Allows stretching of various structures.
5. Reticular Consists of a network of interlacing fibers; cells are thin and flat and wrapped around the fibers.	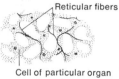Reticular fibers Cell of particular organ	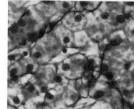	Liver, spleen, and lymph nodes. Forms framework of organs; binds together smooth muscle tissue cells.
B. Cartilage Dense network of collagenous fibers and some elastic fibers embedded in a gel-like matrix; cells are called chondrocytes and occur in spaces termed lacunae; surface of cartilage is covered by a membrane called perichondrium.			
1. Hyaline Also called gristle; appears as a bluish white, glossy, homogeneous mass; contains numerous chondrocytes; is the most abundant type of cartilage.	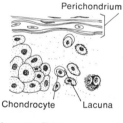Perichondrium Chondrocyte Lacuna	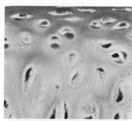	Ends of long bones; ends of ribs, nose, larynx, trachea, bronchi, bronchial tubes, and embryonic skeleton. Provides movement at joints, flexibility, and support.
2. Fibrocartilage Consists of chondrocytes scattered among bundles of collagenous fibers.	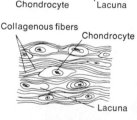Collagenous fibers Chondrocyte Lacuna		Symphysis pubis and intervertebral discs. Support and fusion.
3. Elastic Consists of chondrocytes located in a threadlike network of elastic fibers.	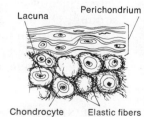Lacuna Perichondrium Chondrocyte Elastic fibers	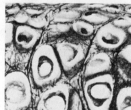	Epiglottis, larynx, external ear, and eustachian tube. Gives support and maintains shape.

C. Bone   Refer to Chapter 7.

D. Vascular   Refer to Chapter 16.

Photomicrograph sources: Embryonal mesenchyme (x1,000) and mucous (x100); adult loose or areolar (x450), collagenous (x430), and reticular (x430) connective tissue; and fibrocartilage (x430) courtesy of Dr. Donald I. Patt, from *Comparative Vertebrate Histology,* Donald I. Patt and Gail R. Patt, Harper & Row, Publishers, Inc., New York, 1969. Adult adipose and elastic connective tissue, and hyaline and elastic cartilege courtesy of Carolina Biological Supply Company.

tract. Another kind of cell in loose connective tissue is called the *mast cell*. This cell is found in abundance along blood vessels. It forms an anticoagulant, a substance that prevents blood from clotting within the vessels, called heparin. Mast cells are also believed to produce histamine, a chemical that dilates, or enlarges, small blood vessels. Other cells in loose connective tissue include *melanocytes*, or pigment cells, fat cells, and white blood cells.

Loose connective tissue is continuous throughout the body. It is present in all mucous membranes and around all blood vessels and nerves. And it occurs around body organs. Combined with adipose tissue, it forms the subcutaneous layer, or the layer of tissue that attaches the skin to underlying tissues and organs.

ADIPOSE TISSUE. This kind of tissue is basically a form of loose connective tissue in which the fibroblasts are specialized for fat storage. Adipose cells have the shape of a "signet ring" because the cytoplasm and nucleus are pushed to the edge of the cell by a large droplet of fat (Exhibit 4–2). In general, adipose tissue is found throughout the body wherever loose connective tissue is located. Specifically, it is found in the subcutaneous layer below the skin, around the kidneys, at the base and on the surface of the heart, in the marrow of long bones, and as a padding around joints. Adipose tissue is a poor conductor of heat and therefore reduces heat loss through the skin. It is also an important food reserve and generally supports and protects various organs.

COLLAGENOUS CONNECTIVE. This kind of tissue has a predominance of collagenous fibers, or white fibers, arranged in bundles (Exhibit 4–2). The cells found in collagenous connective tissue are fibroblasts, which are placed in rows between the bundles. The tissue is silvery white in appearance. It is tough, yet somewhat pliable. Because of the great strength of this tissue, collagenous connective tissue is the principal component of (1) tendons, which attach muscles to bones; (2) ligaments, which hold bones together at joints; (3) aponeuroses, which are flat bands connecting one muscle with another or with a bone; (4) membranes around various organs such as the kidneys and heart; and (5) fasciae, which are sheets of connective tissue wrapped around muscles to hold them in place.

ELASTIC CONNECTIVE. Unlike collagenous connective tissue, elastic connective tissue has a predominance of elastic fibers that branch freely (Exhibit 4–2). These fibers give the tissue a yellowish color. Fibroblasts are present only in the spaces between the fibers. Elastic connective tissue can be stretched. As the name implies, it is elastic and will snap back into shape. It is a component of the cartilages of the voice box, the walls of arteries, the windpipe, bronchial tubes to the lungs, and the lungs themselves. It is also found between the vertebrae of the backbone. Elastic connective tissue provides "stretch," allowing the structures in which it is found to perform their functions more efficiently.

RETICULAR CONNECTIVE. This kind of connective tissue consists of interlacing reticular fibers (Exhibit 4–2). It helps to form the framework, or body, of many organs, including the liver, spleen, and lymph nodes. Reticular connective tissue also helps to bind together the cells of smooth muscle tissue. It is especially adapted to providing strength and support in the structures in which it is found.

As a whole, connective tissue proper is characterized by a more or less fluid intercellular material. Moreover, the typical cell is a fibroblast. Cartilage, the next type of adult connective tissue to be studied, has a more solid intercellular material and a different type of cell.

Cartilage

This type of connective tissue is capable of enduring considerably more stress than the connective tissues just discussed. *Cartilage* consists of a fairly dense network of collagenous fibers and some elastic fibers firmly embedded in a gel-like substance. The cells of cartilage, called *chondrocytes*, occur singly or in groups in spaces called *lacunae* in the intercellular substance. The surface of cartilage is surrounded by a connective tissue covering called the *perichondrium*. The combining form *chondro*, which you will see often, means cartilage, and the form *peri* means around. On the basis of the texture of the intercellular substance, three kinds of cartilage are recognized. These are, in the order of discussion, hyaline cartilage, fibrocartilage, and elastic cartilage.

HYALINE CARTILAGE. This cartilage, also called gristle, appears as a bluish white, glossy, homogenous mass. The collagenous fibers, although present, are not visible, and the prominent chondrocytes are found in lacunae (Exhibit 4–2). Hyaline cartilage is the most abundant kind of cartilage in the body. It is found at joints, where it is called *articular cartilage*. And it forms the ventral ends of the ribs, where it is referred to as *costal cartilage*. Hyaline cartilage also helps to form the nose, voice box, windpipe, and the bronchi and bronchial tubes leading to the lungs. Most of the embryonic skeleton consists of hyaline cartilage. This kind of cartilage affords flexibility and support.

FIBROCARTILAGE. Chondrocytes scattered through many bundles of visible collagenous fibers are found in this type of cartilage (Exhibit 4–2). Fibrocartilage is found at the symphysis pubis, or the point where the pelvic bones fuse just in front of the external opening of the urinary tract. It is also in the discs that lie between each of the vertebrae. This tissue combines the properties of strength and rigidity.

ELASTIC CARTILAGE. In this tissue, chondrocytes are located in a threadlike network of elastic fibers (Exhibit 4–2). Elastic cartilage provides strength and maintains the shape of certain organs. Among these are the voice box, the external part of the ear, and the internal tubes that connect the nose to the ears.

Bone

The details of bone tissue, another kind of connective tissue, will be discussed in Chapter 7 as part of the skeletal system.

Vascular tissue

This kind of connective tissue, also known as blood, will be treated in Chapter 16 as a component of the circulatory system.

MUSCLE TISSUE AND NERVOUS TISSUE

Epithelial and connective tissue can take a variety of forms to provide a variety of services for the body. In a sense, they are all-purpose tissues. By contrast, the third major type of tissue, called muscle tissue, consists of highly modified cells that perform one basic function. This function is to contract. We shall look at muscle tissue in Chapter 10, when we discuss how the body moves. The fourth major type, called nervous tissue, is specialized to perform the function of sending electrical impulses. Nerve cells and their tissue will be examined in Chapter 12, where we explain how these electrical messages are sent.

Meanwhile, let us look at some structures that are composed of epithelial and connective tissue. We shall start here with the membranes of the body. In Chapter 5, we shall study the skin.

MEMBRANES

The combination of an epithelial layer and an underlying connective tissue layer constitutes a **membrane.** The principal membranes of the body are: (1) mucous membranes, (2) serous membranes, (3) synovial membranes, and (4) the cutaneous membrane.

1. *Mucous membranes.* These membranes line the body cavities that open to the exterior. Examples include the membranes lining the mouth and the entire digestive tract, the respiratory passages, the reproductive system, and the urinary system. The epithelial layer of a mucous membrane secretes mucus, which keeps the cavities from being dried out by air. It also traps dust in the respiratory passageways and lubricates food as it moves through the digestive tract. In addition, the epithelial layer is responsible for functions such as the secretion of digestive enzymes and the absorption of food. The connective tissue layer of a mucous membrane binds the epithelium to the underlying structures. It holds the blood vessels in place. And it provides a thick, tough covering that protects underlying muscles from abrasion or puncture.

2. *Serous membranes.* These membranes are found lining body cavities that do not open to the exterior, and they cover the organs that lie within those cavities. Serous membranes consist of two portions. The part that is attached to the wall of a cavity is called the *parietal* portion. The part that covers the organs lying inside these cavities is called the *visceral* portion. The serous membrane lining the thoracic cavity and covering the lungs is called the pleura. The membrane lining the heart cavity and covering the heart is referred to as the pericardium. The root word *cardio* means heart. Last, the serous membrane lining the abdominal cavity and covering the abdominal organs and some pelvic organs is called the peritoneum.

The epithelial layer of a serous membrane secretes a lubricating fluid, which allows the organs to glide easily over each other or along the walls of the cavities. The connective tissue layer ties the organs to the cavity walls and keeps them from falling to the bottom of the chest or abdomen.

3. *Synovial membranes.* Synovial membranes

line the cavities of the joints. Like serous membranes, they line structures that do not open to the exterior. The epithelial layer of a synovial membrane secretes synovial fluid, a liquid that lubricates the ends of the bones as they move in the joints. The connective tissue layer consists mostly of hyaline cartilage, a very tough and flexible substance. Hyaline cartilage keeps the membrane from rupturing under the pressure of joint movement. It also lines the ends of the bones so that they do not grate against each other.

4. *The cutaneous membrane.* The cutaneous membrane, or skin, constitutes an organ of the body and will be discussed in the next chapter.

All tissues of the body are subjected to a certain degree of wear and tear. Moreover, tissues may become damaged or diseased. When this occurs, the tissue must be repaired, if it is to function properly and prevent further damage to the body. We shall now examine how tissues respond to injury and the factors involved in tissue repair.

TISSUE INFLAMMATION

When cells are damaged, the injury sets off an *inflammatory response.* The injury, which may be viewed as a stress, can have various causes. It could result from mechanical means, such as a clean knife wound during surgery. Bacteria that give off poisonous chemicals could enter through the nose, pores in skin, or by way of a splinter or nail. Injury to cells can also occur if the blood supply is cut off, which causes the cells to "starve."

Inflammation is usually characterized by four fundamental symptoms: *redness, pain, heat,* and *swelling.* A fifth symptom can be the loss of function of the injured area. Whether loss of function occurs depends on the site and extent of the injury. In addition to these effects on the body, the inflammatory response, almost as a contradiction, serves a protective and defensive purpose. It attempts to neutralize and destroy toxic or poisonous agents at the site of injury and to prevent their spread to other organs. In other words, the inflammatory response is an attempt to restore tissue homeostasis.

The immediate inflammatory response to tissue injury consists of a complicated sequence of physiological and anatomical adjustments. Various parts of the body are involved in the initial response. These include the blood vessels, intercellular fluid mixed with parts of injured cells, called the exudate, the cellular components of the blood, and the surrounding epithelial and connective tissues. Other factors that affect the inflammatory response are the individual's age and general state of health. Healing processes of all types exert a great demand on the body's stores of all nutrients. Thus, nutrition, in addition to adequate circulation and tissue drainage, also plays an essential role in healing.

The following discussion outlines how the inflammatory response operates. We shall see that the inflammatory response is one of the body's internal systems of defense (Figure 4–1a). In the example presented here, a tissue is wounded by a rusty nail. The same response occurs, however, when bacterial invasion gives you a sore throat.

1. The injury stimulates tissue in the damaged area to release a chemical called histamine. *Histamine* is believed to increase the diameter of the blood vessels, or dilate them. It also increases the permeability of the plasma membranes of the blood vessel cells. Dilatation increases the amount of blood that can enter the area. And the increase in permeability allows defensive substances in the blood to pass through the vessel walls and into the injured tissue. Such defensive substances include white blood cells, antibodies, oxygen, and scab-forming chemicals. The increased blood supply also carries off poisonous products and dead cells, preventing them from complicating the injury. These poisonous, or toxic, substances include waste products given off by invading microorganisms.

2. The body may also respond by increasing its metabolic rate and by quickening the heartbeat so that more blood circulates to the injured area per minute. Within minutes after the injury, the quickened metabolism and circulation and especially the dilatation and increased permeability of capillaries produce heat, redness, and swelling. The heat comes from the large amount of warm blood that accumulates in the area and, to some extent, from the heat energy given off by the metabolic reactions. The large amounts of blood in the area are also responsible for the redness. The increased permeability of the capillary walls allows quantities of fluid to move out of the blood and into the intercellular spaces in the tissue. Because the fluid moves into the intercellular spaces faster than it can be drained off, it accumulates in the tissue, causing it to swell or puff up. The swelling is called *edema.*

3. Pain, either immediate or delayed, is a cardinal symptom of inflammation. It can result from an injury to nerve fibers or from an irritation

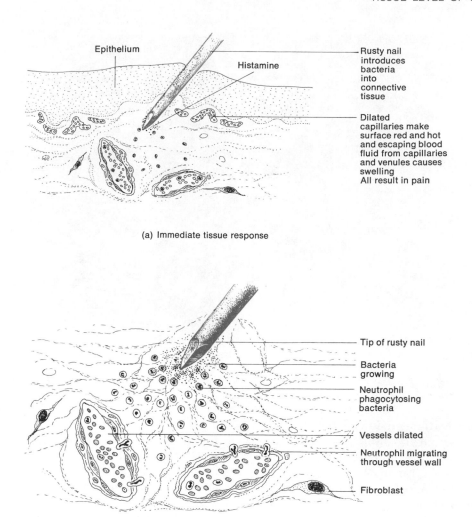

(a) Immediate tissue response

(b) Specific cell response

Figure 4-1. (a) Immediate tissue response and (b) specific cell response as a result of injury.

caused by the release of poisonous chemicals from microorganisms. Pain can also be due to increased pressure from an accumulation of extracellular fluid within the tissues.

4. Very soon after the injury occurs, white blood cells are mysteriously mobilized and rushed to the site from all over the body. The white blood cell, or leucocyte, is the body's first line of defense against the effects of the injury. The first white blood cells to arrive at the site of injury are neutrophils (Figure 4-1b). These are actively phagocytic cells that can engulf foreign particles equal to their own size (Figure 4-2). When neutrophils die, their lysosomes release valuable enzymes that cause the decomposition of injured cells and bacteria in the affected area. Additional neutrophils then ingest the resulting bacterial debris and other refuse that the body cannot use. Other white blood cells, called lymphocytes, are

believed to be involved either in the formation or release of *antibodies*, which are proteins that render invading bacteria and their chemicals harmless.

5. Nutrients stored in the body are used to support the defensive cells. They are also used in the increased metabolic reactions of the cells under attack.

6. The blood brings a soluble protein called fibrinogen to the site of injury. Fibrinogen is then converted to an insoluble, thick network called fibrin, which localizes and traps the invaders, thereby preventing their spread. This network eventually forms a blood clot, which prevents hemorrhage.

In all but very mild inflammations, a substance called pus is discharged. **Pus** is a thick fluid that contains living as well as nonliving white blood cells plus debris from other dead tissue. If the

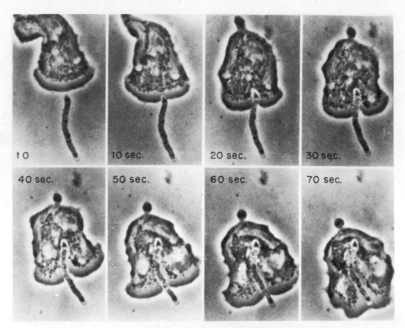

Figure 4–2. Phagocytosis. In this time-lapse photography sequence, a bacterial cell (the rodlike structure) is ingested and destroyed by a human neutrophil. *(From Robert M. Dowben, Cell Biology, Harper & Row, Publishers, Inc., New York, 1971. Photographs courtesy of Dr. James G. Hirsch.)*

pus cannot drain to the outside of the body, an **abscess** occurs, which is simply an excessive accumulation of pus in a confined space. When inflamed tissue is shed many times, it produces an open sore, called an **ulcer,** on the surface of an organ or tissue. Ulcers may result from a prolonged inflammatory response to a continuously injured tissue. For instance, overproduction of digestive acids in the stomach may cause a steady erosion of parts of the epithelial tissue lining the stomach. Elderly people with poor circulation are susceptible to ulcers in the tissues of their legs. The ulcers develop when the tissues are continuously damaged by a shortage of oxygen and nutrients.

As we have seen, white blood cells are the body's first line of defense against injury. The blood supply drains off and blocks the spread of harmful materials. This removal cannot restore the tissues to normalcy, however, especially if the cells have been damaged. The body has some other method to replace the injured areas with new tissue. And the body also has to restore cells that are worn out through normal usage. The process by which tissues replace dead or damaged cells is called repair, our next topic for discussion.

TISSUE REPAIR

Tissue repair begins during the active phase of inflammation. But it can be completed only after all harmful substances have been neutralized or removed from the site of injury. New cells originate by cell reproduction from either the connective tissue stroma of an organ or from the parenchymal cells. *Parenchymal cells* are the cells that make up the functioning part of an organ. For instance, the epithelial cells that secrete and absorb are the parenchymal cells of the intestine. The connective tissue that supports, protects, and binds together the parenchymal cells is called the *stroma*. The restoration of an injured organ or tissue to normal structure and function depends entirely upon which type of cells—parenchymal or stromal—are active in the repair. If only parenchymal elements accomplish the repair, a perfect or near-perfect reconstruction of the injured tissue may occur. However, if the fibroblast cells of the stroma do the repair work, the tissue will be replaced with new connective tissue called scar tissue. This condition is known as **fibrosis.** Since scar tissue is not specialized to carry out the functions of the parenchymal tissue, function is impaired. An example of repair by scar tissue is an inflammatory disease of the small intestine called *enteritis.* Enteritis results in progressive fibrosis of the intestinal wall. The function of the parenchymal tissue in the small intestine is to move foods along and to break down and absorb nutrients. The scar tissue resulting from enteritis is not capable of performing these functions, so the patient suffers from malnutrition, diarrhea, and constipation.

The single most important factor in tissue repair lies in the capacity of parenchymal tissue to regenerate. This, in turn, depends on the ability of the parenchymal cells to reproduce, and to reproduce quickly.

Regenerative capacity of cells

The various types of tissues in the body exhibit vast differences in regenerative capacities. Some parenchymal cells continue to multiply throughout life. They replace worn-out or destroyed cells under normal physiological conditions. The various epithelial surfaces such as the skin and the linings of the mouth, uterus, vagina, respiratory tract, and urinary tract are examples of this type of tissue. In these areas, the surface cells are continually lost, but they are constantly being replaced with new ones. The restoration of tissue is perfect. A good example of the ability of epithelium to repair itself is the replacement of cells lining the uterus after they are shed during menstruation. Immature red and white blood cells also carry on cell reproduction and constantly replace the old, dead, and dying mature cells. When you donate a pint of blood, the body quickly replaces the blood cells as well as the water and dissolved chemicals.

Nerve and muscle cells are incapable of mitosis and cytokinesis and, therefore, cannot replace themselves. Nerve cells have long projections of the cytoplasm, which are the telephone wires over which messages are sent. The projections that lie outside the brain and spinal cord can be repaired as long as the rest of the nerve cell is alive and healthy. Projections that lie within the brain and spinal cord cannot be replaced. With the exception, then, of the cytoplasmic projections of certain nerve cells, the repair of muscle and nerve tissue is by scar tissue, with a resulting loss of function.

Some glandular tissue cells are capable of vigorous reproduction only under the appropriate stimulation, such as that produced by hormones. Examples of tissues that can regenerate under hormonal stimulation are cells of the liver, salivary glands, endocrine glands, kidney tubules, and the oil and sweat glands of the skin. Other cells in this category are bone cells and the chondroblasts and fibroblasts of connective tissue.

The repair process

If injury to a tissue is slight or superficial, repair may sometimes be accomplished with the drainage and reabsorption of pus, followed by parenchymal regeneration. When the area of skin loss is great, fluid moves out of the capillaries, and the area becomes dry. Fibrin seals the open tissue by hardening into a *scab*.

When tissue and cell damage are extensive and severe, as in large, open wounds, both the connective tissue stroma and the parenchymal cells are active in repair. This kind of repair involves the rapid cell division of many fibroblasts, the manufacture of new collagenous fibers, which provide strength, and an increase by cell division of the number of small blood vessels in the area. All these processes create an actively growing, connective tissue called **granulation tissue.** This new granulation tissue forms across a wound or surgical incision to provide a framework, or stroma. The framework supports the epithelial cells that move in and fill the open area. The newly formed granulation tissue also secretes a fluid that kills bacteria.

Conditions affecting repair

Three factors that affect tissue repair are nutrition, good blood circulation, and age. Nutrition is extremely important in the healing process since a great demand is put on the body's stores of nutrients. Protein-rich diets are important since most of the cell structure is made from proteins. Vitamins also play a direct role in wound healing. The following vitamins are believed to be essential to proper wound repair: (1) Vitamin A is important in the replacement of epithelial tissues, especially in the respiratory tract. (2) The B vitamins, thiamine, nicotinic acid, and riboflavin, are necessary for proper functioning of many enzyme systems in cells. They are needed especially for the enzymes involved in decomposing carbohydrates to glucose, which is important to both heart and nervous tissue. These vitamins may relieve pain in some cases. (3) Vitamin C directly affects the normal production and maintenance of intercellular substances. It is important to the manufacture of cementing elements of connective tissues, especially collagen. Vitamin C also strengthens and promotes the formation of new blood vessels. With vitamin C deficiency, even superficial wounds fail to heal, and the walls of the blood vessels become fragile and are easily ruptured. (4) Vitamin D is necessary for the proper absorption of calcium from the intestine. Calcium gives bones their hardness and is necessary for the healing of fractures. (5) Vitamin K helps the blood to clot and thus prevents the injured person from bleeding to death.

In tissue repair, the importance of good blood circulation cannot be overestimated. It is the blood that transports oxygen, nutrients, antibodies, and many defensive cells to the site of injury. The blood also plays an important role in carrying away tissue fluid, blood cells that have been depleted of oxygen, bacteria, foreign bodies, and debris. All these elements would otherwise interfere with the healing process.

Generally speaking, tissues heal faster and leave less obvious scars in the young than in the aged. In the young, tissues have a better blood supply. The young body is generally in a much better nutritional state, and the cells of younger people have a faster metabolic rate. This means that the cells can duplicate their materials and divide more quickly.

The tissues of the body perform various roles in attempting to maintain homeostasis. Among these functions are protection, support, filtration, absorption, secretion, movement, transportation, and defense against disease. Damage to a tissue is a stress because it interferes with these vital functions. When cells are injured, many parts of the body work to overcome the stress and return the body to homeostasis. They do this by contributing to the inflammatory response and to the repair process.

Chapter summary in outline

EPITHELIAL TISSUE

1. A tissue is a group of cells specialized for a particular function.

2. Epithelium covers and lines body surfaces and forms glands.

3. Epithelium has many cells, little intercellular material, and no blood vessels. It is attached to connective tissue by a basement membrane. It is also capable of replacing itself.

Covering and lining epithelium

1. Simple epithelium is a single layer of cells adapted for absorption or filtration.

2. Stratified epithelium is several layers of cells adapted for protection.

3. Epithelial cell shapes include squamous (flat), cuboidal (cubelike), columnar (rectangular), and transitional (variable).

4. Simple squamous epithelium is adapted for diffusion and filtration and is found in lungs and kidneys. Endothelium lines the heart and blood vessels. Mesothelium lines body cavities and covers internal organs.

5. Simple cuboidal epithelium is adapted for secretion and absorption in kidneys and glands.

6. Simple columnar epithelium lines the digestive tract. Specialized cells containing microvilli perform absorption. Goblet cells perform secretion. In the respiratory tract, the cells are ciliated to carry dust particles out of the body.

7. Stratified squamous epithelium is protective. It lines the upper digestive tract and forms the outer layer of skin.

8. Stratified cuboidal epithelium is found in adult sweat glands.

9. Stratified columnar epithelium protects and secretes. It is found in the male urethra and excretory ducts.

10. Transitional epithelium is found in the urinary bladder and is capable of stretching.

11. Pseudostratified epithelium is one layer but gives the appearance of having many layers. It lines excretory and respiratory structures where it protects and secretes.

Glandular epithelium

1. A gland is a single cell or a mass of epithelial cells adapted for secretion.

2. Exocrine glands (sweat, oil, and digestive glands) secrete into ducts.

3. Endocrine glands secrete hormones directly into the blood.

CONNECTIVE TISSUE

1. Connective tissue is the most abundant body tissue. It has few cells, an extensive matrix, and a rich blood supply.

2. It functions in protection, support, and binding organs together.

Embryonal connective tissue

1. Mesenchyme, found only in the embryo, forms all other kinds of connective tissue.

2. Mucous connective tissue is found only in the umbilical cord of the fetus, where it gives support.

Connective tissue proper

1. This tissue has a more or less fluid matrix. The fibroblast is the typical cell.

2. Loose connective tissue, or areolar tissue, is widely distributed. It contains three kinds of fibers (collagenous, elastic, and reticular). It also has several kinds of cells (fibroblasts, macrophages, plasma cells, mast cells, and white blood cells). Loose connective tissue forms the subcutaneous layer under the skin. It is present in mucous membranes and around blood vessels, nerves, and body organs. When the fibroblasts in loose connective tissue become loaded with fat, the tissue is then known as adipose tissue.

3. Dense connective tissue forms tendons, ligaments, and fasciae. All three are usually primarily composed of closely packed collagenous fibers. In tendons and ligaments, collagenous fibers are arranged in parallel bundles. In fasciae, these fibers are interwoven at various angles with each other. A few ligaments are composed of closely packed elastic fibers.

Cartilage

1. Cartilage has a gel-like matrix of collagenous and elastic fibers that contain chondrocytes.

2. Hyaline cartilage is found at the ends of bones, in the nose, and in respiratory structures. It is flexible, allows movement, and provides support.

3. Fibrocartilage connects the pelvic bones and the vertebrae. It provides strength.

4. Elastic cartilage maintains the shape of organs such as the ear.

MEMBRANES

1. A membrane is an epithelial layer overlying a connective tissue layer.

2. Mucous membranes line cavities that open to the exterior, such as the digestive tract.

3. Serous membranes (pleura, pericardium, peritoneum) line closed cavities and the organs lying within the cavities. These membranes consist of the parietal and visceral portions.

4. Synovial membranes line joint cavities.

5. The cutaneous membrane is the skin.

TISSUE INFLAMMATION

1. Damage to a tissue causes an inflammatory response characterized by redness, pain, heat, and swelling.

2. The inflammatory response is initiated by histamine released by the damaged tissue.

3. Further cell injury is prevented by phagocytic neutrophils and by antibody-producing lymphocytes.

TISSUE REPAIR

1. Tissue repair is the replacement of damaged or destroyed cells by healthy ones.

2. If replacement is accomplished by parenchymal cells, the repair is perfect or nearly so, and function is restored. If the replacement is accomplished by the stromal fibroblasts, scarring, or fibrosis, results, and function is impaired.

The repair process

1. If the injury is superficial, tissue repair involves pus removal (if pus is present), scab formation, and parenchymal regeneration.

2. If damage is extensive, granulation tissue is involved.

Conditions affecting repair

1. Nutrition is important to tissue repair. Various vitamins and a protein-rich diet are needed.

2. Adequate circulation of blood is needed.

3. The cells of young people repair faster and more efficiently.

Review questions and problems

1. Define a tissue. What are the four basic kinds of human tissues?

2. What characteristics are common to all epithelium? Distinguish covering and lining epithelium from glandular epithelium.

3. Discuss the classification of epithelium using layering and cell type as criteria.

4. For each of the following kinds of epithelium, briefly describe the microscopic appearance, location in the body, and functions: simple squamous, simple cuboidal, simple columnar, stratified squamous, stratified cuboidal, stratified columnar, transitional, and pseudostratified.

5. What is a gland? Distinguish between endocrine and exocrine glands.

6. Enumerate the ways in which connective tissue differs from epithelium. How are connective tissues classified?

7. Compare embryonal connective tissue with adult connective tissue.

8. Describe the following connective tissues with regard to microscopic appearance, location in the body, and function: loose (areolar), adipose, collagenous, elastic, reticular, hyaline cartilage, fibrocartilage, and elastic cartilage.

9. Define the following kinds of membranes: mucous, serous, synovial, and cutaneous. Where is each located in the body? What are their functions?

10. Describe the principal physiological responses associated with inflammation. What is the immediate tissue response to injury? What is the response to tissue injury by neutrophils and lymphocytes?

11. What is meant by tissue repair? How does the repair take place? What are the various kinds of tissue regeneration? What conditions affect tissue repair?

12. In order to increase your understanding of the location and function of various tissues in the body, we suggest the following: Below are some descriptive statements for various tissues. Next to each statement, write the name of the tissue described.
    A stratified epithelium that permits distention. _____
    A single layer of flat cells concerned with filtration and absorption. _____
    Forms all other kinds of connective tissue. _____
    Specialized for fat storage. _____
    An epithelium with waterproofing qualities. _____
    Forms the framework of many organs. _____
    Produces perspiration, wax, oil, and digestive enzymes. _____
    Cartilage that shapes the external ear. _____
    Contains goblet cells and lines the intestine. _____
    Most widely distributed connective tissue. _____
    Forms tendons, ligaments, and aponeuroses. _____
    Specialized for the secretion of hormones. _____
    Provides support in the umbilical cord. _____
    Lines kidney tubules and is specialized for absorption and secretion. _____
    Permits extensibility of lung tissue. _____

CHAPTER 5

THE INTEGUMENTARY SYSTEM: STRUCTURE, PHYSIOLOGY, AND DISORDERS

STUDENT OBJECTIVES

After you have read this chapter, you should be able to:

1. Define the skin as an organ and component of the integumentary system

2. Explain how the skin is structurally divided into an epidermis, dermis, and subcutaneous layer

3. List the structural layers of the epidermis and describe their functions

4. Explain the composition and functions of the dermis

5. Contrast the structure and functions of such accessory organs of the skin as hair, glands, and nails

6. Describe the role of the skin in the maintenance of normal body temperature

7. List the causes and treatment of acne

8. Define and describe the effects of a burn

9. Classify burns into first-, second-, and third-degree types

10. Define the "rule of nines" for estimating the extent of a burn

11. Define medical terminology associated with the integumentary system

An aggregation of tissues that performs a specific function is an organ. Recall that organs represent the next level of organization in the body. In considering the organ level of organization, we shall use the skin as an example. From organs, the next higher level of organization is a system—a group of organs that operate together to perform specialized functions. The skin and the organs derived from it, such as hair, nails, and glands, and several specialized receptors, constitute the **integumentary system** of the body. First, let us consider the skin as an organ.

THE SKIN AS AN ORGAN

The skin is an organ because it consists of tissues that are structurally joined together to perform specific activities. Most people view the skin as a simple thin covering that keeps the body together and gives it protection. A detailed analysis of the skin, however, reveals that it is quite complex in structure and performs several vital functions. In fact, this organ is essential for survival.

Considered by itself, the skin is the largest organ of the body. For the average adult, the skin occupies a surface area of approximately 3,000 square inches. It covers the body and protects the underlying tissues, not only from bacterial invasion but also from drying out and from harmful light rays. In addition to its protective function, the skin helps to control body temperature, prevents excessive loss of inorganic and organic materials, receives stimuli from the environment, stores chemical compounds, excretes water and salts, and synthesizes several important compounds.

Structurally, the skin consists of two principal parts (Figure 5–1). The outer, thinner portion, which is composed of epithelium, is called the *epidermis*. The epidermis is cemented to the inner, thicker, connective tissue part, which is called the *dermis*. Beneath the dermis is a *subcutaneous layer* of tissues. The combining form *sub* means under. The subcutaneous layer is also called the superficial fascia, and it consists of areolar and adipose tissues. Fibers from the dermis extend down into the superficial fascia and anchor the skin to the subcutaneous layer. The superficial fascia, in turn, is firmly attached to underlying tissues and organs.

The epidermis

The **epidermis** is composed of stratified epithelium in four or five cell layers, depending on its location in the body. (See Figure 5–1.) In areas where exposure to friction is greatest, such as the palms and soles, the epidermis has five layers. In all other parts of the body, the epidermis has four layers. The names of these layers from the inside outward are as follows:

1. *Stratum basale.* This is a single layer of columnar cells capable of continued cell division. The epidermis grows by the division of cells in the stratum basale and deep layers of the stratum spinosum, the next higher layer. As these cells multiply, they push up toward the surface. Their nuclei degenerate, and the cells die. Eventually, the cells are shed in the top layer of the epidermis.

2. *Stratum spinosum.* This layer of the epidermis, just above the stratum basale, contains 8 to 10 rows of polygonal (many-sided) cells that fit closely together. Since the surfaces of these cells assume a prickly appearance, the layer is so named. The word *spinosum* means prickly. The stratum spinosum helps in the continual production of new epithelium.

3. *Stratum granulosum.* This third layer of the epidermis consists of two or three rows of flattened cells that contain dark staining granules of a substance called keratohyaline. This compound is involved in the first step of keratin formation. Keratin is a waterproofing protein found in the top layer of the epidermis. The stratum granulosum contains cells whose nuclei are in various stages of degeneration. As these nuclei break down, the cells are no longer capable of carrying out vital metabolic reactions and die.

4. *Stratum lucidum.* This layer exists only in the thick skin of the palms and soles. It consists of three to four rows of clear, flat, dead cells that contain droplets of a translucent substance called

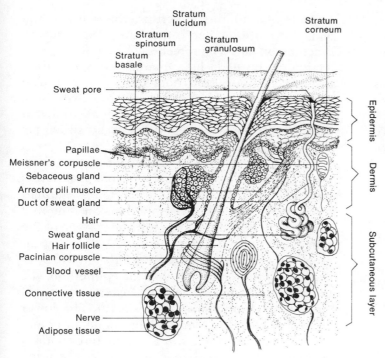

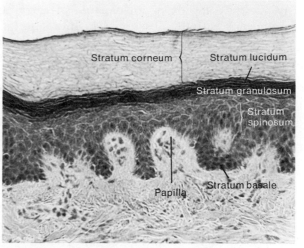

Figure 5–1. The skin. Note the structures in the epidermis and dermis and the underlying subcutaneous layer. *(Photomicrograph courtesy of Dr. Donald I. Patt, from Comparative Vertebrate Histology, Donald I. Patt and Gail R. Patt, Harper & Row, Publishers, Inc., New York, 1969. Magnification of 400X.)*

eleidin. Eleidin is formed from keratohyaline and is eventually transformed to keratin. This layer is so named because of the translucent property of eleidin. The word root *lucidus* means clear.

5. *Stratum corneum.* This layer consists of 25 to 30 rows of flat, dead cells containing the protein *keratin.* The keratin serves as a waterproof covering. These cells are continuously shed and replaced. The stratum corneum serves as an effective barrier against light and heat waves, bacteria, and many chemicals.

The color of the skin is due to a yellow to black pigment called *melanin.* This pigment is found throughout the basale and spinosum layers and in the granulosum of all Caucasian people. In blacks, melanin is found in all epidermal layers. When the skin is exposed to ultraviolet radiation, both the amount and darkness of melanin increase, which causes tanning and further protects the body against radiation. Thus, melanin serves a very important protective function. Another pigment called *carotene* is found in the corneum and fatty areas of the dermis in Oriental people. Together, carotene and melanin account for the yellowish hue of their skin. The pink color of Caucasian skin is due to blood vessels in the dermis. The redness of the vessels is not heavily masked by pigment. The epidermis has no blood vessels.

The dermis

The second principal part of the skin, the dermis, is composed of connective tissue containing collagenous and elastic fibers. (See Figure 5–1.)

Numerous blood vessels, nerves, glands, and hair follicles are also embedded in the dermis. The upper region of the dermis, which is about one-fifth of the total layer, is referred to as the *papillary region.* This part of the dermis is so named because its surface area is greatly increased by small, fingerlike projections called **papillae.** These structures project into the epidermis and contain loops of capillaries. In some cases, the papillae contain Meissner's corpuscles, which are nerve endings that are sensitive to touch. The dermis also contains nerve endings called Pacinian corpuscles, which are sensitive to deep pressure.

The series of ridges marking the external surface of the epidermis are caused by the size and arrangement of the papillae in the dermis. Some of the ridges cross at various angles and can be seen on the back surface of your hand. Other ridges on your palms and fingertips prevent slipping. The ridge patterns on the tips of the fingers and thumbs are different in each individual. Because of this, fingerprints can be taken and accurately used for purposes of identification.

The remaining portion of the dermis is called the *reticular region.* This area of the dermis contains many blood vessels and also contains collagenous and elastic fibers. The spaces between the interlacing fibers are occupied by adipose tissue and sweat glands. The reticular zone is attached to the organs beneath it, such as bone and muscle, by the subcutaneous layer.

It should now be obvious to you that the skin, despite its relatively simple physical appearance, is a very complex organ capable of carrying on numerous activities essential to life. The tissues of the skin are joined to form an organ that performs specific activities. These tissues consist of (1) the various epithelial layers of the epidermis that, together, protect, waterproof, and add new cells, and (2) the connective tissues (areolar and adipose) of the dermis that protect, contain nerve endings for touch and pressure, and connect the epidermis to the subcutaneous layer.

ACCESSORY ORGANS OF THE SKIN

Organs that are derived from the skin, such as hair, glands, and nails, perform functions that are necessary and sometimes vital. Hair and nails offer further protection to the body, whereas the sweat glands perform the vital function of helping to regulate body temperature.

Hair

Certain growths of the epidermis variously distributed over the body are **hairs** or **pili.** Some of the regions of the body not covered by hair are the surfaces of the palms and undersides of the fingers, the back surfaces of the fingers from the second joint to the fingertips, the soles of the feet, the nipples, and the lips. The primary function of hair is protection. Though the protection is limited, hair guards the scalp from mechanical injury and from the sun's rays. The eyebrows and eyelashes protect the eyes from foreign particles. The hair in the nostrils and external ear canal also protects these structures from insects and dust particles.

Hairs are composed of a number of parts (Figure 5–2). Each hair consists of a free shaft and a root. The *shaft* is the visible portion of the hair projecting above the surface of the skin. The *root* is the portion of the hair below the surface of the skin that penetrates deep into the dermis. Surrounding the root is the *hair follicle,* which is made up of an internal zone of epithelium, the *internal root sheath,* and an external zone of epithelium, the *external root sheath.* The lower ends of each follicle are enlarged into an onion-shaped structure, the *bulb.* This structure contains an indentation, the *dermal papilla,* that is filled with loose connective tissue. The papilla contains many blood vessels and provides nourishment for the growing hair. The base of the bulb also contains a region of cells called the *matrix.* The cells

Figure 5–2. Principal parts of a hair and associated structures. *(Photomicrograph courtesy of Dr. Donald I. Patt, from Comparative Vertebrate Histology, Donald I. Patt and Gail R. Patt, Harper & Row, Publishers, Inc., New York, 1969. Magnification of 200X.)*

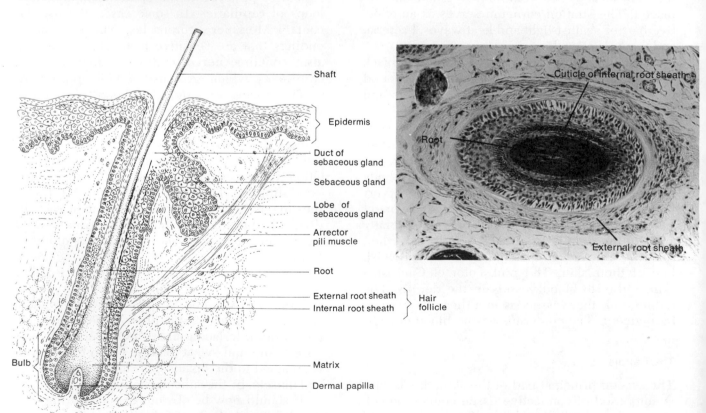

of the matrix produce new hairs by cell division when older hairs are shed. This replacement occurs within the same follicle.

Sebaceous glands, which will be discussed shortly, and bundles of smooth muscle are also associated with hair. These smooth muscles, called *arrector pili*, extend from the dermis of the skin to the side of the hair follicle. In its normal position hair is arranged at an angle to the skin. The arrector pili muscles, which are not voluntarily controlled, contract under stresses of fright and cold and pull the hairs into a more vertical position. This contraction results in "goosebumps" or "gooseflesh" because the skin around the shaft forms slight elevations.

Glands

Two principal kinds of glands associated with the skin are sebaceous glands and sweat glands. **Sebaceous glands,** with only few exceptions, are connected to hair follicles. (See Figure 5–2.) They are multilobed structures connected directly to the follicle by a short duct. These glands are absent in the palms and soles and differ in size and shape in other regions of the body. For example, they are fairly small in most areas of the trunk and extremities but are relatively large in the skin of the face, neck, and upper chest. The sebaceous glands secrete an oily substance called *sebum*, which is a mixture of fats, cholesterol, proteins, and inorganic salts. The functions of the sebaceous glands are to keep the hair from drying and becoming brittle and to form a protective film that prevents excessive evaporation of water from the skin. The sebum also keeps the skin soft and pliable. When sebaceous glands of the face become enlarged because of accumulated sebum, blackheads are produced. Since sebum is fine food for certain bacteria, this frequently results in pimples or boils.

Sweat (sudoriferous) glands are distributed throughout the skin except on the nail beds of the fingers and toes, margins of the lips, eardrum, and tip of the penis. In contrast to sebaceous glands, sweat glands are most numerous in the skin of the palms and the soles. They are also found in abundance in the armpits and forehead. Each gland consists of a coiled end embedded in the subcutaneous tissue and a single tube that projects upward through the dermis and epidermis. This tube, actually the excretory duct, terminates in a pore at the surface of the epidermis. (See Figure 5–1.) The base of each sweat gland is surrounded by a network of small blood vessels.

Perspiration, or *sweat*, is the substance produced by sweat glands. It is a mixture of water, salts (mostly NaCl), urea, uric acid, amino acids, ammonia, sugar, lactic acid, and ascorbic acid. Although perspiration helps eliminate waste materials from the body, its principal value is to help regulate body temperature. This vital function will be discussed subsequently.

Nails

Modified horny cells of the epidermis are the **nails,** or **ungues.** The cells form a clear, solid covering over the back surfaces of the terminal portions of the fingers. Structurally, each nail consists of a body, root, and lunule. The *body* is the visible portion of the nail, and the *root* is hidden under the skin. The *lunule* is the white, half-moon-shaped area at the base of the nail that is the actively growing region. The nails appear pink, except at the lunule, because of the underlying capillaries. These capillaries do not show through the lunule.

Just as the epidermis and dermis form the organ called the skin, the skin together with its accessory organs of hair, glands, and nails make up a system called the integumentary system. As a system, the skin and its accessory organs protect underlying tissues from physical injury, harmful light rays, bacterial invasion, and drying out. The system receives stimuli from the environment, eliminates water and salts in the form of perspiration, and helps regulate body temperature. This last function is vital to survival and will now be considered as a homeostatic mechanism.

THE INTEGUMENTARY SYSTEM AND HOMEOSTASIS

One of the best examples of homeostasis in man is the regulation of body temperature. Why is the integument important in the homeostasis of body temperature? Man, like other mammals, is a *homeotherm*, or warm-blooded organism. This means that man is able to maintain a remarkably constant internal body temperature (98.6°F), even though the environmental temperature may vary over a broad range. Let us assume that a person is participating in an athletic event where the temperature is 100°F. A sequence of events is set into operation to counteract this above-normal temperature, which may be considered a stress. Sensing devices in the skin called receptors pick up the stimulus—in this case, heat—and activate nerves that send the message to the brain. A temperature-regulating area of the brain then sends a nerve impulse to the sweat glands in the skin. When the sweat glands are stimulated by the impulse from the brain, they produce more perspiration. As the perspiration evaporates from the surface of the skin, the skin is cooled, and body temperature is lowered. This sequence of events is shown in Figure 5–3.

Note that temperature regulation by the skin involves a feedback system because the output (cooling of the skin) is fed back to the skin receptors and becomes part of a new stimulus-response cycle. In other words, after the sweat glands are activated, the skin receptors continue to keep the brain informed about the external temperature. The brain, in turn, continues to send messages to the sweat glands until the temperature returns to 98.6°F. Like most of the body's feedback systems, temperature regulation is a negative feedback system. That is, the output, cooling, is the opposite of the original condition, overheating.

DISORDERS OF THE INTEGUMENTARY SYSTEM

Two of the most common disorders of the skin are acne and burns. Acne involves just the sebaceous glands of the skin. Burns can involve the whole integumentary system and the mucous membranes that are in contact with external stresses such as fire, electricity, radioactivity, and chemicals.

What causes acne?

A common inflammatory disease of the sebaceous glands is **acne**. It occurs in areas where the glands are largest, most numerous, and most active.

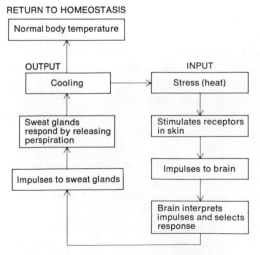

Figure 5–3. The role of the skin in the homeostasis of body temperature.

Common acne is called *acne vulgaris* and is found mostly in individuals between the ages of 14 and 25. It affects more than 80 percent of teenagers at puberty since the glands that control the oil secretions are very active at this time.

Acne can start with a blackhead called a *comedo*, which is a plugged duct of a sebaceous gland. The comedo can become infected and form a *cyst*, or sac of connective tissue cells, that could destroy and displace epidermal cells resulting in permanent scarring. Care must be taken to avoid squeezing, pinching, and scratching of the lesions.

Certain foods seem to aggravate acne in some people. A change in diet, in addition to applying antiseptic ointments, sometimes helps clear up the condition. Frequent, thorough cleansing of the skin with warm water and soap may also help. Recently, vitamin A has been reported to be an effective agent in treating acne. Severe acne may need x-ray, antibacterial (antibiotic), or hormone therapy.

Burns

Tissues may be damaged by thermal (heat), electrical, radioactive, or chemical agents. These agents can destroy the proteins in the exposed cells and thereby cause cell injury or death. Such damage results in a **burn.** The tissues that are directly or indirectly in contact with the environment, such as the skin or the linings of the respiratory and digestive tracts, are affected. Generally, however, the systemic effects of a burn are a greater threat to life than the local effects. Systemic effects are those that occur throughout the body—the term **systemic** referring to the whole body. **Local** effects are changes that occur in one area of the body. The systemic effects of burn include: (1) a large loss of water, plasma, and plasma proteins, which causes shock; (2) bacterial infection; (3) a slower circulation of blood; and (4) a decrease in urine production.

Burns are classified into three types: first-degree, second-degree, and third-degree. In *first-degree* burns, the damage is restricted to the epidermal layers of the skin, and symptoms are limited to local effects such as redness, tenderness, pain, and edema—the cardinal signs of inflammation. In *second-degree* burns, both the epidermal and dermal layers of the skin are damaged, but rapid regeneration of epithelium is still possible. Blisters containing elements of blood and lymph form on the skin surface or beneath the epidermis. Blisters beneath or within the epidermis are called *bullae*. In *third-degree* burns, both the epidermis and dermis are destroyed. The skin surface may be charred or white, or have patches of both. It is lifeless and insensitive to touch. Here, the regeneration of epithelium originates from the wound edges. Regeneration is slow, and much granulation tissue forms before being covered by epithelium. Even if skin grafting is started quickly, these wounds frequently contract and produce disfiguring or disabling scars.

An accurate means for estimating the extent of a burn is the *"rule of nines"*: (1) if the head and neck are affected, the burn covers 9 percent of the body surface; (2) each arm and hand, including the shoulder muscle, also comprise 9 percent of the body surface; (3) each foot and leg as far up as the bottom of the buttocks comprise 18 percent; (4) the front and back portions of the trunk including the buttocks make up 18 percent each; and (5) the perineum makes up 1 percent. The perineum includes the anal region and the urogenital region (Figure 5–4).

A severely burned individual should be moved as quickly as possible to a hospital. Treatment then may include: (1) cleansing the burn wounds thoroughly, (2) removing all dead tissue, so that antibacterial agents can directly contact the wound surface and thereby prevent infection, (3) replacing lost body fluids, and (4) covering wounds with grafts as soon as possible.

A *reminder:* as we mentioned in the preface, each chapter in this text which discusses a major system of the body will be followed by a glossary of medical terminology associated with that system. Both normal and pathological conditions of the system are included in these glossaries. You should familiarize yourself with the terms since they will play an essential role in your medical vocabulary. We would also like to alert you to the fact that some of these disorders, as well as disorders discussed in the text, will be referred to as local or systemic. A **local disease** is one that affects one part or a limited area of the body. A **systemic disease,** on the other hand, affects either the entire body or several of its parts.

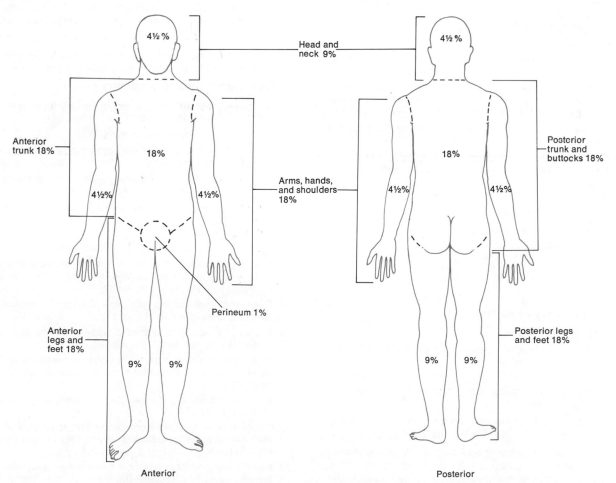

Figure 5–4. The "rule of nines" for estimating the extent of burns.

Medical terminology associated with the integumentary system

Albinism (*alb* = white; *ism* = condition) Congenital (existing at birth) absence of pigment from the skin, hair, and parts of the eye.

Anhidrosis (*an* = without; *hidr* = sweating; *osis* = condition) Inability to sweat.

Callus An area of hardened and thickened skin that is usually seen in palms and soles and is due to pressure and friction.

Carbuncle A hard, round, deep, painful inflammation of the subcutaneous tissue that causes necrosis (deadness) and pus formation (abscess).

Comedo A collection of sebaceous material and dead cells in the hair follicle and excretory duct of the sebaceous gland. Usually found over the face, chest, and back, and more commonly during adolescence. Also called *blackhead* or *whitehead.*

Cyst (*cyst* = sac containing fluid) A sac with a distinct connective tissue wall, containing a fluid or other material.

Decubitus ulcer A bedsore. An ulcer formed due to continual pressure over the skin.

Epidermophytosis (*epi* = upon; *derm* = skin; *phyto* = pertaining to plants; *osis* = condition) Any fungus infection of the skin producing scaliness with itching. Called *athlete's foot* when it affects the feet.

Furuncle (boil) An abscess resulting from infection of a hair follicle.

Hypodermic (*hypo* = under; *derm* = skin) The area beneath the skin.

Intradermal (intra = within) Within the skin. Also called *intracutaneous.*

Melanoma (*melano* = dark colored; *oma* = tumor) A cancerous tumor consisting of melanocytes that produce skin pigment.

Nevi Round, pigmented, flat, or raised skin areas that may be present at birth or develop later. Varying in color from yellow-brown to black. Also called *moles* or *birthmarks.*

Nodule A large cluster of cells raised above the skin but extending deep into the tissues.

Papule A small round skin elevation varying in size from a pin point to that of a split pea. One example is a pimple.

Polyp A tumor on a stem found especially on mucous membranes.

Pustule A small round elevation of the skin containing pus.

Subcutaneous (*sub* = under; *cutis* = skin) Beneath the skin.

Wart A common, contagious, noncancerous epithelial tumor caused by a virus.

Chapter summary in outline

THE SKIN AS AN ORGAN

1. The skin and its accessory organs of hair, glands, and nails constitute the integumentary system.

2. The skin is the largest body organ. It performs the functions of protection, maintaining body temperature, picking up stimuli, and excretion.

3. The principal parts of the skin are the outer epidermis and inner dermis. The dermis overlies the subcutaneous layer.

4. The epidermal layers, from the inside outward, are the stratum basale, spinosum, granulosum, lucidum, and corneum. The basale and spinosum undergo continuous cell division and produce all outer layers.

5. The dermis consists of a papillary region and a reticular region. The papillary region is connective tissue containing blood vessels, nerves, oil glands, hair follicles, and papillae. The reticular region is connective tissue containing fat and sweat glands.

ACCESSORY ORGANS OF THE SKIN

1. The accessory organs of the skin are hair, sebaceous glands, sweat glands, and nails.

2. Hairs consist of a shaft above the surface, a root anchored in the dermis, and a hair follicle.

3. Sebaceous glands are connected to hair follicles. They secrete sebum, which moistens hair and waterproofs the skin.

4. Sweat glands produce perspiration, which carries wastes to the surface and helps regulate body temperature.

5. Nails are modified epidermal cells.

THE INTEGUMENTARY SYSTEM AND HOMEOSTASIS

1. Normal body temperature is 98.6°F.

2. If environmental temperature is high, skin receptors pick up the stimulus (heat) and convey the message to the brain. The brain then causes the sweat glands to produce sweat. As the sweat evaporates, the skin is cooled.

3. The skin-cooling response is a negative feedback mechanism.

DISORDERS OF THE INTEGUMENTARY SYSTEM

1. Acne is an inflammation of sebaceous glands.

2. Skin burns are classified as first-degree (damage to the epidermis), second-degree (damage to both epidermis and dermis but rapid regeneration of epidermis is still possible), and third-degree (damage to both epidermis and dermis).

3. The extent of burns can be found by using the "rule of nines."

4. Burn treatment may involve cleansing, removing dead tissues, replacing body fluids, and performing skin grafts.

Review questions and problems

1. Define an organ. In what respect is the skin an organ? What is the integumentary system?

2. List the principal functions of the skin.

3. Compare the structures of the epidermis and dermis. What is the subcutaneous layer?

4. List and describe the epidermal layers from the inside outward. What is the importance of each layer?

5. How is the dermis adapted to receive stimuli for touch or pressure?

6. Describe the structure of a hair. How are hairs moistened? What produces "goosebumps" or "gooseflesh"?

7. Contrast the locations and functions of sebaceous glands and sweat glands. What are the name and chemical components of the secretions of each?

8. From what layer of the skin do nails form? Describe the principal parts of a nail.

9. Explain, by way of a labeled diagram, how the skin helps to maintain normal body temperature.

10. What is a feedback system? A negative feedback system? Relate these definitions to the maintenance of normal body temperature.

11. What causes acne? How may the condition be aggravated? How is it treated?

12. Define a burn. Classify burns according to degree. How is the "rule of nines" used? What steps may be employed in treating burns?

13. Refer to the glossary of medical terminology associated with the integumentary system. Be sure you can define each term.

CHAPTER 6

THE INTEGRATED BODY: STRUCTURAL PLAN

STUDENT OBJECTIVES

After you have read this chapter, you should be able to:

1. List by name and location the principal body cavities and their major organs

2. Define the anatomical position

3. Compare common and anatomical terms used to describe the external features of the body

4. Define directional terms used in association with the body

5. Describe the common anatomical planes of the body

The human organism possesses certain anatomical characteristics that are easily identifiable and can, therefore, serve as reference landmarks. For example, man has a *backbone*, a characteristic that places him in a large group of organisms called vertebrates. In terms of body form, man is said to be bilaterally symmetric. *Bilateral sym-* *metry* means that the left and right sides of the body are mirror images. Another characteristic of the body's organization is that it resembles a *tube within a tube*. The outer tube is formed by the body wall, whereas the inner tube is the digestive tract. In the chapters that follow, you will need to know such general characteristics as well

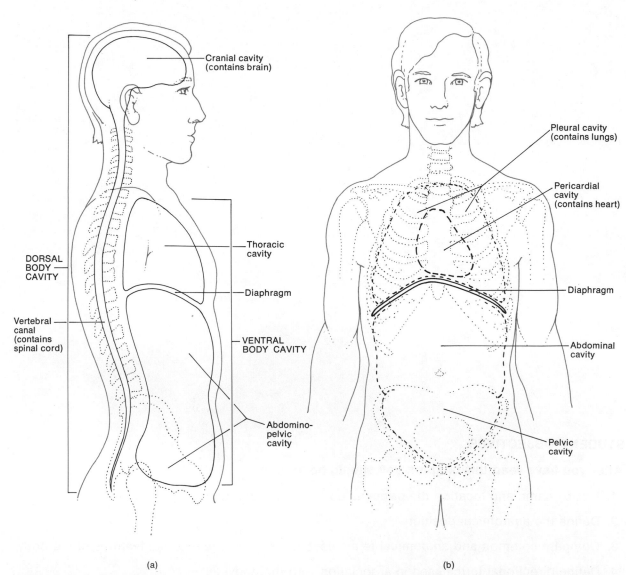

(a)

(b)

Figure 6–1. Body cavities. (a) Median section through the body to indicate the location of the dorsal and ventral body cavities. (b) Subdivisions of the ventral body cavity.

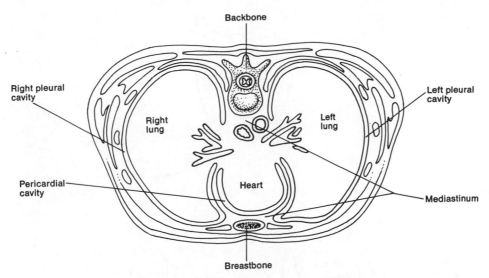

Figure 6-2. The mediastinum. Shown here is a cross section of the body through the thoracic region. A cross section is a section made perpendicular to the long axis of a part. The mediastinum is the space between the lungs and extends from the breastbone to the backbone.

BODY CAVITIES

Spaces within the body that contain various internal organs are called body cavities. Specific cavities may be distinguished if the body is viewed after making a *median,* or *sagittal, section*—that is, after cutting it into right and left sides. Figure 6-1a shows the two principal body cavities. The first of these, which may be called the *dorsal body cavity,* is located near the dorsal, or back, surface of the body. It is a bony cavity formed by the skull bones and the bones that enclose the spinal cord. These are referred to as the cranial bones and the vertebrae, respectively. The dorsal body cavity is further subdivided into a *cranial cavity,* which contains the brain, and a *vertebral canal,* which contains the spinal cord.

The second principal body cavity may be called the *ventral body cavity.* This cavity, also called the *celom,* is located inside the ventral, or front, surface of the body. The organs inside the ventral body cavity are collectively called the **viscera.** Its walls are composed of skin, connective tissue, bone, muscles, and serous membrane. Like the dorsal body cavity, the ventral body cavity has two principal subdivisions. These are the upper portion, called the *thoracic cavity,* or chest cavity, and the lower portion, called the *abdominopelvic cavity.* The anatomical landmark that divides the ventral body cavity into the thoracic and abdominopelvic cavities is the muscular diaphragm. The thoracic cavity, in turn, contains several divisions. These are two *pleural cavities,* one around each lung; the *pericardial cavity,* around the heart (Figure 6-1b); and the **mediastinum,** a space between the pleura of the lungs that extends from the breast bone to the backbone. The mediastinum contains the thymus gland, windpipe, gullet, heart and large blood vessels, and nerves (Figure 6-2). The abdominopelvic cavity, as the name suggests, is divided into two portions, although no wall lies between them (Figure 6-1b). The upper portion, called the *abdominal cavity,* contains the stomach, spleen, liver, gallbladder, pancreas, small intestine, most of the large intestine, the kidneys, and the ureters. The lower portion, called the *pelvic cavity,* contains the urinary bladder, sigmoid colon, rectum, and, in the female, reproductive organs. The abdominal cavity is arbitrarily separated from the pelvic cavity by drawing an imaginary line across the tops of the hipbones. In order to describe the location of organs more easily, the abdominopelvic area is sometimes also divided into the nine regions shown in Figure 6-3.

THE ANATOMICAL POSITION AND REGIONAL NAMES

When a region of the body is described in an anatomical text or chart, we assume that the body is in the anatomical position. The body in the **anatomical position** is erect and facing the observer. The arms are at the sides, and the palms

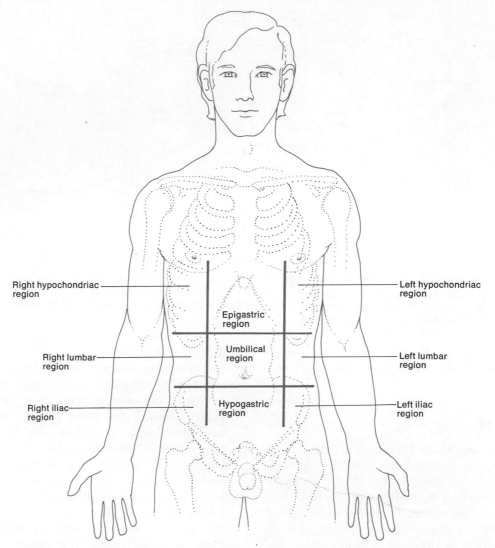

Right hypochondriac region

Left hypochondriac region

Epigastric region

Right lumbar region

Umbilical region

Left lumbar region

Right iliac region

Hypogastric region

Left iliac region

Figure 6–3. Nine regions of the abdominopelvic area. The top horizontal line is drawn near the bottom of the rib cage. The bottom horizontal line is drawn along the tops of the hipbones. The two vertical lines are drawn just to the inside of the nipples. These lines divide the area into a larger middle section and smaller left and right sections. The names of the regions are labeled.

of the hands are turned forward, as illustrated in Figure 6–4. The common terms and the anatomical terms, in parentheses, of certain body regions have been included in Figure 6–4. The anatomical names may not mean much to you at this point, but gaining some familiarity with them will assist you in later chapters.

DIRECTIONAL TERMS

In order to explain exactly where a structure of the body is located, anatomists must use certain *directional terms*. For instance, if you want to point out the breast bone to someone who knows

where the collar bone is, you can say that the breast bone is inferior (farther away from the head) and to the medial (toward the middle of the body) part of the collar bone. As you can see, using the terms *inferior* and *medial* avoids a great deal of complicated description. Since the directional terms listed in Exhibit 6–1 will be used throughout this book and in your laboratory work, we suggest that you learn them. Many of the terms defined in the Exhibit may be understood by referring to Figure 6–5. Essential parts of the figure are labeled so that you can see the directional relationships among parts. Refer to the figure after you read the definition for the directional term.

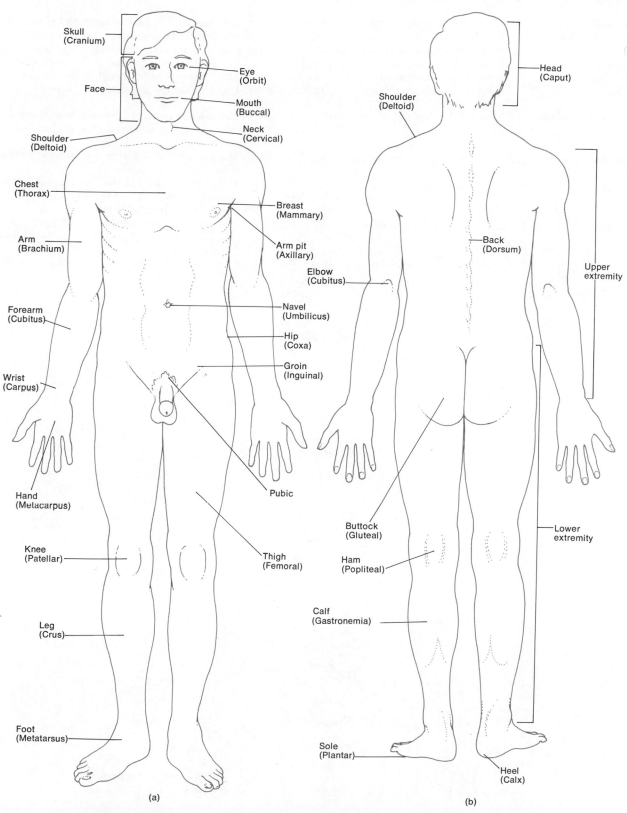

Skull
(Cranium)

Face

Shoulder
(Deltoid)

Chest
(Thorax)

Arm
(Brachium)

Forearm
(Cubitus)

Wrist
(Carpus)

Hand
(Metacarpus)

Knee
(Patellar)

Leg
(Crus)

Foot
(Metatarsus)

Eye
(Orbit)

Mouth
(Buccal)

Neck
(Cervical)

Breast
(Mammary)

Arm pit
(Axillary)

Elbow
(Cubitus)

Navel
(Umbilicus)

Hip
(Coxa)

Groin
(Inguinal)

Pubic

Thigh
(Femoral)

(a)

Shoulder
(Deltoid)

Back
(Dorsum)

Buttock
(Gluteal)

Ham
(Popliteal)

Calf
(Gastronemia)

Sole
(Plantar)

Head
(Caput)

Upper
extremity

Lower
extremity

Heel
(Calx)

(b)

Figure 6–4. The anatomical position. (a) Anterior view. (b) Posterior view. Both common terms and anatomical terms, in parentheses, are indicated for various regions of the body.

Exhibit 6–1. SUMMARY OF DIRECTIONAL TERMS

TERM	DEFINITION	EXAMPLE
1. Superior (cephalic)	Toward the head; toward the upper part of a structure	The heart is superior to the liver
2. Inferior (caudal)	Away from the head; toward the lower part of a structure	The rectum is inferior to the liver
3. Anterior (ventral)	Nearer to or at the front of the body	The urinary bladder is anterior to the rectum
4. Posterior (dorsal)	Nearer to or at the back of the body	The lungs are posterior to the breastbone
5. Medial	Nearer the midline of the body	The ulna is on the medial side of the forearm
6. Lateral	Farther from the midline of the body	The kidneys are lateral to the backbone
7. Proximal	Nearer the attachment of a limb to the trunk	The humerus is proximal to the radius
8. Distal	Away from the attachment of a limb to the trunk	The phalanges are distal to the wrist
9. External (superficial)	Toward the surface of the body	The muscles of the abdominal wall are external to the viscera in the abdominal cavity
10. Internal (deep)	Away from the surface of the body	The muscles of the arm are internal to the skin of the arm
11. Parietal	The walls of a cavity	The parietal layer of the serous pericardium helps form the wall of the pericardial cavity (see Figure 17–1)
12. Visceral	The covering of an organ	The visceral layer of the serous pericardium covers the external aspect of the heart (see Figure 17–1)

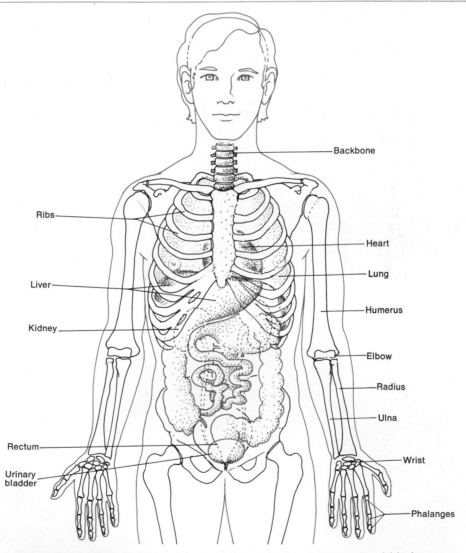

Figure 6–5. Directional terms. The various organs shown should help you understand the meanings of the terms superior, inferior, anterior, posterior, medial, lateral, proximal, and distal.

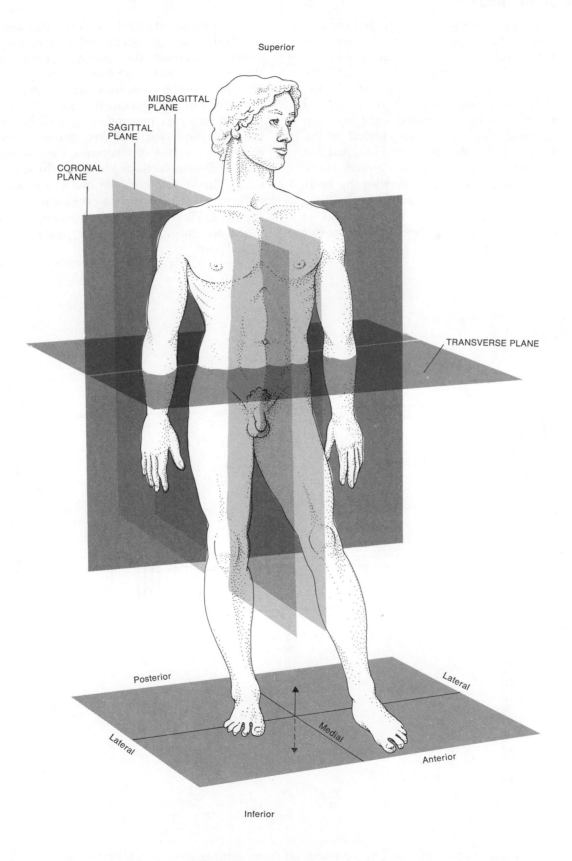

Figure 6–6. Planes of the body.

PLANES OF THE BODY

The structural plan of the human body may also be discussed with respect to planes that pass through it. Several of these planes are used commonly. Figure 6–6 illustrates the planes described below. A *midsagittal plane* through the midline of the body runs vertical to the ground and divides the body into equal right and left sides. A *sagittal plane* also runs vertical to the ground, but it divides the body into unequal left and right portions. A *frontal* or *coronal plane* runs vertical to the ground and divides the body into anterior and posterior portions. Finally, a *horizontal* or *transverse plane* runs parallel to the ground and divides the body into superior and inferior portions.

In subsequent chapters, you will view parts of the body that are sectioned in various ways in order to illustrate their most important features. When you examine sections of organs, it is important to understand how the section is made so that you can understand the anatomical relationship of one part to another. It will be helpful to illustrate how sections are made. Figure 6–7a presents a series of diagrams that indicate how three different sections are made through a simple tube. The sections shown are a *cross section*, an *oblique section*, and a *longitudinal section*. Figure 6–7b is a series of diagrams showing how the same three sections are made through the spinal cord.

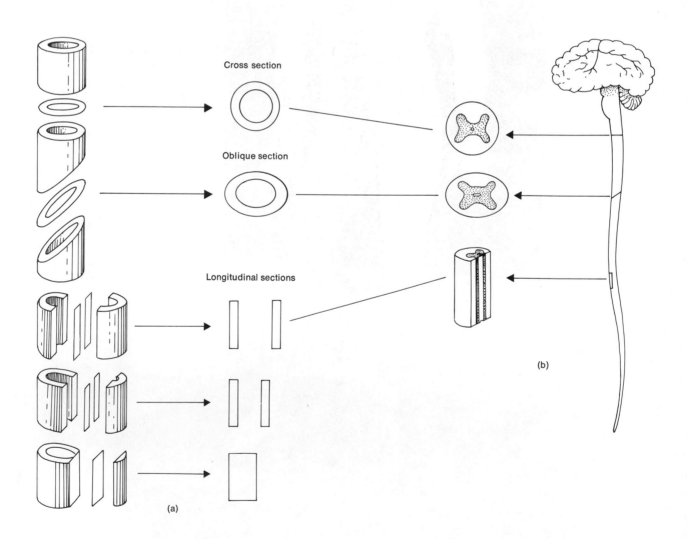

Figure 6–7. How sections are made. (a) Three sections through a tube. (b) Same three sections through the spinal cord.

Chapter summary in outline

BODY CAVITIES

1. Spaces in the body that contain internal organs are called cavities.

2. Dorsal and ventral cavities are the two principal body cavities. The dorsal cavity contains the brain and spinal cord. The organs of the ventral cavity are collectively called the viscera.

3. Dorsal cavity is subdivided into the cranial cavity and vertebral canal.

4. Ventral body cavity is subdivided by the diaphragm into an upper thoracic cavity and a lower abdominopelvic cavity.

5. Thoracic cavity contains two pleural cavities, pericardial cavity, and the mediastinum.

6. Abdominopelvic cavity, which is actually an upper abdominal cavity and a lower pelvic cavity, is divided into nine anatomical regions.

THE ANATOMICAL POSITION AND REGIONAL NAMES

1. The position in which the body is studied and described is the anatomical position. The subject stands erect and faces the observer, with arms at sides and palms turned forward.

2. Regional names are terms given to specific regions of the body for reference purposes. Examples of regional names include cranium (skull), thorax (chest), brachium (arm), patellar (knee), caput (head), and gluteal (buttock).

DIRECTIONAL TERMS

1. These are names given to indicate the relationship of one part of the body to another.

2. Some examples of directional terms are superior (toward the head), anterior (near the front), medial (nearer the midline), distal (farther from the attachment of a limb), and external (toward the surface).

PLANES OF THE BODY

1. Planes of the body are flat surfaces that divide the body into definite areas. The midsagittal plane divides the body into equal right and left sides; the sagittal plane, into unequal right and left sides; the frontal (coronal) plane, into anterior and posterior portions; the horizontal (transverse) plane, into superior and inferior portions.

2. Sections of organs include cross sections, oblique sections, and longitudinal sections.

Review questions and problems

1. What does bilateral symmetry mean? Why is the body considered to be a tube within a tube?

2. Define a body cavity. List the body cavities discussed, and tell which major organs are located in each. What landmarks separate the various body cavities from each other?

3. Discuss how the abdominopelvic area is subdivided into regions. Name and locate each region.

4. When is the body in the anatomical position? Why is the anatomical position used?

5. Review Figure 6–4. See if you can locate each region on your own body, and name each by its common and anatomical term.

6. What is a directional term? Why are these terms important? Can you use each of the directional terms listed in Exhibit 6–1 in a complete sentence?

7. Construct a series of figures of the human body, and indicate each directional term used in Exhibit 6–1 by labeling the relationship of organs to each other.

8. Describe the various planes that may be passed through the body. Explain how each plane divides the body.

9. What is meant by the phrase, "A part of the body has been sectioned?" Given an orange, can you make a cross section, oblique section, and longitudinal section with a knife?

Bibliography

Unit 1

Allison, A.: "Lysosomes and Disease," *Scientific American,* **217**:62, November 1967.

Arey, L. B.: *Human Histology,* 3d ed., W. B. Saunders Company, Philadelphia, 1968.

Atlas of Human Anatomy, College Outline Series, Barnes & Noble Books, div. of Harper & Row, Publishers, Inc., New York, 1970.

Berns, M. W. and D. E. Rounds: "Cell Surgery by Laser," *Scientific American,* February 1970.

Best, C. H. and N. B. Taylor: *Physiological Basis of Medical Practice,* The Williams & Wilkins Company, Baltimore, 1966.

Beveridge, G. W. and E. W. Powell: "The Problem of Acne," *The Practitioner,* 204, May 1970.

Bickers, D. R. and A. Kappas: "Metabolic and Pharmacologic Properties of the Skin," *Hospital Practice,* May 1974.

Bourne, G. H.: *Division of Labor in Cells,* Academic Press, Inc., New York, 1970.

Burgoon, C. F.: "Acne," *Modern Medicine,* Apr. 17, 1972.

"The Cell Wall," *Medical World News,* Aug. 13, 1971.

DuPraw, E. J.: *DNA and Chromosomes,* Holt, Rinehart and Winston, Inc., New York, 1970.

"Facets of Genetics," *Readings from Scientific American,* W. H. Freeman and Company, San Francisco, 1970.

Frenay, A. C., Sr.: *Understanding Medical Terminology,* 4th ed., The Catholic Hospital Association, Publishers, St. Louis, 1970.

"From Cell to Organism," *Readings from Scientific American,* W. H. Freeman and Company, San Francisco, 1967.

Goldberg, N. D.: "Cyclic Nucleotides and Cell Function," *Hospital Practice,* May 1974.

Good, R. A. and D. W. Fisher (eds.): *Immunobiology,* Sinauer Associates, Inc., Stamford, Conn., 1971.

Green, D. E.: "The Mitochondrion," *Scientific American,* **210**:63, January 1964.

Greep, R. O. and L. Weiss: *Histology,* 3d ed., W. B. Saunders Company, Philadelphia, 1968.

Gross, V. E.: *Mastering Medical Terminology,* Halls of Ivy Press, North Hollywood, Calif., 1969.

Gurdon, J. B.: "Transplanted Nuclei and Cell Differentiation," *Scientific American*, December 1968.

Guyton, A. C.: *Textbook of Medical Physiology*, 4th ed., W. B. Saunders Company, Philadelphia, 1971.

Ham, A. W.: *Histology*, 6th ed., J. B. Lippincott Company, Philadelphia, 1969.

Hendler, R. W.: "Biological Membrane Ultrastructure," *Physiol. Rev.*, **51**:66, 1971.

Hoerr, N. L. and A. Osol (eds.): *Blakiston's New Gould Medical Dictionary*, 2d ed., The Blakiston Division, McGraw-Hill Book Company, New York, 1956.

Hollingshead, W. H.: *Textbook of Anatomy*, 2d ed., Harper & Row, Publishers, Inc., New York, 1967.

Holtzman, E.: "The Biogenesis of Organelles," *Hospital Practice*, March 1974.

Holvey, D. N. and J. H. Talbott (eds.): *The Merck Manual*, 12th ed., Merck Sharp & Dohme Research Laboratories, Publishers, Rahway, N.J., 1972.

Kornberg, A.: "The Synthesis of DNA," *Scientific American*, October 1968.

Langley, L. L.: *Homeostasis*, Reinhold Publishing Corporation, New York, 1965.

Larson, D. and R. Gaston: "Current Trends in the Care of Burned Patients," *Amer. J. Nurs.*, **67**:2, 319, February 1967.

"The Living Cell," *Readings from Scientific American*, W. H. Freeman and Company, San Francisco, 1965.

Markert, C. L. and H. Ursprung: *Developmental Genetics*, Prentice-Hall, Inc., Englewood Cliffs, N.J., 1971.

Marples, M. J.: "Life on the Human Skin," *Scientific American*, January 1969.

Memmler, R. L. and R. B. Rada: *The Human Body in Health and Disease*, 3d ed., J. B. Lippincott Company, Philadelphia, 1970.

Mikal, S.: *Homeostasis in Man*, Little, Brown and Company, Boston, 1967.

Montagna, W.: *Advances in Biology of Skin*, vols. 1–12, Appleton-Century-Crofts, New York, 1972.

——: "The Skin," *Scientific American*, **212**:56, February 1965.

Neutra, M. and C. P. LeBlond: "The Golgi Apparatus," *Scientific American*, **220**:100, February 1969.

Nicoll, P. A. et al.: "The Physiology of the Skin," *Ann. Rev. Physiol.*, **34**:177, 1972.

Nomura, M.: "Ribosomes," *Scientific American*, **221**:28, October 1969.

Reith, E. J. and M. H. Ross: *Atlas of Descriptive Histology*, 2d ed., Harper & Row, Publishers, Inc., New York, 1970.

Rosenberg, E.: *Cell and Molecular Biology*, Holt, Rinehart and Winston, Inc., New York, 1971.

Ross, R.: "Wound Healing," *Scientific American*, June 1969.

—— and P. Bornstein: "Elastic Fibers in the Body," *Scientific American*, **224**:44, June 1971.

Satir, P.: "How Cilia Move," *Scientific American*, October 1974.

"Scanning Electron Microscopy," *Medical World News*, Dec. 7, 1973.

Sharon, N.: "The Bacterial Cell Wall," *Scientific American*, **220**:92, May 1969.

Solomon, A. K.: "The State of Water in Red Cells," *Scientific American*, **224**:89, February 1971.

Steen, E. B. and M. F. Montagu: *Anatomy & Physiology*, vols. 1 and 2, College Outline Series, Barnes & Noble Books, div. of Harper & Row, Publishers, Inc., New York, 1959.

Taber, C. W.: *Taber's Cyclopedic Medical Dictionary*, 8th ed., F. A. Davis Company, Philadelphia, 1959.

Temin, H. M.: "RNA-Directed Synthesis," *Scientific American*, **226**:24, January 1972.

Toner, P. G. and K. E. Carr: *Cell Structure: An Introduction to Biological Electron Microscopy*, 2d ed., The Williams & Wilkins Company, Baltimore, 1971.

Wessells, N. K. and W. J. Rutter: "Phases in Cell Differentiation," *Scientific American*, **220**:36, March 1969.

Winchester, A. M.: *Human Genetics*, Charles E. Merrill Company, New York, 1971.

Young, C. C. and J. D. Barger: *Introduction to Medical Science*, 2d ed., The C. V. Mosby Company, St. Louis, 1969.

—— and ——: *Learning Medical Terminology Step by Step*, 3d ed., The C. V. Mosby Company, St. Louis, 1969.

PRINCIPLES OF SUPPORT AND MOVEMENT

CHAPTER 7

SKELETAL TISSUE: STRUCTURE, PHYSIOLOGY, AND DISORDERS

STUDENT OBJECTIVES

After you have read this chapter, you should be able to:

1. Describe the components of the skeletal system

2. Describe the functions of the skeletal system

3. Describe the gross features of a long bone

4. Describe the histological features of dense bone tissue

5. Compare the histological characteristics of cancellous and dense bone

6. Define ossification

7. Contrast the steps involved in intramembranous and endochondral ossification

8. Interpret roentgenograms of normal ossification

9. Describe bone construction and destruction as a homeostatic mechanism

10. Describe the conditions necessary for normal bone growth

11. Define rickets, osteomalacia, and scurvy as vitamin deficiency disorders

12. Contrast the causes and clinical symptoms associated with osteoporosis, Paget's disease, and osteomyelitis

13. Compare the origin of tumors of the skeletal system

14. Define medical terminology related to the skeletal system

Without the skeletal system, man would be little more than a slug. We would be unable to make coordinated movements, such as walking or grasping. The slightest knock on the head or chest could damage the brain or heart. It would even be impossible for us to chew our food. The framework of bones that protects our organs and allows us to move is called the **skeletal system.** In addition to bone, the skeletal system consists of cartilage, which is found in the nose, larynx, outer ear, and where bones attach to one another. The points at which the surfaces of bones attach to each other are called *joints,* or *articulations.*

The skeletal system performs several important, basic functions. First, it *supports* the soft tissues of the body so that the form of the body and an erect posture can be maintained. Second, the system *protects* delicate structures, such as the brain, the spinal cord, the lungs, the heart, and the major blood vessels in the thoracic cavity. Third, the bones serve as *levers* to which the muscles of the body are attached. When the muscles contract, the bones acting as levers produce *movement.* Fourth, the bones serve as *storage areas* for mineral salts — especially calcium and phosphorus. One last important feature of the skeletal system is *blood-cell production,* which occurs in the red marrow of the bones.

In this chapter, we shall discuss the microscopic structure of bone, the formation of bone, and disorders of bone. In Chapter 8, we shall begin a study of specific bones of the body.

A MICROSCOPIC VIEW

Structurally, the skeletal system consists of two types of connective tissue: cartilage and bone. In Chapter 4, we discussed the microscopic structure of cartilage. You may wish to review that discussion as you study the skeletal system. Here, our attention will be directed to discussing the microscopic structure of bone tissue.

Like other connective tissues, **bone,** or **osseous tissue,** contains a great deal of intercellular substance surrounding widely separated cells.

Unlike other connective tissues, however, the intercellular substance of bone contains mineral salts, primarily calcium phosphate and calcium carbonate. These salts are responsible for the hardness of bone, which is thus said to be ossified. Embedded in the intercellular substance are collagenous fibers, which further reinforce the tissue.

The microscopic structure of bone may be analyzed by considering the anatomy of a long bone such as the humerus, or arm bone. As shown in Figure 7–1a, b, a typical long bone consists of the following parts: (1) the **diaphysis,** which is the shaft or main, long portion; (2) the **epiphyses,** which are the extremities or ends of the bone; (3) the **articular cartilage,** a thin layer of hyaline cartilage covering the epiphysis where the bone forms a joint with another bone; and (4) the **periosteum,** a dense, white, fibrous membrane covering the remaining surface of the bone. The periosteum, coming from the terms *peri,* meaning around, and *osteo,* meaning bone, consists of two layers. The outer, fibrous layer is composed of connective tissue containing blood vessels, lymphatic vessels, and nerves that run into the bone. The inner layer contains elastic fibers, blood vessels, and **osteoblasts,** which are cells responsible for forming new bone during growth and repair (Figure 7–1c). The word *blast,* which is a part of many terms, means a germ or bud. It denotes an immature cell or tissue that develops into a more specialized form later on. The periosteum is essential for bone growth, repair, and nutrition. It also serves as a point of attachment for ligaments and tendons. In addition to the four parts listed, a typical long bone also has: (5) a **medullary cavity,** or **marrow cavity,** which is the space within the diaphysis that contains the fatty *yellow marrow;* and (6) an **endosteum,** which is a membrane that lines the medullary cavity and contains osteoblasts.

Despite its macroscopic appearance, bone is not a solid, homogeneous substance. In fact, all bone is porous. In other words, it is full of pores or holes. As you will see, the pores contain living

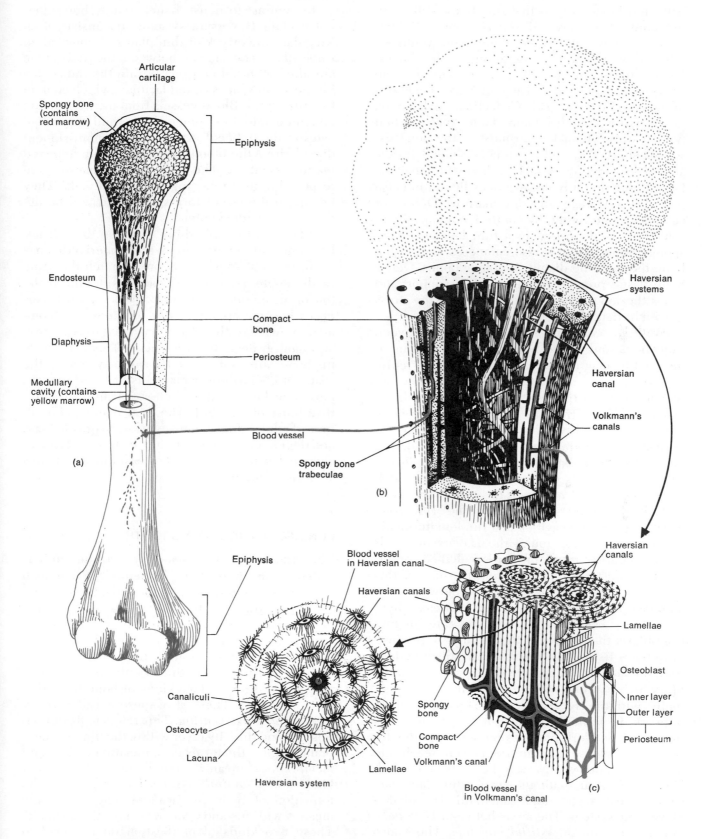

Figure 7–1. Osseous tissue. (a) Macroscopic appearance of a long bone that has been sectioned lengthwise. (b) Histological structure of bone. (c) Enlarged aspect of Haversian systems.

cells and blood vessels that supply the cells with nutrients. The pores also make bones lighter. Think of how much more energy you would expend if you had to drag around solid bones. Depending upon the degree of porosity, the regions of a bone may be categorized as cancellous or dense (Figure 7–1b). **Cancellous,** or **spongy,** bone tissue contains many large spaces filled with marrow. It makes up most of the bone tissue of short, flat, and irregularly shaped bones, and of the epiphyses of long bones. Cancellous bone tissue not only makes bones lighter but also provides a storage area for marrow. **Dense,** or **compact,** bone tissue, by contrast, contains fewer spaces. It is deposited in a thin layer over the spongy bone tissue. It also composes most of the bone tissue of the diaphysis of long bones. Dense bone tissue provides protection and support and helps the long bones of the body resist the stress of weight that is placed on them.

We can compare the differences between cancellous and dense bone tissue by looking at the highly magnified, transverse sections represented in Figure 7–1. One main difference between the bone tissues is that dense bone has a concentric-ring structure, whereas spongy bone does not. Blood vessels and nerves from the periosteum penetrate into the compact, or dense, bone through *Volkmann's canals* (Figure 7–1c). The blood vessels of these canals connect with blood vessels and nerves of the medullary cavity and of the *Haversian canals.* The Haversian canals are circular cavities that run longitudinally through the bone. Around the Haversian canals are concentric rings of hard, calcified, intercellular substance. The concentric rings of this substance are called *lamellae.* Between the lamellae are small spaces called *lacunae,* where osteocytes are found. **Osteocytes** are mature osteoblasts that have lost their ability to produce new bone tissue. Radiating out in all directions from the lacunae are minute canals called *canaliculi,* which connect with other lacunae and, eventually, with the Haversian canal. Thus, an intricate network is formed throughout the bone. This branching network provides numerous routes for blood vessels so that nutrients can reach the osteocytes and wastes can be removed. Each Haversian canal, with its surrounding lamellae, lacunae, osteocytes, and canaliculi, is called a **Haversian system.** The areas between Haversian systems contain *interstitial lamellae.* These also possess lacunae with osteocytes and canaliculi, but their lamellae are usually not connected to the Haversian systems.

In contrast to dense bone, spongy bone does not contain Haversian systems. It consists of an irregular latticework of thin plates of bone called *trabeculae.* The spaces between the trabeculae are filled with *red marrow.* Within the trabeculae lie the small spaces called lacunae, which contain the osteocytes. Blood vessels from the periosteum penetrate into the spongy bone, and the osteocytes in the trabeculae receive their nourishment directly from the blood circulating through the red marrow cavities. The cells of red marrow are responsible for producing new blood cells. They belong to the circulatory system and will be described in a later chapter.

When most people think of bones, they visualize some sort of very hard, white material that has a variety of shapes, yet is pretty much the same in all people. However, if you consider the skeleton of an infant or a young child, certain differences become apparent. For example, it is common knowledge that it is very dangerous to drop an infant, especially on his head. This is because his bones are "soft," and the fall may change the shape of his head or damage his brain. Moreover, most of us know that the bones of a child are softer than those of an adult. The final shape and hardness of adult bone take many years to develop and are dependent upon a rather complex series of chemical changes. Let us now examine how bones are formed and how they grow.

BONE FORMATION AND GROWTH

The process by which bone forms in the body is called **ossification.** The "skeleton" of a human embryo is composed of fibrous membranes and hyaline cartilage, both of which are shaped like bones and provide the medium in which ossification occurs (Figure 7–2). The actual process of ossification begins around the sixth week of embryonic life and continues until an individual reaches adulthood. Two kinds of bone formation are recognized. The first of these is called *intramembranous ossification.* This refers to the formation of bone directly on or within the fibrous membranes—thus, the terms *intra* meaning within, and *membranous* meaning membrane. The second kind, *endochondral ossification,* refers to the formation of bone in cartilage. The term *endo* means within, and *chondro* means cartilage. These two kinds of ossification do *not* lead to any differences in the structure of mature bones. They simply indicate different methods of bone formation.

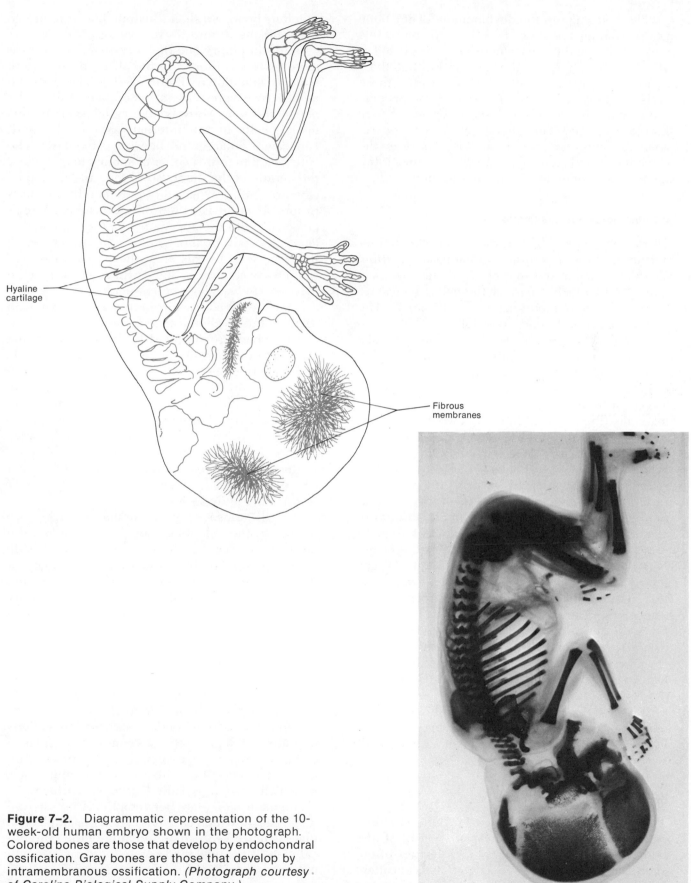

Hyaline
cartilage

Fibrous
membranes

Figure 7–2. Diagrammatic representation of the 10-week-old human embryo shown in the photograph. Colored bones are those that develop by endochondral ossification. Gray bones are those that develop by intramembranous ossification. *(Photograph courtesy of Carolina Biological Supply Company.)*

The first stage in the development of any bone is the coming together of embryonic connective tissue cells in the area where bone formation is about to begin. Next, the area is infiltrated by many small blood vessels. Soon thereafter, these cells increase in number and size. Some become transformed into chondroblasts, and some change into osteoblasts. The chondroblasts will be responsible for cartilage formation, whereas the osteoblasts will form bone tissue by either intramembranous or endochondral ossification.

Intramembranous ossification

Of the two types of bone formation, the simpler and more direct is *intramembranous ossification.* The flat bones of the roof of the skull, lower jaw bone, and probably part of the collar bones are formed by intramembranous ossification. The essentials of this process are as follows.

Osteoblasts formed from mesenchymal cells cluster in the fibrous membrane. The site of such a cluster is called a *center of ossification.* The osteoblasts then secrete intercellular substances. These substances are partly composed of collagenous fibers that form a framework, or matrix, in which calcium salts are quickly deposited. The laying down of calcium salts is called *calcification.* When a cluster of osteoblasts is completely surrounded by the calcified matrix, it is called a *trabecula.* As trabeculae form in nearby ossification centers, they fuse into the open latticework characteristic of spongy bone. With the formation of successive layers of bone, some of the osteoblasts become entrapped in the minute spaces referred to as lacunae. The entrapped osteoblasts lose their ability to form bone and are called osteocytes. The spaces between the trabeculae fill with red marrow. The original connective tissue that surrounds the growing mass of bone then becomes the periosteum. The ossified area has now become true spongy bone. Eventually, the surface layers of the spongy bone will be reconstructed into dense bone. Much of this newly formed bone will be destroyed and re-formed so that the bone may reach its final adult size and shape.

Endochondral ossification

The replacement of cartilage by bone is called *endochondral ossification.* Most bones of the body, including the majority of the bones of the skull, are formed by endochondral ossification. Since this type of ossification is best observed

in a long bone, we shall illustrate how it occurs in the shinbone (Figure 7-3).

Early in embryonic life, a cartilage model that more or less conforms to the shape of the future bone is laid down (Figure 7-3a). The cartilage model is covered by a membrane called the *perichondrium.* Midway along the shaft of this model, cells in the internal layer of the perichondrium enlarge and become osteoblasts. The cells begin to form a collar of spongy bone around the middle of the diaphysis of the cartilage model (Figure 7-3b). Once the perichondrium starts to form bone, it is called the periosteum. Simultaneously with the appearance of the bone collar, changes occur in the cartilage in the center of the diaphysis. In this area, the *primary ossification center* (Figure 7-3c), cartilage cells increase in size. The cartilage cells enlarge because they accumulate glycogen for energy and enzymes to catalyze future chemical reactions. The glycogen and enzymes are supplied by blood vessels that grow into the region. Some of the cartilage cells become osteoblasts. The osteoblasts start laying down bone in the interior of the diaphysis. The remaining cartilage cells in the area degenerate, leaving cavities behind. Blood vessels then grow along the spaces where cartilage cells were (Figure 7-3d) and open up the cavities further. Gradually, these spaces in the middle of the shaft connect with each other, and the marrow cavity is formed (Figure 7-3e).

As all these developmental changes are occurring, the osteoblasts of the periosteum deposit successive layers of bone on the outside so that the collar thickens, becoming thickest at the epiphyses (Figure 7-3f). The cartilage model continues to grow at its ends so that it steadily increases in length. Eventually, secondary ossification centers appear in the epiphyses and also lay down spongy bone. In the shinbone, one secondary ossification center develops in the proximal epiphysis soon after birth. The other center develops in the distal epiphysis during the child's second year (Figure 7-3f, g).

After the two secondary ossification centers have formed, bone tissue has completely replaced cartilage, except in two regions. Cartilage continues to cover the outside of the epiphyses, in which case it is called articular cartilage. It also remains as a plate between the epiphysis and diaphysis, in which case it is called the **epiphyseal plate** (Figure 7-3h). The function of the epiphyseal plate is to allow the bone to increase in length until early adulthood. As the child grows, cartilage cells are produced by mitosis on the epiphyseal

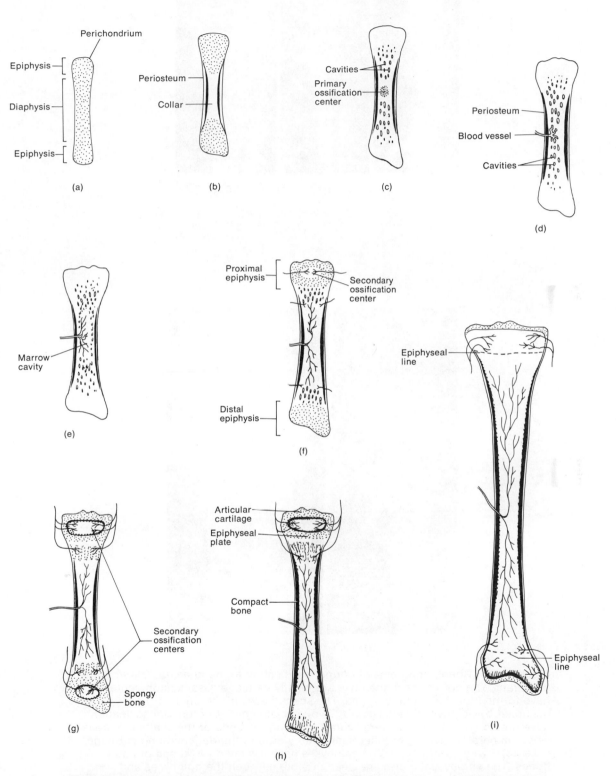

Figure 7–3. Endochondral ossification of the shinbone (tibia). (a) Cartilage model. (b) Collar formation. (c) Development of primary ossification center. (d) Entrance of blood vessels. (e) Marrow-cavity formation. (f) Thickening and lengthening of the collar. (g) Formation of secondary ossification centers. (h) Remains of cartilage as the articular cartilage and epiphyseal plate. (i) Formation of the epiphyseal line.

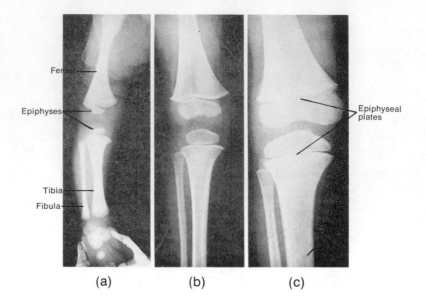

(a)   (b)   (c)

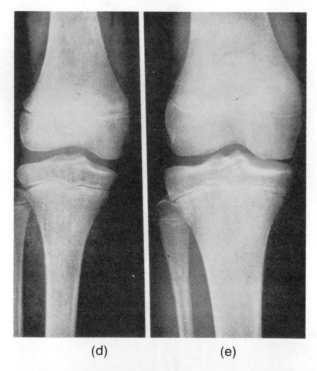

(d)   (e)

Figure 7–4. Roentgenograms of normal ossification at the knee. (a) One-month-old infant. The epiphyses at the knee are mostly cartilage. Ossification centers have formed for the femur and tibia. The space between the ends of the bones is occupied by epiphyseal cartilage. (b) Two-year-old child. Centers of ossification have grown. The white transverse zones marking the ends of the shafts are areas where mineral salts are deposited temporarily around the degenerating cartilage cells. (c) Five-year-old child. The epiphyses have assumed the shape of adult bones. The epiphyseal plates are clearly visible between the epiphyses and diaphyses of all three bones. (d) Eight-year-old child. The epiphyseal plates are still distinct as ossification continues. (e) Twelve-year-old child. The epiphyses have ossified almost completely. The epiphyseal plates are assuming the character of epiphyseal lines. *(From Lester W. Paul and Jonn H. Juhl, The Essentials of Roentgen Interpretation, 3d ed., Harper & Row, Publishers, Inc., New York, 1972.)*

side of the plate. Cells are then destroyed and replaced by bone on the diaphyseal side of the plate. In this way, the thickness of the epiphyseal plate remains fairly constant, but the bone on the diaphyseal side increases in length. Growth in diameter occurs along with growth in length. In this process, the bone lining the marrow cavity is destroyed so that the cavity increases in diameter. At the same time, osteoblasts from the periosteum add new osseous tissue around the outside of the bone. Initially, diaphyseal and epiphyseal ossification produce all spongy bone. Later, by reconstruction, the outer region of spongy bone is reorganized into compact bone.

Around the age of 17, the epiphyseal cartilage cells stop multiplying, and the entire cartilage begins to be replaced by bone. Bone growth now stops. The remnant of the epiphyseal plate is called the *epiphyseal line* (Figure 7–3i). Ossification of all bones is usually completed by age 25. Figure 7–4 consists of a series of **roentgenograms,** or photographs taken with x-rays, that show ossification in the epiphyses of two long bones at the knee. Bones undergoing either intramembranous or endochondral ossification are continually remodeling themselves from the time that initial calcification occurs until the final structure appears. Compact bone is made by the reworking of spongy bone. The diameter of a long bone is increased by the destruction of the bone closest to the marrow cavity and the construction of new bone around the outside of the diaphysis. However, even after bones have reached their adult shapes and sizes, old bone is perpetually destroyed, and new osseous tissue is laid down in its place. We can now take a look at how an equilibrium is reached between the growth and destruction of adult bone.

HOMEOSTASIS OF BONE

Bone shares with the skin the unique feature of replacing itself throughout adult life. This remodeling of adult bone takes place at different rates in different parts of the body. For instance, the distal area of the thighbone replaces itself about every 4 months. By contrast, bone in certain areas of the shaft will not go through a complete turnover in the individual's life. Remodeling allows worn or injured bone to be removed and replaced with new tissue. It also allows bone to serve as the body's storage area for calcium. Many other tissues in the body need small amounts of calcium in order to perform their functions. For instance, muscle needs calcium in order to contract. The muscle cells take their calcium from the blood. However, the blood itself needs calcium in order to clot. The blood continually trades off calcium with the bones, removing calcium when other tissues are not receiving enough of this element and resupplying the bones with calcium to keep them from getting soft.

The cells responsible for the destruction of bone tissue are called **osteoclasts,** the term *clast* coming from *klan,* meaning to break. In the healthy adult, a delicate homeostasis is maintained between the action of the osteoclasts and the removal of calcium, on the one hand, and the action of the bone-making osteoblasts and the deposition of calcium on the other. Should too much new tissue be laid down, the bones will become abnormally thick and heavy. If too much calcium is deposited in the bone, the surplus may form thick bumps, or spurs, on the bone that interfere with movement at joints. A loss of too much tissue or calcium weakens the bones and makes them break easily.

Normal bone growth in the young and bone replacement in the adult depend upon several factors. First, sufficient quantities of calcium and phosphorus, components of the chief salt that makes bone hard, must be included in the diet. Second, the individual must obtain sufficient amounts of vitamins A, C, and D. These substances are particularly responsible for the proper utilization of calcium and phosphorus by the body. Third, the body must manufacture the right amounts of the hormones responsible for bone-tissue activity.

Hormones regulate the growth and remodeling of bones. Certain hormones are responsible for the general growth of bones. Too much or too little of these hormones during childhood causes the person to be abnormally tall or short. Other hormones specialize in regulating the osteoclasts. And still others, especially the sex hormones, aid osteoblastic activity and thus promote the growth of new bone. The sex hormones act as a double-edged sword. They aid in the growth of new bone, but they also bring about the degeneration of all the cartilage cells in the epiphyseal plates. Due to the sex hormones, the typical adolescent goes through a spurt of growth around puberty, when sex-hormone production starts to increase. The individual then quickly finishes his or her growth as the epiphyseal cartilage disappears. Premature puberty can actually prevent the individual from reaching an average adult height because it brings with it premature degeneration of the plates. Still another kind of hormone, which is produced by the parathyroid glands in the neck, determines whether the blood will deposit calcium and phosphorus in osseous tissue or whether it will remove these elements from the bones. The names and activities of these hormones probably do not mean much to you now. For the time being, simply remember that hormones control the homeostasis of bone growth and maintenance. In a later chapter, we shall explain how the hormones accomplish these tasks.

BONE DISORDERS

Disorders always result from some disruption in the homeostasis. You have learned how important certain vitamins, minerals, and hormones are to bone growth and maintenance. Many bone disorders result from deficiencies in vitamins or minerals or from too much or too little of the hormones that regulate bone homeostasis. Infections and tumors, which can interfere with the homeostasis of any tissue, are also responsible for some bone disorders.

Vitamin deficiencies

Vitamin D is important to normal bone growth and maintenance. Vitamin D is essential for the synthesis of a protein that transports the calcium obtained from foods across the lining of the intestine and into the extracellular fluid. When the body lacks this vitamin, it is unable to absorb calcium and phosphorus from foods. A deficiency of vitamin D produces rickets in children and a condition called osteomalacia in adults.

In the condition called **rickets**, epiphyseal cartilage cells cease to degenerate, and new cartilage continues to be produced. This causes the width of the epiphyseal cartilage to become wider than normal. At the same time, the soft matrix laid down by the osteoblasts in the diaphysis fails to calcify. As a result, the bones stay soft. When the child walks, the weight of the body causes the bones in the legs to bow. Malformations of the head, chest, and pelvis also occur.

The cure and prevention of rickets consists of adding generous amounts of calcium, phosphorus, and vitamin D to the diet. Exposing the skin to the ultraviolet rays of sunlight also helps the body manufacture additional vitamin D.

A deficiency of vitamin D in an adult means that the bones give up excessive amounts of calcium and phosphorus. This loss, called *demineralization,* is especially heavy in the bones of the pelvis, legs, and spine. Demineralization caused by vitamin D deficiency is called **osteomalacia,** the term *malacia* meaning softness. After the bones soften, the weight of the body produces a bowing of the leg bones, a shortening of the backbone, and a flattening of the pelvic bones. Osteomalacia mainly affects women who live on poor cereal diets devoid of milk, are seldom exposed to the sun, and have repeated pregnancies that deplete the body of calcium. The condition responds to the same treatment as rickets: if the disease is severe enough and threatens life, large doses of vitamin D should be given.

Osteoblast and osteoclast regulation

Osteoporosis is a bone disorder affecting middle-aged and elderly people. The majority of those affected are women over the age of 60 years. Between puberty and the middle years, the sex hormones maintain strong, healthy, osseous tissue by stimulating the osteoblasts to make new bone. However, after the "change of life," women produce less of the sex hormones. During old age, both men and women produce smaller amounts of these hormones. As a result, the osteoblasts become less active, and the bones become less dense and more porous—hence, the name of the disorder. Most elderly people suffer from some degree of osteoporosis. But it is apt to be most severe in women who experience a drastic cutback in female hormones after the menopause and in men and women who do not have enough calcium in their diets. Osteoporosis most often affects the spine. The thighbone is also a frequently affected area. As the spine fails to maintain itself, the individual shrinks in height and may develop a "dowager's hump." The porous bones also become too weak to stand much strain and fracture easily. Most physicians suggest that middle-aged and elderly people include more calcium in their diets, as well as increased amounts of vitamin A, C, and D. Hormone therapy is also sometimes a part of treatment or prevention.

Another disorder, called **Paget's disease,** is characterized by an irregular thickening and softening of the bones. It is rarely seen in people under the age of 50. The cause, or *etiology,* of the disease is unknown. But the bone-producing osteoblasts and the bone-destroying osteoclasts apparently become uncoordinated. In some areas, too much bone is produced, whereas in other areas too much old bone is removed. This is an excellent example of a homeostatic disturbance. That is, the balance between bone formation and bone destruction is altered. Paget's disease affects the skull, the pelvis, and the bones of the limbs. Very little can be done to alter the course of the disease.

Stress of infection

Osteomyelitis is the term that includes all the infectious diseases of bone. These diseases may be localized, or they may affect many bones. The infections may also involve the periosteum, marrow, and cartilage. Quite a few microorganisms may give rise to bone infection. But the most frequent are bacteria known as *Staphylococcus aureus*, which are commonly called "staph." These bacteria may reach the bone by various means: the bloodstream; an injury such as a fracture; or an infection, such as a soft-tissue abscess, a sinus infection, or an abscess of the tooth.

Before antibiotic treatment became available, osteomyelitis often became a long-lasting condition. It can destroy extensive areas of bone, spread to nearby joints, and, in rare cases, lead to death by producing abscesses in many parts of the body. Penicillin and other antibiotics have been effective in treating the disease and in preventing it from spreading through extensive areas of the bone.

Cancer of bone

When cells in some area of the body reproduce unusually quickly so that an excess of tissue develops, the excess is called a growth, or **tumor.** Tumors may be cancerous and fatal, or they can be quite harmless. A cancerous growth is called a **malignant tumor,** or **malignancy.** A noncancerous growth is called a **benign growth.**

The cells of malignant growths all have one thing in common: they reproduce continuously and often very quickly. This growth continues until the victim dies or until every malignant cell is removed through surgery or destroyed by other means. As the cancer grows, it runs out of room and begins to compete with normal tissues for space and nutrients. Eventually, the normal tissues will lose the fight and die. The organ will function less and less efficiently, until it finally ceases to function altogether. The cancer cells may also spread to regions of the body that are far away from the original, or *primary*, growth. Cancer of the breast, for instance, has a tendency to spread to the lungs. The spread of the cancer to other regions of the body is called **metastasis.** Metastasis occurs when a malignant cell breaks away from the growth, enters the bloodstream, and is carried through the body. Wherever the cell finally comes to rest, it will set up another tumor, called a *secondary* growth. Usually death is caused when a vital organ loses its fight against the cancer cells for room and nutrients. Pain develops when the growth impinges on nerves or blocks a passageway so that secretions build up pressure behind the blockage.

The name of the cancer is derived from the type of tissue in which it originally develops. *Sarcoma* is a general term for any cancer arising from connective tissue. For example, *osteogenic sarcomas* are malignant growths of osteoblasts. *Osteo*, of course, means bone; *genic* comes from the same words as *genesis* and *generate*, meaning origin. This is the most frequent type of childhood cancer, but it is rare in adults. Osteogenic sarcomas destroy normal bone tissue and eventually spread to other areas of the body. If the affected bone is not removed before metastasis occurs, death is inevitable. *Myelomas* are malignant tumors occurring in the bone marrow of middle-aged and older people. The term *myelos* means marrow, and *oma* is the term for tumor. These tumors interfere with the blood-cell-producing function of bone marrow and cause anemia. They also destroy normal bone tissue and eventually spread throughout the body. *Chondrosarcoma* is a cancerous growth of the cartilage.

Benign tumors are composed of cells that do not metastasize. That is, the growth remains in one small area of the body and does not spread to other organs. All or part of a benign tumor may be removed so that cells can be examined under a microscope to make sure they are not malignant. This procedure is called a biopsy. A benign tumor is also removed if it gets in the way of a normal body function or if it causes disfiguration. Like malignancies, benign tumors are named after the type of cell from which they originate. The names of these tumors usually contain the combining form *oma*. For instance, an osteoblastoma is a benign growth of osteoblasts.

Medical terminology associated with the skeletal system

Achondroplasia   (a = without; chondro = cartilage; plasia = growth)   Imperfect ossification within cartilage of long bones during fetal life; also called fetal rickets.

Brodie's abscess   Infection in the spongy tissue of a long bone, with a small inflammatory area.

Craniotomy   (cranium = skull; tome = a cutting)   Any surgery that requires cutting through the bones surrounding the brain.

Necrosis   (necros = death; osis = condition)   Death of tissues or organs; in the case of bone, necrosis results from deprivation of blood supply; could result from fracture, extensive removal of periosteum in surgery, exposure to radioactive substances, and other causes.

Osteitis   Inflammation or infection of bone.

Osteoarthritis   (arthro = joint)   A degenerative condition of bone and also the joint.

Osteoblastoma   (oma = tumor)   A benign tumor of the osteoblasts.

Osteochondroma   A benign tumor of the bone and cartilage.

Osteoma   A benign bone tumor.

Osteomyelitis   Infection that involves bone marrow.

Osteosarcoma   (sarcoma = connective-tissue tumor)   A malignant tumor composed of osseous tissue.

Pott's disease   Inflammation of the backbone, caused by the microorganism that produces tuberculosis.

Chapter summary in outline

1. The skeletal system consists of all bones attached at joints, cartilage between joints, and cartilage found elsewhere (nose, larynx, and outer ear).

2. The functions of the skeletal system include support, protection, movement, mineral storage, and blood-cell formation.

A MICROSCOPIC VIEW

1. Parts of a typical long bone are the diaphysis (shaft), epiphyses (ends), articular cartilage, periosteum, medullary (marrow) cavity, and endosteum.

2. Cancellous, or spongy, bone has many marrow-filled pores and does not contain Haversian systems. It consists of trabeculae containing osteocytes and lacunae. It lies under dense bone.

3. Dense, or compact, bone has fewer pores and structural units called Haversian systems. Dense bone lies over spongy bone and composes most of the bone tissue of the diaphyses.

BONE FORMATION AND GROWTH

1. Bone forms by a process called ossification, which begins when mesenchymal cells become transformed into osteoblasts.

2. Intramembranous ossification occurs within fibrous membranes of the embryo.

3. Endochondral ossification occurs within a cartilage model.

4. The primary ossification center of a long bone is in the diaphysis. Osteoblasts lay down bone, and cartilage degenerates, leaving cavities that merge to form the marrow cavity. Next ossification occurs in the epiphyses, where bone replaces cartilage, except for the epiphyseal plate.

5. In both types of ossification, spongy bone is laid down first. Dense bone is later reconstructed from spongy bone.

HOMEOSTASIS OF BONE

1. The homeostasis of bone growth is dependent upon a balance between bone formation and destruction.

2. Old bone is constantly destroyed by osteoclasts, while new bone is constructed by osteoblasts. This process is called remodeling.

3. Normal growth depends on calcium, phosphorus, and vitamins (A, C, and D) and is controlled by hormones.

BONE DISORDERS

1. Rickets is a vitamin D deficiency in children in which the body does not absorb calcium and phosphorus. The bones are soft and bend under the body's weight.

2. Osteomalacia is a vitamin D deficiency in adults that leads to demineralization.

3. With osteoporosis, the amount and strength of bone tissue decreases due to decreases in hormone output.

4. Paget's disease is the irregular thickening and softening of bones in which osteoclast and osteoblast activities are imbalanced.

5. Osteomyelitis is a term for the infectious diseases of bones, marrow, and periosteum. It is frequently caused by "staph" bacteria.

6. Tumors may be classified as malignant (cancerous) or benign (noncancerous). Tumors of the skeletal system include osteogenic sarcomas (arise in bone), chondrosarcomas (arise in cartilage), and myelomas (arise in marrow).

Review Questions and Problems

1. Define the skeletal system. What are the five principal functions of the system?

2. Diagram the parts of a long bone, and list the functions of each part. What is the difference between cancellous and spongy bone tissue? Diagram the microscopic structure of bone.

3. What is meant by ossification? When does the process begin and end?

4. Distinguish between the two principal kinds of ossification.

5. Outline the major events involved in intramembranous and endochondral ossification.

6. What is a roentgenogram?

7. List the primary factors involved in bone growth.

8. How does osteoblast activity in balance with osteoclast activity demonstrate the homeostasis of bone?

9. Define rickets and osteomalacia in terms of symptoms, cause, and treatment. What do these two diseases have in common?

10. What are the principal symptoms of osteoporosis, osteomyelitis, and Paget's disease? What is the etiology of each?

11. Define the tumors associated with the skeletal system according to origin. Contrast malignant and benign tumors. What is metastasis?

12. Refer to the glossary of medical terminology associated with the skeletal system. Be sure that you can define each term.

CHAPTER 8

THE SKELETAL SYSTEM: STRUCTURAL FEATURES

STUDENT OBJECTIVES

After you have read this chapter, you should be able to:

1. Define the four principal types of bones in the skeleton

2. Note the relationship between bone structure and function

3. Describe the various kinds of markings on the surfaces of bones

4. Relate the structure of the marking to the function it performs

5. Describe the components of the axial and appendicular skeleton

6. Identify the bones of the skull and the major markings associated with each

7. Identify the sutures and fontanels of the skull

8. Identify the paranasal sinuses of the skull in projection diagrams and roentgenograms

9. Identify the bones of the vertebral column

10. List the defining characteristics and curves of each region of the vertebral column

11. Identify the bones of the thorax and their principal markings

12. Identify the bones of the shoulder girdle and their major markings

13. Identify the upper extremity and its component bones and their markings

14. Identify the components of the pelvic girdle and their principal markings

15. Identify the lower extremity and its component bones and their markings

16. Define the structural features and importance of the arches of the foot

17. Compare the principal structural differences between male and female skeletons

18. Define a fracture and list eight kinds of fractures

19. Describe the sequence of events involved in fracture repair

20. Interpret sequential roentgenograms of fracture repair

The skeletal system forms the framework of the body. For this reason, a familiarity with the names, shapes, and positions of individual bones will help you to understand some of the anatomy and physiology of the other organ systems. For example, movements such as throwing a ball, typing, and walking require the coordinated use of bones and muscles. In order to understand how muscles produce different movements, you need to learn the parts of the bones to which the muscles are attached. Efficient functioning of the respiratory system is also highly dependent on normal bone structure. The bones in the nasal cavity form a series of passageways that help clean, moisten, and warm inhaled air. Furthermore, the bones of the thorax are specially shaped and positioned so that the chest can expand during inhalation. Many bones also serve as reference landmarks to students of anatomy as well as to surgeons. Blood vessels and nerves often run along bones. These organs can be located more easily if the bone is identified first. The tops of the lungs are found just below the collar bone. The bottom of the rib cage can be used as a landmark in finding the diaphragm and liver.

We shall study bones by examining the various regions of the body. For instance, we shall look at the skull first and see how the bones of the skull relate to each other. Then we shall move on to the chest. This regional approach will allow you to see how the many bones of the body all fit together.

TYPES OF BONES

The bones of the body may be classified into one of four principal types on the basis of their shape: (1) long, (2) short, (3) flat, and (4) irregular. *Long bones* have greater length than width and consist of a diaphysis and two epiphyses. They are more or less curved for greater strength. Curvature of these bones is rather important for body support. A curved bone is structurally designed to absorb the stress of body weight at several different points. In other words, the stress is evenly distributed. If such bones were straight, the weight of the body would be unevenly distributed, and the bone would fracture very easily. Long bones have considerably more compact bone than spongy bone and thus have a further structural adaptation to their weight-bearing function. Examples of long bones include bones of the legs, arms, fingers, and toes. Refer to Figure 7–1 for a description of the parts of a long bone. *Short bones* are somewhat cube-shaped, and differences in length and width are not important. Their texture is spongy throughout except at the surface, where there is a thin layer of compact bone. Examples of short bones are the wrist and ankle bones. *Flat bones* are generally thin and flat and are composed of two more or less parallel plates of compact bone enclosing a layer of spongy bone. The term *diploe* is applied to the spongy bone of the cranial bones. Flat bones afford considerable protection and provide extensive areas for muscle attachment. Examples of flat bones include the cranial bones, which protect the brain, the breastbone and ribs, which protect organs in the thorax, and the shoulder blades. *Irregular bones* have very complex shapes and cannot be grouped into any of the three categories just described. They also vary in the amount of spongy and compact bone present. Such bones are the vertebrae and some facial bones.

In addition to these four principal types of bones, two other types are recognized. *Wormian*, or *sutural*, *bones* are small clusters of bones between the joints of certain cranial bones. Their number is quite variable among different individuals. *Sesamoid bones* are small bones found in various tendons where a lot of pressure develops—for instance, in the wrist. These bones, like the Wormian bones, are also variable in number. Two sesamoid bones, the knee caps, are present in all individuals.

113

Exhibit 8-1. BONE MARKINGS

MARKING	DESCRIPTION	EXAMPLE
A. Depressions and openings		
1. Fissure	A narrow, cleftlike opening between adjacent parts of bones through which blood vessels and nerves pass	Superior orbital fissure of the sphenoid bone (Exhibit 8-3)
2. Foramen (*foramen* = hole)	A rounded opening through which blood vessels, nerves, and ligaments pass	Infraorbital foramen of the maxilla (Exhibit 8-3)
3. Meatus (canal)	A tubelike passageway running within a bone	External auditory meatus of the temporal bone (Exhibit 8-3)
4. Paranasal sinus (*sin* = cavity)	An air-filled cavity within a bone connected to the nasal cavity	Frontal sinus of frontal bone (Exhibit 8-14)
5. Groove or sulcus (*sulcus* = ditchlike groove)	A furrow or groove that accommodates a soft structure such as a blood vessel, nerve, or tendon	Intertubecular groove of humerus (Exhibit 8-35)
6. Fossa (*fossa* = basinlike depression)	A depression in or on a bone	Mandibular fossa of the temporal bone (Exhibit 8-9)
B. Process	Any prominent, roughened projection	The mastoid process of temporal bone. (Exhibit 8-3)
Processes that form joints		
1. Condyle	A relatively large, convex prominence (knucklelike)	Medial condyle of the femur (Exhibit 8-41)
2. Head	A rounded projection supported on the constricted portion (neck) of a bone	Head of the femur (Exhibit 8-41)
3. Facet	A flat or shallow surface	Articular facet for tubercle of rib on a vertebra (Exhibit 8-25)
Processes to which tendons, ligaments, and other connective tissues attach		
1. Tubercle (*tuber* = knob)	A small, rounded process	Greater tubercle of humerus (Exhibit 8-35)
2. Tuberosity	A large, rounded, usually roughened process	Ischial tuberosity of hipbone (Exhibit 8-39)
3. Trochanter	A very large, blunt projection found only on the femur	Greater trochanter of femur (Exhibit 8-41)
4. Crest	A prominent border or ridge on a bone	Iliac crest of hipbone (Exhibit 8-38)
5. Line	A less prominent ridge	Linea aspera of the femur (Exhibit 8-41)
6. Spinous process (spine)	A sharp, slender process	Spinous process of a vertebra (Exhibit 8-23)
7. Epicondyle (*epi* = above)	A prominence above or on a condyle	Medial epicondyle of femur (Exhibit 8-41)

SURFACE MARKINGS

If you look at the surfaces of bones, you will see various kinds of *markings*. The structure of many of these markings indicates their functions. For instance, long bones that bear a great deal of weight have very large, rounded ends that can form sturdy joints. Other bones have depressions that receive the rounded ends. Roughened areas serve for the attachment of muscles, tendons, and ligaments. Grooves in the surfaces of bones provide a roadbed for the passage of blood vessels, and openings occur where blood vessels and nerves pass in and out of the bone. Exhibit 8–1 describes the different kinds of markings and their functions. As you learn the names and parts of bones in this chapter, see if you can define each bone marking and indicate its function.

DIVISIONS OF THE SKELETAL SYSTEM

The adult human skeleton consists of approximately 206 bones that are grouped in two principal divisions: the **axial** and the **appendicular**. (Exhibit 8–2). The *axis*, or center, of the human body is a straight line that runs along the center of gravity of the body. This imaginary line runs through the head and down to the space between the feet. The midsagittal section is drawn through this line. The axial division of the skeleton (shown in gray) consists of the bones that lie around the axis. These include the ribs and the breastbone and the bones of the skull and backbone. The appendicular division contains the bones of the free *appendages*, which are the upper and lower extremities, plus the bones called girdles, which connect the free appendages to the axial skeleton. The 80 bones of the axial and the 126 bones of the appendicular divisions are typically grouped as shown in the following outline:

I. Axial skeleton
  A. Skull
    1. Cranium — 8
    2. Face — 14
  B. Hyoid (above the voicebox) — 1
  C. Ossicles (ear bones)° — 6
    3 in each ear
  D. Vertebral column — 26
  E. Thorax
    1. Sternum — 1
    2. Ribs — 24
    — 80

° Although the ossicles are not considered part of the axial or appendicular skeleton, but rather as a separate group of bones, they are placed with the axial skeleton for convenience.

II. Appendicular skeleton
  A. Shoulder girdles
    1. Clavicle — 2
    2. Scapula — 2
  B. Free upper extremities
    1. Humerus — 2
    2. Ulna — 2
    3. Radius — 2
    4. Carpals — 16
    5. Metacarpals — 10
    6. Phalanges — 28
  C. Pelvic girdle
    1. Coxal, hip, or pelvic bone — 2
  D. Free lower extremities
    1. Femur — 2
    2. Fibula — 2
    3. Tibia — 2
    4. Patella — 2
    5. Tarsus — 14
    6. Metatarsus — 10
    7. Phalanges — 28
    — 126

Exhibits 8–3 to 8–45, which follow, describe the principal structural features of the individual bones of the body. Two terms that will be useful in studying these Exhibits have been mentioned briefly but will be reviewed again. As you remember, the term *articulation* means joint. Therefore, to say that a bone **articulates** with another bone is to say that the two bones form a joint. The other term, **roentgenogram**, is a film exposed to x-rays. *Roentgenography* is the use of x-rays. Roentgenograms of some normal bones are included in these Exhibits. After studying these representative roentgenograms, you will have a more comprehensive understanding of how they are used by the physician. With roentgenograms, a doctor can determine the location and extent of a fracture, see what progress has been made in fracture repair, and diagnose certain bone diseases.

Exhibit 8-2. DIVISION OF THE SKELETAL SYSTEM

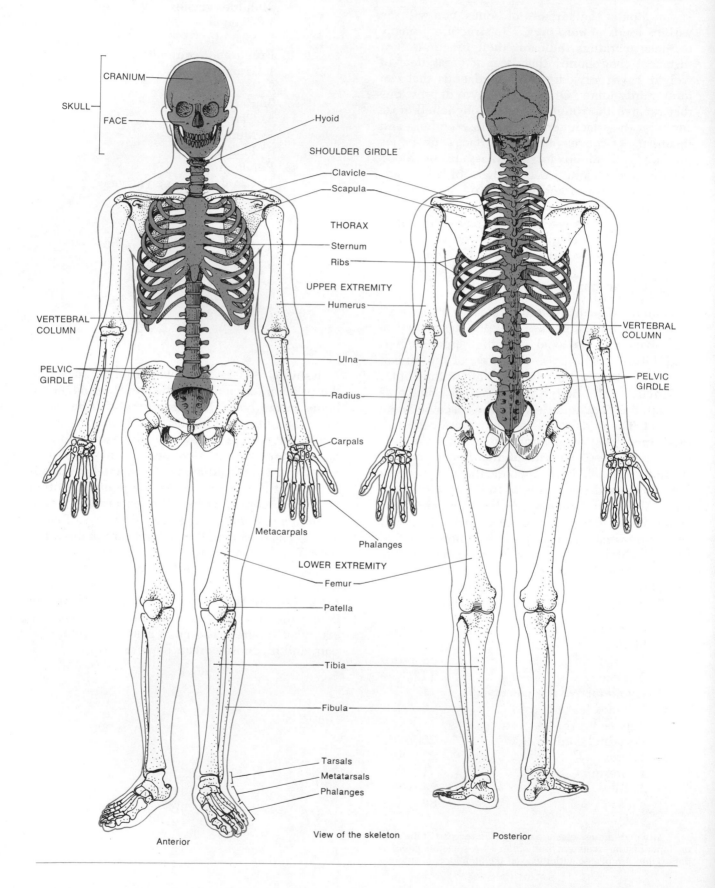

CRANIUM

SKULL

FACE

Hyoid

SHOULDER GIRDLE

Clavicle

Scapula

THORAX

Sternum

Ribs

UPPER EXTREMITY

Humerus

VERTEBRAL COLUMN

VERTEBRAL COLUMN

Ulna

PELVIC GIRDLE

PELVIC GIRDLE

Radius

Carpals

Metacarpals

Phalanges

LOWER EXTREMITY

Femur

Patella

Tibia

Fibula

Tarsals

Metatarsals

Phalanges

Anterior

View of the skeleton

Posterior

Exhibit 8–3. GENERAL DESCRIPTION OF THE SKULL

The **skull**, which contains 22 bones, rests upon the superior end of the vertebral column and is composed of two sets of bones: cranial bones and facial bones. The *cranial bones* enclose and protect the brain and the organs of sight, hearing, and balance. The eight cranial bones are the frontal bone, parietal bones (2), temporal bones (2), the occipital bone, sphenoid, and ethmoid. There are 14 *facial bones*, or bones of the face. These include: the nasal bones (2), maxillae (2), zygomatic bones (2), the mandible, lacrimal bones (2), palatine bones (2), inferior conchae (2), and the vomer. Be sure you can locate all the cranial and facial bones in the various views of the skull shown.

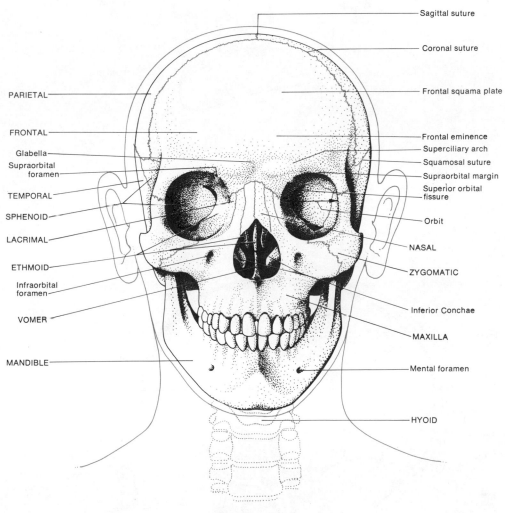

Anterior view of skull

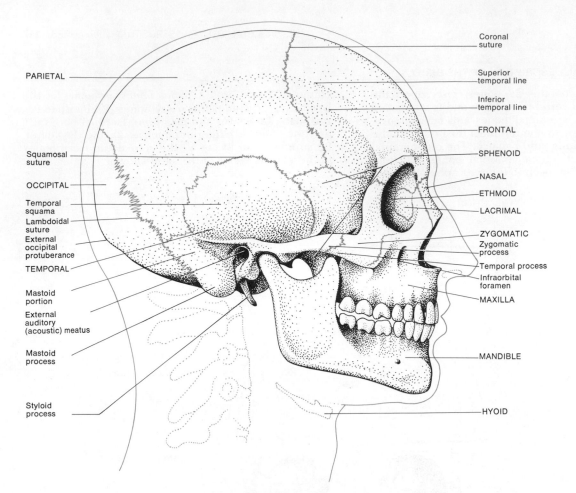

Exhibit 8–3 *(cont'd.)*

PARIETAL

Coronal suture

Superior temporal line

Inferior temporal line

FRONTAL

SPHENOID

Squamosal suture

NASAL

OCCIPITAL

ETHMOID

Temporal squama

LACRIMAL

Lambdoidal suture

ZYGOMATIC

External occipital protuberance

Zygomatic process

TEMPORAL

Temporal process

Infraorbital foramen

Mastoid portion

MAXILLA

External auditory (acoustic) meatus

Mastoid process

MANDIBLE

Styloid process

HYOID

Right lateral view of skull

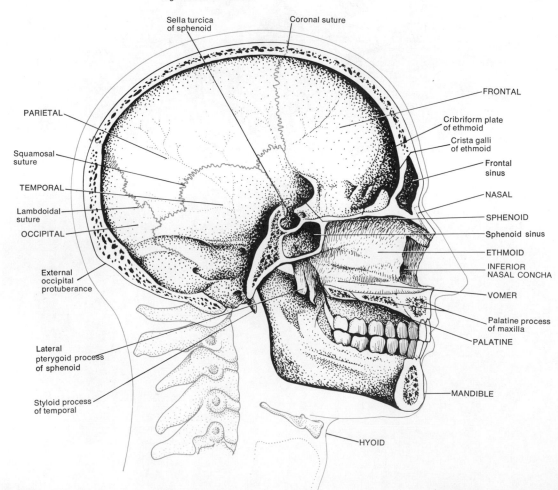

Sella turcica of sphenoid

Coronal suture

FRONTAL

PARIETAL

Cribriform plate of ethmoid

Crista galli of ethmoid

Squamosal suture

Frontal sinus

TEMPORAL

NASAL

Lambdoidal suture

SPHENOID

OCCIPITAL

Sphenoid sinus

ETHMOID

INFERIOR NASAL CONCHA

External occipital protuberance

VOMER

Palatine process of maxilla

Lateral pterygoid process of sphenoid

PALATINE

Styloid process of temporal

MANDIBLE

HYOID

Exhibit 8–4. SUTURES OF THE SKULL

The term **suture** means seam. It is an immovable joint found only between skull bones. Very little connective tissue is found between the bones of the suture. Four prominent skull sutures include: (1) the *coronal suture* between the frontal bone and the two parietal bones, (2) the *sagittal suture* between the two parietal bones, (3) the *lambdoidal suture* between the parietal bones and the occipital bone, and (4) the *squamosal suture* between the parietal bones and the temporal bones. Refer to Exhibits 8–3 and 8–5 for the locations of these sutures.

Exhibit 8–5. FONTANELS OF THE SKULL

The "skeleton" of a newly formed embryo consists of cartilage or fibrous membrane structures shaped like bones. Gradually the cartilage or fibrous membrane is replaced by bone. At birth, membrane-filled spaces called **fontanels**, meaning fountains, are found between cranial bones. These so-called "soft spots" are areas where the bone-making process is not yet complete. They allow the skull to be compressed during birth. Physicians find the fontanels helpful in determining the position of the infant's head prior to delivery. Although an infant may have many fontanels at birth, the form and location of six of them are fairly constant.

1. The *anterior (frontal) fontanel* is located between the angles of the two parietal bones and the two segments of the frontal bone. This fontanel is roughly diamond-shaped, and it is the largest of the six fontanels. It usually closes at 18 to 24 months. In the disorder called *microcephalus*, meaning "small head," the fontanel closes earlier than normal. The brain consequently does not have enough room to grow, and mental retardation results. In *hydrocephalus*, meaning "water on the brain," pressure from excess fluids inside the skull may cause the fontanel to remain open.

2. The *posterior (occipital) fontanel* is situated between the two parietal bones and the occipital bone. This fontanel is considerably smaller than the anterior fontanel. It is diamond-shaped and generally closes about 2 months after birth.

3. The *anterolateral (sphenoidal) fontanels* are two in number. One is located on each side of the skull at the junction of the frontal, parietal, temporal, and occipital bones. These fontanels are quite small and irregular in shape. They normally close by the third month after birth.

4. The *posterolateral (mastoid) fontanels* are also two in number. One is situated on each side of the skull at the junction of the parietal, occipital, and temporal bones. These fontanels are somewhat irregularly shaped. They start to close 1 or 2 months after birth, but closure is not generally complete until the age of 1 year.

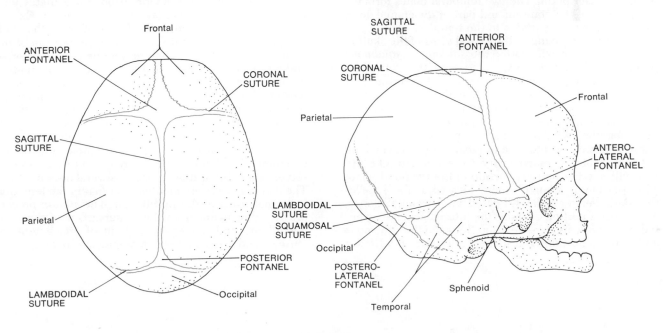

View of skull at birth

Superior   Right lateral

Exhibit 8–6. BONES OF THE SKULL: THE FRONTAL BONE

General Description: The **frontal bone** forms the anterior part of the cranium, or the forehead, the upper portion of the *orbits*, or eye sockets, and most of the anterior part of the cranial floor. At birth, the frontal bone consists of left and right parts that unite soon thereafter.

Markings: If you examine the anterior and lateral views of the skull in Exhibit 8–3, you will note the *squama*, or vertical plate. The term *squam* means scale. This scalelike plate gradually slopes down from the coronal suture, then turns abruptly downward. On either side of the midline of the body, it projects slightly to form the *frontal eminences*. Below each eminence is a horizontal ridge caused by the projec-

tion of the frontal sinuses behind the eyebrow. This is called the *superciliary arch*. Between the eminences and the arches just above the nose is a flattened area, the *glabella*. A thickening of the frontal bone below the superciliary arches is called the *supraorbital margin*. From this margin the frontal bone extends backwards to form the roof of the orbit and part of the floor of the cranial cavity. Within the supraorbital margin, slightly medial to its midpoint, is a hole called the *supraorbital foramen*, the term *foramen* meaning passageway. A nerve and blood vessel pass through this opening. The *frontal sinuses* (see Exhibit 8–13) lie behind the superciliary arches. These mucosa-lined cavities act as sound chambers to provide the voice with resonance.

Exhibit 8–7. BONES OF THE SKULL: THE PARIETAL BONES

General Description: The two **parietal bones**, *paries* meaning wall, form the greater portion of the sides and roof of the cranial cavity.

Markings: The external surface contains two slight ridges that may be observed by looking at the lateral

view of the skull in Exhibit 8–3. These are the *superior temporal line* and a less conspicuous *inferior temporal line*. The internal surface has many eminences and depressions that accommodate the blood vessels supplying the brain.

Exhibit 8–8. BONES OF THE SKULL: THE TEMPORAL BONES

General Description: The two **temporal bones** form the lower sides of the cranium and part of the cranial floor. The term *tempora* pertains to the temples.

Markings: Looking at the lateral view of the skull in Exhibit 8–3, you will notice the *squama* or *squamous portion* — a thin, large, expanded area that forms the anterior and upper part of the temple. Projecting from the lower portion of the squama is the *zygomatic process*, which articulates with the temporal process of the zygomatic bone. The zygomatic process of the temporal bone together with the temporal process of the zygomatic bones constitutes the *zygomatic arch*. If you look at the floor of the cranial cavity, shown in Exhibit 8–10, you will see the *petrous portion* of the temporal bone. This portion is triangular and located at the base of the skull between the sphenoid and occipital bones. The petrous portion contains the internal ear, the essential part of the organ of hearing. Between the squamous and petrous portions is a socket called the *mandibular*

fossa. This part of the temporal bone articulates with the condyle of the lower jaw. It is seen best in Exhibit 8–9. If you now examine the lateral view of the skull in Exhibit 8–3, you will see the *mastoid portion* of the temporal bone, located behind and below the external auditory meatus, or ear canal. In the adult, this portion of the bone contains a number of air spaces called *mastoid "cells."* These spaces are separated from the brain only by thin bony partitions. If *mastoiditis*, or the inflammation of these bony cells, occurs, the infection may spread to the brain or its outer covering. The *mastoid process* is a rounded projection of the temporal bone behind the external auditory meatus. It serves as a point of attachment for several neck muscles. The *external auditory meatus* is a canal in the temporal bone that leads to the middle ear. The *styloid process* projects downward from the undersurface of the temporal bone and serves as a point of attachment for muscles and ligaments of the tongue and neck.

Exhibit 8–9. BONES OF THE SKULL: THE OCCIPITAL BONE

General Description: The **occipital bone** forms the back part and a good portion of the base of the cranium.

Markings: The *foramen magnum* part is a large hole in the inferior part of the bone through which the spinal cord passes. The *occipital condyles* are oval-shaped processes with convex surfaces, one on either side of the foramen magnum, which articulate with depressions on the first cervical vertebra. The *external occipital protuberance* is a prominent projection on the posterior surface of the bone just above the foramen magnum. You can feel this structure as a definite bump on the back of your head, just above your neck. The protuberance is also visible in the lateral view of the skull in Exhibit 8–3.

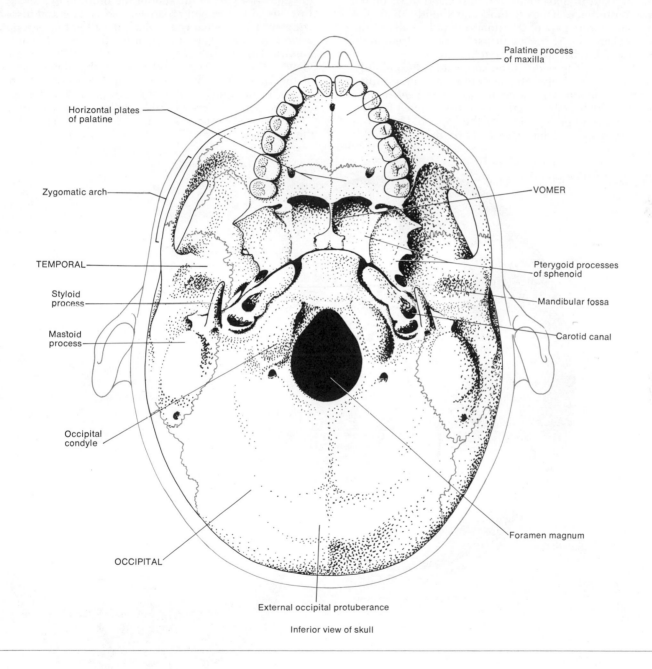

Inferior view of skull

Exhibit 8–10. BONES OF THE SKULL: THE SPHENOID BONE

General Description: The **sphenoid bone** is situated at the anterior part of the base of the skull. The combining form *spheno* means wedge. This bone is referred to as the "keystone" of the cranial floor because it binds the other cranial bones together. If you view the floor of the cranium from above, you will note that the sphenoid articulates with the temporal bones on either side, the ethmoid and frontal bones anteriorly, and the occipital bone posteriorly. It lies behind and slightly above the nose and forms part of the floor and side walls of the eye socket. The shape of the sphenoid is frequently described as a bat with outstretched wings.

Markings: The *body* of the sphenoid is the cubelike central portion between the ethmoid and occipital bones. It contains two large air spaces, the *sphenoidal sinuses,* which drain into the nasal cavity. (See Exhibit 8–14.) On the upper surface of the sphenoid body is a depression called the *sella turcica,* meaning Turk's Saddle. This depression houses the pituitary gland. The *greater wings* of the sphenoid are lateral projections from the body and form the anterolateral floor of the cranium. The greater wings also form part of the sidewall of the skull just anterior to the temporal bone. The *lesser wings* are anterior and superior to the greater wings. They form part of the floor of the cranium and the posterior part of the roof of the orbit, or eye socket. Just lateral to the body between the greater and lesser wings is a somewhat triangular gap called the *superior orbital fissure*. It is an opening for several cranial nerves. This fissure may also be seen in the anterior view of the skull in Exhibit 8–3. On the lower surface of the sphenoid bone you can see the *pterygoid processes,* or *laminae*. These structures project downward from the points where the body and greater wings unite. The pterygoid processes form part of the lateral wall of the nasal cavity.

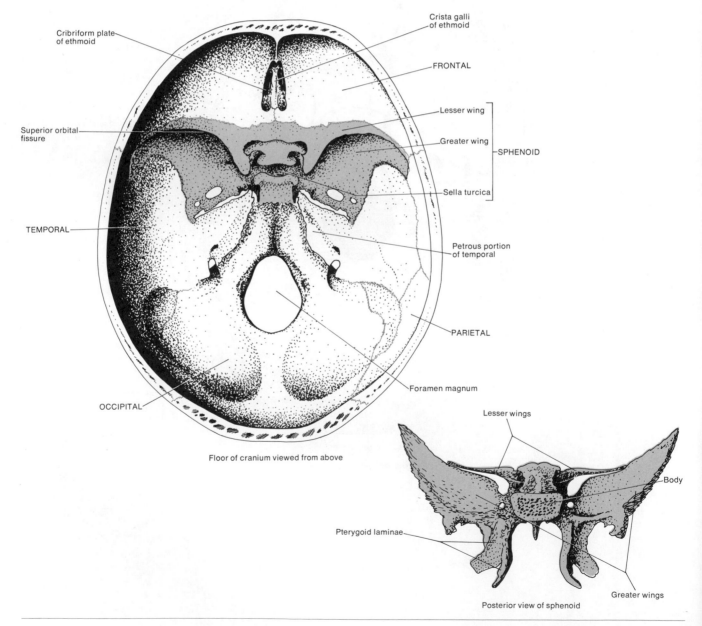

Floor of cranium viewed from above

Posterior view of sphenoid

Exhibit 8–11. BONES OF THE SKULL: THE ETHMOID BONE

General Description: The **ethmoid bone** is a light, cancellous bone located anterior to the sphenoid and posterior to the nasal bones. The combining form *ethmos* means sieve. This bone helps form the anterior portion of the cranial floor, the medial wall of the orbits, the upper portions of the nasal septum, or partition, and most of the sidewalls of the nasal roof. The ethmoid is the principal supporting structure of the nasal cavity.

Markings: Its lateral masses or *labyrinths* compose most of the wall between the nasal cavities and the orbits. They contain several air spaces, or "cells," which together form the *ethmoidal sinuses.* The sinuses are shown in Exhibit 8–14. The *perpendicular plate* (Exhibit 8–13) forms the upper portion of the nasal septum.

The *cribriform plate,* or horizontal plate, lies in the anterior floor of the cranium and forms the roof of the nasal cavity. The nerves that function in smelling pass through the cribriform plate. Projecting upward from the horizontal plate is a triangular process called the *crista galli,* which means Cock's Comb. This structure serves as a point of attachment for the membranes that cover the brain. On either side of the nasal septum, the labyrinths contain two thin scroll-shaped bones. These are called the *superior nasal concha* and the *middle nasal concha.* The conchae allow for the efficient circulation and filtration of inhaled air before it passes into the lungs. See Exhibit 8–3 also.

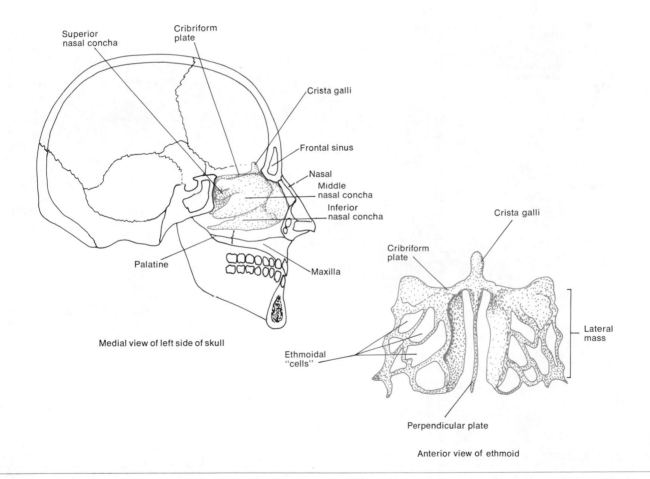

Medial view of left side of skull

Anterior view of ethmoid

Exhibit 8–12. BONES OF THE SKULL: THE NASAL BONES

General Description: The paired **nasal bones** are small, oblong bones located side by side at the middle and upper part of the face. Their fusion forms the upper part of the bridge of the nose. The lower portion of the nose, indeed the major portion, consists of cartilage. See Exhibit 8–13.

Exhibit 8–13. BONES OF THE SKULL: THE MAXILLAE

General Description: The paired maxillary bones unite to form the upper jaw bone. The **maxillae** articulate with every bone of the face except the mandible, or lower jaw bone. They form part of the floor of the orbits, part of the roof of the mouth, and part of the sidewalls and floor of the nose. The two portions of the maxillary bones unite, and the fusion is normally completed before birth. If the palatine processes of the maxillary bones do not unite before birth, a condition called **cleft palate** results. Another form of the condition, called **harelip**, involves a split in the upper lip. Harelip is often associated with cleft palate. Depending on the extent and position of the cleft, activities such as speech and swallowing may be affected.

Markings: A maxillary bone contains a *maxillary sinus* that empties into the nose. See Exhibit 8–14. The *alveolar process*, *alveus* meaning hollow, contains the bony sockets into which the teeth are set. The *palatine process* is a horizontal projection of the maxilla that forms the anterior and larger part of the hard palate, or forward portion of the roof of the mouth. The *infraorbital foramen*, which can be seen in the anterior view of the skull in Exhibit 8–3, is a hole in the maxilla below the orbit. Blood vessels and a large nerve are transmitted through this opening.

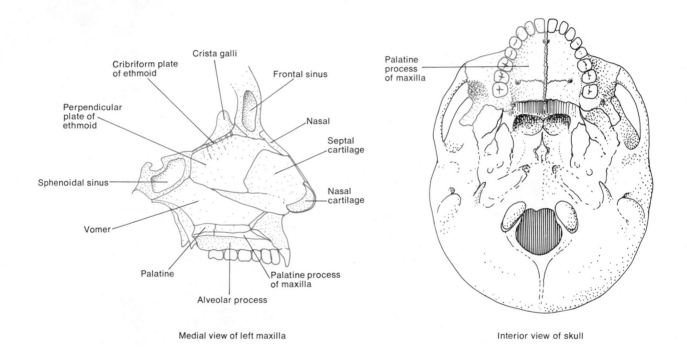

Medial view of left maxilla

Interior view of skull

Exhibit 8–14. SINUSES OF THE SKULL

While discussing several cranial bones, we have referred to sinuses. These cavities, called **paranasal sinuses**, are located within certain bones near the nasal cavity. The paranasal sinuses are lined with mucous membranes that are continuous with the lining of the nasal cavity. Cranial bones containing paranasal sinuses are the frontal bone, the sphenoid, the ethmoid, and the maxillae. The ethmoid sinus consists of a series of small cavities called ethmoid "cells," which range in number from 3 to 18.

Secretions produced by the mucous membranes of the paranasal sinuses drain into the nasal cavity. An inflammation of the membranes due to an allergic reaction or infection is called *sinusitis*. If the membranes swell enough to block drainage into the nasal cavity, fluid pressure builds up in the paranasal sinuses, and a common sinus headache results.

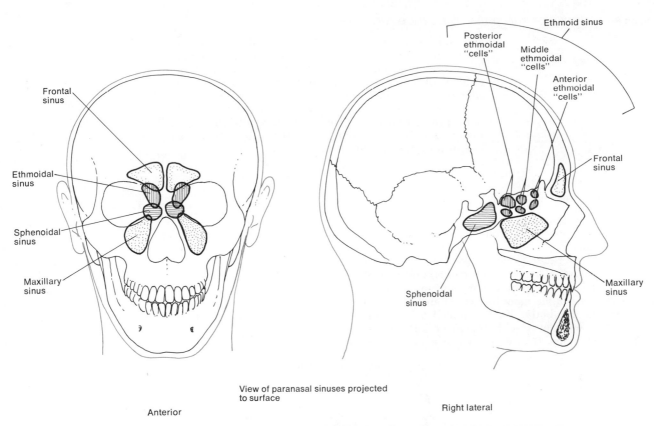

View of paranasal sinuses projected to surface

Anterior

Right lateral

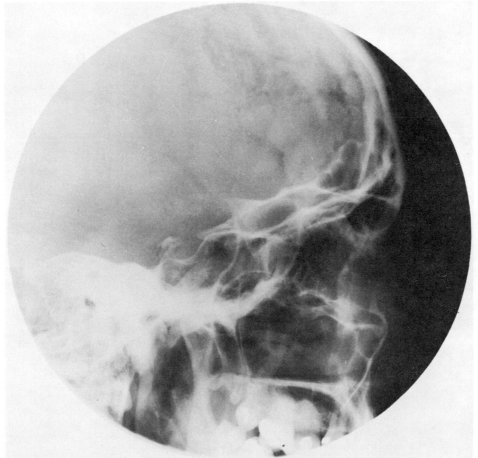

x-ray courtesy of Eastman Kodak Company.

Exhibit 8–15. BONES OF THE SKULL: THE ZYGOMATIC BONES

General Description: The two **zygomatic**, or **malar, bones** are commonly referred to as the cheekbones. They form the prominences of the cheeks and part of the outer wall and floor of the orbits. These bones can be seen in the lateral view of the skull in Exhibit 8–3.

Markings: The *temporal process* of the zygomatic bone projects backward and articulates with the zygomatic process of the temporal bone. These two processes form the *zygomatic arch*.

Exhibit 8–16. BONES OF THE SKULL: THE MANDIBLE

General Description: The **mandible** or lower jawbone is the largest, strongest facial bone. It is the only movable bone in the skull.

Markings: In lateral view you can see that the mandible consists of a curved, horizontal portion called the *body* and two perpendicular portions called the *rami*. The *angle* of the mandible is the area where each ramus meets the body. Each ramus has a *condylar process* that articulates with the mandibular fossa of the temporal bone. It also has a *coronoid process* to

which the temporal muscle attaches. The depression between the coronoid and condylar processes is called the *mandibular notch*. The *mental foramen*, the term *mentum* meaning chin, is approximately below the first molar tooth. A nerve and blood vessels pass through this opening. Dentists inject anesthetics through this foramen. The *alveolar process*, like that of the maxillae, is an arch containing the sockets for the teeth.

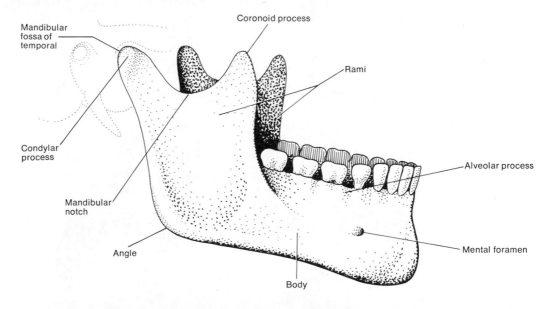

Right lateral view of mandible

Exhibit 8–17. BONES OF THE SKULL: THE LACRIMAL BONES

General Description: The paired **lacrimal bones** are thin bones roughly resembling a fingernail in size and shape. The term *lacrima* means tear. These bones are posterior and lateral to the nasal bones in the medial wall of the orbit. They can be seen in the anterior and lateral views of the skull in Exhibit 8–3. The lacrimal bones help to form the sidewall of the nasal cavity. They also contain part of the canal through which the tear ducts pass into the nasal cavity. The lacrimal bones are the smallest bones of the face.

Exhibit 8–18. BONES OF THE SKULL: THE PALATINE BONES

General Description: The two **palatine bones** are L-shaped and form the posterior portion of the hard palate, part of the floor and outer walls of the nasal cavities, and a small portion of the floor of the orbit. The posterior portion of the hard palate is formed by the *horizontal plates* of the palatine bones. These can be seen in Exhibit 8–9 and the medial view shown in Exhibit 8–3.

Exhibit 8–19. BONES OF THE SKULL: THE INFERIOR CONCHAE

General Description: Refer to the anterior view of the skull in Exhibits 8–3 and 8–11. The two **inferior conchae** are scroll-like bones that project into the nasal cavity below the superior and middle conchae of the ethmoid bone. The name of these bones is derived from *concha*, meaning shell. They serve the same function as the superior and middle conchae. This is, they allow for the circulation and filtration of air before it passes into the lungs. The inferior conchae are separate bones and are not part of the ethmoid.

Exhibit 8–20. BONES OF THE SKULL: THE VOMER

General Description: The **vomer,** which means plowshare, is a roughly triangular bone that forms the lower part of the nasal septum. It is clearly seen in the anterior view of the skull in Exhibit 8–3. Its lower border articulates with the cartilage septum that divides the nose into a right and left nostril. Its upper border articulates with the perpendicular plate of the ethmoid bone. The structures, then, that form the nasal septum or partition are the perpendicular plate of the ethmoid, the septal cartilage, and the vomer. If the vomer is pushed to one side—that is, deviated—the nasal chambers are of unequal size. See the skull viewed from below (Exhibit 8–9) for another view of the vomer.

Exhibit 8–21. THE HYOID BONE

General Description. The single **hyoid bone,** *hyoid* meaning U-shaped, is a unique component of the axial skeleton because it does not articulate with any other bone. Rather, it is suspended from the styloid process of the temporal bone by ligaments. The hyoid is located in the neck between the mandible and larynx. It supports the tongue and affords attachment for some of its muscles. Refer to the anterior and lateral views of the skull in Exhibit 8–3 to see the position of the hyoid bone in the body.

Markings: The hyoid consists of a horizontal *body* and paired projections called the *lesser cornu* and the *greater cornu.* Muscles attach to these paired projections.

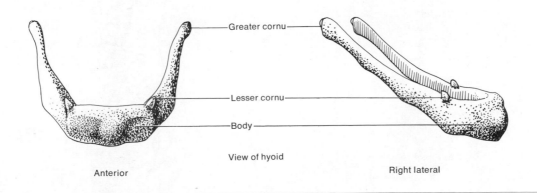

View of hyoid

Anterior   Right lateral

Exhibit 8–22. THE VERTEBRAL COLUMN: GENERAL CONSIDERATIONS

Together with the sternum and ribs, the **vertebral column,** or **spine,** constitutes the skeleton of the **trunk** of the body. The column is composed of a series of bones called **vertebrae.** And in the average adult, the column measures about 28 inches. In effect, the vertebral column is a strong, flexible rod that moves forward, backward, and sideways. It encloses and protects the spinal cord, supports the head, and serves as a point of attachment for the ribs and for the muscles of the back. Between the vertebrae are openings called *intervertebral foramina.* The nerves that connect the spinal cord to various parts of the body run out of these openings.

The adult vertebral column typically contains 26 vertebrae. These are distributed as follows: 7 *cervical vertebrae* in the neck region; 12 *thoracic vertebrae* behind the thoracic cavity; 5 *lumbar vertebrae* supporting the small of the back; 5 *sacral vertebrae* fused into one bone called the *sacrum;* and *coccygeal vertebrae* fused into one or two bones called the *coccyx.* Prior to the fusion of the sacral and coccygeal vertebrae, the total number of vertebrae is 33. Between the vertebrae are fibrocartilaginous **intervertebral discs.** These discs form strong joints and permit various movements of the column. A number of conditions, including injury, disease, and old age, may cause an intervertebral disc to rupture or protrude. Such a "slipped" disc is rather painful because nerves passing over the disc are irritated.

When viewed from the side, the vertebral column shows four **curves.** When seen from the anterior view, these are alternately convex, meaning they curve out toward the viewer, and concave, meaning they curve away from the viewer. The curves of the column, like the curves in a long bone, are very important because they increase its strength, help to maintain balance in the upright position, absorb shocks from walking, and help protect the column from fracture.

In the fetus, the four curves of the vertebrae are not present. There is only a single curve that is anteriorly concave. At about the third postnatal month when an infant begins to hold its head erect, the *cervical curve* develops. Later, when the child stands erect and walks, the *lumbar curve* develops. The cervical and lumbar curves are convex anteriorly. Because they are modifications of the fetal positions, they are called *secondary curves.* The other two curves, the *thoracic curve* and the *sacral curve,* are anteriorly concave. Since they retain the anterior concavity of the fetus, they are referred to as *primary curves.*

As a result of various conditions, the normal curves of the column may become exaggerated. In such cases, they are called **curvatures.** For example, muscular paralysis on either side of the spine, poor posture, or disease may result in **scoliosis,** the term *scolio* meaning bent. With this condition, there is a curvature toward the left or right side of the body. Scoliosis may occur in any region of the backbone. An exaggerated anterior concavity of the thoracic region is called **kyphosis,** or hunchback. **Lordosis,** or swayback, refers to an exaggerated anterior convexity in the lumbar region. Kyphosis and lordosis may be caused by tuberculosis of the vertebrae, rickets, or poor posture.

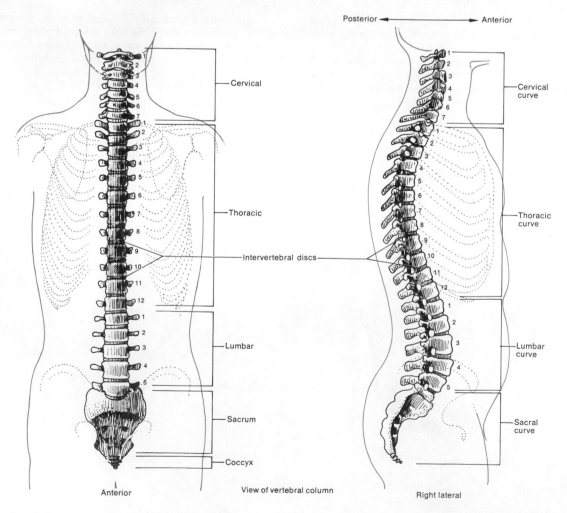

Anterior   View of vertebral column   Right lateral

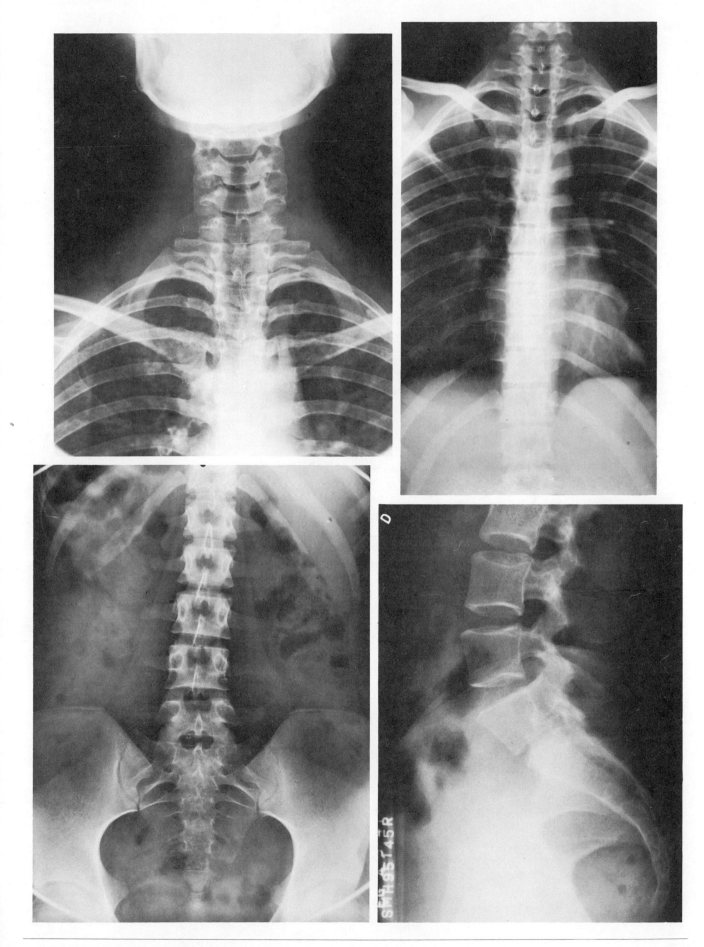

x-rays courtesy of Eastman Kodak Company.

Exhibit 8–23. THE VERTEBRAL COLUMN: STRUCTURE OF A TYPICAL VERTEBRA

All the vertebrae of the column are basically similar in structure. But there are differences in size, shape, and specialized details. A typical vertebra consists of the following components that may be viewed from above and laterally:

1. The *body* is the thick, disc-shaped anterior portion that is the weight-bearing part of a vertebra. Its superior and inferior surfaces are roughened for the attachment of intervertebral discs. The anterior and lateral surfaces contain foramina for blood vessels.

2. The *vertebral,* or *neural, arch* extends posteriorly from the body of the vertebra. Along with the body of the vertebra, it surrounds the spinal cord. It is formed by two short, thick processes, the *pedicles,* which project posteriorly to unite with the *lamina.* The space that lies between the vertebral arch and body contains the spinal cord. This space is known as the *vertebral foramen.* The vertebral foramina of all vertebrae together form the *vertebral canal.* The pedicles are notched above and below in such a way that, when they are arranged in the column, there is an opening between vertebrae on each side of the column. This opening, the *intervertebral foramen,* permits the passage of nerves to and from the spinal cord.

3. Seven *processes* arise from the vertebral arch. At the point where the lamina and each pedicle join, there is a *transverse process* extending laterally on each side. A single *spinous process* projects posteriorly and inferiorly from the lamina, forming the *spine.* These three processes serve as points of muscular attachment. The function of the remaining four processes is to form joints with other vertebrae. The two *superior articular processes* of a vertebra articulate with the vertebra immediately above it. The two *inferior articular processes* of a vertebra articulate with the vertebra below it.

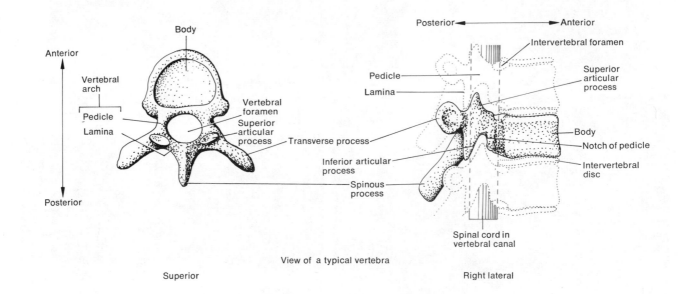

View of a typical vertebra

Superior   Right lateral

Exhibit 8–24. THE VERTEBRAL COLUMN: THE CERVICAL REGION

When viewed from above, it can be seen that the bodies of **cervical vertebrae** are smaller than those of the thoracic vertebrae. The arches, however, are larger. The spinous processes of the second through sixth cervical vertebrae are often *bifid*—that is, cleft in two. Each transverse process contains an opening, the *transverse foramen*. An artery and its accompanying vein and nerve filaments pass through this opening.

The first two cervical vertebrae differ considerably from the others. The first cervical vertebra, the **atlas**, is so named because it supports the head. Essentially, the atlas is a ring of bone with *anterior* and *posterior arches* and large *lateral masses*. It lacks a body and a spinous process. The superior surfaces of the lateral masses, called *superior articular surfaces*, are concave and articulate with the occipital condyles of the occipital bone. This articulation permits the movement seen when shaking the head "yes." The inferior surfaces of the lateral masses, the *inferior articular surfaces*, articulate with the second cervical vertebra. The transverse processes and *transverse foramina* of the atlas are quite large.

The second cervical vertebra, the **axis**, does have a *body*. A peglike process called the *dens*, or odontoid process, projects up through the ring of the atlas. This makes a pivot on which the atlas and head rotate. This arrangement permits movement from side to side as in shaking the head to mean "no."

The third through sixth cervical vertebrae correspond to the structural pattern of the typical cervical vertebra shown. The seventh cervical vertebra, however, is somewhat different. It is called the *vertebra prominens* and is marked by a large, nonbifid spinous process that may be seen and felt at the base of the neck. (See Exhibit 8–22.)

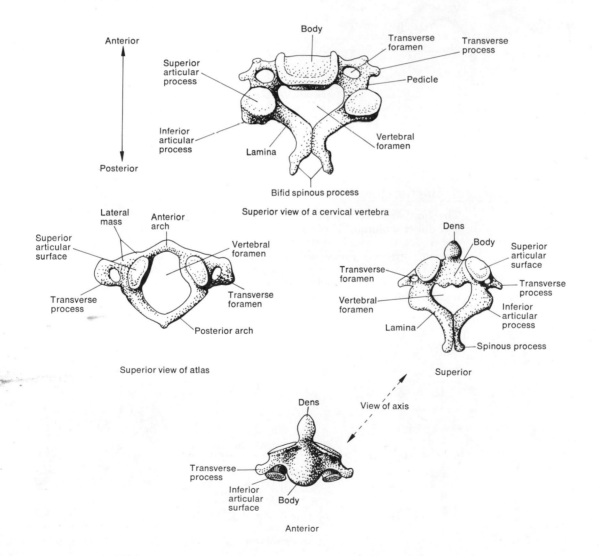

Superior view of a cervical vertebra

Superior view of atlas

View of axis

Exhibit 8–25. THE VERTEBRAL COLUMN: THE THORACIC REGION

Viewing a typical **thoracic vertebra** from above, you can see that it is considerably larger and stronger than a vertebra of the cervical region. In addition, the spinous process on each vertebra is long and pointed and projects inferiorly. Thoracic vertebrae also have longer and heavier transverse processes than cervical vertebrae. Except for the eleventh and twelfth thoracic vertebrae, the transverse processes have facets for articulating with the tubercles of the ribs.

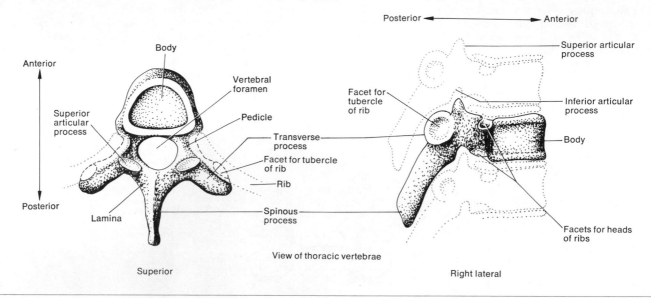

View of thoracic vertebrae

Exhibit 8–26. THE VERTEBRAL COLUMN: THE LUMBAR REGION

The **lumbar vertebrae** are the largest and strongest in the entire column. Their superior articular processes are directed medially instead of superiorly. And their inferior articular processes are directed laterally instead of inferiorly. Their various projections are short and thick, and the spinous process is heavy for attachment of the large back muscles.

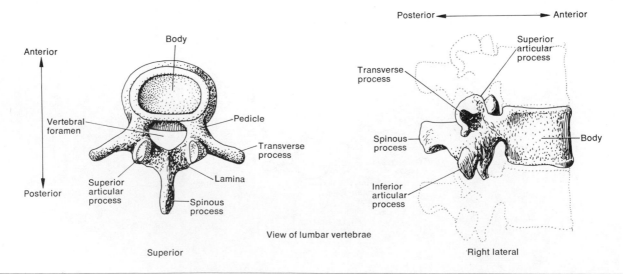

View of lumbar vertebrae

Exhibit 8–27. THE VERTEBRAL COLUMN: THE SACRUM AND COCCYX

The **sacrum** is a triangular bone formed by the union of five sacral vertebrae. It serves as a strong foundation for the pelvic girdle. It is positioned at the back of the pelvic cavity between the two hipbones. Anterior and posterior views of the bone are shown here. The concave anterior side of the sacrum faces the pelvic cavity. It is smooth and contains four *transverse lines* that mark the joining of the bodies of the vertebrae. At the ends of the lines are four pairs of *pelvic foramina*. The convex, posterior surface of the sacrum is irregular. It contains a *median sacral crest*, a *lateral sacral crest*, and four pairs of *dorsal foramina*. These foramina communicate with the pelvic foramina through which nerves and blood vessels pass. The *sacral canal* is a continuation of the vertebral canal. Laterally, the sacrum has a large *articular surface* for articulating with the ilium of the hipbone. Its superior articular process articulates with the fifth lumbar vertebra.

A clinical condition associated with the lamina of the vertebrae, especially those in the lumbosacral region, is referred to as *spina bifida*. In this condition, there is an imperfect union of the two sides of the lamina, leaving a cleft in the arch. The membranes and spinal cord may protrude through the cleft, forming a "tumor" on the back.

The **coccyx** is also triangular in shape and is formed by the fusion of the coccygeal vertebrae, usually the last four. It articulates above with the sacrum. The coccyx is the most rudimentary part of the column, representing the vestige of a tail.

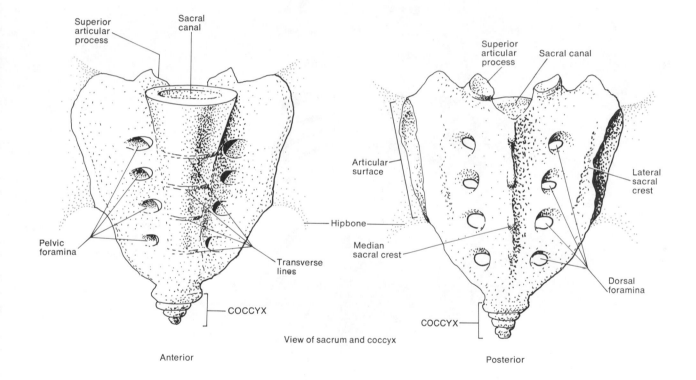

View of sacrum and coccyx

Anterior

Posterior

Exhibit 8–28. THE THORAX: GENERAL CONSIDERATIONS

The term **thorax** refers to the chest. Its skeleton is a bony cage formed by the sternum, costal cartilage, ribs, and the bodies of the thoracic vertebrae. It is shown here in the anterior view. The thoracic cage is roughly cone-shaped, the narrow portion being superior and the broad portion inferior. It is flattened from front to back. The thoracic cage encloses and protects the organs in the thoracic cavity. It also provides support for the bones of the shoulder girdle and upper extremities.

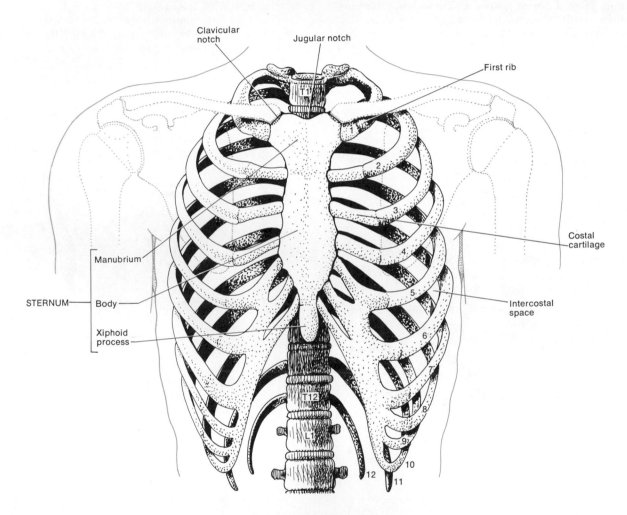

Anterior view of thoracic cage

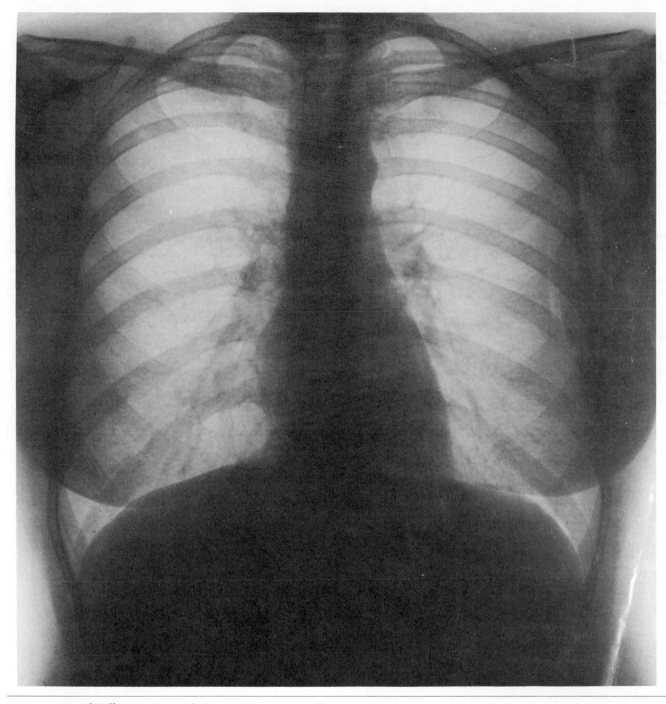

x-ray courtesy of William L. Leonard, Bergen Community College.

Exhibit 8–29. BONES OF THE THORAX: THE STERNUM

General Description: The **sternum**, or breastbone, is a flat, narrow bone measuring about 6 inches in length. It is located in the median line of the anterior thoracic wall. Physicians frequently use a sternal puncture to examine red bone marrow. In this procedure, a large needle is inserted into the sternum and a sample of marrow is withdrawn. Examining the marrow is very important in the diagnosis of blood disorders.

Markings: The sternum (see Exhibit 8–28) consists of three basic portions: (1) the *manubrium*, which is a triangular, superior portion; (2) the *body*, which is the middle, largest portion; and (3) the *xiphoid process*, which is the inferior, smallest portion. The manubrium has a depression on its superior surface called the *jugular notch*. On each side of the jugular notch are *clavicular notches* that articulate with the proximal ends of the clavicles. The manubrium also articulates with the first and second ribs. The body of the sternum articulates with the second through tenth ribs. These bones articulate either directly or indirectly. The xiphoid process has no ribs attached to it but affords attachment for some abdominal muscles.

Exhibit 8–30. BONES OF THE THORAX: THE RIBS

General Description: Twelve pairs of **ribs** are located on either side of the thoracic cavity. (See Exhibit 8–28.) The ribs increase in length from the first through seventh. Then, they decrease in length to the twelfth rib. Each rib articulates posteriorly with its corresponding thoracic vertebra. The first through seventh ribs are also attached directly to the sternum by a strip of hyaline cartilage, called *costal cartilage*. The term *costa* means rib. These ribs are called *true ribs*. The remaining five pairs of ribs are referred to as *false ribs* because their costal cartilages do not attach directly to the sternum. Instead, the cartilages of the eighth, ninth, and tenth ribs attach to each other and then to the cartilage of the seventh rib. The eleventh and twelfth ribs are also designated as *floating ribs* because their anterior ends do not attach even indirectly to the sternum. They attach to the muscles of the body wall instead.

Markings: Although there is some variation in rib structure, we will examine the parts of a typical rib when viewed from the right side from behind. The *head* of a typical rib is a projection at the posterior end of the rib. The *neck* is a constricted portion just below the head. A knoblike structure just below the neck is called a *tubercle*. The *body,* or *shaft,* is the main part of the rib. The inner surface of the rib has a *costal groove* that shelters blood vessels and a nerve. The posterior portion of the rib is connected to a vertebra by its head and tubercle. The head fits into a facet on the body of a vertebra, and the tubercle articulates with the transverse process of the vertebra. Each of the seond through ninth ribs articulates with the bodies of two adjacent vertebrae. The first, tenth, eleventh, and twelfth ribs articulate with only one vertebra each. On the eleventh and twelfth ribs, there is no articulation between the tubercles and the transverse processes of their corresponding vertebrae. Spaces between ribs are called *intercostal spaces.*

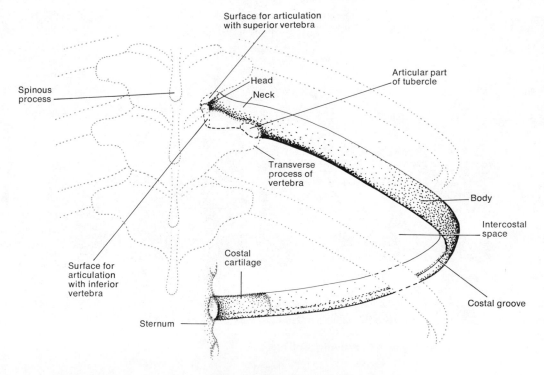

Ribs viewed from above and behind

Exhibit 8–31. THE SHOULDER GIRDLE: GENERAL CONSIDERATIONS

The **shoulder,** or **pectoral,** girdles serve to attach the bones of the free upper extremities to the axial skeleton. Structurally, each of the two shoulder girdles consists of only two bones: a clavicle and a scapula. The shoulder girdles have no articulation with the vertebral column. The clavicle is the anterior component of the shoulder girdle and articulates with the sternum. The posterior component, the scapula, which is positioned freely within a complex musculature, articulates with a clavicle and with the humerus. Although the shoulder joints are weak, they are freely movable and allow movements in many directions.

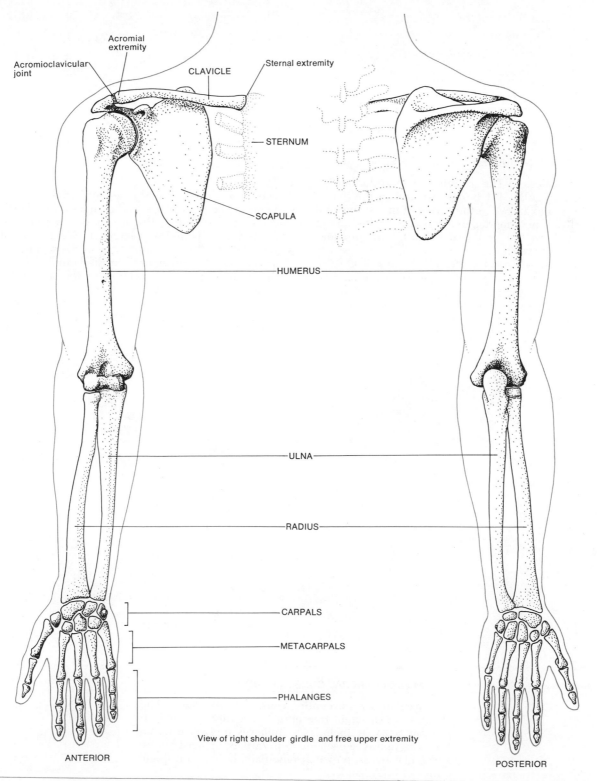

View of right shoulder girdle and free upper extremity

ANTERIOR

POSTERIOR

Exhibit 8–32. THE SHOULDER GIRDLE: THE CLAVICLES

General Description: The **clavicles,** or collarbones, are long slender bones with a double curvature. The two bones lie horizontally in the superior and anterior part of the thorax above the first rib.

Markings: The medial end of a clavicle, the *sternal extremity*, is rounded and articulates with the sternum.

The broad, flat, lateral end, the *acromial extremity*, articulates with the acromion of a scapula. This joint is called the *acromioclavicular joint*. Refer to Exhibit 8–31 for a view of these articulations. The *conoid tubercle* on the inferior surface of the lateral end of the bone serves as a point of attachment for a ligament.

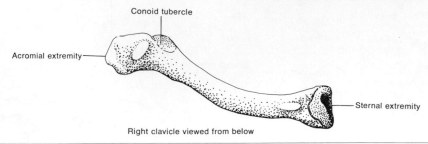

Right clavicle viewed from below

Exhibit 8–33. THE SHOULDER GIRDLE: THE SCAPULAE

General Description: The **scapulae,** or shoulder blades, are large, triangular, flat bones situated in the dorsal part of the thorax between the levels of the second and seventh ribs. Their medial borders are located about two inches from the vertebral column. Anterior and posterior views are shown here.

Markings: A sharp ridge, the *spine*, runs diagonally across the posterior surface of the flattened, triangular *body*. The end of the spine projects as a flattened, expanded process called the *acromion*. This process articulates with the clavicle. Below the acromion is a depression called the *glenoid cavity*. This cavity articulates with the head of the humerus to form the

shoulder socket. The thin edge of the body near the vertebral column is the *medial* or *vertebral border*. The thick edge closer to the arm is the *lateral* or *axillary border*. The medial and lateral borders join at the *inferior angle*. The superior edge of the scapular body is called the *superior border*. At the lateral end of the superior border is a projection of the anterior surface called the *coracoid process* to which muscles attach. Above and below the spine are two fossae: the *supraspinatous fossa* and the *infraspinatous fossa*, respectively. Both serve as surfaces of attachment for shoulder muscles.

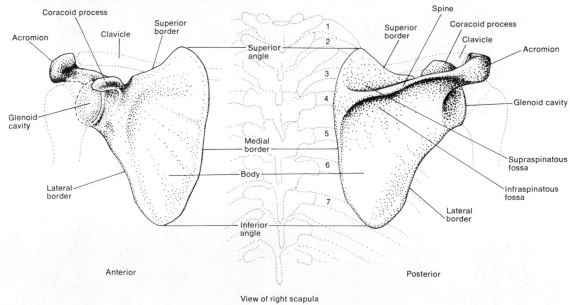

View of right scapula

Exhibit 8–34. THE FREE UPPER EXTREMITIES: GENERAL CONSIDERATIONS

The **free upper extremities** consist of 60 bones. The skeleton of the right free upper extremity is shown in Exhibit 8–31. It includes a humerus in each arm, an ulna and radius in each forearm, carpals, or wrist bones, and metacarpals, which are the palm bones, and finger bones of each hand.

Exhibit 8–35. THE FREE UPPER EXTREMITY: THE HUMERUS

General Description: The **humerus,** or arm bone, is the longest and largest bone of the free upper extremity. It articulates proximally with the scapula and distally at the elbow with both the ulna and radius.

Markings: The proximal end of the humerus consists of a *head* that articulates with the glenoid cavity of the scapula. It also has an *anatomical neck,* which is an oblique groove just below the head. The *greater tubercle* is a lateral projection below the neck. The *lesser tubercle* is an anterior projection. Between these tubercles runs an *intertubercular groove.* The *surgical neck* is a constricted portion just below the tubercles and is so named because of its liability to fracture. The *body* or shaft of the humerus is cylindrical at its proximal end. It gradually becomes triangular and is flattened and broad at its distal end. About midway down the shaft, there is a roughened, V-shaped area called the *deltoid tuberosity.* This serves as a point of attachment for the deltoid muscle. The following parts are found in the distal end of the humerus: The *capitulum* is a rounded knob that receives the head of the radius when the forearm is flexed. The *radial fossa* is a depression that articulates with the head of the radius when the arm is flexed. The *trochlea* is a pulleylike surface that articulates with the ulna. The *coronoid fossa* is an anterior depression that receives part of the ulna when the forearm is flexed. The *olecranon fossa* is a posterior depression that receives the olecranon of the ulna when the forearm is extended. The *medial epicondyle* and *lateral epicondyle* are rough projections on either side of the distal end.

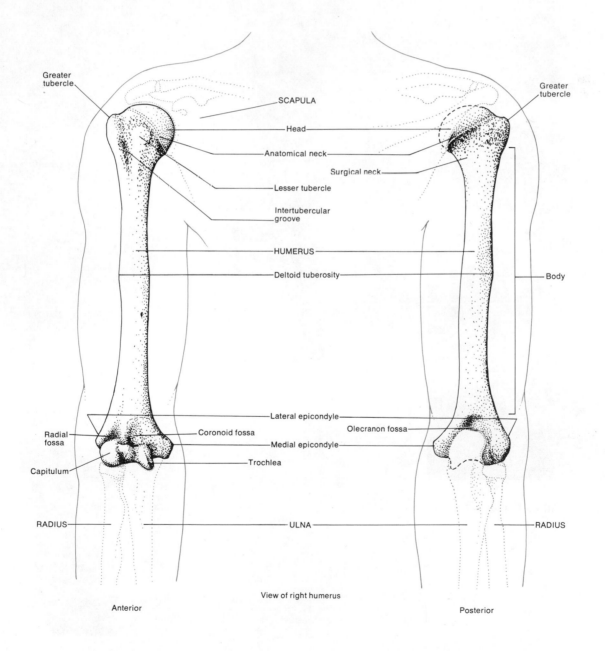

View of right humerus

Anterior

Posterior

Exhibit 8–35 *(cont'd.)*

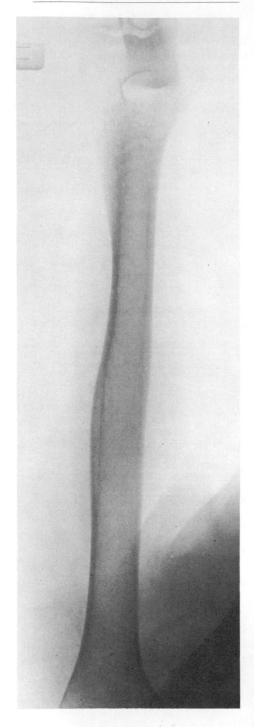

x-ray courtesy of William L. Leonard,
Bergen Community College.

Exhibit 8–36. THE FREE UPPER EXTREMITY: THE ULNA AND RADIUS

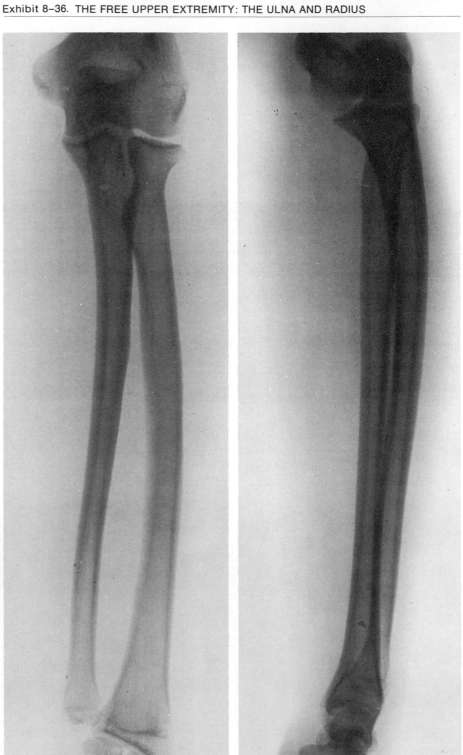

General Description of the Ulna: The **ulna** is the medial bone of the forearm. In other words, it is located on the side of the little finger.

Markings: The proximal end of the ulna contains an *olecranon*, which forms the prominence of the elbow. The *coronoid process* is an anterior projection that, together with the olecranon, receives the trochlea of the humerus. The *trochlear notch* is a curved area between the olecranon and the coronoid processes. The trochlea of the humerus fits into this notch. The *radial notch* is a depression located laterally and inferiorly to the trochlear notch. It receives the head of the radius. The distal end of the ulna consists of a *head* that is separated from the wrist by a fibrocartilage disc. A *styloid process* is on the posterior side of the distal end.

General Description of the Radius: The **radius** is the lateral bone of the forearm. That is, it is situated on the thumb side.

Markings: The proximal end of the radius has a disc-shaped *head* that articulates with the capitulum of the humerus and radial notch of the ulna. It also has a raised, roughened area on the medial side called the *radial tuberosity*. This is a point of attachment for the biceps muscle. The shaft of the radius widens distally to form a concave inferior surface that articulates with two bones of the wrist called the lunate and navicular bones. Also at the distal end is a *styloid process* on the lateral side and a medial, concave *ulnar notch* for articulation with the distal end of the ulna. A very common fracture of the radius, called a Colles' fracture, occurs about 1 inch up from the distal end of the bone.

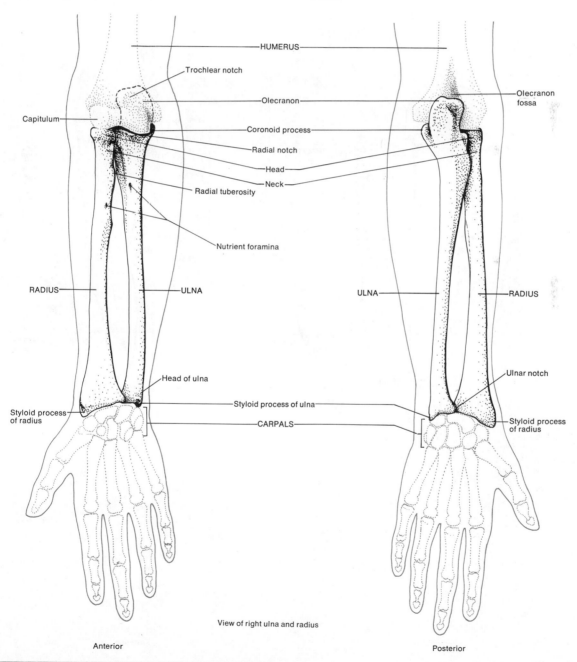

View of right ulna and radius

Anterior

Posterior

x-ray courtesy of **William L. Leonard, Bergen Community College.**

Exhibit 8–37. THE FREE UPPER EXTREMITY: THE WRIST AND HAND

Carpus: The **carpus**, or wrist, consists of eight small bones united to each other by ligaments. The bones are arranged in two transverse rows, with four bones in each row. The proximal row, from the medial to lateral position, consists of the following bones: *pisiform; triquetrum,* or triangular; *lunate;* and *scaphoid,* or navicular. In about 70 percent of the cases involving carpal fractures, only the scaphoid is involved. The distal row of bones, from the medial to lateral position, consists of the following: *hamate; capitate; trapezoid,* or lesser multangular; and *trapezium,* or greater multangular.

Metacarpus: The five bones of the **metacarpus** constitute the palm of the hand. Each metacarpal bone consists of a proximal *base,* a *shaft,* and a distal *head.* The metacarpal bones are numbered one to five, starting with the lateral bone. The bases articulate with the distal row of carpal bones and with one another. The heads articulate with the proximal phalanges of the fingers.

Phalanges: The **phalanges,** or bones of the fingers, number 14 in each hand. There are two phalanges in the first digit, the thumb, and three phalanges in each of the remaining four digits. The first row of phalanges, the *proximal row,* articulates with the metacarpal bones and second row of phalanges. The second row of phalanges, the *middle row,* articulates with the proximal row and the third row. Finally, the third row of phalanges, the *distal row,* articulates with the middle row.

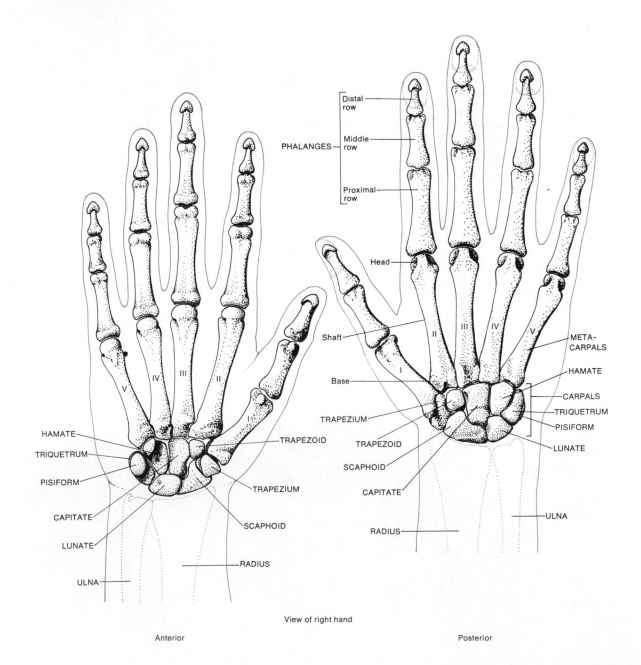

View of right hand

Anterior   Posterior

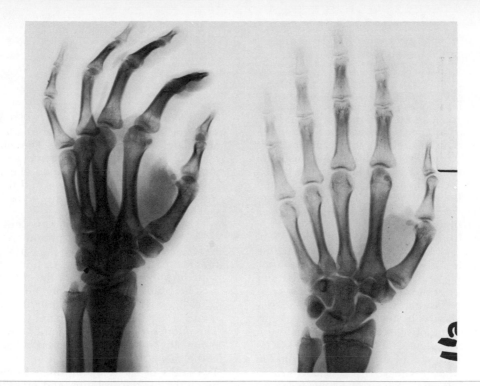

x-ray courtesy of William L. Leonard, **Bergen** Community College.

Exhibit 8–38. THE PELVIC GIRDLES: GENERAL CONSIDERATIONS

The **pelvic girdles** consist of the two **coxal bones,** commonly called the pelvic or hipbones. The pelvic girdles provide a strong and stable support for the free lower extremities on which the weight of the body is carried. The coxal bones are united to each other anteriorly at the pubic symphysis. They unite posteriorly to the sacrum.

Together with the sacrum and coccyx, the pelvic girdles form the basinlike structure called the **pelvis.** The pelvis is divided into a greater pelvis and a lesser pelvis. These are called the false pelvis and true pelvis, respectively. The *greater pelvis* is the expanded portion situated above the narrow bony ring called the *brim of the pelvis.* The greater pelvis consists laterally of the two ilia and posteriorly of the upper portion of the sacrum. There is no bony component in the anterior aspect of the greater pelvis. Rather, the front is formed by the walls of the abdomen. The *lesser pelvis* is below and behind the pelvic brim. It is constructed of parts of the ilium, pubis, sacrum, and coccyx. The lesser pelvis contains an opening above, called the *pelvic inlet,* and an opening below, called the *pelvic outlet. Pelvimetry* is the measurement of the size of the inlet and outlet of the birth canal. Measurement of the pelvic cavity is important to the physician since the baby must pass through the lesser pelvis at birth.

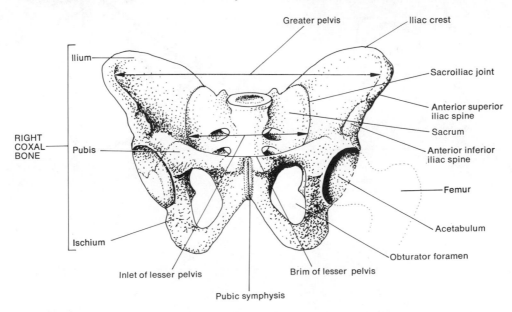

Anterior view of pelvic girdle

Exhibit 8-39. THE PELVIC GIRDLES: THE COXAL BONES

General Description: The **coxal bones** of a newborn consist of three components: a superior *ilium*, an inferior and anterior *pubis*, and an inferior and posterior *ischium*. Eventually, the three separate bones fuse into one. The area of fusion is a deep, lateral fossa called the *acetabulum*. This structure is the socket for the head of the thigh bone. Although the adult coxae are both single bones, it is common to discuss the bones as if they consisted of three portions.

Markings: The ilium is the largest of the three subdivisions of the coxal bone. Its superior border, the *iliac crest*, ends anteriorly in the *anterior superior iliac spine*. The *anterior inferior iliac spine* is located under the anterior superior spine. Posteriorly, the iliac crest ends in the *posterior superior iliac spine*. The *posterior inferior iliac spine* is just below. The spines serve as points of attachment for muscles of the abdominal wall. Just under the inferior iliac spine is the *greater sciatic notch*. The internal surface of the ilium seen from the medial side is the *iliac fossa*. It is a concavity where the iliacus muscle attaches. Posterior to this fossa are

the *iliac tuberosity*, a point of attachment for the sacroiliac ligament, and the *auricular surface*, which articulates with the sacrum. The other conspicuous markings of the ilium are three arched lines on its gluteal surface (buttock) called the *posterior gluteal line*, the *anterior gluteal line*, and the *inferior gluteal line*. Between these lines the gluteal muscles are attached.

The ischium is the lower posterior portion of the coxal bone. It contains a prominent *ischial spine*, a *lesser sciatic notch* under the spine, and an *ischial tuberosity*. The rest of the ischium, the *ramus*, joins with the pubis and surrounds the *obturator foramen*.

The pubis is the anterior and inferior part of the coxal bone. It consists of a *superior ramus*, an *inferior ramus*, and a body that enters into the pubic symphysis. The *pubic symphysis* is the joint between the two coxal bones. It can be seen in Exhibit 8-38. The *acetabulum* is the socket formed by the ilium, ischium, and pubis. Two-fifths of the acetabulum is formed by the ilium, two-fifths by the ischium, and one-fifth by the pubis.

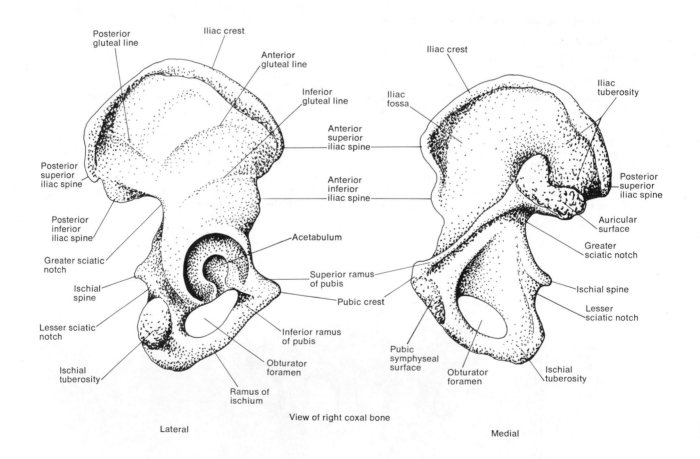

View of right coxal bone

Exhibit 8–40. THE FREE LOWER EXTREMITIES: GENERAL CONSIDERATIONS

The **free lower extremities** are made up of 60 bones. These include the femur of each thigh, each kneecap, the fibula and tibia in each leg, the ankle bones in each leg, and the metatarsals and phalanges of each foot.

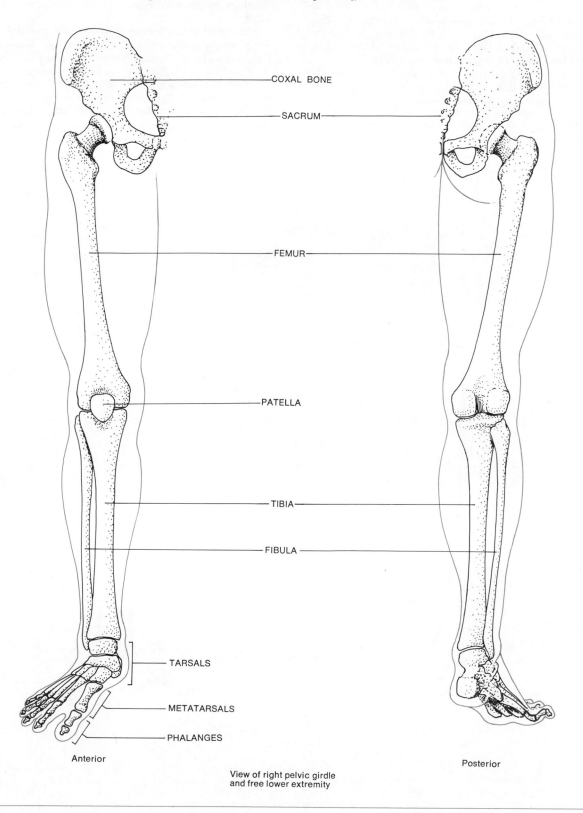

COXAL BONE

SACRUM

FEMUR

PATELLA

TIBIA

FIBULA

TARSALS

METATARSALS

PHALANGES

Anterior

Posterior

View of right pelvic girdle
and free lower extremity

Exhibit 8–41. THE FREE LOWER EXTREMITY: THE FEMUR

General Description: The **femur,** or thigh bone, is the longest and heaviest bone in the body. Its proximal end articulates with the coxal bone. And its distal end articulates with the tibia and patella. The shaft of the femur bows medially so that it approaches the femur of the opposite leg. As a result of this covergence, the knee joints are brought nearer to the line of gravity of the body. The degree of convergence is greater in the female because the female pelvis is broader.

Markings: The proximal end of the femur consists of a rounded *head* that articulates with the acetabulum of the coxal bone. The *neck* of the femur is a constricted region below the head. A fairly common fracture in elderly people occurs at the neck of the femur. Apparently the neck becomes so weak that it fails to support the body weight. The *greater trochanter* and *lesser trochanter* are projections that serve as points of attachment for some of the muscles of the thigh and buttock. Between the trochanters on the anterior surface is a narrow *intertrochanteric line.* Between the trochanters on the posterior surface is an *intertrochanteric crest.*

The shaft of the femur contains a roughened vertical ridge on its posterior surface called the *linea aspera.* This serves for the attachment of several thigh muscles.

The distal end of the femur is expanded and includes the *medial condyle* and *lateral condyle.* These articulate with the tibia. A depressed area between the condyles on the posterior surface is called the *intercondylar fossa.* Lying above the condyles are the *medial epicondyle* and *lateral epicondyle.*

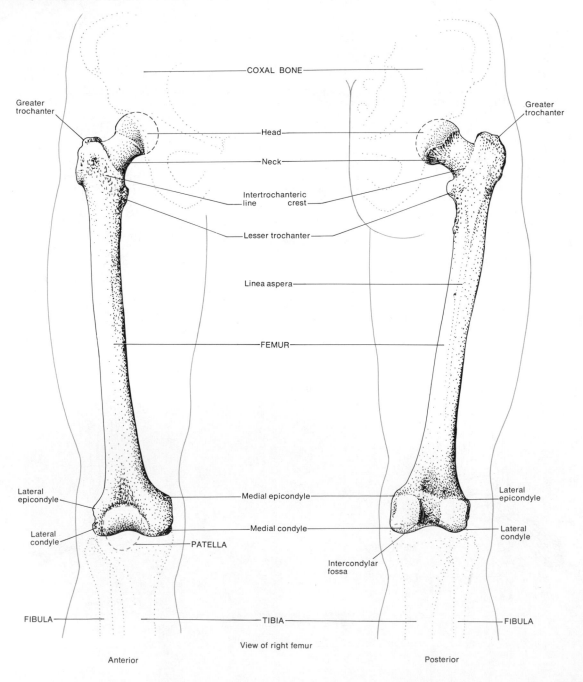

View of right femur

Anterior

Posterior

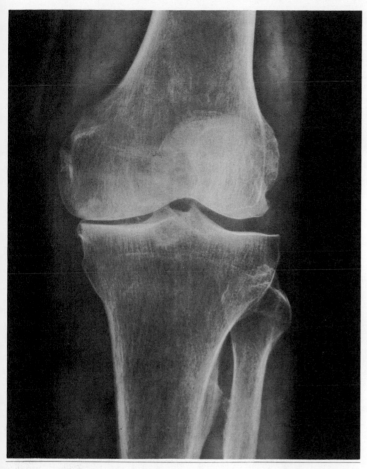

x-ray courtesy of Eastman Kodak Company.

Exhibit 8-42. THE FREE LOWER EXTREMITY: THE PATELLA

The **patella,** or kneecap, is a small, triangular bone in front of the knee joint. It develops in the tendon of the quadriceps fermoris muscle. A bone that forms in a tendon, such as the patella, is called a *sesamoid* bone. The broad superior end of the patella is called the *base*. And the pointed inferior end is referred to as the *apex*. The posterior surface contains two articular surfaces. These are the *articular facets* for the medial and lateral condyles of the femur.

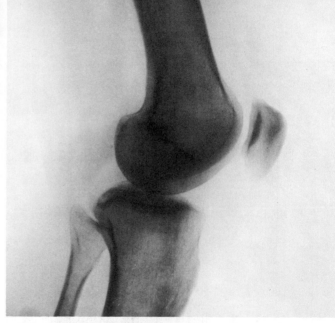

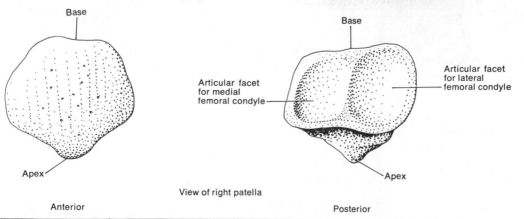

View of right patella

Anterior   Posterior

x-ray courtesy of William L. Leonard, Bergen Community College.

Exhibit 8-43. THE FREE LOWER EXTREMITY: THE TIBIA AND FIBULA

General Description of Tibia: The **tibia,** or shinbone, is the larger, medially placed bone of the lower leg. It bears the brunt of the weight on the leg. The tibia articulates at its proximal end with the femur and at its distal end with the fibula of the leg and talus of the ankle.

Markings: The proximal end of the tibia is expanded into a *lateral condyle* and a *medial condyle*. These articulate with the condyles of the femur. The slightly concave condyles are separated by an upward projection called the *intercondylar eminence*. The *tibial tuberosity* on the anterior surface is a point of attachment for the patellar ligament.

The distal end of the tibia contains a *medial malleolus*. This structure articulates with the talus bone of the ankle and forms the prominence that you can feel on the inner surface of your ankle. The *fibular notch* articulates with the fibula.

General Description of the Fibula: The **fibula** is lateral and runs parallel to the tibia. It is smaller than the tibia and is the slenderest of all long bones.

Markings: The *head* of the fibula, the proximal end, articulates with the tibia. The distal end has a projection called the *lateral malleolus*, which articulates with the talus bone of the ankle. This forms the prominence on the outer surface of the ankle. The lower portion of the fibula also articulates with the tibia at the fibular notch. A fracture of the lower end of the fibula with injury to the tibial articulation is called a Pott's fracture.

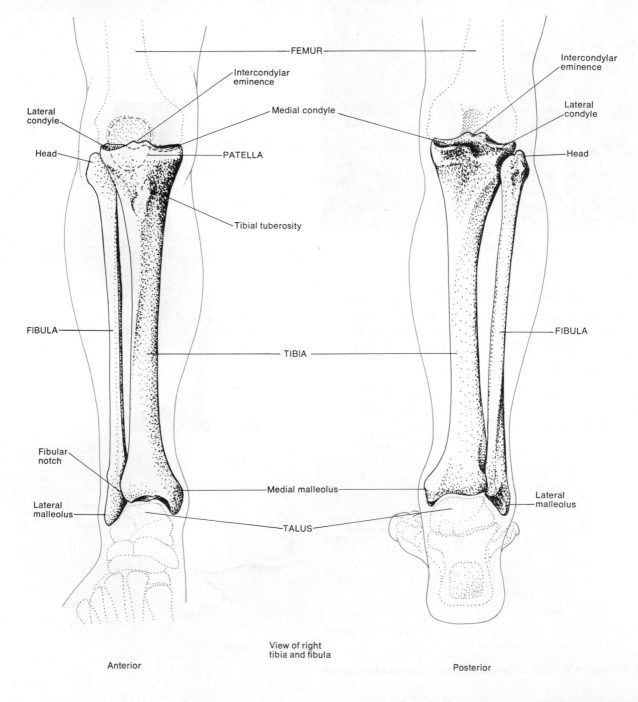

View of right
tibia and fibula

Anterior

Posterior

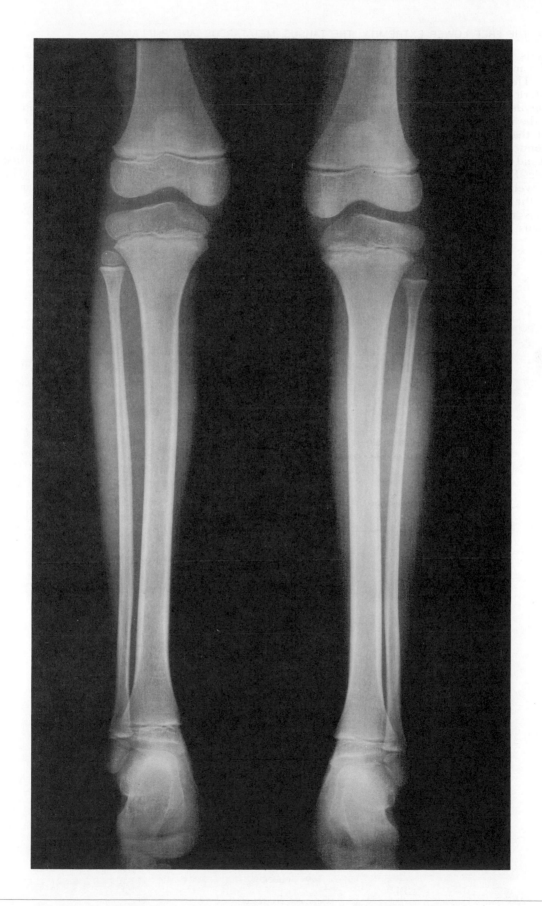

x-ray courtesy of Eastman Kodak Company.

Exhibit 8-44. THE FREE LOWER EXTREMITY: THE TARSUS, METATARSUS, AND PHALANGES

The **tarsus** is a collective designation for the seven bones of the ankle. The term *tarsos* pertains to a broad, flat surface. The *talus* and *calcaneus* bones are located on the posterior part of the foot. The anterior part contains the *cuboid, navicular,* and three *cuneiform* bones. The talus, the uppermost tarsal bone, is the only bone of the foot that articulates with the fibula and tibia. It is surrounded on one side by the medial malleolus of the tibia and on the other side by the lateral malleolus of the fibula. During walking, the talus initially bears the entire weight of its extremity. About half of the weight is then transmitted to the calcaneus. The remainder is transmitted to the other tarsal bones. The calcaneus, or heel bone, is the largest and strongest tarsal bone.

The **metatarsus** consists of five metatarsal bones numbered one to five from the medial to lateral position. The metatarsals articulate proximally with the first, second, and third cuneiform bones and with the cuboid. Distally, they articulate with the proximal row of phalanges. The first metatarsal is thicker than the others because it bears more weight.

The **phalanges** of the foot resemble those of the hand both in number and arrangement. The first toe, or big toe, has two large, heavy phalanges. The other four toes each have three phalanges. These are the proximal, middle, and distal phalanges.

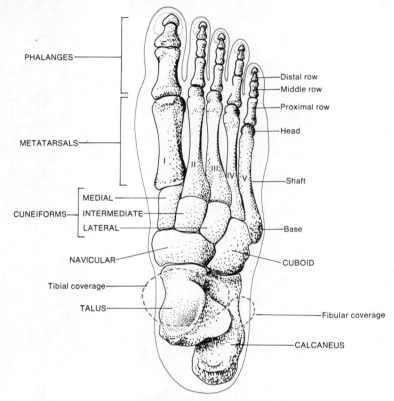

Superior view of right foot

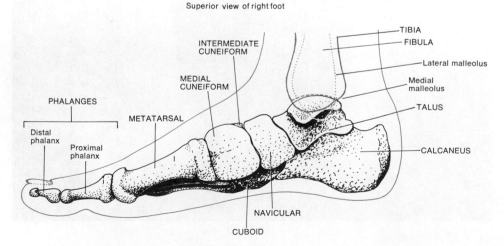

Medial view of right foot

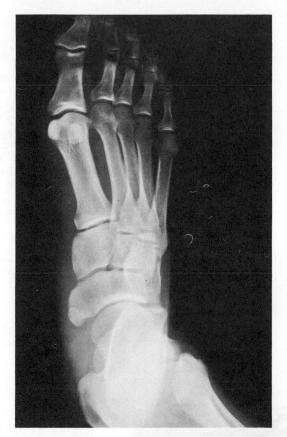

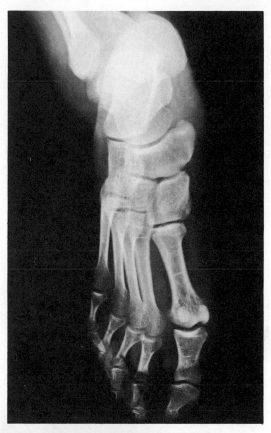

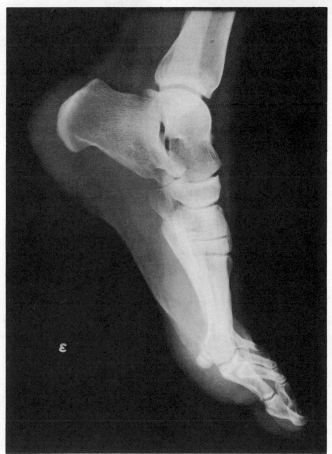

Exhibit 8–45. THE FREE LOWER EXTREMITY: ARCHES OF THE FOOT

The bones of the foot are arranged in two **arches.** These arches enable the foot to support the weight of the body and provide leverage while walking. The arches are not rigid. They yield as weight is applied and spring back when the weight is lifted.

The *longitudinal arch* has two parts. Both consist of tarsal and metatarsal bones arranged to form an arch from the front to the back of the foot. The *medial*, or inner, part of the longitudinal arch originates at the calcaneus. It rises to the talus, and descends forward through the navicular, the three cuneiforms, and the three medial metatarsals. The talus is the keystone of this arch. The *lateral*, or outer, part of the longitudinal

arch also begins at the calcaneus. It rises in the cuboid and descends to the two lateral metatarsals. The cuboid is the keystone of the arch.

The *transverse arch* is formed by the calcaneus, navicular, cuboid, and the posterior parts of the five metatarsals.

The bones comprising the arches are held in position by ligaments and tendons. If these ligaments and tendons are weakened, the height in the longitudinal arch may decrease or "fall." The result is *flatfoot*. Weakening of the transverse arch may cause an enlargement of the joint between the first metatarsal and big toe. This condition is called a *bunion*.

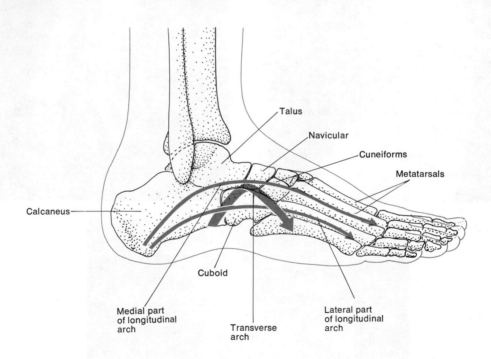

Lateral view of arches of the right foot

COMPARISON BETWEEN MALE AND FEMALE SKELETONS

The bones of the male are generally larger and heavier than those of the female. And the articular ends are thicker when compared with the shafts. In addition, some of the muscles of the male are larger than those of the female. Consequently,

the male skeleton has larger tuberosities, lines, and ridges for the attachment of the heavier muscles.

One marked difference between male and female skeletons can be seen in the pelvis. The main differences between the male pelvis and the female pelvis concern adaptations directly related to the childbearing function (Figure 8–1). The female pelvis is wider, shallower, and lighter in structure than that of the male. The ilia of the female flare out to the sides and give her broader hips. The inlet of the true pelvis of the female is nearly oval, whereas that of the male is triangular. The sacrum of the female is shorter, wider, and less curved than that of the male. And the female coccyx is more movable. The ischial spines and tuberosities of the female turn outward and are further apart than in the male. The pubic arch thus forms an obtuse angle rather than an acute angle as in the male.

All these characteristics contribute to the wider outlet of the true pelvis in the female. This, of course, accommodates the birth of the child. In addition, the ligaments of the sacroiliac joint stretch during pregnancy and childbirth. More room is provided for the developing fetus, and delivery is made easier.

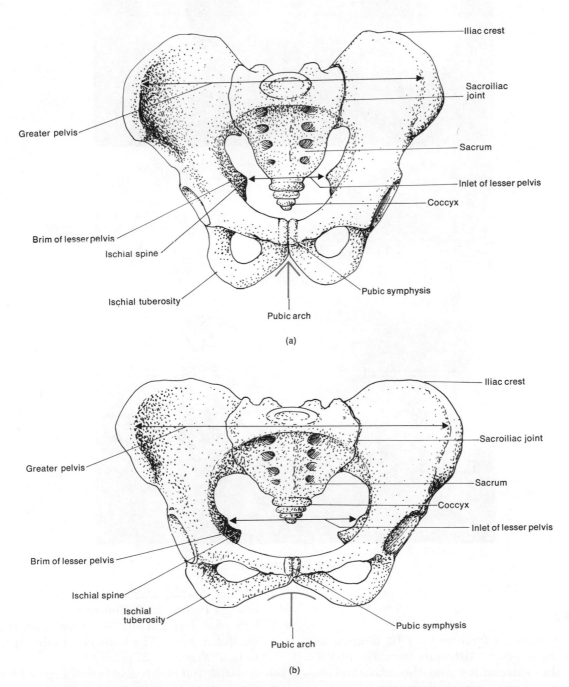

Figure 8–1. Comparison between (a) anterior view of male pelvis and (b) anterior view of female pelvis. [*x-rays (next page) courtesy of Eastman Kodak Company.*]

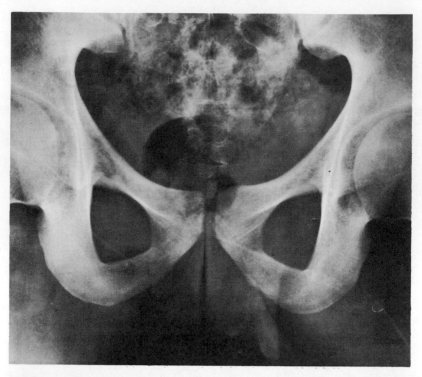

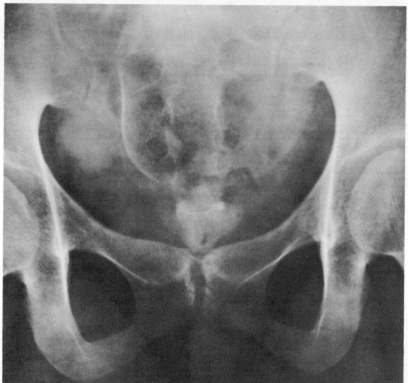

Figure 8–1 (cont'd.)

FRACTURES

In simplest terms, a **fracture** may be defined as any break in a bone. Although fractures of the bones of the extremities may be classified in several different ways, we shall use the following scheme for purposes of our discussion:

1. With a *partial* fracture, the break across the bone is incomplete (Figure 8–2a).
2. In a *complete* fracture, the break occurs across the entire bone. The bone is completely broken in two (Figure 8–2b).
3. A *simple* fracture is also called a *closed* fracture. The fractured bone does not break through the skin (Figure 8–2c).

4. In a *compound*, or *open*, fracture, the broken ends of the fractured bone protrude through the skin (Figure 8–2d).
5. With a *comminuted* fracture, the bone is splintered at the site of impact, and smaller fragments of bone are found between the two main fragments (Figure 8–2e).
6. A *greenstick* fracture is a partial fracture in which a portion of the bone bends (Figure 8–2f).
7. A *displaced* fracture is a fracture in which the anatomical alignment of the bone fragments is not preserved.
8. A *nondisplaced* fracture is a fracture in which the anatomical alignment of the bone fragments has not been disrupted.

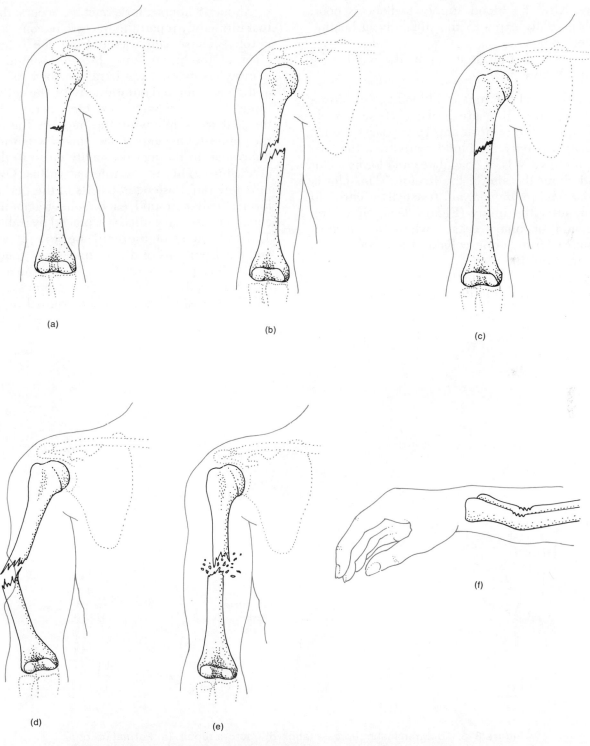

(a)

(b)

(c)

(d)

(e)

(f)

Figure 8–2. Types of fractures. (a) Partial. (b) Complete. (c) Simple. (d) Compound. (e) Comminuted. (f) Greenstick.

Unlike the skin, which may repair itself in days, or muscle, which may mend in weeks, a bone sometimes requires months to heal. A fractured femur, for example, may take 6 months to heal because sufficient calcium to strengthen and harden new bone is deposited very gradually. In addition, bone cells grow and reproduce slowly. Also the blood supply to bone is poor, which partially explains the difficulty in healing an infected bone.

The following steps occur in the repair of a fracture (Figure 8–3):

1. As a result of the fracture, blood vessels crossing the fracture line are broken. These vessels are found in the periosteum, Haversian systems, and marrow cavity. As blood pours from the torn ends of the vessels, it coagulates and forms a clot in and about the site of the fracture. This clot is called a *fracture hematoma*. It usually occurs 6 to 8 hours after the injury (Figure 8–3a). Since the circulation of blood ceases when the fracture hematoma forms, bone cells and periosteal cells at the fracture line die.

2. A growth of new tissue develops in and around the fractured area (Figure 8–3b). This new tissue is called a *callus*. It forms a bridge between separated areas of bone so that they are united. The callus that forms around the outside of the fracture is called an *external callus*. The callus formed between the two ends of bone fragments and between the two marrow cavities is called the *internal callus*.

About 48 hours after a fracture occurs, the cells that ultimately repair the fracture become actively mitotic. These cells come from the osteogenic layer of the periosteum, the endosteum of the marrow cavity, and the bone marrow. As a result of their accelerated mitotic activity, the cells of the three regions grow toward the fracture. During the first week following the fracture, the cells of the endosteum and bone marrow form new trabeculae in the marrow cavity close to the line of fracture. This is the internal callus. Over the next few days, osteogenic cells of the periosteum form a collar around each bone fragment. The collar, or external callus, is replaced by trabeculae. The trabeculae of the calli are joined to living and dead portions of the original bone fragments.

3. The final phase of fracture repair is the *remodeling* of the calli. In the remodeling process (Chapter 7), the dead portions of the original fragments

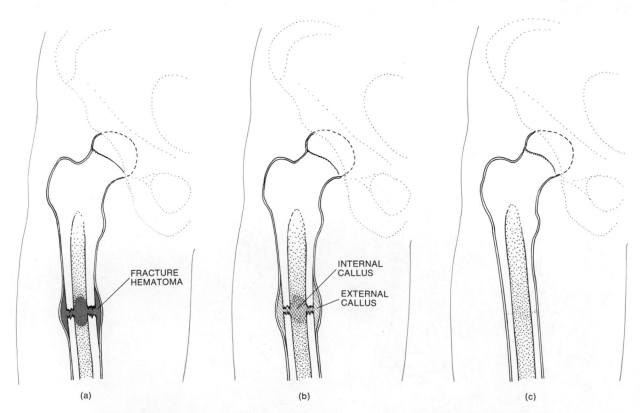

(a) (b) (c)

FRACTURE HEMATOMA

INTERNAL CALLUS

EXTERNAL CALLUS

Figure 8–3. Diagrammatic representation of fracture repair. (a) Formation of fracture hematoma. (b) Formation of external and internal callus. (c) Completely healed fracture.

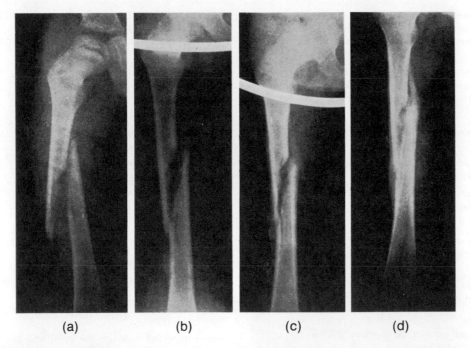

(a)   (b)   (c)   (d)

Figure 8–4.   Roentgenograms of the repair of a fractured femur. (a) Immediately after fracture. (b) Two weeks later, a hazy external callus is visible around the margins of the fracture. (c) About 3 1/2 weeks after the fracture occurs, the internal and external calli begin to bridge the separated fragments. (d) Almost 9 weeks after the fracture, the bridge between fragments is fairly well developed. In time, remodeling will occur, and the fracture will be repaired. *(From Lester W. Paul and John H. Juhl, The Essentials of Roentgen Interpretation, 3d ed., Harper & Row Publishers, Inc., New York, 1972. Photographs courtesy of Dr. Ralph C. Frank, Eau Claire, Wisconsin.)*

are gradually resorbed. Compact bone replaces cancellous bone around the periphery of the fracture (Figure 8–3c). In some cases, the healing is so complete that the fracture line is undetectable, even by x-ray. The healing of a fracture of the femur is illustrated by sequential roentgenograms in Figure 8–4.

Now that we have examined the types of bones that comprise the skeletal system and the relationship of the bones to each other, we are prepared to study the manner in which bones are attached to each other. This attachment, which is called a joint or articulation, determines how bones move.

Chapter summary in outline

TYPES OF BONES
1. On the basis of shape, bones are classified as long, short, flat, or irregular.
2. Wormian or sutural bones are found between the sutures of certain cranial bones. Sesamoid bones develop in tendons or ligaments.

SURFACE MARKINGS
1. Markings are definitive areas on the surfaces of bones.
2. Each marking is structured for a specific function such as joint formation, muscle attachment, or allowing nerves and blood vessels to pass.

3. Examples of markings are fissure, foramen, meatus, fossa, process, condyle, head, facet, tuberosity, crest, and spine.

DIVISIONS OF THE SKELETAL SYSTEM
1. The axial skeleton consists of bones arranged along the longitudinal axis. Parts of the axial skeleton are the skull, hyoid, ossicles, vertebral column, sternum, and ribs.
2. The appendicular skeleton consists of bones of the girdles and the upper and lower extremities. Parts of the appendicular skeleton are the shoulder girdle, bones of the free upper extremities, pelvic girdles, and bones of the free lower extremities.

Sutures and fontanels of the skull

1. Sutures are immovable joints between bones of the skull. Examples are coronal, sagittal, lambdoidal, and squamosal sutures.

2. Fontanels are membrane-filled spaces between the cranial bones of fetuses and infants. Major fontanels are the anterior, posterior, anterolateral, and posterolateral.

Bones of the skull

1. The skull consists of the cranium and the face. It is composed of 22 bones.

2. The eight cranial bones include the frontal, parietal, temporal, occipital, sphenoid, and ethmoid bones.

3. The fourteen facial bones are the nasal, maxilla, zygomatic, mandible, lacrimal, palatine, inferior conchae, and vomer.

Sinuses of the skull

1. Paranasal sinuses are cavities in bones of the skull. They are lined by mucous membranes and communicate with the nasal cavity.

2. The cranial bones containing such sinuses are the frontal, sphenoid, ethmoid, and maxilla.

The vertebral column

1. Much of the vertebral column, the sternum, and the ribs constitute the skeleton of the trunk.

2. The bones of the column are the cervical vertebrae (7), thoracic vertebrae (12), lumbar vertebrae (5), and the sacrum and coccyx.

3. The column contains primary curves (thoracic and sacral) and secondary curves (cervical and lumbar). These curves give strength, support, and balance.

The thorax

1. The thoracic skeleton consists of the sternum, the ribs and costal cartilages, and the thoracic vertebrae.

2. The thorax protects vital organs in the chest area.

The shoulder girdle

1. A shoulder girdle or pectoral girdle consists of a clavicle and scapula.

2. It attaches the free upper extremity to the trunk.

The free upper extremity

1. The bones of a free upper extremity include the humerus, ulna, radius, carpals, metacarpals, and phalanges.

The pelvic girdles

1. The pelvic girdles consist of two coxal bones or hipbones.

2. They attach the free lower extremities to the trunk.

The free lower extremity

1. The bones of a free lower extremity include the femur, tibia, fibula, tarsus, metatarsus, and phalanges.

Arches of the foot

1. The arches of the foot are bones arranged for support and leverage.

2. The two parts of the longitudinal arch are the higher medial and lower lateral arches. The other arch is the transverse arch.

COMPARISON BETWEEN MALE AND FEMALE SKELETONS

1. The bones in a male are generally larger than the female's bones.

2. The female pelvis is adapted for pregnancy and childbirth.

FRACTURES

1. A fracture is any break in a bone.

2. The types of fractures include: partial, complete, simple, compound, comminuted, greenstick, displaced, and nondisplaced.

3. Fracture repair consists of forming a fracture hematoma, forming a callus, and remodeling.

Review questions and problems

1. What are the four principal kinds of bones? Give an example of each.

2. What are surface markings? Describe and give an example of each.

3. Distinguish between the axial and appendicular skeletons. What subdivisions and bones are contained in each?

4. What bones comprise the skull? The cranium? The face?

5. Define a suture. What are the four principal sutures of the skull? Where are they located?

6. What is a fontanel? Describe the location of the six major fontanels.

7. What is a paranasal sinus? Give examples of cranial bones that contain such sinuses.

8. Identify each of the following: sinusitis, cleft palate, harelip, and mastoiditis.

9. What is the hyoid bone? In what respect is it unique? What is its function?

10. What bones form the skeleton of the trunk? Distinguish between the number of nonfused vertebrae found in the adult vertebral column and that of a child.

11. What is a curve in the vertebral column? How are primary and secondary curves differentiated? What is a curvature? Give three examples of curvatures.

12. What bones form the skeleton of the thorax? What are the functions of the thoracic skeleton? What is a sternal puncture?

13. What is a shoulder girdle? What are the bones of the free upper extremity? What is a pelvic girdle? What are the bones of the free lower extremity?

14. Define a Colles' fracture and a Pott's fracture.

15. Define an arch of the foot. What is its function? Distinguish between a longitudinal arch and a transverse arch.

16. How do bunions and flatfeet arise?

17. What are the principal structural differences between male and female skeletons?

18. What is a fracture? Distinguish eight principal kinds.

19. Outline the steps involved in fracture repair.

CHAPTER 9

ARTICULATIONS: STRUCTURE, PHYSIOLOGY, AND DISORDERS

STUDENT OBJECTIVES

After you have read this chapter, you should be able to:

1. Define an articulation and identify the factors that determine the degree of movement at a joint

2. Contrast the structure, kind of movement, and location of fibrous, cartilaginous, and synovial joints

3. Describe the detailed structure of a synovial joint

4. Discuss and compare the movements possible at various kinds of synovial joints

5. Describe the causes and symptoms of common joint disorders, including arthritis, bursitis, tendinitis, and intervertebral disc abnormality

6. Define medical terminology associated with joints

Bones are much too rigid to bend. In fact, if the skeleton of your body were composed of a continuous mass of ossified tissue, you would have less flexibility than a two-by-four. Fortunately, the skeletal system consists of many separate bones, which are held together at joints by flexible types of connective tissue. All movements that change the positions of the bony parts of the body, such as the arms and legs, take place at joints. Joints are necessary for a raft of activities that we take for granted, for example, holding a pencil, turning the head, swinging the arms, driving a car, and dancing. You can understand the importance of joints if you imagine how a cast over the knee joint prevents flexing of the leg or how a splint on a finger limits the ability to manipulate small objects.

The term **articulation** or **joint** refers to a point of contact between bones or between cartilage and bones. In every case, the structure of a joint determines its function. Some joints permit no movement. Others permit a slight degree of movement. And still others afford relatively unrestricted movement. In general, the more closely the bones fit together, the stronger the joint. However, at tightly fitted joints, movement is restricted. The greater the degree of movement, the looser the fit. Unfortunately, loosely fitted joints are more prone to dislocation. Movement at joints is also determined by the flexibility of the connective tissue that binds the bones together and by the position of ligaments, muscles, and tendons. First, we shall look at the various kinds of joints in the body. Later in the chapter, we can discuss joint disorders.

KINDS OF JOINTS

The joints of the body may be classified into three principal kinds on the basis of their anatomy and the type and degree of movement that they allow.

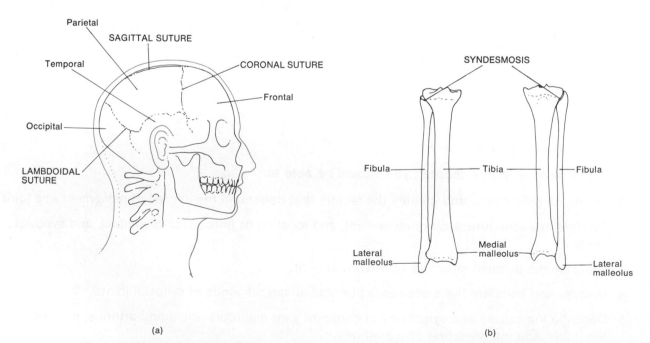

(a)

(b)

Figure 9–1. Fibrous joints. (a) Sutures as seen in a lateral view of the skull. (b) Syndesmosis at the tibiofibular articulation seen in anterior (left) and posterior (right) views.

These classifications are fibrous joints, cartilaginous joints, and synovial joints.

Fibrous joints

Joints that allow little or no movement are called **fibrous joints.** An example is the kind of joint found between bones of the skull. These joints lack a joint cavity; and the articulating bones are held close together by fibrous connective tissue. The two types of fibrous joints are sutures and syndesmoses (Figure 9–1). **Sutures** are found between bones of the skull. Some sutures consist of interlocking, jagged margins of the bone that fit together like a jigsaw puzzle. In other sutures, the margins of the bones overlap. In either case, the bones are very slightly separated by a thin layer of fibrous tissue. As a result of this tight fit, such joints are immovable. The bone surfaces of a **syndesmosis** are united by dense fibrous

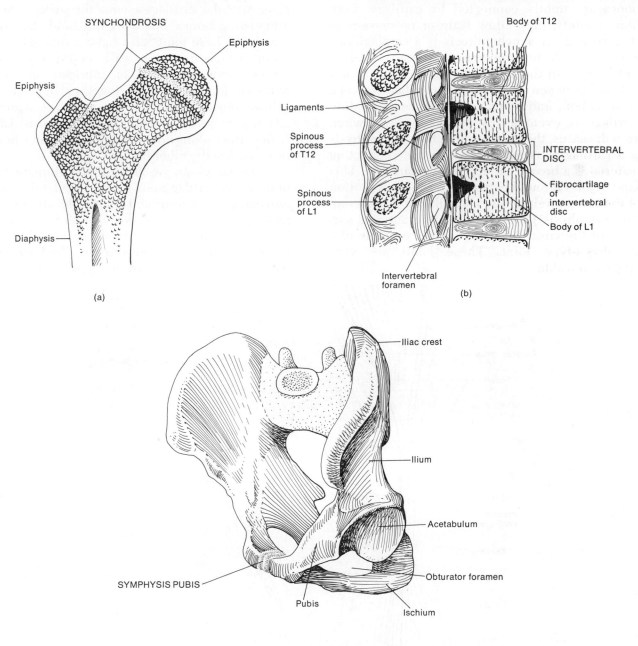

Figure 9–2. Cartilaginous joints. (a) Synchondrosis between the diaphysis and epiphysis of a growing femur. (b) Symphysis joint between the bodies of vertebrae as seen in sagittal section. (c) Symphysis joint (symphysis pubis) between the coxal bones as seen in an oblique view.

tissue. The joint is very slightly movable because the bones are more separated from each other than they are in a suture. An example of a syndesmosis-type joint is the proximal articulation of the tibia with the fibula.

Cartilaginous joints

Another type of joint that has no joint cavity is the **cartilaginous joint.** Here, the articulating bones are tightly connected by cartilage. Like fibrous joints, they allow little or no movement (Figure 9–2). **A synchondrosis** is a cartilaginous joint in which the connecting material is hyaline cartilage, as in the epiphyseal plate. Such a joint is found between the epiphysis and diaphysis of a growing bone and is immovable. Since the hyaline cartilage is eventually replaced by bone when growth ceases, the joint is temporary. **A symphysis** is a cartilaginous joint in which the connecting material is a broad, flat disc of fibrocartilage. This kind of joint is found between vertebrae. A portion of the intervertebral disc is cartilaginous material. The symphysis pubis between the anterior surfaces of the coxal bones is another example of a symphysis-type joint. These joints are very slightly movable.

Synovial joints

When a joint cavity is present, the articulation is called a synovial joint (Figure 9–3). The cavity, called a *synovial cavity,* is a space between the articulating bones. Because of this cavity and because no tissue exists between the articulating surfaces of the bones, synovial joints are freely movable. These joints are also characterized by a layer of hyaline cartilage, called *articular cartilage.* Articular cartilage covers the surfaces of the articulating bones but does not bind the bones together. These joints also have a *synovial membrane,* which lines the walls of the cavity and secretes synovial fluid to lubricate the joint. Synovial joints are held together by bands of collagenous fibers called *ligaments.* A ligament attaches to processes on one of the articulating bones, runs alongside the joint, and attaches to processes on the other bone.

Synovial joints are free of the limitations of fibrous and cartilaginous joints. In synovial joints, movement is determined by the location of the ligaments, by the muscles and their tendons, and by the presence of other bones that might get in the way of particular movements. The knee joint, which is the largest articulation in the body,

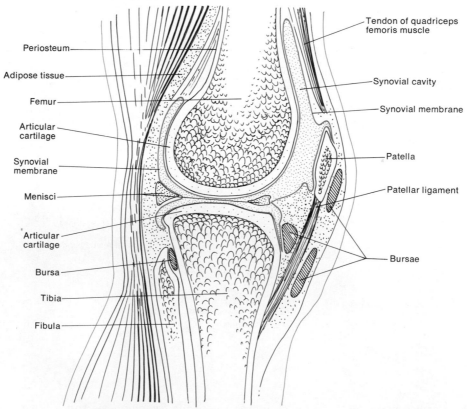

Figure 9–3. Structure of a synovial joint, the knee joint, as seen in sagittal section.

illustrates well the structure of a synovial joint and the restrictions on its movement.

You can examine Figure 9–4 and note the following relationships among ligaments, tendons, and fibrocartilage in the knee joint:

1. Externally the joint is strengthened by muscles and tendons. These include the tendon of the quadriceps femoris muscle in front (Figure 9–4a) and the gastrocnemius muscle in back (Figure 9–4c). The patella lies within the tendon of the quadriceps femoris.

2. The thickened portion of the quadriceps femoris tendon between the top of the patella and the tibia is called the patellar ligament (Figure 9–4a). This ligament strengthens the anterior portion of the joint and prevents the lower leg from being flexed backward too far.

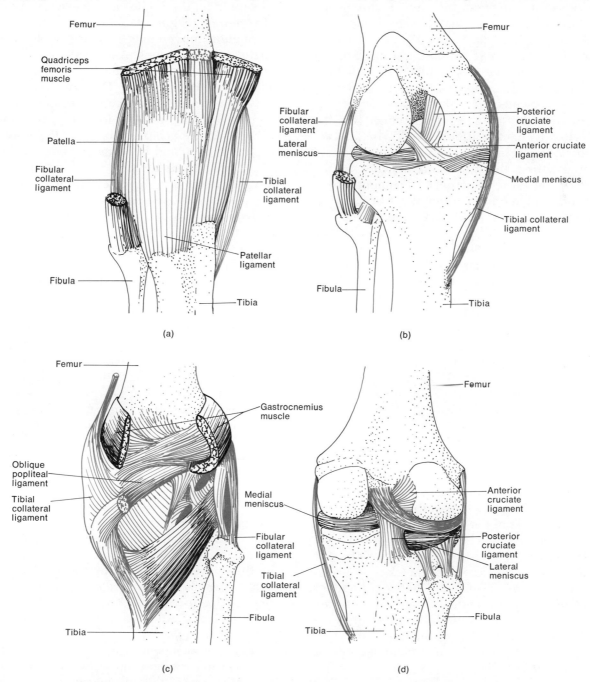

(a)

(b)

(c)

(d)

Figure 9–4. Relationship of ligaments, tendons, and menisci to the knee joint. (a) Anterior, superficial view of the knee. (b) Anterior view of the knee with many superficial structures removed. (c) Posterior, superficial view of the knee. (d) Posterior view of the knee with many superficial structures removed.

3. On either side of the joint are the fibular and tibial collateral ligaments. The fibular collateral ligament, between the femur and fibula, strengthens the lateral side of the joint. The tibial collateral ligament, between the femur and tibia, strengthens the medial side of the joint. Both ligaments prohibit any side-to-side movement at the joint (Figure 9–4a, d).

4. The oblique popliteal ligament is located on the posterior surface of the joint (Figure 9–4c). It starts in a tendon that lies over the tibia and runs upward and laterally to the lateral side of the femur. It supports the back of the knee and prevents hyperextension—that is, bending the knee in the opposite direction it normally bends.

5. Internally, the joint is strengthened by the cruciate ligaments (Figure 9–4b, d). The anterior cruciate ligament passes backward and laterally from the tibia and attaches to the femur. The posterior cruciate ligament passes forward and medially from the tibia and attaches to the femur. Both ligaments are believed to stabilize the knee joint during its movements.

6. Between the articular cartilage of the femur and tibia are **menisci**, concentric wedge-shaped pieces of fibrocartilage (Figure 9–4b, d). These are called the lateral meniscus and the medial meniscus. The menisci afford support for the continuous weight placed on the knee joint. Their surfaces also assist in rotation of the knee. A tearing of the menisci, commonly called torn cartilage, occurs frequently among athletes.

The various movements of the body could cause friction to develop between moving parts. Saclike structures called **bursae** are situated in the body tissues to reduce this friction. (See Figure 9–3.) These sacs resemble joints in that their walls consist of connective tissue lined by a synovial membrane. They are also filled with synovial fluid. Bursae are located between the skin and bone in places where the skin rubs over the bone. They are also found between tendons and bones, muscles and bones, and between ligaments and bones. As fluid-filled sacs, they act as cushions and ease the movement of one part of the body over another. An inflammation of a bursa is called **bursitis**.

Movements at synovial joints

The movements possible at synovial joints may be categorized into five principal kinds: gliding movements, angular movements, rotation, circumduction, and special movements.

A **gliding movement** is the simplest kind that can occur in a joint. One surface moves back and forth and from side to side over another surface without angular or rotary movements. Some joints that glide are those between the carpals, between the tarsals, and between the sacrum and ilium. The heads and tubercles of ribs glide on the bodies and transverse processes of vertebrae (Figure 9–5).

Angular movements increase or decrease the angle between bones. Among the angular movements are flexion, extension, abduction, and adduction (Figure 9–6). **Flexion** involves a decrease in the angle between two bones. Examples include bending the head forward, where the joint is between the occipital bone and the atlas, bending the elbow, and bending the knee. Flexion of the foot at the ankle joint is referred to as **dorsiflexion**. **Extension** involves an increase in the angle between two bones. Extension restores a body part to its anatomical position after it has been flexed. Examples of extension are returning the head to the anatomic position after flexion, straightening out the arm after flexion, and straightening out the leg after flexion. Continuation of extension beyond the anatomic position, as in bending the head backward, is called **hyperextension**. Extension of the foot at the ankle joint is called **plantar flexion**.

Abduction usually means the movement of a bone *away from* the midline of the body. An example of abduction is moving the arms upward and away from the body until they are held straight out at right angles to the chest. In the case of the fingers and toes, however, the midline of the body is not used as the line of reference. Abduction of the fingers is a movement away from an imaginary line drawn through the middle finger—in other words, spreading the hands. Abduction of the toes is relative to an imaginary line drawn through the second toe. **Adduction** is usually the movement of a part *toward* the midline of the body. Moving the arms back to the sides after abduction is adduction. As in abduction, adduction of the fingers is relative to the middle finger. Adduction of the toes is relative to the second toe.

Rotation means moving a bone around its own axis. During rotation, no other motion is permitted. We rotate the atlas around the odontoid process of the axis when we turn the head from side to side to say "no." Turning the arm palm up, then palm down, and then palm up again is an

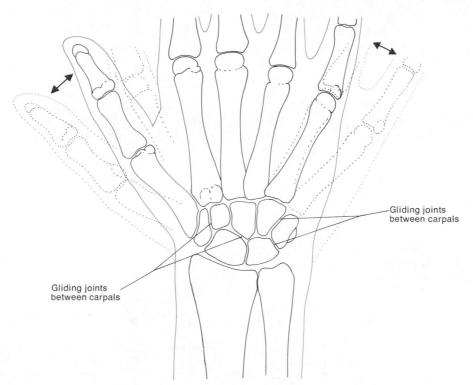

Figure 9–5. Gliding movement between carpal bones.

Gliding joints
between carpals

Gliding joints
between carpals

example of slight rotation of the forearm (Figure 9–7a).

Circumduction is a movement in which the distal end of a bone moves in a circle while the proximal end remains relatively stable. The bone draws a cone in the air. Circumduction involves flexion, abduction, adduction, extension, and rotation. An example is moving the outstretched arm in a circle to wind up for a pitch (Figure 9–7b).

Special movements are those that are found only at the joints indicated in Figure 9–8. They include the following:

Inversion is the movement of the sole of the foot inward at the ankle joint. **Eversion** is the movement of the sole of the foot outward at the ankle joint. **Protraction** is the movement of the mandible or clavicle forward on a plane parallel to the ground. Thrusting out the jaw is protraction of the mandible. Bringing your arms forward until the elbows touch each other requires protraction of the clavicle. **Retraction** is the movement of a protracted part of the body backward on a plane parallel to the ground. Pulling the lower jaw back in line with the upper jaw is retraction. **Supination** is a movement of the forearm in which the palm of the hand is turned forward (anterior). If you wish to practice supina-

tion, flex your arm at the elbow so that rotation of the humerus in the shoulder joint is prevented. **Pronation** is a movement of the flexed forearm in which the palm of the hand is turned backward (posterior).

Types of synovial joints

Even though all synovial joints are basically similar in structure, variations exist in the shape of the articulating surfaces. Accordingly, synovial joints are divided into six subtypes. These include gliding, hinge, pivot, ellipsoidal, saddle, and ball-and-socket joints.

The articulating surfaces of **gliding joints** are usually flat. Only side-to-side and back-and-forth movements are permitted. Since this kind of joint allows movements in two planes, the joint is referred to as *biaxial*. Twisting and rotation are inhibited at gliding joints generally because either the arrangement of the ligaments or the presence of other bones close by restricts the range of movement. Examples are the joints found between carpal bones, tarsal bones, the sacrum and ilium, the sternum and clavicle, and the scapula and clavicle. (See Figure 9–5.)

Hinge joints are characterized by a convex

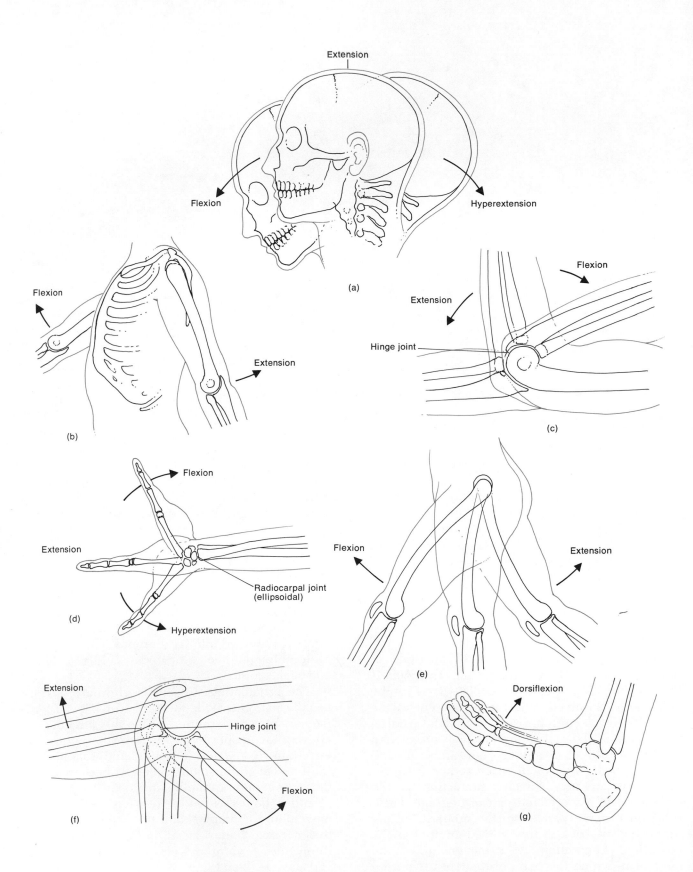

Figure 9–6. Angular movements at synovial joints.

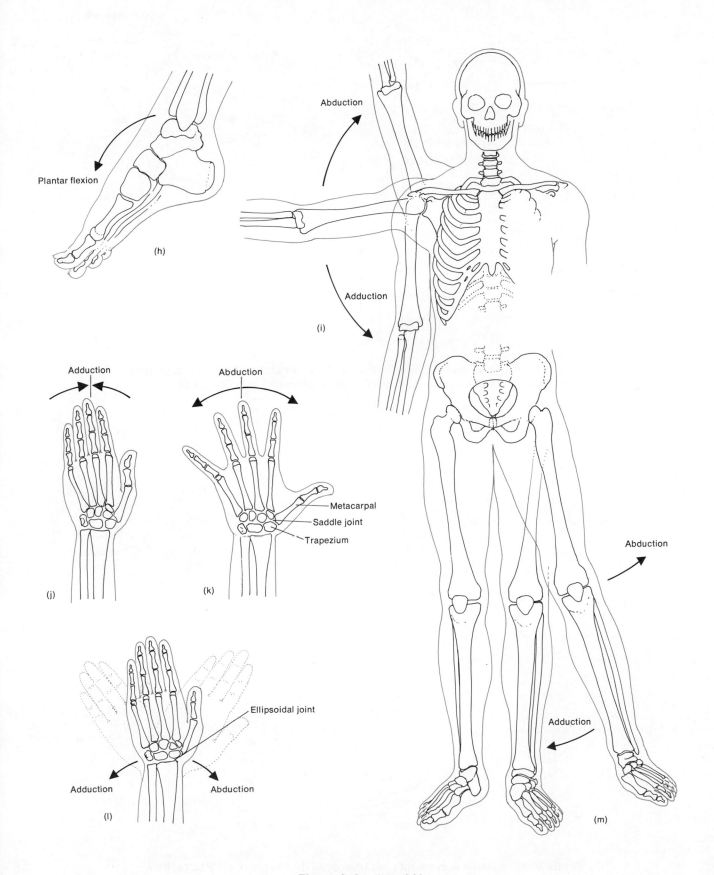

Figure 9–6. *(cont'd.)*

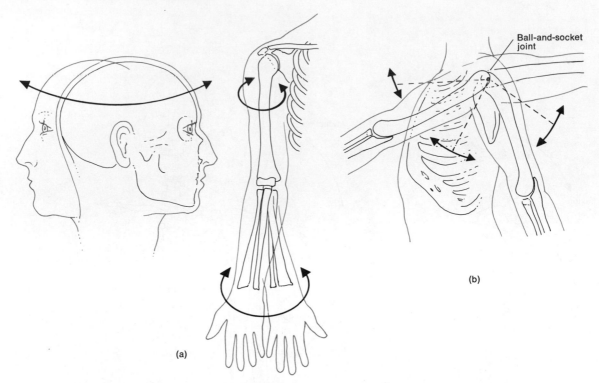

Figure 9–7. Rotation and circumduction. (a) Rotation at the atlas-axis joint (left) and rotation of the humerus (right). (b) Circumduction of the humerus at the shoulder joint.

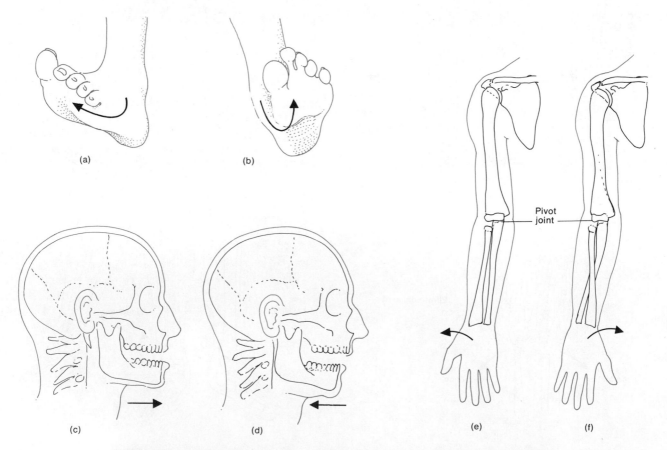

Figure 9–8. Special movements. (a) Inversion. (b) Eversion. (c) Protraction. (d) Retraction. (e) Supination. (f) Pronation.

surface of one bone that fits into a concave surface of another bone. Movement is primarily in a single plane and is usually flexion and extension. The joint is therefore referred to as *monaxial*. The motion is similar to that of a hinged door. Examples of hinge joints include the elbow, knee, ankle, and interphalangeal joints. The movement afforded by a hinge joint is illustrated by flexion and extension at the elbow and knee. (See Figure 9–6c, f.)

In a **pivot joint,** a rounded, pointed, or conical surface of one bone articulates with a shallow depression of another bone. The primary movement permitted is rotation, and the joint is therefore monaxial. Examples include the joints between the atlas and axis and between the proximal ends of the radius and ulna. Movement at a pivot joint is illustrated by supination and pronation of the palms. (See Figure 9–8e, f.)

Ellipsoidal joints are structured so that an oval-shaped condyle of one bone fits into an elliptical cavity of another bone. Since the joint permits side-to-side and back-and-forth movements, it is biaxial. The joint at the wrist between the radius and carpals is ellipsoidal. The movement permitted by such a joint is illustrated when you flex and extend and abduct and adduct the wrist. (See Figure 9–6d, l.)

In a **saddle joint,** the articular surfaces of both of the bones is saddle-shaped—in other words, concave in one direction and convex in the other direction. Essentially, the saddle joint is a modified ellipsoidal joint in which the movement is somewhat freer. Movements at a saddle joint are side to side and back and forth. Thus, the joint is biaxial. The joint between the trapezium and metacarpal of the thumb is an example of a saddle joint. (See Figure 9–6k.)

Ball-and-socket joints consist of a ball-like surface of one bone fitting into a cuplike depression of another bone. Such a joint permits *triaxial* movement. That is, there is movement in three planes of motion: flexion-extension, abduction-adduction, and rotation. Examples of ball-and-socket joints are the shoulder joint and the hip joint. The range of movements afforded at a ball-and-socket joint is illustrated by circumduction of the arm. (See Figure 9–7b.)

A summary of joints is presented in Exhibit 9–1.

DISORDERS OF JOINTS

Various disorders may occur in and around the joints. Structures associated with the joints, such as muscles, tendons, articular cartilage, synovial membranes, and bursae, may be affected.

Arthritis

A disease that has plagued mankind through the ages is **arthritis.** Signs of the disorder have been found in the bones of the Java man and in the mummies of Egypt. The term arthritis actually refers to at least 25 different diseases, the most common being rheumatoid arthritis, osteoarthritis, and gout. However, all these ailments are characterized by inflammation in one or more joints. Inflammation, pain, and stiffness may also be present in neighboring parts of the body, such as the muscles near the joint.

The causes of arthritis are unknown. In some cases, it has followed the stress of sprains, infections, and joint injury. Some researchers believe that the cause is a bacteria or virus, whereas others suspect an allergy. There are those who believe the nervous system or hormones are involved, whereas some suspect a disorder of the metabolic system. Still others believe that certain types of prolonged psychologic stress, such as inhibited hostility, can upset homeostatic balance and bring on arthritic attacks.

Rheumatoid arthritis

Rheumatoid arthritis is a disease that is usually limited to the destruction of joints and their surrounding structures, such as muscles and tendons. In about 80 percent of the cases, the onset of the disease occurs between the ages of 20 and 50 years. Women are affected three times as often as men. About one-third of all arthritic people have the rheumatoid type of arthritis.

The primary symptom of rheumatoid arthritis is inflammation of the synovial membrane (Figure 9–9). The membrane thickens, and synovial fluid accumulates. The resulting pressure causes pain and tenderness. The membrane then produces an abnormal tissue called *pannus*, which grows onto the surface of the articular cartilage. The pannus formation sometimes erodes the cartilage completely. When the cartilage is destroyed, fibrous tissue grows out of the exposed bone ends. The tissue ossifies and fuses the joint so that it is immovable. This is the ultimate crippling effect of rheumatoid arthritis. Most cases do not progress this far. But the range of motion of the joint is greatly inhibited by the severe inflammation and swelling.

Osteoarthritis

A degenerative joint disease that is far more common than rheumatoid arthritis, and usually less

Exhibit 9–1. SUMMARY OF JOINTS

TYPE	DESCRIPTION	MOVEMENT	EXAMPLES
Fibrous	No joint cavity; bones held together by a thin layer of fibrous tissue or dense fibrous tissue		
Suture	Found only between bones of the skull; articulating bones are separated by a thin layer of fibrous tissue	None	Lambdoidal suture between occipital and parietal bones
Syndesmosis	Articulating bones united by dense fibrous tissue	Slight	Distal ends of tibia and fibula
Cartilaginous	No joint cavity; articulating bones united by cartilage		
Synchondrosis	Connecting material is hyaline cartilage	None	Temporary joint between the diaphysis and epiphyses of a long bone
Symphysis	Connecting material is a broad, flat disc of fibrocartilage	Slight	Intervertebral joints and symphysis pubis
Synovial	Joint cavity and articular cartilage present; synovial membrane lines cavity; freely movable		
Gliding	Articulating surfaces usually flat	Biaxial (flexion-extension, abduction-adduction)	Intercarpal and intertarsal joints; sacroiliac joint
Hinge	Spool-like surface fits into a concave surface	Monaxial (flexion-extension)	Elbow, knee, ankle, and interphalangeal joints
Pivot	Rounded, pointed, or concave surface fits into a shallow depression	Monaxial (rotation)	Atlas-axis and radioulnar joints
Ellipsoidal	Oval-shaped condyle fits into an elliptical cavity	Biaxial (flexion-extension, abduction-adduction)	Radiocarpal joint
Saddle	Articular surfaces concave in one direction and convex in opposite direction	Biaxial (flexion-extension, abduction-adduction)	Carpometacarpal joint of thumb
Ball-and-socket	Ball-like surface fits into a cuplike depression	Triaxial (flexion-extension, abduction-adduction, rotation)	Shoulder and hip joints

damaging, is **osteoarthritis.** This type of arthritis is one of the oldest diseases known. It apparently results from a combination of aging, irritation of the joints, and normal wear and tear. Symptoms rarely occur before the age of 40.

Osteoarthritis is a chronic inflammation of the articular cartilage of some of the joints. The inflammation causes swelling, stiffness, and pain. The cartilage slowly degenerates. And as the bone ends become exposed, they lay down little bumps, or spurs, of new osseous tissue. These spurs decrease the space of the joint cavity and put re-

strictions on movement. Unlike rheumatoid arthritis, osteoarthritis usually affects only the articular cartilage. The synovial membrane is very rarely destroyed, and other tissues are unaffected. The disorder does not cause the articulating bones to fuse.

Gouty arthritis

Uric acid is a waste product given off during the synthesis of the nucleic acid purine. Normally, all the acid is quickly excreted in the urine. And, in fact, it gives urine its name. The person who

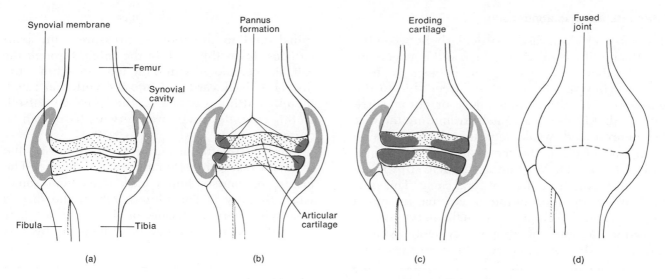

Figure 9–9. Progressive stages of rheumatoid arthritis at the knee joint. (a) Inflammation of the synovial membrane. (b) Early stage of pannus formation and erosion of articular cartilage. (c) Advanced stage of pannus formation and further erosion of cartilage. (d) Obliteration of joint space and fusion of bones.

suffers from *gout* either produces excessive amounts of uric acid or is not able to excrete normal amounts. The result is a buildup of uric acid in the blood. This excess acid then reacts with sodium to form a salt called sodium urate. Crystals of this salt are deposited in soft tissues, some favorite spots being the kidneys and the cartilage of the ears and joints.

In **gouty arthritis,** sodium urate crystals are deposited in the soft tissues of the joints. The crystals irritate the cartilage, causing inflammation, swelling, and acute pain. Eventually, the crystals destroy all the joint tissues. If the disorder is not treated, the ends of the articulating bones fuse, and the joint becomes immovable.

Gout occurs primarily in males of any age. It is believed to be the cause of 2 to 5 percent of all chronic joint diseases. Numerous studies indicate that gout is sometimes caused by an abnormal gene. The gene instructs the body to manufacture purine by a mechanism that gives off unusually large amounts of uric acid. Diet and environmental factors such as stress and climate are also suspected causes of gout. Many famous men have suffered from the disorder—Michelangelo, Martin Luther, Benjamin Franklin, and Charles Darwin, to name a few. Because of this fact, some humorists have related gout to high intelligence and the stresses related to high responsibility.

Although other forms of arthritis cannot be treated with complete success, gout has almost been conquered. The advent of various drugs has been the reason. One chemical called colchine has been used on and off since the sixth century to relieve the pain, swelling, and tissue destruction that occurs during attacks of gouty arthritis. This chemical is derived from the variety of crocus plant from which we also get the spice saffron. Other drugs, which either inhibit uric-acid production or help the kidneys eliminate excess uric acid, are used to prevent further attacks. The drug, allopurinal, is used to treat gout because it prevents the formation of uric acid without interfering with purine synthesis.

Bursitis

An acute or chronic inflammation of a bursa is called **bursitis.** The condition may be caused by *trauma,* which is a wound or injury, by an acute or chronic infection including syphilis and tuberculosis, or by rheumatoid arthritis. It is characterized by pain, swelling, tenderness, and limitation of any motion that involves the inflamed bursa. The patient usually recovers in 1 to 2 weeks, but the condition may become chronic.

Tendinitis

A disorder involving the inflammation of a tendon and synovial membrane at a joint is called **tendinitis** or **tenosynovitis.** The shoulder, the elbow (tennis elbow), the joints in the fingers (trigger finger), and their associated tendons are most often the sites of inflammation.

Intervertebral disc abnormality

The intervertebral discs, particularly between the fourth and fifth lumbar vertebrae and between the fifth lumbar and sacrum, are subject to great forces. They are thus highly susceptible to degeneration resulting from injury or chronic irritation. Should the ligaments surrounding the disc be injured and weakened, the disc begins to "slip" out of place and the result is **intevertebral disc abnormality.** Most often the disc slips backward and away from the vertebral canal. This puts pressure on the roots of the nerves that run out of the spinal cord. The pressure on the nerves causes considerable, and sometimes very acute, pain. If the root of the sciatic nerve, which runs from the spinal cord to the foot, is pressured, the pain radiates down the back of the thigh, through the calf, and occasionally into the foot. Sometimes the disc slips backward and into the vertebral canal, thereby putting pressure on the spinal cord itself. In this case, destruction of nervous tissue can be quite serious.

Surgical treatment may involve removing the disc and fusing together the two vertebrae. The fusion, of course, limits back movement somewhat. Surgery is considered only if the disc is destroying nervous tissue, or if the chronic pain and inflammation cannot be alleviated by less drastic methods.

Medical terminology associated with joints

Ankylosis (*ankyl* = bent; *osis* = condition) Severe or complete loss of movement at a joint.

Arthralgia (*algia* = pain) Pain in a joint.

Arthrosis A disease of a joint; also means articulation or joint.

Bursectomy (*ectomy* = removal of) Removal of a bursa.

Chondritis Inflammation of a cartilage.

Dislocation Displacement of a bone from its natural position in a joint.

Sprain Tearing of tendons and ligaments.

Synovitis Inflammation of a synovial membrane in a joint.

Chapter summary in outline

KINDS OF JOINTS
1. A joint or articulation is a point of contact between two bones.
2. Closely fitting bones are strong but not freely movable. Loosely fitting joints are weaker but freely movable.

Fibrous joints
1. Bones held by fibrous connective tissue, with no joint cavity, are fibrous joints.
2. These joints include immovable sutures (found in the skull) and slightly movable syndesmoses (such as the tibiofibular articulation).

Cartilaginous joints
1. Bones held together by cartilage, with no joint cavity, are cartilaginous joints.
2. These joints include immovable synchondroses united by hyaline cartilage (temporary cartilage between diaphysis and epiphyses) and partially movable symphyses united by fibrocartilage (the symphysis pubis).

Synovial joints
1. These joints contain a joint cavity, articular cartilage, and synovial membranes. They are held together by ligaments and tendons.
2. All synovial joints are freely movable.
3. Types of synovial joints include gliding joints (wrist bones), hinge joints (elbow), pivot joints (radioulnar), ellipsoidal joints (radiocarpal), saddle joints (carpometacarpal), and ball-and-socket joints (shoulder and hip).
4. Planes of movement at synovial joints include the monaxial, biaxial, and triaxial planes.
5. Types of movements at synovial joints include gliding movements, angular movements, rotation, circumduction, and special movements, such as inversion, eversion, protraction, retraction, supination, and pronation.

DISORDERS OF JOINTS
1. Arthritis is the inflammation of a joint. Types of arthritis include rheumatoid arthritis, osteoarthritis, and gouty arthritis.
2. Bursitis is the inflammation of a bursa.
3. Tendinitis is the inflammation of the tendon and synovial membrane at a joint.
4. Intervertebral disc abnormality, called "slipped" disc, involves misplacement of a disc by injury or chronic irritation.

Review questions and problems

1. Define an articulation. What factors determine the degree of movement at joints?
2. Distinguish among the three kinds of joints. List the subtypes of joints. Be sure to include structure, degree of movement, and specific examples.
3. Using the knee as a typical joint, explain the components of a synovial joint. Indicate the relationship of ligaments and tendons to the strength of the joint and restrictions on movement.
4. What are bursae? What is their function? Define bursitis.
5. Define the following principal movements: gliding, angular, rotation, circumduction, and special. Name a joint where each occurs.
6. Have your partner assume the anatomical position and have him execute each of the movements at joints discussed in the text. Reverse roles, and see if you can execute the same moves.
7. Contrast monaxial, biaxial, and triaxial planes of movement. Give examples of each, and name a joint at which each occurs.
8. What is arthritis? Distinguish among the following kinds: rheumatoid arthritis, osteoarthritis, and gouty arthritis. What is a pannus?
9. Define tendinitis and intervertebral disc abnormality. Why is a "slipped" disc usually so painful?
10. Refer to the glossary of medical terminology associated with joints. Be sure that you can define each term.

CHAPTER 10

MUSCULAR TISSUE: STRUCTURE, PHYSIOLOGY, AND DISORDERS

STUDENT OBJECTIVES

After you have read this chapter, you should be able to:

1. List the distinguishing properties of muscle tissue

2. Compare the location, microscopic appearance, nervous control, and functions of the three kinds of muscle tissue

3. Identify the histological characteristics of skeletal muscle

4. Define epimysium, perimysium, endomysium, tendons, and aponeuroses and list their modes of attachment to muscles

5. Describe the relationship of blood vessels and nerves to skeletal muscles

6. Describe the source of energy for muscular contraction

7. Describe the physiological importance of the motor unit

8. Describe the physiology of contraction by listing the events associated with the sliding-filament hypothesis

9. Define the all-or-none principle of muscular contraction

10. Contrast the kinds of normal contractions performed by skeletal muscles

11. Describe the phases of contraction in a typical myogram

12. Compare oxygen debt and heat production as examples of muscle homeostasis

13. Define such common muscular disorders as fatigue, fibrosis, muscular dystrophy, myasthenia gravis, and tumors

14. Compare spasms, cramps, convulsions, tetany, and fibrillation as abnormal muscular contractions

15. Define medical terminology associated with the muscular system

In the preceding chapters on the skeletal system, we discussed how bones are connected in various ways to form joints. Although bones and joints provide leverage and form the framework of the body, they are not capable of moving the body by themselves. Motion is an essential body function that is made possible by the contraction of muscles. Muscle tissue constitutes about 40 to 50 percent of the total body weight and is composed of highly specialized cells having four striking characteristics. One of these features, **irritability**, is the ability of muscle tissue to receive and respond to stimuli. A second characteristic of muscle is **contractility**, the ability to shorten and thicken, or contract, when a sufficient stimulus is received. Muscle tissue also exhibits **extensibility**, which means that it stretches when pulled. You will see later that many skeletal muscles are arranged in opposing pairs. While one is contracting, the other is undergoing extension. The final characteristic of muscle tissue is **elasticity**, the ability of muscle to return to its original shape after contraction or extension. Other tissues of the body also exhibit irritability, extensibility, and elasticity, but only muscle can contract. Through contraction, muscle performs the three important functions of motion, maintaining posture, and producing heat.

The most obvious kinds of motions of the body are walking, running, and moving from one place to another. Other obvious types of movements, such as grasping a pencil or nodding the head, are limited to one or more parts of the body. These kinds of movement rely on the integrated functioning of the bones, joints, and muscles that are attached to the bones. Less noticeable kinds of motion produced by muscles are the beating of the heart, the churning of food in the stomach, the pushing of food through the intestines, the contraction of the gall bladder to release bile, and the contraction of the urinary bladder to expel urine.

In addition to performing the function of movement, muscle tissue also enables the body to maintain posture. The contraction of skeletal muscles holds the body in stationary positions, such as standing and sitting. Completely relax the muscles of your legs and torso, and see how quickly you flop to the floor!

The third function of muscle tissue is heat production. Skeletal muscle contractions produce heat and are thereby important in maintaining normal body temperature. The mechanisms involved in muscle contraction will be discussed later in greater detail.

KINDS OF MUSCLE TISSUE

Three kinds of muscle tissue are recognized: (1) **skeletal,** (2) **smooth,** and (3) **cardiac** muscle. These three types of muscle are further described and categorized on the basis of location, presence of microscopic striations, and nervous control. Skeletal muscle, which is named after its location, is found attached to bones. It is a *striated* muscle because striations, or bandlike structures, are visible when the tissue is examined under a microscope. Finally, it is a *voluntary* muscle because it can be made to contract by conscious, or voluntary, control. Smooth muscle is located in the walls of hollow internal structures, such as blood vessels, the stomach, and the intestines. It is, therefore, described as a *visceral* muscle. Smooth muscle is *nonstriated*, which is the reason why it is called smooth. It is an *involuntary* muscle because its contraction is not under the conscious control of the individual. Cardiac muscle forms the walls of the heart and is named after its location. Cardiac muscle is also a striated and involuntary muscle tissue. In sum, all muscle tissues are classified in the following ways: (1) skeletal or striated voluntary muscle, (2) cardiac or striated involuntary muscle, and (3) smooth or visceral involuntary muscle. We shall now examine each of these groups of muscle tissue in some detail.

SKELETAL MUSCLE

In order to understand the underlying mechanisms of movement by skeletal muscle, you will need to have some knowledge of its histology, or microscopic structure, its connective tissue components, and its nerve and blood supply.

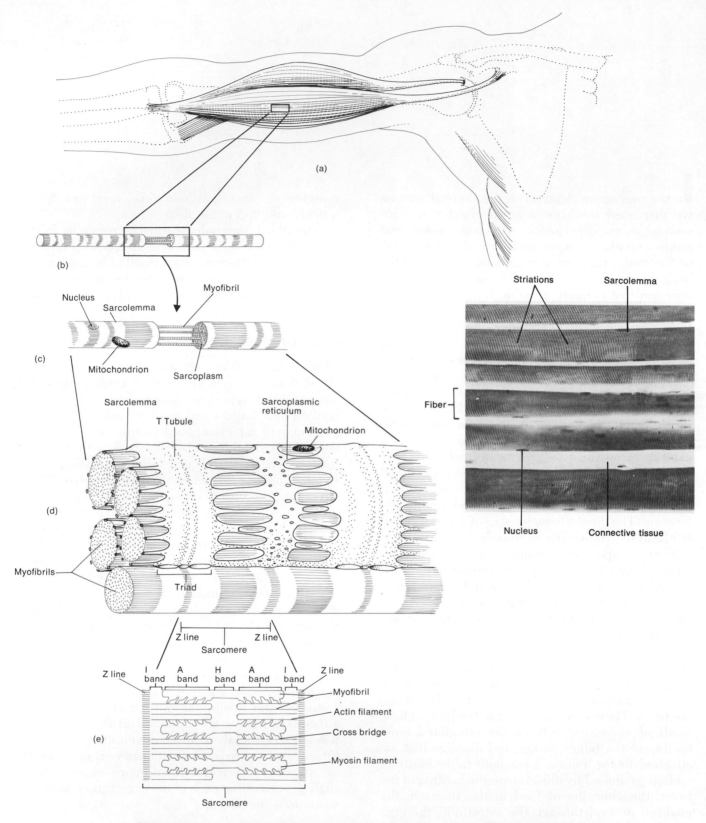

Figure 10-1. Levels of structural organization of a typical skeletal muscle. (a) Entire muscle. (b) Enlarged aspect of a single muscle fiber with a portion removed to show internal structure. (c) Further enlargement of a single muscle fiber showing more detail. (d) Relationship of internal structures to a sarcomere and Z lines. (e) Details of a sarcomere showing myosin and actin filaments and various internal zones. *(Photograph courtesy of Edward J. Reith, from Atlas of Descriptive Histology, Edward J. Reith and Michael H. Ross, Harper & Row, Publishers, Inc., New York, 1970.)*

Histology

Muscle tissue that is attached to and responsible for the movement of bones is generally referred to as *skeletal muscle*. When a typical skeletal muscle (Figure 10–1a) is teased apart and viewed microscopically, it can be seen that the muscle consists of thousands of elongated, cylindrical cells called *muscle fibers* (Figure 10–1b). These fibers lie parallel to each other and range from 10 to 100 μ in diameter. They can be as long as 2½ inches. Each muscle fiber (Figure 10–1c) is surrounded by a membrane called the *sarcolemma*. The combining form *sarco* means flesh, and *lemma* means sheath or husk. The sarcolemma contains a quantity of cytoplasm called *sarcoplasm*. Within the sarcoplasm of a fiber and lying close to the sarcolemma are many nuclei and a number of mitochondria. Also found in a fiber is the *sarcoplasmic reticulum*, a tubelike network of roughly parallel vesicles (Figure 10–1d). Running transversely through the fiber and perpendicularly to the sarcoplasmic reticulum are *T tubules*. The tubules connect with the sarcoplasmic reticulum and open to the outside of the fiber. A *triad* consists of a T tubule and the segments of sarcoplasmic reticulum on both sides of it.

A highly magnified view of a skeletal muscle fiber reveals the presence of threadlike structures, about 1 μ in diameter, called **myofibrils** (Figure 10–1c–e). The prefix *myo* means muscle. These myofibrils (Figure 10–1c, d) run longitudinally through the fiber and consist of two kinds of still smaller filaments. The thin actin filaments are composed of the protein **actin**. The thick myosin filaments consist of the protein **myosin** (Figure 10–1e).

Note that the filaments of a myofibril do not extend the entire length of a muscle fiber; they are stacked in definite compartments called *sarcomeres*. Sarcomeres are partitioned by separations called Z *lines*, which are narrow zones of dense material. Thick and thin filaments overlap for part of their respective lengths and form a dark, dense band called the *A band*. A light-colored, less dense area called the *I band* is composed of thin filaments only. This combination of alternating dark and light bands gives the muscle fiber its striped or striated appearance. A narrow *H band* contains thick filaments only and is located in the middle of the *A band*.

Connective tissue component

Skeletal muscles are protected, strengthened, and attached to other structures by several connective tissue components. The entire muscle is usually wrapped with a substantial quantity of fibrous connective tissue called the *epimysium* (Figure 10–2a). When the muscle is cut in cross section, one can see that extensions of the epimysium divide the muscle into bundles called *fasciculi*. These extensions of the epimysium are called the *perimysium*. In turn, extensions of the perimysium, called *endomysium*, penetrate into the interior of each fasciculus and separate each fiber. The epimysium, perimysium, and endomysium are all continuous with the connective tissue that attaches the muscle to some other structure, such as bone or another muscle (Figure 10–2b). For example, all three elements may be continuous with a *tendon*, which is a cord of connective tissue that attaches a muscle to the periosteum of a bone. In other cases, the connective tissue elements may extend as a broad, flat tendon called an *aponeurosis*. This structure also attaches to the coverings of a bone or another muscle. When a muscle contracts, the tendon or aponeurosis and its corresponding bone or muscle are pulled toward the contracting muscle. It is in this way that skeletal muscles produce movement.

Nerve and blood supply

Skeletal muscles are well supplied with nerves and blood vessels. This heavy infiltration of nervous and circulatory tissues is directly related to contractions, the chief characteristic of muscle. In order for a muscle fiber to contract, it must first be stimulated by an impulse from a nerve cell. A muscle without a functional nerve connection cannot operate. Muscle contraction also requires a great deal of energy. To carry out the reactions that produce this energy, muscle fibers must be supplied with large amounts of nutrients and oxygen. The waste products of these energy-producing reactions must also be eliminated. In fact, muscle action is greatly hindered by any drastic lowering of the blood supply.

Generally, an artery and one or two veins accompany a nerve into a skeletal muscle. The larger branches of the blood vessel accompany the nerve branches through the connective tissue of the muscle (Figure 10–3). Microscopic blood vessels called capillaries are arranged in the endomysium. Each muscle fiber is thus in close contact with one or more capillaries. (See Figure 10–2a.) Each muscle fiber also makes contact with a fiber of a nerve cell.

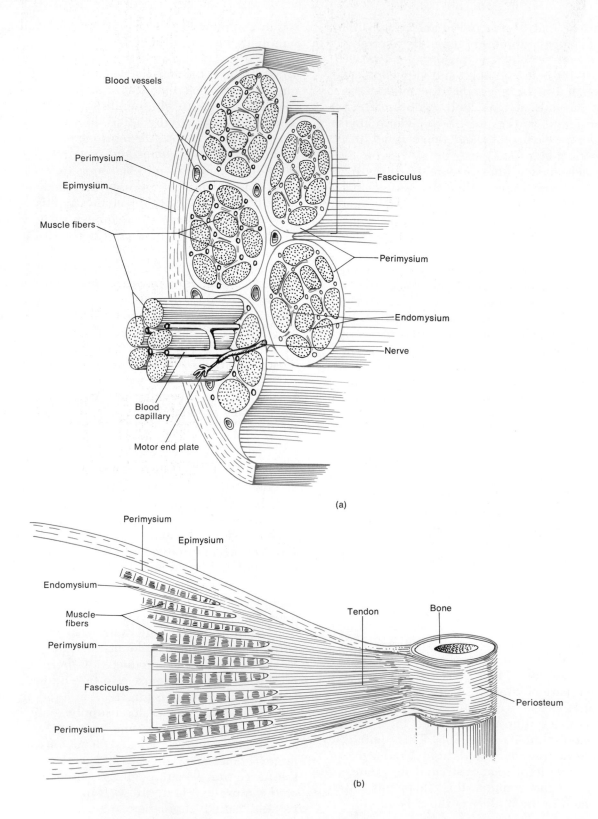

Blood vessels

Perimysium

Epimysium

Muscle fibers

Fasciculus

Perimysium

Endomysium

Nerve

Blood capillary

Motor end plate

(a)

Perimysium

Epimysium

Endomysium

Muscle fibers

Perimysium

Fasciculus

Perimysium

Tendon

Bone

Periosteum

(b)

Figure 10–2. Connective tissue relationships to skeletal muscle. (a) Cross section of a skeletal muscle showing connective tissue coverings. (b) Longitudinal section of a muscle illustrating the attachment of a tendon to the periosteum of bone.

CONTRACTION OF SKELETAL MUSCLE

In order for a muscle to contract, it must have a supply of energy. It must be stimulated by a nerve. And it must be provided with calcium ions. We shall now discuss in detail the contraction of skeletal muscle. As you will learn later, the contractions of cardiac and smooth muscle differ in several ways from the contractions of skeletal muscle. However, the basic mechanisms of muscle contraction are the same—whether the muscle be skeletal, cardiac, or smooth.

The energy for contraction

Contraction of a muscle requires work, and in order for work to be done, energy is needed. Attached to the actin filaments of a muscle fiber is the high-energy compound ATP. (You may wish to reread the section on ATP in Chapter 2.) When a nerve impulse stimulates a muscle fiber, ATP breaks down into ADP + P, and energy is released. As far as we know, ATP is always the immediate source of energy for muscle contraction.

Like the other cells of the body, muscle cells synthesize ATP as follows:

$$ADP + P + energy \rightarrow ATP$$

The energy for replenishing ATP is derived from the breakdown of digested foods. However, unlike most other cells of the body, muscle fibers alternate between a high degree of activity and virtual inactivity. When a muscle is contracting, its energy requirements are high, and the synthesis of ATP is accelerated. If the exercise is strenuous, ATP is used up even faster than it can be manufactured. Thus, muscles must be able to build up a reserve supply of energy. They do this in two ways. A resting muscle needs relatively little energy and produces much more ATP than it can use. At first, the muscle fiber stores the excess ATP on the myosin filaments. When the fiber runs out of storage space for the ATP molecules, it then combines the remainder of the ATP with a substance called *creatine*. Creatine can accept a high-energy phosphate from ATP to become the high-energy compound *creatine phosphate*.

$$ATP + creatine \rightarrow creatine phosphate + ADP$$

Creatine phosphate is produced only when the muscle fibers are resting. During strenuous contraction, the reaction reverses itself, as shown in the following equation:

$$ADP + creatine phosphate \rightarrow ATP + creatine$$

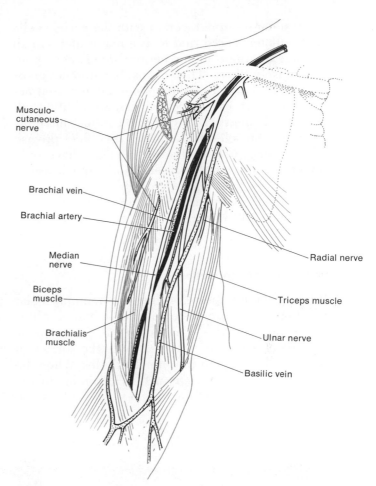

Figure 10–3. Relationship of blood vessels and nerves to skeletal muscles of the right arm seen in anterior view.

The motor unit

In order for muscular contraction to occur, a stimulus must be applied to the muscle tissue. Such a stimulus is normally transmitted by nerve cells called neurons. As you will see in Chapter 12, a neuron has a threadlike process, or fiber, that may run as far as three feet to a muscle. A bundle of such fibers belonging to many different neurons composes a nerve. A neuron that transmits a stimulus to muscle tissue is called a motor neuron. Upon entering a skeletal muscle, the motor neuron branches. These branches then make contact with the individual muscle cells. The area of contact between a neuron and a muscle fiber is called a **motor end plate**, or **neuromuscular junction** (Figure 10–4). When a nerve impulse reaches a motor end plate, small vesicles in the terminal branches of the nerve fiber release a chemical called acetylcholine. This chemical transmits the nerve impulse from the neuron, across the motor end plate, to the muscle fibers, thus stimulating the fibers to contract.

A motor neuron, together with the muscle cells it stimulates, is referred to as a **motor unit.** On an average, a single motor neuron innervates about 150 muscle fibers. This means that stimulation of one neuron will cause the simultaneous contraction of about 150 muscle fibers. In addition, all the muscle fibers of a motor unit contract and relax together. Muscles that control very fine, precise movements, such as the eye muscles, have only about 10 muscle fibers to each motor unit. Muscles of the body that are responsible for gross movements, such as the abdominal muscles, may have as many as 200 or more muscle fibers in each motor unit.

Physiology of contraction

When a nerve impulse reaches the neuromuscular junction, the neuron releases acetylcholine, which causes an electrical change in the sarcolemma. The electrical change travels over the surface of the sarcolemma and into the T tubules. When the impulse is conveyed from the T tubules to the sarcoplasmic reticulum, the reticulum, in some yet unexplained manner, releases calcium ions into the sarcoplasm surrounding the myofibrils. These calcium ions trigger the contractile process. Muscle contraction lasts only as long as calcium ions are present in the sarcoplasm. When the nerve impulse is over, the calcium ions recombine with the reticulum, and muscle contraction ceases.

How calcium ions cause a muscle to contract is still somewhat of a puzzle. The most commonly accepted explanation is the **sliding-filament hypothesis.** We know that the contraction of a muscle fiber is caused by the contraction of the myofibrils in the fiber. We also know that the myofibrils contain a protein called troponin and that contraction requires actin, myosin, calcium ions, and energy. According to the sliding-filament hypothesis, also called the myosin filament hypothesis, the myosin filaments are intersected by a series of cross bridges. When calcium is released by the sarcoplasmic reticulum, the cross bridges connect with actin filaments (Figure 10–5a). The bridges pull the actin filaments of each sarcomere inward toward each other until

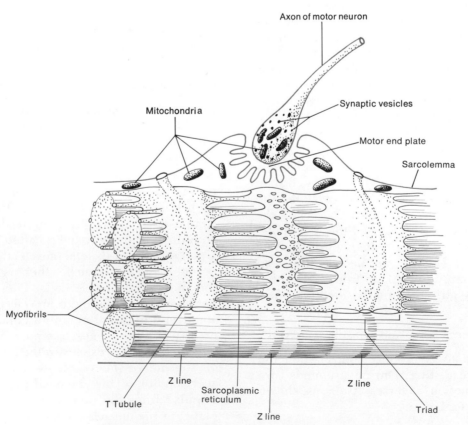

Figure 10–4. Structure of a motor end plate.

their approaching ends overlap (Figure 10–5b). As the actin filaments slide inward, the Z lines are drawn toward the A band, and the sarcomere is shortened. The shortening of the sarcomeres and myofibrils causes the shortening of the muscle fibers.

Researchers believe that the protein troponin, which is found in the filaments, prevents any interaction between actin and myosin in a non-contracting muscle. When a nerve impulse reaches the sarcoplasmic reticulum and triggers the release of calcium ions, the ions combine with the troponin, rendering it inactive. The myosin and actin are then free to interact. At the same time, the nerve impulse somehow stimulates the breakdown of ATP. The energy that is released

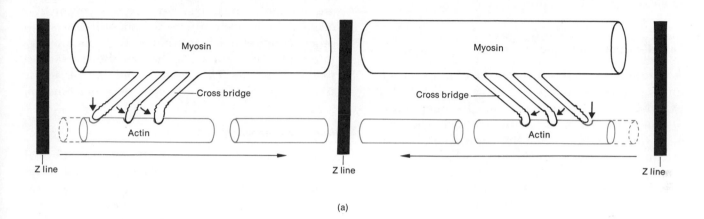

(a)

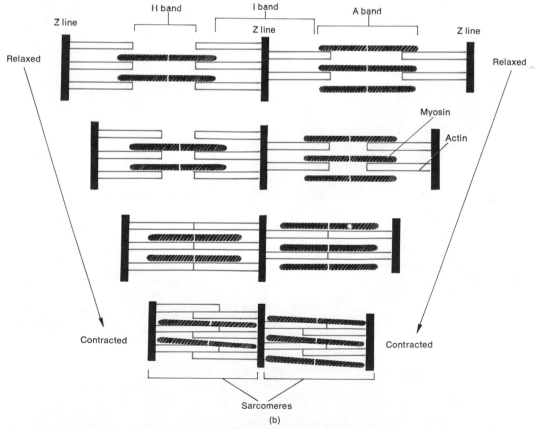

(b)

Figure 10–5. Sliding-filament hypothesis of muscular contraction. (a) Direction of movement of actin filaments during contraction. (b) Position of various zones in relaxed and contracted sarcomeres and illustration of sliding actin filaments.

by ATP breakdown is then used for the attachment of the cross bridges and the sliding of the actin filaments. After the impulse ends, the calcium ions return to the sarcoplasmic reticulum. The troponin breaks the cross-bridge attachments, and the actin filaments slip back outward to their original place. The sarcomeres are thereby returned to their resting lengths, and the muscle resumes its resting shape. A summary of these events is given in Figure 10–6.

The all-or-none principle

According to the **all-or-none principle,** the muscle fibers of a motor unit will contract to their fullest extent or will not contract at all. In other words, *muscle fibers* do not partially contract. The principle does not imply that the strength of the contraction is the same every time the fiber is stimulated. For example, the strength of contraction may be decreased by fatigue, lack of nutrients, or lack of oxygen. The weakest stimulus from a neuron that can initiate a contraction is called a *liminal stimulus.* A stimulus of lesser intensity, or one that cannot initiate contraction, is referred to as a *subliminal stimulus.*

Kinds of contractions

The various skeletal muscles are capable of producing different kinds of contractions, depending on the kind of stimulus applied. We shall now describe several of these.

Isotonic contraction

These contractions are the ones with which you are probably most familiar. As the contraction occurs, the muscle shortens and pulls on some other structure, such as a bone, to produce movement. During such a contraction, the tension remains constant. This is reflected in the terms *iso,* meaning equal, and *tono,* meaning tension.

Isometric contraction

In this kind of contraction, there is a minimal shortening of the muscle. It retains much the same length. But the *tension* in the muscle increases greatly. Isometric contractions do not result in the movement of a part of the body. You can demonstrate such a contraction by carrying a six pack of beer with your arm extended or by pushing the palms of the hands against each other.

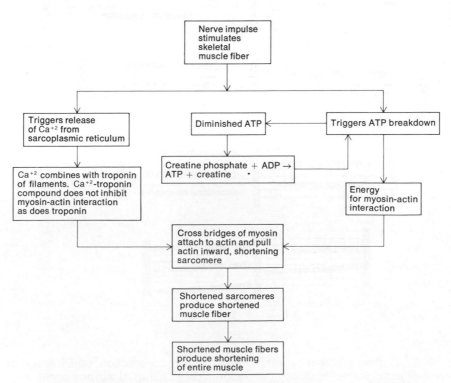

Figure 10–6. Summary of events involved in muscle contraction.

In the first example, the weight of the six pack pushes the arm downward, stretching the shoulder and arm muscles. The isometric contraction of the shoulder and arm muscles work to counteract the stretch. The two forces—contraction and stretching—applied in opposite directions create the tension.

Tonic contraction (tone)

A sustained contraction of some of the fibers in a skeletal muscle is called **tone,** or tonic contraction. At any given time, some of the cells in a muscle are contracted, while others are relaxed. This kind of contraction tightens a muscle, but not enough fibers are contracting at any one time to produce movement. Tone is essential for maintaining posture. When the muscles in the back of the neck are in tonic contraction, they keep the head in the anatomical position and prevent it from slumping forward onto the chest. In tonic contraction, these muscles do not apply enough force to pull the head back into hyperextension. The term *flaccid* is applied to muscles with less than normal tone. Such a loss of tone may be the result of damage or disease of the nerve that conducts a constant flow of impulses to the muscle. If the muscle does not receive impulses for an extended period of time, it may progress from flaccidity to *atrophy*, which is a state of wasting away. Muscles may also become flaccid and atrophied if they are not used. For example, bedridden individuals and people with casts may experience atrophy because the flow of impulses to the inactive muscle is greatly reduced. Upon resumption of normal activities, the atrophied muscles recover.

Twitch contraction

This kind of contraction is a very rapid, jerky response to a single stimulus. Although they probably do not occur in the body, twitch contractions can be artificially brought about in bits of muscle removed from an animal such as a frog. These contractions are worth describing because they show the different phases of muscle contraction quite clearly. Figure 10–7 is a graph of a twitch contraction. This kind of record of a contraction is called a *myogram*. Note that a brief period of time exists between the application of the stimulus and the beginning of contraction. This period of time, about 0.01 second in frog muscle, is called the *latent period*. The second phase, the *contraction period*, lasts about 0.04

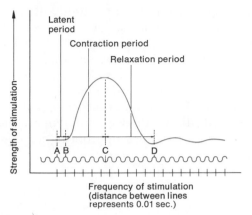

Figure 10–7. Myogram of a twitch contraction.

second and is indicated by the upward tracing. The third phase, the *relaxation period*, lasts about 0.05 second and is indicated by the downward tracing. The duration of these periods varies with the kind of muscle involved. For example, the latent period, contraction period, and relaxation period for muscles that move the eyes are very short. Muscles that move the leg undergo longer periods.

If additional stimuli are applied to the muscle after the initial stimulus, other responses may be noted. For example, if two stimuli are applied one immediately after the other, the muscle will respond to the first stimulus but not to the second stimulus. When a muscle fiber receives enough stimulation to contract, it temporarily loses its irritability and cannot contract again until its responsiveness is regained. This period of lost irritability is referred to as the *refractory period*. The duration of this period also varies with the kind of muscle involved. Skeletal muscle has a short refractory period. Cardiac muscle has a long refractory period. The importance of a long refractory period in cardiac muscle will be noted later.

When two stimuli are applied and the second stimulus is delayed until the refractory period is over, the skeletal muscle will respond to both stimuli. In fact, if the second stimulus is applied after the refractory period but before the muscle has finished contracting, the second contraction will be even stronger than the first. This phenomenon is referred to as **summation.**

Tetanic contractions

If a frog muscle is stimulated at a rate of 20 to 30 stimuli per second, the muscle can only partially relax between stimuli. As a result, the muscle maintains a sustained contraction called **incomplete tetanus.** Stimulation at an increased rate (35 to 50 stimuli per second) results in **complete tetanus,** a sustained contraction that lacks even partial relaxation. Essentially, both kinds of tetanus result from the rapid succession of separate twitches. Relaxation is either partial or does not occur at all. Voluntary contractions, such as contraction of the biceps in order to flex the forearm, are tetanic contractions.

Treppe

This kind of contraction is the condition in which a skeletal muscle contracts more forcefully after it has contracted several times. Assume, for example, that a series of liminal stimuli are introduced into a muscle. Time must be allowed for the muscle to contract and relax, but this takes only 0.1 second in frog muscle. If stimuli are repeated at intervals of 0.5 second, the first few tracings on the myogram will show an increasing height with each contraction. This is treppe, or the staircase phenomenon. It is the principle employed by athletes when warming up. After the first few stimuli, the muscle reaches its peak of performance and undergoes its strongest contraction. After numerous contractions, the muscle does not relax completely, and the relaxation phase finally disappears.

Now that we have discussed some of the structural and functional aspects of skeletal muscle, we will examine the histology and physiology of cardiac tissue. The continued, efficient functioning of this tissue is essential to survival.

CARDIAC MUSCLE

The principal constituent of the wall of the heart is **cardiac muscle.** It has the same striated appearance as skeletal muscle, but, unlike skeletal muscle, it is involuntary. The cells of cardiac muscle are roughly quadrangular in shape (Figure 10–8a). The individual fibers are covered by a thin, poorly defined sarcolemma, and the internal myofibrils produce the characteristic striations. Cardiac muscle cells have the same basic arrangement of actin, myosin, and sarcoplasmic reticulum that is found in skeletal muscle cells. In addition, they contain a system of transverse tubules similar to the T tubules of skeletal muscle. However, the nuclei in cardiac cells are centrally located as compared to the peripheral location of nuclei in skeletal muscle. And whereas the fibers of skeletal muscle are arranged in a parallel fashion, those of

cardiac muscle branch freely with other fibers to form two separate networks. The muscular walls and septum of the upper chambers of the heart compose one network. The muscular walls and septum of the lower chambers of the heart compose the other network. When a single fiber of either network is stimulated, all the fibers in the network become stimulated as well. Thus, each network contracts as a functional unit. The fibers of each network were once thought to be fused together into a multinucleated mass called a syncytium. But it is now known that each fiber in a network is separated from the next fiber by a transverse thickening of the sarcolemma called an **intercalated disc.** These discs strengthen the cardiac muscle tissue and allow a faster spread of nerve impulses across the tissue from one cell to another.

Under normal conditions, cardiac muscle contracts rapidly, continuously, and rhythmically about 72 times a minute, never stopping. This is a major physiological difference between cardiac and skeletal muscle. Another difference between skeletal and cardiac muscle is the source of stimulation. Skeletal muscle ordinarily contracts only when it is stimulated by a nerve impulse. In contrast, cardiac muscle can contract without any nerve stimulation. Its source of stimulation is a specialized conducting tissue that lies only within the heart. About 72 times a minute, this tissue sends out electrical impulses that stimulate cardiac contraction. Nerve stimulation merely causes the conducting tissue to increase or decrease its rate of discharge. A third difference is that cardiac muscle has a long refractory period that extends into part of the relaxation period. As a result, even though the rate of the heartbeat can be drastically increased, the heart cannot undergo a complete or incomplete tetanus.

SMOOTH MUSCLE

Like cardiac muscle, **smooth muscle** is involuntary. Unlike either cardiac or skeletal muscle, however, it is nonstriated. A single fiber of smooth muscle is spindle-shaped, and within the fiber is a single, oval, centrally located nucleus (Figure 10–8b). Smooth muscle cells also contain actin and myosin filaments, but the filaments are not as neatly arranged as in skeletal and cardiac muscle. This is the reason why well-differentiated striations do not occur.

Two kinds of smooth muscle, visceral and multiunit, are recognized. The more common type is called *visceral muscle.* It is found in wrap-

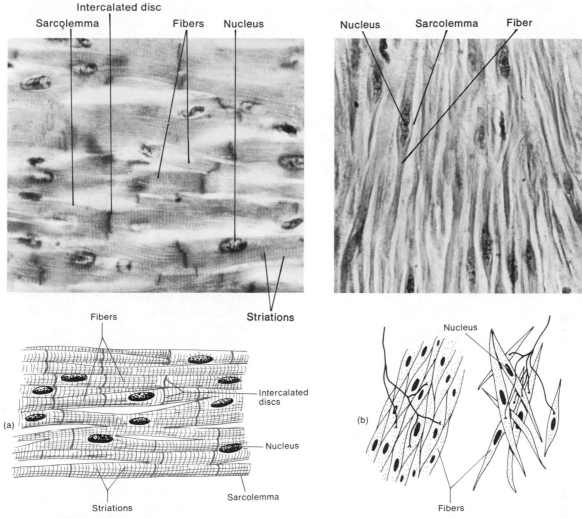

Figure 10–8. Histology of cardiac and smooth muscle. (a) Cardiac muscle. (b) Smooth muscle. Visceral smooth muscle on left; multiunit smooth muscle on right. *(Photographs courtesy of Edward J. Reith, from Atlas of Descriptive Anatomy, Edward J. Reith and Michael H. Ross, Harper & Row, Publishers, Inc., New York, 1970.)*

around sheets that form part of the walls of the hollow viscera such as the stomach, intestines, uterus, and urinary bladder. The terms *smooth muscle* and *visceral muscle* are sometimes used interchangeably. The fibers in a visceral muscle are arranged in a branching pattern and are tightly bound together to form a continuous network (Figure 10–8b). When a neuron stimulates one fiber, the impulse travels over the other fibers so that contraction occurs in a wave over many adjacent fibers.

The second kind of smooth muscle, called *multiunit smooth muscle*, consists of individual fibers each with its own motor-nerve endings (Figure 10–8b). Whereas stimulation of a single visceral muscle fiber causes contraction of many adjacent fibers, stimulation of a single multiunit fiber causes contraction of only that fiber. In

this respect, multiunit muscle is like skeletal muscle. Multiunit smooth muscle is found in the walls of blood vessels, in the arrector muscles attached to hair follicles, and in the intrinsic muscles of the eye — the iris, for example.

Both kinds of smooth muscle contract more slowly and relax more slowly than skeletal muscle. The reason is probably because the actin and myosin filaments of smooth muscle are so poorly arranged. Moreover, whereas skeletal muscle cells contract as individual units, visceral muscle cells contract one after another as the impulse spreads from one to another.

MUSCLE HOMEOSTASIS

Throughout the text we have emphasized the concept of homeostasis, particularly its importance in health and disease. Muscle tissue also has a vital role in maintaining the homeostasis of the body. Two such examples are the relationship of muscle tissue to oxygen and heat production.

Oxygen debt

Earlier we said that the energy required to convert ADP into ATP comes from the breakdown of digested foods. The nutrient that supplies this energy is the sugar glucose, and the reaction proceeds as follows:

$$Glucose \rightarrow pyruvic\ acid + energy$$

When the muscle is resting, the breakdown of glucose is slow enough for the blood to supply sufficient oxygen to change the pyruvic acid to the waste products carbon dioxide and water:

$$Pyruvic\ acid + O_2 \rightarrow CO_2 + H_2O$$

Reactions that involve oxygen, like the one above, are called *aerobic reactions.*

When a muscle is contracting and ATP manufacture is sped up, the breakdown of glucose occurs too rapidly for the blood to supply enough oxygen to convert all the pyruvic acid to carbon dioxide and water. The removal of pyruvic acid must then proceed *anaerobically* (without oxygen) as follows:

$$Pyruvic\ acid \rightarrow lactic\ acid$$

Most of the lactic acid is transported to the liver, where it is eventually resynthesized to glycogen or glucose, which can be reused later on. However, some of the lactic acid accumulates in the muscle tissue. Physiologists suspect that the acid is responsible for the feeling of muscle fatigue. Ultimately, the lactic acid in the muscle must be converted to CO_2 and H_2O, and oxygen is needed for this conversion. The amount of oxygen required is called the **oxygen debt.** When violent activity is over, a person continues to breathe hard or pant in order to pay back the debt. Until this is done, the exercise must be interrupted. Thus, the accumulation of lactic acid causes hard breathing and enough discomfort to stop muscle activity until homeostasis is restored.

Heat production

As noted earlier, the production of heat by skeletal muscle is an important homeostatic mechanism for maintaining normal body temperature. Of the total energy released during muscular contraction,

only 20 to 30 percent is used for mechanical work (contraction). The remainder is released as heat, which is used by the body to help maintain a normal body temperature.

Heat production by muscles may be divided into two phases: (1) *initial heat,* which is produced by the contraction and relaxation of a muscle, and (2) *recovery heat,* which is produced after relaxation. Initial heat is independent of O_2 and is associated with ATP breakdown. Recovery heat is associated with the anaerobic breakdown of glucose to pyruvic acid and the aerobic conversion of lactic acid to CO_2 and H_2O.

DISORDERS OF THE MUSCULAR SYSTEM

Disorders of the muscular system are related to disruptions of homeostasis. The disorders may involve lack of nutrients, buildup of toxic products, disease, injury, disuse, or faulty nervous connections.

Fatigue

If a muscle or group of muscles are continuously stimulated, after a point, the strength of contraction becomes progressively weaker, until the muscle refuses to respond. This condition is referred to as **fatigue.** It results partly from diminished availability of oxygen and partly from the toxic effects of lactic acid and carbon dioxide accumulated during exercise. The significant factors that contribute to muscle fatigue are

1. Excessive activity resulting in the accumulation of toxic products
2. Malnutrition resulting in insufficient supplies of glucose and, therefore, ATP
3. Circulatory disturbances that impair the delivery of useful substances to muscles and the removal of waste products from muscles
4. Respiratory disturbances that interfere with the oxygen supply and increase the oxygen debt.

Fibrosis and fibrositis

The formation of fibrous tissue in places where it normally does not exist is called **fibrosis.** Muscle cells cannot undergo mitosis, and dead muscle cells are normally replaced with fibrous connective tissue. Fibrosis, then, is often a consequence of muscle injury or degeneration.

Fibrositis is an inflammation of fibrous tissue. If it occurs in the lumbar region and low back, it is referred to as *lumbago*. Fibrositis is a common condition characterized by pain, stiffness, or soreness of fibrous tissue, especially in the muscle coverings. It is not destructive or progressive. But it may persist for years, or it may disappear spontaneously. Attacks of fibrositis may follow an injury, repeated muscular strain, or prolonged muscular tension. *Myositis* is an inflammation of muscle cells.

Muscular dystrophy and myasthenia gravis

The term *atrophy* refers to a reduction in the size of an organ. Muscular atrophy may be caused by disuse, either because the individual fails to exercise the muscle or because nerve impulses do not reach the muscle fibers. Muscular atrophy may also be caused by a *myopathy*, which is a general term for the diseases that attack and destroy muscle tissue. Muscular atrophy is always accompanied by some degree of muscular weakness. If the connections with nervous tissue are destroyed, the fibers cease to contract at all.

The term **muscular dystrophy** applies to a number of inherited myopathies. The word *dystrophy* means degeneration. The disease is characterized by degeneration of the individual muscle cells, which leads to a progressive atrophy of the muscle. Usually the voluntary skeletal muscles are weakened equally on both sides of the body, whereas the internal muscles, such as the diaphragm, are not affected. Histologically, the changes that occur include variation in muscle fiber size, degeneration of fibers, and deposition of fat.

The cause of muscular dystrophy has been variously attributed to an embryonic defect, faulty metabolism of potassium, protein deficiency, and inability of the body to utilize creatine.

Muasthenia gravis is an atrophy of the skeletal muscles caused by an abnormality at the motor end plate that prevents the muscle fibers from contracting. Recall that motor neurons stimulate the skeletal muscle fibers to contract by releasing a chemical called acetylcholine. Myasthenia gravis is caused by failure of the neurons to release acetylcholine or by excess amounts of cholinesterase, a chemical that interferes with the action of acetylcholine. As the disease progresses, more motor end plates become affected. The muscle becomes increasingly weaker and may eventually cease to function altogether.

The cause of myasthenia gravis is unknown. It is more common in females, occurring most frequently between the ages of 20 to 50. The muscles of the face and neck are most apt to be involved. Initial symptoms include a weakness of the eye muscles and difficulty in swallowing. Later, the individual has trouble chewing and talking. Eventually, the muscles of the limbs may become involved. Death may result from paralysis of the respiratory muscles, but usually the disorder does not progress to this stage.

Abnormal contractions

One kind of abnormal contraction of a muscle is **spasm,** a sudden, involuntary contraction of short duration. **A cramp** is an involuntary complete tetanic contraction in a muscle. **Convulsions** are involuntary tetanic contractions of a whole group of muscles. Convulsions occur when motor neurons are stimulated by factors such as fever, poisons, hysteria, and changes in body chemistry due to the withdrawal of certain drugs. The stimulated neurons send many seemingly "senseless" impulses to the muscle fibers. **Fibrillation** is the uncoordinated contraction of individual muscle fibers so that the muscle fails to contract smoothly. Cardiac muscle is most prone to this abnormality.

Medical terminology associated with the muscular system

Gangrene   Death of a muscle that results from almost complete interruption of its blood supply.

Myology   Study of muscles.

Myomalacia   (*malaco* = soft) Softening of a muscle.

Myopathy   Any disease of muscle tissue.

Myosclerosis   (*scler* = hard) Hardening of a muscle.

Myospasm   Spasm of a muscle.

Myotonia   Increased muscular irritability and contractility with decreased power of relaxation; tonic spasm of the muscle.

Sprain   The forcible twisting of a joint causing rupture of a ligament or causing the ligament to loosen at one of its attachments.

Trichinosis   A myositis caused by the parasitic worm *Trichinella spiralis,* which may be found in the muscles of men, rats, and pigs. Man contracts the disease by eating infected pork that is insufficiently cooked.

Volkmann's contracture   (*contra* = against) Permanent contraction of a muscle due to replacement of destroyed muscle cells with fibrous tissue that lacks ability to stretch. Destruction may occur from interference with circulation caused by a tight bandage, a piece of elastic, or a cast.

Wryneck or torticollis   Complete tetanus of one of the sternocleidomastoid muscles; produces twisting of the neck and an unnatural position of the head.

Chapter summary in outline

CHARACTERISTICS

1. Irritability is the property of receiving and responding to stimuli.

2. Contractility is the ability to shorten and thicken, or contract.

3. Extensibility is the ability to stretch or extend.

4. Elasticity is the ability to return to original shape after contraction or extension.

KINDS OF MUSCLE TISSUE

1. Skeletal muscle is attached to bones, striated, and voluntary.

2. Smooth muscle is located in viscera. It is nonstriated and involuntary.

3. Cardiac muscle forms the walls of the heart. It is striated and involuntary.

SKELETAL MUSCLE

Histology

1. The muscle consists of fibers covered by a sarcolemma. The fibers contain sarcoplasm, nuclei, sarcoplasmic reticulum, and T tubules.

2. Each fiber contains myofibrils (actin and myosin filaments). The myofibrils are compartmentalized into sarcomeres.

Connective tissue components

1. The entire muscle is covered by the epimysium. Fasciculi are covered by perimysium. Fibers are covered by endomysium.

2. Tendons and aponeuroses attach muscle to bone.

Nerve and blood supply

1. Nerves convey impulses, and blood provides nutrients and oxygen for contraction.

CONTRACTION OF SKELETAL MUSCLE

Energy for contraction

1. The primary source of energy is ATP.

2. When muscles are resting, ATP combines with creatine to form creatine phosphate, which breaks down to produce ATP when muscles contract strenuously.

The motor unit

1. A motor neuron transmits the stimulus to a muscle for contraction.

2. The area of contact between a motor neuron and muscle fiber is a motor end plate.

3. A motor neuron and the muscle fibers it stimulates form a motor unit.

Physiology of contraction

1. A nerve impulse travels over the sarcolemma and enters the T tubules and sarcoplasmic reticulum.

2. Calcium ions released by the reticulum trigger the contractile process.

3. Actual contraction is brought about when the action filaments slide toward each other.

The all-or-none principle

1. Muscle fibers of a motor unit contract to their fullest extent or not at all.

2. The weakest stimulus capable of causing contraction is a liminal stimulus.

3. A stimulus not capable of inducing contraction is a subliminal stimulus.

Kinds of contractions

1. The various kinds of contraction are isotonic, isometric, tonic, twitch, tetanic, and treppe.

2. A record of any contraction is called a myogram. The refractory period is the time when a muscle has temporarily lost irri-

tability. Skeletal muscles have a short refractory period. Cardiac muscle has a long refractory period.

3. Summation is the stronger contraction of a muscle in response to additional stimuli.

CARDIAC MUSCLE

1. This muscle is found only in the heart. It is striated and involuntary.

2. Cells are quadrangular and contain centrally placed nuclei.

3. The fibers form a continuous, branching network that contracts as a functional unit.

4. Intercalated discs provide strength and aid impulse conduction.

SMOOTH MUSCLE

1. Smooth muscle is found in viscera. It is nonstriated and involuntary.

2. Visceral smooth muscle is found in the walls of viscera, and the fibers are arranged in a network.

3. Multiunit smooth muscle is found in blood vessels and the eye. The fibers operate singly rather than as a unit.

MUSCLE HOMEOSTASIS

1. Oxygen debt is the amount of O_2 that is required to convert lactic acid into CO_2 and H_2O. It occurs during strenuous exercise and is paid back by breathing fast. Until it is paid back, the homeostasis between muscular activity and oxygen requirements is not restored.

2. The heat given off during muscular contraction is used to maintain the homeostasis of body temperature.

DISORDERS OF THE MUSCULAR SYSTEM

Fatigue

1. The strength of contraction becomes progressively weaker with fatigue.

2. It is caused by insufficient oxygen that results in a buildup of lactic acid.

Fibrosis and fibrositis

1. Fibrosis is the formation of fibrous tissue where it normally does not exist.

2. Fibrositis is an inflammation of fibrous tissue. If it occurs in the lumbar region, it is called lumbago. Myositis is muscle tissue inflammation.

Muscular dystrophy and myasthenia gravis

1. Muscular dystrophy is a hereditary disease of muscles characterized by degeneration of individual muscle cells.

2. Myasthenia gravis is a disease exhibiting great muscular weakness and fatigability resulting from improper neuromuscular transmission.

Abnormal contractions

1. Abnormal contractions include: spasms, cramps, convulsions, and fibrillation.

Review questions and problems

1. How is the skeletal system related to the muscular system? What are the three kinds of motion that are accomplished by the muscular system?

2. What are the four characteristics of muscle tissue?

3. What criteria are employed for distinguishing the three kinds of muscle tissue?

4. Discuss the microscopic structure of skeletal muscle.

5. Define epimysium, perimysium, endomysium, tendon, and aponeurosis. Describe the nerve and blood supply to a muscle.

6. In considering the contraction of skeletal muscle, describe the following:

   (a) sources of energy

   (b) motor unit

   (c) importance of calcium and troponin

   (d) sliding filament hypothesis

7. What is the all-or-none principle? Relate it to a liminal and subliminal stimulus.

8. Define each of the following contractions and state the importance of each: isotonic, isometric, tonic, twitch, tetanic, and treppe.

9. What is a myogram? How might a muscle become flaccid?

10. Describe the latent period, contraction period, and relaxation period of muscle contraction. Construct a diagram to substantiate your answer.

11. Define the refractory period. How does it differ between skeletal and cardiac muscle? What is summation?

12. Compare cardiac and smooth muscle with regard to microscopic structure, functions, and locations.

13. Discuss each of the following as examples of muscle homeostasis: oxygen debt and heat production.

14. What do you think might be the relationship between shivering (uncontrolled muscular contractions) and body temperature? Can you relate sweating (cooling of the skin) following strenuous muscular activity to the homeostasis of body temperature?

15. What is muscle fatigue? Describe some of the factors that might contribute to the condition.

16. Define fibrosis. What is one of its causes? Define myositis.

17. What is muscular dystrophy?

18. What is myasthenia gravis? In this disease, why do the muscles not contract normally?

19. Define each of the following: abnormal muscular contractions: spasm, cramp, convulsion, and fibrillation.

20. Refer to the glossary of medical terminology associated with the muscular system. Be sure that you can define each term.

CHAPTER 11

THE MUSCULAR SYSTEM: ACTIONS AND GROSS STRUCTURE

STUDENT OBJECTIVES

After you have read this chapter, you should be able to:

1. Describe the relationship between bones and skeletal muscles in producing body movements

2. Define a lever and fulcrum and compare the three classes of levers on the basis of placement of the fulcrum, effort, and resistance

3. Describe most body movements as activities of groups of muscles by explaining the roles of the prime mover, antagonist, and synergist

4. Define the criteria employed in naming skeletal muscles

5. Identify the principal skeletal muscles in different regions of the body by name, origin, insertion, and action

6. Compare the common sites for intramuscular injections

When speaking of muscle tissue, we are referring to all the contractile tissues of the body, that is, skeletal, cardiac, and smooth muscle. But when we speak of the *muscular system,* we refer to the *skeletal* muscle system, which is composed of skeletal muscle and connective tissues that make up individual muscle organs, such as the biceps. Cardiac and smooth muscle tissues are classified with other organ systems. For instance, cardiac muscle tissue is formed in the heart, an organ of the circulatory system. Smooth muscle of the intestine is part of the digestive system. And smooth muscle of the urinary bladder is part of the urinary system. In this chapter, we shall discuss the muscular system only. We shall do this by taking a look at how skeletal muscles produce movement and by describing and locating some of the principal skeletal muscles.

HOW MUSCLES PRODUCE MOVEMENT

Skeletal muscles produce movements by pulling on tendons, which, in turn, pull on bones. The majority of muscles cover at least one joint and are attached to the articulating bones that form the joint (Figure 11-1a). When such a muscle contracts, it pulls on the articulating bones, and this pull draws one of the articulating bones toward the other. The two articulating bones usually do not move equally in response to the contraction. One of them is kept pretty much in its original position because other muscles contract to pull it in the opposite direction or because the structure of the bone makes it less movable. Ordinarily, the attachment of a muscle tendon to the more stationary bone is called the **origin.** The attachment of the other muscle tendon to the more movable bone is referred to as the **insertion.** A good analogy for remembering this is a spring on a door. The part of the spring attached to the door represents the insertion, whereas the part of the spring attached to the door frame represents the origin. The fleshy portion of the muscle between the tendons of

the origin and insertion is referred to as the *belly,* or *gaster.* In the appendages, especially, the origin is proximal, and the insertion is distal. In addition, muscles that move a part of the body generally do not cover the moving part. Note in Figure 11-1a, for example, that although contraction of the biceps muscle moves the forearm, the major portion of the muscle lies over the humerus of the arm.

In bringing about a body movement, bones act as levers, and joints function as fulcrums of these levers. **A lever** may be defined as a rigid rod that moves about on some fixed point or support called a **fulcrum.** A fulcrum may be symbolized as △F. A lever is acted upon at two different points by two different forces: (1) the *resistance,* which can be symbolized as R, and (2) the *effort,* which can be symbolized as E. The resistance may be regarded as something to be overcome or balanced, whereas the effort is exerted to overcome the resistance. The resistance may be the weight of the body that is to be moved, such as an arm or leg, or some object to be lifted, or both. The muscular effort (contraction) is applied to the bone at the insertion of the muscle and brings about the work of motion. As an example, consider the biceps flexing the forearm at the elbow as a weight is lifted (Figure 11-1b). In this case, when the forearm is raised the elbow is the fulcrum. The weight of the forearm plus the weight in the hand is the resistance. And the shortening of the biceps pulling the forearm upward is the effort.

Levers are categorized into three types on the basis of the relative placement of the fulcrum, effort, and resistance. In a *first-class lever,* the fulcrum is placed between the effort and resistance (Figure 11-2a). An example of a first-class lever is a seesaw. Examples of first-class levers in the body are not abundant, however. One example is the head resting on the vertebral column. When the head is raised, the facial portion of the skull is the resistance. The joint between the atlas and occipital bone (atlanto-occipital joint) is the fulcrum. And the muscles of the back in contraction represent the effort.

Second-class levers are similar to the mechanics of a wheelbarrow. In this case, the fulcrum is at one end. The effort is at the opposite end. And the resistance is in between (Figure 11–2b). Most authorities agree that there are no examples of second-class levers in the body. Some, however, consider that raising the body on the toes (resistance) and utilizing the ball of the foot as the fulcrum is an example. Here, the calf muscles pull the heel upward as it shortens (effort).

Third-class levers are the most common kinds of levers in the body and consist of the fulcrum at one end, the resistance at the opposite end, and the effort in between (Figure 11–2c). A common example is flexing the forearm at the elbow. Here, the weight of the forearm is the resistance, the contraction of the biceps is the effort, and the elbow joint is the fulcrum.

In general, most movements are coordinated by the activity of several skeletal muscles. In fact, skeletal muscles usually act in groups rather than individually. As an example, let us examine

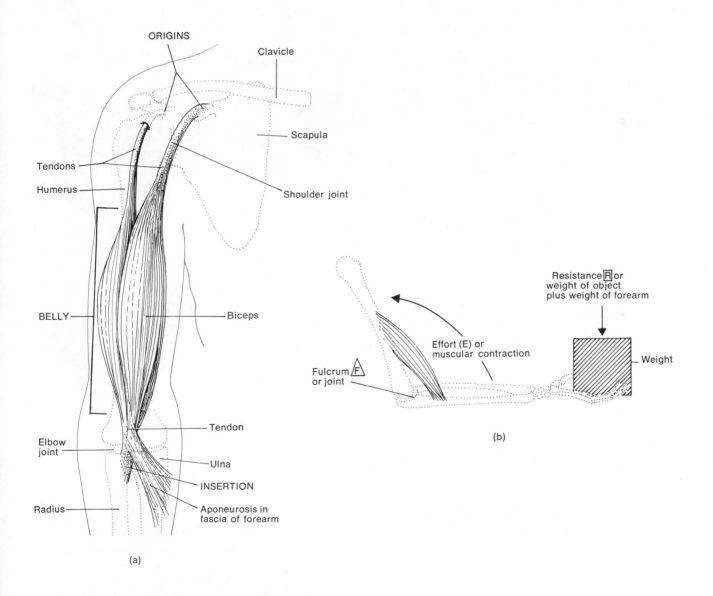

Figure 11–1. Muscle-bone relationships. (a) Skeletal muscles produce movements by pulling on bones. (b) Bones serve as levers, and joints act as fulcrums for the levers. Here the lever-fulcrum principle is illustrated by the movement of the forearm lifting a weight. Note where the resistance and effort are applied in this example.

flexing the forearm at the elbow. A muscle that causes a desired action is referred to as the **agonist** or **prime mover.** In this instance, the biceps is the agonist. Simultaneously with the contraction of the biceps, another muscle, called the **antagonist,** is relaxing. In this movement, the triceps serves as the antagonist. The antagonist has an effect opposite to that of the agonist. That is, the antagonist relaxes and gives way to the movement of the agonist. If we consider the extension of the forearm at the elbow, the triceps would assume the role of the agonist, and the biceps would act as antagonist. It will become obvious to you that most joints are operated by antagonistic groups of muscles. Still other muscles called **synergists,** or **fixators,** assist the agonist by reducing undesired action or unnecessary movements in the less mobile articulating bone. While flexing the forearm, the synergists, in this case the deltoid and pectoralis major muscles, would hold the arm and shoulder in a suitable position for the flexing action. Whereas the deltoid abducts the humerus, the pectoralis major adducts and medially rotates the humerus. Essentially,

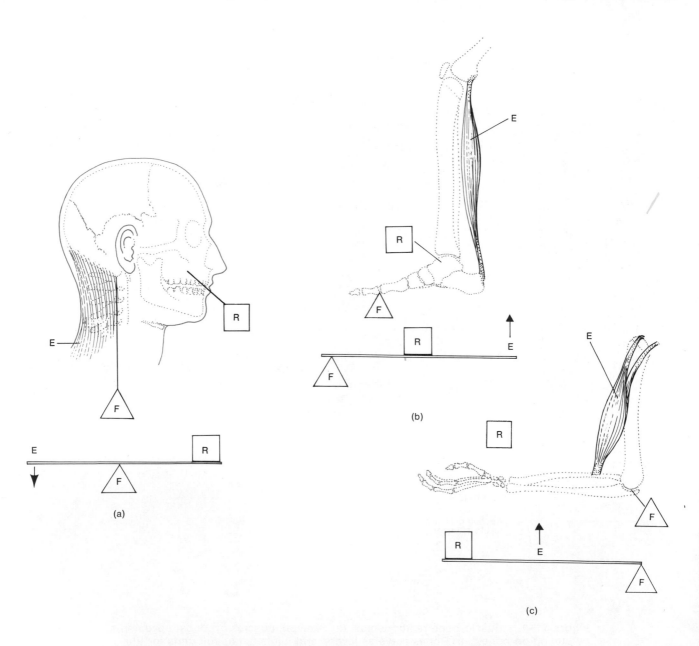

Figure 11–2. Classes of levers. (a) First-class lever. (b) Second-class lever. (c) Third-class lever. Each is defined on the basis of the placement of the fulcrum, effort, and resistance.

synergists contract at the same time as the prime mover and help the prime mover produce a more effective movement.

NAMING MUSCLES

Most of the almost 700 skeletal muscles are named on the basis of one or more distinctive criteria. If you understand these criteria, you will find it much easier to learn and remember the names of individual muscles. Some muscles are named on the basis of the *direction of the muscle fibers*. There are, for example, rectus (straight), transverse, and oblique muscles. Rectus fibers usually run parallel to the midline of the body. Transverse fibers run perpendicular to the midline. And oblique fibers are diagonal to the midline. Muscles named according to these three directions include the rectus abdominis, transversus abdominis, and external oblique, respectively.

Another criterion employed is *location*. For example, the temporalis is so named because of its proximity to the temporal bone. The tibialis anterior is located near the tibia. *Size* is also commonly employed. For instance, the term *maximus* means largest, and *minimus* means smallest. *Longus* means long, and *brevis* means short. Examples include the gluteus maximus, gluteus minimus, adductor longus, and peroneus brevis.

Some muscles such as the biceps, triceps, and quadriceps are named on the basis of the *number of origins* they have. For instance, the biceps has two origins. The triceps has three, and the quadriceps has four. Other muscles are named on the basis of *shape*. Common examples include the deltoid (meaning triangular) and trapezius (meaning trapezoid). Muscles may also be named after their *insertion and origin*. One example is the sternocleidomastoid, which originates on the sternum and clavicle and inserts at the mastoid process of the temporal bone. The stylohyoideus originates on the styloid process of the temporal bone and inserts at the hyoid bone.

Still another criterion used for naming muscles is *action*. Listed here are the principal actions of muscles, their definitions, and examples of muscles that perform the actions. For convenience, the actions are grouped as antagonistic pairs where possible.

Flexor   Decreases the angle at a joint
(flexor carpi radialis)

Extensor   Increases the angle at a joint
(extensor carpi ulnaris)

Abductor   Moves a bone away from the midline
(abductor hallucis)

Adductor   Moves a bone closer to the midline
(adductor longus)

Levator   Produces an upward movement
(levator scapulae)

Depressor   Produces a downward movement
(depressor labii infemoris)

Supinator   Turns the palm upward or to the anterior
(supinator)

Pronator   Turns the palm downward or to the posterior
(pronator teres)

Dorsiflexor   Extends the ankle joint
(extensor digitorum longus)

Plantar flexor   Flexes the ankle joint
(plantaris)

Invertor   Turns the sole of the foot inward
(tibialis anterior)

Evertor   Turns the sole of the foot outward
(peroneus tertius)

Sphincter   Decreases the size of an opening
(pyloric sphincter between stomach and duodenum)

Tensor   Makes a body part more rigid
(tensor fasciae latae)

Rotator   Moves a bone around its longitudinal axis
(obturator)

PRINCIPAL MUSCLES OF THE BODY

In the following pages, a series of Exhibits are provided for you to learn the principal skeletal muscles of the body. The Exhibits contain a listing of the muscles in terms of their origins, insertions, and actions. Diagrams of the muscles under consideration are also included. The second column in all but the first Exhibit is called the Learning Key. It contains a listing of prefixes, roots, suffixes, and definitions that explain the derivation of the muscles' names. You are not expected to memorize the word derivations, but we hope that, by reading them, your task of learning the names of muscles will be easier and more fun. It is strongly suggested that you make frequent reference to Chapter 8 in order to review your knowledge of bone markings since they serve as points of origin and insertion for many of the muscles. By no means have all the muscles of the body been included. Only those considered important for introductory courses will be discussed.

Exhibit 11–1. THE MUSCULAR SYSTEM: AN OVERVIEW

Shown here is a general anterior and posterior view of the muscular system. Do not try to memorize all these muscles yet. As you study groups of muscles in subsequent Exhibits, refer to the figures in this Exhibit to see how each grouping is related to all others.

We have attempted, as much as possible, to indicate whether the muscles are superficial or deep, anterior or posterior, and medial or lateral. We have also tried to show the relationship of the muscles under consideration to other muscles in the area you are studying. If you have mastered the various criteria used in naming muscles, the names and actions of many muscles will have more meaning.

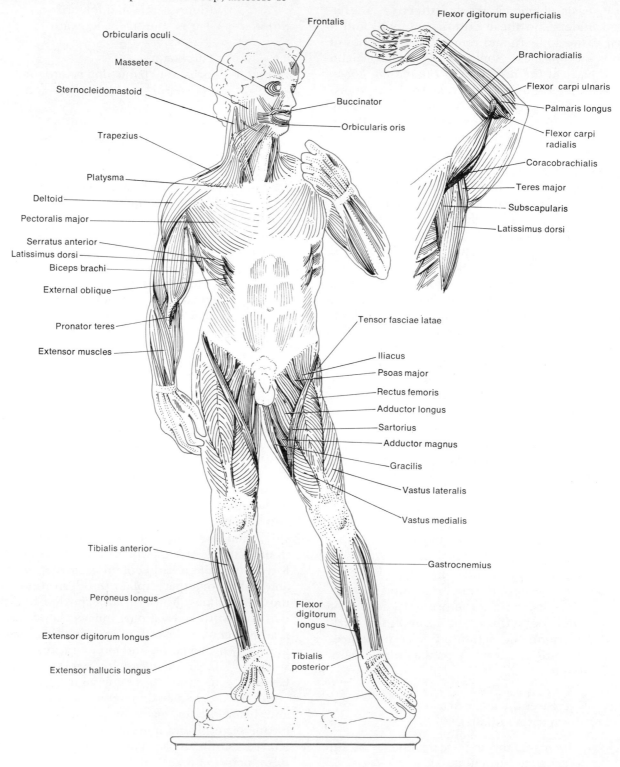

Anterior view of principal superficial muscles

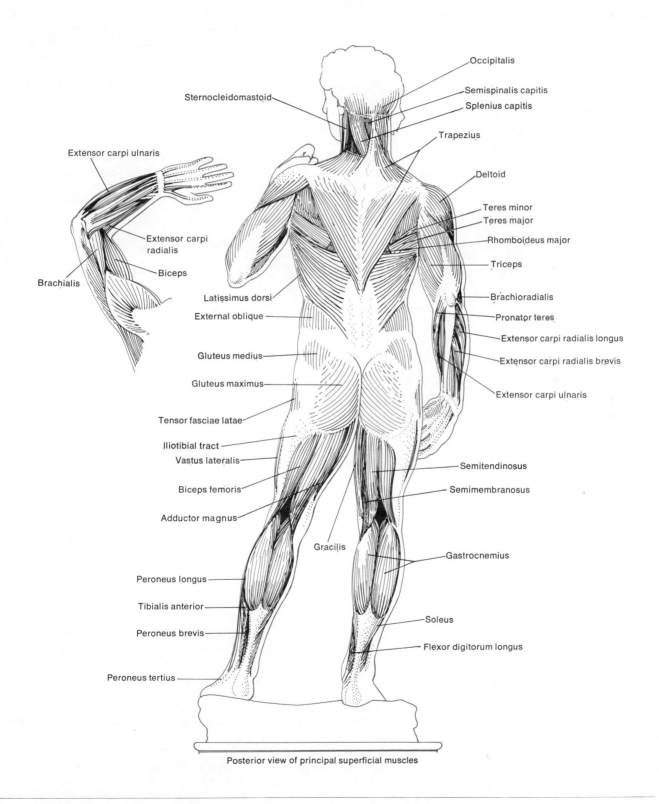

Posterior view of principal superficial muscles

Exhibit 11–2. MUSCLES OF FACIAL EXPRESSION

MUSCLE	LEARNING KEY	ORIGIN	INSERTION	ACTION
Epicranius	*epi* = over *crani* = skull	This muscle is divisible into two portions: the frontalis over the frontal bone and the occipitalis over the occipital bone. The two muscles are united by a strong aponeurosis, the galea aponeurotica, which covers the top and sides of the skull.		
Frontalis	*front* = forehead	Galea aponeurotica	Skin above supraorbital line	Draws scalp forward, raises eyebrows, and wrinkles forehead horizontally
Occipitalis	*occipito* = base of skull	Occipital bone and mastoid process of temporal bone	Galea aponeurotica	Draws scalp backward
Orbicularis oris	*orb* = circular *or* = mouth	Muscle fibers surrounding opening of mouth	Skin at corner of mouth	Closes lips, compresses lips against teeth, protrudes lips, and shapes lips during speech
Zygomaticus major	*zygomatic* = cheek bone *major* = greater	Zygomatic bone	Orbicularis oris	Draws angle of mouth upward and outward as in smiling or laughing
Levator labii superioris	*levator* = raises or elevates *labii* = lip *superioris* = upper	Below infraorbital foramen of maxilla	Orbicularis oris	Elevates (raises) upper lip
Depressor labii inferioris	*depressor* = depresses or lowers *inferioris* = lower	Mandible	Skin of lower lip	Depresses (lowers) lower lip
Buccinator	*bucc* = cheek	Alveolar processes of maxilla and mandible	Orbicularis oris	Major cheek muscle; compresses cheek as in blowing air out of mouth and causes the cheeks to cave in, producing the action of sucking
Mentalis	*mentum* = chin	Mandible	Skin of chin	Elevates and protrudes lower lip and pulls skin of chin up as in pouting
Platysma	*platy* = flat, broad	Fascia over deltoid and pectoralis major muscles	Mandible and muscles around angle of mouth	Draws outer part of lower lip downward and backward as in pouting
Risorius	*risor* = laughter	Fascia over parotid (salivary) gland	Skin at angle of mouth	Draws angle of mouth laterally as in tenseness
Orbicularis oculi	*ocul* = eye	Medial wall of orbit	Circular path around orbit	Closes eye

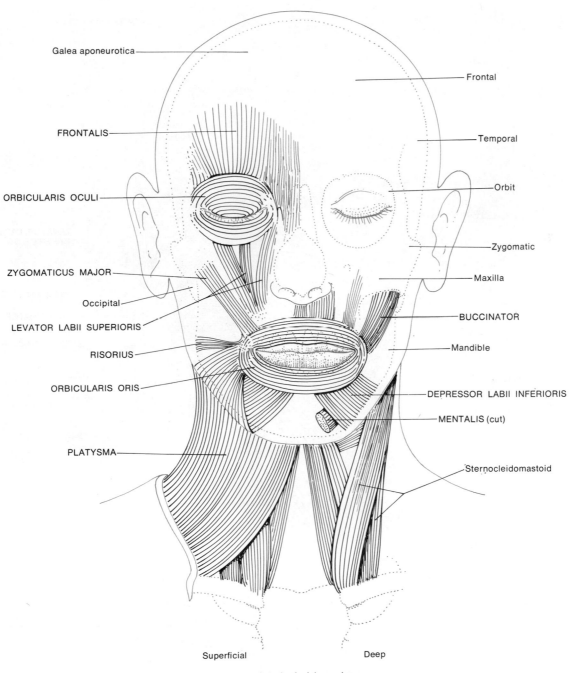

Galea aponeurotica

Frontal

FRONTALIS

Temporal

ORBICULARIS OCULI

Orbit

Zygomatic

ZYGOMATICUS MAJOR

Maxilla

Occipital

LEVATOR LABII SUPERIORIS

BUCCINATOR

RISORIUS

Mandible

ORBICULARIS ORIS

DEPRESSOR LABII INFERIORIS

MENTALIS (cut)

PLATYSMA

Sternocleidomastoid

Superficial

Deep

Anterior facial muscles

Exhibit 11–2 (*cont'd.*)

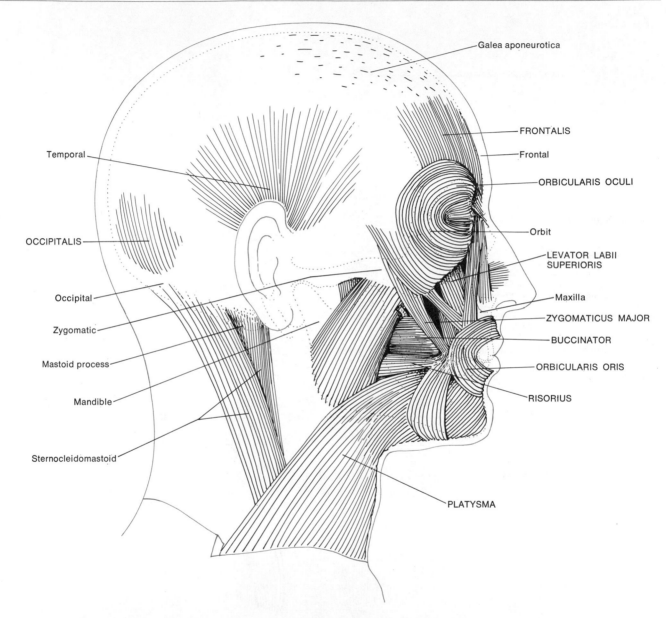

Right lateral view of superficial facial muscles

Exhibit 11-3. MUSCLES THAT MOVE THE MANDIBLE

MUSCLE	LEARNING KEY	ORIGIN	INSERTION	ACTION
Masseter	*maseter* = chewer	Maxilla and zygomatic arch	Angle and ramus of mandible	Elevates mandible as in closing the mouth and protracts (protrudes) mandible
Temporalis	*tempora* = temples	Temporal bone	Coronoid process of mandible	Elevates and retracts mandible
Medial pterygoid	*medial* = closer to midline *pterygoid* = like a wing; pterygoid plate of sphenoid	Medial surface of lateral pterygoid plate of sphenoid; maxilla	Angle and ramus of mandible	Elevates and protracts mandible and moves mandible from side to side
Lateral pterygoid	*lateral* = farther from midline	Great wing and lateral surface of lateral pterygoid plate of sphenoid	Condyle of mandible; temporal-mandibular articulation	Protracts mandible, opens mouth, and moves mandible from side to side

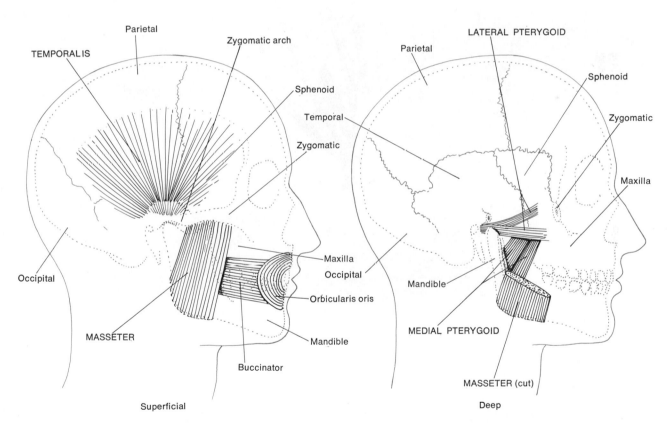

Right lateral view of muscles that move the mandible

Exhibit 11–4. MUSCLES THAT MOVE THE EYEBALLS

MUSCLE	LEARNING KEY	ORIGIN	INSERTION	ACTION
Superior rectus	*superior* = above *rectus* = in this case, muscle fibers running parallel to long axis of eyeball	Orbital cavity	Upper and central part of eyeball	Rolls eyeball upward
Inferior rectus	*inferior* = below	Orbital cavity	Lower and central part of eyeball	Rolls eyeball downward
Lateral rectus		Orbital cavity	Lateral side of eyeball	Rolls eyeball laterally
Medial rectus		Orbital cavity	Medial side of eyeball	Rolls eyeball medially

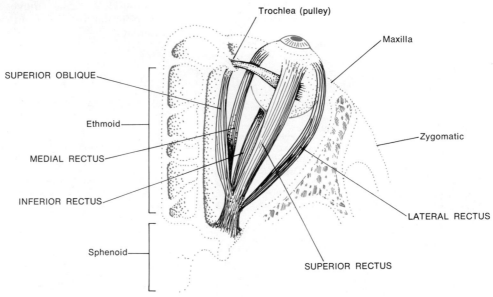

Muscles of the right eyeball seen from above

MUSCLE	LEARNING KEY	ORIGIN	INSERTION	ACTION
Superior oblique	*oblique* = in this case, muscle fibers running diagonally to long axis of eyeball	Orbital cavity	Eyeball between superior and lateral recti	Rotates eyeball on its axis; directs cornea downward and laterally; note that it moves through a ring of fibrocartilagenous tissue called the trochlea *(trochlea-pulley)*
Inferior oblique		Maxilla (front of orbital cavity)	Eyeball between superior and lateral recti	Rotates eyeball on its axis; directs cornea upward and laterally

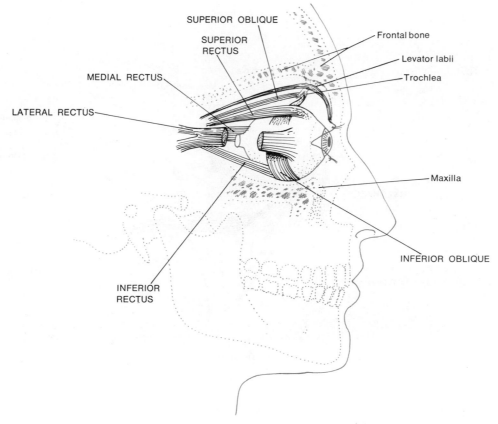

Right lateral view of muscles of the right eyeball

Exhibit 11–5. MUSCLES THAT MOVE THE TONGUE

MUSCLE	LEARNING KEY	ORIGIN	INSERTION	ACTION
Genioglossus	*geneion* = chin *glossus* = tongue	Mandible	Undersurface of tongue and hyoid bone	Depresses and thrusts tongue forward (protraction)
Styloglossus	*stylo* = stake or pole; styloid process of temporal	Styloid process of temporal bone	Side and undersurface of tongue	Elevates tongue and draws it backward (retraction)
Stylohyoid	*hyoeides* = U-shaped; pertaining to hyoid bone	Styloid process of temporal bone	Body of hyoid bone	Elevates and retracts tongue
Hyoglossus		Body of hyoid bone	Side of tongue	Depresses tongue and draws down its sides

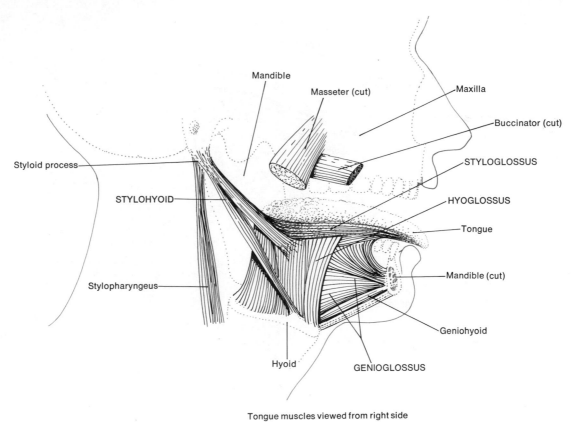

Tongue muscles viewed from right side

Exhibit 11–6. MUSCLES THAT MOVE THE HEAD

MUSCLE	LEARNING KEY	ORIGIN	INSERTION	ACTION
Sternocleido-mastoid (See Exhibit 11–9)	*sternum* = breast-bone *cleido* = clavicle *mastoid* = mastoid process of temporal	Sternum and clavicle	Mastoid process of temporal bone	Contraction of both muscles flexes the head on the chest Contraction of one muscle rotates face toward side opposite contracting muscle
Semispinalis capitis (See Exhibit 11–13)	*semi* = half *spine* = spinous process *caput* = head	Articular process of seventh cervical vertebra and transverse processes of upper six thoracic vertebrae	Occipital bone	Both muscles extend head; contraction of one muscle rotates face toward same side as contracting muscle
Splenius capitis (See Exhibit 11–13)	*splenion* = bandage	Ligamentum nuchae and spines of seventh cervical vertebra and upper four thoracic vertebrae	Occipital bone and mastoid process of temporal bone	Both muscles extend head; contraction of one rotates it to same side as contracting muscle
Longissimus capitis (See Exhibit 11–13)	*longissimus* = long-est	Transverse processes of upper four thoracic vertebrae	Mastoid process of temporal bone	Extends head and rotates face toward side opposite contracting muscle

Exhibit 11–7. MUSCLES THAT ACT ON THE ANTERIOR ABDOMINAL WALL

MUSCLE	LEARNING KEY	ORIGIN	INSERTION	ACTION
Rectus abdominis	*abdomino* = belly	Pubic crest and symphysis pubis	Cartilage of fifth to seventh ribs and xiphoid process	Flexes vertebral column
External oblique	*external* = closer to the surface	Lower eight ribs	Iliac crest; linea alba (midline aponeurosis)	Both muscles compress abdomen; one side alone bends vertebral column laterally
Internal oblique	*internal* = farther from the surface	Iliac crest, inguinal ligament, and vertebral column	Cartilage of last three or four ribs	Compresses abdomen; one side alone bends vertebral column laterally
Transversus abdominis	*transverse* = muscle fibers run transversely to midline	Iliac crest, inguinal ligament, vertebral column, and cartilages of lower six ribs	Xiphoid process, linea alba, and pubis	Compresses abdomen

Exhibit 11–7 (*cont'd.*)

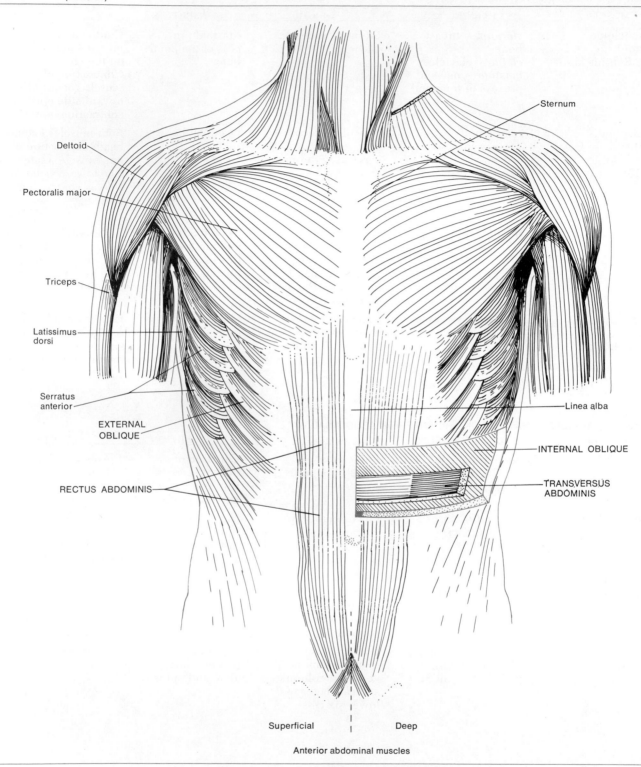

Anterior abdominal muscles

Exhibit 11–8. MUSCLES USED IN BREATHING

MUSCLE	LEARNING KEY	ORIGIN	INSERTION	ACTION
Diaphragm	*dia* = across, between *phragm* = fence	Xiphoid process, costal cartilages, and lumbar vertebrae	Central tendon	Forms floor of thoracic cavity; contraction pulls central tendon downward and increases vertical length of thorax during inspiration
External intercostals	*inter* = between *costa* = rib	Inferior border of rib	Superior border of rib below	Elevate ribs during inspiration and thus increase lateral and anteroposterior dimensions of the thorax
Internal intercostals		Superior border of the rib below	Inferior border of rib above	Draw adjacent ribs together during expiration and thus decrease the lateral and anteroposterior dimensions of the thorax

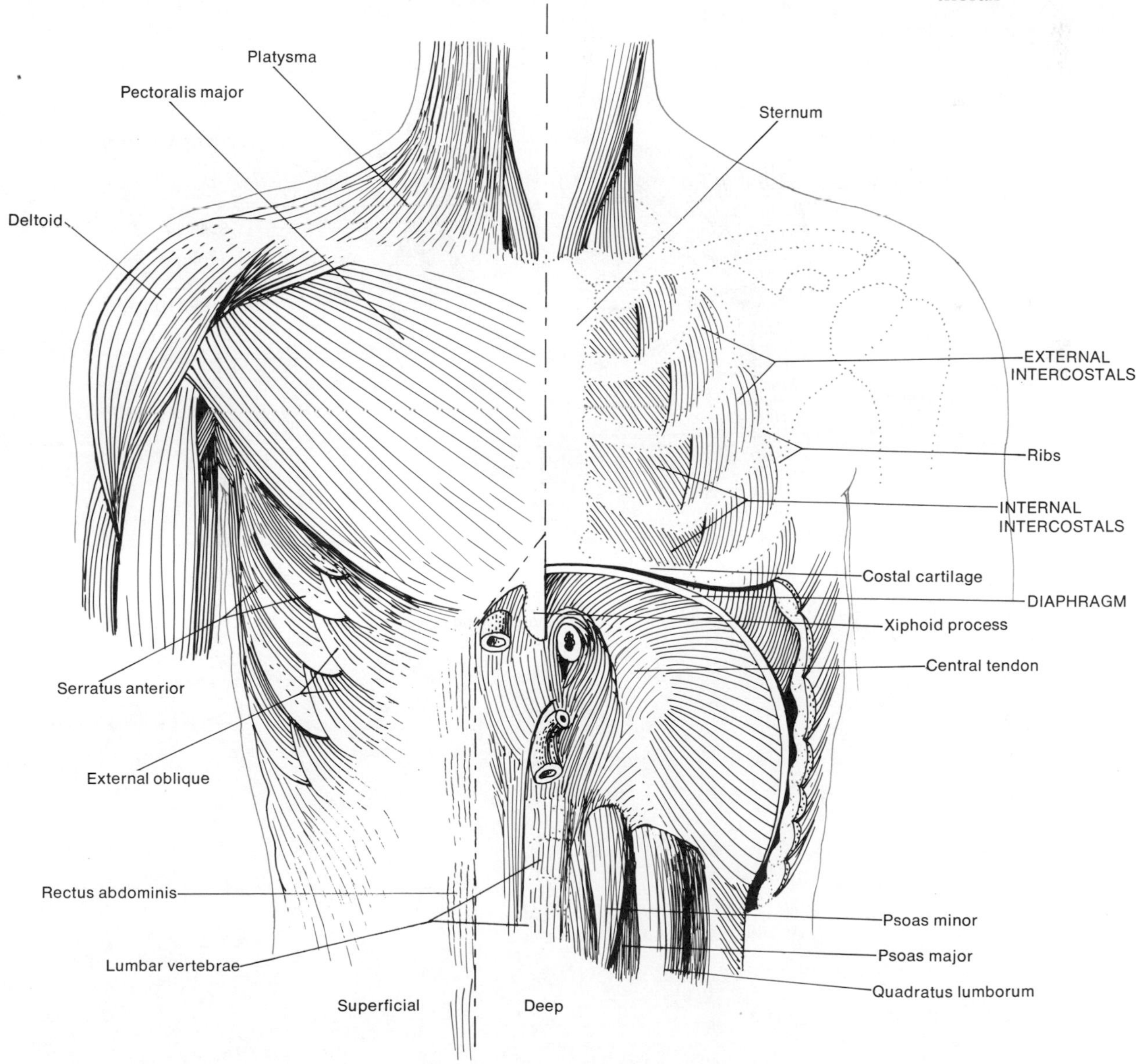

Anterior muscles used in breathing

Exhibit 11–9. MUSCLES THAT MOVE THE SHOULDER GIRDLE

MUSCLE	LEARNING KEY	ORIGIN	INSERTION	ACTION
Subclavius	*sub* = under *clavius* = clavicle	First rib	Clavicle	Draws clavicle downward
Pectoralis minor	*pectus* = breast, chest, thorax *minor* = lesser	Third through fifth ribs	Coracoid process of scapula	Draws scapula downward, rotates shoulder joint upward, and raises third through fifth ribs during forced inspiration when scapula is fixed
Serratus anterior	*serratus* = serrated *anterior* = front	Upper eight or nine ribs	Vertebral border and inferior angle of scapula	Moves scapula forward and rotates scapula upward

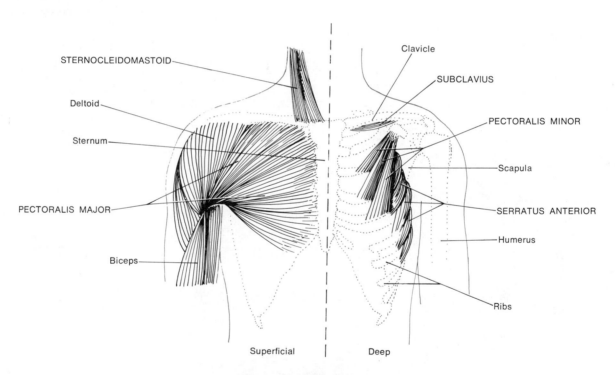

Anterior muscles that move the shoulder girdle

MUSCLE	LEARNING KEY	ORIGIN	INSERTION	ACTION
Trapezius	*trapezoides* = trapezoid-shaped	Occipital bone and spines of seventh cervical and all thoracic vertebrae	Acromion process of clavicle and spine of scapula	Elevates clavicle, adducts scapula, and elevates or depresses scapula
Levator scapulae	*levator* = raises *scapulae* = scapula	Upper four or five cervical vertebrae	Vertebral border of scapula	Elevates scapula
Rhomboideus major	*rhomboides* = rhomboid-shaped, or diamond-shaped	Spines of second to fifth thoracic vertebrae	Vertebral border of scapula	Moves scapula backward and upward and slightly rotates it downward
Rhomboideus minor		Spines of seventh cervical and first thoracic vertebrae	Coracoid process of scapula	Adducts scapula

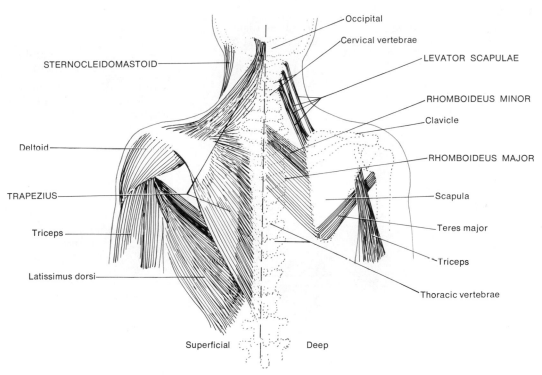

Posterior muscles that move the shoulder girdle

Exhibit 11-10. MUSCLES THAT MOVE THE HUMERUS

MUSCLE	LEARNING KEY	ORIGIN	INSERTION	ACTION
Pectoralis major (See Exhibit 11–9)		Clavicle, sternum, cartilage of second to sixth ribs	Greater tubercle of humerus	Flexes, adducts, and rotates arm medially
Latissimus dorsi	*dorsum* = back	Spines of lower six thoracic vertebrae, lumbar vertebrae, crests of sacrum and ilium, lower four ribs	Intertubercular groove of humerus	Extends, adducts, and rotates arm medially; draws shoulder downward and backward
Deltoid	*delta* = triangular-shaped	Clavicle and acromion process and spine of scapula	Lateral surface of body of humerus	Abducts arm
Supraspinatus	*supra* = above *spinatus* = spine of scapula	Fossa above spine of scapula	Greater tubercle of humerus	Assists deltoid muscle in adducting arm
Infraspinatus	*infra* = below	Fossa below spine of scapula	Greater tubercle of humerus	Rotates humerus laterally
Teres major	*teres* = long and round	Dorsal medial aspect of scapula	Below lesser tubercle of humerus	Extends humerus and draws it down; assists in adduction and medial rotation of arm
Teres minor		Dorsal medial aspect of scapula	Greater tubercle of humerus	Rotates humerus laterally

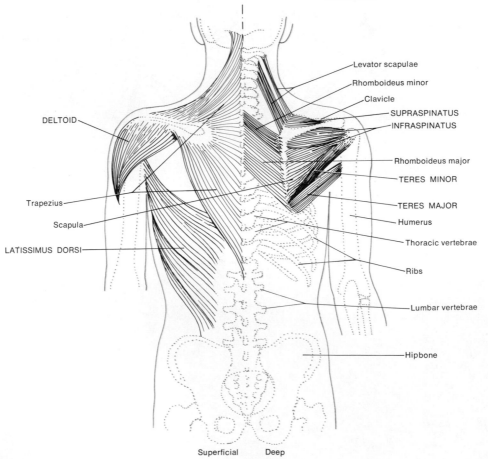

Posterior muscles that move the humerus

Exhibit 11–11. MUSCLES THAT MOVE THE FOREARM

MUSCLE	LEARNING KEY	ORIGIN	INSERTION	ACTION
Biceps brachii	*biceps* = two heads of origin *brachion* = arm	Tuberosity above glenoid cavity and coracoid process of scapula	Radial tuberosity	Flexes forearm and supinates forearm and hand
Brachialis		Anterior surface of humerus	Tuberosity and coronoid process of ulna	Flexes forearm
Brachioradialis (See Exhibit 11–12 also)	*radialis* = radius	Supracondyloid ridge of humerus	Above styloid process of radius	Flexes forearm
Triceps brachii	*triceps* = three heads of origin	Infraglenoid tuberosity of scapula, lateral and posterior surface of humerus above radial groove, posterior surface of humerus below radial groove	Olecranon of ulna	Extends forearm
Supinator	*supination* = turning palm upward or anteriorly	Lateral epicondyle of humerus, ridge on ulna	Oblique line of radius	Supinates the forearm and hand
Pronator teres	*pronation* = turning palm downward or posteriorly	Medial epicondyle of humerus, coronoid process of ulna	Midlateral surface of radius	Pronation of forearm and hand

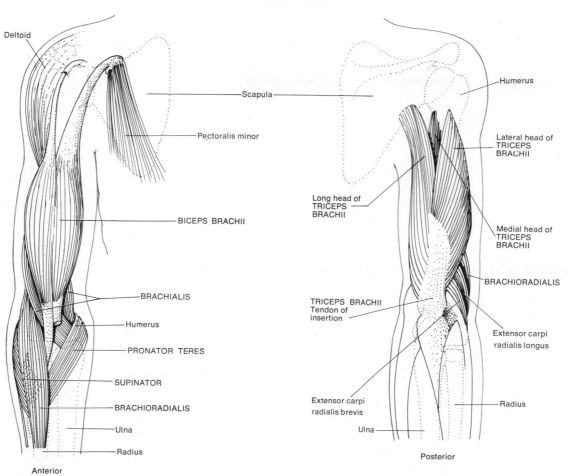

Muscles that move the forearm

211

Exhibit 11–12. MUSCLES THAT MOVE THE WRIST AND FINGERS

MUSCLE	LEARNING KEY	ORIGIN	INSERTION	ACTION
Flexor carpi radialis	*flexor* = decreases angle *carpus* = wrist	Medial epicondyle of humerus	Second and third meta-carpals	Flexes and abducts wrist
Flexor carpi ulnaris	*ulnaris* = ulna	Medial epicondyle of humerus and upper dorsal border of ulna	Pisiform, hamate, and fifth meta-carpal	Flexes and adducts wrist
Extensor carpi radialis longus	*extensor* = increases angle at a joint *longus* = long	Lateral supra-condylar ridge of humerus	Second meta-carpal	Extends and abducts wrist
Extensor carpi ulnaris		Lateral epicondyle of humerus and dorsal border of ulna	Fifth meta-carpal	Extends and adducts wrist
Flexor digitorum profundus	*digit* = finger or toe *profundus* = deep	Anterior medial surface of body of ulna	Bases of distal phalanges	Flexes terminal phalanges of each finger
Flexor digitorum superficialis	*superficialis* = superficial	Medial epicondyle of humerus, coronoid process of ulna, oblique line of radius	Middle phalanges	Flexes middle phalanges of each finger
Extensor digitorum		Lateral epicondyle of humerus	Middle and distal phalanges of each finger	Extends phalanges
Extensor indicis	*indicis* = index	Dorsal surface of ulna	Tendon of extensor digitorum into index finger	Extends index finger

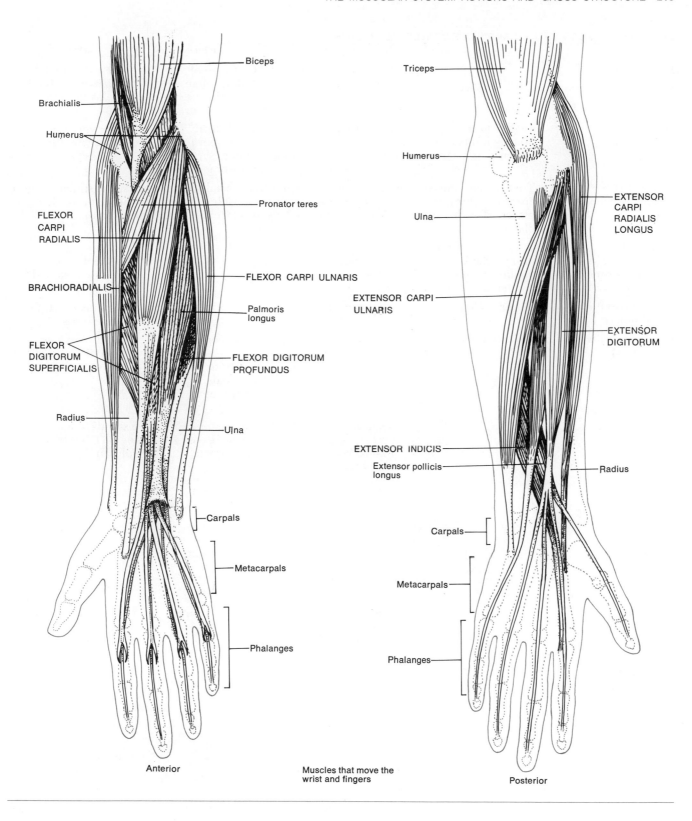

Biceps

Brachialis

Humerus

FLEXOR
CARPI
RADIALIS

BRACHIORADIALIS

FLEXOR
DIGITORUM
SUPERFICIALIS

Radius

Pronator teres

FLEXOR CARPI ULNARIS

Palmoris
longus

FLEXOR DIGITORUM
PROFUNDUS

Ulna

Carpals

Metacarpals

Phalanges

Anterior

Triceps

Humerus

Ulna

EXTENSOR CARPI
ULNARIS

EXTENSOR INDICIS

Extensor pollicis
longus

EXTENSOR
CARPI
RADIALIS
LONGUS

EXTENSOR
DIGITORUM

Radius

Carpals

Metacarpals

Phalanges

Posterior

**Muscles that move the
wrist and fingers**

Exhibit 11–13. MUSCLES THAT MOVE THE VERTEBRAL COLUMN

MUSCLE	LEARNING KEY	ORIGIN	INSERTION	ACTION
Rectus abdominis (See Exhibit 11–7)		Body of pubis of coxal bone	Cartilage of fifth through seventh ribs	Flexes vertebral column at the lumbar spine as in doing a sit up
Quadratus lumborum	*quad* = four *lumb* = lumbar region	Iliac crest	Twelfth rib and upper four lumbar vertebrae	Flexes vertebral column laterally
Sacrospinalis (erector spinae)	This muscle consists of three posterior groupings: iliocostalis, longissimus, and spinalis. These groups, in turn, consist of a series of overlapping muscles. The iliocostalis group is laterally placed. The longissimus group is intermediate in placement. And the spinalis is medially placed.			
Lateral Iliocostalis lumborum	*ilium* = flank *lumbus* = loin	Iliac crest	Lower six ribs	Extends lumbar region of vertebral column
Iliocostalis thoracis	*thorax* = chest	Six lower ribs	Six upper ribs	Maintains erect position of spine
Iliocostalis cervicis	*cervix* = neck	First six ribs	Transverse processes of fourth to sixth cervical vertebrae	Extends cervical region of vertebral column
Intermediate Longissimus thoracis		Transverse processes of lumbar vertebrae	Transverse processes of all thoracic and upper lumbar vertebrae, and the ninth and tenth ribs	Extends thoracic region of vertebral column
Longissimus cervicis		Transverse processes of fourth and fifth thoracic vertebrae	Transverse processes of second to sixth cervical vertebrae	Extends cervical region of vertebral column
Longissimus capitis		Transverse processes of upper four thoracic vertebrae	Mastoid process of temporal bone	Extends head and rotates it to opposite side
Medial Spinalis thoracis		Spines of upper lumbar and lower thoracic vertebrae	Spines of upper thoracic vertebrae	Extends vertebral column

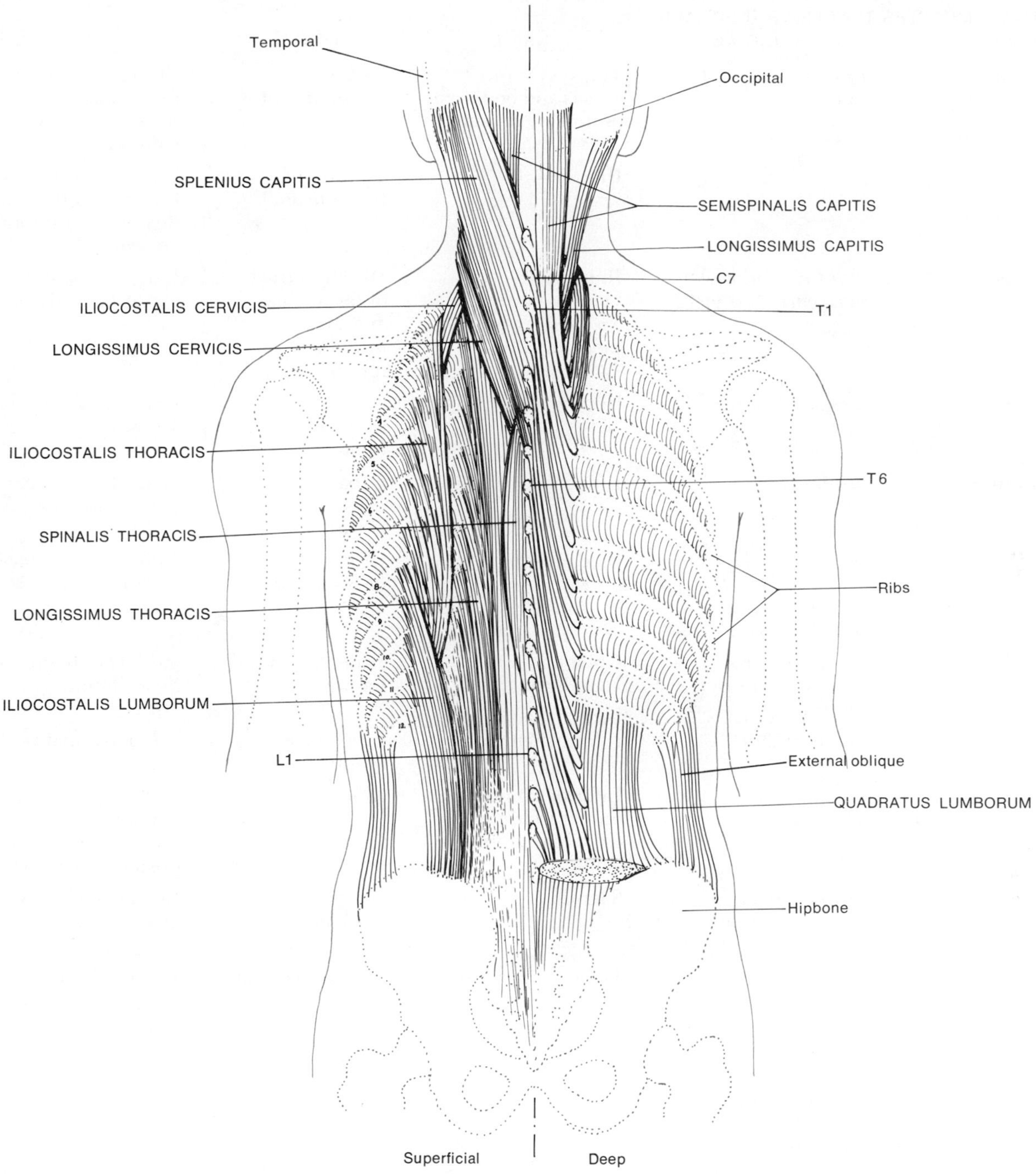

Temporal

Occipital

SPLENIUS CAPITIS

SEMISPINALIS CAPITIS

LONGISSIMUS CAPITIS

C7

ILIOCOSTALIS CERVICIS

T1

LONGISSIMUS CERVICIS

ILIOCOSTALIS THORACIS

T 6

SPINALIS THORACIS

Ribs

LONGISSIMUS THORACIS

ILIOCOSTALIS LUMBORUM

L1

External oblique

QUADRATUS LUMBORUM

Hipbone

Superficial

Deep

Muscles that move the vertebral column

Exhibit 11–14. MUSCLES THAT MOVE THE FEMUR

MUSCLE	LEARNING KEY	ORIGIN	INSERTION	ACTION
Psoas major	*psoa* = muscle of loin	Transverse processes and bodies of lumbar vertebrae	Lesser trochanter of femur	Flexes and rotates femur laterally; flexes vertebral column
Iliacus		Iliac fossa	Tendon of psoas major	Flexes and rotates femur laterally; slight flexion of vertebral column
Gluteus maximus	*gloutos* = buttock *maximus* = largest	Iliac crest, sacrum, coccyx, and aponeurosis of sacrospinalis	Iliotibial tract of fascia lata and gluteal tuberosity of femur	Extends and rotates femur laterally
Gluteus medius	*media* = middle	Ilium	Greater trochanter of femur	Abducts and rotates femur laterally
Gluteus minimus	*minimus* = small	Ilium	Greater trochanter of femur	Abducts and rotates femur medially
Tensor fasciae latae	*tensor* = makes tense *fascia* = band *latus* = broad, wide	Iliac crest	Iliotibial tract of fascia lata	Flexes and abducts femur
Adductor longus	*adductor* = moves a part closer to the midline	Crest and symphysis	Linea aspera of femur	Adducts, rotates, and flexes femur
Adductor brevis	*brevis* = short	Inferior ramus of pubis	Linea aspera of femur	Adducts, rotates, and flexes femur
Adductor magnus	*magnus* = large	Inferior ramus of pubis; ischium to ischial tuberosity	Linea aspera of femur	Adducts, flexes, and extends femur (anterior part flexes, posterior part extends)
Piriformis	*pirum* = pear *forma* = shape	Sacrum	Greater trochanter of femur	Rotates femur laterally and abducts it
Quadratus femoris	*femoris* = femur	Ischial tuberosity	Linea quadrata	Rotates femur laterally
Obturator externus	*obturator* = closed because it arises over the obturator foramen, which is closed by a heavy membrane *external* = outside	Pubis and ischium	Trochanteric fossa of femur	Rotates thigh laterally

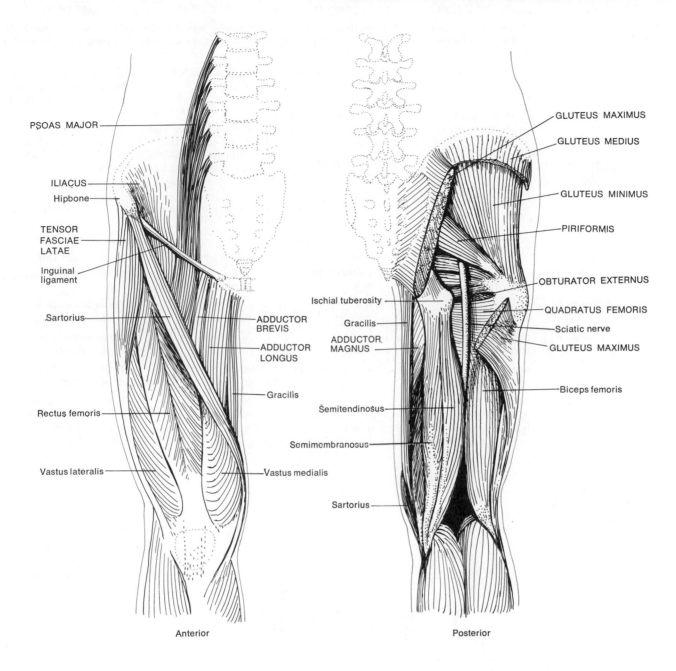

Anterior

Posterior

Muscles that move the femur

Exhibit 11–15. MUSCLES THAT ACT ON THE KNEE JOINT

MUSCLE	LEARNING KEY	ORIGIN	INSERTION	ACTION
Quadriceps femoris	A composite muscle that includes four distinct parts, usually described as four separate muscles. The common tendon from the patella to the tibial tuberosity is known as the patellar ligament.			
Rectus femoris		Anterior, inferior spine of ilium	Upper border of patella	All four heads extend knee; rectus portion alone also flexes thigh
Vastus lateralis	*vastus* = vast, large *lateralis* = lateral	Greater trochanter and linea aspera of femur	Upper border and sides of patella; tibial tuberosity through the patellar ligament (tendon of quadriceps)	
Vastus medialis	*medialis* = medial	Linea aspera of femur		
Vastus intermedius	*intermedius* = middle	Anterior and lateral surfaces of body of femur		
Hamstrings	A collective designation for three separate muscles.			
Biceps femoris		Ischial tuberosity and linea aspera of femur	Head of tibula and lateral condyle of tibia	Flexes knee and extends thigh
Semitendinosus	*semi* = half *tendo* = tendon	Ischial tuberosity	Upper body of tibia	Flexes knee and extends thigh
Semimembranosus	*membran* = membrane	Ischial tuberosity	Medial condyle of tibia	Flexes knee and extends thigh
Gracilis	*gracilis* = slender	Symphysis pubis and pubic arch	Medial surface of body of tibia	Flexes knee and adducts thigh
Sartorius	*sartor* = tailor; refers to cross-legged position in which tailors sit	Anterior superior spine of ilium	Medial surface of body of tibia	Flexes knee; flexes thigh and rotates it laterally, thus crossing leg

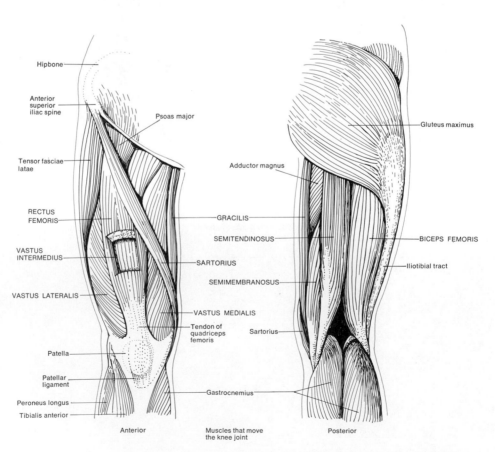

Muscles that move the knee joint

Exhibit 11–16. MUSCLES THAT MOVE FOOT AND TOES

MUSCLE	LEARNING KEY	ORIGIN	INSERTION	ACTION
Gastrocnemius	*gaster* = belly *kneme* = leg	Lateral and medial condyles of femur and capsule of knee	Calcaneus	Plantar flexes foot
Soleus	*soleus* = sole of foot	Head of fibula and medial border of tibia	Calcaneus	Plantar flexes foot
Peroneus longus	*perone* = fibula	Head and body of fibula and lateral condyle of tibia	First metatarsal and first cuneiform bone	Plantar flexes and everts foot
Peroneus brevis		Body of fibula	Fifth metatarsal	Plantar flexes and everts foot
Peroneus tertius	*tertius* = third	Distal third of fibula	Fifth metatarsal	Dorsally flexes and everts foot
Tibialis anterior	*tibialis* = tibia	Lateral condyle and body of tibia	First metatarsal and first cuneiform	Dorsally flexes and inverts foot
Tibialis posterior	*posterior* = back	Interosseus membrane between tibia and fibula	Second, third, and fourth metatarsals; navicular; third cuneiform, cuboid	Plantar flexes and inverts foot
Flexor digitorum longus	*digitorum* = digit, finger, or toe	Tibia	Distal phalanges of four outer toes	Flexes toes and plantar flexes and inverts foot
Extensor digitorum longus		Lateral condyle of tibia and anterior surface of fibula	Middle and distal phalanges of four outer toes	Extends toes and dorsally flexes and everts foot

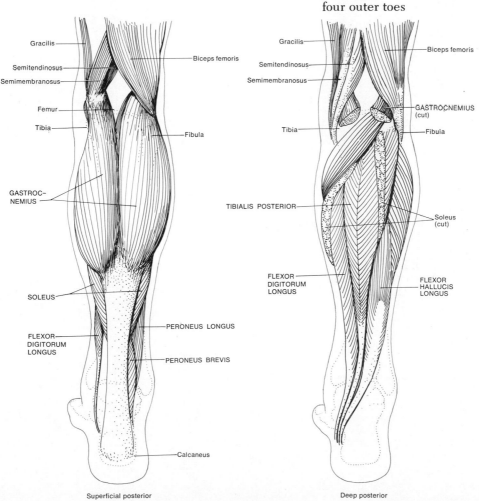

Muscles that move the foot and toes

Superficial posterior   Deep posterior

Exhibit 11–16 (*cont'd.*)

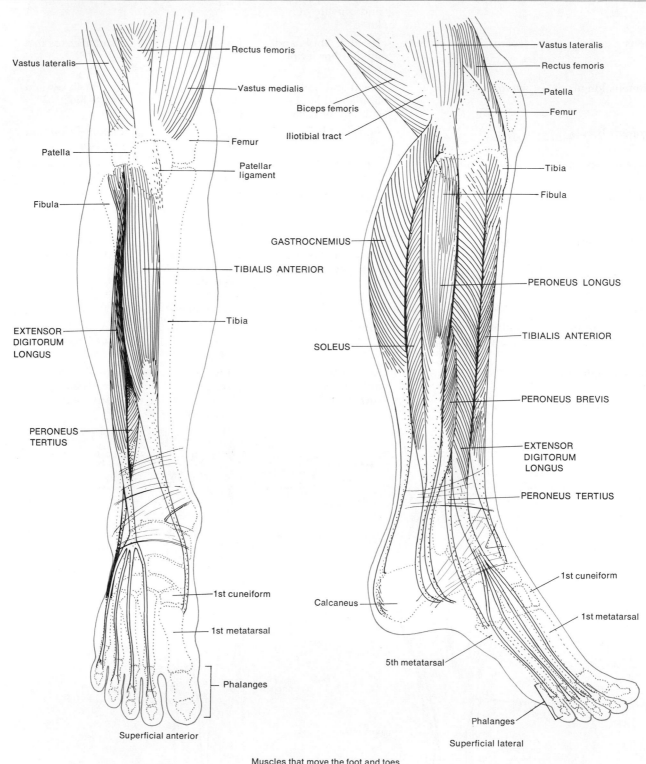

Muscles that move the foot and toes

INTRAMUSCULAR INJECTIONS

The way a particular drug is administered is determined primarily by the physical and chemical properties of the drug, the area of the body in which the drug is supposed to act, and the speed of response required. Generally, drugs are administered for one of two effects: local or systemic. With a *local* effect, the action of the drug is confined to the site of application. With a *systemic* effect, the drug is first absorbed into the blood and then diffuses into body tissues far away from the site of administration. Local effects are achieved by applying a medication to the surface of the tissue being treated. For instance, lotions, ointments, wet dressings, and certain anti-inflammatories may be put on the surface of the skin. Mucous membranes of the eyes, nose, mouth, throat, respiratory tract, reproductive tract, and urinary passageways may be treated locally with sprays, swabs, irrigations, douches, and suppositories.

Drugs designed to produce systemic effects may be administered several ways, depending upon the nature and amount of the drug, the speed of response desired, and the general condition of the patient. Methods of administration include swallowing, placing the drug under the tongue and allowing it to dissolve, inserting the drug rectally, inhaling, and injecting the drug with a needle. All the ways that a drug may be administered with a needle are described by the term **parenteral.** Among the parenteral routes of administration are (1) *intradermal* or *intracutaneous*, in which the needle penetrates the epidermis and is inserted into the dermis; (2) *subcutaneous* or *hypodermic*, in which the needle is inserted into the subcutaneous layer; (3) *intramuscular*, in which the needle is inserted into a muscle; (4) *intravenous*, in which the needle is inserted directly into a vein; and (5) *intraspinal*, in which the needle is inserted into the vertebral canal. At this point, we shall discuss intramuscular

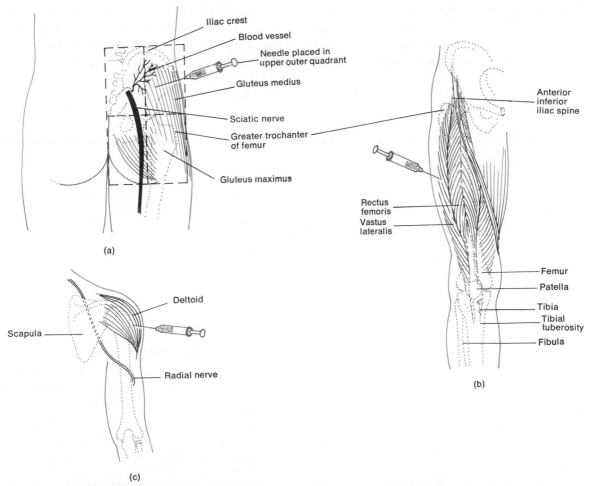

Figure 11–3. Intramuscular injections. Shown are the three common sites for intramuscular injections: (a) buttock, (b) lateral surface of thigh, and (c) deltoid region of the arm.

injections only. Other parenteral routes will be considered later.

When an intramuscular injection is given, it is made through the skin and subcutaneous tissue into the muscle itself. Intramuscular injections are preferred when prompt absorption is desired, when larger doses than can be given cutaneously are indicated, or if the drug is too irritating to give subcutaneously. The common sites for intramuscular injections include the buttock, lateral side of the thigh, and the deltoid region of the arm. Muscles in these areas, especially the gluteal muscles in the buttock, are fairly thick. Because of the large number of muscle fibers and extensive fascia, the drug has a larger surface area for absorption. Absorption is further promoted by the extensive blood supply to muscles. Ideally, intramuscular injections should be given deep within the muscle and away from major nerves and blood vessels.

For many intramuscular injections, the preferred site is the buttock (Figure 11–3a). In giving such an injection, the buttock should be divided into four quadrants and the upper outer quadrant used as the injection site. The iliac crest serves as a good landmark for this quadrant.

The spot for injection should be about 2 to 3 inches below the iliac crest. The upper outer quadrant is chosen because, in this area, the muscle is quite thick with few nerves. For this reason, there is not much chance of injuring the sciatic nerve. Injury to the nerve can cause paralysis of the lower extremity. The probability of injecting the drug into a blood vessel is also remote in this area. After the needle is inserted, the plunger should be pulled up for a few seconds. If the syringe fills with blood, this indicates that the needle is in a blood vessel, and a different injection site on the opposite buttock should be chosen.

Injections given in the lateral side of the thigh are inserted into the midportion of the vastus lateralis muscle (Figure 11–3b). This site is determined by using the knee and greater trochanter of the femur as landmarks. The midportion of the muscle is located by measuring a handbreadth above the knee and a handbreadth below the greater trochanter.

The deltoid injection is given in the midportion of the muscle about two to three fingerbreadths below the acromion of the scapula and lateral to the axilla (Figure 11–3c).

Chapter summary in outline

HOW MUSCLES PRODUCE MOVEMENT
1. Skeletal muscles produce movements by pulling on bones.
2. The stationary attachment is the origin; the movable attachment is the insertion.
3. Bones serve as levers and joints as fulcrums.
4. Levers are acted upon by two different forces: resistance and effort. There are first-class, second-class, and third-class levers.
5. The agonist produces the desired action. The antagonist produces an opposite action. The synergist assists the agonist by reducing unnecessary movements.

NAMING MUSCLES
1. Skeletal muscles are named on the basis of one or more distinctive criteria, including direction of fibers, location, size, number of heads, shape, origin-insertion, and action.

INTRAMUSCULAR INJECTIONS
1. Drugs are generally administered for either local or systemic effects.
2. The term *parenteral* is applied to all ways that drugs are administered with a needle.
3. Parenteral routes may be intradermal, subcutaneous, intramuscular, intravenous, or intraspinal.
4. Advantages of intramuscular injections are prompt absorption, use of larger doses than can be given cutaneously, and minimal irritation.
5. Common sites for intramuscular injections are the buttock, lateral side of the thigh, and deltoid region of the arm.

Review questions and problems
1. What is meant by the muscular system? Explain fully.
2. Using the terms origin, insertion, and belly in your discussion, describe how skeletal muscles produce body movements by pulling on bones.
3. What is a lever? Fulcrum? Apply these terms to the body, and indicate the nature of the forces that act on levers.
4. Describe the three classes of levers, and provide one example for each in the body.
5. Define the role of the agonist, antagonist, and synergist in producing body movements.
6. At the beginning of this chapter, several criteria for naming muscles were discussed. These were direction of fibers,

location, size, number of heads, shape, origin and insertion, and action. Select at random the names of some muscles presented in Exhibits 11–2 through 11–16, and see if you can determine the criterion or criteria employed for each. In addition, refer to the Learning Key in each Exhibit as a guide. Select as many muscles as you wish, as long as you feel you understand the concept involved.

7. Discuss the muscles and their actions involved in facial expression.

8. What muscles would you use to do the following: (a) frown; (b) pout; (c) show surprise; (d) show your upper teeth; (e) pucker your lips; (f) squint; (g) blow up a balloon, (h) smile?

9. What are the principal muscles that move the mandible? Give the function of each.

10. What would happen if you lost tone in the masseter and temporal muscles?

11. What muscles move the eyeballs? In which direction does each muscle move the eyeball?

12. Describe the action of each of the muscles acting on the tongue.

13. What tongue, facial, and mandibular muscles would you use when chewing a piece of gum?

14. What muscles are responsible for moving the head, and how do they move the head?

15. Which of the muscles listed above would you use to signify "yes" and "no" by moving your head?

16. What muscles accomplish compression of the anterior abdominal wall?

17. What are the principal muscles involved in breathing? What are their actions?

18. In what directions is the shoulder girdle drawn? What muscles accomplish these movements?

19. What muscles are used to (a) raise your shoulders, (b) lower your shoulders, (c) join your hands behind your back, (d) join your hands in front of your chest?

20. What movements are possible at the shoulder joint? What muscles accomplish these movements?

21. What muscles move the forearm? In which directions do these movements occur?

22. What muscles move the forearm and what actions are used when lighting a cigarette?

23. Discuss the various movements possible at the wrist and fingers. What muscles accomplish these movements?

24. How many muscles and actions of the wrist and fingers used when writing can you list?

25. Discuss the various muscles and movements of the vertebral column.

26. Can you perform an exercise that would involve the use of each of the muscles listed in Exhibit 11–13?

27. What muscles accomplish movements of the femur? What actions are produced by these muscles?

28. Review in your mind the various movements involved in your favorite kind of dancing. What muscles listed in Exhibit 11–14 would you be using and what actions would you be performing?

29. What muscles act at the knee joint? What kinds of movements do these muscles perform?

30. Determine the muscles and their actions listed in Exhibit 11–15 that you would use in climbing a ladder to a diving board, diving into the water, swimming the length of a pool, and then sitting at pool side.

31. Discuss the muscles that plantar flex, evert, pronate, dorsiflex, and supinate the foot.

32. In which directions are the toes moved? What muscles bring about these movements?

33. Distinguish between local and systemic drug actions. How are drugs applied locally?

34. What routes of administration are used for drugs that produce systemic effects?

35. Define parenteral. What are the various routes of parenteral administration?

36. What are the advantages of intramuscular injections?

37. Describe how you would locate the sites for an intramuscular injection in the buttock, lateral side of the thigh, and deltoid region of the arm.

Bibliography

Unit 2

Aufranc, O. E. and R. H. Turner: "Total Replacement of the Arthritic Hip," *Hospital Practice,* October 1971.

Basmajian, J. F. and M. A. MacConell: *Muscles and Movements: A Basis for Human Kinesiology,* The Williams & Wilkins Company, Baltimore, 1969.

Bethlem, J.: *Muscle Pathology,* American Elsevier Publishing Company, Inc., New York, 1970.

Bones, Joints and Muscles of the Human Body: A Programmed Text for Physical Therapy Aides, Glencoe Press, The Macmillan Company, New York, 1970.

Bourne, G. W.: *The Biochemistry and Physiology of Bone,* 2d ed., Academic Press, Inc., New York, 1972.

Chapman, C. B. and J. H. Mitchell: "The Physiology of Exercise," *Scientific American,* **212:**88, May 1965.

Close, R. I. "Dynamic Properties of Mammalian Skeletal Muscles," *Physiol. Rev.,* **52:**129, 1972.

Evans, F. G. (ed.): *Studies in the Anatomy and Function of Bone and Joints,* Springer-Verlag New York Inc., New York, 1966.

Hall, B. K. "Cellular Differentiation in Skeletal Tissue," *Biol. Rev.,* **45:**455, 1970.

Harris, W. H. and R. P. Heaney: *Skeletal Renewal and Metabolic Bone Disease,* Little, Brown and Company, Boston, 1970.

Herring, G. M.: "The Chemical Structure of Tendon, Cartilage, Dentin and Bone Matrix," *Clin. Orthop.,* **60:**261, 1968.

Hollander, J. L. and D. J. McCarty, Jr.: *Arthritis and Allied Conditions,* 8th ed., Lea & Febiger, Philadelphia, 1972.

Hoyle, G.: "How Is Muscle Turned On and Off?", *Scientific American,* **222:**845, April 1970.

Intramuscular Injections, Wyeth Laboratories, New York, 1972.

"Joints, Rheumatoid Arthritis, Osteoarthritis," *Chemical & Engineering News,* July 22, 1968.

Laki, K.: *Contractible Muscle and Proteins,* Marcel Dekker, Inc., New York, 1971.

Larson, C. B. and M. Gould: *Orthopedic Nursing,* 7th ed., The C. V. Mosby Company, St. Louis, 1970.

McLean, F. C. and N. R. Urist (eds.), *Bone: Fundamentals of the Physiology of Skeletal Tissue,* 3d. ed., University of Chicago Press, Chicago, 1968.

"The Most Exciting Area in Arthritis Research," *Medical World News*, Oct. 8, 1971.

Mouratoff, G. J.: "Pharmacotherapy of Rheumatoid Arthritis," *Modern Medicine*, Aug. 7, 1972.

Murray, J. M. and A. Weber: "The Cooperative Action of Muscle Proteins," *Scientific American*, **230:**58, February 1974.

Porter, K. R. and C. Franzini-Armstrong: "The Sarcoplasmic Reticulum," *Scientific American*, **212:**73, March 1965.

Riiegg, J. C.: "Smooth Muscle Tone," *Physiol. Rev.*, **51:**201, 1971.

Rodnan, G. P. (ed.): "Primer on the Rheumatic Diseases," *J. Amer. Med. Assoc.*, **224:**5 (supplement), April 1973.

Sandow, A.: "Skeletal Muscle," *Ann. Rev. Physiol.*, **32:**87, 1970.

Steinbach, H. L. and R. H. Gold: "Pyogenic Infections of Bone: A Roentgenologic Guide," *Hospital Medicine*, July 1972.

Tonomuna, Y.: *Muscle Protein, Muscle Contraction and Cation Transport*, University Park Press, Baltimore, 1972.

Tronzo, R. G.: "Bone: Self-repairing, Self-renewing," *Consultant*, April 1972.

Trueta, J.: *Studies of the Development and Decay of the Human Frame*, W. B. Saunders Company, Philadelphia, 1968.

Vaughn, J. M.: *The Physiology of Bone*, Clarendon Press, Oxford, 1970.

Weissmann, G.: "The Molecular Basis of Acute Gout," *Hospital Practice*, July 1971.

CONTROL SYSTEMS

CHAPTER 12

NERVOUS TISSUE: STRUCTURE AND PHYSIOLOGY

STUDENT OBJECTIVES

After you have read this chapter, you should be able to:

1. Describe the function of the nervous system in maintaining homeostasis

2. Classify the organs of the nervous system into central and peripheral divisions

3. Contrast the histological characteristics and functions of neuroglia and neurons

4. Categorize neurons on the basis of shape and function

5. Describe the necessary conditions for the regeneration of nervous tissue

6. List the sequence of events involved in the initiation and transmission of a nerve impulse

7. Define the all-or-none principle of impulse transmission by a neuron

8. List the factors that determine the rate of impulse transmission

9. Discuss the components of a reflex arc

10. Contrast the operation of a two-neuron and three-neuron reflex arc

11. Define a reflex

12. List the factors involved in the conduction of an impulse across a synapse

13. Define the roles of acetylcholine and cholinesterase in the transmission of an impulse across a synapse

14. Compare the functions of excitatory and inhibitory impulses in helping to maintain homeostasis

15. List the factors that may inhibit or block nerve impulses

The nervous system is the control tower and communications network of the body. In human beings it performs three broad functions. First, it stimulates movements that are vital to life as well as movements that simply make life easier and more enjoyable. Second, it shares responsibility for the maintenance of homeostasis. And, third, it allows us to express uniquely human traits. Human life simply cannot exist without a functioning nervous system. For instance, skeletal and smooth muscle cells cannot contract until they are stimulated by a nerve impulse. If the intercostal muscles and diaphragm do not contract, we cannot breathe. If the smooth muscles of the digestive system do not contract, food cannot be pushed through the gullet, stomach, and intestines. And if the digestive glands are not stimulated to release their secretions, food cannot be digested. It is obvious, then, that our cells cannot receive nutrients unless the digestive system is connected to a functioning nervous system. But suppose our muscles and glands could stimulate themselves. Even then, we could not live very long without our nervous system. This is because of the second great function of the nervous system—keeping the body in homeostasis. Recall that homeostasis is the maintenance of a constant internal environment. The nervous system *senses* changes that occur inside the body and in the outside environment. It then interprets these changes and decides on a course of action. This property is called *integration*. After deciding which action to take, it elicits a *response* by sending impulses to the appropriate muscles and glands. As you will discover later, the nervous system shares the responsibility for homeostasis with the endocrine system, which sends out chemical messengers. The third broad function of the human nervous system is to provide the uniquely human pleasures of thinking, feeling, and acting upon our thoughts and feelings.

For convenience of study, the entire nervous system may be divided into two principal portions: (1) the central nervous system and (2) the peripheral nervous system (Figure 12–1). The **central nervous system** is the control center for the entire system and consists of the brain and the spinal cord. All body sensations must be relayed to the central nervous system if they are to be felt and acted upon. All the impulses that stimulate muscles to contract and that cause glands to secrete must pass through the central system.

The **peripheral nervous system** contains all remaining nervous tissue—that is, the nerves which connect the central nervous system to all other parts of the body. The peripheral nervous system is frequently divided into a somatic nervous system and an autonomic nervous system. The *somatic nervous system* contains all the nerve fibers that run between the central nervous system and the skeletal muscles and skin. The term *soma* refers to the body. Since the somatic division produces movement only in the skeletal muscles, it is under our conscious control and is therefore said to be *voluntary*.

The *autonomic nervous system* consists of all the nerve fibers that run between the central nervous system and smooth muscle, cardiac muscle, and glands. The autonomic system produces movement only in involuntary muscles and glands. Thus, the system is said to be *involuntary*. With very few exceptions, the viscera receive nerve fibers from two divisions of the autonomic nervous system: the sympathetic division and the parasympathetic division. The particular functions of the sympathetic and parasympathetic divisions will be considered later.

HISTOLOGY OF THE NERVOUS SYSTEM

Despite the organizational complexity of the nervous system, it consists of only two principal kinds of cells. The first of these, the neurons or nerve cells, make up the nervous tissue that forms the structural and functional portion of the system. Neurons are highly specialized for impulse conduction and are responsible for all the special attributes of the nervous system, such as thinking, controlling muscle activity, and regulating glands.

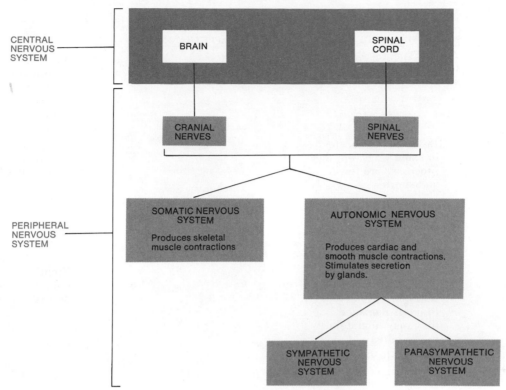

Figure 12–1. Organization of the nervous system.

The second type of cell, the neuroglia, forms a special kind of connective tissue component of the central nervous system. Neuroglia generally perform the less specialized activity of binding together nervous tissue. They do not transmit impulses.

Neuroglia

The connective tissue cells of the central nervous system are called **neuroglia** or **glial cells.** The combining form *neuro* means nerve, and *glia* means glue. Nervous tissue lacks the compactness of muscle and rigidity of bone and must be heavily supported by thick connective tissue. Many of the glial cells form a supporting network by twining around the nerve cells in the brain and spinal cord. Other glial cells bind nervous tissue to other supporting structures and attach the neurons to their blood vessels. A few of the glial cells also serve very specialized functions. For example, many nerve fibers are coated with a thick, fatty sheath that is produced by a particular type of neuroglia. Certain small glial cells are phagocytotic. They protect the central nervous system from disease by engulfing invading microbes and clearing away debris. All these duties of the neural connective tissue are divided among several different kinds of glial cells, which are classified according to their size and shape. Neuroglia are of special clinical interest because they are a very common source of tumors of the nervous system. Exhibit 12–1 lists the cells and summarizes their functions.

Neurons

Nerve cells, called **neurons,** are responsible for the conduction of impulses from one part of the body to another.

Structure

A neuron consists of three structurally and functionally distinct portions: (1) the cell body, (2) dendrites, and (3) an axon (Figure 12–2a). The *cell body* or *perikaryon* contains a well-defined nucleus and nucleolus surrounded by a granular cytoplasm. Within the cytoplasm are found typical organelles such as mitochondria, Golgi apparatus, and endoplasmic reticulum. The cell body receives a stimulus from the

Exhibit 12–1. NEUROGLIA

TYPE	DESCRIPTION	MICROSCOPIC APPEARANCE	FUNCTION
Astrocytes (*astro* = star; *cyte* = cell)	Star-shaped cells with numerous processes		Twine around nerve cells to form a supporting network in the brain and spinal cord; attach neurons to their blood vessels
Oligodendrocytes (*oligo* = few; *dendro* = tree)	Resemble astrocytes in some ways, but the processes are fewer and shorter		Give support by forming semirigid connective tissue rows between neurons in brain and spinal cord; produce a thick, fatty sheath called the myelin sheath on neurons of the central nervous system
Microglia (*micro* = small)	Small cells with few processes; normally stationary; if nervous tissue is damaged, they may migrate to injured area		Engulf and destroy microbes and cellular debris

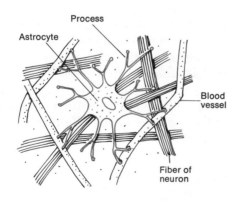

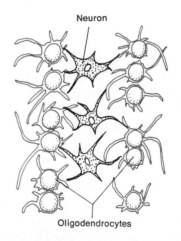

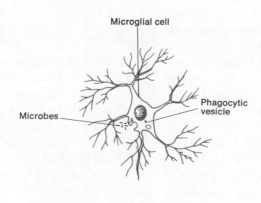

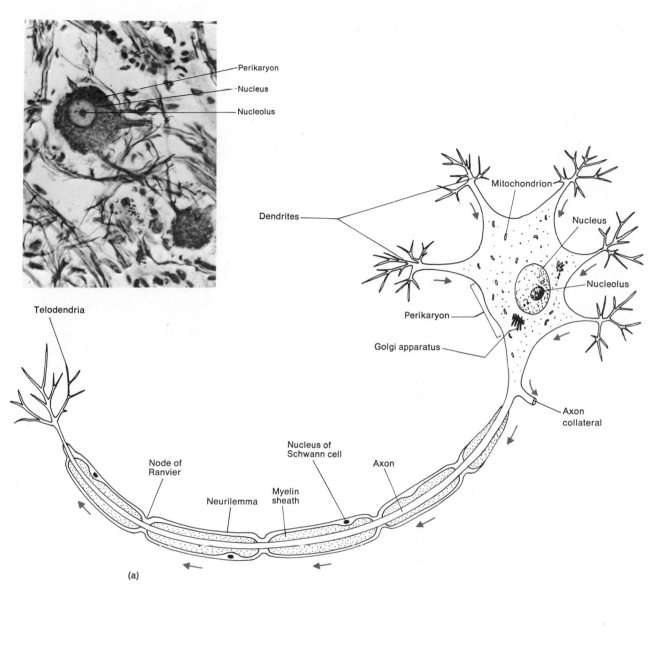

(a)

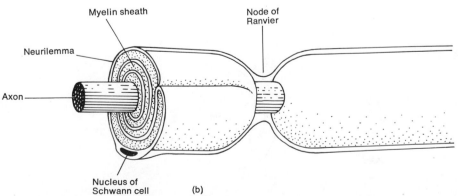

(b)

Figure 12–2. Structure of a neuron. (a) Shown in an entire neuron from the ventral gray area of the spinal cord. The arrows indicate the direction in which a nerve impulse passes. (b) Cross section through a myelinated axon. *(Photograph courtesy of Edward J. Reith, from Atlas of Descriptive Histology, Edward J. Reith and Michael H. Ross, Harper & Row, Publishers, Inc., New York, 1970.)*

dendrites, which are short, highly branched extensions of the cytoplasm of the perikaryon. Their essential function is to pick up a stimulus and transmit it *to the perikaryon.* The **axon,** or **nerve fiber,** by contrast, is a single, highly specialized and relatively long process that conducts impulses *away from the perikaryon* to another neuron or to an organ of the body. Axons vary tremendously in length. Some of the axons inside the brain may be a fraction of an inch long. Those that run between the spinal cord and toes may be over 3 feet long. Axons also vary in diameter, and, as you will see later, this variation is related to the rate of impulse conduction. Axons contain one or more side branches called *axon collaterals.* The axon and its collaterals each terminate by branching into many fine filaments called *telodendria.*

Figure 12–2b shows a cross section of a nerve fiber. The axons of many neurons are surrounded by a white, lipid, segmented covering called the *myelin sheath.* The myelin sheath is formed by flattened cells, called *Schwann cells,* located along the nerve fiber. Schwann cells produce the myelin sheath around axons outside the central nervous system. Oligodendrocytes produce a similar sheath on some of the axons inside the central nervous system. Each Schwann cell produces a segment of the sheath by first encircling an area of the axon until the ends of the encircling cell meet and overlap (Figure 12–3). The cell then winds itself around the axon a second or third time. The cytoplasm of the Schwann cell is gradually squeezed out of the inside wrap-around layers and into the outside layer. The result is an axon surrounded by several double layers of cell membrane. Between the segments of sheath created by the Schwann cells are unmyelinated gaps called the *nodes of Ranvier.* (See Figure 12–2b.) An electron micrograph of the myelin sheath is shown in Figure 3–7c. The function of the myelin layer is to insulate the axon. Nerve fibers that are wrapped in the sheath are called *myelinated axons.* Those that are not protected by the sheath are called *unmyelinated axons.* The *neurilemma* is a delicate, continuous sheath that encloses the myelin of nerve fibers that lie outside the brain and spinal cord. The word *lemma* means husk or sheath. The essential function of the neurilemma is to assist in the regeneration of injured axons.

Classification

The many different kinds of neurons found in the body may be classified according to two criteria: structure and function. The structural classification is based upon the number of processes extending from the cell body. For example, *multipolar neurons* have several dendrites and one axon. Most neurons in the brain and spinal cord are of this type. *Bipolar neurons* have one dendrite and one axon and are found in the retina of the eye and in the inner ear. The third structural type of neuron is called a *unipolar neuron.* These have only one process extending from the cell body. The single process divides into a central branch, which functions as an axon, and a peripheral branch, which is an axon in structure but functions as a dendrite. Unipolar neurons originate in the embryo as bipolar ·neurons. During development, the axon and dendrite fuse into a single process.

The functional classification of neurons is based upon the direction in which they carry impulses. **Sensory neurons,** called **afferent neurons,** carry impulses from receptors in the skin and sense organs to the brain and spinal cord. Sensory neurons are usually unipolar. **Motor neurons,** called **efferent neurons,** convey impulses from the brain and spinal cord to effectors, which may be either muscles or glands. Other neurons, called **association** or **internuncial**

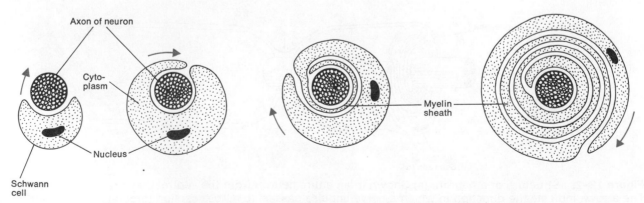

Axon of neuron

Cytoplasm

Nucleus

Schwann cell

Myelin sheath

Figure 12–3. Stages in the formation of the myelin sheath around an axon.

neurons, carry impulses from sensory neurons to motor neurons and are located inside the brain and spinal cord.

PHYSIOLOGY OF NERVOUS TISSUE

Two striking features of nervous tissue are (1) its very limited ability to regenerate, and (2) its highly developed activity to send electrical messages called nerve impulses.

Regeneration of nervous tissue

Unlike the cells of epithelial tissue, neurons have very limited powers for regeneration. Before or around the time of birth, the perikarya of most developing nerve cells lose their mitotic apparatus and thus their ability to reproduce. This means that if the cell body of a neuron is destroyed, the neuron cannot be replaced by other reproducing cells in the tissue. The function of the neuron is permanently lost. However, axons that have a neurilemma can be repaired, as long as the perikaryon is intact. Most nerves that lie outside of the brain and spinal cord consist of axons that are covered with a neurilemma. A person who injures a nerve in his arm has a good chance of regaining nerve function. Axons that lie within the brain and spinal cord do not have a neurilemma. This means that any injury to the brain or spinal cord is permanent.

The nerve impulse

Nerve cells have the ability to respond to stimuli and convert them into nerve impulses. This property is called **irritability.** Nerve cells also have **conductivity,** which is the ability to transmit the impulse to another neuron or to another tissue.

Irritability

You may recall from Chapter 3 that molecules or ions diffuse from areas where their concentration is high to areas where their concentration is low. Also recall that the cell can actively transport ions in the reverse direction—that is, from an area of low concentration to an area of high concentration. As you read the following discussion on nerve impulses, keep in mind that impulses revolve around one simple phenomenon that you already know—the passive and active transport of sodium and potassium ions.

When a nerve cell is supposedly resting or not being stimulated, it is nevertheless busy transporting ions across its membrane. Sodium ions (Na^+) do not easily diffuse into the cell, and the few ions that do make it across the cell membrane are actively transported back out of the cell. This is called the "sodium pump." Potassium ions (K^+) diffuse in and out of the cell quite easily, but the cell also has a "potassium pump" in operation. The pump actively transports potassium ions inward so that many more K^+ lie inside the cell than out. The neuron also contains a large number of negative ions that cannot diffuse to the outside or that diffuse very poorly. The net result is that a great many sodium ions are dumped around the outside of the cell. Since the sodium ions are positive, a positive charge exists on the outside of the cell membrane. A large number of potassium ions lie inside the cell. The potassium ions are also positive, but there are not enough of them to equalize the even larger numbers of negative ions that are trapped inside the cell. The inside of the cell membrane thereby has a negative charge. This difference in charge on either side of the membrane of a resting neuron is referred to as the *resting potential*, and such a membrane is said to be *polarized* (Figure 12–4).

If a stimulus of adequate strength is applied to the polarized membrane of a neuron (Figure 12–5a), the membrane's permeability to Na^+ ions at the point of stimulation is greatly increased, and the membrane's permeability to K^+ remains the same. This causes Na^+ to move into the cell by diffusion. The electrical difference across the membrane starts to decrease, going from the resting potential level down to zero. At zero point, the membrane is *depolarized*. However, even after depolarization is reached, the Na^+ ions continue to rush inside the membrane, and a membrane potential develops again. But this time, the inside is positively charged, and the outside is negatively charged. In other words, the influx of Na^+ ions is so great that a reversal in the membrane potential occurs in the stimulated area (Figure 12–5b).

Once these events have occurred, an action potential, or **nerve impulse,** is initiated. The stimulated negatively charged point on the outside of the membrane sends out an electrical current to the positive point (still polarized) adjacent to it (Figure 12–5c). This local current acts as a stimulus and causes the adjacent point on the membrane to reverse its potential from positive to negative. The reversal of polarization repeats itself over and over until the nerve impulse travels the length of the neuron.

By the time the impulse has traveled from one point on the membrane to the next, the previous point becomes *repolarized.* That is, its resting potential is restored. Repolarization results from a new series of changes in membrane permeability (Figure 12–5d). The membrane becomes permeable to K⁺ and impermeable to Na⁺. The outward movement of K⁺ ions causes the outer surface to become electrically positive, and the heavy loss of positive ions leaves the inner surface negative again. Finally, a recovery period occurs during which any ions that have moved into or out of the nerve cell are restored to their original site. This means that Na⁺ ions are actively transported outside, and K⁺ ions move back into the cell. The repolarization period returns the cell to its resting potential. The recovery period restores the original concentrations of Na⁺ and K⁺ inside and outside the cell. The neuron is now ready to receive another stimulus and to transmit it in the same manner.

THE ALL-OR-NONE PRINCIPLE. Any stimulus strong enough to initiate impulse transmission is referred to as a *threshold* or *liminal stimulus.* In this regard, a nerve cell transmits an impulse according to the **all-or-none principle.** The principle states that if a stimulus is strong enough to bring about an action potential, the impulse is transmitted along the entire neuron at a constant and maximum strength. The transmission is independent of any further intensity of the stimulus. Any stimulus weaker than a threshold stimulus is referred to as a *subthreshold stimulus.* Such a stimulus is incapable of initiating a nerve impulse. If, however, a second stimulus or a series of subthreshold stimuli is quickly applied to the neuron, the cumulative effect may be sufficient to initiate an impulse. This phenomenon is called *summation of inadequate stimuli.*

SPEED OF NERVE IMPULSES. The speed of a nerve impulse is also independent of the strength of the stimulus. It is normally determined by the size, type, and physiological condition of the fiber. Fibers with larger diameters transmit impulses faster than those with smaller diameters. Fibers with the greatest diameter are called *A fibers.* They are all myelinated, and they transmit impulses at speeds up to about 394 to 410 feet/second. A fibers are found in the larger sensory nerves that relay impulses about touch, pressure, position of joints, heat, and cold. They are also found in all the motor nerves that convey impulses to the skeletal muscles. Sensory A fibers generally connect the brain and spinal cord with sensors

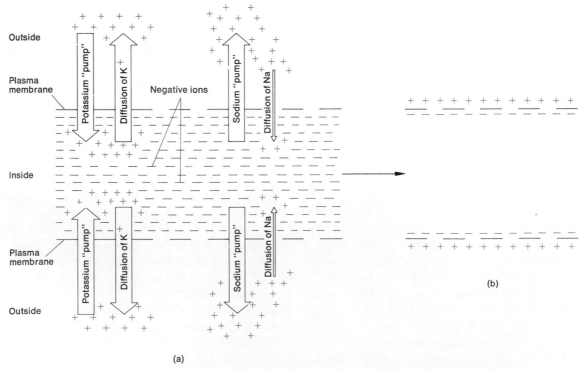

(a)

(b)

Figure 12–4. Development of the resting potential. (a) Schematic representation of the movement and distribution of sodium ions, potassium ions, and negative ions. Consult text for details. (b) Simplified representation of the polarized membrane.

that detect potentially dangerous situations in the outside environment. Motor A fibers innervate the muscles that can do something about the situation. For instance, if you pick up a hot frying pan, information about the heat passes over sensory A fibers to the spinal cord. There, it is relayed to motor A fibers that stimulate the muscles of the hand to drop the pan instantaneously. A fibers are located where split-second reaction may be important to survival. Other fibers, called B and C fibers, transmit impulses more slowly, and they are generally found where instantaneous response is not a life-and-death matter. *B fibers* have a middle-sized diameter. They are myelinated, and they transmit impulses at speeds of about 10 to 46 feet/second. B fibers make up some of the sensory nerves that transmit impulses for pain from the skin and viscera to the brain and cord. They also comprise all the motor nerves that leave the lower part of the brain and spinal cord and terminate in relay stations called ganglia. The ganglia ultimately link up with other fibers that stimulate the smooth muscle and glands of the viscera. *C fibers* have the smallest diameter and transmit impulses at the rate of about 2 to 8 feet/second. C fibers are unmyelinated and are located in the skin and visceral nerves. These fibers also make up some of the sensory nerves that transmit impulses for pain and perhaps some impulses for touch, pressure, heat, and cold. C fibers are found in all motor nerves that lead from the relay stations and stimulate the smooth muscle and glands of the viscera. Some examples of the operation of B and C fibers are constricting and dilating the pupils of the eyes, speeding up and slowing down heartbeat, and contracting and relaxing the urinary bladder.

Conductivity

Conductivity is the ability to transmit an impulse to another neuron or to another tissue, such as a muscle cell. The simplest conduction path is a reflex arc, which we shall look at now.

REFLEX ARCS. Once an impulse is initiated in the dendrites or cell body of a neuron, it moves through the cell body and then along the axon to the telodendria. At this point several alternatives exist. The impulse may be passed on to another neuron. Or, if the axon terminates in a muscle or gland, the impulse may stimulate contraction or secretion. The third possibility is that the impulse may not pass to a neuron, muscle, or gland. In such cases, the impulse is "lost" and produces no response.

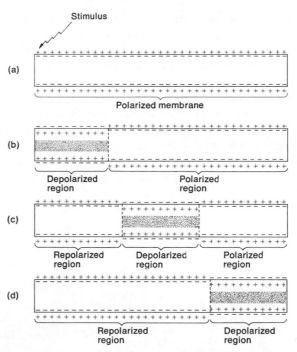

Figure 12–5. Initiation and transmission of a nerve impulse. The colored area represents the region of the membrane that has initiated and is transmitting the nerve impulse. The direction in which the impulse is moving is from left to right.

The route that an impulse takes from its origin in the dendrites or cell body to its termination somewhere else is called its *conduction pathway.* The most basic conduction pathway is the **reflex arc,** the functional unit of the nervous system. A reflex arc connects a receptor and an effector and consists of the following basic components: (1) The **receptor** is the distal end of a dendrite of a sensory neuron. Its role in the reflex arc is to respond to a change (stimulus) in the internal or external environment by producing a nerve impulse. (2) The **sensory neuron** passes the impulse generated by the receptor to the central nervous system. (3) The **center** is a region usually within the central nervous system. At this point, the incoming impulse may be blocked, transmitted, or rerouted. The majority of reflex arcs contain an *association neuron* between the sensory neuron and the neuron leading to a muscle or gland. However, in some reflex arcs, the sensory neuron delivers the impulse directly to the dendrites of the motor neuron. (4) The **motor neuron** transmits the impulse from the sensory or association neuron in the center to the organ that will respond. (5) The **effector** is the organ of the body, either a muscle or a gland, that responds to the motor neuron impulse. It should be noted that the axons of one neuron in the pathway never quite touch the dendrites of the next neuron in the pathway. Rather, the impulse must travel across a minute gap called a synapse.

Let us consider the operation of a *two-neuron reflex arc*. It is also called a *monosynaptic* arc because there is only one synapse in the pathway. The patellar reflex is one example (Figure 12–6a). When a quick tap (stimulus) is applied to the tendon below the patella, the dendrites of the sensory neurons ending in the muscle and tendon attached to the patella receive the stimulus and convert it into a nerve impulse. The impulse is transmitted along the sensory neuron toward the spinal cord. Inside the cord, the impulse passes from the axon of the sensory neuron, across the synapse, to the dendrites or cell body of the motor neuron. From here, the impulse is conveyed along the motor axon to the muscle on the front of the thigh (effector). When the impulse reaches the muscle, it contracts, and a jerking response occurs. This is the reflex that the physician is checking when he taps your knee.

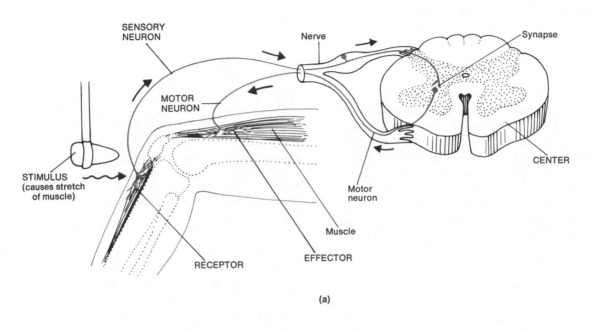

(a)

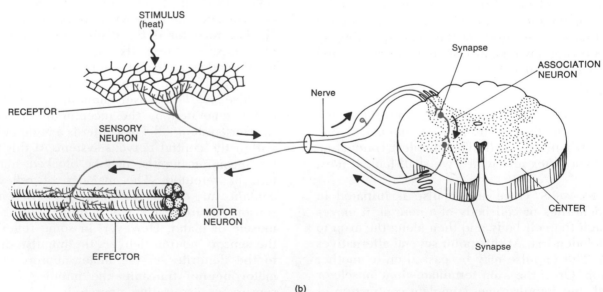

(b)

Figure 12–6. Reflex arcs. (a) Two-neuron (monosynaptic) reflex arc such as the patellar reflex. (b) Three-neuron (bisynaptic) reflex arc as exemplified by the withdrawal of the hand from a hot object. Note especially the essential components of the reflex arc.

Very few reflex arcs, however, involve only two neurons. Most also involve one or more association neurons. A common example of a *three-neuron reflex arc*, also called a *bisynaptic arc*, is the withdrawal of the hand from a hot object. In this case, the impulse must travel over two synapses from receptor to effector. Since the same principles apply to the three-neuron reflex arc and the two-neuron reflex arc, the details will not be discussed. You may refer to Figure 12–6b. It should be noted, however, that when the hand is stimulated by heat, other responses also occur. For example, the person might look in the direction of the stimulus, exhibit facial and vocal expressions of pain, and probably experience a series of emotions such as fear, apprehension, or anger. All these responses indicate that a single stimulus can elicit a complex response involving many components of the nervous system. This is explained by the fact that axons of sensory neurons often have a number of branches, each of which may synapse with other association neurons. Thus, the sensory neuron for pain in the hand can relay impulses to motor neurons that stimulate neck and facial muscles and to neurons in the brain which are involved in feeling and emotion.

A **reflex** is a quick, involuntary response to a stimulus transmitted over a reflex arc. Reflexes are conducted over relatively simple reflex arcs. That is, they do not involve the association neurons of the brain that are responsible for thought and decision making.

CONDUCTION ACROSS SYNAPSES. The axons of a neuron never quite touch the dendrites of the next neuron in a conduction pathway. In order for an impulse to move from one neuron to the next, it must bridge a small gap called a **synapse.** The term *synapsis* means a connection.

Note in Figure 12–7 that the telodendria at the terminal end of an axon contain rounded or oval expansions referred to as *synaptic knobs* or *end feet.* Electron micrograph studies reveal that the end feet contain numerous granular structures called *synaptic vesicles.* The vesicles, in turn, contain any one of a number of chemical transmitter substances. The nature of the substance is dependent upon the location of the synapse in the nervous system. When the impulse arrives at the end of the axon, it causes the synaptic vesicles to rupture and release the chemical transmitter. The chemical diffuses from the end feet, into the synapse, and makes contact with the dendritic portion or perikaryon of the next neuron. Such a neuron may be called a *postsynaptic neuron* since it is located after the synapse. When the chemical transmitter reaches the postsynaptic neuron, it stimulates the membrane of the neuron to become more permeable to Na^+. If enough sodium ions flow inward, the impulse is transmitted along the postsynaptic neuron.

CHEMICAL TRANSMITTERS. The chemical transmitters are responsible for several important properties of the nervous system. When a nerve cell is stimulated, the impulse may spread back into other dendrites at the same time as it is moving toward the axon. However, only axons liberate the transmitter substance. This means that there is only one-way impulse conduction at a synapse (axon → synapse → dendrite). Im-

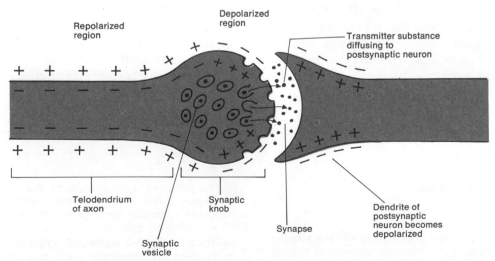

Figure 12–7. Diagrammatic representation of impulse conduction from the axon of a presynaptic neuron to the dendrite of a postsynaptic neuron.

pulses must move forward over their pathway; they cannot back up into another presynaptic neuron. Such a mechanism is very important in preventing impulse conduction along wrong pathways. Imagine the result if impulses transmitted along the motor neurons that move your hand could move back and stimulate the sensory neuron which relays information about heat. You would feel heat, cry in pain, and go through all the emotions of being burned every time you simply wanted to move your hand.

Chemical transmitters are responsible for the highly selective nature and appropriateness of responses in still other ways. The best-known transmitter at synapses is **acetylcholine,** the chemical that also stimulates skeletal muscle contraction. Acetylcholine can stimulate a postsynaptic neuron almost indefinitely. What, then, stops the stimulation and allows us to relax a contracting muscle? The answer is a chemical called *cholinesterase,* which is also released at a synapse. Cholinesterase destroys acetylcholine. In a normal person, cholinesterase inactivates all of the acetylcholine by the time the postsynaptic neuron is returned to its resting state and ready to transmit a second impulse. If we desire to continue contracting a muscle, a second impulse must flow over the presynaptic neuron and more acetylcholine must be released at the synapse.

The body has other types of chemical transmitters besides acetylcholine. One type includes the inhibitory transmitters. Inhibitory transmitters are their own worst enemies, but at the same time they protect the body in an important way. For example, some presynaptic neurons release an inhibitory transmitter that does two things to the postsynaptic neuron: (1) it stimulates the neuron, and (2) it makes the membrane of the neuron more permeable to the inward flow of negative ions. As a result, whenever an impulse tries to cross the synapse, the inside of the postsynaptic neuron at first becomes more negative than a normal resting neuron. More Na^+ must flow into the stimulated neuron in order to depolarize it and generate an impulse. If enough Na^+ flows inward, the impulse will be conducted across the synapse. If insufficient Na^+ flows inward, the impulse will stop at the synapse and never reach the brain or spinal cord. Postsynaptic neurons that synapse with inhibitory neurons also synapse with the more common noninhibitory type of neuron. Impulses coming in over the noninhibitory neuron cross the synapse, but those coming in over the inhibitory neurons pass the synapse only with difficulty. This means that the noninhibitory neuron impulses always have the right of way.

Inhibitory transmitters are important in maintaining homeostasis. It is believed that impulses which are important to life and health run over the noninhibitory neurons. Less essential impulses run over the inhibitory neurons. The nervous system is thus able to give priority to the more important situations and avoid responding in two contradictory ways.

OTHER FACTORS THAT AFFECT IMPULSE TRANSMISSION. Impulse transmission and conduction may also be inhibited or blocked by a decrease in temperature, by pressure, and by certain drugs. For example, chemical transmitters become inactive at temperatures of 32°F or lower. If excessive or prolonged pressure is applied to a nerve, impulse transmission is interrupted, and a part of the body may go to sleep. Removal of the pressure results in the "prickly" feeling when impulse transmission is reestablished. Among the drugs that may alter impulse transmission are anesthetics. *Anesthetics* are chemical substances that produce partial or complete loss of sensation. They probably act by altering the permeability of the nerve cell membrane so that it is unable to generate an impulse.

Chapter summary in outline

ORGANIZATION

1. The nervous system controls and integrates all body activities by sensing changes, interpreting them, and reacting to them.

2. The central nervous system consists of the brain and spinal cord.

3. The peripheral nervous system consists of the somatic nervous system, which innervates skeletal muscles, and the autonomic nervous system, which innervates cardiac muscle, smooth muscle, and glands.

HISTOLOGY OF THE NERVOUS SYSTEM

1. Neuroglia are connective tissue cells that support, attach neurons to blood vessels, produce the myelin sheath, and carry out phagocytosis.

2. Neurons, or nerve cells, consist of a perikaryon, which is the cell body, dendrites, which pick up stimuli and convey them to the perikaryon, and usually a single axon, which transmits impulses from the perikaryon to the dendrites or perikaryon of another neuron or to an organ of the body.

3. On the basis of structure, neurons are multipolar, bipolar, and unipolar.

4. On the basis of function, sensory (afferent) neurons transmit impulses to the central nervous system; association neurons transmit impulses to other neurons including motor ones; and motor (efferent) neurons transmit impulses to effectors.

PHYSIOLOGY OF NERVOUS TISSUE
Regeneration of nervous tissue

1. Around the time of birth, the perikaryon loses its mitotic apparatus and is no longer able to divide.

2. Axons that have a neurilemma are capable of regeneration.
Nerve impulse
IRRITABILITY

1. Irritability is the ability of neurons to respond to stimuli and convert them into impulses.

2. The membrane of a neuron is positive on the outside and negative on the inside. This difference in charge is called a resting potential, and the membrane is said to be polarized.

3. When a stimulus causes the inside to become positive and the outside negative, the membrane is said to have an action potential, which travels from point to point along the membrane. The traveling action potential is a nerve impulse.

4. A stimulus strong enough to initiate impulse transmission is called a threshold stimulus. It travels at a constant and maximum rate, which is called the all-or-none principle.

5. Several subthreshold stimuli may together initiate impulse transmission (summation of inadequate stimuli).

6. A fibers have a large diameter. B fibers have an intermediate-sized diameter, and C fibers have a small diameter. The larger the diameter, the faster the impulse is transmitted.
CONDUCTIVITY

1. Conductivity is the ability of neurons to transmit an impulse to another neuron or to a muscle or gland.

2. A reflex arc is the shortest route that can be taken by an impulse from a receptor to an effector.

3. A two-neuron reflex arc contains one sensory and one motor neuron; the patellar reflex is an example.

4. A three-neuron reflex arc contains a sensory, association, and motor neuron; an example is pulling the hand away from a hot object.

5. A reflex is a quick, involuntary response to a stimulus that travels over a reflex arc.

6. Conduction involves transmission of an impulse across a synapse. In many instances, acetylcholine transmits the impulse across the synapse. Further conduction is prevented by cholinesterase, which inactivates acetylcholine.

7. Inhibitory chemical transmitters are released at some synapses. Both excitatory and inhibitory neurons function in the nervous system so that the brain can select important impulses for appropriate action.

8. Impulse conduction and transmission may be inhibited or blocked by temperature changes, pressure changes, and certain drugs, such as anesthetics.

Review questions and problems

1. How does the nervous system maintain homeostasis? Distinguish between the central and peripheral nervous systems. Relate the terms voluntary and involuntary to the nervous system.

2. What are neuroglia? List their principal functions.

3. Define a neuron. Diagram and label a neuron, and next to each part list its function.

4. Discuss how neurons are classified on the basis of structure and function.

5. What factors determine neuron regeneration?

6. Define irritability and conductivity.

7. Outline the principal steps involved in the origin and transmission of a nerve impulse. What determines the speed of a nerve impulse?

8. Define resting potential, polarized membrane, action potential, depolarized membrane, and repolarized membrane.

9. What is the all-or-none principle? Relate the terms threshold stimulus, subthreshold stimulus, and summation of inadequate stimuli to the principle.

10. Define a reflex arc. What are the components of a reflex arc? Distinguish between a monosynaptic and bisynaptic reflex arc.

11. What events are involved in the transmission of a nerve impulse across a synapse?

12. How are nerve impulses inhibited? What advantage is this to the body?

13. Support the statement: "The nerve impulse is the body's best means for rapid correction of a deviation that tends to disrupt homeostasis."

CHAPTER 13

THE NERVOUS SYSTEM: STRUCTURE, PHYSIOLOGY, AND DISORDERS

STUDENT OBJECTIVES

After you have read this chapter, you should be able to:

1. Define white matter, gray matter, nerves, ganglia, tracts, and nuclei

2. Identify the three principal parts of the brain

3. Identify the structural features of the cerebrum

4. Compare the motor, association, and sensory functions of the cerebrum

5. Describe the principle of an electroencephalograph and its significance in the diagnosis of certain disorders

6. Identify the anatomical characteristics and functions of the cerebellum

7. Compare the components of the brain stem with regard to structure and function

8. Describe the principal structural features of the spinal cord

9. Identify the conducting and reflex activities of the spinal cord

10. Contrast the functions of ascending and descending tracts of the spinal cord

11. Define spinal cord injury and list the immediate and long-range effects

12. Identify the factors responsible for the maintenance and protection of the central nervous system

13. Describe the formation and circulation of cerebrospinal fluid

14. Define the anatomical subdivisions of the somatic portion of the peripheral nervous system

15. Identify by number and name the 12 pairs of cranial nerves

16. List the 12 cranial nerve pairs with respect to type, origin, distribution, and function

17. List the distribution of the 31 pairs of spinal nerves

18. Describe the structure of a typical spinal nerve

19. Define a plexus

20. Note the name, composition, and functions of the principal plexuses

21. Describe the conditions necessary for peripheral nerve regeneration

22. Compare the structural and functional differences between the somatic and autonomic portions of the nervous system

23. Identify the structural features of the autonomic nervous system

24. Compare the sympathetic and parasympathetic divisions of the autonomic nervous system in terms of structure, physiology, and chemical transmitters released

25. Identify the relationship between reflexes and the maintenance of homeostasis

26. Classify reflexes on the basis of organs stimulated and location of receptors

27. List several clinically important reflexes

28. List the clinical symptoms of disorders of the nervous system, including poliomyelitis, syphilis, cerebral palsy, Parkinsonism, epilepsy, multiple sclerosis, cerebral vascular accidents, and tumors

29. Define medical terminology associated with the nervous system

The neurons of the body are collected together to form a highly organized neural tissue. Neural and connective tissues together make up the major organs of the system: the brain, the spinal cord, and the somatic and autonomic structures running between the central nervous system and the other parts of the body.

THE GROUPING OF NEURAL TISSUE

Neurons are not strung out in a maze of separate, criss-crossing, tangled fibers. Rather, the axons of a group of neurons are usually found neatly bundled together and headed in the same direction. The cell bodies and dendrites of such a bundle of neurons are usually collected together in specific areas. The term **white matter** refers to aggregations of myelinated axons from many neurons. The fatty substance, myelin, has a whitish color that gives white matter its name. The gray colored areas of the nervous system are called, obviously enough, **gray matter.** They contain either nerve cell bodies and dendrites or bundles of unmyelinated axons.

A **nerve** is a bundle of fibers located outside the central nervous system. Since the axons of most of the neurons of the peripheral nervous system are myelinated, most nerves are white matter. Nerve cell bodies that lie outside the central nervous system are generally grouped together with other nerve cell bodies to form **ganglia.** The term *ganglion* means knot. Ganglia, of course, are masses of gray matter.

A **tract** is a bundle of fibers located inside the central nervous system. Tracts may run long distances up and down the spinal cord. Short tracts exist inside the brain and connect parts of the brain with each other and with the spinal cord. The chief tracts that conduct impulses up the cord are concerned with sensory impulses and are called *ascending tracts.* By contrast, bundles of fibers that carry impulses down the cord are motor tracts and are called *descending tracts.* The major tracts consist of myelinated fibers, and they are therefore white matter. A **nucleus** is a mass of nerve cell bodies and dendrites in-

side the brain. It consists of gray matter. **Horns** are the chief areas of gray matter in the spinal cord.

THE CENTRAL NERVOUS SYSTEM

For purposes of study and discussion, the white matter, gray matter, ganglia, tracts, and nuclei of the body's nervous system are categorized into two major divisions: the central nervous system, abbreviated CNS, and the peripheral nervous system, abbreviated PNS. These two major divisions each have further divisions and areas that will be covered in the following sections. In this section, we shall view the two main parts of the central nervous system: the brain and the spinal cord. Both the brain and spinal cord will be examined in terms of their structure and physiology. We shall also take a look at how the body protects and maintains these organs of the central nervous system. As we discuss the brain, you will want to keep in mind the structural and functional relationships that exist among its various parts.

The brain

The **brain** of an average adult is one of the largest organs of the body, weighing about 3 pounds. Look at Figure 13–1: you will see that the brain resembles a rather stocky mushroom. It is divided into three principal areas: the cerebrum, cerebellum, and brain stem. The brain stem, the stalk of the mushroom, consists of the medulla oblongata, pons, midbrain, thalamus, and hypothalamus. The lower end of the brain stem is a continuation of the spinal cord. The upper end supports the cap of the mushroom, the large cerebrum. The cerebrum constitutes about seven-eights of the total weight of the brain and occupies most of the skull. Below the cerebrum is the cerebellum.

The cerebrum

Supported on the brain stem and forming the bulk of the brain is the **cerebrum** (Figure 13–1).

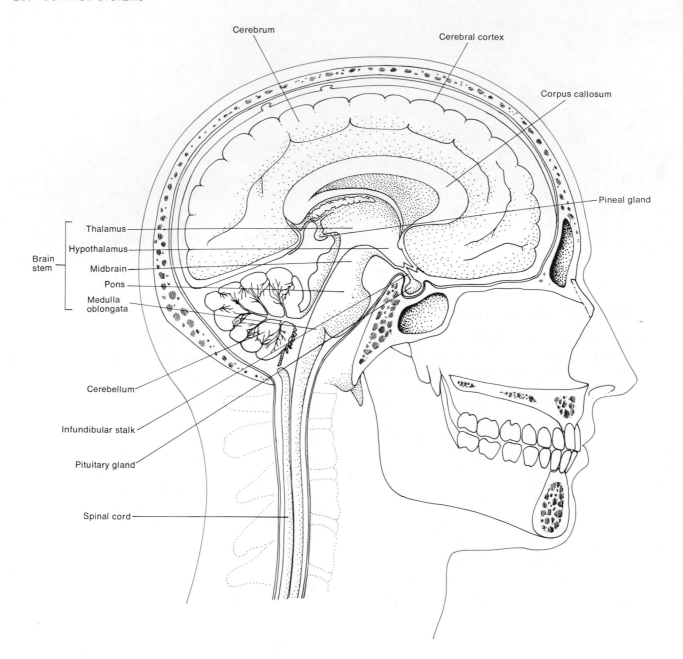

Cerebrum

Cerebral cortex

Corpus callosum

Pineal gland

Thalamus

Hypothalamus

Brain
stem

Midbrain

Pons

Medulla
oblongata

Cerebellum

Infundibular stalk

Pituitary gland

Spinal cord

Figure 13–1. Principal portions of the brain as seen in sagittal section.

The surface of the cerebrum is composed of gray matter (2 to 4 millimeters thick) and is referred to as the *cerebral cortex*. The word *cortex* means rind or bark. The cortex, containing millions and millions of cells, consists of six layers of nerve cell bodies. Underneath the cortex lies the cerebral white matter.

During embryonic development when there is a rapid increase in brain size, the gray matter of the cortex enlarges out of all proportion to the underlying white matter. As a result, the cortical region rolls and folds upon itself. The upfolds are called *gyri* or *convolutions* (Figure 13–2a).

The deep downfolds are referred to as *fissures*, whereas the shallow downfolds are termed *sulci*. The most prominent fissure, the *longitudinal fissure*, almost completely separates the cerebrum into right and left halves, or *hemispheres* (Figure 13–2b). The hemispheres, however, are connected internally by a large bundle of transverse fibers composed of white matter and called the *corpus callosum* (Figure 13–1).

Each cerebral hemisphere is further subdivided into four lobes by other sulci or fissures (Figure 13–2). The *central sulcus*, or *fissure of Rolando*, separates the *frontal lobe* from the

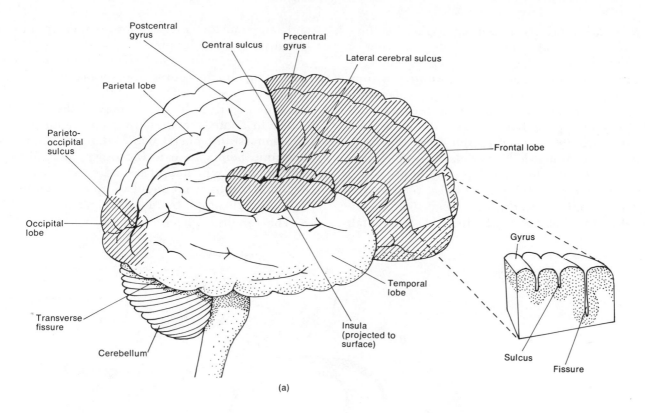

(a)

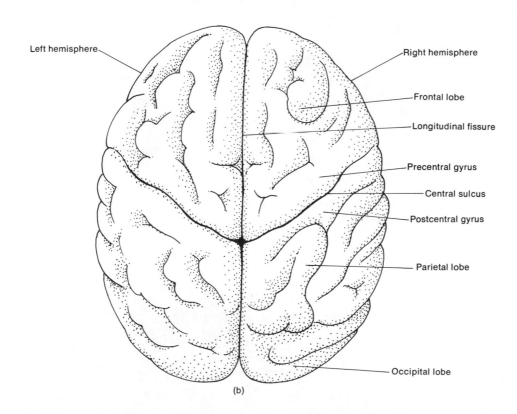

(b)

Figure 13–2.   Lobes and fissures of the cerebrum. (a) Right lateral view.
(b) Superior view. Since the insula cannot be seen externally, it has been projected
to the surface. The insert in (a) indicates the relative differences between a gyrus,
sulcus, and fissure.

parietal lobe. A very important gyrus, the precentral gyrus, is located immediately in front of the central sulcus. The *lateral cerebral sulcus,* or *fissure of Sylvius,* separates the *frontal lobe* from the *temporal lobe.* The *parietooccipital sulcus* separates the *parietal lobe* from the *occipital lobe.* Another prominent fissure, the *transverse fissure,* separates the cerebrum from the cerebellum. The frontal lobe, parietal lobe, temporal lobe, and occipital lobe are named after the bones that cover them. A fifth part of the cerebrum, the *insula (island of Reil),* lies deep within the lateral cerebral fissure, under the parietal, frontal,

and temporal lobes. It cannot be seen in an external view of the brain.

The white matter underlying the cortex consists of myelinated nerve fibers arranged in three principal directions: (1) *association fibers,* which transmit impulses from one part of the cerebral cortex to another within the same hemisphere; (2) *commissural fibers,* which transmit impulses from one hemisphere to the other; and (3) *projection fibers,* which form ascending and descending tracts that transmit impulses from the cerebrum to other parts of the brain and spinal cord.

The *cerebral nuclei* are paired masses of gray

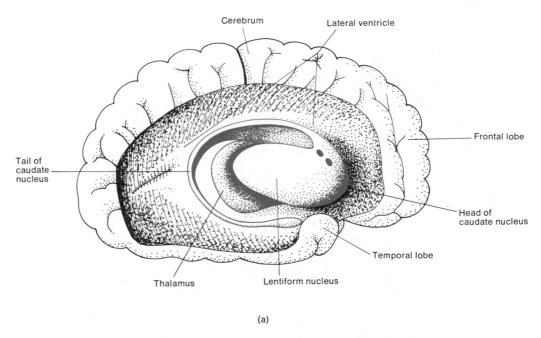

(a)

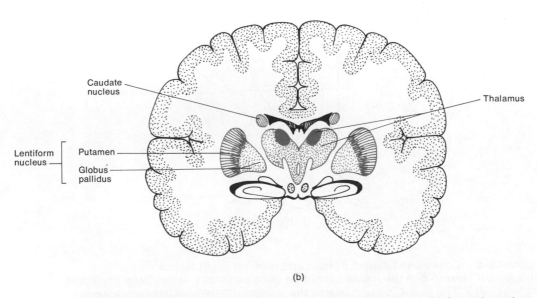

(b)

Figure 13–3. Cerebral nuclei viewed in (a) sagittal section and (b) frontal section.

matter located deep inside the white matter of the cerebral hemispheres. Two of the more prominent cerebral nuclei are the *caudate nucleus* and the *lentiform nucleus* (Figure 13–3). The lentiform nucleus consists of two portions, the *putamen* and the *globus pallidus*. The cerebral nuclei are interconnected by many fibers. In addition, they are connected to the cerebral cortex and to the thalamus and hypothalamus of the brain stem. The caudate nucleus and the putamen control large unconscious movements of the skeletal muscles. An example of this is swinging the arms while walking. Such gross movements are also consciously controlled by the cerebral cortex. The globus pallidus is concerned with the regulation of muscle tone required for specific body movements. For example, if you wished to perform a very specific function with one of your hands, you might first position your body appropriately and then tense the muscles of the upper arm. Damage to the nuclei results in abnormal body movements, such as uncontrollable shaking, called tremors, and involuntary movements of skeletal muscle. Moreover, destruction of a substantial portion of the caudate nucleus almost totally paralyzes the opposite part of the body. It will be seen shortly that damage to the left side of the brain affects the right side of the body and vice versa.

The functions of the cerebrum are numerous and complex. In a very general way, the cerebral cortex is divided into motor, sensory, and association areas. The *motor areas* are regions that govern muscular movement. The *sensory areas* are concerned with the interpretation of sensory impulses. And the *association areas* are concerned with emotional and intellectual processes.

The three areas of the cortex are subdivided into the small, numbered areas shown in Figure 13–4. Each of these small areas is responsible for a particular function, such as speech or smell. If one of the areas is damaged, the person usually has a great deal of trouble performing the function for which it is responsible. Or, he may not be able to perform the function at all. The control of a function is not, however, limited to a single center in the cortex. All portions of the cerebrum are interconnected, and every cerebral activity results from the interaction of many parts.

The *primary motor area* of the cortex is located mostly in the precentral gyrus of the frontal lobe, that is, in the gray matter immediately in front of the central sulcus (Figure 13–4a). This region, also called area 4, transmits voluntary nervous impulses to skeletal muscles. Area 6, a gyrus im-

mediately anterior to area 4, transmits impulses concerned with more complex motor activities, such as postural movements and skilled movements. Area 8, just anterior to area 6, is primarily concerned with the control of eye movements. Area 44, *Broca's speech area*, is inferior to the primary motor area. This area, as well as regions in the frontal, temporal, and parietal lobes, is concerned with speaking and writing. The motor speech area is usually located in the left hemisphere of right-handed individuals and in the right hemisphere of left-handed individuals. Injury to the speech areas may result in *aphasia*, which is an inability to speak; *agraphia*, an inability to write; *word deafness*, an inability to understand spoken words; or *word blindness*, an inability to understand written words.

Sensory areas are located behind the central sulcus (Figure 13–4a). The *general* or *somesthetic sensory area* is located mainly in the parietal lobe on the postcentral gyrus (areas 3, 1, and 2). The functions of this and adjoining areas are (1) detecting the position of the body in space (proprioception); (2) recognizing the size, shape, weight, temperature, and texture of objects; (3) comparing stimuli as to intensity and location; and (4) evaluating and integrating stimuli. The *primary auditory area* in the superior part of the temporal lobe (areas 41, 42) receives impulses from the ears. The *primary visual area* in the occipital lobe (area 17) receives impulses from the eyes. The *primary olfactory area* in the temporal lobe is the area concerned with the sense of smell. And the *primary gustatory area* (area 43) at the base of the postcentral gyrus is concerned with the sensation of taste.

The *association areas* of the cortex are made up of association fibers that connect motor and sensory areas (Figure 13–4b). These areas are also concerned with memory, emotions, reasoning, will, judgment, personality traits, and intelligence. The association region of the cortex occupies the greater portion of the lateral surfaces of the occipital, parietal, and temporal lobes and the frontal lobes in front of the motor areas.

Brain waves: electroencephalogram

Brain cells have the capacity to generate electrical potentials called brain waves, which indicate activity of the cerebral cortex. Brain waves pass easily through the skull, and they can be detected by sensors called electrodes. A record of such waves is called an **electroencephalogram,** or **EEG.** An EEG is obtained by placing electrodes on the

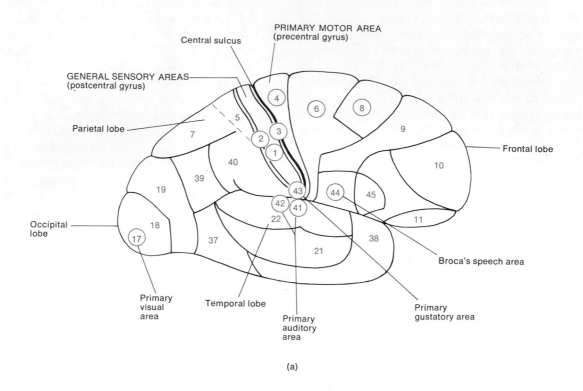

(a)

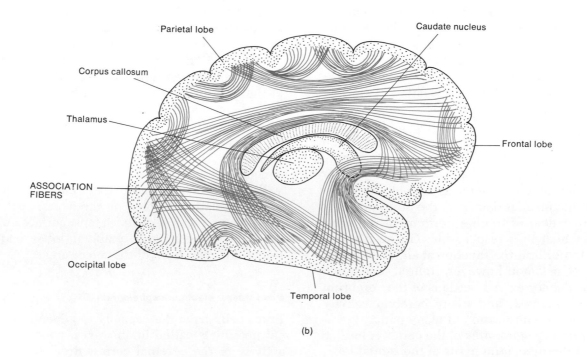

(b)

Figure 13–4. Functional areas of the cerebrum. The (a) lateral view indicates the sensory and motor areas. The (b) sagittal section shows the association areas.

head of an individual and amplifying the waves by using an instrument called an electroencephalograph. As indicated in Figure 13–5, four kinds of waves are produced by normal individuals:

1. *Alpha waves.* These rhythmic waves occur at a frequency of about 8 to 13 cycles/second. They are found in the EEG's of almost all normal individuals when they are awake and in the resting state. These waves disappear entirely during sleep.
2. *Beta waves.* The frequency of these waves is more than 14 cycles/second and may occur as high as 50 cycles. Beta waves generally appear when the nervous system is active.
3. *Theta waves.* These waves have frequencies of 4 to 7 cycles/second. They are normal in children but also occur in EEG's of adults who are undergoing a great deal of emotional stress.
4. *Delta waves.* The frequency of these waves is below 3½ cycles/second and may be as low as 1 cycle every 2 to 3 seconds. Delta waves appear during deep sleep. They are normal in an awake infant. But when they are produced by an awake adult, they indicate certain types of brain damage.

Distinct EEG patterns appear in certain abnormalities. In fact, the EEG is used clinically in the diagnosis of epilepsy, infectious diseases, tumors, trauma, and hematomas. Some of these are discussed later in this chapter.

The cerebellum

The **cerebellum** is the second largest portion of the brain and occupies the inferior and posterior aspects of the cranial cavity. Specifically, it is below the posterior portion of the cerebrum and is separated from it by the *transverse fissure*. (See Figure 13–2.) The cerebellum is shaped somewhat like a butterfly. The central constricted area is called the *vermis*, which means worm-shaped, and the lateral "wings" are referred to as *hemispheres* (Figure 13–6).

The surface of the cerebellum consists of gray matter that is thrown into a series of slender, parallel *sulci*. These sulci are less prominent than those seen on the cerebral cortex. Beneath the gray matter are white matter tracts (*arbor vitae*) that resemble branches of a tree. Deep within the white matter are masses of gray matter, the *cerebellar nuclei*.

The cerebellum is attached to the brain stem by three paired bundles of fibers called *cerebellar peduncles*. These are as follows:

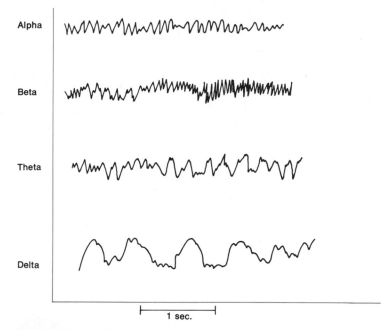

Figure 13–5. Kinds of waves recorded in an electroencephalogram.

1. *Inferior cerebellar peduncles,* which connect the cerebellum with the medulla at the base of the brain stem and with the spinal cord
2. *Middle cerebellar peduncles,* which connect the cerebellum with part of the brain stem called the pons
3. *Superior cerebellar peduncles,* which connect the cerebellum with the midbrain, also a part of the brain stem

The cerebellum is a motor area of the brain that produces certain unconscious movements in the skeletal muscles. These movements are required for coordination, for maintenance of posture, and for keeping the body balanced on its center of gravity. The cerebellar peduncles are the telephone lines that allow the cerebellum to carry out its functions.

Let us now see how the cerebellum produces coordinated movement. Motor areas of the cerebral cortex voluntarily initiate muscle contraction. Once the movement has begun, the sensory areas of the cortex receive impulses from nerves in the joints. The impulses provide information about the extent of muscle contraction and the amount of joint movement. The term *proprioception* is applied to this sense of the position of one body part relative to another. The cortex uses the proprioceptive sensations to decide what muscles need to contract next and how much contraction is needed in order to continue moving in the desired direction. The cortex sends its

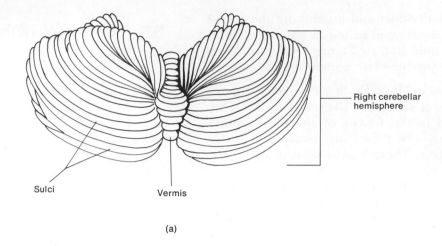

Right cerebellar
hemisphere

Sulci

Vermis

(a)

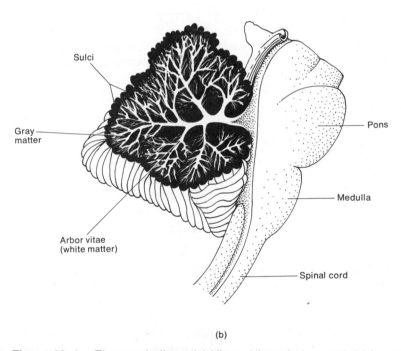

Sulci

Gray
matter

Arbor vitae
(white matter)

Pons

Medulla

Spinal cord

(b)

Figure 13–6. The cerebellum. (a) Viewed from below and (b) in sagittal section.

decision over tracts to the pons and midbrain, which relay the impulses over the middle and superior cerebellar peduncles to the cerebellum. The cerebellum is now ready to act. It sends unconscious motor impulses over the inferior cerebellar peduncles to the medulla and spinal cord. The impulses pass out of the spinal cord and over nerves that stimulate the prime movers and synergists to contract and that interfere with the contraction of the antagonists. The result is smooth, coordinated movement. A well-functioning cerebellum is particularly essential for skilled, delicate movements such as playing the piano or threading a needle.

The cerebellum also transmits impulses that control postural muscles. In other words, the cerebellum is required for maintaining normal muscle tone. Finally, the cerebellum maintains body equilibrium. The inner ear contains structures that sense balance. Information such as whether the body is leaning to the left or to the right or about to tip over frontward is probably transmitted from the inner ear to the cerebellum. The cerebellum then discharges impulses that bring about the contraction of the muscles necessary for maintaining equilibrium.

Damage to the cerebellum through trauma or disease is characterized by certain symptoms involving skeletal muscles. For example, there may be muscle uncoordination, called *ataxia.* A sign of ataxia is the inability to accurately touch an indicated point on the body without looking. For instance, people with ataxia cannot touch the tip of their nose with their finger because they cannot coordinate movement with their sense of where a body part is located. Another sign of ataxia is a change in the speech pattern due to uncoordination of speech muscles. Cerebellar damage may also result in disturbances of gait in which the subject staggers or cannot coordinate normal walking movements.

The brain stem

The **brain stem,** the third principal portion of the brain, consists of the medulla oblongata, pons, midbrain, thalamus, and hypothalamus.

MEDULLA OBLONGATA. The **medulla oblongata** is a continuation of the upper portion of the spinal cord and forms the most inferior portion of the brain stem. (See Figure 13–1.) It lies just above the level of the foramen magnum and extends upward to the lower portion of the pons. The medulla is composed of white fiber tracts. Among these tracts are scattered nuclei that serve as controlling centers for various activities or that contain the cell bodies of cranial nerves. Cranial nerves are nerve fibers that connect the brain with many of the muscles and glands of the body. They are part of the peripheral nervous system and will be explained more fully later.

On the ventral side of the medulla are two roughly triangular structures called the *pyramids.* The pyramids are composed of the largest motor tracts that run from the cortex to the spinal cord. At the junction of the medulla with the cord, most of the fibers in the left pyramid cross to the right side, and most of the fibers in the right pyramid cross to the left. The crossing is referred to as the *decussation* (crossing) *of the pyramids.* The principal fibers that undergo decussation belong to the lateral corticospinal tracts, so named because they originate in the cortex and run down the lateral sides of the spinal cord. Fibers in the left lateral tract of the spinal cord synapse with spinal nerves that stimulate skeletal muscles on the left side of the body. Likewise, the right lateral tract relays motor impulses that go to the right side of the body. The phenomenon of decussation explains why the motor areas of the right cerebral cortex control voluntary movement in the left side of the body, and the left

cerebral cortex controls voluntary movement on the right side (Figure 13–7).

On the dorsal surface of the medulla are two prominent nuclei, the nucleus gracilis and nucleus cuneatus. These nuclei receive sensory impulses from some ascending tracts of the cord and relay them to the opposite side of the medulla, from which they pass to the sensory areas of the cortex. Nearly all sensory impulses received on one side of the body are registered on the opposite side of the brain.

In addition to its function as a conduction pathway for sensory and motor impulses between the brain and cord, the medulla also contains the nuclei, that is, the cell bodies and dendrites, of four pairs of cranial nerves. Three of the pairs exit on the lateral surfaces of the medulla (Figure 13–7a). These are the glossopharyngeal nerve (IX), which relays impulses related to swallowing, salivation, and taste; the vagus (X), which relays impulses related to the functioning of many thoracic and abdominal viscera and also controls voluntary muscle; and the accessory (XI), which conveys impulses related to head and shoulder movements. The hypoglossal (XII) pair, which controls tongue movements, emerges more ventrally.

Three vital reflex centers are also located in the nuclei of the medulla. These are the *cardiac center,* which regulates heart beat, the *respiratory center,* which adjusts the rate and depth of breathing, and the *vasoconstrictor center,* which regulates the diameter of the blood vessels. Other centers in the medulla mediate the swallowing, vomiting, coughing, sneezing, hiccoughing, and blinking reflexes.

In view of the many vital activities controlled by the medulla, it is not surprising that a hard blow to the base of the skull can be fatal. Nonfatal medullary injury may be indicated by cranial nerve malfunctions on the same side of the body as the area of medullary injury (cranial nerves do not decussate), paralysis and loss of sensation on the opposite side of the body, and irregularities in respiratory control.

PONS. The **pons,** which means bridge, lies directly above the medulla and anterior to the cerebellum. (See Figure 13–1.) As its name implies, one of the chief functions of the pons is to serve as a bridge connecting the spinal cord with the brain and parts of the brain with each other. Like the medulla, the pons consists of interlaced white fibers with nuclei scattered throughout. The transverse fibers carry information about skeletal

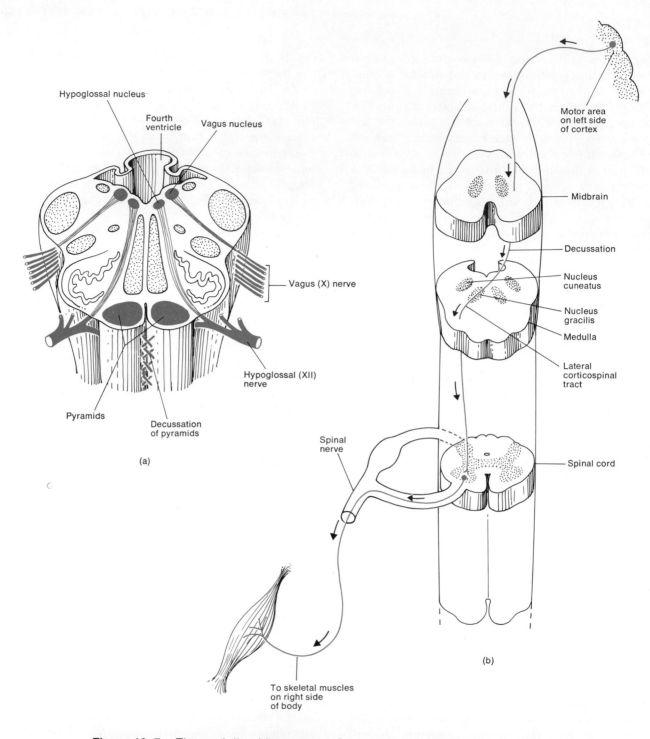

Figure 13-7. The medulla oblongata. (a) Cross section indicating location of pyramids, decussation of pyramids, nuclei, and exit of cranial nerves X and XII. (b) Path of lateral corticospinal tract from the motor cortex on the left side of the brain to skeletal muscles on the right side of the body.

muscle movement from the cortex and relay the impulses over the middle peduncles of the cerebellum to the opposite sides of the cerebellum. Some of the longitudinal fibers of the pons belong to the corticospinal tract that relays motor impulses from the cortex to the spinal cord. The remainder of the longitudinal fibers belong to motor and sensory tracts that connect the spinal cord or medulla with the upper parts of the brain stem.

The nuclei for certain cranial nerves are also contained in the pons. These include the trigeminal nerve (V), which relays impulses for chewing and for sensations of the head and face; the abducens (VI), which regulates eye movements; the facial nerve (VII), which conducts impulses related to taste, salivation, and facial expression; and the vestibulocochlear nerve (VIII), which controls hearing and equilibrium.

MIDBRAIN. The **midbrain** is a short, constricted portion of the brain that connects the pons and cerebellum with the cerebral hemispheres. (See Figure 13–1.) Structurally, the midbrain consists of an internal cavity (cerebral aqueduct), a ventral *basilar* portion, and a dorsal *tegmental* portion. The internal cavity will be discussed later. The ventral portion contains a pair of fiber bundles, the *cerebral peduncles*. These are composed of motor fibers that convey impulses from the cerebral cortex to the pons and to the spinal cord. Large bundles of sensory fibers also pass through the midbrain on their way from the cord to the thalamus. The dorsal portion of the midbrain contains four rounded eminences, the *corpora quadrigemina*, which are concerned with auditory and visual reflexes. For example, these structures send out impulses that control turning the head to hear sounds coming from a given direction. They also control eyeball movements in response to changes in the position of the head.

The midbrain, like the pons, also contains the nuclei of some of the cranial nerves. These are the oculomotor nerve (III), which mediates movements of the eyeball and changes in the sizes of the pupils, and the trochlear (IV), which moves the eyeball.

THALAMUS. The **thalamus** is a large, oval structure located above the midbrain. (See Figure 13–1.) It consists of two masses of gray matter covered by a thin layer of white matter. The thalamus is both a relay station and a center for the interpretation of certain impulses. As a relay station, it sends all sensory impulses, except those for smell, to the cerebral cortex. As an interpretation center, the thalamus produces conscious recognition of the sensation of pain.

HYPOTHALAMUS. The **hypothalamus** is composed of several nuclei and is located below the thalamus. (See Figure 13–1.) Despite its relatively small size, the hypothalamus controls many body activities, most of which are related to homeostasis. A summary of its more important functions follows:

1. The hypothalamus controls and integrates the autonomic nervous system, which stimulates smooth muscle, regulates the rate of contraction of cardiac muscle, and controls the secretions of many of the body's glands. Through the autonomic system, the hypothalamus is the chief regulator of visceral activites. For instance, it controls the heartbeat, the movement of food through the digestive tract, and contraction of the urinary bladder.

2. The hypothalamus receives and interprets sensory impulses from the viscera.

3. The hypothalamus is the principal intermediary between the nervous system and endocrine system—the two great control systems of the body. The hypothalamus lies just above the pituitary, which is the major endocrine gland. When the hypothalamus detects certain changes in the body, it releases chemicals that stimulate the pituitary. The pituitary then releases hormones that determine the fate of carbohydrates, fats, and proteins, regulate the concentrations of some ions, and stimulate the sex organs. The hypothalamus also produces two hormones of its own, which will be described in Chapter 15 (the endocrine system).

4. The hypothalamus is the center for the mind-over-body phenomenon. And, believe it or not, the mind can have a great deal of influence over how the body functions. When the cerebral cortex interprets strong emotions, it often sends impulses over tracts that connect the cortex with the hypothalamus. The hypothalamus then sends impulses over the autonomic system and releases chemicals that stimulate the pituitary gland. The result can be a wide range of changes in body activities. For instance, when you panic, impulses go out from the hypothalamus to stimulate your heart to beat faster. Likewise, continued psychological stress can produce long-term abnormalities in body function that can make a person quite ill. These are the so-called psychosomatic disorders. Psychosomatic disorders are real and not imaginary.

5. The hypothalamus may be the area of the brain that feels rage.

6. It controls normal body temperature. Certain cells of the hypothalamus serve as a thermostat—a mechanism sensitive to changes in temperature. If blood flowing through the hypothalamus is above normal temperature, the hypothalamus sends impulses over the autonomic system to stimulate activities that promote heat loss. Heat can be lost through relaxation of the smooth muscle in the blood vessels and by sweat-

ing. Conversely, if the temperature of the blood is below normal, the hypothalamus sends out impulses that promote heat retention by the body. Heat can be retained through the contraction of cutaneous blood vessels, cessation of sweating, and shivering.

7. The hypothalamus regulates the amount of food intake through two centers. The feeding center is stimulated by hunger sensations from an empty stomach. When enough food has been ingested, the satiety center becomes stimulated and sends out impulses that inhibit the feeding center.

8. The hypothalamus also contains a thirst center. Certain cells in the hypothalamus become stimulated when the water content of the blood is low. The stimulated cells produce the sensation of thirst in the hypothalamus.

9. The hypothalamus serves as one of the centers that maintains the waking state and sleep patterns.

The spinal cord

The **spinal cord** originates as a continuation of the medulla and extends from the foramen magnum to the level of the second lumbar vertebra (Figure 13–8a).

Structure

The cord is somewhat oval in shape and decreases in size as it descends. The average length of the cord is about 17 to 18 inches. When the cord is viewed externally, two conspicuous enlargements can be seen. The upper enlargement, the *cervical enlargement*, extends from the fourth cervical to about the first thoracic vertebra. It contains the nerves that supply the upper extremities. The lower enlargement, the *lumbar enlargement*, contains nerves that supply the lower extremities. It extends downward from the eleventh thoracic segment. The lumbar enlargement is widest at the twelfth thoracic vertebra and thereafter tapers to a conical portion, the *conus medullaris.* Some of the spinal nerves that arise from the lower portion of the cord do not leave the spine immediately. Instead, they run farther downward in the vertebral canal and look like wisps of coarse hair flowing from the end of the cord. They are appropriately named the *cauda equina,* which means "horse's tail."

The cord is essentially a series of 31 segments,

each of which gives rise to a pair of spinal nerves. Each spinal nerve is connected to the cord by two roots: a **dorsal root,** which contains sensory fibers, and a **ventral root,** which contains motor fibers. (See Figure 13–8b.) In cross section it can be seen that the cord is partially divided into right and left halves by two grooves: the anterior median fissure and the posterior median sulcus. The *anterior median fissure* is a deep groove in the ventral surface. The *posterior median sulcus* is a shallower groove on the dorsal surface. In the center of the cord is a small round space called the *central canal.*

The spinal cord consists of both gray and white matter. Looking at the cross section of the cord illustrated in Figure 13–8b, you will see that the gray matter lies in an area that is shaped like the letter H. The central canal is located in the middle of the H. The transverse bar of the H is called the *gray commissure* and serves to connect the upright portions of the letter H. The upright portions may be further anatomically divided into *anterior horns (columns), lateral horns (columns),* and *posterior horns (columns)* of gray matter. The term *horn* or *column* simply designates a mass of cell bodies of association and motor neurons.

The white matter consists of myelinated fibers that are arranged around and between the columns of gray matter. The white matter in each half of the cord is divided into three *funiculi* (columns): *anterior funiculus, lateral funiculus,* and *posterior funiculus.* Each funiculus is, in turn, divided into a number of tracts called *fasciculi.* The long ascending tracts consist of sensory fibers that conduct impulses from sensory spinal nerves to the brain. The long descending tracts convey impulses from the brain to the motor neurons of spinal nerves in the spinal cord. The shorter tracts contain ascending and descending fibers that convey impulses from one level of the cord to another. (See Figure 13–9.)

Functions

The spinal cord has two functions: (1) it serves as a two-way conduction system between the brain and the periphery, and (2) it controls all reflexes except those mediated by the cranial nerves. Let us first consider relay functions of the cord. The long tracts within the cord transmit sensory and motor impulses between the brain and the periphery (Figure 13–9). The names of these tracts are generally descriptive enough to indicate the location of the point of origin, the point of termination, and the direction of impulse conduction.

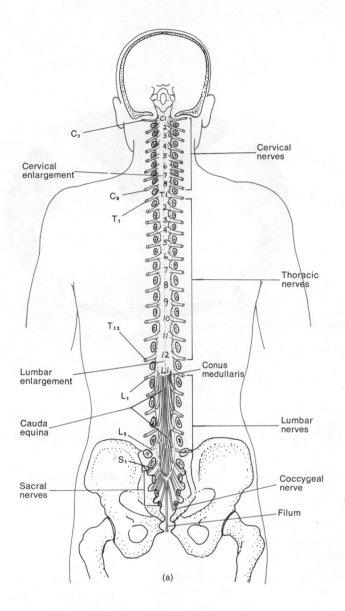

(a)

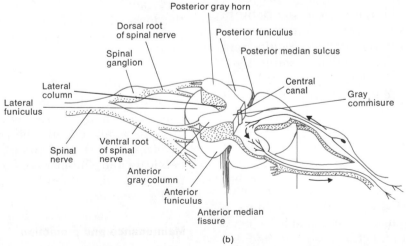

(b)

Figure 13-8. The spinal cord. (a) Seen in dorsal view in relation to surrounding structures. (b) Viewed in cross section.

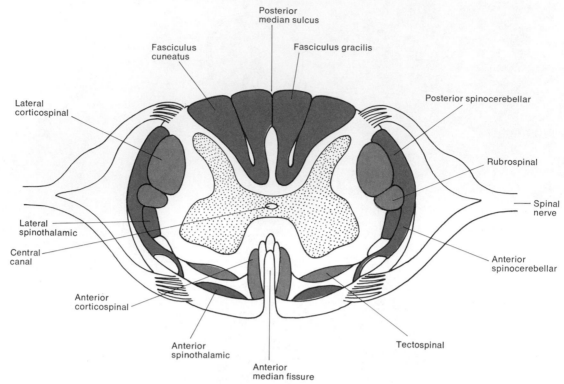

Figure 13–9. Selected tracts of the spinal cord. Ascending (sensory) tracts are indicated in color; descending (motor) tracts are shown in dark gray. See Figure 13–7b for a general idea as to how tracts communicate between the brain and the cord.

The point of origin is the location of the nerve cell bodies from which axons of the tract arise. The point of termination is the structure in which the axons end. For example, the spinothalamic tract originates in the spinal cord and terminates in the thalamus of the brain. Since it conveys impulses from the cord to the brain, it is an ascending (sensory) tract. The corticospinal tract originates in the cortex and terminates in the spinal cord. Because it conducts impulses from the brain to the cord, it is a descending (motor) tract. A summary of some clinically important ascending and descending tracts is presented in Exhibit 13–1.

The second function of the spinal cord is to form a pathway for reflexes. Within the spinal cord are located the association neurons and the sensory-motor synapses for all reflexes that involve skeletal muscles innervated by the spinal nerves. The short ascending and descending tracts that convey impulses from one level of the cord to another form pathways for these reflex arcs. They allow a single stimulus to produce a variety of motor responses. For example, when a person touches a hot object, the sensory impulse not only synapses with motor neurons that move the hand, but may also travel up an ascending tract to a spinal nerve which stimulates neck muscles to turn the head toward the object.

Spinal cord injury

The spinal cord may be damaged by fracture or dislocation of the vertebrae enclosing it or by wounds, such as those from bullets or shell fragments. All can result in *transection*, that is, partial or complete severing of the spinal cord. Complete transection means that all ascending and descending pathways are cut. It results in loss of all sensations and voluntary muscular movements below the level of transection. In fact, individuals with complete cervical transections close to the base of the skull usually die of asphyxiation before treatment can be administered. This happens because impulses from the medulla to the breathing muscles are interrupted. If the upper cervical cord is partially transected, both the arms and legs are paralyzed, and the patient is classified as *quadriplegic*. Partial transection between the cervical and lumbar enlargements results in paralysis of the legs only, and the patient is classified as *paraplegic*.

Maintenance and protection

Since the brain and spinal cord are crucial to our survival, the body must constantly nourish these organs and provide for their protection. Just as is the case with other organs of the body, the brain

Exhibit 13–1. SELECTED ASCENDING AND DESCENDING TRACTS OF THE SPINAL CORD

TRACT	LOCATION	ORIGIN	TERMINATION	FUNCTION
A. Ascending tracts				
1. **Anterior spinothalamic**	Anterior funiculus	Posterior gray column on opposite side of the cord	Mostly in thalamus of the brain	Conveys sensations of touch to the brain
2. **Lateral spinothalamic**	Lateral funiculus	Posterior gray column on opposite side of the cord	Thalamus of the brain	Conveys impulses to the brain that give rise to sensations of pain and temperature
3. **Fasciculus gracilis** and **fasciculus cuneatus**	Posterior funiculi	Spinal ganglia on the same side of the cord	Medulla	Convey impulses to the brain from kinesthetic receptors — that is, receptors on muscles, tendons, and joints which give rise to sensations of movement and position. Other impulses conducted are for touch and pressure, and two-point discrimination, the ability to feel the two points of a compass placed on the skin as a single or double prick
4. **Posterior spinocerebellar** and **anterior spinocerebellar**	Lateral funiculi	Posterior gray columns of the cord	Cerebellum	Convey impulses to the brain from kinesthetic receptors
B. Descending tracts				
1. **Lateral corticospinal**	Lateral funiculi	Cerebral cortex	Anterior gray column of opposite side of cord	Conveys motor impulses to spinal nerves that terminate in skeletal muscles; involves impulses which coordinate skilled movements on the opposite side of the body
2. **Anterior corticospinal**	Lateral funiculi	Cerebral cortex	Anterior gray column on same side of cord	Conveys motor impulses to spinal nerves that terminate in skeletal muscles; conducts impulses which coordinate skilled movements
3. **Rubrospinal**	Lateral funiculus	Midbrain	Anterior gray column of spinal cord	Conveys excitatory or inhibitory impulses to the cord; impulses reach skeletal muscles via spinal nerves
4. **Tectospinal**	Anterior funiculus	Midbrain	Anterior gray column of the cervical spinal cord	Conveys impulses to the cord that mediate reflex activities of the head and neck muscles in response to visual stimuli

receives oxygen and nutrients from its blood supply. Both the brain and the spinal cord are protected by two kinds of coverings and by a shock-absorbing substance that is called cerebrospinal fluid.

Blood supply to the brain

The brain is well supplied with blood vessels, which bring it oxygen and glucose. Although the brain actually consumes less oxygen than do most other organs of the body, it must receive its allotment of oxygen continuously. If the blood supply to the brain is interrupted for only a few moments, unconsciousness may result. An interruption that lasts a minute or two can weaken cells by starving them of oxygen. If the cells are totally deprived of oxygen for about 4 minutes, many of them are permanently injured. Sometimes babies are cut off from the oxygen supply of their mothers' blood before they leave the birth canal and can breathe. Many of these children are born dead or suffer throughout their lives from brain damage that may cause mental retardation, epileptic seizures, and paralysis.

Blood supplying the brain also contains glucose, the principal source of energy for brain cells. Because carbohydrate storage in the brain is limited, the blood must also transport a continuous supply of sugar. If blood entering the brain has a low glucose level, conditions such as mental confusion, dizziness, convulsions, and even loss of consciousness may occur. It may interest you to know that these are also the symptoms of insulin shock, a condition that occurs when a diabetic gives himself too much insulin. Recall from Chapter 1 that insulin decreases the amount of glucose in the blood.

Coverings

The brain and spinal cord are extremely delicate and vital organs. They are therefore provided with two kinds of coverings. The first of these, the outer covering, is bone. The brain is enclosed and protected by the cranium, and the spinal cord is protected by the bones of the vertebral column. (See Chapter 8.) The inner covering is formed by three membranes, collectively called **meninges.** These membranes run continuously around the brain and down around the spinal cord (Figure 13–10). The outer membrane, the *dura mater*, is a tough fibrous tissue. A fold of the dura mater, the *tentorium cerebelli*, separates the cerebellum from the cerebrum. At the termination of the cord, an extension of the dura mater, the *filum*, continues downward to the first coccygeal vertebra. (See Figure 13–8a.) The middle membrane, the *arachnoid*, is a delicate fibrous tissue between the dura mater and the innermost membrane. The inner membrane, the *pia mater*, is a transparent layer that adheres to the surfaces of the brain and cord and contains blood vessels. *Meningitis*, or inflammation of the meninges, usually involves the middle and inner layers.

Cerebrospinal fluid

The central nervous system is further protected against injury by a substance called **cerebrospinal fluid.** This fluid circulates through the subarachnoid space around the brain and cord and through the ventricles of the brain. The subarachnoid space is the area between the arachnoid and pia mater. The **ventricles** are cavities within the brain that communicate with each other and with the central canal of the spinal cord. The two *lateral ventricles* are located one in each cerebral hemisphere under the corpus callosum. (See Figure 13–10.) The *third ventricle* is a slitlike cavity between the right and left halves of the thalamus and between the lateral ventricles. Each lateral ventricle communicates with the third ventricle by a narrow, oval opening, the *interventricular foramen*. The *fourth ventricle* lies between the cerebellum, medulla, and pons. It communicates with the third ventricle via the *cerebral aqueduct*. The roof of the fourth ventricle has three openings: a *median aperture* and two *lateral apertures*. Through these openings, the fourth ventricle also communicates with the subarachnoid space of the brain and cord.

Cerebrospinal fluid is a clear, colorless fluid of watery consistency. Chemically, it contains proteins, glucose, urea, and salts. It also contains some white blood cells. The fluid serves as a shock absorber for the central nervous system. It also circulates nutritive substances filtered from the blood. Cerebrospinal fluid is formed primarily by filtration from networks of capillaries, called *choroid plexuses*, in the ventricles. The fluid formed in the lateral ventricles circulates through the interventricular foramina to the third ventricle, where more fluid is added. It then flows through the cerebral aqueduct to the fourth ventricle. Here, further additions occur, and the fluid circulates through the apertures of the fourth ventricle and into the subarachnoid space. It passes downward to the subarachnoid space around the cord and upward around the brain. From here, it is gradually reabsorbed into veins. Removal of the fluid from the subarachnoid space in the lumbar region of the cord is referred to as a *lumbar puncture*.

Cerebrospinal fluid is continuously formed in the ventricles. Its production never stops. If an obstruction, such as a tumor, arises in the brain and interferes with the drainage of the fluid from the ventricles into the subarachnoid spaces, large amounts of fluid start building up in the ventricles. Fluid pressure inside the brain increases, and, if the fontanels have not yet closed, the head bulges to relieve the pressure. This condition is called *internal hydrocephalus*. The term *hydro* means water, whereas *cephalo* means head. If an obstruction interferes with drainage somewhere in

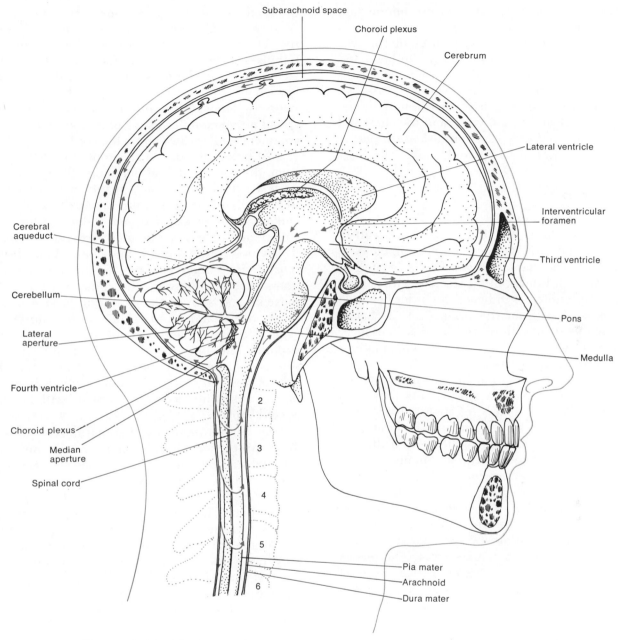

Figure 13–10. Meninges and ventricles seen in sagittal section. The direction of flow of cerebrospinal fluid is indicated by the colored arrows.

the subarachnoid space and cerebrospinal fluid accumulates inside the subarachnoid space, the condition is referred to as *external hydrocephalus*.

SOMATIC PORTION OF THE PERIPHERAL NERVOUS SYSTEM

The structures that connect the central nervous system to the other parts of the body constitute the **peripheral nervous system.** It is synonymous with all the neural tissue that lies outside the brain and spinal cord. The peripheral nervous system is subdivided into two portions: the autonomic nervous system and the somatic

nervous system. The autonomic nervous system, which will be discussed later, is responsible for unconscious movements, such as the constriction of blood vessels. It does not produce conscious movements, nor does it transmit sensory impulses. It also has some rather special features that cause it to be classified separately. The **somatic nervous system** consists of all the motor fibers that run from the central nervous system to the skeletal muscles and all the sensory fibers that run from the skeletal muscles, skin, and viscera to the central nervous system. The somatic nervous system is responsible for all movements in which there is some conscious control. It is also responsible for transmitting sensory information from all parts of the body. The somatic nervous system consists of the 12 pairs of cranial nerves that originate from various areas of the brain and the 31 pairs of spinal nerves that originate at various points along the length of the spinal cord. The cranial nerves are primarily distributed to the head and neck and to the viscera of the chest and abdomen. Spinal nerves, by contrast, are primarily distributed to the arms, legs, and trunk.

Cranial nerves

Of the 12 pairs of **cranial nerves,** 10 pairs originate from the brain stem (Figure 13–11), but all 12 pairs leave the skull through foramina in the base of the skull. The cranial nerves are designated in two ways, with Roman numerals and with names. The Roman numerals indicate the order in which the nerves arise from the brain (front to back), whereas the names indicate the distribution or function of the nerves. Some of the cranial nerves are referred to as *mixed nerves.* That is, they contain both sensory and motor fibers. Other cranial nerves contain sensory fibers only, and still others consist mainly of motor fibers. The cell bodies of the sensory fibers are located in ganglia outside the brain. The cell bodies of motor fibers lie in nuclei within the brain.

Relevant information concerning the cranial nerves is presented in Exhibit 13–2. As you study the information, you will notice that some of the motor fibers control unconscious movements, whereas the somatic nervous system has been defined as a conscious system. The reason for this seeming contradiction is that some of the fibers of the autonomic system leave the brain bundled together with the somatic fibers of the cranial nerves. Damage to a cranial nerve can easily include damage to the autonomic fibers that travel with it. Therefore, unconscious functions transmitted by the autonomic fibers are described along with the conscious functions of the somatic fibers of the cranial nerves.

Spinal nerves

Thirty-one pairs of **spinal nerves** originate on the spinal cord. These nerves, unlike the cranial nerves, are named for the region of the vertebral column from which they emerge. There are 8 pairs of *cervical* spinal nerves, 12 pairs of *thoracic,* 5 pairs of *lumbar,* 5 pairs of *sacral,* and 1 pair of *coccygeal* spinal nerves. (See Figure 13–8a.)

Each of the 31 pairs of spinal nerves attaches indirectly to the cord by means of two short roots (Figure 13–12). The *dorsal (posterior* or *sensory) root* contains afferent nerve fibers and conducts impulses to the spinal cord. Each dorsal root also has a swelling, the *dorsal root (spinal) ganglion,* formed by the cell bodies of peripheral sensory neurons. The *ventral (anterior* or *motor) root* of a spinal nerve contains axons of motor neurons that conduct impulses away from the cord.

Just before the dorsal and ventral roots emerge from the intervertebral foramen, they unite to form a spinal nerve. As a result of this convergence, all spinal nerves are mixed. That is, they contain both sensory and motor fibers. After each spinal nerve leaves the intervertebral foramen, it divides into several branches. The *dorsal (posterior) branch* contains motor and sensory fibers that run between the spinal cord and the muscles and skin of the posterior surface of the head, neck, and trunk. The *ventral (anterior) branch* contains sensory and motor fibers and supplies the muscles and skin of the anterior surface of the head, neck, and trunk and the extremities. The *visceral branch,* which belongs to the autonomic nervous system, will be discussed shortly.

The ventral branches of spinal nerves (except those of eleven thoracic nerves) are not distributed directly to skin and muscle. Instead, they combine and form complex networks of nerve fibers called **plexuses.** The term *plexus* means braid. Emerging from these plexuses are nerves bearing names that are often descriptive of the general regions they supply or the course they take. Each of these nerves, in turn, may have several branches named for the specific structures they supply (Figure 13–13). Pertinent information concerning the principal plexuses is provided in Exhibit 13–3.

Regeneration of peripheral nerves

As mentioned previously, once nerve cell bodies of the central nervous system are destroyed, the cells are not replaced. The same is true of nerve cell bodies of the peripheral nervous system. However, fibers of the peripheral nervous system may regenerate if the cell body is still intact and functioning and a neurilemma is present around the axon. After the axon of a peripheral nerve is severed, the distal portion of the axon degenerates so that an empty neurilemma remains. Next, the proximal portion of the axon begins to produce several fibrils that extend into the empty neurilemma. The fibrils grow until one

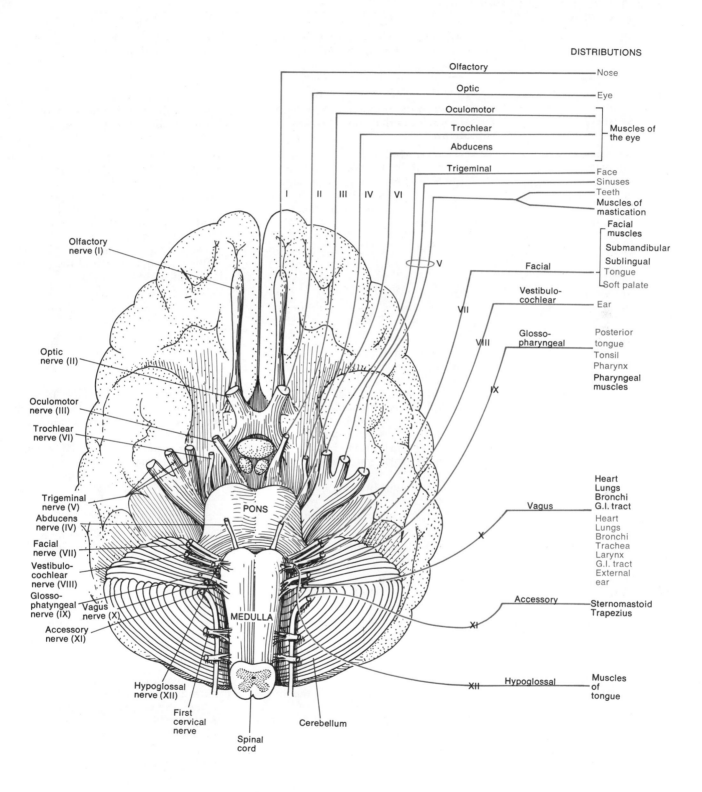

Figure 13–11. The location of the 12 pairs of cranial nerves. Sensory nerve distributions are indicated in color. Motor nerve distributions are shown in black.

Exhibit 13–2. SUMMARY OF CRANIAL NERVES

NUMBER, NAME, AND (TYPE)	LOCATION	FUNCTION AND COMMENTS
I. Olfactory (Sensory)	Arise in the nasal mucosa and end in the olfactory bulb (relay to olfactory cortex), where they synapse with sensory neurons that lead to olfactory area of cerebrum	Smell. Injury results in loss of the sense of smell (*anosmia*)
II. Optic (Sensory)	Arise in the retina of eye and end in nuclei in thalamus and mid-brain	Vision. Injury to one nerve may result in blindness of one eye
III. Oculomotor (Motor)	Origin°: Midbrain Distribution: Levator palpebrae (muscle of eyelid); extrinsic eye muscles except superior oblique and lateral rectus, ciliary muscle of the eye	Movement of the eyelid and eyeball. Symptoms of damage are drooping (*ptosis*) of eyelid, turning of the eyeball downward and outward, seeing two objects instead of one (*diplopia*), a dilated pupil, and blurred vision Accommodation of the lens for near vision and constriction of the pupil for bright light
IV. Trochlear (Motor)	Origin: Midbrain Distribution: Superior oblique eye muscle	Movement of the eye. The trochlears are the smallest of the cranial nerves
V. Trigeminal (Mixed)	Origin: Pons Distribution of motor portion: Muscles of mastication (mandibular branch) Distribution of sensory portions: 1. Ophthalmic branch: cornea of eye, skin of forehead and scalp 2. Maxillary branch: skin of upper cheek, roof of oral cavity, nasal mucosa, upper teeth 3. Mandibular branch: skin of chin, temporal region; mouth, cheek, tongue, lower teeth	Chewing. Injury results in paralysis of the muscles of mastication Sensations of head and face. Injury produces a loss of sensation of touch and temperature in the face, mucous membranes, and tongue; the trigeminal nerves are the largest of the cranial nerves
VI. Abducens (Motor)	Origin: Medulla Distribution: Lateral rectus eye muscle	Eye movement. Injury results in turning of the eyeball inward
VII. Facial (Mixed)	Origin: Medulla Distribution of motor portion: Muscles of facial expression and submaxillary and sublingual salivary glands Distribution of sensory portion: Taste buds of tongue	Facial expression and secretion of saliva. Injury produces paralysis of the muscles of facial expression (Bell's palsy) and reduced salivary secretion Taste. Injury produces loss of taste on part of tongue
VIII. Vestibulocochlear (Sensory)	Origin: Medulla Distribution: 1. Cochlear branch: spiral organ (of Corti) in ear 2. Vestibular branch: semicircular canals and vestibule of ear	Hearing and equilibrium. Injury to the cochlear branch results in deafness; injury to the vestibular branch results in lack of balance and dizziness
IX. Glossopharyngeal (Mixed)	Origin: Medulla Distribution of motor portion: a small muscle of pharynx; parotid salivary gland Distribution of sensory portion: pharynx and taste buds	Swallowing movements and secretion of saliva. Injury results in reduced production of saliva. Taste. Injury results in loss of sensation in the throat and posterior part of tongue

NUMBER, NAME, AND (TYPE)	LOCATION	FUNCTION AND COMMENTS
X. Vagus (Mixed)	Origin: Medulla Distribution of motor portion: voluntary muscles of pharynx, larynx (voice box); also respiratory passageways, heart, stomach, small intestine, most of large intestine, and gallbladder	Movement of organs supplied
	Distribution of sensory portion: structures innervated by motor portion	Sensations from many of the organs supplied Severing of both nerves in the upper body seriously interferes with swallowing, paralyzes vocal cords, and destroys sensations from many organs. Injury to both nerves in abdominal area has little effect on functioning of abdominal organs since they are also supplied by autonomic fibers that emerge from spinal cord
XI. Accessory (Motor)	Origin: Medulla and spinal cord Distribution: Sternocleidomastoid, trapezius, muscles of thoracic and abdominal viscera, larynx, and pharynx	Head and shoulder movements. Injury results in an inability to rotate the head and raise the shoulder
XII. Hypoglossal (Motor)	Origin: Medulla Distribution: Muscles of the tongue	Movement of the tongue. Injury results in difficulty in chewing, speaking, and swallowing

Note: For cranial nerves III through XII, the term *origin* refers to their origin from the brain stem, and the term *distribution* refers to their termination in a part of the body.

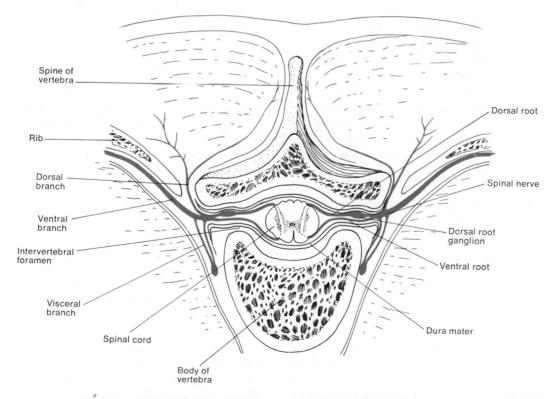

Figure 13–12. Structure of a typical spinal nerve.

of them reaches the end organ originally supplied by the severed nerve. After contact is made between the fibril and the end organ, the fibril becomes the functional axon. The remaining fibrils degenerate. The process of regeneration may take from several months to a year or so.

THE AUTONOMIC NERVOUS SYSTEM

The portion of the nervous system that controls the activities of smooth muscle, cardiac muscle, and glands is the **autonomic nervous system.** Structurally, the system includes nerves, ganglia, and plexuses. Functionally, it operates without any conscious control. Biologists originally thought that the autonomic system functioned autonomously, without any control from the central nervous system—hence its name, the autonomic nervous system. In truth, however, the autonomic system is neither structurally nor functionally independent of the central nervous system. It is regulated by centers in the brain, in particular by the cerebral cortex, the hypothalamus, and the medulla oblongata. However, the autonomic nervous system does differ markedly from the somatic nervous system in some ways. For convenience of study, then, the two must be separated.

Whereas the somatic system produces conscious movement in skeletal muscles, the autonomic nervous system regulates visceral activities. And it does so involuntarily and automatically. Examples of visceral activities that are regulated by the autonomic nervous system are changes in the size of the pupil of the eye, accommodation for near vision, dilatation and constriction of blood vessels, rate and force of the heartbeat, emptying of the urinary bladder, muscular movements of the gastrointestinal tract, formation of gooseflesh, and secretion by most glands. Note again that these activities lie beyond conscious control. They are automatic activities.

Also unlike the somatic system, the autonomic nervous system is entirely motor. All of its axons are efferent fibers, which transmit impulses from the central nervous system to visceral effectors. Autonomic fibers are called *visceral efferent fibers. Visceral effectors* include cardiac muscle, smooth muscle, and glandular epithelium. It should not be implied that there are no afferent (sensory) impulses from visceral effectors, however. Visceral sensations pass over visceral afferent fibers that structurally and functionally belong to the somatic nervous system. Some of the functions of these fibers were described with the cranial and spinal nerves. However, the hypothalamus, which largely controls the autonomic nervous system, also receives impulses from the visceral sensory fibers.

The somatic and autonomic systems also differ in the way they bring about effector relaxation. The autonomic system consists of two principal divisions: the sympathetic and the parasympathetic. Many organs innervated by the autonomic system receive visceral efferent neurons from both components of the autonomic system—one set from the sympathetic division, another from the parasympathetic division. In general, impulses transmitted by the fibers of one division stimulate the organ to become active, whereas impulses from the other division bring about a decrease or halt in the activity of the organ. In the somatic nervous system, only one kind of motor neuron innervates an organ. When the somatic neurons stimulate the cells of the organ, the organ becomes active. When the neuron ceases to stimulate the organ, contraction stops altogether. Recall from Chapter 10 that the skeletal muscle cells of each motor unit contract only when they are stimulated by their motor neuron. When the impulse stops, contraction stops.

Structure of the autonomic system

Visceral efferent neurons

Autonomic pathways always consist of two neurons. One runs from the central nervous system to a ganglion. The other runs directly from the ganglion to the effector.

The first of these visceral efferent neurons in an autonomic pathway is called a *preganglionic neuron* (Figure 13–14). Preganglionic neurons have their cell bodies in the brain or spinal cord. Their myelinated axons, called *preganglionic fibers*, pass out of the central nervous system as part of a cranial or spinal nerve. At some point, they leave their somatic nerves and run to autonomic ganglia where they synapse with the dendrites or cell bodies of postganglionic neurons.

Postganglionic neurons, the second visceral efferent neurons in an autonomic pathway, lie entirely outside the central nervous system. (See Figure 13–14.) Their cell bodies and dendrites (if they have dendrites) are located in the autonomic ganglia, where they synapse with the preganglionic fibers. The axons of postganglionic neurons are called *postganglionic fibers*. Post-

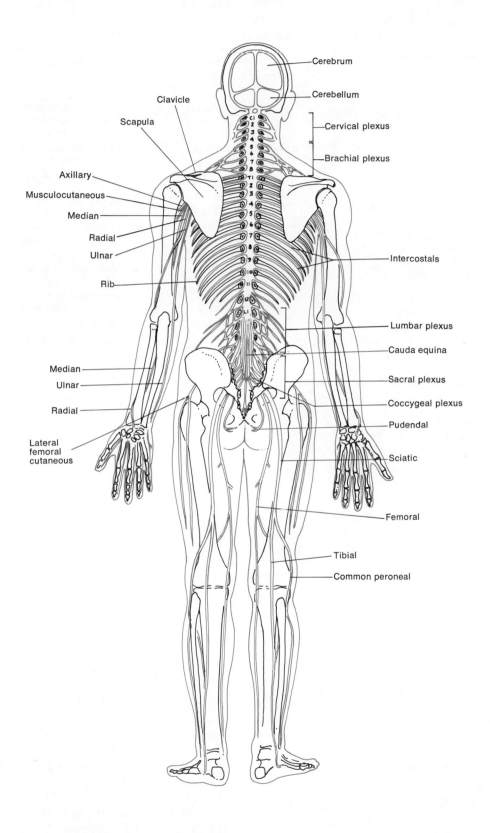

Figure 13–13. Principal plexuses and their distributions.

Exhibit 13–3. PLEXUSES OF SPINAL NERVES

SPINAL NERVE[1]	PLEXUS FORMED	PRINCIPAL BRANCHES FROM PLEXUS	INNERVATION (PARTS OF THE BODY SUPPLIED)
C1 C2 C3 C4	**Cervical plexus**	Lesser occipital Great auricular Transverse cervical Supraclavicular Phrenic	Motor fibers to muscles and sensory fibers to skin of head, neck, and upper part of shoulders Motor fibers to the diaphragm
C5 C6 C7 C8 T1	**Brachial plexus**	Musculocutaneous Median Ulnar Radial	Motor fibers to anterior arm muscles and sensory fibers to skin of lateral forearm Motor fibers to most of the anterior muscles of forearm and some in the palm, and sensory fibers to skin on radial half of palm Motor fibers to anteromedial muscles of forearm and most muscles of palm, and sensory fibers to skin of the ulnar side of the hand Motor fibers to muscles on posterior aspect of arm and forearm, and sensory fibers to the skin of the same regions and posterior aspect of the hand
T2 T3 T4 T5 T6 T7 T8 T9 T10 T11 T12	The anterior branches of spinal nerves T2 through T12 do not form a plexus. They are called *intercostal (thoracic) nerves* because they pass between the ribs		Motor fibers to intercostal muscles and muscles of the abdominal wall and sensory fibers to the overlying skin of the wall
L1 L2 L3 L4	**Lumbar plexus**[2]	Iliohypogastric Ilioinguinal Lateral femoral cutaneous Femoral Obturator	Sensory fibers to anterior abdominal wall Sensory fibers to same area as above plus external genitals, and motor fibers to muscles of abdominal wall Sensory fibers to lateral aspect of thigh Motor fibers to muscles on anterior and lateral aspects of thigh and leg and sensory fibers from skin on anteromedial aspect of thigh Motor fibers to anteromedial muscles of thigh

ganglionic fibers are nonmyelinated, and they terminate in visceral effectors.

If we review this organization, then, it can be seen that preganglionic neurons convey efferent impulses from the central nervous system to autonomic ganglia. Postganglionic neurons relay the impulses from the autonomic ganglia to visceral effectors.

Autonomic ganglia

Synapses between visceral efferent neurons occur in *autonomic ganglia*. These ganglia may

be divided into three general groups (Figure 13–15). The *sympathetic trunk,* or *vertebral chain, ganglia* are a series of ganglia that lie in a horizontal row on either side of the vertebral column close to the bodies of the vertebrae. These ganglia are connected to each other by short fibers and look like beads on a chain, one chain on each side of the spinal column. The sympathetic trunk extends downward through the neck, thorax, and abdomen to the coccyx. Typically, there are 22 ganglia in each chain: 3 cervical, 11 thoracic, 4 lumbar, and 4 sacral. However, even though the sympathetic trunk ganglia extend the length

SPINAL NERVE[1]	PLEXUS FORMED	PRINCIPAL BRANCHES FROM PLEXUS	INNERVATION (PARTS OF THE BODY SUPPLIED)
L4 L5 S1 S2 S3	Sacral plexus	Sciatic—divides into two branches:	
		1. Tibial	Motor fibers to most posterior muscles of the thigh and to muscles of calf of leg; and sensory fibers to skin of the calf and sole of foot
		2. Common peroneal	Motor fibers to dorsiflexor and evertor muscles of foot, and sensory fibers to dorsum (top surface) of foot
		Superior and inferior gluteal	Motor fibers to gluteal muscles and tensor fascia latae
		Posterior femoral cutaneous	Sensory fibers to skin of buttocks, and posterior aspect of thigh and leg
		Pudendal	Motor fibers to muscles of anus and external genitals, and sensory fibers to skin around anus, scrotum, labia majora, and external genitals
S4 S5 Col	Coccygeal plexus	Coccygeal	Sensory fibers to skin in region of coccyx

[1] C = cervical, T = thoracic, L = lumbar, S = sacral, Co = coccygeal

[2] The lumbar, sacral, and coccygeal plexuses are frequently considered as a single plexus, called the lumbosacral plexus. They have been considered separately here for convenience of study.

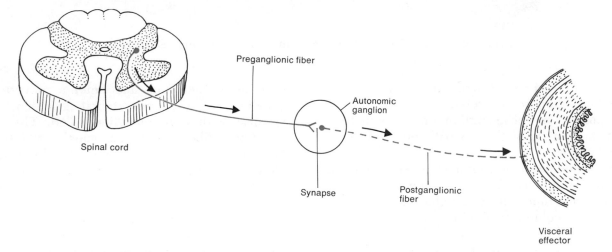

Figure 13–14. Relationship between preganglionic and postganglionic neurons of the autonomic nervous system.

of the vertebral column, they receive preganglionic fibers only from the thoracic and lumbar regions of the spinal cord.

The preganglionic fibers of the sympathetic trunk ganglia leave the vertebral canal as part of the spinal nerve. They then branch away from the somatic fibers of the nerve and enter the nearest sympathetic trunk ganglia. The term *white ramus* refers to this portion of the preganglionic axon. It connects the spinal cord with the ganglia of the sympathetic trunk. Some axons synapse in the first ganglion they enter. Others pass through the ganglion and run upward or downward to form the fibers on which the ganglia are strung. These latter axons may not synapse until they reach a ganglion in the cervical or sacral area. Some of the postganglionic fibers leaving the sympathetic trunk ganglia pass directly to visceral effectors of the head, neck, chest, and abdomen. Most rejoin the spinal nerves before supplying peripheral visceral effectors such as sweat glands and the smooth

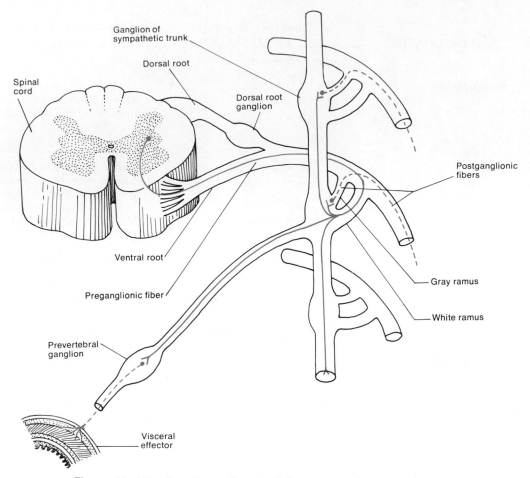

Figure 13–15. Ganglia and rami of the autonomic nervous system.

muscle in blood vessels and around hair follicles. The term *gray ramus* refers to the portions of the axons of the postganglionic neurons that run from the ganglia of the sympathetic trunk to the spinal nerves.

A second kind of ganglion of the autonomic nervous system is called a *prevertebral*, or *collateral, ganglion.* These ganglia lie anterior to the spinal column and close to the large abdominal arteries from which their names are derived. For example, the *celiac ganglia* are on either side of the celiac artery just below the diaphragm. The *superior mesenteric ganglion* is located near the beginning of the superior mesenteric artery in the upper abdomen. And the *inferior mesenteric ganglion* is located near the beginning of the inferior mesenteric artery in the middle of the abdomen. (See Figure 13–16.) Prevertebral ganglia receive preganglionic fibers from the thoracic and lumbar regions of the spinal cord. The preganglionic fibers leave the spinal cord along with a spinal nerve, branch into a white ramus, enter the sympathetic trunk, and

then leave the trunk and enter a prevertebral ganglion. In the prevertebral ganglion, they synapse with a postganglionic cell. The postganglionic fibers leave the ganglia and follow the course of various arteries to the visceral effectors in the abdomen and pelvis.

The third kind of autonomic ganglion is referred to as a *terminal* or *intramural, ganglion.* (See Figure 13–16.) These ganglia, as the name implies, are located at the end of the visceral efferent pathway. Some terminal ganglia lie very close to visceral effectors that are supplied by their postganglionic fibers. Others are located within the walls of visceral effectors. Terminal ganglia receive preganglionic fibers from the brain and from the sacral region of the spinal cord. The preganglionic fibers do not pass through the sympathetic trunk ganglia.

The two autonomic divisions

As mentioned, the autonomic nervous system is divided into the sympathetic and parasympathetic

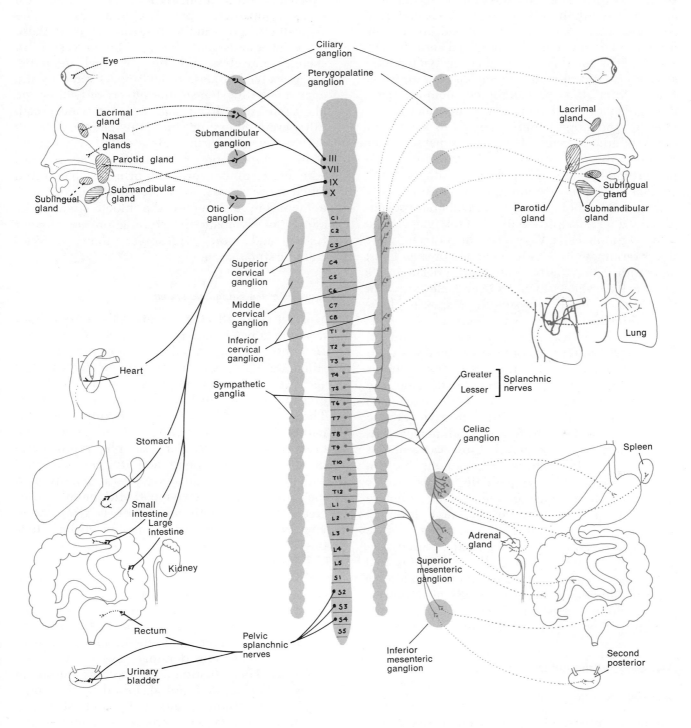

Figure 13–16. Structure of the autonomic nervous system. Sympathetic fibers are indicated in color, and parasympathetic fibers in black. Preganglionic fibers are shown as solid lines, and postganglionic fibers are indicated as broken lines.

divisions. These are also referred to as the thoracolumbar and craniosacral divisions, respectively. In the **sympathetic division,** the preganglionic neurons have their cell bodies in the lateral gray columns of the thoracic segment and first two or three lumbar segments of the spinal cord (Figure 13–16). It is for this reason that the sympathetic division is also called the *thoracolumbar division.* Axons from the preganglionic neurons pass through the ventral roots and white rami of the thoracic and upper lumbar nerves to ganglia of the sympathetic trunk. They then synapse either in one of the sympathetic trunk ganglia or in the prevertebral ganglia. Note that sympathetic trunk ganglia and prevertebral ganglia receive only sympathetic fibers. Each sympathetic preganglionic fiber synapses with several sympathetic postganglionic cells. Very often the postganglionic fibers terminate in widely separated organs. Thus, an impulse that starts in a single preganglionic neuron may affect several visceral effectors. For this reason, most sympathetic responses have very widespread effects on the body.

The cell bodies of preganglionic neurons of the **parasympathetic division** are located in nuclei in the brain stem and in the lateral gray columns of the sacral portion of the spinal cord. Thus, the synonymous term, *craniosacral division.* The preganglionic fibers of the parasympathetic neurons pass through four cranial nerves (III, VII, IX, and X) and generally three sacral nerves (2d, 3d, and 4th). Unlike sympathetic preganglionic fibers, parasympathetic preganglionic fibers travel a considerable distance before synapsing. The synapse occurs in a terminal ganglion, near or within a visceral effector. Since each parasympathetic preganglionic neuron usually synapses with postganglionic neurons that innervate a common organ, impulses along a parasympathetic pathway generally affect only a single visceral effector. Each parasympathetic response, then, produces a very specific effect.

Physiology of the autonomic system

Chemical transmitters

Autonomic fibers, like other axons of the body, release chemical transmitters at synapses and at points of contact between autonomic fibers and visceral effectors. These points are called **neuroeffector junctions.** On the basis of the particular chemical transmitter produced, autonomic fibers may be classified as either cholinergic or adrenergic. **Cholinergic fibers** produce *acetycholine*

and include the following: (1) both sympathetic and parasympathetic preganglionic axons, (2) parasympathetic postganglionic axons, and (3) some sympathetic postganglionic axons. The sympathetic postganglionic axons include those to sweat glands and those to blood vessels in skeletal muscles and the external genitalia. Since acetylcholine is quickly inactivated by the enzyme cholinesterase, the effects of cholinergic fibers are short-lived and are not widespread. **Adrenergic fibers** produce the chemical transmitter *norepinephrine,* also called *sympathin.* Most sympathetic postganglionic axons are adrenergic. Since norepinephrine is inactivated much more slowly than acetylcholine and norepinephrine may enter the blood stream, the effects of sympathetic stimulation are longer lasting and more widespread than parasympathetic stimulation.

Activities of the autonomic system

Most visceral effectors receive fibers from both the sympathetic and parasympathetic divisions. In these cases, impulses from one division stimulate the organ to initiate or increase its activities, whereas impulses from the other division decrease or inhibit the organ's activities. The stimulating division may be either the sympathetic or the parasympathetic, depending on the organ involved. For example, sympathetic impulses increase heart activity, but parasympathetic impulses decrease heart activity. On the other hand, parasympathetic impulses increase digestive activities, whereas sympathetic impulses inhibit them. A summary of the activities of the autonomic system is presented in Exhibit 13–4.

The parasympathetic division is primarily concerned with activities that restore and conserve body energy. For instance, under normal body conditions, parasympathetic impulses to the digestive glands and the smooth muscle of the digestive system dominate over sympathetic impulses. This situation allows energy-supplying foods to be digested and absorbed by the body.

The sympathetic division, by contrast, is primarily concerned with processes involving the expenditure of energy. When the body is in homeostasis, the main function of the sympathetic division is to counteract the parasympathetic effects just enough to carry out normal processes requiring energy. However, during a time of extreme stress, the sympathetic dominates the parasympathetic. For example, when people are confronted with a dangerous situation, their bodies become very alert, and they sometimes perform feats of unusual strength. This is because fear stimulates the sympathetic division. It, in turn, produces the following effects: (1) the pupils of the eyes dilate; (2) the heart rate increases; (3) the blood vessels of the skin and viscera constrict; and (4) the remainder of the blood vessels dilate. This causes a rise in blood pressure and a faster flow of blood into the dilated blood vessels of skeletal muscles, cardiac muscle, the lungs, and brain—organs that are involved in fighting off the danger. Rapid breathing occurs as the bronchioles dilate to allow movement of air in and out of the lungs at a faster rate. Blood sugar level rises as liver glycogen is converted to glucose to supply the additional energy needs of the body. The sympathetic division also stimulates the adrenal gland to produce epinephrine, a hormone that intensifies and prolongs the sympathetic effects noted above. During this period of stress, the sympathetic effects inhibit other processes that are not essential for meeting the situation. For example, muscular movements of the gastrointestinal tract and digestive secretions are slowed down or even stopped.

Exhibit 13–4. ACTIVITIES OF THE AUTONOMIC NERVOUS SYSTEM

VISCERAL EFFECTOR	EFFECT OF SYMPATHETIC STIMULATION	EFFECT OF PARASYMPATHETIC STIMULATION
Eye		
Iris	Contracts dilator pupillae muscle and brings about dilatation of the pupil	Contracts sphincter pupillae muscle and brings about constriction of the pupil
Ciliary muscle	Accommodates the lens for far vision	Contracts the muscle and accommodates the lens for near vision
Glands		
Sweat	Stimulates secretion	No innervation
Lacrimal (tear)	No innervation	Normal or excessive secretion
Salivary	Decreases secretion and saliva is mucus-rich	Increases secretion of thin, watery saliva
Gastric	No known effect	Secretion stimulated
Intestinal	No known effect	Secretion stimulated
Adrenal medulla	Promotes epinephrine secretion	No innervation
Lungs (bronchial tubes)	Dilatation	Constriction
Heart	Increases rate and strength of contraction; dilates coronary vessels that supply blood to heart muscle cells	Decreases rate and strength of contraction; constricts coronary vessels
Blood vessels		
Skin	Constriction	No innervation for most
Skeletal muscle	Dilatation	No innervation
Visceral organs (except heart and lungs)	Constriction	No innervation for most
Stomach	Decreases motility	Increases motility
Intestines	Decreases motility	Increases motility
Kidney	Constriction that results in decreased urine volume	No effect
Spleen	Contraction and discharge of stored blood into general circulation	No innervation
Urinary bladder	Relaxes muscular wall	Contracts muscular wall
Hair follicles	Contraction results in erection of hairs ("goose pimples")	No innervation

HOMEOSTASIS THROUGH REFLEXES

One vital role of both the nervous and endocrine systems is to respond to stimuli that upset homeostasis. The nervous system reacts to stimuli quickly and produces responses by almost instantaneous electrical conduction over its pathways. The endocrine system reacts more slowly and produces changes by releasing chemicals into the bloodstream. As you will see in Chapter 15, the nervous and endocrine systems are structurally and functionally interconnected, and their responses to stimuli are highly coordinated.

The parasympathetic division dominates when the body is in relative homeostasis, and the sympathetic division dominates during extreme stress. We shall describe extreme stress and its effects in Chapter 15. For the time being, let us look at how the nervous system helps to maintain homeostasis by responding to the more usual types of stress.

The reflex is the major mechanism of the nervous system for responding quickly to changes in the internal or external environment. As you recall from Chapter 12, a reflex is a quick, involuntary response that occurs when a nervous impulse passes over a reflex arc. Reflexes may be categorized as visceral or somatic reflexes, depending upon the kind of organ stimulated. Visceral reflexes pass over visceral sensory fibers and then pass back over the autonomic motor fibers to adjust activities such as heartbeat, breathing, and blood pressure. Somatic reflexes are conducted over somatic neurons and stimulate skeletal muscle to contract.

Reflexes may also be classified on the basis of the location of receptors that pick up the impulse. These are summarized in Exhibit 13–5 and will be discussed further in Chapter 14.

Exhibit 13–5. KINDS OF REFLEXES

CRITERION AND KIND	DESCRIPTION
I. Nature of organ stimulated	
A. Somatic reflex	Contraction of skeletal muscle
B. Visceral (autonomic) reflex	Contraction of smooth or cardiac muscle or secretion by glands
II. Location of receptors	
A. Exteroceptive reflex	The receptor is located at or near the external body surface
B. Visceroceptive reflex	The receptor is within a visceral organ or blood vessel
C. Proprioceptive reflex	The receptor is within a muscle or tendon

Reflexes can be important tools for diagnosing certain disorders of the nervous system and for locating the area of an injured tissue. If a reflex ceases to function or functions abnormally, the physician may suspect that the damage lies somewhere along the conduction pathway. Visceral reflexes, however, are usually not practical tools for diagnosis. For one thing, it may be difficult to stimulate visceral receptors without opening up the body. In contrast, many somatic reflexes can be tested simply by tapping or stroking external areas of the body. Among the more important somatic reflexes that have clinical significance are the following:

1. *Patellar reflex* (knee jerk). This reflex involves the extension of the lower leg by contraction of the quadriceps in response to tapping the tendon over the patella. The reflex is abolished by disease or injury of afferent or efferent nerves to the muscle or reflex centers in the lumbar region of the spinal cord. The reflex is exaggerated in disease or injury involving the corticospinal tracts descending from the cortex to the cord.
2. *Achilles reflex* (ankle jerk). This reflex involves the extension of the foot in response to tapping the Achilles tendon. Abolition of the ankle jerk indicates disease or injury of the nerves to the leg muscles or of the nerve cells of the lumbosacral region of the cord. An exaggerated Achilles reflex indicates a lesion of the motor tracts of the cord at this level.
3. *Babinski reflex*. Stimulation of the outer margin of the sole of the foot resulting in extension of the great toe, with or without fanning of the other toes. This phenomenon occurs in normal children under $1\frac{1}{2}$ years of age due to incomplete development of the nervous system. That is, myelination of the fibers in corticospinal tract has yet to occur. A positive Babinski reflex up to this age is considered normal. After that age, it is viewed as abnormal. A positive Babinski reflex after age $1\frac{1}{2}$ indicates an interruption of the corticospinal tract as the result of a lesion in the tract, usually in the upper portion. The normal response to stimulation of the outer edge of the sole after $1\frac{1}{2}$ years is the *plantar reflex*. This response is a curling under of all the toes accompanied by a slight turning in and flexion of the anterior part of the foot.
4. *Abdominal reflex*. This reflex involves pulling in the abdominal wall in response to stroking the side of the abdomen. This reflex, if absent,

is also associated with lesions of the cortico-spinal system. In addition, it may be abolished in lesions of the peripheral nerves or reflex centers in the thoracic part of the cord.

DISORDERS OF THE NERVOUS SYSTEM

Many disorders can affect the various parts of the nervous system. Some of the diseases are known to be caused by viruses or bacteria. Other conditions are caused by damage to the nervous system during birth. The origins of many conditions, however, are unknown. Here, we shall discuss the origins and describe the symptoms of some common nervous system disorders.

Poliomyelitis

Poliomyelitis, also known as **infantile paralysis,** is a viral infection that is most common during childhood. The onset of the disease is marked by fever, severe headache, a stiff neck and back, deep muscle pain and weakness, and loss of some somatic reflexes. The virus may spread by means of the blood and respiratory passages to the central nervous system. In the central nervous system, it destroys the motor nerve cells, specifically those in the anterior horns of the spinal cord and in the nuclei of the cranial nerves. The injury to the spinal gray matter gives the disease its name—*polio*, meaning gray matter, and *myel*, meaning spinal cord. Destruction of the anterior horns produces paralysis of one or more limbs. Poliomyelitis can cause death from respiratory or heart failure, if the virus invades the brain cells of the vital medullary centers.

Syphilis

Syphilis is a venereal disease caused by the *Treponema pallidum* bacterium. Venereal diseases are infectious disorders that are spread through sexual contact of any sort. The disease process of syphilis goes through several stages: primary, secondary, latent, and sometimes tertiary. During the primary stage, the chief symptom is an open sore, called a chancre, at the point of contact. The chancre eventually heals. About 6 weeks later, a range of generalized symptoms, such as a skin rash, fever, and aches in the joints and muscles, ushers in the secondary stage. At this stage, syphilis can usually be cured with penicillin. However, even if the person does not undergo treatment, his symptoms will eventually disappear. Within a few years, he will usually cease to be infectious. During this later "symptomless" period, called the latent

stage, one of two things happen: Either the patient develops an immunity to the bacteria and becomes cured, or the bacteria may invade and slowly destroy any of the body organs. This is why untreated syphilis is so dangerous. When symptoms of organ degeneration appear, the disease is said to be in its fourth, or tertiary, stage. If the syphilis bacteria attack the organs of the nervous system, the tertiary stage is called *neurosyphilis*. Neurosyphilis may take any one of several forms, depending on the tissue involved. For instance, about 2 years after the onset of the disease, the bacteria may attack the meninges, producing meningitis. Or the blood vessels that supply the brain may become infected. In this case, symptoms depend on the parts of the brain that are destroyed by oxygen and glucose starvation. Over the years, the bacteria may spread through the nerve cells of the brain. As one nerve cell after another is destroyed, the patient experiences corresponding symptoms. Cerebellar damage is manifested by uncoordinated movements. For instance, the person may have trouble with skilled activities such as writing. As the motor areas become extensively damaged, the victim may be unable to control urine and bowel movements. Eventually, he may be bedridden, without even the motor control necessary to feed himself. Damage to the cerebral cortex produces memory loss and personality changes that range from irritability to confusion to hallucinations.

A common form of neurosyphilis is *tabes dorsalis*, a progressive degeneration of the posterior columns of the spinal cord. The sensory ganglia and the sensory nerve roots are affected. Tabes dorsalis forms an interesting contrast with polio. The polio virus attacks the anterior columns and destroys motor neurons. The polio victim is unable to move the affected muscles voluntarily, but he retains all his sensory functions. The person with tabes dorsalis suffers from just the opposite problem. He retains motor control, but he loses many of his sensory functions. He often has tingling or numbness in his limbs and trunk. Normally, receptors in the joints are able to tell the central nervous system how much a joint is flexed and where one part of the body is in relation to another part. This information is necessary for coordinating movement and maintaining posture and balance. When the sensory nerve roots are destroyed, however, this information cannot pass from the receptors to the brain. Consequently, the person with tabes dorsalis must use his eyes in order to carry out motor activities successfully. He has trouble walking in the dark because he does not know whether his legs are flexed or where his feet are. In lighted areas, he may walk with a characteristic shuffle, which consists of jerking the knee up and then letting the leg extend abruptly to the ground.

Syphilis can be treated with penicillin during the primary, secondary, and latent periods. Some forms of neurosyphilis can also be successfully treated; however, the prognosis for tabes dorsalis is very poor. Unfortunately, not everyone with syphilis shows noticeable symptoms during the first two stages of the disease. But syphilis can be diagnosed through a blood test. The importance of these blood tests and follow-up treatment cannot be overemphasized.

Cerebral palsy

The term **cerebral palsy** refers to a group of motor disorders caused by damage to the motor areas of the brain during fetal life, birth, or infancy. One cause is infection of the mother with German measles during the first 3 months of pregnancy. During early pregnancy, certain cells in the fetus are dividing and changing in order to lay down the basic structures of the brain. These cells can be abnormally changed by toxin from the measles virus. Radiation during fetal life, temporary oxygen starvation during birth, and hydrocephalus during infancy can also damage brain cells.

Cases of cerebral palsy are categorized into three groups depending on whether the cortex, the cerebral nuclei of the cerebrum, or the cerebellum is affected most severely. Most cerebral palsy victims have at least some damage in all three areas. The location and extent of motor damage determine the symptoms. For instance, the cerebral palsy victim may have a contorted face caused by partial facial paralysis. If his tongue is paralyzed, he may be able to make only glutteral sounds. Extensive damage to the cerebellum causes very uncoordinated movements. Although cerebral palsy refers only to motor damage, sensory and association areas of the brain may be affected as well. The person may be deaf or partially blind. About 70 percent of cerebral palsy victims appear to be mentally retarded. However, the apparent mental slowness is often due to the person's inability to speak or hear well. These people are often much brighter than they seem.

Cerebral palsy is not a progressive disease. In other words, it does not get worse as time elapses. But, once the damage is done, it is irreversible.

Parkinsonism

This disorder, also called **Parkinson's disease,** is a progressive malfunction of the cerebral nuclei of the cerebrum. Recall that the cerebral nuclei regulate unconscious contractions of skeletal muscles that aid activities desired by the motor areas of the cerebral cortex. Examples of movement produced by the cerebral nuclei are swinging the arms when walking and making facial expressions when talking. In Parkinsonism, the cerebral nuclei produce useless skeletal movements that often interfere with voluntary movement. For instance, the muscles of the arms and hands may alternately contract and relax so that the patient's hands shake. This type of shaking is called *tremors*. Other muscles may contract continuously and make the involved part of the body rigid. *Rigidity* of the facial muscles gives the face a masklike appearance. The expression is characterized by a wide-eyed, unblinking stare and a slightly open mouth with saliva drooling from the corners. Vision, hearing, and intelligence are unaffected by the disorder, indicating that Parkinsonism does not attack the cerebral cortex.

Parkinsonism seems to be caused by a malfunction at the neuron synapses. The motor neurons of the cerebral nuclei release the chemical transmitter acetylcholine. In normal people, the cerebral nuclei also produce an enzyme called dopamine, which quickly inactivates the acetylcholine and prevents continuous conduction across the synapse. People with Parkinsonism do not manufacture enough dopamine in their bodies. As a result, stimulated cerebral nuclei neurons do not easily stop conducting impulses. Unfortunately, injections of dopamine are useless because the drug is not able to diffuse from the blood into the brain. However, a few years ago researchers developed a drug called levodopa, which is very similar to dopamine. Levodopa is able to diffuse into the brain, where it is converted by a chemical reaction to dopamine.

Epilepsy

Epilepsy is a disorder characterized by short, periodic attacks of motor, sensory, and/or psychological malfunction. The attacks, called **epileptic seizures,** are brought on by abnormal and irregular discharges of electricity by millions of neurons in the brain. The discharges stimulate many of the neurons to send impulses over their conduction pathways. As a result, a person undergoing an attack may contract skeletal muscles involuntarily. He may sense lights, noise, or smells when the eyes, ears, and nose actually have not been stimulated. The electrical discharges may also inhibit certain brain centers. For instance, the awake center in the hypothalamus may be depressed so that the person loses consciousness.

Many different types of epileptic seizures exist. The particular type of seizure depends on the area of the brain that is electrically stimulated and whether the stimulation is restricted to a small area or spreads throughout the brain. *Grand mal* seizures are brought on by a burst of electrical discharges that travel throughout the motor areas and spread to the areas of consciousness in the hypothalamus. The person loses consciousness, has spasms of his voluntary muscles, and may also lose urinary and bowel control. Sensory and intellectual areas may also be involved. For instance, just as the attack begins, the person may sense a peculiar taste in his mouth or see flashes of light. The unconsciousness and motor activity last a few minutes. Then the muscles relax, and the person awakens. Afterward, he may be mentally confused for a short period of time. Studies with EEGs show that grand mal attacks are characterized by rapid brain waves occurring at the rate of 25 to 30 per second (Figure 13–17). The normal adult rate is 10 waves/second.

Many epileptics suffer from electrical discharges that are restricted to one or several relatively small areas of the brain. An example is the *petit mal* form, which apparently involves the thalamus and hypothalamus. Petit mal seizures are characterized by an abnormally slow brain wave pattern occurring at the rate of 3 waves/second (see Figure 13–17). The person loses consciousness for about 5 to 30 seconds, but he does not undergo the embarrassing loss of motor control that is typical of a grand mal seizure. The victim merely seems to be daydreaming. A few people experience several hundred petit mal seizures each day. For them, the chief problems are a loss of productivity in school or work and periodic inattentiveness while driving a car.

Some epileptics experience motor seizures that are restricted to the precentral motor area of one cerebral hemisphere. These attacks consist of spasms that pass up or down one side of the body. People who suffer from sensory seizures may see lights or distorted objects if the discharge occurs in the occipital lobe. They may hear voices or a roaring in their ears if the discharge is located in the temporal lobe. Or, they may taste or smell unpleasant substances if the discharge is in the parietal lobe. People undergoing attacks of localized motor or sensory disturbances may or may not lose consciousness. A form of epilepsy that is sometimes confused with mental illness is *psychomotor epilepsy*. The electrical outburst occurs in the temporal lobe, where it causes the person to lose contact with reality. It may spread to some of the motor areas and produce mild spasms in some of the voluntary muscles. The person may stare into space and involuntarily smack his lips or clap his hands during an attack. If the motor areas are not involved, he may simply walk aimlessly. When he comes back to reality, he is surprised to find himself in a strange or different place.

The causes of epilepsy are varied. Many conditions can cause nerve cells to produce periodic bursts of impulses. Head injuries, tumors and abscesses of the brain, and childhood infections, such as mumps, whooping cough, and measles, are some of the causes. It should be noted, however, that epilepsy almost never affects intelligence. If frequent severe seizures are allowed to occur over a long period of time, some cerebral damage may occasionally result. However, damage can be prevented by controlling the seizures with drug therapy.

Epileptic seizures can be eliminated or alleviated by drugs that make neurons more difficult to stimulate. Many of these drugs change the permeability of the nerve cell membrane so that it does not depolarize as easily.

Multiple sclerosis

This disorder causes a progressive destruction of the myelin sheaths of neurons in the central nervous system. The sheaths deteriorate to *scleroses*, which are hardened scars or plaques, in multiple regions. Hence, the disorder is given the name **multiple sclerosis.** The destruction of the myelin sheaths interferes with the transmission of impulses from one neuron to another, literally "short-circuiting" conduction pathways. Multiple sclerosis is one of the most common disorders of the central nervous system. Usually the first symptoms occur between the ages of 20 and 40. The early symptoms are generally produced by the formation of just a few plaques

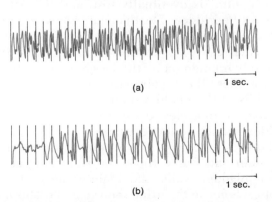

(a) 1 sec.

(b) 1 sec.

Figure 13–17. EEG of (a) grand mal seizure and (b) petit mal seizure.

and are, consequently, mild. For instance, plaque formation in the cerebellum may produce some incoordination in one hand. The patient complains that his handwriting has become sloppy. A short-circuiting of some pathways in the corticospinal tract may partially paralyze the leg muscles so that the patient drags his foot when walking. The diseased sheaths often heal, and the symptoms disappear for a while. Later, a new series of plaques develop, and the victim suffers a second attack. One attack follows another over the years. Each time the plaques form, some of the neurons are damaged by the hardening of their sheaths. Other neurons are uninjured by their plaques. The result is a progressive loss of function interspersed with healing periods during which the undamaged neurons regain their ability to transmit impulses.

The symptoms of multiple sclerosis depend on the areas of the central nervous system most heavily infested with plaques. Sclerosis of the white matter of the spinal cord is common. As the sheaths of the neurons in the corticospinal tract deteriorate, the patient loses the ability to contract skeletal muscles. Damage to the ascending tracts produces numbness and short-circuits information about the position of body parts and the flexion of joints. Damage to either set of tracts also destroys spinal cord reflexes. Very often the white matter of the brain stem and cerebellum deteriorates. In this case, the patient loses muscular coordination, cannot control movement of the eyeballs, and pronounces words slowly and hesitatingly. Damage to the tracts in the cerebral white matter can short-circuit impulses running to and from the cerebral cortex. In such cases, the person is unable to carry out motor activities initiated in the cortex or cannot express himself in speech. Incoming sensory impulses, such as those for vision, may be interrupted.

As the disease progresses, most voluntary motor control is eventually lost, and the patient becomes bedridden. Death occurs anywhere from 7 to 30 years after the first symptoms appear. The usual cause of death is a severe infection indirectly brought on by the loss of motor activity. For instance, the involuntary muscles that form the walls of the bladder often become paralyzed. When these muscles are unable to contract, the bladder cannot squeeze urine to the outside of the body. A tube may be inserted so that the urine can drain out. But without the squeezing action of the bladder walls, the bladder never totally empties. Some of the urine then stagnates, forming a good environment for bacteria. The patient undergoes bouts of bladder infection that easily spreads to the kidney and destroys the kidney cells.

The etiology of multiple sclerosis is unknown. Many researchers suspect that the disease is caused by a common virus that does not affect the myelin sheaths of most people. Other possible causes are metal poisons, accidental injury to the central nervous system, and diseases of the blood vessels that supply the central nervous system.

Cerebral vascular accidents (CVA)

By far the most common type of brain disorder is *stroke,* also called **cerebral apoplexy** and **cerebrovascular accident.** Stroke is the destruction of brain tissue resulting from any of several disorders in the vessels that supply the brain. Common causes of stroke are intracerebral hemorrhage, embolism, and arteriosclerosis of the cerebral arteries. An *intracerebral hemorrhage* is a rupture of one of the vessels within the pia mater or brain. Blood oozes into the brain and can damage neurons by increasing intracranial fluid pressure in all or part of the brain. An *embolus* is a blood clot or bit of foreign material, most often debris from an inflammation, that becomes lodged in an artery and blocks circulation. *Arteriosclerosis,* also called hardening of the arteries, is the formation of plaques on the walls of the arteries. The plaques may slow down circulation by partially closing off the vessel. Both thrombosis and arteriosclerosis cause brain damage by reducing or shutting off the supply of oxygen and glucose needed by brain cells.

Many elderly people suffer "little strokes." These are short periods of reduced blood supply usually caused by arteriosclerosis. Damage is generally either undetectable or very mild. During these mild strokes the person may have a short blackout, blurred vision, or dizziness, and does not realize that anything serious has occurred. However, a stroke can also cause sudden, massive damage. Severe stroke causes about 21 percent of all deaths from heart and blood vessel disease. If the person recovers, he may suffer partial paralysis, speech difficulty, mental disturbances, severe headaches, visual disturbances, deafness, and dizziness.

The type of malfunction following a stroke depends on the parts of the brain that were injured. Vascular disorders are more common after the age of 40, and thus stroke is primarily a disease of the middle aged and elderly.

Tumors of the nervous system

Brain tumors are generally divided into different classes that present a variety of symptoms, depending on their location and the extent of impairment. Classes include tumors of the skull, meninges, neuroglia, cranial nerves, pituitary, and pineal body. These tumors may be congenital, or they may occur at any age. They are most common, however, in early adult or middle life. Males and females are affected about equally. Malignant tumors of the neuroglia are called *gliomas,* and the most common of these are *astrocytomas.* Another common malignant tumor is the *medulloblastoma,* which occurs in the cerebellum of children and young adults. A *meningioma* is generally a benign tumor that arises in the meninges.

Symptoms arise when the tumor damages neurons by invading them directly, by putting pressure on other parts of the brain as the tumor enlarges, and by obstructing cerebrospinal flow.

In order to determine whether a patient has a brain tumor, the physician first tests the patient's reflexes, coordination, muscular strength, and ability to interpret stimuli correctly, and the functioning of certain involuntary muscles such as the muscles that control dilation of the pupil of the eye. In this way, the physician can determine what parts of the nervous system seem to be diseased. An EEG may be given to determine whether the brain is involved. If the cause of malfunction is a brain tumor, the tumor may, finally, be located through various techniques that provide pictures of the brain and its blood vessels.

Medical terminology associated with the nervous system

Analgesia   (*an* = without; *algia* = painful condition) Insensibility to pain.

Anesthesia   (*esthesia* = feeling) Loss of feeling.

Aphasia   (*a* = without; *phasis* = speech; *ia* = condition) Diminished or complete loss of ability to comprehend and/or express spoken or written words, due to injury or disease of the brain centers; most common cause is stroke.

Bacterial meningitis   Acute inflammation of the meninges caused by bacteria.

Coma   Abnormally deep unconsciousness with an absence of voluntary response to stimuli; varying degrees of reflex activity remain. May be due to illness or to an injury.

Neuralgia   (*neur* = nerve) Attacks of pain along the entire course or branch of a peripheral sensory nerve; one common type involves one or more branches of the trigeminal nerve and is called trigeminal neuralgia (tic douloureux).

Neuritis   Inflammation of a nerve; can result from irritation to the nerve produced by trauma, bone fractures, nutritional deficiency (usually thiamine), poisons such as carbon monoxide and carbon tetrachloride, heavy metals such as lead, and some drugs. Neuritis of the facial nerve that results in paralysis of facial muscles is called Bell's palsy.

Paralysis   Diminished or total loss of motor function resulting from damage to nervous tissue or a muscle.

Sciatica   Severe pain along the sciatic nerve and its branches. Usually due to rupture of an intervertebral disc or to osteoarthritis of the lower spinal column: the disc or arthritic joint puts pressure on the nerve root supplying the sciatic nerve and thereby causes the pain.

Shingles   Acute inflammation caused by a virus that attacks sensory cell bodies of dorsal root ganglia. Inflammation spreads peripherally along a spinal nerve and infiltrates dermis and epidermis over the nerve, producing a characteristic line of skin blisters.

Spastic   (*spas* = draw or pull) Resembling spasms or convulsions.

Spina bifida   (*bifid* = into two parts) An abnormality in one or many vertebral arches. The arches fail to fuse during embryonic development so that part of the spinal cord may be exposed.

Torpor   Abnormal inactivity or no response to normal stimuli.

Viral encephalitis   Acute inflammation of the brain caused by any of a number of viruses.

Chapter summary in outline

THE GROUPING OF NEURAL TISSUE

1. White matter is an aggregation of myelinated axons.
2. Gray matter is a collection of nerve cell bodies and dendrites or unmyelinated axons.
3. A nerve is a bundle of fibers outside the central nervous system.
4. A ganglion is a collection of cell bodies outside the central nervous system.
5. A bundle of fibers of similar function inside the central nervous system forms a tract.
6. A mass of nerve cell bodies and dendrites inside the brain forms a nucleus.
7. A horn is an area of gray matter in the spinal cord.

THE CENTRAL NERVOUS SYSTEM

The brain

1. Principal parts are the cerebrum, cerebellum, and brain stem, which contains the medulla, pons, midbrain, thalamus, and hypothalamus.
2. The cerebrum is the largest portion. Its surface (cortex) contains gyri, fissures, and sulci.
3. The functions of the cerebrum are motor (voluntary muscular movement), sensory (interpreting sensory impulses), and associational (emotional and intellectual processes).
4. Brain waves generated by the cerebral cortex are recorded on an EEG. They may be used to diagnose epilepsy, infections, and tumors.
5. The cerebellum is below and behind the cerebrum. It functions in the coordination of skeletal muscles and in posture and balance.
6. The medulla oblongata is continuous with the upper part of the spinal cord. It contains nuclei that are reflex centers for heartbeat, respiration, vasoconstriction, and dilatation.
7. The pons is above the medulla. It relays impulses concerned with voluntary skeletal movements from the cortex to the cerebellum.
8. The midbrain connects the pons and cerebellum with the cerebrum. It conveys motor impulses from the cerebrum to the cerebellum and cord, and conveys sensory impulses from cord to thalamus. It regulates auditory and visual reflexes.
9. The thalamus is located above the midbrain and serves as a relay station for impulses that govern proprioception.
10. The hypothalamus is under the thalamus. It controls the autonomic nervous system, regulates body temperature, food and fluid intake, waking state, and sleep.

The spinal cord

1. The gray matter in the spinal cord is divided into horns and the white matter into funiculi.
2. The cord serves as a two-way conduction system between brain and periphery and controls reflexes not mediated by cranial nerves.

Maintenance and protection

1. Interruption of the oxygen supply to the brain could result in paralysis, mental retardation, epilepsy, or death. Glucose deficiency may produce dizziness, convulsions, and unconsciousness.
2. The central nervous system is protected by bones and the meninges (dura mater, arachnoid, and pia mater).
3. Cerebrospinal fluid protects by serving as a shock absorber.

SOMATIC PORTION OF PERIPHERAL NERVOUS SYSTEM

1. The peripheral nervous system connects the central nervous system to other parts of the body.
2. The somatic portion consists of all neurons that relay motor impulses from the central nervous system to skeletal muscle and all neurons that relay sensory impulses to the central nervous system.

Cranial nerves

1. Twelve pairs of cranial nerves originate from the brain.
2. The pairs are named primarily on the basis of distribution and are numbered on the basis of order of origin.

Spinal nerves

1. Thirty-one pairs originate from the spinal cord. They are attached to the cord by a dorsal and a ventral root. All are mixed nerves.
2. Principal branches are dorsal, ventral, and visceral.
3. Ventral branches form plexuses (networks of fibers) before they innervate various parts of the body.
4. Peripheral nerve regeneration depends upon the presence of a neurilemma.

THE AUTONOMIC NERVOUS SYSTEM

1. This system automatically controls activities of smooth muscle, cardiac muscle, and glands (visceral effectors). All autonomic axons are efferent.
2. Efferent neurons are preganglionic (with myelinated axons) and postganglionic (with unmyelinated axons).
3. Autonomic ganglia are classified as sympathetic trunk ganglia (on sides of spinal column), prevertebral ganglia (anterior to spinal column), and terminal ganglia (near or inside visceral effectors).
4. Sympathetic responses are widespread and, in general, are concerned with energy expenditure. Parasympathetic responses are restricted and are typically concerned with energy restoration and conservation.

HOMEOSTASIS THROUGH REFLEXES

1. Reflexes represent principal mechanisms of the body for responding to changes in the internal and external environment.
2. Among clinically important reflexes are patellar reflex, Achilles reflex, Babinski reflex, and abdominal reflex.

DISORDERS OF THE NERVOUS SYSTEM

1. Poliomyelitis is a viral infection that results in paralysis.
2. Syphilis is caused by the bacterium *Treponema pallidum* and may result in blindness, memory defects, abnormal behavior, and loss of sensory functions in trunk and limbs.
3. Cerebral palsy includes a group of central nervous system disorders that primarily involve the cerebral cortex, cerebellum, and basal ganglia. The disorders damage motor centers.
4. Parkinsonism is a malfunction of the basal ganglia of the cerebrum caused by insufficient dopamine.
5. With epilepsy, the victim experiences convulsive seizures. It results from irregular electrical discharges of brain cells and may be diagnosed by an EEG.
6. Multiple sclerosis is the destruction of myelin sheaths of the neurons of the central nervous system. Impulse transmission is interrupted.
7. Cerebral vascular accidents are also called strokes. Brain tissue is destroyed due to hemorrhage, thrombosis, and arteriosclerosis.
8. Tumors may involve the neuroglia (gliomas), the cerebellum (medulloblastomas), and the meninges (meningiomas).

Review questions and problems

1. Define the following terms: white matter, gray matter, nerve, ganglion, tract, and nucleus.

2. How is the central nervous system differentiated from the peripheral nervous system? What are the components of the peripheral nervous system?

3. Categorize the brain into its three principal subdivisions.

4. What are the principal sulci and fissures of the cerebrum, and how are they related to the lobes of the cerebrum? What is the cerebral cortex?

5. What relationship do the names of the cranial bones have to the names of the lobes of the cerebrum?

6. Compare the fibers in the white matter of the cerebrum with regard to direction and function.

7. What are basal ganglia? List two and describe their possible functions.

8. Discuss the motor, sensory, and association functions of the cerebrum. What is aphasia? Agraphia? Word deafness? Word blindness?

9. What is an electroencephalograph? What is its diagnostic value?

10. Describe the structure and functions of the cerebellum. What are cerebellar peduncles? With which body system is the cerebellum most closely associated in terms of coordination?

11. How would damage to the cerebellum affect skeletal muscles?

12. What is ataxia? List several signs of ataxia.

13. What is the brain stem? Compare the medulla oblongata, pons, and midbrain with regard to structure and function.

14. Define decussation. Give an example.

15. What is the function of the thalamus?

16. Summarize the major functions of the hypothalamus.

17. Where is the spinal cord located? Describe its gross, external appearance.

18. Diagram and label a cross section of the spinal cord and explain the function of each area. What are the functions of the spinal cord?

19. Compare the ascending and descending tracts of the spinal cord with respect to location, origin, termination, and function.

20. Discuss the importance of a constant supply of oxygen and glucose to brain functioning.

21. How are the brain and cord protected?

22. Describe the composition, formation, and circulation of cerebrospinal fluid.

23. What is the difference between internal and external hydrocephalus? What is a lumbar puncture?

24. What is the somatic nervous system? How is it related to the peripheral nervous system?

25. What is a cranial nerve? Why are some cranial nerves classified as mixed nerves? How are cranial nerves named and numbered?

26. List the location and function of each of the cranial nerves.

27. Describe some of the abnormalities associated with damage to cranial nerves I, II, III, V, VI, VII, VIII, IX, X, XI, and XII.

28. What is a spinal nerve? How is a spinal nerve related to the spinal cord? How are spinal nerves named?

29. What are the three branches of a spinal nerve? Describe the distribution of each.

30. Define a plexus. What are the principal plexuses and what organs do each of the plexuses innervate?

31. What factors are related to the regeneration of peripheral nerves?

32. What are the principal components of the autonomic nervous system? What is its general function? Why is it called involuntary?

33. What would you say is the principal anatomical difference between the voluntary nervous system and the autonomic nervous system?

34. Relate the role of visceral efferent fibers and visceral effectors to the autonomic nervous system.

35. Distinguish the following with respect to location and function: preganglionic neurons and postganglionic neurons.

36. What is an autonomic ganglion? Describe the location of the three types. Define white ramus and gray ramus.

37. On what basis are the sympathetic and parasympathetic divisions of the autonomic nervous system differentiated anatomically and functionally?

38. Discuss the distinction between cholinergic and andrenergic fibers of the autonomic nervous system.

39. Give several examples of the antagonistic effects of the sympathetic and parasympathetic divisions of the autonomic nervous system.

40. Can you summarize the principal functional differences between the voluntary nervous system and the autonomic nervous system?

41. Below is a diagram of the human body as it might appear in a fear situation. Specific parts of the body have been labeled for you. Write the *sympathetic response* for each labeled part.

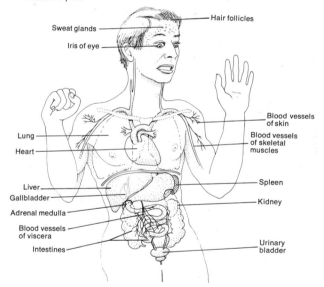

42. Define a reflex. How are reflexes related to the maintenance of homeostasis?

43. Distinguish the following kinds of reflexes: somatic, visceral, exteroceptive, visceroceptive, and proprioceptive.

44. Indicate the clinical importance of the following reflexes: patellar, achilles, Babinski, and abdominal.

45. Summarize the ways in which homeostasis is disrupted by the following conditions: poliomyelitis, syphilis, cerebral palsy, Parkinsonism, epilepsy, multiple sclerosis, and cerebral vascular accidents. Knowledge of the clinical symptoms will help you to formulate your response.

46. What is the diagnostic usefulness of the EEG for epilepsy?

47. Contrast the various kinds of epilepsy with regard to clinical symptoms.

48. What are the principal kinds of tumors of the nervous system?

49. Refer to the glossary of medical terminology associated with the nervous system. Be sure that you can define each term.

CHAPTER 14

THE PERCEPTION OF SENSATIONS

STUDENT OBJECTIVES

After you have read this chapter, you should be able to:

1. Define a sensation and list the four prerequisites necessary for its transmission

2. Define projection, adaptation, after images, and modality as characteristics of sensations

3. Compare the location and function of exteroceptors, visceroceptors, and proprioceptors

4. Describe the distribution of cutaneous receptors by interpreting the results of the two-point discrimination test

5. List the location and function of the receptors for touch, pressure, cold, heat, pain, and proprioception

6. Distinguish between somatic, visceral, referred, and phantom pain

7. Define acupuncture and discuss its possible use as a pain reliever

8. Locate the receptors for olfaction and describe the neural pathway for smell

9. Identify the gustatory receptors and describe the neural pathway for taste

10. Describe the structure and physiology of the accessory visual organs

11. List the structural divisions of the eye

12. Discuss retinal image formation by describing refraction, accommodation, constriction of the pupil, convergence, and inverted image formation

13. Define emmetropia, myopia, hypermetropia, and astigmatism

14. Diagram and discuss the rhodopsin cycle responsible for light sensitivity of rods

15. Describe the afferent pathway of light impulses to the brain

16. Define the anatomical subdivisions of the ear

17. List or describe the principal events involved in the physiology of hearing

18. Identify the receptor organs for equilibrium

19. Discuss the receptor organs' roles in the maintenance of dynamic and static equilibrium

20. Contrast the causes and symptoms of cataracts, glaucoma, conjunctivitis, trachoma, Ménière's disease, and impacted cerumen

21. Define medical terminology associated with the sense organs

Your ability to "sense" stimuli is vital to your survival. If you could not "sense" pain, you would probably not remove your hand quickly from a hot pot handle. You would not be alerted to the fact that your appendix was about to burst, your heart was malfunctioning, you were developing a cavity in a tooth, or an ulcer was forming in your stomach. In all these cases a great deal of damage could be done to your body before you did something about the situation. If you could not see, you might run the risk of physical injury whenever you faced an unfamiliar obstacle. If you could not smell, you might breathe in a harmful gas. If you could not hear, you might not notice verbal warnings, a car honking, or an aggressor sneaking up. If you could not taste, you might swallow some kind of poisonous material. In short, if you could not "sense" your environment and make the necessary homeostatic adjustments, you could not survive on your own.

As vital as the ability to "sense" stimuli is to your survival, it also provides us with many pleasures. Think of the pleasures involved in holding the hand of a loved one, or listening to your favorite music, or enjoying the taste of your favorite food, or smelling a flower, or seeing a ballet.

In this chapter we shall look at the different kinds of sensations, how we detect sensations, how they are transmitted to and interpreted by the nervous system, and what happens if sensory receptors are damaged or diseased.

In its broadest context, the term **sensation** refers to a state of awareness of external or internal conditions of the body. In order for a sensation to occur, four prerequisites must be met: (1) A *stimulus,* or change in the environment, capable of initiating a response by the nervous system must be present. (2) A *receptor* or sense organ must pick up the stimulus and convert it to a nervous impulse. (3) The impulse must be *conducted* along a nervous pathway from the receptor or sense organ to the brain. (4) A region of the brain must *translate* the impulse into a sensation. **A sense receptor** or **sense organ** may be viewed as specialized nervous tissue that exhibits a high degree of sensitivity to internal or external conditions.

A receptor may be quite simple. For example, it might be just the dendrites of a single neuron that picks up the sensation of pain in the skin. Or, it may be a complex organ, such as the eye, that contains highly specialized neurons, epithelium, and connective tissues. Regardless of complexity, though, all sense receptors contain the dendrites of sensory neurons. The dendrites occur either alone or in close association with specialized cells belonging to other types of tissues. Receptors exhibit a very high degree of excitability, and their threshold stimulus is low. Except for receptors associated with pain, each is specialized for receiving a particular kind of stimulus.

The majority of sensory impulses are conducted to the sensory areas of the cerebral cortex because it is only in this region of the body that a stimulus can produce conscious feeling. Sensory impulses that terminate in the spinal cord or brain stem can initiate motor activities, but they can never produce conscious sensations. Once a stimulus is received by a receptor and converted into an impulse, the impulse is conducted along an afferent pathway that enters either the spinal cord or the brain.

SENSATIONS

Before studying the receptors and neural pathways of various senses, you will want to know something about the general characteristics of sensations and their basic classifications.

Characteristics

Although it may seem contrary to personal experience, conscious sensations occur in the cortical regions of the brain. In other words, you see, hear, and feel pain in the brain. We seem to see with our eyes, hear with our ears, and feel pain in an injured part of our body only because the cortex interprets the sensation as coming from the stimulated sense receptor. The term *projection* describes this process by which the brain refers sensations to their point of stimulation.

A second characteristic of many sensations is *adaptation*. According to this phenomenon, a sensation may disappear even though a stimulus is still being applied. For example, when you first get into a tub of hot water, you might feel an intense burning sensation. But, after a brief period of time, the sensation decreases to one of comfortable warmth, even though the stimulus (hot water) is still present. Other examples of adaptation include placing a ring on your finger, putting on your shoes or hat, and sitting on a chair. Initially, you are conscious of the sensations involved, but they are lost soon thereafter.

Sensations may also be characterized by *afterimages*. That is, some sensations tend to persist even though the stimulus has been removed. This phenomenon is just the opposite of adaptation. One common example of afterimage occurs when you look at a bright light and then look away or close your eyes. You still see the light for several seconds or minutes afterward.

A fourth characteristic of sensations is *modality*, which refers to the specific kind of sensation felt. The sensation may be one of pain, pressure, touch, body position, equilibrium, hearing, vision, smell, or taste. In other words, the distinct properties by which one sensation may be distinguished from another is its modality. As sensory signals enter the brain, they are sorted according to their modality and transferred to the appropriate region where a specific sensation is produced.

Classification

One convenient method of classifying sensations is to categorize them according to the location of the receptor. On this basis, receptors may be classified as exteroceptors, visceroceptors, and proprioceptors. **Exteroceptors** provide information about the external environment. They pick up stimuli outside the body and transmit sensations of hearing, sight, touch, pressure, temperature, and pain on the skin. Exteroceptors are located near the surface of the body.

Visceroceptors pick up information about the internal environment. These sensations arise from within the body and may be felt as pain, taste, fatigue, hunger, thirst, and nausea. Visceroceptors are located in blood vessels and viscera.

Proprioceptors allow us to feel position and movement. Such sensations give us information about muscle tension, the position and tension of our joints, and equilibrium. These receptors provide information about body position and movements.

Sensations may also be classified according to the simplicity or complexity of the receptor and the neural pathway involved. *General senses* are those that involve a relatively simple receptor and neural pathway. In addition, the receptors for general sensations are numerous and are found throughout widespread areas of the body. Examples include cutaneous sensations, such as touch, pressure, heat, cold, and pain. *Special senses*, by contrast, involve rather complex receptors and neural pathways. The receptors for each special sense are found in only one or two specific areas of the body. Among the special senses are smell, taste, sight, and hearing. In the following sections, we shall cover both general and special senses.

GENERAL SENSES

Your skin contains the receptor organs for many general senses. Receptor organs are also located in your muscles, tendons, joints, subcutaneous tissue, and viscera.

Cutaneous sensations

Touch, pressure, cold, heat, and pain are known as the **cutaneous sensations.** The receptors for these sensations are located in the skin, connective tissue, and the ends of the gastrointestinal tract. Inasmuch as the sensation of pain is not limited to cutaneous receptors, we shall consider pain under a separate heading. The cutaneous receptors are randomly distributed over the body surface so that some parts of the skin are densely populated with receptors and other parts contain only a few, widely separated ones. Areas of the body that have few cutaneous receptors are relatively insensitive, whereas those regions containing large numbers of cutaneous receptors are very sensitive. This can be demonstrated by using the *two-point discrimination test* for touch (Figure 14–1). In this test, a compass is applied to the skin, and the distance in millimeters between the two points of the compass is varied. The subject then indicates when he feels two points and when he feels only one.

The compass may be placed on the tip of the tongue, an area where receptors are very densely

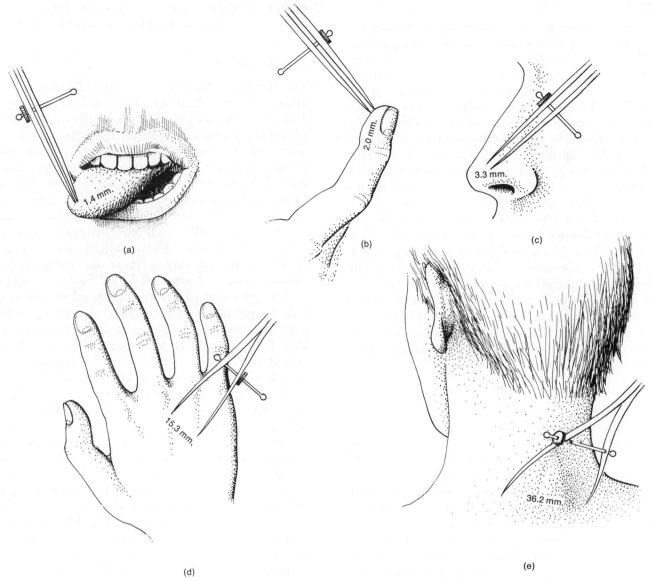

Figure 14-1. Two-point discrimination test for touch. (a) Tip of tongue. (b) Tip of finger. (c) Side of nose. (d) Back of hand. (e) Back of neck.

packed. The distance between the two points can then be narrowed to 1.4 millimeters. At this distance, the points are able to stimulate two different receptors, and the subject feels that he is being touched by two objects. However, if the distance is decreased below 1.4 millimeters, he feels only one point, even though both points are touching him. This is because the points are so close together that they reach only one receptor. The compass can then be placed on the back of the neck, where receptors are relatively few and far between. In this case, the subject feels two distinctly different points only if the distance between them is 36.2 millimeters or more.

The results of this test indicate that the more sensitive the area, the closer the compass points may be placed and still be felt separately. The following order, from greatest sensitivity to least, has been established from the test: tip of tongue, tip of finger, side of nose, back of head, and back of neck.

Cutaneous receptors have relatively simple structures. Basically, they consist of the dendrites of one or several sensory neurons that may or may not be enclosed in a capsule of epithelial or connective tissue. Impulses received by cutaneous touch receptors pass into spinal and cranial nerves, through the thalamus, to the general sensory area of the parietal lobe of the cortex.

Touch

Cutaneous receptors for **touch** include Meissner's corpuscles, Merkel's discs, and root hair plexuses (Figure 14–2). *Meissner's corpuscles* are egg-shaped receptors in which a mass of dendrites is enclosed by epithelium. They are found in the papillae of the skin and enable us to detect when two points on the skin are being touched. Meissner's corpuscles are most numerous in the fingertips, palms of the hand, and soles of the feet. They are also abundant in the eyelids, tip of the tongue, lips, nipples, clitoris, and tip of the penis. *Merkel's discs* are receptors for touch that consist of disclike formations of dendrites attached to deeper layers of epidermal cells. They are distributed in many of the same locations as Meissner's corpuscles. *Root hair plexuses* are dendrites arranged in networks around the roots of hairs. If a hair shaft is moved, the dendrites are stimulated. Since the root hair plexuses are not surrounded by any supportive or protective structures, they are called *free*, or *naked, nerve endings.*

Pressure

Sensations of **pressure** are longer lasting and have less variation in intensity than do sensations of touch. Moreover, whereas touch is felt in a small, "pinprick" area, pressure is felt over a much larger area. The pressure receptors are oval-shaped structures called *Pacinian corpuscles* (Figure 14–2). They are composed of a capsule resembling an onion made from layers of connective tissue enclosing dendrites. Pacinian corpuscles are found in the deep subcutaneous tissues that lie under mucous membranes, in serous membranes of the abdominal cavity, around joints and tendons, and in some viscera.

Cold

The cutaneous receptors for the sensation of **cold** are called *end bulbs of Krause* (Figure 14–2). The commonest form of these receptors is an oval-shaped connective tissue capsule containing dendrites. They are widely distributed in the dermis and subcutaneous connective tissue and are also located in the cornea of the eye, the tip of the tongue, and external genitals.

Heat

The cutaneous receptors for **heat** are referred to as *end organs of Ruffini* (Figure 14–2). The end organs of Ruffini are deeply embedded in the dermis and are less abundant than cold receptors.

Pain sensations

The receptors for **pain** are simply the branching ends of the dendrites of certain sensory neurons. Pain receptors are found in practically every tissue of the body and adapt only slightly or not at all. They may be stimulated by any type of stimulus. Excessive stimulation of any sense organ causes pain. For example, when stimuli for other sensations such as touch, pressure, heat, and cold reach a certain threshold, they stimulate pain receptors as well. Additional stimuli for pain receptors include excessive distention or dilation of an organ, prolonged muscular contractions, muscle spasms, inadequate blood flow to an organ, or the presence of certain chemical substances. Pain receptors, because of their sensitivity to all stimuli, have a general protective function in that they inform us of changes that could be potentially dangerous to health or life. Adaptation to pain does not readily occur. This is rather important since pain indicates disorder or disease. If we became used to it and ignored it, irreparable damage could result.

Sensory impulses for pain are conducted through spinal and cranial nerves. The lateral spinothalamic tract of the cord relays impulses to the thalamus. From here, the impulses are relayed to the postcentral convolution of the parietal lobe. Recognition of the kind and intensity of pain is then localized in the cerebral cortex.

In general, pain may be divided into two types: somatic and visceral. **Somatic pain** arises from stimulation of the skin receptors. In this case, it is called superficial somatic pain. It may also arise from stimulation of receptors in skeletal muscles, joints, tendons, and fascia, in which case it is called deep somatic pain. **Visceral pain** results from stimulation of receptors in the viscera.

In most cases of somatic pain and in some instances of visceral pain, the cortex accurately projects the pain back to the stimulated area. For example, if you burn your finger, you feel the pain in your finger. Also, an individual with inflammation of the lining of the pleural cavity experiences pain in the affected area. In most instances of visceral pain, however, the sensation of pain is not projected back to the point of stimulation. Rather, the pain may be felt in or just under the skin that overlies the stimulated organ. Or, the pain may be felt in a surface area of the body

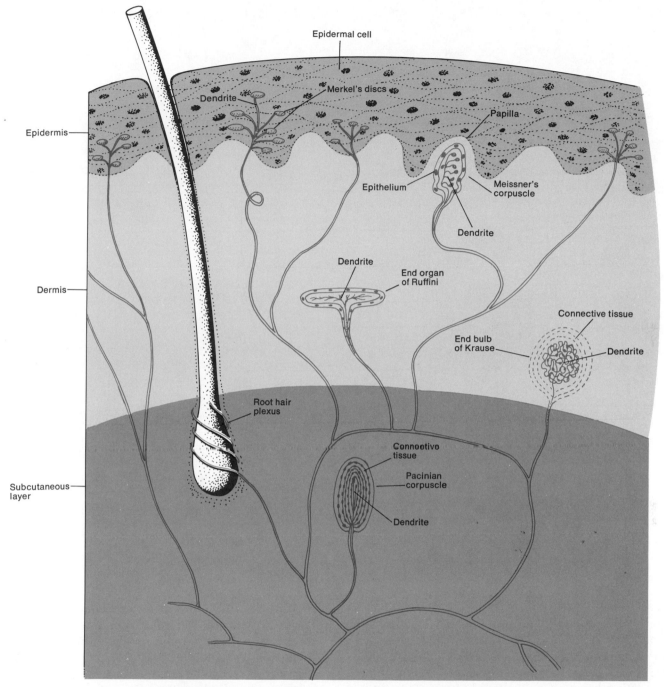

Figure 14–2. Structure and location of cutaneous receptors.

that is quite far removed from the stimulated organ. This phenomenon is called **referred pain.** In general, the area to which the pain is referred and the visceral organ involved receive their innervation from the same segment of the spinal cord. Consider the following example: Afferent fibers from the heart enter segments T1 to T4 of the spinal cord as do afferent fibers from the skin over the heart and the skin over the medial surface of the left arm. Thus, the pain of a heart attack is typically felt in the skin over the heart and down the left arm. Figure 14–3 illustrates cutaneous regions to which visceral pain may be referred.

A kind of pain frequently experienced by amputees is called **phantom pain.** In this instance,

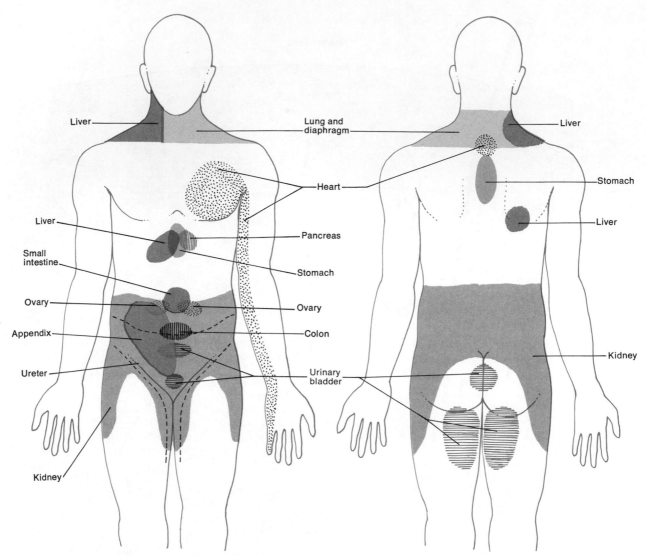

Figure 14–3. Referred pain. The shaded parts of the diagram indicate cutaneous areas to which visceral pain is referred.

the person experiences pain in a limb or part of a limb after it has been amputated. Here is how phantom pain occurs: Let us say that a foot has been amputated. A sensory nerve that originally terminated in the foot is severed during the operation but repairs itself and returns to function within the remaining leg. From past experience the brain has always projected stimulation of the neuron back to the foot. So when the distal end of this neuron is now stimulated, the brain continues to project the sensation back to the missing part. Thus, even though the foot has been amputated, the patient still "feels" pain in his toes.

Most pain sensations respond to pain-killing drugs. In a few individuals, however, pain may be controlled only by surgery. The aim of surgical treatment is to interrupt the pain impulse somewhere between the receptors and the interpreta-

tion centers of the brain. This is done by severing the sensory nerve, its spinal root, or certain tracts in the spinal cord or brain. Examples are sympathectomy, which is excision of some of the neural tissue of the autonomic nervous system, cordotomy, which is severing of a spinal cord tract, and prefrontal lobotomy, which is the destruction of tracts that connect the thalamus with the prefrontal and frontal lobes of the cerebral cortex. In each instance, the pathway for pain from a given receptor is severed so that the pain impulses are no longer conducted to the cortex.

Acupuncture and pain

Another method of inhibiting pain impulses is called **acupuncture.** The word comes from the terms *acus*, meaning needle, and *pungere*, meaning to sting. Here is how the procedure is performed: Needles are inserted through selected areas of the skin. The needles are then twirled by the acupuncturist or by a battery-operated device. In about 20 to 30 minutes after the twirling starts, pain is deadened for 6 to 8 hours. The location of needle insertion varies with the part of the body the acupuncturist desires to anesthetize. To pull a tooth, one needle is inserted in the web between the thumb and the index finger. For a tonsillectomy, one needle is inserted about 2 inches above the wrist. For removal of the lung, one needle is placed in the forearm, midway between the wrist and the elbow.

There is no complete or totally satisfactory explanation of why acupuncture works. According to the "gate control" theory, the twirling of the acupuncture needle stimulates two sets of nerves that eventually enter the spinal cord and synapse with the same association neurons. One very fine nerve is the nerve for pain and the other, a much thicker nerve, is the nerve for touch. The speed of the impulse passing along the touch nerve is faster than that passing along the pain nerve. Recall that fibers with larger diameters conduct impulses faster than those with smaller diameters. Because the touch impulse reaches the dorsal horn of the cord first, it has right of way over the pain impulse. It thus "closes the gate" to the brain before the pain impulse reaches the cord. Since the pain impulse does not pass to the brain, no pain is felt. We should mention, however, that before acupuncture can be used as a routine procedure by American physicians, for the well-being of the public additional research and understanding of the process must take place.

Proprioceptive sensations

An awareness of the activities of muscles, tendons, and joints is provided by the **proprioceptive** or **kinesthetic sense.** It informs us of the degree to which tendons are tensed and muscles are contracted. The proprioceptive sense enables us to recognize the location and rate of movement of one part of the body in relation to other parts. It also allows us to estimate weight and to determine the muscular work necessary to perform a task. With the proprioceptive sense, we can judge the position and movements of our limbs without using our eyes when we walk, type, play a musical instrument, or dress in the dark.

Proprioceptive receptors are located in mus-

cles, tendons, and joints and in the connective tissue that surrounds muscle fibers. The receptors are fairly simple in structure and similar to cutaneous receptors. Among the proprioceptors are two types of sensory neurons that originate on the sarcolemma of skeletal muscle fibers. When the fiber is stretched, one kind of neuron is stimulated. When the fiber reaches maximum contraction, the other type of neuron is stimulated. Changes in the shape of a muscle stimulate a different kind of receptor, which consists of naked nerve endings in the connective tissue between the fibers and in the fascia surrounding the muscle. When a muscle is partially contracted, the resulting moderate stretch on its tendon stimulates still other sensory neurons that are wrapped around the fibers of the tendon.

Proprioceptors adapt only slightly. This is beneficial since the brain must be appraised of the status of different parts of the body at all times so that adjustments can be made to ensure coordination.

The afferent pathway for muscle sense consists of impulses sent from proprioceptors into cranial and spinal nerves. Impulses for conscious proprioception travel in ascending tracts in the cord, where they are relayed to the thalamus and cerebral cortex. The sensation is registered in the general sensory area in the parietal lobe of the cortex posterior to the central fissure. Proprioceptive impulses that result in reflex action travel to the cerebellum through spinocerebellar tracts.

SPECIAL SENSES

In contrast to the general senses, the four special senses of smell, taste, sight, and hearing have receptor organs that are highly complex. The sense of smell is the least specialized, as opposed to the sense of sight, which is the most specialized. Like the general senses, however, the special senses allow us to detect changes in our environment. First, you will read about the anatomy of each special sense. Then, you will be able to understand the physiology of their respective functions.

Olfactory sensations

The receptors for the **olfactory sense** are located in the nasal epithelium at the upper portion of each nasal cavity on either side of the nasal septum (Figure 14–4a). The nasal epithelium consists of two principal kinds of cells: supporting and olfactory (Figure 14–4b). The *supporting cells*

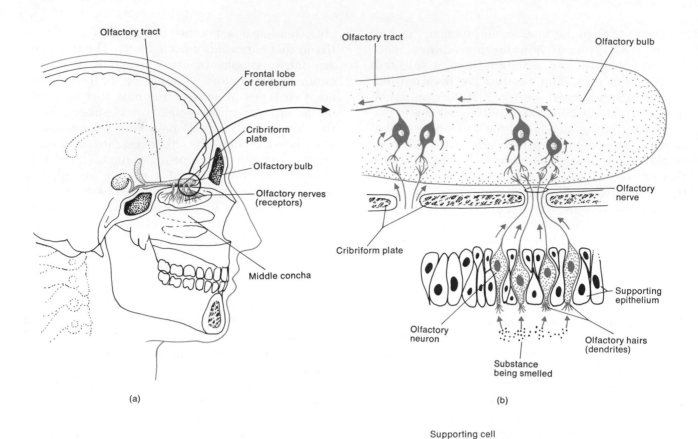

Figure 14-4. Receptors for olfaction. (a) Location of receptors in nasal cavity. (b) Enlarged aspect of olfactory receptors. *(Photograph courtesy of Donald I. Patt, from Comparative Vertebrate Histology, Donald I. Patt and Gail R. Patt, Harper & Row, Publishers, Inc., New York, 1969. Magnification × 450.)*

are columnar epithelial cells of the mucous membrane lining the nose. The *olfactory cells* are bipolar neurons whose cell bodies lie between the supporting cells. The distal (free) end of each olfactory cell contains six to eight dendrites, called *olfactory hairs*. The axons of the olfactory cells unite to form several nerves, which pass through openings in the cribriform plate. These olfactory nerves terminate in masses of gray matter called the *olfactory bulbs*. The olfactory bulbs lie beneath the frontal lobes of the cerebrum on either side of the crista galli of the ethmoid bone. The first synapse of the olfactory neural pathway occurs in the olfactory bulbs between the axons of the olfactory nerves and the dendrites of neurons

inside the olfactory bulbs. Axons of these neurons run posteriorly to form the *olfactory tract*. From here, impulses are conveyed to the olfactory portion of the cortex. In the cortex, the impulses are interpreted as odor and give rise to the sensation of smell.

The mechanism by which the stimulus for smell is converted to a nerve impulse is explained by three widely accepted theories. One theory holds that substances capable of producing odors emit gaseous particles. On entering the nasal cavities, these particles become dissolved in the mucus secretions of the nasal membrane. This fluid then acts chemically on the olfactory hairs to create a nerve impulse.

According to the second theory, radiant energy given off by the molecules of the stimulating substance is the stimulus rather than the molecules themselves. The third theory purports that substances detected by smell are usually soluble in fat. Since the membrane of an olfactory hair is largely fat, it is assumed that molecules of substances to be smelled are dissolved in the membrane where they initiate a nerve impulse.

The sensation of smell happens quickly, but adaptation to odors also occurs rapidly. For this reason, we become accustomed to some odors and are also able to endure unpleasant ones. Rapid adaptation also accounts for the failure of a person to detect a gas that accumulates slowly in a room. The cortex stores memories of odors quite well. Once you have smelled a substance, you generally recognize its odor if you smell it again.

The supporting cells of the nasal epithelium and tear ducts are innervated by branches of the trigeminal nerve, or cranial nerve V. These neurons receive stimuli of pain, cold, heat, tickling, and pressure. Olfactory stimuli, such as pepper, onions, ammonia, ether, and chloroform, are irritating and may cause tearing because they stimulate the receptors of the trigeminal nerve as well as the olfactory neurons.

Gustatory sensations

The receptors for **gustatory sensations,** or sensations of taste, are located in the taste buds (Figure 14–5b). Taste buds are most numerous on the tongue, but they are also found on the soft palate and in the throat. The *taste buds* are oval-shaped bodies consisting of two kinds of cells. The *supporting cells* are specialized epithelium that forms a capsule. Inside each capsule are four to twenty *gustatory cells*, which are the sensory neurons for taste. Each gustatory cell contains a dendrite that projects to the external surface through an opening in the taste bud called the *taste pore*. Gustatory cells make contact with taste stimuli through the taste pore.

Taste buds are found in some connective tissue elevations on the tongue called **papillae** (Figure 14–5a). The papillae give the upper surface of the tongue its characteristic rough appearance. *Circumvallate papillae,* the largest type, are circular and form an inverted V-shaped row at the posterior portion of the tongue. *Fungiform papillae* are knoblike elevations and are found primarily on the tip and sides of the tongue. All circumvallate and most fungiform papillae contain taste buds. *Filiform papillae* are threadlike

structures that cover the anterior two-thirds of the tongue. (See Figure 20–5.)

In order for gustatory cells to be stimulated, the substances we taste must be in solution in the saliva so that they can enter the taste pores in the taste buds. Despite the many substances we taste, there are basically only four taste sensations: sour, salt, bitter, and sweet. Each taste is due to a different response to different chemicals. Certain regions of the tongue react more strongly than other regions to particular taste sensations. For example, although the tip of the tongue reacts to all four taste sensations, it is highly sensitive to sweet and salty substances. The posterior portion of the tongue is highly sensitive to bitter substances, and the lateral edges of the tongue are more sensitive to sour substances. (See Figure 20–5.)

The nerves that supply afferent fibers to taste buds are the facial nerve (VII), which supplies the anterior two-thirds of the tongue; the glossopharyngeal (IX), which supplies the posterior one-third of the tongue; and the vagus (X), which supplies the epiglottis area of the throat. Taste impulses are conveyed from the gustatory cells in taste buds to the three nerves just cited. From these, the impulses enter the medulla, pass through the thalamus, and terminate in the parietal lobe of the cortex.

Visual sensations

The structures related to **vision** are the eyeball, which is the receptor organ for visual sensations, the optic nerve, the brain, and a number of accessory structures.

Accessory structures of the eye

Among the *accessory organs* are the eyebrows, eyelids, eyelashes, and the lacrimal apparatus, which allows us to tear (Figure 14–6). The *eyebrows* protect the eyeball from falling objects, prevent perspiration from getting into the eye, and shade the eye from the direct rays of the sun. The *eyelids,* or *palpebrae,* consist primarily of smooth muscle that is covered externally by skin. The underside of the muscle is lined with a mucous membrane called the *conjunctiva,* which also covers the surface of the eyeball. Inflammation of this membrane is known as *conjunctivitis* or pinkeye. The borders of the eyelids, which contain the roots of the eyelashes, are thickened folds of connective tissue, called the *tarsal plates.* Attached to the upper portion of the lid

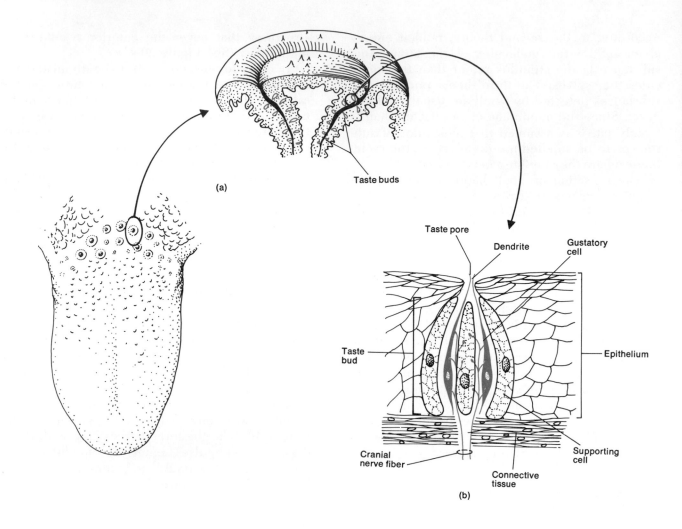

(a)

Taste buds

Taste pore

Dendrite

Gustatory cell

Taste bud

Epithelium

Cranial nerve fiber

Connective tissue

Supporting cell

(b)

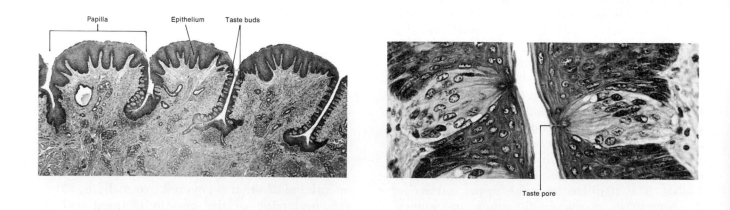

Papilla

Epithelium

Taste buds

Taste pore

Figure 14–5. Taste receptors. (a) Location of taste buds relative to papillae. (b) Structure of a taste bud. *(Photographs courtesy of Edward J. Reith, from Atlas of Descriptive Histology, Edward J. Reith and Michael H. Ross, Harper & Row, Publishers, Inc., New York, 1970.)*

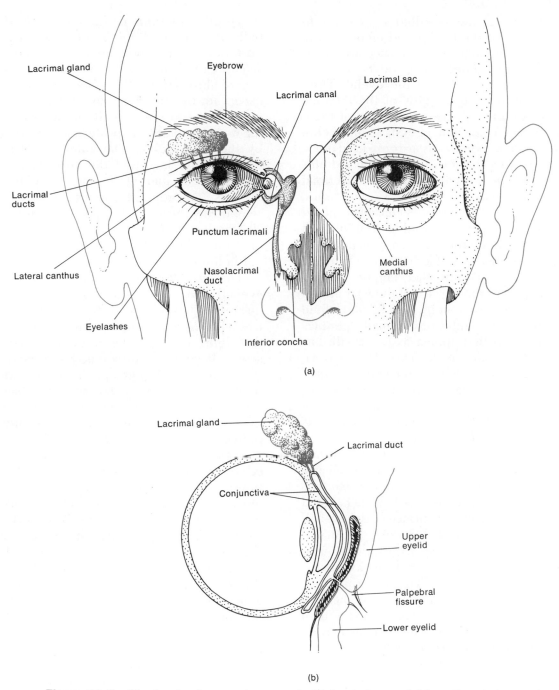

(a)

(b)

Figure 14–6. The lacrimal apparatus seen in (a) front view and (b) sectional view.

is the levator palpebrae superioris, a striated muscle which raises the lid. Another striated muscle, the *orbicularis* oculi, is arranged as a sphincter around both lids. Its function is to close the eyelids. The gap between the upper and lower eyelids that exposes the eyeball is called the *palpebral fissure*. Each corner of the eye is referred to as a *canthus*. The lateral canthus is closer to the temporal bones, whereas the

medial canthus is nearer the nasal bones. The eyelids shade the eye during sleep, protect the eye from light rays and foreign objects, and spread lubricating secretions over the surface of the eyeball.

Projecting from the border of each eyelid is a row of short, thick hairs, the *eyelashes*. Sebaceous glands at the base of the hair follicles of the eyelashes pour a lubricating fluid into the follicles.

Infection of these glands is called a *sty*. Close to the inner margin of the eyelid and embedded in the tarsal plates of each lid are sebaceous glands called *meibomian glands*. The ducts of these glands open onto the edge of the eyelid. Their secretion lubricates the margins of the lids and prevents the overflow of tears. Infection of these glands produces a chalazion, or cyst, on the eyelid.

The *lacrimal apparatus* is a term used for a group of structures that manufactures and drains away tears. These structures are the lacrimal glands, the lacrimal ducts, the lacrimal sacs, and the nasolacrimal ducts. A *lacrimal gland* is located at the superior lateral portion of both orbits. Each is about the size and shape of an almond. Leading from the lacrimal glands are six to twelve excretory ducts that empty lacrimal fluid, or tears, on to the surface of the conjunctiva of the upper lid. From here, the lacrimal fluid passes medially and enters two small openings (puncta lacrimalia) that appear as two small dots at the medial canthus of the eye. The lacrimal secretion then passes into two ducts, the *lacrimal canals*, and is next conveyed into the lacrimal sac. The *lacrimal sac* is the superior expanded portion of the *nasolacrimal duct*, a canal that transports the lacrimal secretion into the nose.

The *lacrimal secretion* is a watery solution containing salts and some mucus. It cleans, lubricates, and moistens the external surface of the eyeball. After being secreted by the lacrimal glands, it is spread over the surface of the eyeball by the blinking action of the eyelids. Normally, the secretion is carried away by evaporation or by passing into the lacrimal canals as fast as it is produced. If, however, an irritating substance makes contact with the conjunctiva, the lacrimal glands are stimulated to oversecrete. Tears then accumulate more rapidly than they can be carried away. This is a very important protective mechanism since the tears dilute and wash away the irritating substance. "Watery" eyes also occur when an inflammation of the nasal mucosa, such as a cold, obstructs the nasolacrimal ducts so that drainage of the tears is blocked.

Structure of the eye

The adult **eye** measures about 1 inch in diameter. Of its total surface area, only the anterior one-sixth is exposed. The remainder is recessed and protected by the orbit into which it fits. Anatomically, the eyeball can be divided into three principal layers: (1) fibrous tunic, (2) vascular tunic, and (3) retina (Figure 14–7a).

The *fibrous tunic* is the outer coat of the eyeball. It can be divided into two regions, the posterior sclera and the anterior cornea. The *sclera*, called the "white of the eye," is a white coat of fibrous tissue that covers the entire eyeball, except for the anterior colored portion. The sclera gives shape to the eyeball and affords protection for its inner parts. Its posterior surface is pierced by the optic nerve. The anterior portion of the fibrous tunic is called the *cornea*. It is a nonvascular, transparent fibrous coat that covers the iris, the colored part of the eye. Like the sclera, the cornea is composed of fibrous tissue. The outer surface of the cornea contains an epithelial layer that is continuous with the epithelium of the conjunctiva.

The *vascular tunic* is the middle layer of the eyeball and is composed of three portions: the posterior choroid, the anterior ciliary body, and iris. The *choroid* is a thin, dark brown membrane that lines most of the internal surface of the sclera. It contains numerous blood vessels and a large amount of pigment. The choroid absorbs light rays so they are not reflected back on the lens. Through its blood supply, it also maintains the nutrition of the retina. The optic nerve pierces the choroid at the back of the eyeball. The anterior portion of the choroid becomes the *ciliary body*. This second division of the vascular tunic contains the ciliary muscle—a smooth muscle that alters the shape of the lens for near or far vision. The *iris* is the third portion of the vascular tunic. It consists of circular and radial smooth muscle fibers arranged to form a doughnut-shaped structure. The black hole in the center of the iris is the *pupil*, the area through which light enters the eye. The iris is suspended between the cornea and the lens and is attached at its outer margin to the ciliary body. One of the principal functions of the iris is to regulate the amount of light entering the eye. For example, when the eye is stimulated by bright light, the circular muscles of the iris contract and decrease the size of the pupil. When the eye must adjust to dim light, the radial muscles of the iris contract and increase the size of the pupil. Muscles of the eye, such as the ciliary, radial, and circular muscles, originate and insert within the eyeball and adjust the eye internally for vision. They are referred to as *intrinsic eye muscles*. Other muscles, such as the recti and oblique muscles, originate in the orbit and insert on the outside surface of the eyeball. (See Exhibit 11–4.) They move the eyeball in various directions and are called *extrinsic eye muscles*.

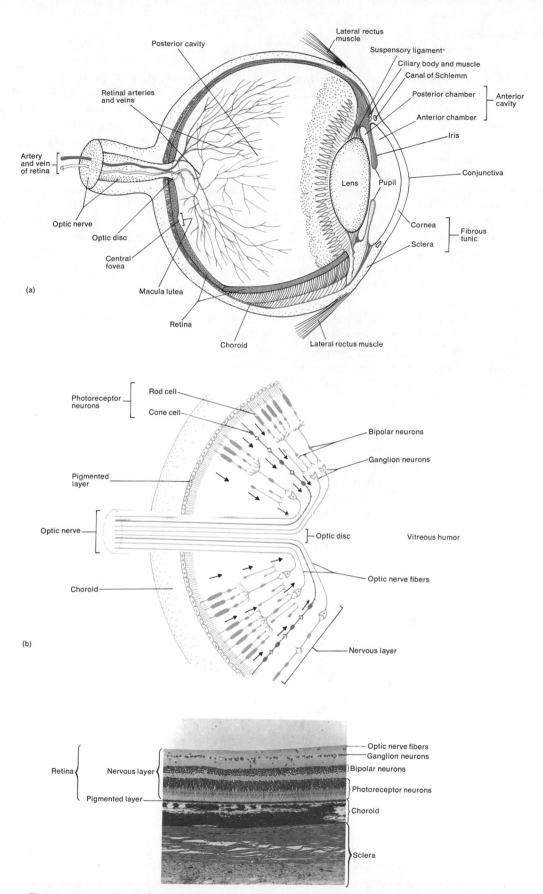

Figure 14–7. Structure of the eye. (a) Gross anatomy. (b) Microscopic structure of retina. [*(b) courtesy of Donald I. Patt, from Comparative Vertebrate Histology, Donald I. Patt and Gail R. Patt, Harper and Row, Publishers, Inc., New York, 1969. Magnification ×100.*]

The third and inner coat of the eye, the *retina*, lies only in the posterior portion of the eye. Its primary function is image formation. It consists of a nervous tissue layer and pigmented layer (Figure 14–7a, b). The outer pigmented layer is comprised of epithelial cells lying in contact with the choroid. The inner nervous layer contains three zones of neurons. These three zones, named in the order in which they conduct impulses, are the photoreceptor neurons, bipolar neurons, and ganglion neurons.

The dendrites of the photoreceptor neurons are called rods and cones because of their respective shapes. They are visual receptors highly specialized for stimulation by light rays. **Rods** are specialized for vision in dim light. (See Figure 3–7b.) They also allow us to discriminate between different shades of dark and light and permit us to see shapes and movement. **Cones,** by contrast, are specialized for color vision and for sharpness of vision, called *visual acuity.* Cones are stimulated only by bright light. This is why we cannot see color by moonlight. It is estimated that there are 7 million cones and somewhere between ten and twenty times as many rods. Cones are most densely concentrated in the *central fovea*, a small depression in the center of the macula lutea. (See Figure 14–7a.) The *macula lutea*, or yellow spot, is situated in the exact center of the retina. The fovea is the area of sharpest vision because of the high concentration of cones. Rods are absent from the fovea and macula, but they increase in density toward the periphery of the retina.

When impulses for sight have passed through the photoreceptor neurons, they are conducted across synapses to the bipolar neurons in the intermediate zone of the nervous layer of the retina. From here, the impulses are passed to the ganglion neurons.

The axons of the ganglion neurons extend posteriorly to a small area of the retina called the *optic disc*, or *blind spot*. This region contains openings through which the fibers of the ganglion neurons emerge as the optic nerve. Since this area contains neither rods nor cones, and only nerve fibers, no image is formed on it. Thus, it is called the blind spot. (See Figure 14–7a.)

In addition to the fibrous tunic, vascular tunic, and retina, the eyeball itself contains the lens, just behind the pupil. The *lens* is constructed of numerous layers of protein fibers arranged like the layers of an onion. Normally, the lens is perfectly transparent. It is enclosed by a clear plastic capsule and held in position by the *suspensory ligaments.* An opacity of the lens is known as a *cataract.*

The interior part of the eyeball contains a large cavity that is divided into two smaller cavities. (See Figure 14–7a.) These are called the anterior cavity and the posterior cavity. They are separated from each other by the lens. The *anterior cavity*, in turn, has two subdivisions referred to as the anterior chamber and the posterior chamber. The *anterior chamber* lies behind the cornea and in front of the iris and lens. The *posterior chamber* lies behind the iris and in front of the suspensory ligament. The anterior cavity is filled with a clear, watery fluid called the *aqueous humor.* The fluid is believed to be secreted in the posterior chamber by the epithelium of the ciliary bodies behind the iris. From the posterior chamber, the fluid permeates the posterior cavity and then passes forward between the iris and the lens, through the pupil into the anterior chamber. After passing into the anterior chamber, the aqueous humor, which is continually produced, is drained off into the *canal of Schlemm* and then into the blood. The pressure in the eye, called *intraocular pressure,* is produced mainly by the aqueous humor. It keeps the retina smoothly applied to the choroid so that the retina may form clear images. Normal intraocular pressure (about 24 millimeters Hg) is maintained by the drainage of the aqueous humor through the canal of Schlemm and into the blood. Abnormal elevation of intraocular pressure, called *glaucoma,* results in degeneration of the retina and blindness. In addition to maintaining normal intraocular pressure, the aqueous humor is also the principal link between the circulatory system and the lens and cornea. Neither the lens nor the cornea has blood vessels.

The second, and larger, cavity of the eyeball is the *posterior cavity.* It lies between the lens and the retina and contains a soft, jellylike substance called the *vitreous humor.* This substance also contributes to intraocular pressure and helps to prevent the eyeball from collapsing. However, the vitreous humor, unlike the aqueous humor, does not undergo constant replacement. It is formed during embryonic life and is not replaced thereafter.

Physiology of vision

Before light can reach the rods and cones of the retina, it must first pass through the cornea, aqueous humor, lens, and vitreous humor. Moreover, for vision to occur, light reaching the rods and cones must form an image on the retina. The resulting nerve impulses must then be conducted to the visual areas of the cerebral cortex. In discussing the physiology of vision, let us first consider retinal image formation.

RETINAL IMAGE FORMATION. The formation of an image on the retina requires four basic processes, all concerned with focusing light rays. These basic processes are (1) refraction of light rays, (2) accommodation of the lens, (3) constriction of the pupil, and (4) convergence of the eyes.

Refraction and accommodation. When light rays traveling through a transparent medium (such as air) pass into a second transparent medium with a different density (such as water), the rays bend at the surface of the two media. This is called *refraction* (Figure 14–8a). The eye has four such media of refraction: the cornea, aqueous humor, lens, and vitreous humor (Figure 14–8b). Light rays entering the eye from the air are refracted at the following points: (1) the anterior surface of the cornea as they pass from the lighter air into the denser cornea; (2) the posterior surface of the cornea as they pass into the less dense aqueous humor; (3) the anterior surface of the lens as they pass from the aqueous humor into the denser lens; and (4) the posterior surface of the lens as they pass from the lens into the less dense vitreous humor.

When an object is 20 feet or more away from the viewer, the light rays that are reflected from the object are nearly parallel to one another. Nature has carefully calibrated the degree of refraction that takes place at each surface in the eye. Because of this, the parallel rays are sufficiently bent to fall exactly on the central fovea, where vision is sharpest. However, light rays that are reflected from close-by objects are divergent rather than parallel. As a result, they must be refracted toward each other to a greater extent. This change in refraction ability is the responsibility of the lens of the eye.

If the surfaces of a lens curve outward, as in a biconvex lens, the lens will refract the rays toward each other so that they eventually intersect (Figure 14–8d). The more the lens curves outward, the more acutely it bends the rays toward each other. Conversely, when the surfaces of a lens curve inward, as in a biconcave lens, the rays bend away from each other (Figure 14–8c). The lens of the eye is biconvex. Furthermore, it has the unique ability to change the focusing power of the eye by becoming moderately curved at one moment and greatly curved the next. When the eye is focusing on a close object, the lens curves greatly in order to bend the rays toward the central fovea. This increase in the curvature of the lens is called *accommodation* (Figure 14–9). During accommodation, the ciliary muscle contracts, pulling the ciliary body and choroid forward toward the lens. This releases the tension on the lens and suspensory ligament. Due to its elasticity, the lens shortens, thickens, and bulges. In near vision, the ciliary muscle is contracted, and the lens is bulging. In far vision, the ciliary muscle is relaxed, and the lens is flatter. With aging, the lens loses elasticity and, therefore, its ability to accommodate.

The normal eye, referred to as an *emmetropic eye,* can sufficiently refract light rays from an object 20 feet away to focus a clear object on the retina. Many individuals, however, do not have this ability because of abnormalities related to improper refraction. Among these are *myopia* (nearsightedness), *hypermetropia* (farsightedness), and *astigmatism* (irregularities in the surface of the lens or cornea). The conditions are illustrated and explained in Figure 14–8c, e. Why do you think nearsightedness can be corrected with glasses containing biconcave lenses? How would you correct farsightedness?

Constriction of the pupil. The muscles of the iris also assume a function in the formation of clear retinal images. Part of the accommodation mechanism consists of the contraction of the circular muscle fibers of the iris to constrict the pupil. Constricting the pupil means narrowing the diameter of the hole through which light enters the eye. This occurs simultaneously with accommodation of the lens and prevents light rays from entering the eye through the periphery of the lens. Light rays entering at the periphery would not be brought to focus on the retina and would result in blurred vision. The pupil, as noted earlier, also constricts in bright light to protect the retina from sudden or intense stimulation.

Convergence. When birds use their eyes, they see a set of objects off to the left through one eye and an entirely different set of objects off to the right through the other eye. This characteristic effectively doubles their field of vision and allows them to detect a predator slinking up from behind. In human beings, both eyes focus on only one set of objects—a characteristic called *single binocular vision.*

Single binocular vision occurs when light rays from an object are directed toward corresponding points on the two retinas. When we stare straight ahead at a distant object, the incoming light rays are aimed directly at both pupils and are refracted to identical spots on the retinas of both eyes. But as we move close to the object, our eyes must rotate medially—that is, become "crossed"—in order for the light rays from the object to hit the same points on both retinas. The term *convergence* refers to this medial movement of the two eyeballs so that they are both directed toward

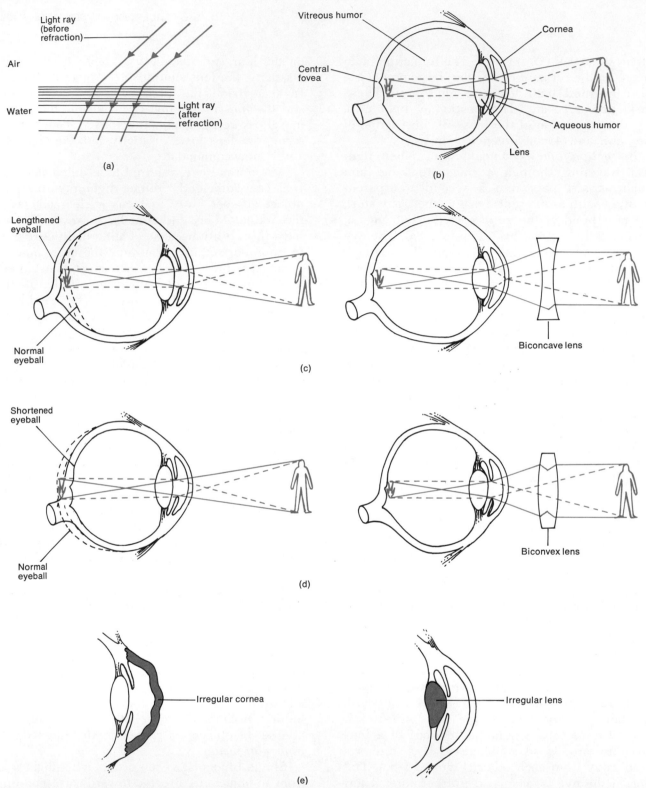

Figure 14-8. Refraction. (a) Refraction of light rays passing from air into water. Normal and abnormal refraction in the eye. (b) In the normal or emmetropic eye, light rays from an object are bent sufficiently by the four refracting media and converged on the central fovea. A clear image is formed. (c) In the myopic eye, the image is focused in front of the retina. The condition may result from an elongated eyeball or a thickened lens. Correction is by use of a concave lens. (d) In the hypermetropic eye, the image is focused behind the retina. The condition may be the result of the eyeball being too short or the lens being too thin. Correction is by a convex lens. (e) Astigmatism. This is a condition in which the curvature of the cornea or lens is uneven. As a result, horizontal and vertical rays are focused at two different points on the retina. Suitable glasses correct the refraction of an astigmatic eye. On the left, astigmatism resulting from an irregular cornea. On the right, astigmatism resulting from an irregular lens. The image is not focused on the area of sharpest vision of the retina. This results in blurred or distorted vision.

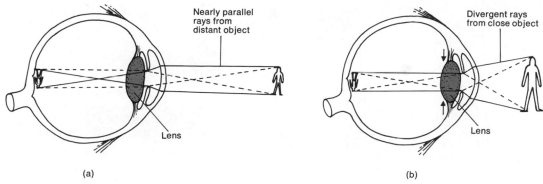

Figure 14–9. Accommodation for (a) objects 20 feet or more away and (b) objects nearer than 20 feet.

the object being viewed. The nearer the object, the greater the degree of convergence necessary to maintain single binocular vision. Convergence is brought about by the coordinated action of the extrinsic eye muscles.

Inverted image. Images are actually focused upside down on the retina. They also undergo mirror reversal. That is, light reflected from the right side of an object hits the left side of the retina and vice versa. Note in Figure 14–8b that reflected light from the top of the object crosses light from the bottom of the object and strikes the retina below the central fovea. Reflected light from the bottom of the object crosses light from the top of the object and strikes the retina above the central fovea. The reason why we do not see a topsy-turvy world is that the brain learns early in life to coordinate visual images with the exact locations of objects. The brain stores memories of reaching and touching objects and automatically turns visual images right-side-up and right-side-around.

STIMULATION OF PHOTORECEPTORS. After an image is formed on the retina by refraction, accommodation, constriction of the pupil, and convergence, light impulses must be converted into nerve impulses by the rods and cones. The exact mechanism by which light acts as a stimulus to initiate nerve impulses that result in the sensation of sight is not entirely clear. The following data, however, are known. Rods contain a reddish pigmented compound called *rhodopsin,* or *visual purple.* This substance consists of the protein scotopsin plus retinene, a derivative of vitamin A. When light rays strike a rod, rhodopsin rapidly breaks down. In some manner, this chemical breakdown stimulates impulse conduction by the rods (Figure 14–10).

Rhodopsin is highly light sensitive—so much

so that even the light rays from the moon or from a candle will break down some of it and thereby allow us to see. The rods, then, are uniquely specialized for night vision. However, they are of only limited help for daylight vision. This is because rhodopsin breaks down extremely quickly in bright light and is rebuilt slowly. In bright light, the rhodopsin is destroyed faster than it can be manufactured. In dim light, production is able to keep pace with a slower rate of breakdown. These characteristics of rhodopsin are responsible for the experience of having to adjust to a dark room after walking in from the sunshine. The period of adjustment is the time it takes for the completely destroyed rhodopsin to reform. The adjustment period is normal. *Night blindness* is the lack of normal night vision following the adjustment period. Night blindness is not considered normal. It is most often caused by vitamin A deficiency.

Cones, which are the receptors for daylight and color, contain photosensitive chemicals that require brighter light for their breakdown. Unlike rhodopsin, the photosensitive chemicals of the cones reform quickly.

AFFERENT PATHWAY TO THE BRAIN. From the rods and cones, impulses are transmitted through bipolar neurons to ganglion cells. The cell bodies of the ganglion cells lie in the retina and their axons leave the eye via the optic nerve (Figure 14–11). The axons pass through the *optic chiasma,* a crossing point of the optic nerves. Some fibers cross to the opposite side. Others remain uncrossed. Upon passing through the optic chiasma, the fibers, now part of the *optic tract,* enter the brain and terminate in the thalamus. Here the fibers synapse with neurons whose axons pass to the visual centers located in the occipital lobes of the cerebral cortex.

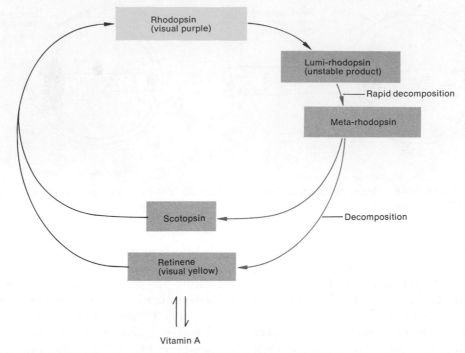

Figure 14–10. The rhodopsin cycle. Color indicates breakdown reactions in light. Black indicates reformation of rhodopsin in darkness.

Analysis of the afferent pathway to the brain reveals that the visual field of each eye is divided into two regions. These are referred to as the *medial*, or *nasal, half* and the *lateral*, or *temporal*, *half*. For each eye, light rays from an object in the nasal half of the visual field fall on the temporal half of the retina. Light rays from an object in the temporal half of the vision field fall on the nasal half of the retina. Note that in the optic chiasma nerve fibers from the nasal halves of the retinas cross and continue on to the thalamus. Also note that nerve fibers from the temporal halves of the retinas do not cross but continue directly on to the thalamus. As a result, the visual center in the cortex of the right occipital lobe "sees" the left side of an object via impulses from the temporal half of the retina of the right eye and the nasal half of the retina of the left eye. The cortex of the left occipital lobe interprets visual sensations from the right side of an object via impulses from the nasal half of the right eye and the temporal half of the left eye. Blind spots in the field of vision may indicate a brain tumor along one of the afferent pathways. For instance, a symptom of tumor in the right optic tract might be an inability to see the left side of a normal field of vision without moving the eyeball.

Auditory sensations and equilibrium

In addition to containing receptors for sound waves, the **ear** also contains receptors for equilibrium. After reviewing the anatomy of the ear, we shall turn to the physiology of hearing and equilibrium. Anatomically, the ear is subdivided into three principal regions: (1) the outer ear, (2) the middle ear, and (3) the inner ear.

The outer ear

The *outer ear* is structurally designed to collect sound waves and then direct them inward (Figure 14–12a). It consists of the pinna, the external auditory canal, and the tympanic membrane, also called the eardrum. The *pinna*, or *auricle*, is a trumpet-shaped flap of elastic cartilage covered by thick skin that has relatively few cutaneous receptors. The rim of the pinna is called the helix, whereas the inferior portion is referred to as the lobe. The pinna is attached to the head by ligaments and muscles. The *external auditory canal* is a tube, about 1 inch in length, that lies in the external auditory meatus of the temporal bone. It leads from the pinna to the eardrum. The walls of the canal consist of bone lined with cartilage that is continuous with the cartilage of the pinna. The cartilage in the external auditory

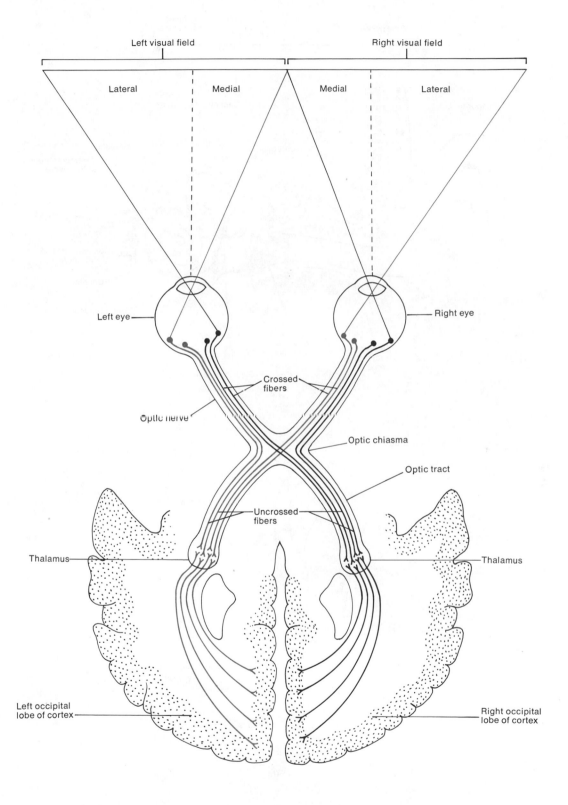

Figure 14-11. Afferent pathway for visual impulses.

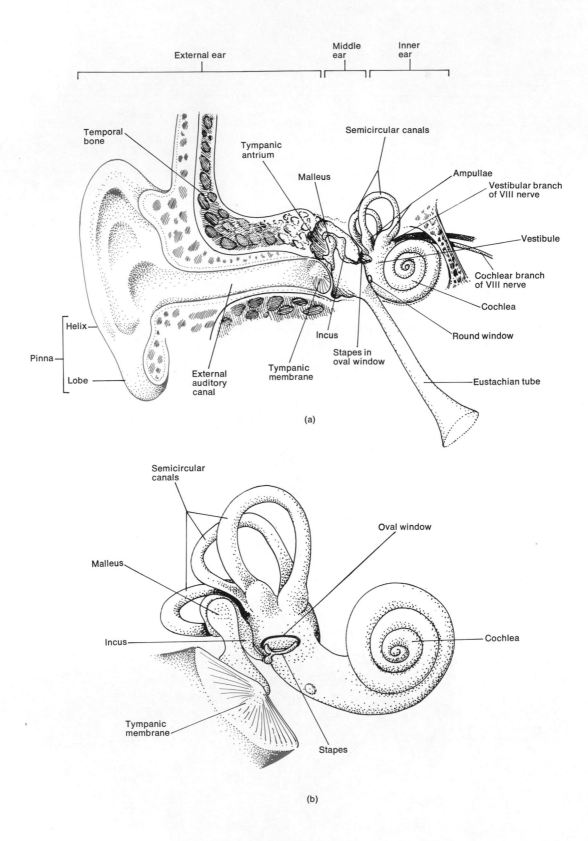

Figure 14–12. Structure of the auditory apparatus. (a) Divisions of the ear into external, middle, and inner portions. (b) Details of the middle ear.

canal is covered with thin, highly sensitive skin. Near the exterior opening, the canal contains a few hairs and *ceruminous glands,* which secrete *cerumen* (earwax). The combination of hairs and cerumen prevents foreign objects from entering the ear. The *tympanic membrane* is a thin, semi-transparent partition of fibrous connective tissue between the external auditory meatus and the middle ear. Its external surface is concave and is covered with skin. Its internal surface is convex and is covered with a mucous membrane.

The middle ear

Also called the *tympanic cavity,* the *middle ear* is a small, epithelial-lined cavity hollowed out of the temporal bone (Figure 14–12a, b). The cavity is separated from the external ear by the eardrum and from the internal ear by a very thin bony partition that contains two small openings, called the oval window and the round window. The posterior wall of the cavity communicates with the mastoid cells of the temporal bone through a chamber called the *tympanic antrum.* This anatomical fact explains why a middle ear infection may spread to the temporal bone, causing mastoiditis, or even spread to the brain. The anterior wall of the cavity contains an opening that leads into the *Eustachian tube,* also called the auditory tube. The Eustachian tube connects the middle ear with the nose and throat. Through this passageway, infections may travel from the throat and nose to the ear. The function of the tube is to equalize pressure on both sides of the tympanic membrane. Abrupt changes in external pressure might otherwise cause the eardrum to rupture. Since the tube opens during swallowing and yawning, these activities allow atmospheric air to enter or leave the middle ear until the internal pressure equals the external pressure. Any sudden pressures on the eardrum may be equalized by deliberately swallowing.

Extending across the middle ear are three exceedingly small bones referred to as **auditory ossicles.** These are called the malleus, incus, and stapes. According to their shapes, they are commonly named the hammer, anvil, and stirrup, respectively. The "handle" of the **malleus** is attached to the inner portion of the tympanic membrane. Its head articulates with the base of the incus. The **incus** is the intermediate bone in the series and articulates with the stapes. The **stapes** fits into a small opening between the middle and inner ear called the *fenestra vestibuli,* or *oval window.* Directly below the oval window

is another opening, the *fenestora chochleae,* or *round window.* This opening separates the middle and inner ears. The auditory ossicles are attached to the tympanic membrane, to each other, and to the oval window by means of ligaments and muscles.

The inner ear

This structure is also called the *labyrinth* (Figure 14–13a) because of its complicated series of canals. Structurally, it consists of two main divisions: (1) a bony labyrinth and (2) a membranous labyrinth that fits within the bony portion. The *bony labyrinth* is a series of cavities within the petrous portion of the temporal bone. It can be divided into three areas named on the basis of shape. These areas are the vestibule, cochlea, and semicircular canals. The bony labyrinth is lined with periosteum and contains a fluid called the perilymph. This fluid surrounds the *membranous labyrinth,* a series of sacs and tubes lying inside and having the same general form as the bony labyrinth. Epithelium lines the membranous labyrinth, and it also contains a fluid called *endolymph.*

The *vestibule* constitutes the oval central portion of the bony labyrinth. The membranous labyrinth within the vestibule consists of two sacs called the *utricle* and *saccule.* These sacs are connected to each other by a small duct.

Projecting upward and posteriorly from the vestibule are the three *semicircular canals.* Each of the semicircular canals is arranged at approximately right angles to the other two. On the basis of their positions, they are called the superior, posterior, and lateral canals. One end of each canal enlarges into a swelling called the *ampula.* Inside the semicircular canals lie portions of the membranous labyrinth, the *semicircular ducts.* These structures are almost identical in shape to the semicircular canals and communicate with the utricle of the vestibule.

Lying in front of the vestibule is the *cochlea,* so designated because of its resemblance to a snail's shell. The cochlea consists of a bony spiral canal that makes about two and one-half turns around a central bony core called the *modiolus.* A cross section through the cochlea shows that the canal is divided by partitions into three separate channels resembling the letter Y lying on its side (Figure 14–13b). The stem of the Y is a bony shelf that protrudes into the canal. The wings of the letter are composed of the membranous labyrinth. The channel above the bony parition is called the

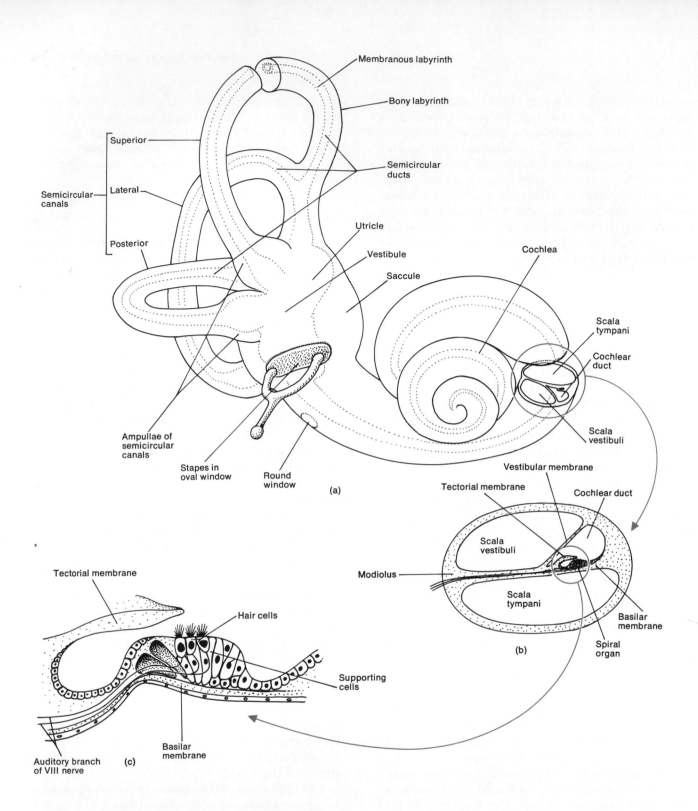

Figure 14–13. Details of the inner ear. (a) The inner ear. The outer, black lines belong to the bony labyrinth. The inner, colored lines, to the membranous labyrinth. (b) Cross section through the cochlea. (c) Enlargement of spiral organ. *(Photograph courtesy of Edward J. Reith, Atlas of Descriptive Histology, Edward J. Reith and Michael H. Ross, Harper & Row, Publishers, Inc., New York, 1970.)*

scala vestibuli, and the channel below, the *scala tympani.* The cochlea adjoins the wall of the vestibule, into which the scala vestibuli opens. The scala tympani terminates at the round window. The perilymph of the vestibule is continuous with that of the scala vestibuli. The third channel (between the wings of the Y) is the membranous labyrinth, the *cochlea duct.* The cochlea duct is separated from the scala vestibuli by the *vestibular membrane,* also called *Reissner's membrane.* It is separated from the scala tympani by the *basilar membrane.* Resting on the basilar membrane is the *spinal organ,* the organ of hearing (Figure 14–13c). This organ consists of supporting cells and hair cells that extend into the endolymph of the cochlea duct. The basal ends of the hair cells are in contact with the fibers of the auditory nerve. Projecting over and in contact with the hair cells is a gelatinous membrane, the *tectorial membrane.*

Physiology of hearing

Sound waves result from the alternate compression and decompression of air. They originate from a vibrating object and travel through air in much the same way that waves travel over the surface of water. The events involved in the physiology of hearing sound waves are listed below. While reading this succession of events, you will want to make constant reference to Figure 14–14.

1. Sound waves that reach the ear are directed by the pinna into the external auditory canal.
2. When the waves strike the tympanic membrane, the alternate compression and decompression of the air cause the membrane to vibrate.
3. The central area of the tympanic membrane is connected to the malleus, which also starts to vibrate. The vibration is then picked up by the incus, which transmits the vibration to the stapes.
4. As the stapes moves back and forth, it pushes the oval window in and out.
5. The movement of the oval window sets up waves in the perilymph.

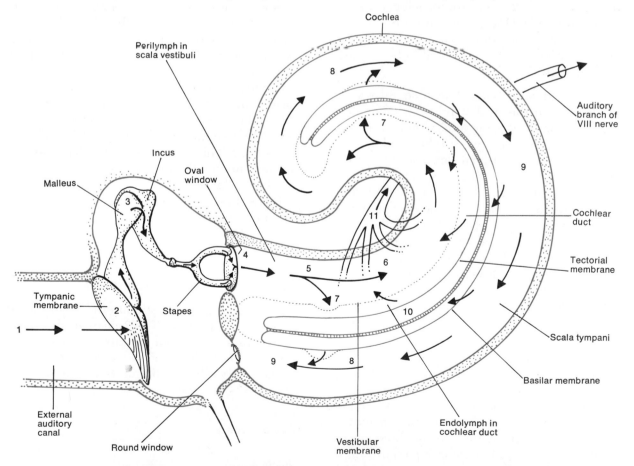

Figure 14–14. The physiology of hearing. Follow the text very carefully to understand the meaning of the indicated numbers.

6. As the window bulges inward, it pushes the perilymph of the scala vestibuli up into the cochlea.

7. This pressure pushes the vestibular membrane inward and increases the pressure of the endolymph inside the cochlea duct.

8. The basilar membrane gives under the pressure and bulges out into the scala tympani.

9. The sudden pressure in the scala tympani pushes the perilymph toward the round window, causing it to bulge back into the middle ear. Conversely, as the sound wave subsides, the stapes moves backward, and the procedure is reversed. That is, the fluid moves in the opposite direction along the same pathway, and the basilar membrane bulges into the cochlea duct.

10. When the basilar membrane vibrates, the hair cells of the spiral organ are moved against the tectorial membrane. In some unknown manner, the movement of the hairs stimulates the dendrites of neurons at their base, and sound waves are converted into nerve impulses.

11. The impulses are then passed on to the auditory branch of the vestibulocochlear nerve and the medulla. Here, some impulses cross to the opposite side and finally travel to the auditory area of the temporal lobe of the cerebral cortex.

It has been demonstrated that if sound waves passed directly to the oval window without first passing through the tympanic membrane and auditory bones, hearing would be inadequate. This is because a minimal amount of sound energy is required to transmit sound waves through the perilymph of the cochlea. Since the tympanic membrane has a surface area about twenty-two times larger than that of the oval window, it can collect about twenty-two times more sound energy. This energy is sufficient to transmit sound waves through the perilymph.

Physiology of equilibrium

The term *equilibrium* has two meanings. One kind of equilibrium, called *static equilibrium,* refers to the orientation of the body (mainly the head) relative to the ground. The second kind of equilibrium, called *dynamic equilibrium,* is the maintenance of the position of the body (mainly the head) in response to sudden movements or to a change in the rate or direction of movement. The receptor organs for equilibrium are the saccule, utricle, and semicircular ducts.

The utricle and saccule each contain within their walls sensory hair cells that project into the cavity of the membranous labyrinth (Figure 14–15). The hairs are coated with a gelatinous layer in which particles of calcium carbonate, called *otoliths,* are embedded. When the head tips downward, the otoliths slide with gravity in the direction of the ground. As the particles move, they exert a downward pull on the gelatinous mass, which in turn, exerts a downward pull on the hairs and makes them bend. The movement of the hairs stimulates the dendrites at the bases of the hair cells. The impulse is then transmitted to the temporal lobe of the brain through the vestibular branch of the vestibulocochlear nerve. The utricle and saccule are considered to be sense organs of static equilibrium. They provide information regarding the orientation of the head in space and are essential for the maintenance of posture.

Let us now consider the role of the semicircular ducts in maintaining dynamic equilibrium. The three semicircular ducts are positioned at right angles to each other in three planes: frontal (the superior duct), sagittal (the posterior duct), and lateral (the lateral duct). This positioning permits correction of an imbalance in three planes. In the ampulla, the dilated portion of each duct, there is a small elevation called the *crista* (Figure 14–16a). Each crista is composed of a group of hair cells covered by a mass of gelatinous material called the *cupula.* When the head moves, the endolymph in the semicircular ducts flows over the hairs and bends them as water in a stream bends the plant life growing at its bottom. The movement of the hairs stimulates sensory neurons, and the impulses pass over the vestibular branch of the vestibulocochlear nerve (Figure 14–16b). The impulses then reach the temporal lobe of the cerebrum. Impulses are then sent to the muscles that must contract in order to maintain body balance in the new position.

DISORDERS OF THE SENSE ORGANS

The sense organs can be altered or damaged by numerous disorders. The causes of disorder can range from congenital origins to bacterial infections to the effects of old age. Here, we shall discuss only a few of the more common disorders of the eyes and ears.

Eye disorders

Cataract

Disorders most commonly resulting in blindness are those that primarily affect elderly people.

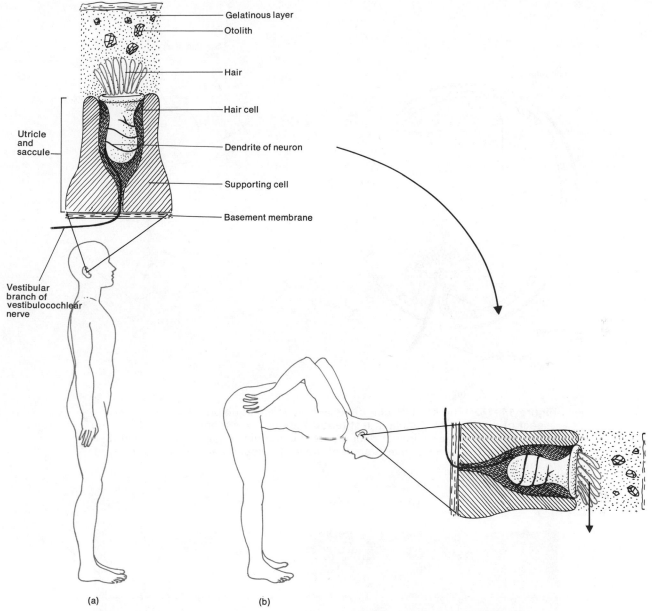

Figure 14–15. The utricle and saccule and static equilibrium. As the person bends, the otoliths drop downward in the direction of the ground, pulling on the gelatinous mass, which slides over the hairs. The bending of the hairs causes changes in the hair cells that stimulate the neurons.

The most prevalent one is **cataract** formation. This disorder causes the lens or its capsule to lose its transparency (Figure 14–17). Cataracts can occur at any age, but we shall discuss only the type that develops with old age. Quite often as a person gets older the cells in the lenses of the eyes degenerate and are replaced with nontransparent fibrous protein. Or the lenses may start to manufacture nontransparent protein. The main symptom of cataract is a progressive, painless loss of vision. The degree of loss depends upon the location and extent of the opacity. If vision loss is very gradual, frequent changes in glasses may help to maintain useful vision for a while. This is because the changes initially affect the density of the lens and thus change the refraction of light rays. Eventually, though, the changes may be so extensive that light rays are blocked out altogether. At this point, surgery is indicated. Essentially, the surgical procedure consists of removing the opaque lens and substituting an artificial lens by means of eyeglasses.

Glaucoma

The second most common cause of blindness, especially in the elderly, is **glaucoma.** This disorder is characterized by an abnormally high

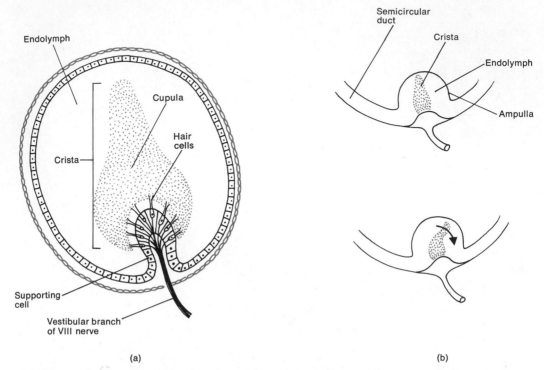

(a) (b)

Figure 14–16. The semicircular ducts and dynamic equilibrium. (a) Enlarged aspect of a crista. (b) Cupula at rest (above) and in movement (below). When the endolymph in the semicircular duct moves, the cupula is displaced. The impulse is picked up by the vestibular branch of the vestibulocochlear nerve and relayed to the brain.

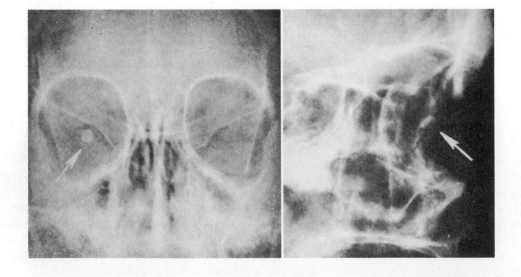

Figure 14–17. Roentgenogram of a cataract (arrow) in the lens of the right eye. *(From Lester W. Paul and John H. Juhl, The Essentials of Roentgen Interpretation, Harper & Row, Publishers, Inc., New York, 1972.)*

pressure of the fluid inside the eyeball. The aqueous humor does not return into the bloodstream through the canal of Schlemm as quickly as it is formed. (See Figure 14–7a.) The fluid accumulates and puts pressure on the neurons of the retina. If the pressure continues over a long period of time, it destroys the neurons and brings about blindness. It can affect a person of any age, but 95 percent of the victims are past 40 years of age. Glaucoma affects the eyesight of more than 1 million people in this country.

Treatment is through drugs or surgery. Drug treatment involves reducing the pressure in the eye by giving compounds that cause the sphincter muscles of the iris and ciliary body to contract. These contractions prevent the obstruction of the anterior chamber and the canal of Schlemm. This results in improved fluid outflow from the eye. Surgery consists of the removal of a small piece of the iris (iridectomy) to allow drainage and lessen the interior pressure.

Conjunctivitis (pinkeye)

Many different kinds of eye inflammations exist. But the most common type is **conjunctivitis,** an inflammation of the membrane that lines the insides of the eyelids and the cornea. (See Figure 14–6.)

Conjunctivitis can be caused by microorganisms — most often the pneumococci or staphylococci bacteria. In such cases, the inflammation is very contagious. It can also be caused by a number of irritants, in which case the inflammation is not contagious. The irritants include dust, smoke, wind, air pollution, and excessive glare from intense light on water, sand, or snow. The condition may be acute or chronic. The epidemic type in children is extremely contagious, but normally it is not serious.

Trachoma

This is a chronic contagious conjunctivitis caused by an organism called the TRIC agent. The organism has characteristics of both viruses and bacteria. **Trachoma** is characterized by many granulations or fleshy projections on the eyelids. These projections can irritate and inflame the cornea if untreated and reduce vision. The disease produces an excessive growth of subconjunctival tissue and the invasion of blood vessels into the upper half of the front of the cornea. The disease progresses until it covers the entire cornea, bringing about a loss of vision because of corneal opacity. Antibiotics, such as tetracycline and the sulfa drugs, kill the organisms that cause trachoma and have reduced the seriousness of this infection.

Ear disorders

Ménière's disease

An important cause of deafness and loss of equilibrium in adults is **Ménière's disease** of the inner ear. It is a disturbance or malfunction of any part of the inner ear. It can be the result of many causes. These include (1) infection of the middle ear; (2) trauma from brain concussion producing hemorrhage or a splitting of the labyrinth; (3) cardiovascular diseases, such as arteriosclerosis and blood vessel disturbances; (4) congenital malformation of the labyrinth; (5) an excessive formation of endolymph; and (6) an allergy. The last two causes can both produce an increase in pressure in the cochlear duct and vestibular system. This, in turn, causes a progressive atrophy of the hair cells of the cochlea or semicircular ducts.

If the cochlea is injured, typical symptoms are hissing, roaring, or ringing in the ears and/or deafness. If the semicircular ducts are involved, the person has a feeling of dizziness or motion sickness, often producing nausea and vomiting. The dizzy spells can be mild or severe, lasting from a few minutes to several days. Ménière's disease is a chronic disorder. It affects both sexes equally and usually begins in late middle life. In severe cases, surgical treatment should be considered. Treatment with ultrasound, which consists of beaming ultrasound waves into the labyrinth, selectively destroys the semicircular ducts (equilibrium), while preserving the cochlear duct (hearing). This has produced excellent results. Another successful surgical procedure involves electrical coagulation of the labyrinth, which means changing the fluid to a jelly or solid. Of all the symptoms of Ménière's disease, the tendency of the patient to fall in the direction of the affected ear is probably the most incapacitating and potentially dangerous.

Impacted cerumen

Some people produce an abnormal amount of cerumen, or earwax, in the external auditory canal of the ear. Here, it becomes impacted, or pressed so firmly together that it becomes immovable, and prevents sound waves from reaching the tympanic membrane. The treatment for **impacted cerumen** is usually periodic ear irrigation or removal of wax with a blunt instrument.

Medical terminology associated with the sense organs

Achromatopsia (*a* = without; *chrom* = color) Complete color blindness.

Ametropia (*ametro* = disproportionate; *ops* = eye; *ia* = condition) Refractive defect of the eye resulting in an inability to focus images properly on the retina.

Blepharitis (*blepharo* = eyelid) An eyelid inflammation.

Eustachitis Eustachian tube infection or inflammation.

Keratitis (*kerato* = cornea) An inflammation or infection of the cornea.

Keratoplasty (*plasty* = mold, shape, form) Corneal graft or transplant; an opaque cornea is removed and replaced with a normal transparent cornea to restore vision.

Labyrinthitis Inner ear or labyrinth inflammation.

Myringitis (*myringa* = eardrum) Inflammation of the eardrum; also called tympanitis.

Nystagmus A constant, rapid involuntary eyeball movement, possibly caused by a disease of the central nervous system.

Otalgia (*oto* = ear; *algia* = pain) Earache.

Otitis Inflammation of the ear.

Presbyopia (*presby* = old) Inability to focus clearly on nearby objects due to weakness of the ciliary muscle. Weakness usually caused by aging.

Ptosis (*ptosis* = fall) Falling or drooping of the eyelid. (This expression is also used for the slipping of any organ below its normal position.)

Retinoblastoma (*blast* = bud; *oma* = tumor) A relatively common tumor arising from immature retinal cells and accounting for 2 percent of childhood malignancies.

Strabismus An eye muscle disorder, commonly called "crossed eyes." The eyeballs do not move in unison. May be caused by lack of coordination of the extrinsic eye muscles.

Tinnitus A ringing in the ears.

Chapter summary in outline

SENSATIONS
1. Awareness of conditions and changes in these conditions inside and outside the body is sensation.
2. The prerequisites are receiving a stimulus, converting it into an impulse, conducting the impulse to the brain, and translating the impulse into a sensation.
3. A receptor picks up a stimulus for a sensation.

Characteristics
1. Projection occurs when the brain refers a sensation to the point of stimulation.
2. Adaptation is the loss of sensation even though the stimulus is still applied.
3. An afterimage is the persistence of a sensation even though the stimulus is removed.
4. The modality is the property by which one sensation is distinguished from another.

Classification
1. Exteroceptors receive stimuli from the external environment.
2. Visceroceptors receive stimuli from blood vessels and viscera.
3. Proprioceptors receive stimuli from muscles, tendons, and joints for body position and movement.

GENERAL SENSES
Cutaneous sensations
1. Touch, pressure, cold, and heat sensations from the skin, connective tissue, and gastrointestinal tract are called the cutaneous sensations.

2. Receptors for these sensations are Meissner's corpuscles, Merkel's discs, root hair plexuses, Pacinian corpuscles, end bulbs of Krause, and end organs of Ruffini.

Pain sensations
1. Receptors are found in almost every body tissue.
2. Two general kinds of pain, recognized in the parietal lobe of the cortex, are somatic and visceral.
3. Referred pain is felt in the skin near or away from the organ sending pain impulses.
4. With phantom pain, a person "feels" pain in a limb that has been amputated.
5. The possible use of acupuncture to relieve pain is based on the theory that the stimulus provided by a needle passes over a touch nerve faster than over a pain nerve. As a result, pain is inhibited.

Proprioceptive sensations
1. Receptors, found in muscles, tendons, and joints, inform us of muscle tone, movement of body parts, and body position.

SPECIAL SENSES
Olfactory sensations
1. Receptor cells in the nasal epithelium send impulses to the olfactory bulbs, to olfactory tracts, to the cortex.

Gustatory sensations
1. Receptors in the taste buds send impulses to the cranial nerves, thalamus, and cortex.

Visual sensations
1. The eye is constructed of three coats: (*a*) fibrous tunic (sclera and cornea); (*b*) vascular tunic (choroid, ciliary body, and iris); and (*c*) retina, which contains rods and cones.
2. The anterior cavity contains aqueous humor, and the posterior cavity contains vitreous humor.
3. Retinal image formation involves refraction of light, accommodation of lens, constriction of pupil, convergence, and inverted image formation.
4. Improper refraction may result from myopia (nearsightedness), hypermetropia (farsightedness), and astigmatism (corneal or lens abnormalities).
5. Rods and cones convert light rays into visual nerve impulses; rhodopsin is necessary for the conversion.
6. Impulses from rods and cones are conveyed through retina to optic nerve, optic chiasma, optic tract, thalamus, and cortex.

Auditory sensations and equilibrium
1. The ear consists of three anatomical subdivisions: (*a*) the outer ear (pinna, external auditory canal, and tympanic membrane); (*b*) the middle ear (Eustachian tube, ossicles, oval window, and round window); and (*c*) the inner ear (bony labyrinth and membranous labyrinth).
2. Sound waves are caused by the alternate compression and decompression of air.
3. Waves enter the external auditory canal, strike the tympanic membrane, pass through the ossicles, strike the oval window, set up waves in the perilymph, strike the vestibular membrane and scala tympani, increase pressure in the endolymph, strike the basilar membrane, and stimulate hairs on the spiral organ. A sound impulse is then initiated.
4. Static equilibrium is the relationship of the body relative to the pull of gravity.
5. Dynamic equilibrium is equilibrium in response to movement of the body.

DISORDERS OF SENSE ORGANS
1. Cataract is the loss of transparency of the lens or capsule.

2. Glaucoma is abnormally high intraocular pressure, which destroys neurons of the retina.

3. Conjunctivitis is an inflammation of the conjunctiva.

4. Trachoma is a chronic, contagious inflammation of the conjunctiva.

5. Ménière's disease is the malfunction of the inner ear that may cause deafness and loss of equilibrium.

6. Impacted cerumen is an abnormal amount of earwax in the external auditory canal.

Review questions and problems

1. Define a sensation and a sense receptor. What prerequisites are necessary for the perception of a sensation?

2. Describe the following characteristics of a sensation: projection, adaptation, after image, and modality.

3. Can you think of any examples of adaptation not discussed in the text?

4. Compare the location and function of exteroceptors, visceroceptors, and proprioceptors.

5. Distinguish between a general sense and a special sense.

6. What is a cutaneous sensation? How are cutaneous receptors distributed over the body? Relate your response to the two-point discrimination test.

7. For each of the following cutaneous sensations, describe the receptor involved in terms of structure, function, and location: touch, pressure, cold, and heat.

8. How do cutaneous sensations help you to maintain homeostasis?

9. Why are pain receptors important? Differentiate somatic pain, visceral pain, referred pain, and phantom pain. How is surgery employed to stop pain?

10. Why is the concept of referred pain exceedingly useful to the physician in diagnosing various internal disorders?

11. Define acupuncture. How is the procedure performed? Describe the "gate control" theory of acupuncture.

12. What is the proprioceptive sense? Where are the receptors for this sense located?

13. Can you relate proprioception to the maintenance of homeostasis?

14. Discuss the origin and path of an impulse that results in smelling.

15. How are papillae related to taste buds? Describe the structure and location of the papillae. Discuss how an impulse for taste travels from a taste bud to the brain.

16. Describe the structure and importance of the following accessory structures of the eye: eyelids, eyelashes, eyebrows, and lacrimal apparatus.

17. By means of a labeled diagram, indicate the principal anatomical structures of the eye. How is the retina adapted to its function?

18. Distinguish a sty from a chalazion.

19. How do extrinsic eye muscles differ from intrinsic eye muscles?

20. Describe the location and contents of the chambers of the eye. What is intraocular pressure? How is the canal of Schlemm related to this pressure?

21. Explain how each of the following events is related to the physiology of vision: (a) refraction of light, (b) accommodation, (c) constriction of the pupil, (d) convergence, and (e) inverted image formation.

22. Distinguish emmetropia, myopia, hypermetropia, and astigmatism by means of a diagram.

23. How is a light stimulus converted into an impulse? Relate your discussion to the rhodopsin cycle by means of a diagram.

24. What is night blindness? What causes it?

25. Describe the path of a visual impulse from the optic nerve to the brain.

26. Define visual field. Relate the visual field to image formation on the retina.

27. Diagram the principal parts of the outer, middle, and inner ear. Describe the function of each part labeled.

28. Explain the events involved in the transmission of sound from the pinna to the spiral organ.

29. What is the afferent pathway for sound impulses from the vestibulocochlear nerve to the brain?

30. Compare the function of the semicircular ducts in maintaining dynamic equilibrium with the role of the saccule and utricle in maintaining static equilibrium.

31. Define each of the following: cataract, glaucoma, conjunctivitis, trachoma, Ménière's disease, and impacted cerumen.

32. Refer to the glossary of medical terminology associated with the sense organs. Be sure that you can define each medical term listed.

CHAPTER 15

THE ENDOCRINE SYSTEM: CHEMICAL REGULATION OF HOMEOSTASIS

STUDENT OBJECTIVES

After you have read this chapter, you should be able to:

1. Discuss the function of the endocrine system in maintaining homeostasis

2. Define an endocrine gland

3. Identify the relationship between an endocrine gland and a target organ

4. Define the anatomical and physiological relationship between the pituitary gland and hypothalamus

5. List the six hormones of the adenohypophysis, their target organs, and their functions

6. Define the source of hormones stored by the neurohypophysis, their target organs, and functions

7. Define a negative feedback mechanism

8. Relate a negative feedback mechanism to the regulation of hormones secreted by the pituitary

9. Describe pituitary dwarfism, and Simmond's disease, giantism, and acromegaly and list the clinical symptoms of each

10. Discuss how thyroxin is synthesized, stored, and transported by thyroid follicles

11. Identify the physiological effects and regulation of secretion of thyroxin and thyrocalcitonin

12. Name four abnormalities of thyroid secretion and list the clinical symptoms of each

13. Describe the physiological effects and regulation of the parathyroid hormone

14. Identify the principal effects of abnormal secretion of the parathyroid hormone on calcium metabolism

15. Distinguish the effects of adrenal cortical mineralocorticoids, glucocorticoids, and gonadocorticoids on physiological activities

16. Identify the function of the adrenal medullary secretions as supplements to sympathetic responses

17. Compare the effects of hypo- and hypersecretions of adrenocortical hormones

18. Compare the roles of glucagon and insulin in the control of blood sugar level

19. Identify the physiological effects of the hormones secreted by the pineal gland

20. Define the role of the thymus in antibody production

21. Define the general stress syndrome and compare homeostatic responses and stress responses

22. Identify the body reactions during the alarm, resistance, and exhaustion stages of stress

23. Define medical terminology associated with the endocrine system

In the previous three chapters, you learned how the nervous system controls the body through electrical impulses that are delivered over neurons. Now it is time to look at the body's other great control system, the endocrine system. The endocrine organs affect bodily activities by releasing chemical messengers, called hormones, into the bloodstream. Obviously, bodily activities would become counterproductive and ineffective if the two great control systems were to pull in opposite directions. The nervous and endocrine systems, therefore, coordinate their activities like an interlocking supersystem. Certain parts of the nervous system routinely stimulate or inhibit the release of hormones. The hormones, in turn, are quite capable of stimulating or inhibiting the flow of particular nerve impulses. Like the proverbial horse and carriage, the two systems go together. In this chapter, you will study the various endocrine glands, the organs that make up the body's means of chemical control.

ENDOCRINE GLANDS

The body contains two different kinds of glands: exocrine and endocrine. **Exocrine glands,** mentioned in Chapter 4, secrete their products into ducts. The ducts then carry the secretions into body cavities or to the external surface of the body. So far, we have talked only about exocrine glands. They include sweat, sebaceous, mucous, and digestive glands. **Endocrine glands,** by contrast, secrete their products into the extracellular space around the secretory cells. Since they secrete internally, the term *endo,* meaning within, is used. The secretion passes through the membranes of cells lining blood vessels and into the blood. Since they have no ducts, endocrine glands are also referred to as *ductless glands.* The endocrine glands of the body are the pituitary, thyroid, parathyroids, adrenals, pancreas, ovaries, testes, pineal, and thymus. The placenta or "afterbirth" is, in some ways, a temporary endocrine gland, but this will not be discussed until Chapter

25. The endocrine glands are organs that together make up the **endocrine system.** The location of the organs of the endocrine system is illustrated in Figure 15–1.

The secretions of endocrine glands are called **hormones,** the term *hormone* meaning to set in motion. A hormone may be a protein, an amine, or a steroid. Amines, like proteins, contain carbon, hydrogen,˙ and nitrogen. Unlike proteins, they lack oxygen, and they do not contain peptide bonds. A steroid is a type of lipid. The one thing that all hormones have in common—whether protein, amine, or steroid—is maintaining homeostasis by changing the physiological activities of cells. A hormone may stimulate changes in the cells of an organ or in groups of organs. These are called the *target organs* of a hormone. Or, the hormone may directly affect the activities of all the cells in the body. Some of the biological activities regulated by hormones are growth, including the closure of the epiphyseal plates in long bones; concentration of ions in the extracellular and intracellular fluid; blood-sugar level; and sexual and reproductive functions.

PITUITARY GLAND

The hormones of the **pituitary gland,** also called the **hypophysis,** regulate so many body activities that the pituitary has been nicknamed the "master gland." Surprisingly, the hypophysis is a rather small round structure measuring about 0.5 inch in diameter and weighing about 0.5 gram. It lies in the sella turcica of the sphenoid bone and is attached to the hypothalamus of the brain via a stalklike structure. This structure, called the *infundibular stalk,* penetrates the dura mater. (See Figure 13–1.)

The pituitary is divided structurally and functionally into an anterior lobe and a posterior lobe, both of which are connected to the hypothalamus. The *anterior lobe* contains many glandular epithelium cells and forms the glandular part of the pituitary. A system of blood vessels

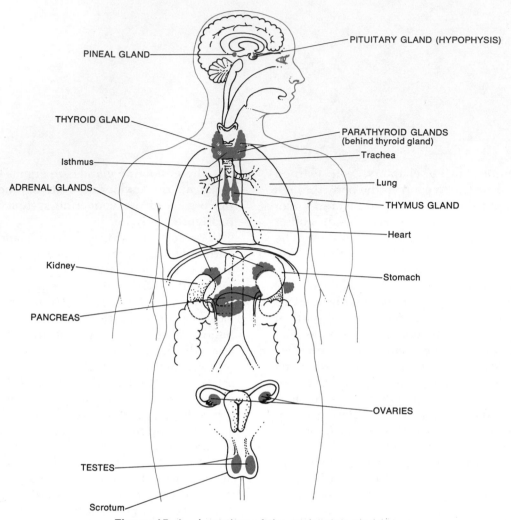

Figure 15–1. Location of the endocrine glands.

connects the anterior lobe with the hypothalamus. The *posterior lobe* contains neurons, which form the neural part of the pituitary. Other nerve fibers connect the neurohypophysis directly with the hypothalamus.

The adenohypophysis

The anterior lobe of the pituitary is also called the **adenohypophysis.** It releases hormones that regulate a whole range of bodily activities, from growth to reproduction. However, the release of these hormones is either stimulated or inhibited by chemical secretions that come from the hypothalamus of the brain. This is one hookup between the nervous system and the endocrine system that we mentioned at the beginning of the chapter. The hypothalamic secretions are delivered to the adenohypophysis in the following way. The hypothalamus lies just above the infundibular

stalk, which receives its blood supply through the *superior hypophysial arteries* (Figure 15–2). As soon as they reach the infundibular stalk, the arteries break up into a network of capillaries. The hypothalamic secretions diffuse into these capillaries. The capillaries then unite to form *portal veins,* which carry the blood and secretions into the anterior lobe.

When the anterior lobe receives the proper chemical stimulation from the hypothalamus, its glandular cells secrete any one of six hormones. The glandular cells, themselves, are called acidophils, basophils, or chromophobes, depending on the way their cytoplasm reacts to laboratory stains. The acidophils, which stain pink, secrete two hormones: growth hormone, which controls bodily growth; and prolactin, which initiates milk secretion from the breasts. The basophils stain darkly and release the other four hormones. These are: thyroid stimulating hormone, which

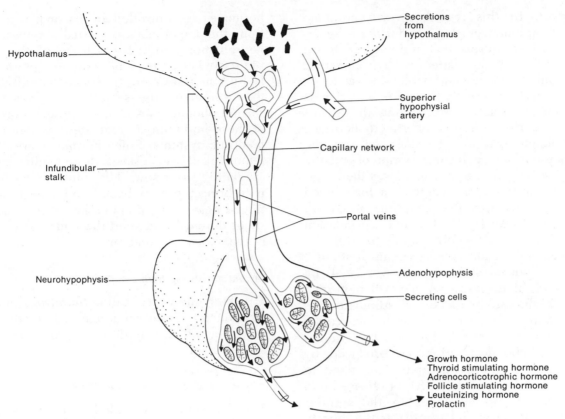

Figure 15–2. Blood supply of the adenohypophysis. Note that chemicals from the hypothalamus pass into the blood which is circulating through the adenohypophysis.

controls the thyroid gland; adrenocorticotrophic hormone, which regulates the adrenal glands; follicle stimulating hormone, which stimulates the production of egg and sperm in the reproductive organs; and leuteinizing hormone, which stimulates other sexual and reproductive activities. The chromophobes may also be involved in the secretion of the adrenocorticotrophic hormone.

Except for the growth hormone, all the secretions are referred to as *trophic hormones*, which means that their target organs are other endocrine glands. Prolactin, follicle stimulating hormone, and leuteinizing hormone are also called *gonadotrophic hormones* because they regulate the functions of the gonads. The gonads are the endocrine glands that produce sex hormones.

Growth hormone (STH)

The *growth hormone* is also referred to as *somatotropin* and as the *somatotrophic hormone (STH)*. The word root *soma* means body, whereas *troph* means nourishment. Its principal function is to act upon the hard and soft tissues of the body to increase their rate of growth and to maintain

their size once growth is attained. Growth hormone causes cells to grow and multiply by increasing the rate at which amino acids enter cells and are built up into proteins. The building processes are called *anabolism.* Thus, growth hormone is considered to be a protein *anabolism* hormone. Growth hormone has two other functions as well. It causes cells to switch from burning carbohydrates to burning fats for energy. For example, it stimulates adipose tissue to release some of its fat. And it stimulates the other cells of the body to break down the released fat molecules. As you remember, when chemical bonds are broken, energy is released. Since energy-releasing processes are referred to as *catabolism,* we can say that growth hormone promotes fat catabolism. At the same time, growth hormone accelerates the rate at which glycogen stored in the liver is converted into glucose and released into the blood. However, since the cells are using fats for energy, they do not consume much glucose. The end result is an increase in the level of blood sugar, and long continued excessive amounts of the hormone may lead to diabetes. This is called the *diabetogenic effect.* Another hormone of the body, insulin, decreases blood sugar level. In this regard, growth hormone and insulin are antagonistic. The former is *hyperglycemic,* which means that it increases blood sugar level, and the latter is *hypoglycemic,* which means that it decreases blood sugar level.

The control of growth hormone secretion is not yet clearly established. Among the factors that increase the secretion of growth hormone are low blood sugar and stress. Exercise, which increases the muscle cell consumption of glucose, lowers blood sugar and also increases the secretion of the hormone. By contrast, a high blood sugar level decreases the secretion of growth hormone (Figure 15–3). Current speculation suggests that the blood circulating through the hypothalamus contains various chemical stimuli. One such chemical stimulus may very well be the level of blood sugar. A lower-than-normal amount of glucose in the blood stimulates the hypothalamus to secrete a neurohumor called *somatotropin releasing factor (SRF).* A **neurohumor** is a chemical substance produced by nerve cells that causes an endocrine gland to secrete its hormones. When SRF is released into the bloodstream, it circulates to the anterior lobe of the pituitary, and stimulates the lobe to secrete growth hormone. As soon as blood sugar level returns to normal, SRF secretion shuts off. In fact, it is believed that an abnormally high level of blood sugar inhibits the release of SRF. Since the pituitary is no longer stimulated, it ceases to secrete growth hormone.

The regulation of growth hormone illustrates two phenomena that are typical of all the secretions of the adenohypophysis. First, each of these

hormones is controlled by its own neurohumor from the hypothalamus. Usually release of the neurohumor stimulates release of the anterior pituitary hormone. The one exception is prolactin's neurohumor. It works to inhibit, rather than stimulate, the secretion of prolactin. The second phenomenon is that the secretion of most of these hormones is regulated through negative feedback systems. Since hormones are chemical regulators of homeostasis, these feedback systems are hardly surprising. The purpose of a hormone is to bring the body back into homeostasis. Continued heavy secretion of a hormone would overshoot the goal and send the body out of balance in the opposite direction.

Thyroid stimulating hormone (TSH)

This hormone is also called *thyrotropin* and *TSH.* Its function is to stimulate the synthesis and secretion of the hormones produced by the thyroid gland.

Adrenocorticotrophic hormone (ACTH)

This hormone, also called *adrenocorticotrophin* and *ACTH,* has a dual function. Its trophic function is to control the production and secretion of the adrenal cortex hormones. The adrenal glands will be discussed later in this chapter. Like growth hormone, ACTH also acts directly on all body cells by increasing their catabolism of fats. Unlike growth hormone, it decreases blood sugar level because it stimulates the liver to remove glucose from the blood and store it in the form of glycogen.

Follicle stimulating hormone (FSH)

In the female, *follicle stimulating hormone,* or *FSH,* is transported to the ovaries where it stimulates the development of an egg each month. FSH also stimulates cells in the ovaries to secrete estrogens, or female sex hormones. In the male, FSH stimulates the testes to produce sperm and to secrete testosterone, a male sex hormone.

Leuteinizing hormone (LH or ICSH)

The *leuteinizing hormone* is called *leuteotropin* and *LH* in the female and *interstitial cell stimulating hormone (ICSH)* in the male. LH assumes a role in the development of the egg following the action of FSH. Namely, it stimulates the ovary to release the developed egg and prepares the uterus for implantation of a fertilized egg. It also stimu-

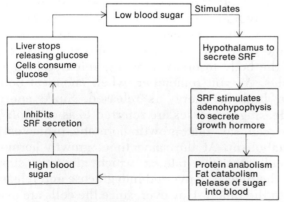

Figure 15–3. Regulation of the secretion of the growth hormone. Like other hormones of the adenohypophysis, the secretion of growth hormone is controlled by a neurohumor. Like most hormones of the body, growth hormone secretion and inhibition involve a negative feedback system. Please refer to Chapter 1 for the details of a negative feedback system.

lates the secretion of progesterone (another female sex hormone) and readies the mammary glands for milk secretion. In the male, ICSH stimulates the interstitial cells in the testes to develop and secrete testosterone.

Prolactin

Prolactin, or *lactogenic hormone*, together with other hormones, initiates milk secretion by the mammary glands.

The neurohypophysis

In a strict sense, the posterior lobe, or **neurohypophysis,** is not an endocrine gland since it does not synthesize hormones. The posterior lobe consists of cells called pituicytes, which are similar in appearance to the connective tissue neuroglia of the nervous system (Figure 15–4). Running from the hypothalamus, down the infundibular stalk, to the posterior lobe are neurons that are vital to the functioning of the neurohypophysis. The hypothalamus produces two hormones, oxytocin and vasopressin-ADH (antidiurectic hormone). These hormones travel along

the outside surface of these neurons to the posterior lobe where they are stored. Later, when the hypothalamus is properly stimulated, it sends impulses over the neurons that stimulate the neurohypophysis to release the hormones into the blood.

Oxytocin

This hormone stimulates the contraction of the smooth muscle cells in the pregnant uterus and the contractile cells around the ducts of the mammary glands (Figure 15–5). It is released in large quantities just prior to giving birth. When labor begins, the uterus and vagina distend. This distension initiates afferent impulses to the hypothalamus. These impulses stimulate the secretion of more oxytocin by the hypothalamus. The oxytocin migrates along the nerve fibers of the hypothalamus to the neurohypophysis. The impulses also cause the neurohypophysis to release oxytocin into the blood. It is then carried to the uterus to reinforce uterine contractions. The effect of oxytocin on milk ejection is as follows. Milk formed by the glandular cells of the breasts is stored until the baby begins active

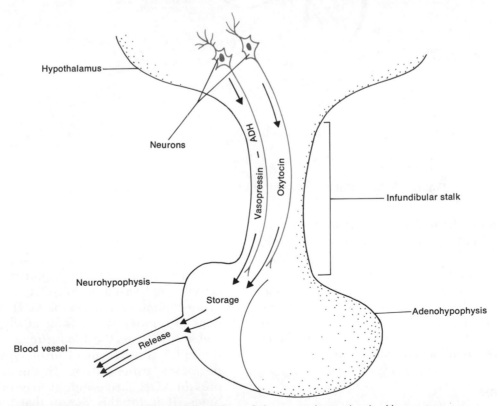

Figure 15–4. Innervation and function of the neurohypophysis. Hormones produced by the hypothalamus travel to the neurohypophysis where they are stored. At the appropriate time, impulses conducted over the neurons stimulate the neurohypophysis to release the stored hormones into the blood.

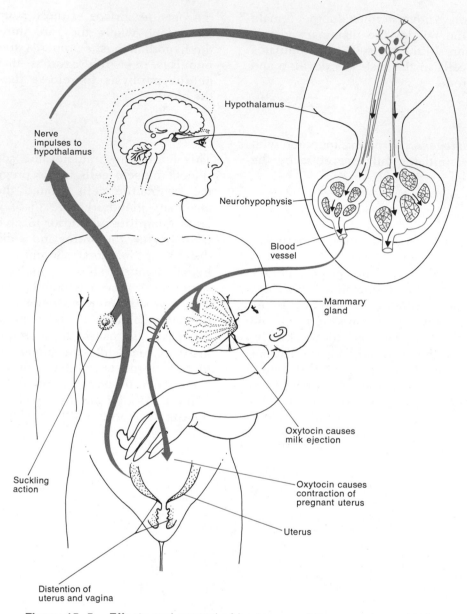

Figure 15–5. Effects and control of hormones of the neurohypophysis.

suckling. From about 30 seconds to 1 minute after nursing begins, the baby receives no milk. During this latent period, nervous impulses from the nipple are transmitted to the hypothalamus. The hypothalamus stimulates the posterior lobe to release stored oxytocin. Oxytocin then flows from the neurohypophysis via the blood to the breasts, where it stimulates the cells to contract and eject milk out of the mammary glands.

Vasopressin-ADH

This hormone has two principal physiological activities. One of these is to cause a rise in blood pressure by bringing about constriction of arterioles. This effect is noted if large quantities

of the purified hormone are injected. Only in rare instances, however, does the body secrete enough hormone to significantly affect blood pressure. The more important physiological activity of vasopressin-ADH is its effect on urine volume. Vasopressin-ADH causes the kidneys to remove water from newly forming urine and return it to the bloodstream. Since water is the chief constituent of urine, vasopressin-ADH decreases urine volume. In the absence of vasopressin-ADH, urine output may be increased ten times. It is for this reason that the abbreviation ADH, meaning antidiuretic hormone, is used. An *antidiuretic* is any chemical substance that prevents excessive urine production.

The amount of vasopressin-ADH normally

secreted varies with the needs of the body. For example, when the body is dehydrated, the concentration of water in the blood falls below normal limits. Receptors in the hypothalamus detect the low water concentration in the plasma and stimulate the hypothalamus to produce vasopressin-ADH. The hormone travels down nerve fibers to the neurohypophysis. It is then released into the bloodstream and transported to the kidneys. The kidneys respond by decreasing urine output, and water is conserved. Vasopressin-ADH also decreases the rate at which perspiration is produced during dehydration. By contrast, if the blood contains a higher-than-normal water concentration, the receptors detect the increase, and secretion of the hormone is stopped. The kidneys can then release large quantities of urine, and the volume of body fluid is brought down to normal.

Secretion of vasopressin-ADH can also be altered by a number of special conditions. Pain, stress, acetylcholine, and nicotine all stimulate secretion of the hormone. Alcohol inhibits secretion and thereby increases urine output. This is why thirst is one of the symptoms of a hangover. The exact mechanism by which vasopressin-ADH actually regulates water volume of the kidneys is discussed in Chapter 22.

Disorders of the pituitary

Disorders of the endocrine system, in general, are based upon under- or overproduction of hormones. The term **hyposecretion** describes an underproduction, whereas the term **hypersecretion** means an oversecretion. The pituitary gland produces many hormones. All these hormones, with the exception of the growth hormone, directly control the activities of other endocrine glands. It is hardly surprising, then, that hypo- or hypersecretion of a pituitary hormone produces widespread and complicated abnormalities.

Among the clinically interesting disorders related to the adenohypophysis are those involving the growth hormone. Growth hormone builds up cells, particularly those of bone tissue. If the growth hormone is hyposecreted during the growth years, bone growth is slow, and the epiphyseal plates close before normal height is reached. This is the condition called **pituitary dwarfism.** Other organs of the body also fail to grow, and the pituitary dwarf is childlike in many physical respects. Treating the condition requires administration of human growth hormone during childhood, before the epiphyseal plates close.

If secretion of growth hormone is normal during childhood, but lower-than-normal during adult life, a rare condition called **pituitary cachexia (Simmond's disease)** occurs. The tissues of a person with Simmond's disease waste away, or **atrophy.** The victim becomes quite thin and shows signs of premature aging. For instance, as his connective tissue degenerates, it loses its elasticity, and the skin hangs and becomes wrinkled. The atrophy occurs because the person is not receiving enough growth hormone to stimulate the protein-building activities that are required for replacing cells and cell parts.

Hypersecretion of the growth hormone produces completely different disorders. For example, hyperactivity during childhood years results in **giantism,** which is an abnormal increase in the length of long bones. Hypersecretion during adulthood is called **acromegaly.** Acromegaly cannot produce further lengthening of the long bones because the epiphyseal plates are already closed. Instead, the bones of the hands, feet, cheeks, and jaws thicken. Other tissues also grow. For instance, the eyelids, lips, tongue, and nose enlarge, the skin thickens and furrows, especially on the forehead and soles of the feet.

The principal abnormality associated with dysfunction of the neurohypophysis is **diabetes insipidus.** This disorder should not be confused with diabetes mellitus, which is a disorder of the pancreas and is characterized by sugar in the urine. Diabetes insipidus is the result of a hyposecretion of vasopressin-ADH, usually caused by damage to the neurohypophysis or to the hypothalamus. Symptoms of the disorder include excretion of large amounts of urine and subsequent thirst. Diabetes insipidus is treated by administering vasopressin-ADH.

THYROID GLAND

The double-lobed organ located just below the voicebox is called the **thyroid gland.** The two lobes lie one on either side of the windpipe and are connected by a mass of tissue called an *isthmus.* (See Figure 15–1.) The gland weighs about 25 grams and has a very rich blood supply, receiving about 80 to 120 milliliters of blood per minute. Histologically, the thyroid is composed of spherical-shaped sacs called *thyroid follicles* (Figure 15–6 top). The walls of each follicle are formed by a layer of simple cuboidal epithelium. This layer manufactures the two hormones produced by the gland, thyroxin and thyrocalcitonin. The interior of each sac is filled with

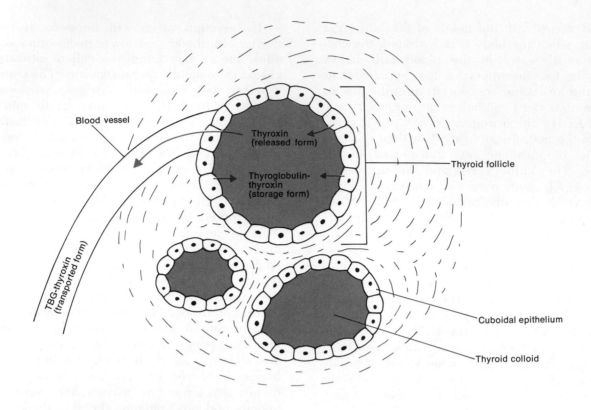

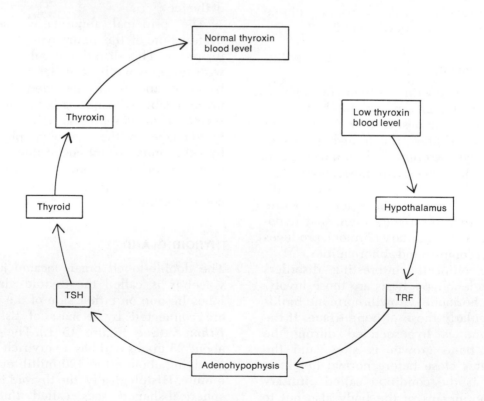

Figure 15–6. The thyroid gland. (Top) Histology of thyroid tissue and the storage, release, and transportation of thyroxin. (Bottom) Regulation of thyroxin secretion.

thyroid colloid, a stored form of the thyroid hormones.

Physiology of the thyroid

One of the unique features of the thyroid gland is its ability to store its hormones and release them in a steady flow over a long period of time. For example, the principal hormone, *thyroxin,* is synthesized from iodine and an amino acid called tyrosine. Synthesis usually occurs on a fairly continuous basis. However, if the body has no immediate need for the hormone, it combines with *thyroglobulin,* a protein secreted by the follicle cells, and is stored in the colloid. When demand for thyroxin occurs, the hormone splits apart from the thyroglobulin and is released into the blood. There, it combines with a plasma protein called *thyroid-binding globulin* or *TBG.* The thyroxin is released from TBG as it enters tissue cells.

The synthesis and transport of thyrocalcitonin, the other thyroid hormone, occurs in basically the same manner. The iodine in thyroxin and thyrocalcitonin that is bound to TBG is called *protein-bound iodine* or *PBI.* Under normal circumstances, the amount of PBI in the blood is fairly constant—4 to 8 micrograms PBI/100 milliliters blood. This amount can easily be measured. For this reason, PBI is a good index of thyroid hormone secretion and is often used as a tool to diagnose suspected thyroid malfunction.

Thyroxin action and control

The major function of thyroxin is to control the rate of metabolism. In other words, it regulates the catabolism, or energy-releasing processes, and the anabolism, or building-up processes. Thyroxin increases the rate at which carbohydrates are burned. And it stimulates cells to break down proteins for energy instead of using them for building processes. At the same time, thyroxin decreases the breakdown of fats. The overall effect, though, is to increase catabolism. It thus produces energy and raises body temperature as heat energy is given off. This is called the *calorigenic effect.* Thyroxin is also an important factor in the regulation of tissue growth and in the development of tissues. Hyposecretion of thyroxin during the early years of life causes some organs to fail to develop. Finally, thyroxin acts as a diuretic, increases the reactivity of the nervous system, and increases the heart rate.

The secretion of thyroxin seems to be brought on by any of several factors (Figure 15–6 bottom).

For instance, if levels of thyroxin in the blood fall below normal, or if the metabolic rate decreases, chemical sensors in the hypothalamus detect the change in the blood chemistry and stimulate the hypothalamus to release a neurohumor called thyrotropin-releasing factor (TRF). TRF stimulates the adenohypophysis to secrete thyroid-stimulating hormone (TSH). TSH acts upon the thyroid, stimulating it to release thyroxin, until the metabolic rate is brought back up to normal. Conditions that increase the body's need for energy, such as a cold environment, high altitude, and pregnancy, also trigger this feedback system and increase the secretion of thyroxin.

When the amount of thyroxin in the blood is brought up to the required level, the secretion of TSH stops, and thyroxin release is cut back. Thyroid activity can also be slowed down by a number of other factors. For instance, when very large amounts of certain sex hormones are circulating with the blood, TSH secretion is decreased. Aging slows down the activities of most glands. Thus thyroid production decreases as an individual gets older.

Thyrocalcitonin action and control

The second thyroid hormone produced by the thyroid gland is *thyrocalcitonin.* It is involved in the homeostasis of blood calcium level. As you remember, bones are continually remolded during adult life. Part of this process consists of the breakdown of osseous tissue and the release of calcium into the blood. The other part of the process is the deposition of calcium in the bones and the subsequent laying down of new ossified tissue. Thyrocalcitonin lowers the amount of calcium in the blood by inhibiting bone breakdown and by accelerating the absorption of calcium by the bones. If thyrocalcitonin is administered to a person with normal blood calcium levels, it causes *hypocalcemia,* or low blood calcium level. If thyrocalcitonin is given to an individual with *hypercalcemia* (high blood calcium level), the level returns to normal. It is suspected that the blood calcium level directly controls the secretion of thyrocalcitonin in a simple negative feedback system which does not involve the pituitary gland. (See Figure 15–7.)

Disorders of the thyroid

Hyposecretion of the thyroid hormone during the growth years results in a condition called **cretinism.** Two outstanding clinical symptoms of the cretin are dwarfism and mental retardation.

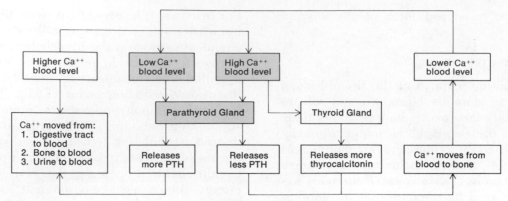

Figure 15–7. Regulation of the secretion of the parathyroid hormone and thyrocalcitonin.

The first is caused by failure of the skeleton to grow and mature. The second is caused by failure of the brain to develop fully. Recall that one of the functions of thyroid hormone is to control tissue growth and development. Cretins also exhibit retarded sexual development and a yellowish skin color. Flat pads of fat develop, giving the cretin the characteristic round face and thick nose; a large, thick, protruding tongue; and protruding abdomen. Because the energy-producing metabolic reactions are so slow, the cretin has a low body temperature and general lethargy. Carbohydrates are stored rather than utilized. Heart rate is also slow. If the condition is diagnosed early, the symptoms are reversible following administration of the thyroid hormone.

Hypothyroidism during the adult years produces a disorder called **myxedema.** The name refers to the fact that thyroxin is a diurectic. Lack of thyroxin causes the body to retain water. And one of the hallmarks of myxedema is an edema that causes the facial tissues to swell and look puffy. Another symptom caused by the retention of water is an increase in blood volume that frequently causes high blood pressure. Like the cretin, the person with myxedema also suffers from slow heart rate, low body temperature, muscular weakness, general lethargy, and a tendency to gain weight easily. The long-term combination of a slow heart rate and high blood pressure may overwork the heart muscles, causing the heart to enlarge. Because the brain has already reached maturity, the person with myxedema does not experience mental retardation. However, in moderately severe cases nerve reactivity may be dulled so that the person lacks mental alertness. Myxedema occurs eight times more frequently in females than in males. Its symptoms are abolished by the administration of thyroxin.

Hypersecretion of thyroid hormone gives rise to a condition called **exophthalmic goiter.** This disease, like myxedema, is also more frequent in females, affecting eight females to every one male. One of its primary symptoms is an enlarged thyroid, called a *goiter,* which may be two to three times its original size. The two other characteristic symptoms are an edema behind the eye, which causes the eyes to "pop out" (**exophthalmos**), and an abnormally high metabolic rate. The high metabolic rate produces a range of effects that are generally opposite to those of myxedema. The person has an increased pulse. The body temperature is high, and the skin is warm, moist, and flushed. Weight loss occurs, and the person is usually full of "nervous" energy. The thyroxin also increases the responsiveness of the nervous system. Thus a person with exophthalmic goiter may become irritable and may exhibit tremors of the fingers when they are extended. The usual methods for treating hyperthyroidism are administering drugs that suppress thyroxin synthesis or surgically removing a part of the gland.

The term **goiter** simply means an enlargement of the thyroid gland. It is a symptom of many thyroid disorders. It may also occur if the gland does not receive enough iodine to produce sufficient thyroxin for the body's needs. The follicle cells then enlarge in a futile attempt to produce more thyroxin, and they secrete large quantities of colloid. This is called *simple goiter.* Simple goiter is most often caused by a lower-than-average amount of iodine in the diet. However, it may also develop if iodine intake is not increased during certain conditions that put a high demand on the body for thyroxin. Such conditions are frequent exposure to cold and high fat and protein diets.

PARATHYROIDS

Embedded on the posterior surfaces of the lateral lobes of the thyroid are small, round masses of tissue called the **parathyroid glands.** Typically, two parathyroids are attached to each thyroid lobe. (See Figure 15–1.) The parathyroids are abundantly supplied with blood and are innervated by the autonomic nervous system. Histologically, the parathyroids contain two kinds of epithelial cells. The first kind, called *principal* or *chief cells*, are believed to be the major synthesizers of the parathyroid hormone. Some researchers believe that the other kind of cell, called an *oxyphil cell*, synthesizes a reserve capacity of hormone.

Physiology of the parathyroids

Parathyroid hormone (PTH) controls the homeostasis of ions in the blood, especially the homeostasis of calcium and phosphate ions. First, if adequate amounts of vitamin D are present, PTH increases the rate at which calcium is absorbed from the intestine into the blood. Second, PTH increases the number of osteoclasts, or bone-destroying cells. As a result, bone tissue is broken down, and calcium and phosphate ions are released into the blood. Recall that thyrocalcitonin secreted by the thyroid has the opposite effect. Finally, PTH produces two changes in the kidneys: (1) It increases the rate at which the kidneys remove calcium ions from the urine and return them to the blood; (2) it accelerates the transportation of phosphate ions from the blood into the urine for elimination. More phosphate is lost through the urine than is gained from the bones. The overall effect of PTH, then, is to decrease blood phosphate level and increase blood calcium level. As far as blood calcium level is concerned, PTH and thyrocalcitonin are antagonists.

PTH secretion is not controlled by the pituitary gland. When the calcium ion level of the blood falls, more PTH is released (Figure 15–7). Conversely, when the calcium ion level of the blood rises, less PTH (and more thyrocalcitonin) is secreted. Once again, this is an example of a negative feedback control system.

Disorders of the parathyroids

A normal amount of calcium in the extracellular fluid is necessary to maintain the resting state of neurons. A deficiency of calcium caused by **hypoparathyroidism** causes neurons to depolarize without the usual stimulus. As a result, nervous impulses increase and result in muscle twitches, spasms, and convulsions. This condition is called **tetany.** (See Chapter 10.) The effects of hypocalcemic tetany are observed in the **Trousseau** and **Chvostek signs.** Trousseau sign is observed when the binding of a blood pressure cuff around the upper arm produces contraction of the fingers and inability to open the hand. The Chvostek sign is a contracture of the facial muscles elicited by tapping the facial nerves at the angle of the jaw. Hypoparathyroidism results from surgical removal of the parathyroids or from parathyroid damage caused by parathyroid disease, infection, hemorrhage, or mechanical injury.

Hyperparathyroidism causes demineralization of bone. This condition is called **osteitis fibrosa cystica** because the areas of destroyed bone tissue are replaced by cavities that fill with fibrous tissue. The bones thus become deformed and are highly susceptible to fracture. Hyperparathyroidism is usually caused by a tumor in the parathyroids.

THE ADRENALS

The body has two **adrenal glands,** and they are located one on top of each kidney. (See Figure 15–1.) Each adrenal gland is structurally and functionally differentiated into two sections: the outer *adrenal cortex*, which makes up the bulk of the gland, and the inner *adrenal medulla* (Figure 15–8a). Covering the gland is a thick layer of fatty connective tissue and an outer, thin fibrous capsule.

The adrenal cortex

Histologically, the cortex is subdivided into three zones. Each zone has a different cellular arrangement and secretes different hormones. The outer zone, directly underneath the connective tissue covering, is referred to as the *zona glomerulosa.* Its cells are arranged in arched loops or round balls, and they secrete a group of hormones called mineralocorticoids. The middle zone, or *zona fasciculata*, is the widest of the three zones and consists of cells arranged in long, straight cords. The zona fasciculata secretes glucocorticoid hormones. The inner zone, the *zona reticularis*, contains cords of cells that branch freely. This zone synthesizes sex hormones, chiefly male hormones called androgens.

Mineralocorticoids

These hormones help control electrolyte homeostasis, particularly the concentrations of sodium

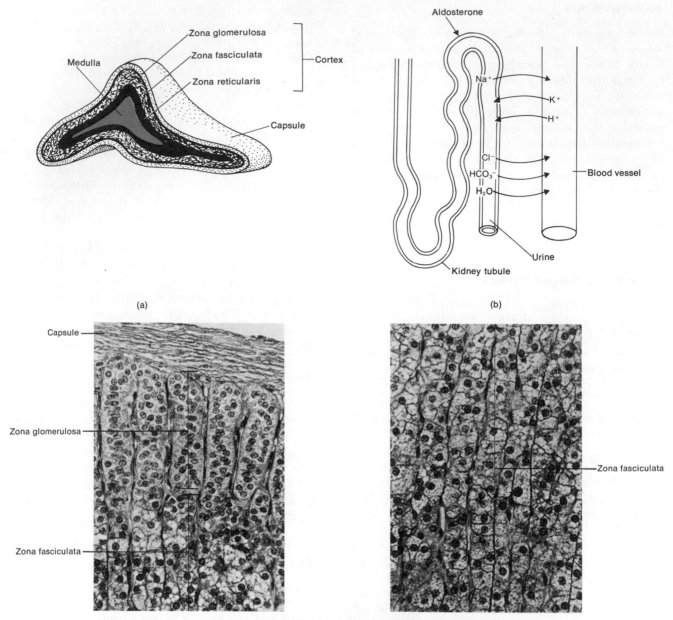

Figure 15–8. The adrenal gland. (a) Cross section of the gland showing the various zones. (b) Effects of aldosterone. *(Photographs courtesy of Edward J. Reith, Atlas of Descriptive Histology, Edward J. Reith and Michael H. Ross, Harper & Row, Publishers, Inc., New York, 1970.)*

and potassium. Although the adrenal cortex secretes at least three different hormones classified as mineralocorticoids, one of these hormones is responsible for about 95 percent of the mineralocorticoid activity. The name of this hormone is *aldosterone*. Aldosterone acts on the tubule cells in the kidneys and causes them to increase their reabsorption of sodium (Figure 15–8b). In other words, sodium ions are removed from the urine and returned to the blood. In this manner, aldosterone prevents rapid depletion of sodium

in the body. On the other hand, aldosterone decreases reabsorption of potassium. Large amounts of potassium are moved from the blood into the urine. These two basic functions—conservation of sodium and elimination of potassium—cause a number of secondary effects. For example, a large proportion of the sodium reabsorption occurs through an exchange reaction whereby positive hydrogen ions pass into the urine to take the place of the positive sodium ions. Since this mechanism removes hydrogen ions, it makes the blood less acidic and prevents acidosis. More will be said about acidosis and the opposite condition, alkalosis, in Chapter 23. The movement of Na^+ ions also sets up a positively charged field in the blood vessels around the kidney tubules. As a result, negatively charged chloride and bicarbonate ions are drawn out of the urine and back into the blood. Finally, the increase in sodium-ion concentration in the blood vessels causes water to move by osmosis from the urine into the blood. In summary, aldosterone causes potassium excretion and sodium reabsorption. The sodium reabsorption leads to the elimination of H^+ ions, the retention of Na^+, Cl^-, and HCO_3^-, and the retention of water.

The control of aldosterone secretion is rather complex and poorly understood. It seems that several different mechanisms operate simultaneously, and they are believed to occur as follows (Figure 15–9). When the body is dehydrated, the blood contains less water. This means that a smaller volume of blood is circulating through the vessels and that blood pressure is lowered. Receptors in the hypothalamus detect the low pressure, and the hypothalamus secretes a neurohumor that circulates through the blood to the adrenal cortex. In the adrenal cortex, the neurohumor stimulates secretion of aldosterone. However, as the low blood pressure stimulates the hypothalamus, it also stimulates

certain kidney cells to secrete into the blood an enzyme called *renin*. Renin converts angiotensinogen, a plasma protein produced by the liver, into *angiotensin I*, which is then converted into *angiotensin II*. Angiotensin II also stimulates the adrenal cortex to produce more aldosterone. Aldosterone brings about Na^+ and water reabsorption. This leads to an increase in extracellular fluid volume and a restoration of blood pressure to normal. The adrenal cortex can also be stimulated simply by abnormal concentrations of Na^+ or K^+. A decrease in Na^+ or an increase of K^+ in the blood acts directly on the cortex to increase aldosterone secretion.

Glucocorticoids

The *glucocorticoids* are a group of hormones that are largely concerned with normal metabolism and the ability of the body to resist stress. Three examples of glucocorticoids are *hydrocortisone (cortisol)*, *corticosterone*, and *cortisone*. Of the three, hydrocortisone is the most abundant. The glucocorticoids have the following effects on the body:

1. Glucocorticoids work with other hormones in promoting normal metabolism. Their role is to make sure that enough energy is provided. They increase the rate at which amino acids are

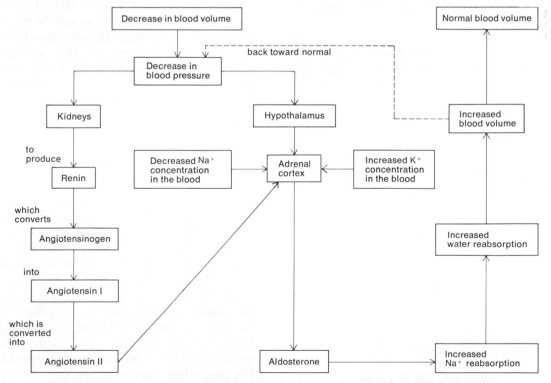

Figure 15–9. Proposed mechanism for the regulation of aldosterone secretion.

removed from cells and transported to the liver. The amino acids may be synthesized into new proteins, such as the enzymes that are needed for the metabolic reactions. Or, if the body's reserves of glycogen and fat are low, the liver may convert the amino acids to glucose. Glucocorticoids also promote the breakdown of carbohydrates to glucose. Both processes increase blood sugar level. Glucocorticoids are therefore hyperglycemic. In addition, the hormones encourage the movement of fats from storage depots to all the cells, where they are catabolized for energy.

2. Glucocorticoids work in many ways to provide resistance to stress. One of the more obvious ways is their hyperglycemic effect. A sudden increase in available glucose makes the body more alert. Additional glucose gives the body energy for combating a range of stressors, such as fright, temperature extremes, high altitude, bleeding, and infection. Glucocorticoids also make the blood vessels more sensitive to vessel-constricting chemicals. They thereby raise blood pressure. This is advantageous if the stressor happens to be blood loss, which causes a drop in blood pressure.

3. Glucocorticoids decrease the blood vessel dilatation and edema associated with inflammations. They are thus anti-inflammatories. Unfortunately, they also decrease connective-tissue regeneration and are thereby responsible for slow wound healing.

The control of glucocorticoid secretion is another example of a typical negative feedback mechanism (Figure 15–10). The two stimuli are extreme stress and low blood levels of glucocorticoids. Such stress could include emotional stress or physical damage, such as that produced by contusions, broken bones, disease, or tissue destruction. The stress may directly stimulate

the hypothalamus. For example, the hypothalamus could be stimulated by low blood pressure resulting from excessive bleeding. Or the stimulus may be relayed to the hypothalamus from other parts of the nervous system. In any case, either extreme stress or abnormally low levels of glucocorticoids stimulate the hypothalamus to secrete a neurohumor called *corticotrophin-releasing factor (CRF)*. This neurohumor initiates the release of ACTH from the anterior lobe of the pituitary. ACTH is carried through the blood to the adrenal cortex, where it then stimulates glucocorticoid secretion.

Gonadocorticoids

The adrenal cortices secrete both male and female *gonadocorticoids*, or *sex hormones*. But the amount of sex hormones secreted by the adrenals is usually so small that it is insignificant. The exception is hypersecretion—an abnormality that will be described shortly.

The adrenal medulla

The adrenal medulla consists of hormone-producing cells, called *chromaffin cells*, which surround large blood-containing sinuses. (See Figure 15–8a.) Dispersed among the chromaffin cells are ganglia of the sympathetic division of the autonomic nervous system. In all other visceral effectors, preganglionic sympathetic fibers first synapse with postganglionic neurons before innervating the effector. In the adrenal medulla, however, the preganglionic fibers pass directly into the gland. The secretion of hormones from the medulla is directly controlled by the autonomic nervous system, and innervation by the preganglionic fibers allows the gland to respond very rapidly to a stimulus.

The two principal hormones synthesized by the adrenal medulla are epinephrine and norepinephrine. Epinephrine constitutes about 80 percent of the total secretion of the gland and is more potent in its action than norepinephrine. Both hormones are *sympathomimetic*. That is, they produce effects similar to those brought about by the sympathetic division of the autonomic nervous system. And, to a large extent, they are responsible for the "fight-or-flight" response. Like the glucocorticoids of the adrenal cortices, these hormones help the body resist stress situations. However, unlike the cortical hormones, the medullary hormones are not essential for life. Under stress conditions, impulses received by

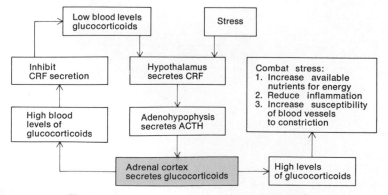

Figure 15–10. Regulation of the secretion of glucocorticoids.

the hypothalamus are conveyed to sympathetic preganglionic neurons, which cause the chromaffin cells to increase their output of epinephrine and norepinephrine. Epinephrine increases blood pressure by increasing the heart rate and by constricting the blood vessels. It accelerates the rate of respiration, dilates respiratory passageways, decreases the rate of digestion, increases the efficiency of muscular contractions, increases blood sugar level, and stimulates cellular metabolism.

Disorders of the adrenals

Hypersecretion of the mineralocorticoid aldosterone results in a decrease in the body's potassium concentration. As you remember, potassium movement is involved in the transmission of nerve impulses. Consequently, if potassium depletion is great enough, neurons cannot depolarize and muscular paralysis results. Hypersecretion also brings about excessive retention of sodium and water. The water increases the volume of the blood and causes high blood pressure. It also increases the volume of the interstitial fluid, producing edema.

Disorders associated with glucocorticoids include Addison's disease and Cushing's syndrome. Hyposecretion of glucocorticoids results in the condition called **Addison's disease.** Clinical symptoms include hypoglycemia, which leads to muscular weakness, mental lethargy, and weight loss. In addition, increased potassium blood levels and decreased sodium blood levels lead to low blood pressure and dehydration. **Cushing's syndrome** is a hypersecretion of glucocorticoids, especially hydrocortisone and cortisone. The condition is characterized by the redistribution of fat. This results in spindly legs accompanied by a characteristic "moon face," "buffalo hump" on the back, and pendulous abdomen. The facial skin is flushed, and the skin covering the abdomen develops stretch marks. The individual also bruises easily, and wound healing is poor.

The **adrenogenital syndrome** results from overproduction of sex hormones, particularly the male androgens, by the adrenal cortex. Hypersecretion in male infants and young male children results in an enlarged penis. In young boys, it also causes premature development of male sexual characteristics. Hypersecretion in adult males is characterized by overgrowth of body hair, enlargement of the penis, and increased sexual drive. Hypersecretion in young girls results in premature sexual development. Hypersecretion in both girls and women usually produces a receding hairline, baldness, an increase in body hair, deepening of the voice, muscular arms and legs, small breasts, and an enlarged clitoris.

Tumors of the chromaffin cells, called **pheochromocytomas,** cause hypersecretion of the medullary hormones. The oversecretion causes high blood pressure, high levels of sugar in the blood and urine, an elevated basal metabolism rate, nervousness, and sweating. Since the medullary hormones create the same effects as does sympathetic nervous stimulation, hypersecretion puts the individual into a prolonged version of the "fight-or-flight" response. Needless to say, this eventually wears out the body, and the individual eventually suffers from general weakness.

PANCREAS

Because of its functions, the **pancreas** can be classified as both an endocrine and an exocrine gland. Since the exocrine functions of the gland will be discussed in the chapter on the digestive system, we shall treat only its endocrine functions at this point. The pancreas is a flattened organ located behind and slightly below the stomach. (See Figure 15–1.) The endocrine portion of the pancreas consists of clusters of cells called *islets of Langerhans.* Two kinds of cells are found in these clusters: (1) *alpha cells,* which comprise about 25 percent of the islet cells and secrete the hormone glucagon; and (2) *beta cells,* which constitute about 75 percent of the islet cells and secrete the hormone insulin. The islets are surrounded by blood capillaries and by the cells that form the exocrine part of the gland.

Physiology of the pancreas

The endocrine secretions of the pancreas—glucagon and insulin—are concerned with regulation of the blood sugar level. Let us now examine how this takes place.

Glucagon

The product of the alpha cells is *glucagon,* a hormone whose principal physiological activity is to increase the blood glucose level (Figure 15–11). Glucagon does this by accelerating the conversion of liver glycogen into glucose. The liver then releases the glucose into the blood,

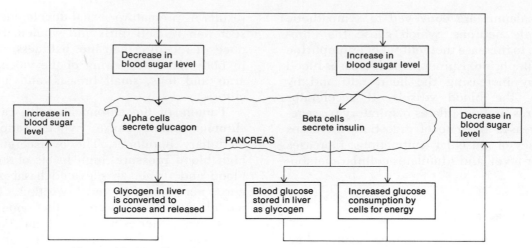

Figure 15–11. Regulation of the secretion of glucagon and insulin.

and the blood sugar level rises. Secretion of glucagon is directly controlled by the level of blood sugar. For example, when the blood sugar level falls below normal, chemical sensors in the alpha cells of the islets stimulate the cells to secrete glucagon. When blood sugar rises, the cells are no longer stimulated and production slackens. If for some reason the self-regulating device fails and the alpha cells secrete glucagon continuously, hyperglycemia may result.

Insulin

The beta cells of the islets produce a hormone called *insulin*. This hormone increases the build-up of proteins in cells. But its chief physiological action is opposite that of glucagon. Insulin decreases blood sugar level (Figure 15–11). This is accomplished in two ways. First, insulin accelerates the transport of glucose from the blood into body cells, especially into the cells of the liver and muscles. Second, insulin accelerates the conversion of glucose into glycogen. The regulation of insulin secretion, like that of glucagon secretion, is directly determined by the level of sugar in the blood. However, other hormones can indirectly affect insulin production. For instance, growth hormone raises blood glucose level, and the rise in glucose level triggers insulin secretion. ACTH, by stimulating the secretion of glucocorticoids, brings about hyperglycemia and also indirectly stimulates the release of insulin. Epinephrine is also an insulin antagonist.

Endocrine disorders of the pancreas

Hyposecretion of insulin results in a number of clinical symptoms referred to as **diabetes mellitus.** Typically an inherited disease, diabetes mellitus

is caused by the destruction or malfunction of the beta cells. Among the symptoms are hyperglycemia and excretion of glucose in the urine as hyperglycemia increases. There is also an inability to reabsorb water, resulting in increased urine production, dehydration, loss of sodium, and thirst. Although the cells need glucose for energy-releasing reactions, glucose cannot enter the cells without the help of insulin. The cells start breaking down large quantities of fats and proteins into glucose. When the fats are decomposed, organic acids called ketone bodies are formed as side products. Excessive decomposition of fats produces more ketone bodies than the body can neutralize through its buffer systems. As a result, the blood pH falls. This form of acidosis is called **ketosis.** The catabolism of stored fats and proteins also causes weight loss. As lipids are transported by the blood from storage depots to hungry cells, lipid particles are deposited on the walls of blood vessels. The deposition leads to hardening of the arteries and a multitude of circulatory problems.

Hyperinsulinism is much rarer than hyposecretion and is generally the result of a malignant tumor in an islet. The principal symptom is a decreased blood glucose level, which stimulates the secretion of epinephrine, glucagon, and the growth hormone. As a consequence, anxiety, sweating, tremor, increased heart rate, and weakness occur. Moreover, brain cells do not have enough glucose to function efficiently. This leads to mental disorientation, convulsions, unconsciousness, shock, and eventual death as the vital centers in the medulla are affected.

OVARIES AND TESTES

The female gonads, called the **ovaries,** are paired oval-shaped bodies located in the pelvic cavity.

(See Figure 15–1.) The ovaries produce female sex hormones that are responsible for the development and maintenance of the female sexual characteristics. Along with the gonadotrophic hormones of the pituitary, the sex hormones also regulate the menstrual cycle, maintain pregnancy, and ready the mammary glands for lactation. The male has two oval-shaped glands, called **testes,** that lie in the scrotum. The testes produce the male sex hormones that stimulate the development and maintenance of the male sexual characteristics. In Chapter 24, more will be said about the sex hormones — their names, how they function, and how they are regulated by the pituitary.

PINEAL GLAND

The cone-shaped gland located in the roof of the third ventricle is known as the **pineal gland,** or **epiphysis cerebri.** (See Figure 13–1.) The gland consists of masses of specialized cells. Around the cells are scattered preganglionic sympathetic fibers. The pineal gland starts to degenerate at about age 7, and in the adult it is largely fibrous tissue.

Although many anatomical facts concerning the pineal gland have been known for years, its physiology is still somewhat obscure. One hormone secreted by the pineal gland is *melatonin,* which appears to affect the secretion of hormones by the ovaries. It has been known for years that light stimulates the sexual endocrine glands. Researchers have also discovered that blood levels of melatonin are low during the day and high at night. Putting these observations together, some investigators now believe that melatonin inhibits the activities of the ovaries. During daylight hours, light entering the eye stimulates neurons to transmit impulses to the pineal that inhibit melatonin secretion. Without melatonin interference, the ovaries are free to step up their hormone production. But at night, the pineal gland is able to release melatonin, and ovarian function is slowed down. One of the functions of the pineal gland might very well be regulation of the activities of the sexual endocrine glands, particularly the menstrual cycle.

Some evidence also exists that the pineal secretes a second hormone called *adrenoglomerulotropin.* This hormone may stimulate the adrenal cortex to secrete aldosterone. Still other functions attributed to the pineal gland are the secretion of a growth-inhibiting factor and the secretion of a hormone called *serotonin* that is involved in normal brain physiology.

THYMUS GLAND

Usually a bilobed organ, the **thymus gland** is located in the upper mediastinum behind the sternum and between the lungs. (See Figure 15–1.) The gland is conspicuous in the infant, and at about 2 years of age it attains its largest relative size. Around puberty, the thymic tissue, which consists primarily of a type of white blood cell, is replaced by fat. By the time the person reaches maturity, the gland has atrophied. The thymus is believed to secrete a hormone that enables the white blood cells to produce antibodies for the defense of the body.

A summary of the principal endocrine activities is presented in Exhibit 15–1.

STRESS AND HOMEOSTASIS

Throughout this text, we have repeatedly emphasized the concept of homeostasis — the maintenance of the body's internal physiological environment in response to stresses that originate within the body or in the external environment. Essentially, homeostasis may be viewed as specific responses of the body to specific stimuli. For instance, when blood calcium goes up, the rise stimulates the thyroid gland to release thyrocalcitonin. When blood calcium falls, thyrocalcitonin secretion is inhibited, and parathyroid hormone secretion is stimulated. Homeostatic mechanisms "fine tune" the body. If the mechanisms are successful, the internal environment maintains a uniform chemistry, temperature, and pressure.

Homeostatic mechanisms are geared toward counteracting the everyday stresses of living. However, if a stress is extreme or unusual, the normal ways of keeping the body in balance may not be sufficient. In this case, the stress triggers a wide-ranging set of bodily changes called the **general stress syndrome.** Unlike the homeostatic mechanisms, the general stress syndrome does not maintain a constant internal environment. In fact, it does just the opposite. For instance, during the general stress syndrome, blood pressure and blood sugar level are raised above normal. The purpose of these changes in the internal environment is to gear up the body to meeting an emergency. Here is how it works.

The hypothalamus can be called the watch dog of the physical and psychological state of the body. It has sensors that detect changes in the chemistry, temperature, and pressure of the blood. It is informed of strong emotions through tracts that connect it with the emotional centers of the cerebral cortex. When the hypothalamus realizes

Exhibit 15–1. SUMMARY OF ENDOCRINE ACTIVITIES

SOURCE	HORMONE	SITE OF ACTION	EFFECT
Pituitary			
1. Adenohypophysis	Growth hormone (STH)	General	Stimulates hard and soft tissue growth; accelerates protein synthesis; promotes fat catabolism; increases blood sugar level
	Thyroid stimulating hormone (TSH)	Thyroid follicles	Stimulates synthesis and secretion of thyroid hormones
	Adrenocorticotrophic hormone (ACTH)	Adrenal cortex	Stimulates synthesis and secretion of glucocorticoid hormones of the adrenal cortex
	Follicle stimulating hormone (FSH)	Ovaries or testes	In female, stimulates growth and development of egg and secretion of estrogens; in male, induces development of sperm and secretion of testosterone
	Leuteinizing hormone Leuteotropin (LH)	Ovaries	Progesterone secretion, preparation of uterus for fertilized egg, mammary gland development
	Interstitial cell stimulating hormone (ICSH)	Testes	Stimulates interstitial cells to secrete testosterone
	Prolactin	Mammary glands	Stimulates milk secretion
2. Neurohypophysis	Oxytocin	Smooth muscle of uterus and mammary glands	Uterine contraction and milk ejection
	Vasopressin-ADH	Arterioles and kidneys	Arteriole constriction; decreases urine volume through reabsorption
Thyroid	Thyroxin	General	Increases metabolic rate
	Thyrocalcitonin	Bones	Decreases blood calcium level
Parathyroids	Parathyroid hormone (PTH)	Bones, intestines, and kidneys	Increases blood calcium level; decreases blood phosphate level

that a severe or unusual stress is occurring, it initiates a chain of reactions that produce the general stress syndrome. The severe or unusual stresses that produce the syndrome are called **stressors.** A stressor may be almost any severe disturbance such as extreme heat or cold, environmental poisons, poisons given off by bacteria during a raging infection, heavy bleeding from a wound or surgical procedure, or a strong emotional reaction.

When a stressor appears on the scene, it stimulates the hypothalamus to initiate the syndrome through two pathways. The first pathway is stimulation of the sympathetic nervous system and adrenal medulla. This produces an immediate set of responses called the alarm reaction. The second pathway, called the resistance reaction, involves the anterior pituitary gland and adrenal cortex. The resistance reaction is slower to start, but its effects are longer lasting.

Alarm reaction

The *alarm reaction,* or "fight-or-flight" response, is the initial reaction of the body to any stressor (Figure 15–12a). It is actually a complex of reactions initiated by the hypothalamic stimulation of the sympathetic nervous system and the adrenal medulla. The responses of the visceral effectors are immediate, though short-lived. And they are designed to counteract a danger by mobilizing the body's resources for immediate physical activity. In essence, the alarm reaction brings tremendous amounts of glucose and oxygen to the organs that are most active in warding off danger. These are the brain, which must become highly alert; the skeletal muscles, which may have to fight off a strong attacker; and the heart, which must work furiously to pump enough materials to the brain and muscles. Among the stress responses that characterize the alarm stage are the following:

SOURCE	HORMONE	SITE OF ACTION	EFFECT
Adrenals			
1. Cortex	Mineralocorticoids (Aldosterone)	General	Increases blood sodium level; decreases blood potassium level; removes hydrogen ions; conserves chloride and bicarbonate ions; increases body fluid
	Glucocorticoids (Hydrocortisone)	General, but primarily the kidneys	Promotes normal carbohydrate, protein, and fat catabolism; resistance to stress; anti-inflammatory
	Gonadocorticoids (Sex hormones)	General	See description under specific sex hormones
2. Medulla	Epinephrine	General	Typical sympathetic actions under stress
Pancreas	Glucagon	General	Increases blood sugar level; mobilizes conversion of proteins and fats to glucose; elevates potassium and phosphate levels in blood
	Insulin	General	Decreases blood sugar level; promotes glucose storage as fat and protein; decreases potassium and phosphate levels in blood
Ovaries	Estrogens	General, but notably mammary glands and reproductive organs	Maturation of female sex organs and development of secondary sex characteristics
	Progesterone	Sexual organs and mammary glands	Prepares uterus for implantation; mammary gland development
Testes	Testosterone	General	Maturation of male sex organs and development of secondary sex characteristics
Pineal	Melatonin	Retina of eye and ovaries	Inhibits ovarian function
	Adrenoglomerulotropin	Adrenal cortex	Stimulates aldosterone secretion
Thymus	Thymus hormone	Liver and lymphoid tissue	Antibody production

1. The heart rate and the strength of cardiac muscle contraction increase. This circulates substances in the blood very quickly to areas where they are needed to combat the stress.
2. Blood vessels supplying the skin and viscera, except the heart and lungs, undergo constriction. At the same time, blood vessels supplying the skeletal muscles and brain undergo dilation. These responses route more blood to organs that are active in the stress responses while decreasing blood supply to organs that do not assume an immediate, active role.
3. The spleen contracts and discharges stored blood into the general circulation to provide additional blood. Moreover, red blood cell production is accelerated, and the ability of the blood to clot is increased. These preparations are made by the body for combating possible bleeding.
4. The liver transforms large amounts of stored glycogen into glucose and releases it into the bloodstream. The glucose is broken down by the active cells to provide the energy needed to meet the stressor.
5. An increase in sweat production also occurs. This response helps to lower body temperature, which is elevated as circulation increases and body catabolism increases. Profuse sweating

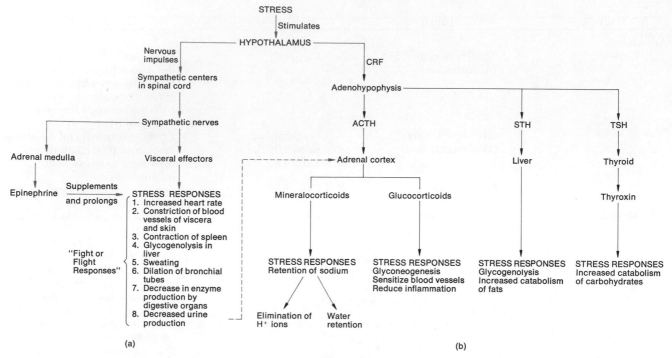

Figure 15–12. Stress responses during the (a) alarm stage and (b) resistance stage. Colored arrows indicate immediate reactions. Black arrows indicate long-term reactions.

also helps to eliminate wastes produced as a result of accelerated catabolism.

6. The rate of breathing increases, and the respiratory passageways widen in order to accommodate larger volumes of air. This response enables the body to acquire more oxygen, which is needed in the decomposition reactions of catabolism. It also allows the body to eliminate more carbon dioxide, which is produced as a side product during catabolism.

7. A decrease in the production of saliva, stomach enzymes, and intestinal enzymes occurs. This takes place since digestive activity is not essential for counteracting the stress.

8. Sympathetic impulses to the adrenal medulla increase its secretion of epinephrine. This hormone supplements and prolongs many sympathetic responses, such as increasing heart rate and strength, constricting blood vessels, accelerating the rate of breathing, widening respiratory passageways, increasing blood sugar level, increasing the rate of catabolism, and decreasing the rate of digestion.

If the stress responses of the alarm stage are grouped according to general functions, you will note that they are designed to rapidly increase circulation, promote catabolism for energy production, and decrease nonessential activities. Exhibit

13–4 will provide you with some additional sympathetic responses that occur during stress.

Resistance reaction

The second stage in the stress responses is the *resistance reaction* (Figure 15–12b). Unlike the short-lived alarm reaction that is initiated by nervous impulses from the hypothalamus, the resistance reaction is initiated by neurohumors secreted by the hypothalamus and is a long-term reaction. The neurohumors are CRF, TSH, and STH.

CRF, or corticotropin releasing factor, stimulates the adenohypophysis to increase its secretion of ACTH. As you know, ACTH stimulates the adrenal cortex to secrete its hormones. The adrenal cortex is also indirectly stimulated by the alarm reaction. During the alarm reaction, kidney activity is cut back because it is not essential for meeting sudden danger. The resultant decrease in urine production provides a stimulus for the secretion of the mineralocorticoids.

The mineralocorticoids secreted by the adrenal cortex bring about the conservation of sodium ions by the body. A secondary effect of sodium conservation is the elimination of H^+ ions. The H^+ ions build up in high concentrations as a result of increased catabolism and tend to make the blood more acidic. Thus, during stress, a lowering of body pH is prevented. Sodium retention also leads to water retention. This maintains the high blood pressure that is typical of the alarm reaction. It also helps to make up for fluid that is lost during the stress of severe bleeding.

The glucocorticoids, which are produced in high concentrations during stress, bring about the following reactions:

1. They accelerate the conversion of fats into glucose so that the body has a large supply of energy long after the immediate stores of glucose have been used up. The glucocorticoids also stimulate the removal of proteins from cell structures and stimulate the liver to break them down into amino acids. The amino acids can then be rebuilt into enzymes that are needed to catalyze the increased chemical activities of the cells. If the body runs out of fat to burn, the amino acids are then converted to glucose. This is a last-resort attempt by the body to provide glucose when fats are depleted.
2. The glucocorticoids make blood vessels more sensitive to stimuli that bring about their constriction. This response counteracts a drop in blood pressure caused by bleeding.
3. The glucocorticoids also inhibit the formation of fibroblasts. Fibroblasts, as you remember, develop into connective tissue cells. In addition, injured fibroblasts release chemicals that play a role in stimulating the inflammatory response. Thus, the glucocorticoids reduce inflammation and prevent it from becoming disruptive rather than protective. Unfortunately, through their effect on fibroblasts, the glucocorticoids also discourage connective tissue formation. Wound healing is therefore slow during a prolonged resistance stage.

The two other neurohumors secreted by the hypothalamus in response to a stressor are TSH and STH. TSH stimulates the thyroid to secrete thyroxin, which increases the catabolism of carbohydrates. STH stimulates the catabolism of fats and the conversion of glycogen to glucose. The combined actions of TSH and STH increase catabolism and thereby supply additional energy for the body.

The resistance stage of the general stress syndrome allows the body to continue fighting a stressor long after the effects of the alarm reaction have fizzled out. It increases the rate at which life processes occur in the body. It also provides the energy, functional proteins, and circulatory changes that are required for meeting emotional crises, performing extremely strenuous tasks, fighting infection, or resisting the threat of bleeding to death. During the resistance stage, blood chemistry returns to nearly normal. The cells use glucose at the same rate that it is dumped into the bloodstream. Thus, blood sugar level returns to normal. Blood pH is brought under control by the kidneys as they excrete more hydrogen ions. However, blood pressure remains abnormally high. This is because the retention of water increases the volume of the blood.

All of us are confronted by stressors from time to time, and we have all experienced the resistance stage. Generally, this stage is successful in seeing us through a very stressful situation, and our bodies then return to normal. Occasionally, the resistance stage fails to combat the stressor, however, and the body "gives up." In this case, the general stress syndrome moves into the stage of exhaustion.

Exhaustion

One of the primary causes of exhaustion is loss of potassium ions. When the mineralocorticoids stimulate the kidney to retain sodium ions, potassium and hydrogen ions are traded off for sodium ions and are secreted in the urine. As the chief positive ion within cells, potassium is partially responsible for controlling the water concentration of the cytoplasm. As the cells lose more and more potassium, they function less and less effectively, until they finally start to die. This is called the *stage of exhaustion.* Unless the condition is rapidly reversed, vital organs cease functioning, and the patient dies. Another cause of exhaustion is depletion of the adrenal cortex hormones. In this case, blood glucose level suddenly falls, and the cells do not receive enough nutrients. A final cause is a weak organ. A long-term or strong resistance reaction puts heavy demands on the body, particularly on the heart, blood vessels, and adrenal cortex. They may not be up to handling the demands, or they may suddenly fail under the strain. In this respect, ability to handle stressors is determined to a large degree by previous health.

In summary, the general stress syndrome is a mechanism for adapting the body to severe stress. Unlike the homeostatic mechanisms, it changes the internal environment. In addition, the general stress syndrome is not a specific response to a specific stimulus. Instead, it produces wide-ranging changes in the body. And the same set of changes occur, no matter what the particular stressor happens to be. Generally, this is advantageous. Occasionally, however, it is not. For instance, just as fear of an immediate danger can produce the general stress syndrome, so can intense worry, tension, and resentment. Some investigators say that people who live under constant psychological stress seem to be in a continual state of mild resistance response. It has been suggested that this often causes chronic high blood pressure and contributes to heart disease by increasing wear and tear on the circulatory system.

Medical terminology associated with the endocrine system

Acromegaly (*acro* = extremity; *megaly* = large) Oversecretion of the pituitary growth hormone during adulthood resulting in enlarged bones of the hands, feet, and face.

Addison's disease Caused by hyposecretion of glucocorticoids by the adrenal cortex; amino acids are not catabolized, and blood sugar is low.

Aldosteronism A disorder caused by hypersecretion of adrenal mineralocorticoids; potassium depletion occurs, sometimes causing paralysis; sodium and water are retained, causing high blood pressure and edema.

Cretinism Hyposecretion of the thyroid, resulting in dwarfism and mental retardation.

Cushing's disease A disease of hypersecretion of the adrenal glucocorticoids; amino acid and fat catabolism increase; fat is redeposited in face and trunk.

Diabetes insipidus A disorder caused by hyposecretion of the antidiuretic hormone (ADH) of the neurohypophysis; basic clinical symptoms are high urine production and thirst.

Diabetes mellitus (*meli* = sugar) A chronic hereditary disease characterized by high blood sugar and due to an absolute or relative insufficiency of insulin; symptoms may appear at any age.

Feminizing adenoma Malignant tumors of the adrenal gland that secrete abnormally high amounts of female sex hormones and produce female secondary sexual characteristics in the male.

Goiter (*gutter* = throat) An enlargement of the thyroid gland with typical swelling of the front of the neck; *simple goiter* most often caused by insufficient iodine, and person suffers symptoms of thyroxin hyposecretion. *Exophthalmic goiter* (Graves' disease) caused by overactivity of thyroid gland; body's metabolic reactions increased. *Exophthalmos* means protrusion of eyeballs.

Hyperplasia Excessive development of any tissue.

Hypoplasia The defective development of any tissue.

Myxedema Hyposecretion of the thyroid resulting in puffy features due to edema.

Neuroblastoma Malignant tumor arising from the adrenal medulla associated with metastases to bones.

Pituitary cachexia (Simmond's disease) Hyposecretion of pituitary growth hormone during adult years; reduced protein anabolism, among other metabolic abnormalities, causes wasting away of tissue.

Pituitary dwarfism Unusual shortness and general underdevelopment of all organs caused by hyposecretion of the growth hormone of the anterior pituitary gland during childhood.

Pituitary giantism A hypersecretion of the growth hormone of the anterior pituitary gland during childhood resulting in abnormal growth, particularly of long bones.

Thyroid storm An aggravation of all symptoms of hyperthyroidism resulting from trauma, surgery, unusual emotional stress, or labor.

Virilism Masculinization.

Virilizing adenoma Malignant tumors of the adrenal gland that secrete high amounts of male sex hormones and produce male secondary sexual characteristics in the female.

Chapter summary in outline

ENDOCRINE GLANDS

1. Exocrine glands (sweat, sebaceous, digestive) secrete their products through ducts into body cavities or onto body surfaces.

2. Endocrine glands are ductless and secrete hormones into the blood.

3. Hormones are proteins, amines, or steroids that change the physiological activities of cells in order to maintain homeostasis.

4. Organs that exhibit changes in response to hormones are called target organs.

PITUITARY GLAND

1. This gland is differentiated into the adenohypophysis (the anterior lobe and glandular portion) and the neurohypophysis (the posterior lobe and nervous portion).

2. The adenohypophysis secretes trophic hormones and gonadotrophic hormones. These hormones are regulated by neurohumors and, like most hormones of the body, are involved in negative feedback systems.

3. Hormones of the adenohypophysis are: *(a)* growth hormone (regulates growth) and is controlled by SRF; *(b)* thyroid stimulating hormone (regulates activities of thyroid); *(c)* adrenocorticotrophic hormone (regulates adrenal cortex); *(d)* follicle

stimulating hormone (regulates ovaries and testes); *(e)* leuteinizing hormone (regulates female and male reproductive activities); and *(f)* prolactin (initiates milk secretion).

4. Hormones of the neurohypophysis are oxytocin (stimulates contraction of uterus and ejection of milk) and vasopressin-ADH (stimulates arteriole constriction and water reabsorption by the kidneys).

THYROID GLAND

1. The gland synthesizes thyroxin, which controls the rate of metabolism by increasing the catabolism of carbohydrates and proteins and serves as a diuretic.

2. Thyrocalcitonin regulates the homeostasis of blood calcium.

PARATHYROIDS

1. Parathyroid hormone regulates the homeostasis of calcium and phosphate.

ADRENALS

1. These glands consist of an outer cortex and inner medulla.

2. Cortical secretions are mineralocorticoids (regulate sodium reabsorption and potassium excretion); glucocorticoids (normal metabolism and resistance to stress); and gonadocorticoids (male and female sex hormones).

3. Medullary secretions are epinephrine and norepinephrine, which produce effects similar to sympathetic responses.

PANCREAS

1. Alpha cells of the pancreas secrete glucagon (increases blood glucose level), and beta cells secrete insulin (decreases blood glucose level).

OVARIES AND TESTES

1. Ovaries are located in the pelvic cavity and produce sex hormones related to development and maintenance of female sexual characteristics.

2. Testes lie inside the scrotum and produce sex hormones related to the development and maintenance of male sexual characteristics.

PINEAL GLAND

1. This gland secretes melatonin (possibly regulates menstrual cycle), adrenoglomerulotropin (may stimulate adrenal cortex), and serotonin (involved in normal brain physiology).

THYMUS GLAND

1. This gland is believed to secrete a hormone related to antibody production.

STRESS AND HOMEOSTASIS

1. A condition of the body usually produced in response to extreme stimuli.

2. Such stimuli are called stressors and include surgical operations, poisons, infections, fever, and strong emotional responses.

3. The alarm reaction is initiated by nervous impulses from the hypothalamus to the sympathetic division of the autonomic nervous system and adrenal medulla. Responses are immediate and short-lived. They are "fight-or-flight" responses which increase circulation, promote catabolism for energy production, and decrease nonessential activities.

4. The resistance reaction is initiated by a neurohumor (CRF) from the hypothalamus to the adenohypophysis, which then causes the secretion of ACTH. ACTH stimulates adrenal cortex to secrete its hormones. Resistance reactions are long term and accelerate catabolism to provide energy to counteract stress.

5. Stage of exhaustion results from dramatic changes that occur during alarm and resistance reactions. If stress is too great, exhaustion may lead to death.

Review questions and problems

1. Distinguish between an endocrine gland and an exocrine gland. What is the relationship between an endocrine gland and a target organ?

2. What is a hormone? Distinguish between trophic and gonadotrophic hormones.

3. In what respect is the pituitary gland actually two separate glands? Describe the histology of the adenohypophysis. Why does the anterior lobe of the gland have such an abundant blood supply?

4. What hormones are produced by the adenohypophysis, and what are their functions?

5. Relate the importance of neurohumors to secretions of the adenohypophysis. How are negative feedback systems related to hormonal regulation?

6. Discuss the histology of the neurohypophysis and the function and regulation of the hormones produced by the neurohypophysis.

7. Distinguish between hyposecretion and hypersecretion. What are the principal clinical symptoms of pituitary dwarfism, Simmond's disease, giantism, and acromegaly?

8. In diabetes insipidus, why does the patient exhibit high urine production and thirst?

9. Describe the location and histology of the thyroid gland. How is the thyroid hormone made, stored, and secreted?

10. Discuss the physiological effects of thyroxin and thyrocalcitonin. How are these hormones regulated?

11. What clinical symptoms are present in cretinism, myxedema, exophthalmic goiter, and simple goiter? Relate these symptoms to the normal activity of thyroxin.

12. Where are the parathyroids located? What is their histology? What are the functions of the parathyroid hormone, and how is it regulated?

13. Distinguish between the cause and symptoms of tetany and the cause and symptoms of osteitis fibrosa cystica.

14. Compare the adrenal cortex and adrenal medulla with regard to location and histology.

15. Describe the hormones produced by the adrenal cortex in terms of type, normal function, and control.

16. What relationship does the adrenal medulla have to the autonomic nervous system? What is the action of adrenal medullary hormones?

17. What are the effects of hypersecretion of aldosterone? Describe Addison's disease, Cushing's syndrome, and the adrenogenital syndrome. What is a pheochromocytoma?

18. Describe the location of the pancreas and the histology of the islets of Langerhans. What are the actions of glucagon and insulin?

19. What are the principal effects of hypoinsulinism and hyperinsulinism?

20. Where is the pineal gland located? What are its assumed functions?

21. Describe the location of the thymus gland. What is its proposed function?

22. Define the general stress syndrome. What is a stressor?

23. How do homeostatic responses differ from stress responses?

24. Outline the reactions of the body during the alarm stage, resistance stage, and stage of exhaustion when placed under stress. What is the central role of the hypothalamus during stress?

25. Correlate your response from Question 41 in Chapter 13 with the knowledge you have acquired in this chapter and respond to the following statement: "The combined activities of the nervous and endocrine systems are essential for the

purposes of maintaining homeostasis and overcoming stress."

26. Refer to the glossary of medical terminology associated with the endocrine system. Be sure that you can define each term.

27. Following is a chart listing the endocrine glands discussed in this chapter. The glands are listed from four points of reference: top, bottom, left, and right. Draw arrows to indicate how two or more glands are related, and write the name of the relationship near the arrow. If a gland bears no relationship to any others, do not draw an arrow. One example is shown in the chart.

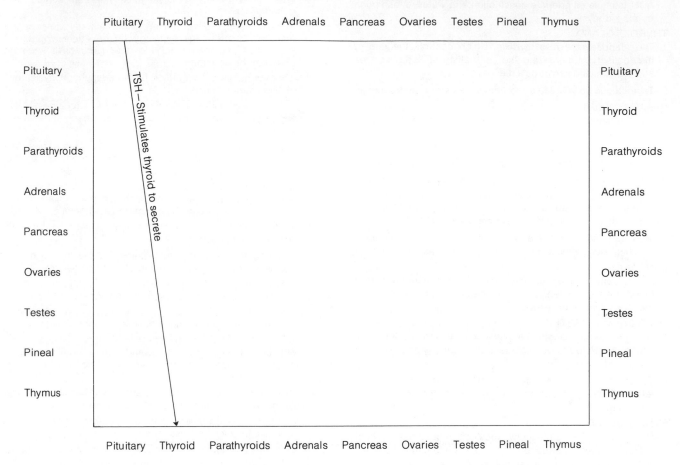

Pituitary Thyroid Parathyroids Adrenals Pancreas Ovaries Testes Pineal Thymus

TSH—Stimulates thyroid to secrete

Pituitary	Pituitary
Thyroid	Thyroid
Parathyroids	Parathyroids
Adrenals	Adrenals
Pancreas	Pancreas
Ovaries	Ovaries
Testes	Testes
Pineal	Pineal
Thymus	Thymus

Pituitary Thyroid Parathyroids Adrenals Pancreas Ovaries Testes Pineal Thymus

Bibliography

Unit 3

Baker, P. F.: "The Nerve Axon," *Scientific American,* **214**:74, 1966.

Barber, G. W.: "Physiological Chemistry of the Eye," *Arch. Ophthalmol.,* **87**:72, 1972.

Barr, M. L.: *The Human Nervous System,* Harper & Row, Publishers, Inc., New York, 1972.

Benzinger, T. H.: "The Human Thermostat," *Scientific American,* **204**:134, 1961.

Best, C. H. and N. B. Taylor: *The Physiological Basis of Medical Practice,* 8th ed., The Williams & Wilkins Company, Baltimore, 1966.

Botelho, S. Y.: "Tears and the Lacrimal Gland," *Scientific American,* **211**:78, October 1964.

"Building Blocks and Stumbling Blocks in Diabetes Management" (in consultation), *Medical World News,* Jan. 18, 1974.

Davis, H. and S. R. Silverman: *Hearing and Deafness,* Holt, Rinehart and Winston, Inc., New York, 1970.

Dean, G.: "The Multiple Sclerosis Problem," *Scientific American,* **223**:40, 1970.

Dey, F. L.: "Auditory Fatigue and Predicted Permanent Hearing Defects from Rock-and-Roll Music," *New England J. Med.,* **282**:467, Feb. 26, 1970.

Diassi, P. A. et al.: "Endocrine Hormones," *Ann. Rev. Pharmacol.,* **10**:219, 1970.

Eccles, J.: "The Synapse," *Scientific American,* **212**:56, January 1965.

Epstein, B. S.: *The Spine, A Radiological Text and Atlas,* Lea & Febiger, Philadelphia, 1969.

"Exploring the Frontiers of the Mind," *Time,* medicine section, p. 50, Jan. 14, 1974.

Field, J. (ed.): "Endocrinology," sec. 7, *Handbook of Physiology,* The Williams & Wilkins Company, Baltimore, 1972.

Guillemin, R. and R. Burgus: "The Hormones of the Hypothalamus," *Scientific American,* **227**:24, November 1972.

Hamwi, G. J.: "Nutrition and Diseases of the Endocrine Glands," *Amer. J. Clin. Nutr.,* **23**:311, 1970.

Harpen, R.: *Human Senses in Action,* The Williams & Wilkins Company, Baltimore, 1972.

Heimer, L.: "Pathways in the Brain," *Scientific American,* **225**:48, July 1971.

Jackson, R. L.: "The Child with Diabetes," *Nutrition Today,* March/April, 1971.

Johnston, R. and B. Roots: *Nerve Membranes,* Pergamon Press, New York, 1973.

Kandel, E. R.: "Nerve Cells and Behavior," *Scientific American,* **223**:57, July 1970.

Kaufman, H. E.: "Corneal Transplantation: A Progress Report," *Hospital Practice,* July 1973.

Kiely, W. F.: "Stress and Somatic Disease" (editorial), *J. Amer. Med. Assoc.,* **224**:521, Apr. 23, 1973.

Kolansky, H. and W. T. Moore: "Toxic Effects of Chronic Marihuana Use," *J. Amer. Med. Assoc.,* **225**:35, Oct. 2, 1972.

Krieger, D. T.: "The Hypothalamus and Neuroendocrinology," *Hospital Practice,* September 1971.

Laros, R. K., Jr., et al.: "Prostaglandins," *Amer. J. Nurs.,* **73**:1001, June 1973.

Levey, R. H.: "The Thymus Hormone," *Scientific American,* **211**:66, July 1964.

Lim, R. K. S.: "Pain," *Ann. Rev. Physiol.,* **32**:269, 1970.

Lipscomb, D. M.: "Ear Damage from Exposure to Rock and Roll Music," *Arch. Otolarying.,* **90**:545, November 1969.

Locke, W. et al. (eds.): *Hypothalamus and Pituitary in Health and Disease,* Charles C Thomas, Publisher, Springfield, Ill., 1972.

Luria, A. R.: "The Functional Organization of the Brain," *Scientific American,* **222**:66, March 1970.

MacNichol, E. F., Jr.: "Three-pigment Color Vision," *Scientific American,* **211**:48, December 1964.

Melzack, R.: *The Puzzle of Pain,* Basic Books, Inc., Publishers, New York, 1973.

Monif, G. R. G.: "Can Diabetes Mellitus Result from an Infectious Disease?," *Hospital Practice,* March 1973.

Moore, R. Y.: "Parkinsonism: An Insidious Disease Yields to Therapy," *Modern Medicine,* Oct. 29, 1973.

Netter, F.: "The Ciba Collection of Medical Diseases," vol. IV, *Endocrine System and Selected Metabolic Diseases,* Ciba Pharmaceutical Co., Summit, N.J., 1965.

Noback, C. R. and R. J. Demarest: *The Nervous System: Introduction and Review,* McGraw-Hill Book Company, New York, 1972.

Nuttall, F. Q.: "The Therapeutic Promise of Prostaglandins," *Modern Medicine,* Aug. 20, 1973.

Perlo, V. P.: "Answers to Questions on the Physical Examination in Central Nervous System Disease," *Hospital Medicine,* May 1971.

Peterson, L. R.: "Short-term Memory," *Scientific American,* **215**:90, July 1966.

Pettigrew, J. D.: "The Neurophysiology of Binocular Vision," *Scientific American,* **227**:84, August 1972.

Pribam, K. H.: "The Neurophysiology of Remembering," *Scientific American,* **220**:73, January 1969.

Rassmussen, H. and M. M. Pechet: "Calcitonin," *Scientific American,* **223**:42, October 1970.

Robertson, D. M. and H. B. Dinsdale: *The Nervous System: Structure and Function in Disease,* The Williams & Wilkins Company, Baltimore, 1972.

Selye, H.: "Stress, It's a G. A. S.," *Psychology Today,* September 1969.

Siegel, M.: "Optics and Visual Physiology," *Arch. Ophthalmol.,* **86**:100, 1971.

Simeone, F. A.: "Shock: Its Nature and Treatment," *Amer. J. Nurs.,* **66**:6, June 1966.

Stent, G. S.: "Cellular Communication," *Scientific American,* **227**:43, September 1972.

Sunderland, S.: *Nerves and Nerve Injuries,* The Williams & Wilkins Company, Baltimore, 1968.

Tepperman, J.: *Metabolic and Endocrine Physiology,* 2d ed., Year Book Medical Publishers, Inc., Chicago, 1968.

Thomas, R. C.: "Electrogenic Sodium Pump in Nerve and Muscle Cells," *Physiol. Rev.,* **52**:563, 1972.

Tudhope, G. R.: *Thyroid and the Blood,* Charles C Thomas, Publisher, Springfield, Ill., 1969.

Werblin, F. S.: "The Control of Sensitivity in the Retina," *Scientific American,* **228**:71, October 1970.

Whittaker, V. P.: "Membranes in Synaptic Function," *Hospital Practice,* April 1974.

Wilentz, J. S.: *Senses of Man,* Apollo Editions, Inc., New York, 1971.

Wilson, V. J.: "Inhibition in the Central Nervous System," *Scientific American,* **214**:102, May 1966.

Wolf, S. and H. Goodell: *Stress and Disease,* 2d ed., Charles C Thomas, Publisher, Springfield, Ill., 1968.

Yahr, M. D.: "Brain Tumors," *Hospital Medicine,* September 1973.

Young, R. W.: "Visual Cells," *Scientific American,* **223**:80, October 1970.

UNIT 4

MAINTENANCE
OF THE
HUMAN BODY

CHAPTER 16

THE CIRCULATORY SYSTEM: BLOOD, INTERSTITIAL FLUID, AND LYMPH

STUDENT OBJECTIVES

After you have read this chapter, you should be able to:

1. Contrast the general roles of blood, lymph, and interstitial fluid in maintaining homeostasis

2. Define the principal physical characteristics of blood and its functions in the body

3. Identify the plasma and formed element constituents of blood

4. Compare the origins of the formed elements in blood and the reticuloendothelial cells

5. Describe the structure of erythrocytes and their function in the carriage of oxygen and carbon dioxide

6. Define erythropoiesis and identify the factors related to erythrocyte production and destruction

7. Describe the importance of a reticulocyte count in the diagnosis of abnormal rates of erythrocyte production

8. List the structural features and types of leucocytes

9. Describe the importance of a differential count

10. Discuss the role of leucocytes in phagocytosis and antibody production

11. Discuss the structure of thrombocytes and explain their role in blood clotting

12. List the components of plasma and identify their importance

13. Describe the stages involved in blood clotting

14. Identify the factors that promote and inhibit blood clotting

15. Contrast a thrombus and an embolus

16. Define clotting and bleeding times

17. Define the ABO and Rh blood grouping classifications

18. Define the antigen-antibody reaction as the basis for ABO blood grouping

19. Define the antigen-antibody reaction of the Rh blood grouping system

20. Define erythroblastosis fetalis as a harmful antigen-antibody reaction

21. Compare the location, composition, and function of interstitial fluid and lymph

22. Contrast the causes of hemorrhagic, hemolytic, aplastic, and sickle cell anemia

23. Compare the clinical symptoms of polycythemia and leukemia

24. Identify the clinical symptoms of infectious mononucleosis

25. Define medical terminology associated with blood

The cells of the human body have developed an ability to perform highly specialized functions. This specialization is usually accompanied by the loss of other, frequently vital functions. Thus, the more specialized a cell becomes, the less capable it is of carrying on an independent existence. For instance, a specialized cell is less capable of protecting itself from extreme temperatures, toxic chemicals, and changes in pH. It often cannot go looking for food or devour whole bits of food. And, if it is firmly implanted in a tissue, it cannot move away from its own wastes. These vital functions must be performed for the cell. The substance that bathes the cell and carries out these functions is called interstitial fluid.

The interstitial fluid, in turn, must be serviced by the blood and lymph. The blood picks up oxygen from the lungs, nutrients from the digestive tract, hormones from the endocrine glands, and enzymes from still other parts of the body. The blood then transports these substances to all the tissues and releases them into the interstitial fluid. In the interstitial fluid, the substances are passed on to the cells and exchanged for wastes.

The blood must service all the tissues of the body. This means that it is an ideal medium for the transport of disease-causing organisms throughout the body. To protect itself from such disease spread, however, the body has a lymphatic system. This system is a collection of vessels containing a fluid called lymph. The lymph picks up wastes from the interstitial fluid, cleanses it of bacteria, and returns the wastes to the blood. The blood then carries the wastes to the lungs, kidneys, and sweat glands, where they are eliminated from the body. The blood also takes wastes to the liver, where they are detoxified and recycled.

Blood inside blood vessels, interstitial fluid around body cells, and lymph inside lymph vessels constitute the *internal environment* of the human organism. Because the cells are too specialized to adjust to more than very limited changes in their environment, the internal environment must be kept as constant as possible. This condition we have called homeostasis. In preceding chapters, we have discussed how the internal environment

was kept in homeostasis. Now we shall take a look at that environment itself.

The blood, heart, and blood vessels constitute the *blood vascular system*. The lymph, lymph vessels, and lymph glands make up the *lymph vascular system*. Together, the two systems are called the **circulatory system.** Let us now take a look at those substances called blood, interstitial fluid, and lymph.

BLOOD

The red body fluid that flows through all the vessels except the lymph vessels is called **blood.** Arterial blood, leaving the heart, tends to be bright red. Venous blood, however, is returning to the heart and is somewhat darker. Blood is a viscous fluid, which means that it is thicker and more adhesive than water. As you know from practical experience, thick fluids flow more slowly than thin, watery ones. Water is considered to have a viscosity of 1. The viscosity of blood, by comparison, ranges from 4.5 to 5.5. This means that it flows $4\frac{1}{2}$ to $5\frac{1}{2}$ times more slowly than water. The adhesive quality of blood, or its stickiness, may be felt by touching it. Blood is also slightly heavier than water. Other physical characteristics of blood include a necessary temperature of about 100.4°F, a pH range of 7.35 to 7.45, and a 3.5 percent concentration of NaCl—about the same concentration as seawater. Blood constitutes about 8 percent of the total body weight. The blood volume of an average-sized man is between 5 and 6 quarts.

Despite its rather simple physical appearance, blood is an exceedingly complex liquid that performs a number of functions vital to the maintenance of homeostasis. Among the important functions are:

1. The transportation of oxygen from the lungs to all cells of the body
2. The transportation of carbon dioxide from the lymph vascular system to the lungs
3. The transportation of nutrients from the digestive organs to the cells of the body

4. The transportation of waste products from the lymph vascular system to the kidneys
5. The transportation of hormones from endocrine glands to cells of the body
6. The transportation of enzymes to various cells of the body
7. The regulation of body pH through buffers dissolved in the blood
8. The regulation of normal body temperature because it contains such a large volume of water (an excellent heat absorber and coolant)
9. The regulation of the water content of cells, principally through dissolved sodium ions
10. The prevention of body fluid loss through the clotting mechanism
11. Protection against toxins and foreign microbes through special combat-unit cells in the blood and lymph

Microscopically, blood is composed of two portions: plasma, which is a liquid containing dissolved substances, and formed elements, which are cells and cell-like bodies suspended in the plasma. We shall look first at the formed elements and then discuss plasma.

Formed elements

In clinical practice, the most common classification of the **formed elements** of the blood is the following:

   I. Erythrocytes, or red blood cells
  II. Leucocytes, or white blood cells
     A. Granular leucocytes (granulocytes)
       1. Neutrophils
       2. Eosinophils
       3. Basophils
     B. Agranular leucocytes (agranulocytes)
       1. Lymphocytes
       2. Monocytes
 III. Thrombocytes, or platelets

We shall now consider some of the more important facts about each of the formed elements.

Where do these formed elements come from? The process by which blood cells are formed is called *hematopoiesis.* During embryonic and fetal life, there are no clear-cut centers for blood cell production. For example, the yolk sac, liver, spleen, thymus gland, lymph nodes, and bone marrow all participate at various times in producing the formed elements. In the adult, however, we can pinpoint the production process precisely. Bone marrow (myeloid tissue) is responsible for the production of red blood cells, granular leucocytes, and platelets. On the other

hand, lymphatic tissue—which includes spleen, tonsils, and lymph nodes—is responsible for the production of agranular leucocytes.

During embryonic life, certain epithelial cells, called *primitive reticular cells,* wander throughout the body and later become trapped in certain organs. Some of the primitive reticular cells move into red bone marrow and are transformed into *hemocytoblasts,* immature cells that are eventually capable of developing into mature blood cells (Figure 16–1). The hemocytoblasts in red bone marrow first develop into distinct cells that are later transformed into specific mature cells. For example, the hemocytoblasts develop into (1) *rubriblasts* that go on to form mature red blood cells, (2) *myeloblasts* that go on to form mature neutrophils, eosinophils, and basophils, and (3) *megakaryoblasts* that go on to form mature platelets. Other hemocytoblasts become entrapped in lymphatic tissue. The fate of these hemocytoblasts differs from that of the hemocytoblasts in the red bone marrow. Hemocytoblasts in lymphatic tissue develop into two distinct cells that are later transformed into other mature blood cells. These are (1) *lymphoblasts* that eventually form lymphocytes and (2) *monoblasts* that eventually form monocytes.

Mature blood cells do not live in the bloodstream for very long. Their disintegrating bodies pose the danger of clogging small blood vessels, so certain cells clear away their bodies after they die. These cells are called *reticuloendothelial cells* and are also formed from primitive reticular cells. However, unlike the hemocytoblasts in bone marrow or lymphatic tissue that produce mature blood cells, reticuloendothelial cells have a different destiny. They enter the spleen, tonsils, lymph nodes, liver, and other organs and become highly specialized for phagocytosis. The reticuloendothelial cells that lie in the lymph nodes are particularly active in destroying microbes and their toxins. The reticuloendothelial cells in the liver and spleen concentrate more on ingesting dead blood cells.

Erythrocytes

Microscopically, *red blood cells,* or **erythrocytes,** appear as biconcave discs averaging about 7.7 microns in diameter (see Figure 16–1). Mature red blood cells are quite simple in structure. They lack a nucleus and can neither reproduce nor carry on extensive metabolic activities. The interior of the cell contains some cytoplasm; protein; lipid substances, including cholesterol; and a red pigment called hemoglobin. *Hemoglobin,* which constitutes about 33 percent of the cell volume, is responsible for the red color of blood.

The function of the erythrocytes is to combine with oxygen and carbon dioxide and transport them through the blood vessels. Red blood cells are highly specialized for this purpose. The hemoglobin molecule consists of a protein called globin and a pigment called heme, which contains iron. As the erythrocyte passes through the lungs, each of the four iron atoms in the hemoglobin molecule combines with a molecule of oxygen. The oxygen is transported in this state to other tissues of the body. In the tissues, the iron-oxygen reaction reverses, and the oxygen is released to diffuse into the interstitial fluid. Can you guess why iron is sometimes prescribed for "tired blood?" On the return trip, the globin portion combines with a molecule of carbon dioxide from the lymph. This complex is transported and released in the lungs. Red blood cells are jampacked with hemoglobin molecules in order to increase their carrying capacity. The shape of a red blood cell also increases its carrying capacity. A biconcave structure has a much greater surface area than, say, a sphere or cube. The erythrocyte thus provides the maximum surface area for the hemoglobin to make contact with the gas molecules.

As we mentioned, a red blood cell does not live long. For some reason, its cell membrane becomes fragile, and the cell is nonfunctional in about 120 days. The blood, however, contains inordinate numbers of these cells. A healthy male has 5.5 to 7 million red blood cells per cubic millimeter of blood. A healthy female has 4.5 to 6 million red blood cells per cubic millimeter. To maintain normal quantities of erythrocytes, the body must produce new mature cells at the astonishing rate of 2 million per second. In the adult production takes place in the red bone marrow in the spongy bone of the cranium, ribs, sternum, bodies of vertebrae, and epiphyses of the humerus and femur. The process by which erythrocytes are formed is called **erythropoiesis**.

Erythropoiesis starts with the transformation of a hemocytoblast into a rubriblast. Subsequently, other intermediate cells are formed until the final stage of red blood cell formation is reached (see Figure 16–1). Hemoglobin synthesis and disintegration of the nucleus are involved in the development of a mature red blood cell. At first, the hemocytoblast is transformed into a rubriblast and then a prorubricyte, the cell that starts hemoglobin synthesis. During the next developmental stage, the nucleus shrinks, more hemo-

globin is synthesized and the cell is now a rubricyte. Next, the cytoplasm of the rubricyte becomes filled with even more hemoglobin, and the nucleus begins to disintegrate and undergo absorption. Now the cell is a normoblast. During absorption of the nucleus, the remaining endoplasmic reticulum in the cytoplasm can still be seen as a network in the cell. For this reason, the cell at this point is called a reticulocyte. Once all the reticulum is absorbed, the reticulocyte becomes a mature red blood cell. This red blood cell contains hemoglobin but no nucleus. Once the erythrocyte is formed, it circulates through the blood vessels until its life span comes to an end. Aged erythrocytes are destroyed by reticuloendothelial cells in the liver and spleen. The hemoglobin molecules are split apart; the iron is reused and the rest of the molecule is converted into other substances for reuse or elimination.

Normally erythropoiesis and red cell destruction proceed at the same pace. But if the body suddenly needs more erythrocytes, or if erythropoiesis is not keeping up with red blood cell destruction, a homeostatic mechanism steps up erythrocyte production. The stimulus for the homeostatic mechanism is oxygen deficiency within the cells of the kidney. This is not surprising because the chief function of the erythrocytes is to deliver oxygen. As soon as the kidney cells become oxygen deficient, they release a hormone called erythropoietin. This hormone circulates through the blood to the red bone marrow, where it stimulates hemocytoblasts to develop into red blood cells.

Cellular oxygen deficiency may occur if you do not breathe in enough oxygen. This commonly happens at high altitudes where the air is "thin." Oxygen deficiency also occurs if you become anemic. The term *anemia* means that the number of functional red blood cells or their hemoglobin content is below normal. Consequently, the erythrocytes are unable to transport enough oxygen from the lungs to the cells. Anemia has many causes. The most common are a lack of iron, lack of certain amino acids, or the lack of vitamin B_{12}. Iron is needed for the oxygen-carrying part of the hemoglobin molecule. And the amino acids are needed for the protein, or globin, part of hemoglobin. Vitamin B_{12} does not actually become part of the new blood cell. Its function is to stimulate the red bone marrow to produce erythrocytes. B_{12} is obtained through meat, especially liver, but it cannot be absorbed through the walls of the digestive tract without the help of another substance. This substance is called intrinsic factor and is produced by the mucous cells that line the stomach. The inability to produce intrinsic factor is the cause of a disorder called *pernicious anemia*. An anemia that arises simply from an inadequate diet is called *nutritional anemia*.

A diagnostic test that informs the physician about the rate of erythropoiesis is the *reticulocyte count*. Some of the normoblasts and reticulocytes are normally released into the bloodstream before they become mature red blood cells. (See Figure 16–1.) If the number of reticulocytes in a sample of blood is less than 0.5 percent of the number of mature red blood cells in the sample, erythropoiesis is occurring too slowly. A low reticulocyte count might confirm a diagnosis of nutritional or pernicious anemia. Or, it might indicate a kidney disease that prevents the kidney cells from pro-

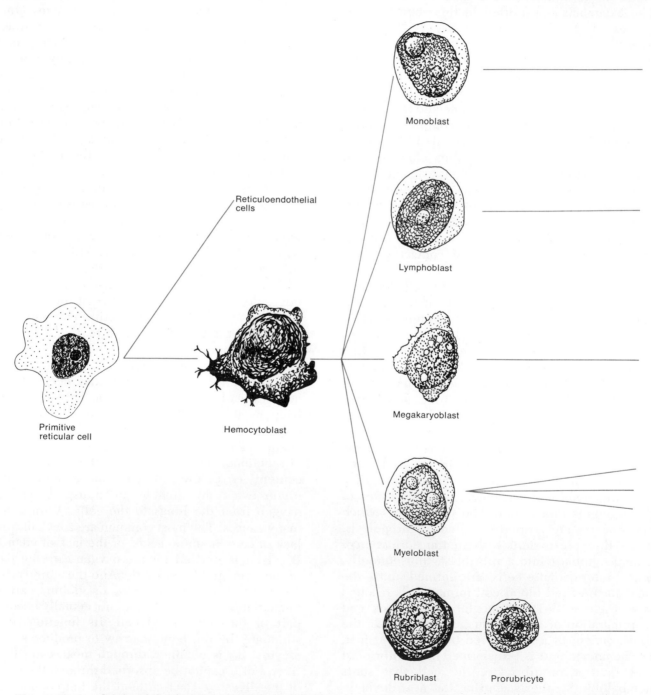

Primitive reticular cell

Reticuloendothelial cells

Hemocytoblast

Monoblast

Lymphoblast

Megakaryoblast

Myeloblast

Rubriblast

Prorubricyte

Figure 16–1. Origin of the formed elements in blood and reticuloendothelial cells. Consult text for details.

ducing erythropoietin. If the reticulocyte number is more than 1.5 percent of the mature red blood cells, erythropoiesis is abnormally rapid. A raft of problems may be responsible for a high reticulocyte count. Among these are most types of anemia, oxygen deficiency, and uncontrolled red blood cell production caused by a cancer in the bone marrow. If the individual has been suffering from a nutritional or pernicious anemia, the high count may indicate that treatment has been effective, and the bone marrow is making up for lost time.

Leucocytes

Unlike red blood cells, **leucocytes,** or *white blood cells,* have nuclei and do not contain hemoglobin. In addition, they are far less numerous, averaging

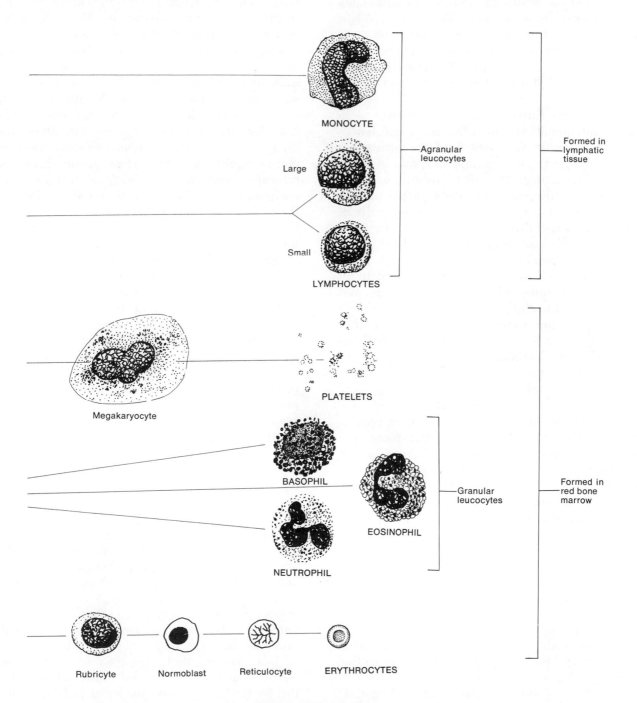

Figure 16–1 (cont'd.)

from 5,000 to 9,000 cells per cubic millimeter. Red blood cells, therefore, outnumber white blood cells about 700 to 1. Leucocytes fall into two major groups (see Figure 16–1). The first group contains the *granular leucocytes*. These develop from hemocytoblasts in the bone marrow. They have granules in the cytoplasm and possess lobed nuclei. Three kinds of granular leucocytes exist. These are the neutrophils, the eosinophils, and the basophils. The second principal group of leucocytes, called the agranular leucocytes, originates from hemocytoblasts in lymphatic tissue. When they are placed under a light microscope, no cytoplasmic granules can be seen. Their nuclei are more or less spherical. The two kinds of agranular leucocytes are lymphocytes and monocytes.

The general function of the leucocytes is to combat inflammation and infection. Some leucocytes are actively *phagocytotic*. This means that they can ingest bacteria and dispose of dead matter. Most leucocytes also possess, to some degree, the ability to crawl through the minute spaces between the cells that form the walls of blood vessels and through connective and epithelial tissue. This movement, the same kind that is exhibited by amoebas, is called *diapedesis*. First, a part of the cell membrane stretches out in an armlike projection. Then the cytoplasm and nucleus flow into the projection. Finally, the rest of the membrane snaps up into place. Another projection is made, and so on, until the cell has "crawled" to its destination.

If a tissue of the body becomes injured or infected, blood vessels in the immediate vicinity dilate and bring more blood to the affected area. This results in inflammation, which is characterized by redness, heat, swelling, pain, and often loss of function. Leucocytes become active and migrate by diapedesis in large numbers through capillary walls into the affected tissue. As the leucocytes engulf the bacteria, some *suppuration*, or pus formation, may occur. Essentially, pus consists of dead and living bacteria, white blood cells, cellular debris, and blood fluid. If the leucocytes destroy the invading bacteria effectively, the affected area returns to normal, and the process of repair takes over. If the leucocytes are ineffective, suppuration increases, and the infection may also spread.

The diagnosis of an injury or infection within the body may involve a *differential count*. In this procedure, the number of each kind of white cell in 100 white blood cells is counted. A normal differential count might appear as follows:

Neutrophils	60–70%
Eosinophils	2–4%
Basophils	0.5–1%
Lymphocytes	20–25%
Monocytes	3–8%
Total	100%

In interpreting the results of a differential count, particular attention is paid to the neutrophils. The neutrophils are the most active in response to tissue destruction. More often than not, a high neutrophil count indicates that the damage is caused by invading bacteria. An increase in the number of monocytes generally indicates a chronic (of long duration) infection such as tuberculosis. It is hypothesized that monocytes take longer to reach the site of infection than neutrophils, but once they arrive they do so in larger numbers and destroy more microbes than neutrophils. High eosinophil counts typically indicate allergic conditions since eosinophils are believed to combat allergens, the causative agents of allergies.

The term *leucocytosis* refers to an increase in the number of white blood cells. If the increase exceeds 10,000, a pathological condition is usually indicated. A decrease below normal in the number of white blood cells is termed *leucopenia*.

Some leucocytes fight infection in another, extremely important way. These are the lymphocytes, which are involved in the production of antibodies. **Antibodies** are special proteins that inactivate antigens. An **antigen** is any type of protein that the body is not capable of synthesizing. Antigens, in other words, are "foreign" proteins that are introduced into the body by any number of ways. For example, many of the proteins that make up the cell structures and enzymes of bacteria are antigens. The toxins released by bacteria are also antigens. When antigens enter the body, they react chemically with substances in the lymphocytes and stimulate some of the lymphocytes to become *plasma cells* (Figure 16–2). The plasma cells then produce **antibodies,** which are globulin-type proteins that attach to antigens, much like enzymes attach to substrates. Like enzymes, a particular type of antibody will generally attach only to a particular type of antigen. However, unlike enzymes, which enhance the reactivity of the substrate, antibodies "cover" their antigens so the antigens cannot come in contact with other chemicals in the body. In this way, bacterial poisons can be "sealed up" and rendered harmless. The bacteria

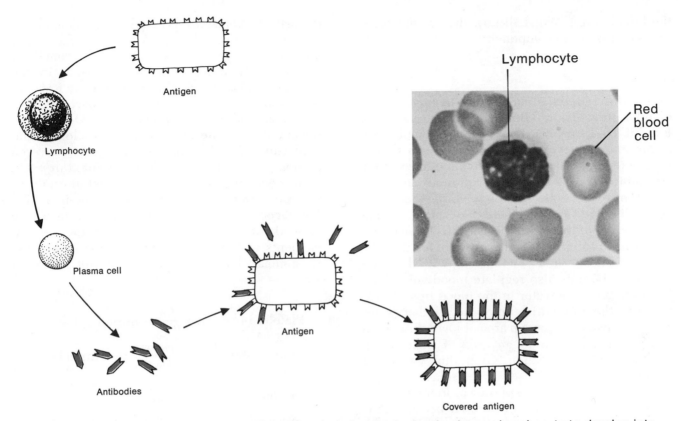

Figure 16–2. Antigen-antibody response. An antigen entering the body stimulates a lymphocyte to develop into an antibody-producing plasma cell. The antibodies attach to the antigen, cover it, and render it harmless. Photomicrograph of lymphocyte ×1800. *(Photomicrograph courtesy of Edward J. Reith, from Atlas of Descriptive Histology, Edward J. Reith and Michael H. Ross, Harper & Row, Publishers; Inc., New York, 1970.)*

themselves are destroyed as the antibodies "walk off" with the proteins in their cell membranes. This process is called the *antigen-antibody response*. The antigen-antibody response allows us to combat infection. It gives us immunity to some diseases. And, as you will find out, it is responsible for blood types, for allergies, and for the body's rejection of organs transplanted from an individual with a different genetic makeup.

Foreign bacteria exist everywhere in the environment and have continuous access to the body through the mouth, nose, and pores of the skin. Furthermore, many cells, especially those of epithelial tissue, age and die, and their remains must be disposed of daily. Even when the body is healthy, the leucocytes actively ingest bacteria and debris. However, a leucocyte can phagocytose only a certain number of substances before the substances interfere with the normal metabolic activities of the leucocyte and bring on its death. Consequently, the life span of a leucocyte is very short. During times of health, some white blood cells will live a couple of days. During a period of infection they may live for only a few hours.

Thrombocytes

If a hemocytoblast does not become an erythrocyte or granular leucocyte, it may develop into still another kind of cell, called a megakaryoblast (see Figure 16–1), that is transformed into a megakaryocyte. Megakaryocytes are large cells whose cytoplasm breaks up into fragments. Each fragment becomes enclosed by a piece of the cell membrane and is called a **thrombocyte** or *platelet*. Platelets are disc-shaped cells without a nucleus. They average from 2 to 4 μ in diameter. Between 250,000 and 400,000 platelets appear in each cubic millimeter of blood.

The function of the platelets is to prevent fluid loss by initiating a chain of reactions that results in blood clotting. Like the other formed elements of the blood, platelets have a short life span, probably only about 1 week. This short life span is due to the facts that platelets are "used up" in clotting and that their cells are too simple to carry on much metabolic activity. When the formed elements—erythrocytes, leucocytes, and thrombocytes—are removed, the liquid portion of the blood is left. We shall now discuss the properties of this liquid component.

Plasma

When the formed elements are removed from blood, a straw-colored liquid called **plasma** is left. Exhibit 16–1 outlines the chemical composition of plasma. However, a few of the solutes should be pointed out. About 7 to 9 percent of the solutes are proteins. Some of these proteins are also found elsewhere in the body, but when they occur dissolved in blood, they are called *plasma proteins*. Albumins, which constitute the majority of plasma proteins, are responsible for the viscosity or thickness of blood. Along with the electrolytes, albumins also regulate blood volume by preventing all the water in the blood from diffusing into the interstitial fluid. Recall that water moves by osmosis from an area of low solute (high water) concentration to an area of high solute (low water) concentration. Globulins, which are antibody proteins released by plasma cells, form a small component of the plasma proteins. Gamma globulin is especially well known because it is able to form an antigen-antibody complex with the proteins of the hepatitis, measles, and tetanus viruses. Fibrinogen, a third plasma protein, takes part in the blood-clotting mechanism, along with the platelets.

Blood clotting

Three aspects of blood clotting that should be known are how clots form, how the formation of clots is prevented, and how clotting tests are performed.

Clot formation

Under normal circumstances, blood maintains its liquid state as long as it remains in the vessels. If, however, it is drawn from the body, it first becomes very thick and then forms a soft jelly. The gel eventually separates from the liquid component. The straw-colored liquid component, called **serum,** is simply plasma minus its clotting proteins. The gel is called a **clot** and consists of a network of insoluble fibers in which the cellular components of blood are trapped.

The clotting process may be initiated when blood comes in contact with a rough surface. Drops of blood, like drops of any fluid, will stick to a rough surface. In the case of blood, the roughened area may be a cut end of a vessel or a

deposit of cholesterol-type lipids. It is assumed that when the platelets stick to the surface, their membranes rupture, and chemicals contained within the platelet cells are released. In the presence of Ca^{2+} ions, the platelet chemicals are transformed through a series of reactions to a substance called *thromboplastin*. During the second step in the formation of the clot, thromboplastin and Ca^{2+} serve as catalysts to convert a plasma protein called *prothrombin* into *thrombin*, an enzyme. In the final step, thrombin catalyzes the conversion of another plasma protein, called *fibrinogen*, into *fibrin*. Fibrin is the insoluble network of fine threads in which the cellular elements of blood are trapped. This is the clot. In summary, the three phases of clot formation are

$$\text{Platelet breakdown} \xrightarrow{\text{Ca}^{2+}} \text{thromboplastin}$$

$$\text{Prothrombin} \xrightarrow[\text{other factors}]{\text{thromboplastin, Ca}^{2+}} \text{thrombin}$$

$$\text{Fibrinogen} \xrightarrow{\text{thrombin}} \text{fibrin}$$

Clot formation is a vital mechanism that controls excessive loss of blood from the body. In order to form clots, the body needs calcium and vitamin K. Vitamin K is not involved in the actual clot formation, but it is required for the synthesis of prothrombin, which occurs in the liver. The vitamin is normally produced by bacteria that live in the intestine. It is also fat-soluble, which means that it can be absorbed through the walls of the intestine and into the blood only if it is attached to fat. People suffering from disorders that prevent absorption of fat often suffer uncontrolled bleeding. Clotting may also be encouraged by applying a thrombin or fibrin spray, a rough surface such as gauze, or heat.

Clot prevention

Unwanted clotting may be brought on by the formation of cholesterol-containing plaques on the walls of the blood vessels. These are the plaques of arteriosclerosis, or "hardening of the arteries." They supply a rough surface that is perfect for the adhesion of platelets and are often the sites of clotting. Clotting within an unbroken blood vessel is referred to as **thrombosis.** The clot itself is called a **thrombus.** A thrombus may dissolve, or it may remain intact and interfere with circulation. In the latter case, it may cause damage to tissues by cutting off the

Exhibit 16–1. CHEMICAL COMPOSITION AND DESCRIPTION OF THE SUBSTANCES IN PLASMA

CONSTITUENT	DESCRIPTION	CONSTITUENT	DESCRIPTION
A. Water	Constitutes about 92 percent of plasma and is the liquid portion of blood. Ninety percent of the water is derived from absorption from the digestive tract, and ten percent from the metabolism of nutrients. The functions of the water are to act as a solvent and suspending medium for the solid components of blood and to absorb heat.	2. Nonprotein nitrogen (NPN) sub-stances	These compounds are the substances in plasma that contain nitrogen but are not proteins. Among such compounds are urea, uric acid, creatine, creatinine, and ammonium salts. They represent breakdown products of protein metabolism and are carried by the blood to the organs of excretion.
B. Solutes 1. Proteins	Plasma proteins constitute about 7 to 9 percent of the solutes in plasma.	3. Food substances	Once foods are broken down in the digestive tract, the products of digestion are passed into the blood for distribution to all cells of the body. These products include amino acids (from proteins), glucose (from carbohydrates), and fats (from lipids).
Albumins	These comprise 55 to 64 percent of the plasma proteins and are the smallest of the plasma proteins. Albumins are produced by the liver and provide the blood with viscosity, a factor related to the maintenance and regulation of blood pressure. Albumins also exert a considerable osmotic pressure. This helps maintain the water balance between the blood and tissues and helps regulate the volume of blood.	4. Regulatory substances	The two principal regulatory substances are enzymes and hormones. Enzymes are produced by the cells of the body and catalyze chemical reactions. Hormones, produced by endocrine glands, regulate growth and developmental processes in the body.
Globulins	These proteins constitute about 15 percent of the plasma proteins. They are the protein group to which antibodies belong. One of the more important sub-groups is the gamma globulins, which attack measles, hepatitis, tetanus, and possibly poliomyelitis viruses.	5. Respiratory gases	The respiratory gases, oxygen and carbon dioxide, are carried by the blood. It should be noted, however, that these gases are more closely associated with the hemoglobin of red blood cells than the plasma itself.
Fibrinogen	This protein represents only a small fraction of plasma proteins (4 percent). It is also produced by the liver and plays an essential role in the blood-clotting mechanisms.	6. Electrolytes	A number of ions constitute the inorganic salts of plasma. The cations include Na^+, K^+, Ca^{2+}, and Mg^{2+}. The anions include Cl^-, PO_4^{3-}, SO_4^{2-}, and HCO_3^-. The salts in plasma function in the maintenance of proper osmotic pressure, the regulation of normal pH, and the maintenance of the proper physiological balance between the tissues and the blood.

oxygen supply. Equally serious is the possibility that the thrombus will become dislodged and be carried with the blood to a smaller vessel. In a smaller vessel, the clot will get stuck and may block the circulation to a vital organ. A blood clot, bubble of air, or piece of debris that is transported by the blood stream is called an **embolus.** When an embolus becomes lodged in a vessel and cuts off circulation, the condition is called an **embolism.**

No matter how healthy the body is, occasional rough spots do appear on uncut vessel walls. In fact, it is believed that blood clotting is a continuous process inside blood vessels and that it is continually combated by clot prevention and clot dissolving mechanisms. Blood contains antithrombic substances—that is, substances that prevent thrombin formation. One of these is *heparin,* which inhibits the conversion of prothrombin to thrombin and prevents most thrombus formation. Heparin is used in open heart surgery to prevent clotting. Should the thrombus form after all, an enzyme that breaks down fibrin may dissolve the clot. The dissolving of the clot is called **fibrinolysis.** Recall that the term *lysis* means to break apart.

In general, any chemical substance that prevents clotting is an **anticoagulant.** Examples of anticoagulants are heparin, dicumarol, and the citrates and oxalates. Heparin is a quick-acting anticoagulant that is extracted from donated human blood. The pharmaceutical preparation *dicumarol* may be given to patients who are thrombosis prone. Dicumarol is isolated from sweet clover and acts as an antagonist to vitamin K. Dicumarol is slower acting than heparin, and it is used primarily as a preventative. The citrates and oxalates are used by laboratories and blood banks to prevent blood samples from clotting. These substances react with calcium to form insoluble compounds. In this way, the blood calcium is tied up and is no longer free to catalyze the conversion of prothrombin to thrombin.

Clotting tests

The time required for blood to coagulate, usually from 5 to 15 minutes, is known as *clotting time.* This time is used as an index of a person's blood-clotting properties. One method for determining clotting time involves taking a sample of blood from a vein and placing 1 milliliter of the blood into each of three Pyrex tubes. The tubes are then submerged in a water bath at 37°C (98.6°F). Every 30 seconds they are examined for the formation of a clot. The clotting process is initiated when the platelets break up upon coming into contact with the glass tubing. When the clot adheres to the walls of the tube, the end point is reached, and the time is recorded. Blood taken from individuals with hemophilia clots very slowly or not at all.

Bleeding time is the time required for the cessation of bleeding from a small skin puncture. This procedure is usually accomplished by puncturing the ear lobe. As the droplets of blood escape, they are dried by gently touching the wound with filter paper. When the paper is no longer stained, the bleeding has stopped. Normally, bleeding time varies from 1 to 4 minutes. Unlike coagulation time, which involves only the breakdown of platelets, bleeding time also involves constriction of injured blood vessels and all three steps of clot formation.

Blood grouping or typing

The surfaces of erythrocytes contain genetically determined antigens called **agglutinogens.** These proteins are responsible for the two major blood-group classifications: the ABO group and the Rh system.

ABO group

The *ABO blood grouping* is based upon two agglutinogens, which are symbolized as A and B (Figure 16–3). Individuals whose erythrocytes manufacture only agglutinogen A are said to have blood type A. Those who manufacture only agglutinogen B are type B. Individuals who manufacture both A and B are typed AB. Others, who manufacture neither, are called type O.

AGGLUTINOGEN ON ERYTHROCYTE MEMBRANE	BLOOD TYPE
A	A
B	B
AB	AB
Neither A nor B	O

The percentages of individuals possessing these four blood types are not equally distributed. The incidence of the various types in the Caucasian population in the United States is as follows: type A (41 percent), type B (10 percent), type AB (4 percent), and type O (45 percent).

The blood plasma of many people contains genetically determined antibodies referred to as **agglutinins.** These are antibody a (Anti-A), which attacks agglutinogen A, and antibody b (Anti-B) which attacks B. The antibodies formed by each of the four blood types are shown in Figure 16–3. Note that an individual does not have antibodies that attack the antigens of his own erythrocytes. For instance, a person with blood type A does not have antibody a. But every person has an antibody against any agglutinogen that he himself does not synthesize. For example, suppose type A blood is accidentally given to a person who does not have A agglutinogens. The individual's body recognizes that the A protein is foreign and therefore treats it as an antigen. Antibody a's rush to the foreign erythrocytes, attack them, and cause them to *agglutinate* (clump). Hence the names agglutinogen and agglutinins. This is another example of an antigen-antibody response.

When blood is given to a patient, care must be taken to ensure that the individual's antibodies will not attack the donated erythrocytes and cause clumping. The destruction of the donated cells will not only undo the work of the transfusion, but the clumps can block vessels and cause serious problems that may lead to death. Referring again to the agglutinogen-agglutinin-chart in Figure 16–3, note that a person can receive blood from others in his blood group. Type AB blood obviously does not have a or b antibodies. This means that AB individuals can receive all four types of blood without any danger of clumping. Thus, AB blood is called the *universal recipient*. Type O blood is called the *universal donor* because it does not have agglutinogens to act as antigens in another person's body.

The Rh system

When blood is transfused, the technician must make sure that the donor and recipient blood types are safely matched—not only for ABO group type, but also for Rh type.

The *Rh system* is so named because it was first worked out in the blood of the Rhesus monkey. Like the ABO grouping, the Rh system is based upon agglutinogens that lie on the surfaces of erythrocytes. Individuals whose erythrocytes contain the Rh agglutinogens are designated as Rh^+. Those who lack Rh agglutinogens are designated as Rh^-. It is estimated that 90 percent of all citizens of the United States are Rh^+, whereas 10 percent are Rh^-.

Under normal circumstances, human plasma does not contain Anti-Rh antibodies. However, if an Rh^- person receives Rh^+ blood, his body starts to make Anti-Rh antibodies that will remain in his blood. If a second transfusion of Rh^+ blood is given later, the previously formed Anti-Rh antibodies will react against the donated blood, and a severe reaction may occur. One of the most common problems with Rh incompatibility arises from pregnancy. During delivery, some of the fetus' blood is apt to leak from the afterbirth into the mother's bloodstream. If the fetus is Rh^+ and the woman is Rh^-, the mother will make Anti-Rh antibodies. If the woman becomes pregnant again, her Anti-Rh antibodies will make their way into the bloodstream of the baby. If the baby is Rh^-, no problem will occur since he does not have the Rh antigen. If he is

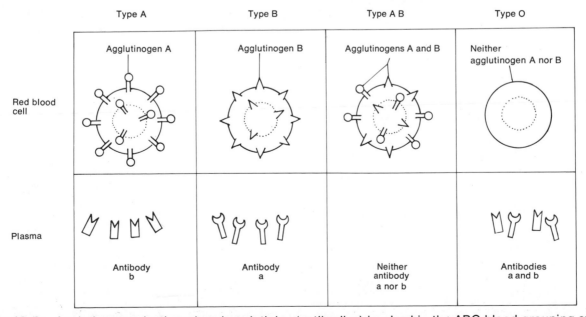

Figure 16–3. Agglutinogens (antigens) and agglutinins (antibodies) involved in the ABO blood grouping system.

Rh+, an antigen-antibody response called *hemolysis* may occur in his blood. Hemolysis means a breakage of erythrocytes resulting in the liberation of hemoglobin. The hemolysis brought on by fetal-maternal incompatibility is called **erythroblastosis fetalis.** When a baby is born with erythroblastosis, all his blood is slowly drained and replaced with antibody-free blood. Moreover, modern technology has made it possible to transfuse blood into the unborn child if erythroblastosis is suspected. More important, though, is the fact that erythroblastosis can be prevented with RhoGAM Rh$_0$(D) Immune Globulin (Human), a commercially available drug that is administered to Rh$^-$ mothers right after each delivery. RhoGAM prevents the mother's blood from forming antibodies against the fetal Rh antigens that are released during delivery. Thus the fetus of the next pregnancy is protected. In the case of an Rh$^+$ mother and an Rh$^-$ child, there are no complications since the fetus cannot make antibodies.

INTERSTITAL FLUID AND LYMPH

For all practical purposes interstitial fluid and lymph are the same. The major difference between the two is location. When the fluid bathes the cells, it is called **interstitial fluid,** or **tissue fluid.** When it flows through the lymphatic vessels, it is called **lymph** (Figure 16–4). Both fluids are similar in composition to plasma. The principal chemical difference is that they contain less protein. This is because the larger protein molecules do not diffuse easily through the cells that form the walls of the blood vessels. Keep in mind that whole blood does not flow into the tissue spaces; it remains within closed vessels. Certain constituents of the plasma do move, however, and once they move out of the blood, they are collectively called interstitial fluid. The transfer of materials between blood and interstitial fluid occurs by osmosis and diffusion across the cells that make up the vessel walls. Both interstitial fluid and lymph contain variable numbers of leucocytes. Leucocytes can enter the tissue fluid by diapedesis, and the lymph tissue itself produces nongranular leucocytes. However, interstitial fluid and lymph both lack erythrocytes and platelets.

Interstitial fluid differs from lymph in one respect. Interstitial fluid contains carbon dioxide, but lymph does not. As you remember, the blood erythrocytes are responsible for the removal of this gas so that it never reaches the lymphatic vessels. Other substances, especially organic molecules, in interstitial fluid and lymph vary in kinds and amounts in relation to the location of the sample analyzed. The lymph vessels of the digestive tract, for example, contain a great deal of lipid that has been absorbed from food.

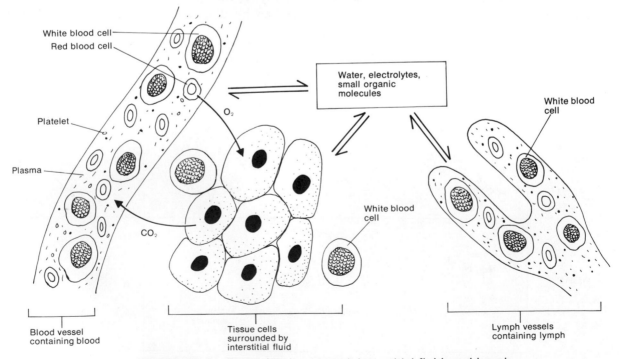

Figure 16–4. Composition of blood, interstitial fluid, and lymph.

BLOOD DISORDERS

The various blood disorders that can arise affect differing portions of the blood. With anemia, for example, the patient may have an abnormally low number of red blood cells, whereas with mononucleosis white blood cells are affected. There are numerous blood disorders, and all may have wide-ranging effects.

Anemia

Anemia is a sign rather than a diagnosis. Many different kinds of anemia exist, and all are characterized by less than normal numbers of erythrocytes or by a below normal hemoglobin content. These conditions lead to fatigue and intolerance to cold, both of which are related to lack of oxygen needed for energy and heat production and paleness due to low hemoglobin content. But diagnosis cannot be made nor can treatment begin until the cause of the anemia is discovered.

Hemorrhagic anemia

An excessive loss of erythrocytes through bleeding is called **hemorrhagic anemia.** Common causes are large wounds, stomach ulcers, and excessive menstrual bleeding. If bleeding is extraordinarily heavy, the anemia is termed acute. Excessive loss of blood fluid may endanger the individual's life. Slow, prolonged bleeding is more apt to produce a chronic anemia whose chief symptom is fatigue.

Hemolytic anemia

The term **hemolytic anemia** comes from the word *hemolysis*, the rupturing of erythrocyte cell membranes. The cell is destroyed, and its hemoglobin pours out into the plasma. A characteristic sign includes distortions in the shapes of erythrocytes that are progressing toward hemolysis. There may also be a sharp increase in the number of reticulocytes. The reticulocytes increase since the destruction of red blood cells stimulates erythropoiesis. Agents that may cause hemolytic anemia are parasites, toxins, and antibodies from incompatible blood, such as occurs with an Rh⁻ mother and Rh⁺ fetus. Erythroblastosis fetalis is an example of a hemolytic anemia.

Aplastic anemia

Destruction or inhibition of the red bone marrow results in **aplastic anemia.** Typically, the marrow is replaced by fatty tissue, fibrous tissue, or tumor cells. Toxins and certain medications are causes. Many of the medications inhibit the enzymes involved in hemopoiesis.

Sickle cell anemia

The erythrocytes of a person with **sickle cell anemia** manufacture an abnormal kind of hemoglobin. When the erythrocyte gives up its oxygen to the interstitial fluid, it tends to lose its integrity in places of low oxygen tension and form long, stiff, rodlike structures that bend the erythrocyte into a sickle shape (Figure 16–5), which gives the disorder its name. The sickled cells rupture easily. And even though erythropoiesis is stimulated by the loss of the cells, it cannot keep pace with the hemolysis. The individual consequently suffers from a hemolytic anemia that reduces the amount of oxygen which can be supplied to his tissues. Prolonged oxygen reduction may eventually cause extensive tissue damage. Furthermore, because of the shape of the sickled cells, they tend to get stuck in blood vessels—a situation that can cut off blood supply to an organ altogether.

Sickle cell anemia is inherited. The gene that is responsible for the disorder also seems to give the individual immunity against malaria. This theory is corroborated by the fact that sickle cell genes are found primarily among populations, or descendants of populations, which live in the malaria belt around the world. This includes parts of Mediterranean Europe and subtropical Africa and Asia. A person with only one of the sickling genes is said to have sickle cell trait. He has a high resistance to malaria—a factor that may have tremendous survival value for him—but he does not develop the anemia. Only people who inherit a sickling gene from both parents experience sickle cell anemia.

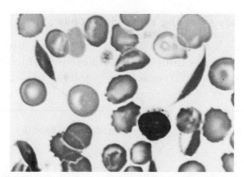

Figure 16–5. Microscopic appearance of erythrocytes in sickle cell anemia. *(Courtesy of Carolina Biological Supply Company.)*

Polycythemia

The term **polycythemia** refers to a condition characterized by an abnormal increase in the number of red blood cells. Increases of 2 to 3 million cells per cubic millimeter are considered to be polycythemic. The disorder is harmful because the thickness of the blood (viscosity) is greatly increased due to the extra red blood cells, and viscosity contributes to a tendency to thrombosis and hemorrhage. It also causes a rise in blood pressure. The tendency to thrombosis results from too many red blood cells piling up as they try to enter smaller vessels. The tendency to hemorrhage is because hyperemia (unusually large amount of blood in an organ part) is observed in all organs.

There are two basic types of polycythemia; *primary* and *secondary*. The primary type is characterized by an overactivity of the red bone marrow and by an enlarged liver and spleen. Its cause is unknown. The secondary type is the result of a lack of oxygen in the arteries of people suffering from chronic cardiac or pulmonary disease. Other causes include kidney tumors or cysts, which could increase secretion of the hormone aldosterone resulting in increasing blood volume; liver cancer, which could interfere with the livers' normal function of disposing of millions of old red blood cells daily; and very high altitudes.

The percentage of the blood that is made up of red blood cells is called the *hematocrit*. It is determined by centrifuging blood and noting the ratio of red blood cells to plasma. The average hematocrit is 45 percent. This means that in 100 milliliters of blood there are 45 milliliters of cells and 55 milliliters of plasma. Anemic blood may have a hematocrit of 15 percent, whereas polycythemic blood may have a hematocrit of 65 percent.

Disorders involving white blood cells

Two well-known disorders affecting white blood cells are infectious mononucleosis and leukemia. Although both disorders are seen fairly frequently, the causes of the conditions are not known.

Infectious mononucleosis

Infectious mononucleosis is a contagious disease of unknown cause that occurs mainly in children and young adults. The trademark of the disease is an elevated white count with an abnormally high percentage of lymphocytes and mononucleocytes —hence, the name mononucleosis. As mentioned earlier in the chapter, an increase in the number of monocytes usually indicates a chronic infection. The various signs and symptoms include slight fever, sore throat, brilliant red throat and soft palate, stiff neck, cough, and malaise. The spleen may enlarge; and secondary complications involving the liver, heart, kidneys, and nervous system may develop.

There is no cure for mononucleosis, and treatment consists of watching for and treating any complications. Usually the disease runs its course in a few weeks, and the individual suffers no permanent ill effects.

Leukemia

This disorder is also called "cancer of the blood." **Leukemia** is an uncontrolled, greatly accelerated production of white cells. Many of the cells fail to reach maturity. As with most cancers, the symptoms and the cause of death do not result so much from the cancer cells themselves as from the interference of the cancer cells with normal body processes. The accumulation of cells leads to abnormalities in organ functions. For example, the anemia and bleeding problems commonly seen in leukemia result from the "crowding out" of normal bone marrow cells. This interferes with the normal production of red blood cells and platelets. The most common cause of death from leukemia is internal hemorrhaging, especially cerebral hemorrhage that destroys the vital centers in the brain. The second most frequent cause of death is uncontrolled infection. This happens because there is a lack of mature or normal white blood cells available to fight infection.

Therapy may temporarily stop the pathologic process. The abnormal accumulation of leucocytes may be reduced or even eliminated by using x-ray and antileukemic drugs. Partial or complete remissions may be induced, with some lasting as long as 15 years.

Medical terminology associated with blood

Corpuscles   Cellular elements in the blood such as red and white cells.

Embolus   Any foreign or abnormal particle that is transported through the blood. It may be solid, liquid, or gaseous. Obstruction occurs if embolus becomes stuck in small vessels.

Hem, Hemo, Hema, Hemato   (*heme* = iron) Various combining forms meaning blood.

Hemolysis   (laking) A swelling and subsequent rupture of erythrocytes with the liberation of meoglobin into the surrounding fluid.

Hemorrhage   (*rrhage* = bursting forth) Bleeding; either internal (from blood vessels into tissues) or external (from blood vessels directly to surface of body).

Plasmolysis (crenation) (*plas* = mold, shape, form; *lysis* = dissolve) Shrinking of the cytoplasm in any cell (in this case, erythrocytes) producing knobbed, starry shapes. Results from a loss of water by osmosis, as when blood is mixed with a 5 percent salt solution.

Septicemia (*sep* = decay, putrefaction; *emia* = condition of blood) Toxins or disease-causing bacteria in the blood. Also called "blood poisoning."

Thrombus (*thrombo* = clot, lump) A blood clot that is lodged in a vessel.

Venesection Opening of a vein for withdrawal of blood.

Medical terminology associated with transfusion

Transfusion The transfer of whole blood, individual blood components (red blood cells only or plasma only), or bone marrow directly into the blood stream of a recipient.

    Direct (immediate) Transfer of blood directly from one person to another without exposing the blood to air.

    Indirect (mediate) Transfer of blood from a donor to a flask or other container and then to the recipient. Permits blood to be stored for an emergency. Also permits breaking down blood into its components so that patient receives only a needed part of the blood.

    Exchange Removing and discarding blood from the recipient while simultaneously replacing it with donor blood. This method is used for erythroblastosis fetalis and for poisoning.

    Reciprocal Blood is transferred from a person who has recovered from a contagious infection into the vessels of a patient suffering with the same infection. An equal amount of blood is returned from the patient to the well person. This method allows the patient to receive antibody-bearing lymphocytes from the recovered person.

Whole blood Blood containing all formed elements, plasma, and plasma solutes in natural concentration.

    Citrated whole blood Whole blood protected from coagulation by a citrate.

    Heparinized whole blood Whole blood in a heparin solution to prevent coagulation.

Fractionated blood (*fract* = break) Blood that has been separated into its components. Only the part of the blood needed by the patient is given.

    Blood cells
        Packed red cells Erythrocytes separated from citrated, whole human blood plasma, which has been identified with the donor's blood group and Rh type. May be used in the treatment of any type of severe anemia, except that produced by excessive bleeding. In the latter case, whole blood is given in order to bring the fluid volume up to normal.
        Platelet concentrates Platelets obtained from freshly drawn whole blood. Used for platelet-deficiency disorders such as hemophilia.

    Blood plasma
        Normal plasma Cell-free plasma that contains normal concentrations of all solutes. Used to bring blood volume up to normal when excessive numbers of blood cells have not been lost. For instance, when extensive areas of the skin have been burned, much fluid is lost through evaporation until new tissue forms. Plasma transfusions are vital in replacing the fluid. Also used for excessive hemorrhaging until patient's blood has been typed.
        Fibrinogen A sterile freeze-dried component of normal plasma, which in solution can be converted to insoluble fibrin by adding thrombin. Can be applied to wounds to stop bleeding. Also indicated in hemorrhagic disorders caused by fibrinogen deficiency.

Normal serum albumin Albumin prepared in a solution that is osmotically equivalent to normal human plasma. Albumin and blood electrolytes are responsible for blood solute concentration and prevent abnormally large quantities of blood water from moving into tissue fluid. In kidney disease, much albumin is excreted into urine. When the liver is diseased, it fails to synthesize albumin. Serum albumin is administered to reduce the resulting edema of either disease.

    Gamma globulin (Immune serum globulin) Solution of globulins from nonhuman blood consisting of antibodies that react with measles, epidemic hepatitis, tetanus, and possibly poliomyelitis viruses. Prepared by injecting these viruses into animals, removing blood from animals after antibodies have accumulated, isolating antibodies, and injecting them into a human for short-term immunity.

Blood plasma substitute A substance that mimics the characteristics of plasma. Used to maintain blood volume during emergency conditions, such as hemorrhage, until blood can be matched, or to prevent dehydration, if a patient cannot swallow liquids. Also used to replace fluid and electrolytes after loss of blood during surgery. Patient will usually replace formed elements within a short time.

Chapter summary in outline

BLOOD

1. The principal functions of blood are the transportation of O_2 and CO_2, nutrients and wastes, and hormones and enzymes. It regulates pH, normal body temperature, and water content of cells; and protects against disease. Blood consists of plasma and formed elements.

2. Formed elements are erythrocytes, leucocytes, and thrombocytes. Wandering epithelial cells become hemocytoblasts. Hemocytoblasts in red bone marrow develop into erythrocytes, granular leucocytes, and thrombocytes. Hemocytoblasts entrapped in lymphatic tissue develop into agranular leucocytes.

3. Erythrocytes, or red blood cells, are biconcave discs without nuclei that contain hemoglobin. Erythrocyte formation is called erythropoiesis and occurs in adult red marrow of certain bones. A reticulocyte count is a diagnostic test that indicates the rate of erythropoiesis.

4. Leucocytes, or white blood cells, are nucleated cells. Two principal types are granular (neutrophils, eosinophils, and basophils) and agranular (lymphocytes and monocytes).

5. One function of leucocytes, especially neutrophils, is to combat inflammation and infection through phagocytosis.

6. In response to the presence of foreign proteins called antigens, lymphocytes are changed into plasma cells. Plasma cells produce antibodies, which cover antigens and render them harmless. This is called the antigen-antibody response and is important in combating infection and providing immunities.

7. Thrombocytes, pr platelets, are disc-shaped structures without nuclei. They are formed from megakaryocytes and initiate clotting.

Plasma

1. The liquid portion of blood, called plasma, consists of 92 percent water and 8 percent solutes. Important solutes include proteins (albumins, globulins, and fibrinogen), foods, enzymes and hormones, gases, and electrolytes.

Blood clotting

1. A clot is a network of insoluble protein (fibrin) in which formed elements of blood are trapped.

2. Clotting in a blood vessel is called thrombosis. A thrombus that moves from its site of origin is called an embolus.

3. Clinically important clotting tests are clotting time (time required for blood to coagulate) and bleeding time (time required for the cessation of bleeding from a small skin puncture).

Blood grouping

1. ABO and Rh systems are based upon antigen-antibody responses.

2. In the ABO system, agglutinogens (antigens) A and B determine blood type. Plasma contains agglutinins (antibodies) that clump agglutinogens that are foreign to the individual.

3. In the Rh system, individuals whose erythrocytes contain Rh agglutinogens are classified as Rh+; those who lack the antigen are classified as Rh−.

INTERSTITIAL FLUID AND LYMPH

1. Interstitial fluid bathes body cells, whereas lymph is found in lymphatic vessels.

2. These fluids are basically similar in chemical composition. They differ chemically from plasma in that both contain less protein, a variable number of leucocytes, and no platelets or erythrocytes.

BLOOD DISORDERS

1. Anemia is indicated by a decreased erythrocyte count or hemoglobin deficiency. Kinds of anemia include hemorrhagic, hemolytic, aplastic, and sickle cell anemia.

2. Polycythemia is an abnormal increase in the number of erythrocytes.

3. Infectious mononucleosis is characterized by an elevated white cell count, especially the monocytes. The cause is unknown.

4. Leukemia is the uncontrolled production of white blood cells that interferes with normal clotting and vital body activities.

Review questions and problems

1. How are blood, interstitial fluid, and lymph related to the maintenance of homeostasis?

2. Distinguish between the blood vascular system and lymph vascular system.

3. Define the principal physical characteristics of blood. List the functions of blood and their relationship to other systems of the body.

4. Distinguish between plasma and formed elements. Where are the formed elements produced?

5. What are reticuloendothelial cells? Describe their function.

6. Describe the microscopic appearance of erythrocytes. What is the essential function of erythrocytes?

7. Define erythropoiesis. Relate erythropoiesis to the homeostasis of red blood cell count. What factors accelerate and decelerate erythropoiesis?

8. What is a reticulocyte count? What is its diagnostic significance?

9. Describe the classification of leucocytes. What are their functions?

10. What is the importance of diapedesis and phagocytosis in fighting bacterial invasion?

11. What is a differential count? What is its significance?

12. Distinguish between leucocytosis and leucopenia.

13. Describe the antigen-antibody response. How is the response protective?

14. What are the major chemicals in plasma? What do they do?

15. What is the difference between plasma and serum?

16. Briefly describe the process of clot formation. What is fibrinolysis? Why does blood usually not remain clotted in vessels?

17. Define the following: thrombus, embolus, anticoagulant, clotting time, and bleeding time.

18. What is the basis for ABO blood grouping? What are agglutinogens and agglutinins?

19. What is the basis for the Rh system? How does erythroblastosis fetalis occur? How may it be prevented?

20. Compare interstitial fluid and lymph with regard to location, chemical composition, and function.

21. Define anemia. Contrast the causes of hemorrhagic, hemolytic, aplastic, and sickle cell anemias.

22. What is leukemia, and what is the cause of some of its symptoms?

23. Define hematocrit. Compare the hematocrits of polycythemic and anemic blood.

24. What is infectious mononucleosis?

25. Refer to the glossary of medical terminology associated with blood and transfusion. Be sure that you can define each term.

CHAPTER 17

THE CIRCULATORY SYSTEM: ORGANS AND VESSELS

STUDENT OBJECTIVES

After you have read this chapter, you should be able to:

1. List the functions of the circulatory system

2. Describe the location of the heart in the mediastinum

3. Distinguish between the structure and location of fibrous and serous pericardium

4. Contrast the structure of the epicardium, myocardium, and endocardium

5. Identify the blood vessels, chambers, and valves of the heart

6. Contrast between the structure and function of veins, capillaries, and arteries

7. Compare systemic, coronary, pulmonary, portal, and fetal circulation

8. Identify the principal arteries and veins of systemic circulation

9. Describe the route of blood in coronary circulation

10. Compare angina pectoris and myocardial infarction as abnormalities of coronary circulation

11. Describe the importance and route of blood involved in portal circulation

12. Identify the major blood vessels of pulmonary circulation

13. Contrast fetal and adult circulation

14. Explain the fate of fetal circulation structures once postnatal circulation is established

15. Define patent ductus arteriosus as an abnormality related to fetal circulation

16. Identify the components and functions of the lymph vascular system

17. Compare the structure of veins and lymphatics

18. Describe the structure and function of lymph nodes

19. Identify the clinically important groups of lymph nodes

20. Contrast the functions of the tonsils, spleen, and thymus gland as lymphatic organs

21. Define immunity

22. Identify the kinds of nonspecific and specific defenses against infection

23. Describe the function of an antibody in the antigen-antibody reaction

24. Describe the role of the thymus gland in antibody production

25. Describe the effects of an allergic reaction on the body

26. Explain the rejection phenomenon in transplantation

27. Contrast the various kinds of transplants

28. Discuss immunosuppresive techniques used to control rejection of transplants

29. Define medical terminology associated with circulatory organs

The circulatory system can be likened to an intricate transportation network within the body. Actually, the circulatory system consists of two principal divisions, the blood vascular system and the lymph vascular system. The first consists of the blood, heart, and blood vessels, and the second consists of lymph, lymph vessels, lymph nodes, and lymphatic organs.

The overall function of the circulatory system is to keep things moving from one part of the body to another. This movement depends on the circulation of blood and lymph within a maze of vessels. The architecture of these vessels is analogous to an interstate highway system. Just as a highway system is designed to route vehicles from one destination to another, the vessels of the circulatory system are designed to transport blood and lymph to and from various parts of the body.

Let us first examine the center of your body's highway system, the heart. This is the pump that maintains circulation in vessels.

THE HEART

The **heart** is a hollow, muscular organ that pumps the blood through the vessels. It is situated between the lungs in the mediastinum, and about two-thirds of its mass lies to the left of the midline of the body (Figure 17–1a). The heart is shaped like a blunt cone about the size of a closed fist. A roentgenogram of a normal heart is provided in Figure 17–1b. Compare it with the adjacent diagram. Its pointed end, called the apex, projects downward, forward, and to the left, and lies above the central depression of the diaphragm. Its broad end, or base, projects upward, backward, and to the right, and lies just below the second rib. The major parts of the heart to be considered here are the pericardium, the walls and chambers, and the valves.

The pericardium

The heart is enclosed in a loose-fitting serous membrane called the **pericardium** (Figure 17–1a, c). The pericardium consists of two principal layers: an external fibrous layer that binds the heart in place and an internal serous layer that secretes a lubricating fluid. The *fibrous layer* is composed of tough fibrous tissue. Its upper surface attaches to the large blood vessels that emerge from the heart. Its lower end attaches the heart to the diaphragm, and its anterior surface binds the heart to the sternum. The internal *serous layer* contains two subdivisions: the *parietal layer,* which lines the inside of the fibrous pericardium, and *visceral layer,* which adheres to the outside of the heart muscle. The visceral layer is also the outermost layer of the heart. Between the parietal and visceral layers of the serous pericardium is a small space, called the *pericardial cavity.* This cavity contains pericardial fluid. As the heart beats, its surface continually moves against the outer layers of the pericardium. The fluid prevents friction between the membranes. An inflammation of the pericardium is called *pericarditis.*

Walls and chambers

The wall of the heart (Figure 17–1c) is divided into three portions: (1) the epicardium, or external layer; (2) the myocardium, or middle layer; and (3) the endocardium, or inner layer. The *epicardium* is the same as the visceral layer of the pericardium, which has just been described. The *myocardium,* which is cardiac muscle tissue, comprises the bulk of the heart. As you will recall from Chapter 10, cardiac muscle fibers are involuntary, striated, and branched, and the tissue is arranged in interlacing bundles of fibers. The myocardium is responsible for the actual contraction of the heart. Inflammation of the myocardium is referred to as *myocarditis.* The *endocardium* is a thin layer of endothelium pierced by tiny blood vessels and some bundles of smooth muscle. It lines the inside of the myocardium and covers the valves of the heart and the tendons that hold the valves open. It is continuous with the endothelium of the large blood vessels of the heart. Inflammation of the endocardium is called *endocarditis.*

The interior of the heart is divided into four spaces or chambers, which receive the circulating blood (Figure 17–2). The two upper chambers are called the right and left **atria.** They are separated by a partition called the *interatrial septum.* The two lower chambers, the right and left **ventricles,**

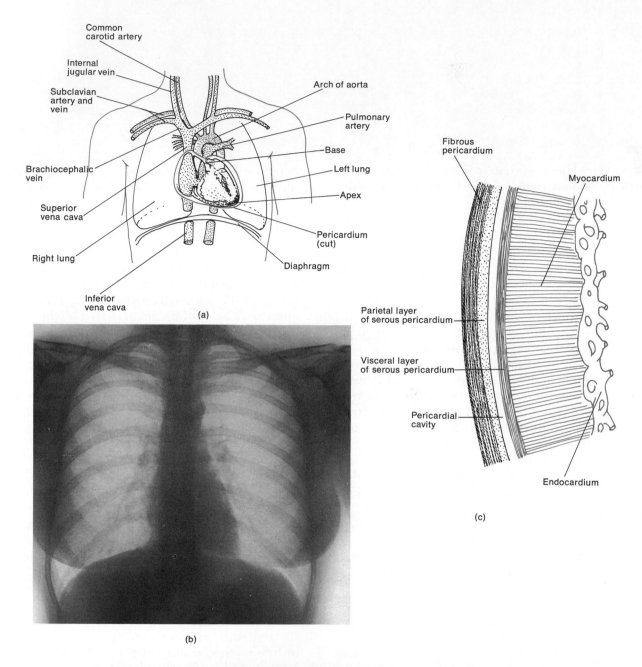

Figure 17–1. The heart. (a) Position of the heart in the thoracic cavity. (b) Roentgenogram of a normal heart. (c) Structure of the pericardium and heart wall. *(x-ray courtesy of William I. Leonard, Bergen Community College.)*

are separated by an *interventricular septum.* The right atrium drains blood from all parts of the body except the lungs. It receives the blood through three veins. One of these veins is the *superior vena cava,* which brings blood from the upper portion of the body. Another of the veins is the *inferior vena cava,* which brings blood from the lower portions of the body. The third vein is the *coronary sinus,* which drains blood from the vessels supplying the walls of the heart. The right

atrium then squeezes the blood into the right ventricle, which pumps it into the *pulmonary artery.* The pulmonary artery then carries the blood to the lungs. In the lungs, the blood releases its carbon dioxide and takes on oxygen. It returns to the heart via four *pulmonary veins* that empty into the left atrium. The blood is then squeezed into the left ventricle and exits from the heart through the *aorta.* This large artery transports the blood to all body parts except the lungs.

If you look closely at the sectioned heart in Figure 17–2, you will see that the sizes of the four chambers vary according to their functions. The right atrium, which must collect blood coming from almost all parts of the body, is slightly larger than the left atrium, which receives only the blood coming from the lungs. The thickness of the walls of the chambers varies too. The atria are relatively thin-walled because they need only enough cardiac muscle tissue to squeeze the fluid into the ventricles. The right ventricle has a much thicker layer of myocardium since it must send blood to the lungs and around back to the left atrium. The left ventricle has the thickest walls since it must pump blood through literally miles of vessels in the head, trunk, arms, and legs.

Valves

As each chamber of the heart constricts, it pushes a portion of blood into a ventricle or out of the heart through an artery. But as the walls of the chambers relax, some structure must prevent the blood from flowing back into the chamber. That structure is a valve.

Atrioventricular valves lie between the atria and their ventricles (Figure 17–3). The atrioventricular valve between the right atrium and right ventricle is called the *tricuspid valve* because it consists of three flaps, or cusps. These flaps are fibrous tissues that grow out of the walls of the heart and are covered with endocardium. The pointed ends of the cusps project into the ventricle. Cords called *chordae tendineae* connect the pointed ends to small *papillary muscles* (muscular columns) that are located on the inner surface of the ventricles. The chordae tendineae and their muscles keep the flaps pointing in the direction of the blood flow. As the atrium relaxes and the ventricle squeezes the blood out of the heart, any blood that is driven back toward the atrium is pushed between the flaps and the walls of the ventricle. This drives the cusps upward until their edges meet and close the opening. The atrioventricular valve between the left atrium and left ventricle is called the *bicuspid* or *mitral valve*. It has two cusps that work in the same way as the cusps of the tricuspid valve.

Each of the arteries that leave the heart has a

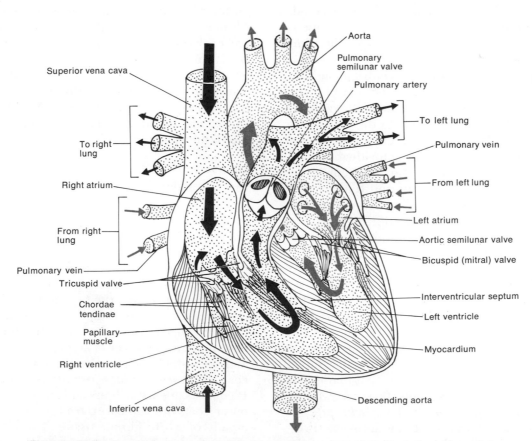

Figure 17–2. Internal anatomy of the heart. Shown here is a frontal section of the heart indicating the principal internal structures.

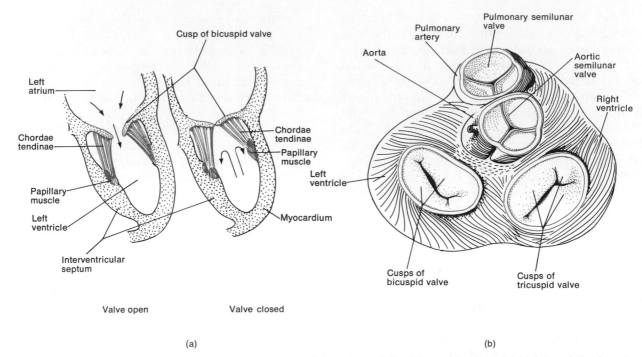

Figure 17–3. Valves of the heart. (a) Structure and function of the bicuspid valve. (b) Valves of the heart viewed from above. The atria have been removed to expose the tricuspid and bicuspid valves.

valve that prevents blood from flowing back into the heart. These are the **semilunar valves**—*semilunar* meaning half-moon or crescent-shaped. The *pulmonary semilunar valve* lies in the opening where the pulmonary artery leaves the right ventricle. The *aortic semilunar valve* is situated at the opening between the left ventricle and the aorta (Figure 17–2). Both valves consist of three semilunar cusps. Each cusp is attached by its convex or inwardly curved margin to the wall of its artery. The free borders of the cusps curve outward and project into the opening inside the blood vessel (Figure 17–3b). Like the atrioventricular valves, the semilunar valves permit the blood to flow in only one direction. In this case, the flow is from the ventricles into the arteries. Next, we shall look at the structure and function of the arteries, as well as the capillaries and veins.

BLOOD VESSELS

The blood vessels form a network of tubes that carry blood away from the heart, transport it to the tissues of the body, and then return it to the heart. If you examine the general plan of these vessels in Figure 17–4, you will note that blood vessels are called either arteries, capillaries, or veins. **Arteries** are the vessels that carry blood from the heart to the tissues. Two large arteries leave the heart and divide into medium-sized vessels that head toward the various regions of the body. The medium-sized arteries, in turn, divide

into small vessels called *arterioles*. As the arterioles enter a tissue, they branch into countless numbers of microscopic vessels called **capillaries**. Through the walls of the capillaries, substances are exchanged between the blood and body tissues. Before leaving the tissue, groups of capillaries reunite to form small veins called *venules*. These, in turn, merge to form progressively larger tubes—the veins themselves. **Veins,** in other words, are blood vessels that convey blood from the tissues back to the heart.

Arteries

Arteries and veins are fairly similar in construction (Figure 17–4a). They both have walls constructed of three coats and a hollow inner core, called a *lumen,* through which the blood flows. Arteries, however, are considerably thicker and stronger than veins because the pressure in an artery is always greater than in a vein. The inner coat of an arterial wall is called the *tunica interna.* It is composed of a lining of endothelium (squamous epithelium) that is in contact with the blood. It also has an overlying layer of areolar connective tissue and an outer layer of elastic tissue. The middle coat, or *tunica media,* is usually the thickest layer. It consists of elastic fibers and smooth muscle. The outer coat, the *tunica externa,* is composed principally of white fibrous tissue. The fibrous tissue is tough and firm and prevents the artery from collapsing if it is cut.

As a result of the structure of the middle coat, especially, arteries have two very important properties: elasticity and contractility. When the ventricles of the heart contract and eject blood into the large arteries, the arteries expand to contain the extra blood volume. Then, as the ventricles relax, the elastic recoil of the arteries forces the blood onward. The contractility of an artery is a function of its smooth muscle. The smooth muscle is arranged in rings around the lumen, resembling somewhat the shape of a donut. As the muscle contracts, it squeezes the wall more tightly around the lumen and consequently narrows the area through which the blood flows. Such a decrease in the size of the lumen is called *vasoconstriction.* The nerves responsible for the action are termed *vasoconstrictor nerves.* Conversely, if all the mus-

cle fibers relax, the size of the arterial lumen increases. This is called *vasodilation*, and the nerves mediating the response are called *vasodilator nerves.*

The contractility of arteries also serves a minor function in stopping bleeding. The blood flowing through an artery is under a great deal of pressure. If you accidentally cut an artery, you will notice that the blood flows out in rapid spurts. Great quantities of blood can be quickly lost from a broken artery. When an artery is cut, its walls constrict so that blood does not flow out of it quite so rapidly. However, there is a limit to how much vasoconstriction can help. For this reason, the arteries lie protected deep in the muscles, close to the bones.

Most parts of the body receive branches from

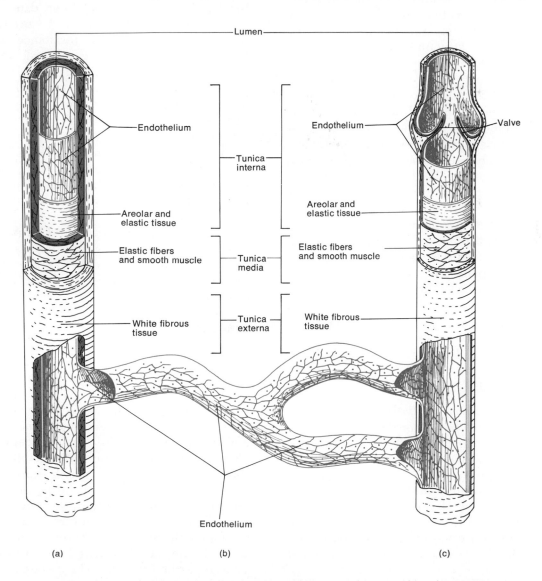

Figure 17–4. Structure of blood vessels. (a) Artery. (b) Capillary. (c) Vein. The relative size of the capillary is enlarged for emphasis.

more than one artery. In such areas the distal ends of the vessels unite and the junction of two or more vessels supplying the same body region is called an **anastomosis**. Anastomoses may also occur between the origins of veins and between arterioles and venules. Anastomoses between arteries provide alternate routes by which blood can reach a given tissue or organ. Thus, if a vessel is occluded by disease, injury, or surgery, circulation to a part of the body is not necessarily stopped. The alternate route of blood to a body part through an anastomosis is known as *collateral circulation*.

Capillaries

Capillaries are microscopic vessels measuring about 0.01 millimeters in diameter. They connect arterioles with venules (Figure 17–4b). The function of the capillaries is to permit the exchange of nutrients and gases between the blood and interstitial fluid. The structure of the capillaries is admirably suited for this purpose. First, the capillary walls are composed of only a single layer of cells (endothelium). This means that a substance in the blood must diffuse through the plasma membranes of just one cell in order to reach the interstitial fluid. It should be noted again that this vital exchange of materials occurs only through capillary walls. The thick walls of arteries and veins present too great a barrier for diffusion to occur. Capillaries are also well suited to their function in that they branch to form an extensive *capillary network* throughout the tissue. The network increases the surface area through which diffusion can take place and thereby allows a rapid exchange of large quantities of materials.

Though capillaries lack the elastic connective fibers of arteries, their walls are still quite elastic. This elasticity permits them to adjust to the amount and force of blood flowing through them.

Veins

Veins are composed of essentially the same three coats as arteries, but they have considerably less elastic tissue and smooth muscle (Figure 17–4c). However, veins do contain more white fibrous tissue. They are also elastic enough to adapt to variations in the volume and pressure of blood passing through them. If you cut a vein, you will notice that the blood leaves the vessel in an even flow rather than in the rapid spurts characteristic of arteries. This is because by the time the blood leaves the capillaries and moves into the veins, it has lost a great deal of its pressure. Most of the structural differences between arteries and veins

reflect this pressure difference. For example, veins do not need to have walls that are as strong as the walls of their corresponding arteries. The low pressure in veins, however, has its disadvantages. For instance, when you stand, the pressure pushing blood up the veins in your legs is barely enough to balance the force of gravity pushing it back down. For this reason, many veins, especially those in the limbs, contain valves that prevent any backflow.

Some people inherit weak valves. When the valves in a vein are weak, large quantities of blood are forced by gravity back down into distal parts of the vein. This overloads the vein and pushes the walls outward. After repeated overloading, the walls lose their elasticity and become permanently stretched and flabby like an overused rubber band. A vein that has been damaged in this way is called a *varicose vein*. A varicosed wall is not able to exert a firm resistance against the blood. Instead of moving upward, some of the blood tends to accumulate in the pouched-out area of the vein, causing it to swell and forcing fluid out into the surrounding tissue. The veins that lie close to the external surfaces of the legs are highly susceptible to varicosities. Veins that lie deeper are not as vulnerable because surrounding skeletal muscles prevent their walls from overstretching. Varicosities are also common in the veins that lie in the walls of the rectum. These are called *hemorrhoids*.

The arteries, capillaries, and veins are organized into definite routes in order to circulate the blood throughout the body. We can now look at the basic routes the blood takes as it is transported through its vessels.

CIRCULATORY ROUTES

Looking at Figure 17–5, you will see that there are a number of basic *circulatory routes* through which the blood travels. The largest route by far is the *systemic circulation*. This includes all the oxygenated blood that leaves the left ventricle through the aorta and returns to the right atrium after traveling to all the organs except the lungs. Two subdivisions of the systemic circulation are the *coronary circulation*, which supplies the myocardium of the heart, and the *portal circulation*, which runs from the digestive tract to the liver. Blood leaving the aorta and traveling through the systemic arteries is a bright red color. As it moves through the capillaries, it loses its oxygen and takes on carbon dioxide. The carbon dioxide gives the blood in the systemic veins its dark color. When blood returns to the heart from the systemic route, it then goes out of the right ventricle through the *pulmonary circulation* to the lungs. In the lungs, it loses its carbon dioxide and takes on oxygen. It is now a bright red color again. It returns to the left atrium of the heart and reenters the systemic circulation. A third major route is one that exists only in the fetus and contains spe-

cial structures that allow the developing human to exchange materials with its mother. This is the *fetal circulation,* which will be described later.

Systemic circulation

The flow of blood from the left ventricle to all parts of the body except the lungs and back to the

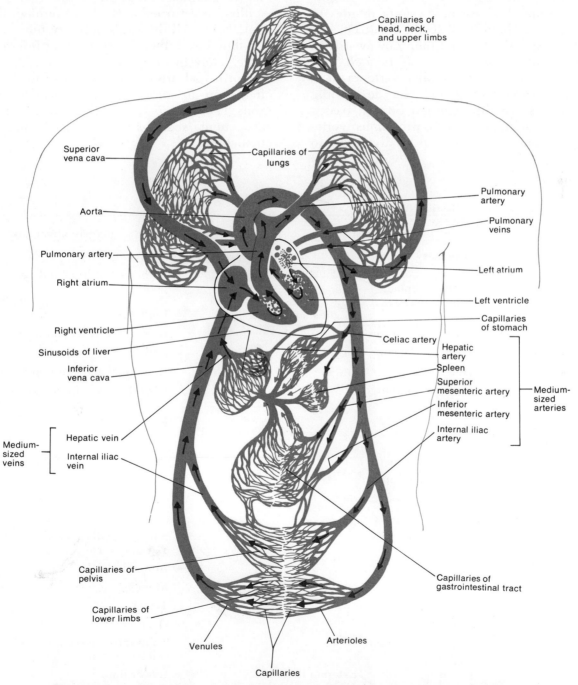

Figure 17-5. Circulatory routes. Systemic circulation is indicated by heavy black arrows; pulmonary circulation by thin black arrows; and portal circulation by thin colored arrows. Refer to Figure 17–6 for the details of coronary circulation and Figure 17–9 for the details of fetal circulation.

right atrium is called the *systemic circulation.* The purpose of systemic circulation is to carry oxygen and nutrients to body tissues and to remove carbon dioxide and other wastes from the tissues. All systemic arteries branch from the *aorta,* which arises from the left ventricle of the heart. (See Exhibit 17–1.) As the aorta emerges from the left ventricle, it passes upward underneath the pulmonary artery. At this point, it is called the *ascending aorta.* The ascending aorta divides to send a coronary branch off to the heart muscle. Then it turns to the left and then downward. As it makes the turn, it is called the *aortic arch.* As it runs down to the fourth lumbar vertebra, it is called the *descending aorta.* The descending aorta lies close to the vertebral bodies, passes through the diaphragm to the fourth lumbar vertebra, and terminates at this level by dividing into

two *common iliac arteries,* which carry blood to the lower extremities. The section of the descending aorta between the aortic arch and the diaphragm is also referred to as the *thoracic aorta.* The section between the diaphragm and the common iliac arteries is termed the *abdominal aorta.* Each section of the aorta gives off arteries that continue to branch into smaller vessels and finally into the capillaries that pierce the tissues.

Blood is returned to the heart through the systemic veins. All the veins of the systemic circulation flow into either the *superior* or *inferior vena cavae.* They in turn empty into the right atrium. The principal arteries and veins of the systemic circulation are described and illustrated in Exhibits 17–1 to 17–12, which follow, and in the subsequent sections on coronary and portal circulation.

Exhibit 17–1. THE AORTA AND ITS BRANCHES

DIVISION OF AORTA	ARTERIAL BRANCH	REGION SUPPLIED
Ascending aorta	Coronary	Heart
Aortic arch	Brachiocephalic (innominate) → Right common carotid	Right side of head and neck
	Brachiocephalic (innominate) → Right subclavian	Right arm
	Left common carotid	Left side of head and neck
	Left subclavian	Left arm
Thoracic aorta	Intercostals	Intercostal and chest muscles, plurae
	Superior phrenics	Posterior and superior surfaces of diaphragm
	Bronchials	Bronchi of lungs
	Esophageals	Esophagus
	Inferior phrenics	Inferior surface of diaphragm
Abdominal aorta	Celiac → Hepatic	Liver
	Celiac → Left gastric	Stomach and esophagus
	Celiac → Splenic	Spleen, pancreas, and stomach
	Superior mesenteric	Small intestines, cecum, ascending and transverse colons
	Suprarenals	Adrenal glands
	Renals	Kidneys
	Testiculars	Testes
	Ovarians	Ovaries
	Inferior mesenteric	Transverse, descending, and sigmoid colons; rectum
	Common iliacs → External iliacs	Lower limbs
	Common iliacs → Internal iliacs	Uterus, prostate, muscles of buttocks

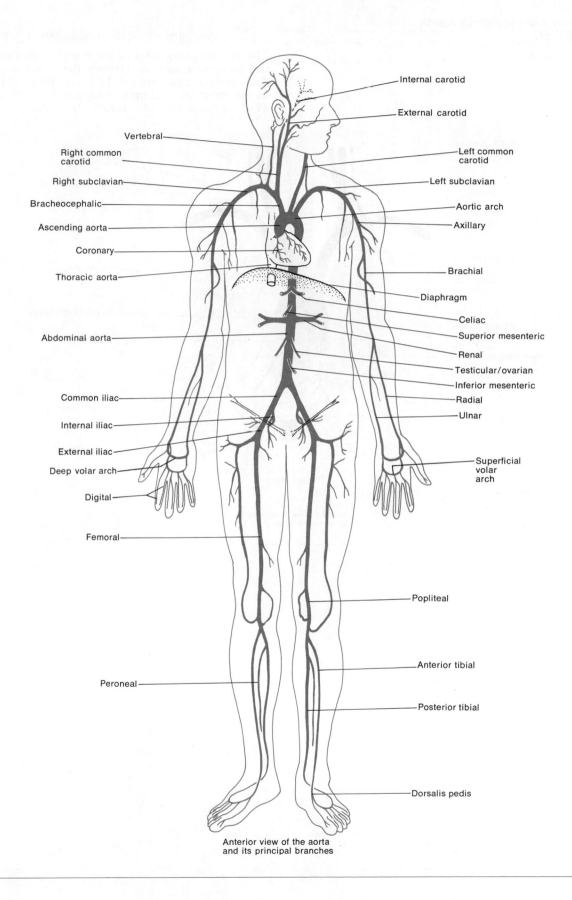

Internal carotid

External carotid

Vertebral

Right common carotid

Right subclavian

Bracheocephalic

Ascending aorta

Coronary

Thoracic aorta

Left common carotid

Left subclavian

Aortic arch

Axillary

Brachial

Diaphragm

Celiac

Superior mesenteric

Renal

Testicular/ovarian

Inferior mesenteric

Radial

Ulnar

Abdominal aorta

Common iliac

Internal iliac

External iliac

Deep volar arch

Digital

Femoral

Superficial volar arch

Popliteal

Anterior tibial

Peroneal

Posterior tibial

Dorsalis pedis

Anterior view of the aorta
and its principal branches

Exhibit 17–2. THE ASCENDING AORTA

BRANCH	DESCRIPTION AND REGION SUPPLIED
Coronary arteries	The two *coronary arteries* are branches that arise from the ascending aorta just above the semilunar valves. They form a crown around the heart giving off branches to the atrial and ventricular myocardium. (See Figure 17–6 for the details of coronary circulation.)

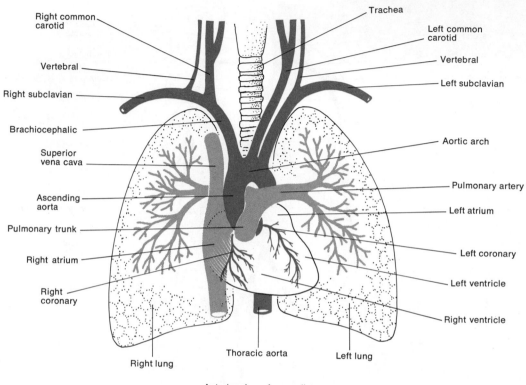

Anterior view of ascending aorta and its branches

Exhibit 17–3. THE AORTIC ARCH

BRANCH	DESCRIPTION AND REGION SUPPLIED
Brachiocephalic	The *brachiocephalic artery* is the first branch off the aortic arch. It subdivides to form the right subclavian artery and right common carotid artery. (See illustration in Exhibit 17–2.) The *right subclavian artery* extends from the brachiocephalic to the first rib, passes into the armpit, or *axilla,* and supplies the arm and hand. See (a) next page. This artery is a good example of the convention of giving the same vessel different names as it passes through different regions. The continuation of the right subclavian into the axilla is called the *axillary artery.* From here, it continues into the upper arm as the *brachial artery.* At the bend of the elbow, the brachial artery divides into the *ulnar* and *radial arteries.* These vessels pass down to the palm, one on each side of the forearm. In the palm, branches of the two arteries anastomose to form two palmar arches—the *superficial volar arch* and the *deep volar arch.* From these arches arise the *digital arteries,* which supply the fingers and the thumb. Before passing into the axilla, the right subclavian gives off an important branch to the brain called the *vertebral artery.* See (b) next page. The right vertebral artery passes through the foramina of the transverse processes of the cervical vertebrae to the undersurface of the brain. Here, it unites with the left vertebral artery to form the *basilar artery.* Anastomoses of the left and right internal carotids along with the basilar artery form an arterial circle at the base of the brain called the *circle of Willis.* See (c) next page. From these vessels arise the arteries supplying the brain. Essentially, the circle of Willis is formed by the union of the *anterior cerebral arteries* (branches of the internal carotids) and the *posterior cerebral arteries* (branches of the basilar artery). The posterior cerebral arteries are connected with the internal carotids by the *posterior communicating arteries.* The anterior cerebral arteries are connected by the *anterior communicating arteries.* The circle of Willis is important because it equalizes blood pressure to the brain and provides alternate routes for blood to the brain should arteries become diseased or damaged.
Right common carotid	The *right common carotid artery* passes upward. At the upper level of the larynx, it divides into the *right external* and *right internal carotid arteries.* See (b) next page. The external carotid supplies the right sides of the thyroid gland, tongue, throat, face, ear, scalp, and dura mater. The internal carotid supplies the brain, right eye, and right sides of the forehead and nose.
Left common carotid	The *left common carotid* branches directly from the arch of the aorta. (See illustration in Exhibit 17–2.) Corresponding to the right common carotid, it divides into basically the same branches with the same names—except that the arteries are now labeled "left" instead of "right."
Left subclavian	The *left subclavian artery* is the third branch off the aortic arch. (See Exhibit 17–2.) It distributes blood to the left vertebral artery and to the vessels of the left arm. The arteries branching from the left subclavian are named like those of the right subclavian.

Exhibit 17–3 (cont'd.)

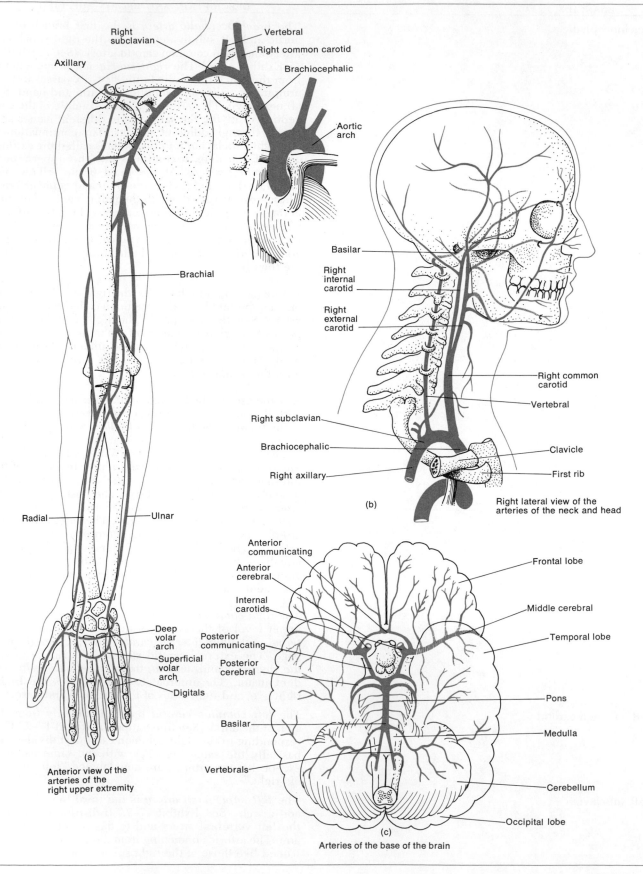

Right subclavian

Vertebral

Axillary

Right common carotid

Brachiocephalic

Aortic arch

Brachial

Radial

Ulnar

Deep volar arch

Superficial volar arch

Digitals

(a)
Anterior view of the arteries of the right upper extremity

Basilar

Right internal carotid

Right external carotid

Right common carotid

Vertebral

Right subclavian

Brachiocephalic

Clavicle

Right axillary

First rib

(b)
Right lateral view of the arteries of the neck and head

Anterior communicating

Anterior cerebral

Internal carotids

Posterior communicating

Posterior cerebral

Basilar

Vertebrals

Frontal lobe

Middle cerebral

Temporal lobe

Pons

Medulla

Cerebellum

Occipital lobe

(c)
Arteries of the base of the brain

Exhibit 17–4. THE THORACIC AORTA

The *thoracic aorta* runs from the fourth to twelfth thoracic vertebrae. See Exhibit 17–5. Along its course, it sends off numerous small arteries to the viscera and skeletal muscles of the chest. The *visceral branches* supply the pericardium around the heart, the bronchial tubes that lead from the windpipe to the lungs, the cells of the lungs (but not the areas of the lungs that oxygenate blood), the gullet, and the tissue lining the mediastinum. The *parietal branches* supply the chest muscles, diaphragm, and mammary glands.

Exhibit 17–5. THE ABDOMINAL AORTA

BRANCH	DESCRIPTION AND REGION SUPPLIED
Visceral	
Celiac	The *celiac artery* (trunk) is the first aortic branch below the diaphragm. It has three branches: (1) *hepatic artery*, which supplies the tissues of the liver; (2) *left gastric artery*, which supplies the stomach; and (3) *splenic artery*, which supplies the spleen, pancreas, and stomach.
Superior mesenteric	The *superior mesenteric artery* distributes blood to the small intestines and to the first half of the large intestine.
Suprarenals	The right and left *suprarenal arteries* supply blood to the adrenal glands.
Renals	The right and left *renal arteries* carry blood to the kidneys.
Testiculars	The right and left *testicular arteries* extend into the scrotum and terminate in the testes.
Ovarians	The right and left *ovarian arteries* are distributed to the ovaries.
Inferior mesenteric	The *inferior mesenteric artery* supplies the last two-thirds of the large intestine and the rectum.
Parietal	
Inferior phrenics	The *inferior phrenics* are distributed to the undersurface of the diaphragm.
Lumbars	The *lumbar arteries*, usually four pairs, supply the spinal cord and its meninges and the muscles and skin of the lumbar region of the back.
Middle sacral	The *middle sacral artery* supplies the sacrum, coccyx, gluteus maximus muscle, and rectum.

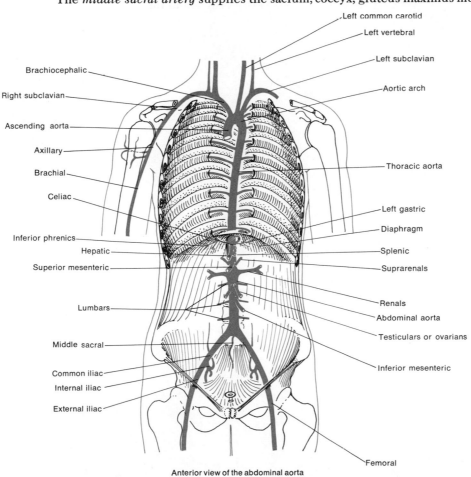

Anterior view of the abdominal aorta

Exhibit 17–6. ARTERIES OF THE PELVIS AND LOWER EXTREMITIES

BRANCH	DESCRIPTION AND REGION SUPPLIED
	At about the level of the fourth lumbar vertebra, the abdominal aorta divides into the right and left *common iliac arteries*. Each of these, in turn, passes downward about two inches and gives rise to two branches: the internal iliac and external iliac.
Internal iliacs	The *internal iliacs* form branches that supply the gluteal muscles, medial side of each thigh, urinary bladder, rectum, prostate gland, uterus, and vagina.
External iliacs	The *external iliac arteries* diverge through the pelvis, enter the thighs, and here become the right and left *femoral arteries*. Both femorals send branches back up to the genitals and to the wall of the abdomen. Other branches run to the muscles of the thigh. The femoral continues down the medial and posterior side of the thigh at the back of the knee joint, where it becomes the *popliteal artery*. Between the knee and ankle, the popliteal runs down the back of the leg and is called the *posterior tibial artery*. Below the knee, the *peroneal artery* branches off of the posterior tibial to supply the structures on the medial side of the fibula and calcaneus. In the calf, another artery, the *anterior tibial artery*, branches off of the popliteal and runs along the front of the legs. At the ankles, it becomes the *dorsalis pedis artery*. At the ankle, the posterior tibial divides into the *medial* and *lateral plantar arteries*. These arteries anastomose with the dorsalis pedis and supply blood to the foot.

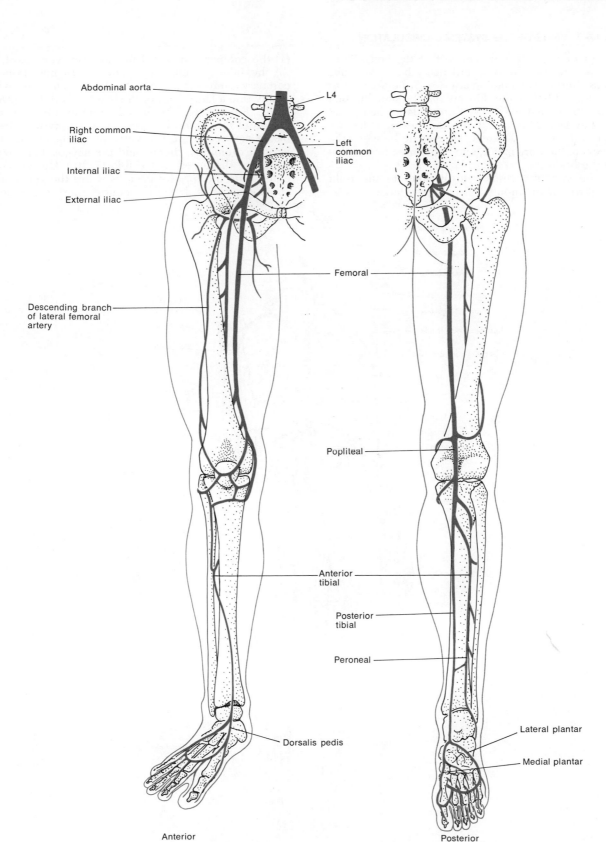

View of the arteries of the right lower extremity

Exhibit 17-7. VEINS OF THE SYSTEMIC CIRCULATION

Deep veins are located deep within the body. They usually accompany arteries, and many have the same names as their corresponding arteries. *Superficial veins* are located just below the skin and may be seen through the skin. Veins that have very thin walls in the cranial cavity (they lack a tunica externa and tunica media) are called *venous sinuses*. Venous sinuses lie between the two layers of the dura mater, one of the membranes that covers the brain.

All the systemic veins return blood to the right atrium of the heart through one of three large vessels:

(1) the coronary sinus, (2) the superior vena cava, and (3) the inferior vena cava. The return flow from the coronary arteries is taken up by the *cardiac veins*, which empty into a large vein of the heart called the *coronary sinus*. From here, the blood empties into the right atrium of the heart. The veins that empty into the *superior vena cava* are the veins of the head and neck, upper extremities, thorax, and the azygos. (The azygos veins are discussed in Exhibit 17–10.) The veins that empty into the *inferior vena cava* are the veins of the abdomen, pelvis, lower extremities, and the azygos.

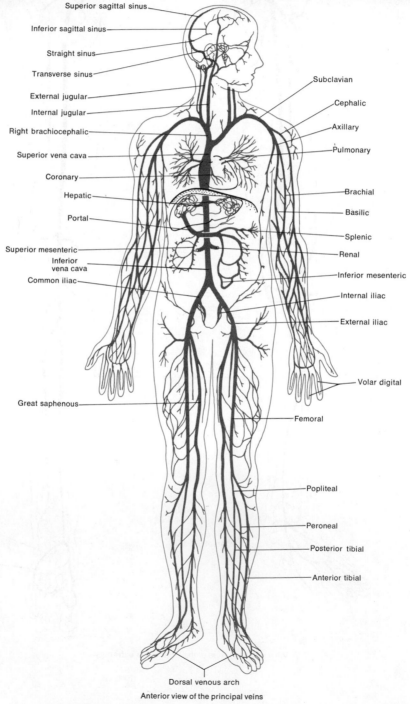

Superior sagittal sinus

Inferior sagittal sinus

Straight sinus

Transverse sinus

External jugular

Internal jugular

Right brachiocephalic

Superior vena cava

Coronary

Hepatic

Portal

Superior mesenteric

Inferior vena cava

Common iliac

Great saphenous

Subclavian

Cephalic

Axillary

Pulmonary

Brachial

Basilic

Splenic

Renal

Inferior mesenteric

Internal iliac

External iliac

Volar digital

Femoral

Popliteal

Peroneal

Posterior tibial

Anterior tibial

Dorsal venous arch

Anterior view of the principal veins

Exhibit 17–8. VEINS OF THE HEAD AND NECK

VEIN	DESCRIPTION AND REGION DRAINED
Internal jugulars	The right and left *internal jugular veins* arise as a continuation of the *transverse (lateral) sinuses* at the base of the skull. The sinuses are located between the layers of the dura mater and receive blood from the brain. Other sinuses that drain into the internal jugular include the *superior sagittal sinus*, the *inferior sagittal sinus*, the *straight sinus*, and *sigmoid sinus*. The internal jugulars descend on either side of the neck. They receive blood from the superior part of the face and neck and pass behind the clavicles, where they join with the right and left *subclavian veins*. The unions of the internal jugulars and subclavians form the right and left *brachiocephalic (innominate) veins*. From here the blood flows into the *superior vena cava*.
External jugulars	The left and right *external jugular veins* run down the neck, along the outside of the internal jugulars. They drain blood from the parotid (salivary) glands, facial muscles, scalp, and other superficial structures into the *subclavian veins*.

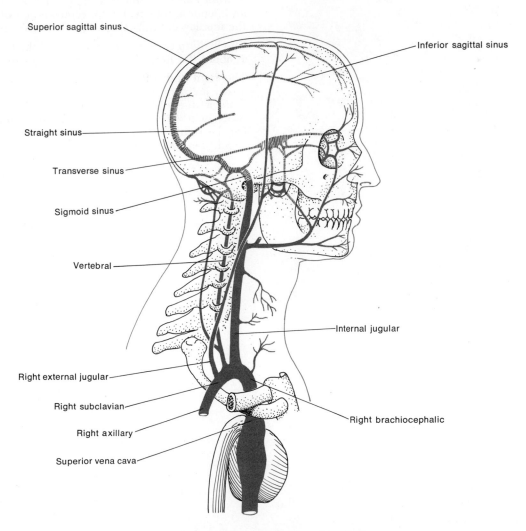

Right lateral view of the veins of the neck and head

Exhibit 17–9. VEINS OF THE UPPER EXTREMITIES

VEIN	DESCRIPTION AND REGION DRAINED
	Blood from the upper extremity is returned to the heart by deep and superficial veins. Both sets of veins contain valves.
Deep veins	The deep veins run alongside arteries and are called the *radial, ulnar, brachial, axillary,* and *subclavian veins.*
Superficial veins	The superficial veins anastomose extensively with each other and with the deep veins. They include the following:
Cephalics	The *cephalic vein* of each arm begins in the *dorsal arch* and winds upward around the radial border of the forearm. At a point just below the elbow, it unites with the accessory cephalic vein to form the cephalic vein of the upper arm. It eventually empties into the subclavian vein.
Basilics	The *basilic vein* of each arm originates in the ulnar part of the dorsal arch. It extends along the posterior surface of the ulna to a point below the elbow where it joins the *median cubital vein.* If a vein must be punctured for an injection, transfusion, or removal of a blood sample, the median cubitals are the preferred veins.
Axillaries	The *axillary vein* is a continuation of the basilic. It ends at about the first rib, where it becomes the subclavian.
Subclavians	The right and left *subclavian veins* unite with the internal jugulars to form the *brachiocephalic veins.* The thoracic duct of the lymphatic system flows into the left subclavian vein at its junction with the internal jugular. On the right, the right lymphatic duct enters the right subclavian vein at the corresponding junction.

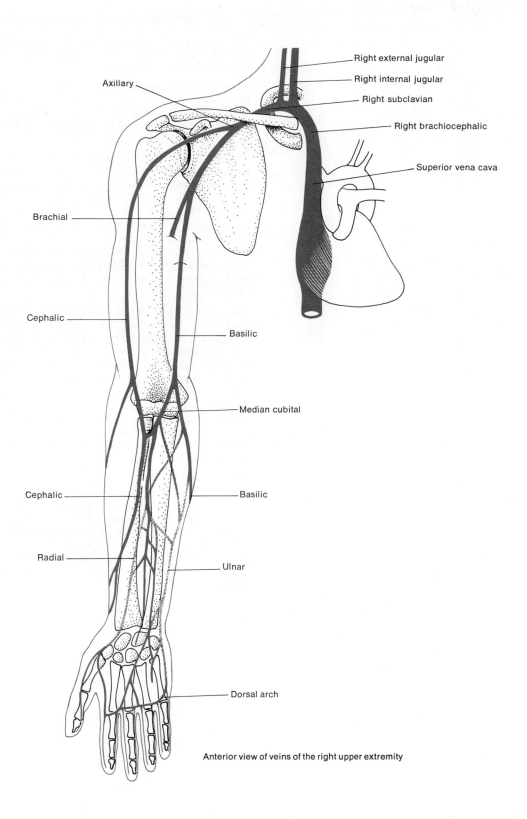

Anterior view of veins of the right upper extremity

Exhibit 17–10. VEINS OF THE THORAX

VEIN	DESCRIPTION AND REGION DRAINED
	The principal thoracic vessels that pour their contents into the superior vena cava are the brachiocephalic veins and the azygos veins.
Brachiocephalic	The right and left *brachiocephalic veins,* formed by the union of the subclavians and internal jugulars, drain blood from the head, neck, mammary glands, and upper thorax. The brachiocephalics unite to form the *superior vena cava.*
Azygos	The *azygos veins,* in addition to collecting blood from various parts of the thorax, serve as a bypass for the inferior vena cava that drains blood from the lower part of the body. Several small veins directly link the azygos veins with the vena cava. And the large veins that drain the legs and abdomen dump some of their blood into the azygos. If the inferior vena cava or hepatic portal vein becomes obstructed, the azygos veins can return blood from the lower part of the body to the superior vena cava. The three azygos veins are as follows:
Azygos vein	The azygos vein lies in front of the vertebral column, slightly to the right of the midline. It begins as a continuation of the right ascending lumbar vein. And it has connections with the inferior vena cava, right common iliac, and lumbar veins. The azygos receives blood from the right intercostal veins that drain the chest muscles; from the hemiazygos and accessory hemiazygos veins (discussed below); from several esophageal, mediastinal, and pericardial veins; and from the right bronchial vein. The vein ascends to the fourth thoracic vertebra, arches over the right lung, and empties into the superior vena cava.
Hemiazygos vein	The hemiazygos vein is in front of the vertebral column and slightly to the left of the midline. It begins as a continuation of the left ascending lumbar vein. It receives blood from the lower four or five intercostal veins and from some esophageal and mediastinal veins. At about the level of the ninth thoracic vertebra, it connects with the azygos vein.
Accessory hemiazygos vein	The accessory hemiazygos vein is also in front and to the left of the vertebral column. It receives blood from three or four intercostal veins and from the left bronchial vein. It joins the azygos at the level of the eighth thoracic vertebra.

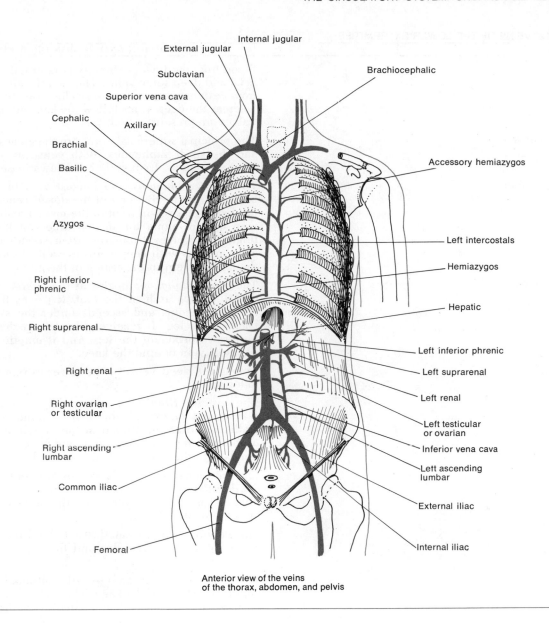

Internal jugular

External jugular

Subclavian

Superior vena cava

Brachiocephalic

Cephalic

Axillary

Brachial

Accessory hemiazygos

Basilic

Azygos

Left intercostals

Hemiazygos

Right inferior
phrenic

Hepatic

Right suprarenal

Left inferior phrenic

Right renal

Left suprarenal

Right ovarian
or testicular

Left renal

Left testicular
or ovarian

Right ascending
lumbar

Inferior vena cava

Lett ascending
lumbar

Common iliac

External iliac

Internal iliac

Femoral

Anterior view of the veins
of the thorax, abdomen, and pelvis

Exhibit 17–11. VEINS OF THE ABDOMEN AND PELVIS

VEIN	DESCRIPTION AND REGION DRAINED
Inferior vena cava	The *inferior vena cava* is the largest vein of the body. (See Exhibit 17–10.) It is formed by the union of the two *common iliac veins* that drain the legs and abdomen. It extends upward through the abdomen and thorax to the right atrium. Numerous small veins enter the inferior vena cava. For the most part, they carry return flow from branches of the abdominal aorta, and their names correspond to the names given to arteries. Among these are: the left and right *renal veins* from the kidneys and uterus; the right *testicular vein* from the testes (the left testicular vein empties into the left renal vein); the right *ovarian vein* from the ovaries (the left ovarian vein also empties into the left renal vein); the right *suprarenal veins* from the adrenal glands (the left suprarenal vein empties into the left renal vein); the right *inferior phrenic vein* from the diaphragm (the left inferior phrenic vein sends a tributary to the left renal vein); and the *hepatic veins* from the liver. In addition, a series of roughly parallel *lumbar veins* drain blood from both sides of the posterior abdominal wall. The lumbars connect at right angles with the right and left ascending lumbar veins, which form the origin of the corresponding azygos or hemiazygos vein. The lumbars drain some of their blood into the ascending lumbars and then run to the inferior vena cava, where they release the remainder of their flow.

Exhibit 17–12. VEINS OF THE LOWER EXTREMITIES

VEIN	DESCRIPTION AND REGION DRAINED
	Blood from the lower extremity is returned by a superficial and deep set of veins. The superficials are formed from extensive anastomoses close to the surface, whereas the deep veins follow the large arterial trunks. Valves are present in both sets.
Superficial veins	The principal superficial veins are the great saphenous and the small saphenous. Both veins, especially the great saphenous, frequently become varicosed.
Great saphenous	The *great saphenous,* the longest vein in the body, begins at the medial end of the *dorsal venous arch* of the foot. It passes in front of the medial malleolus and then upward along the medial aspect of the leg and thigh. It receives tributaries from superficial tissues and connects with deep veins as well. It empties into the femoral vein in the region of the groin.
Small saphenous	The *small saphenous* begins at the lateral end of the dorsal venous arch of the foot. It passes behind the lateral malleolus and ascends under the skin of the back of the leg. It receives blood from the foot and posterior portion of the leg. And it empties into the popliteal vein behind the knee.
Deep veins	Among the deep veins of the lower extremity are the following:
Posterior tibial	The *posterior tibial* vein is formed by the union of the *medial* and *lateral plantar veins* behind the medial malleolus. It ascends deep in the muscle at the back of the legs, receives blood from the *peroneal vein,* and unites with the anterior tibial vein just below the knee.
Anterior tibial	The *anterior tibial* vein is an upward continuation of the *dorsalis pedis* veins in the foot. It runs between the tibia and fibula and unites with the posterior tibial to form the popliteal vein.
Popliteal	The *popliteal vein,* located just behind the knee, receives blood from muscles and the small saphenous vein.
Femoral	The *femoral vein* is an upward continuation of the popliteal just above the knee. The femorals run up the posterior of the leg and drain deep structures of the thighs. After receiving the great saphenous veins in the region of the groin, they continue as the right and left *external iliac veins.* The right and left *internal iliac veins* receive blood from the pelvic wall and viscera, external genitals, buttocks, and medial aspect of the thigh. The right and left *common iliac veins* are formed by the union of the internal and external iliacs. The common iliacs unite to form the inferior vena cava.

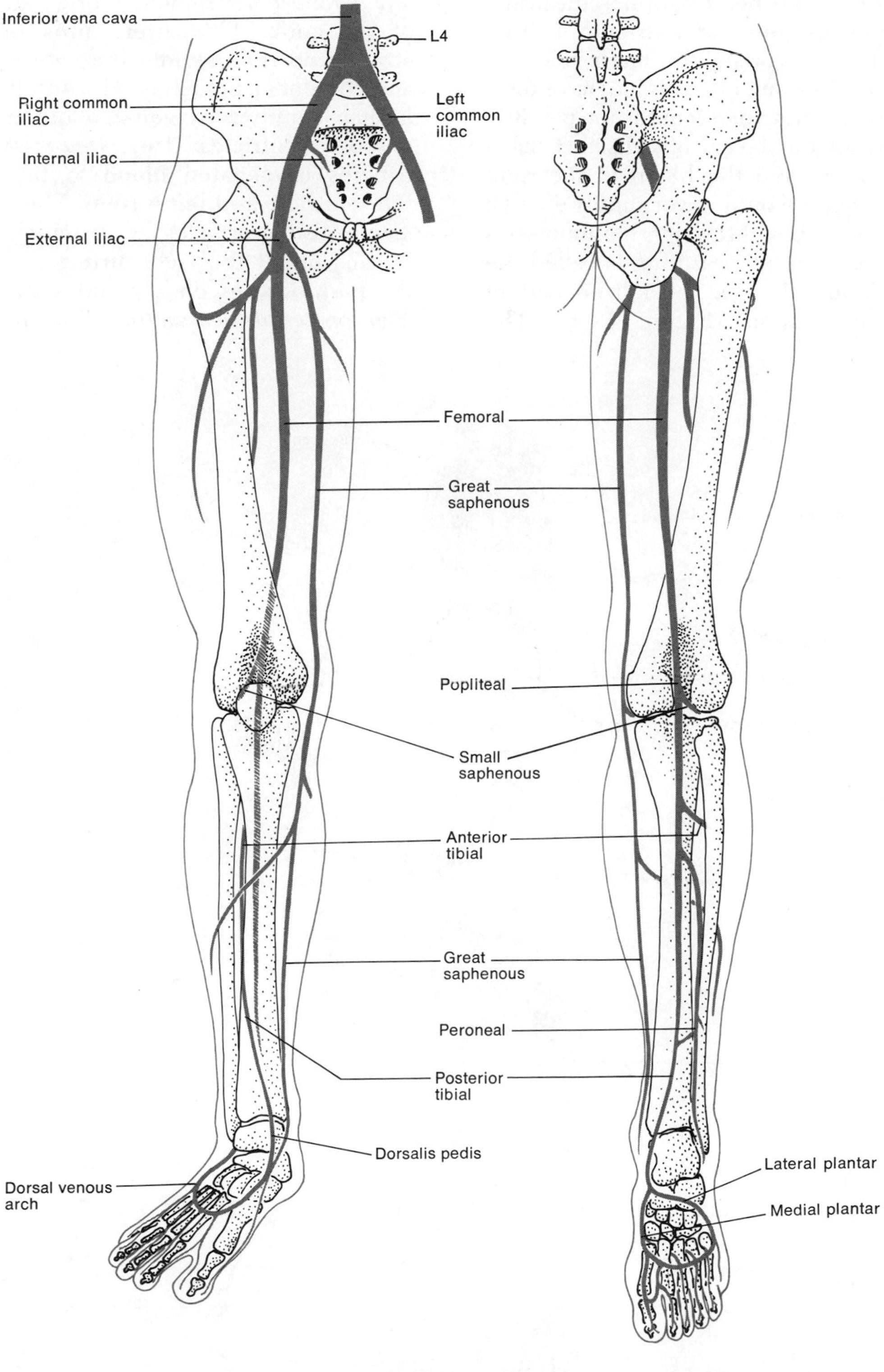

Inferior vena cava

L4

Right common iliac

Left common iliac

Internal iliac

External iliac

Femoral

Great saphenous

Popliteal

Small saphenous

Anterior tibial

Great saphenous

Peroneal

Posterior tibial

Dorsalis pedis

Dorsal venous arch

Lateral plantar

Medial plantar

Anterior

View of the veins of the right lower extremity

Posterior

Coronary circulation

You may have assumed that the blood collected in the chambers of the heart supplies the walls of the heart with oxygen and nutrients. In that case, you will be surprised to learn that the walls of the heart, like any other tissue, have their own blood vessels. Nutrients could not possibly diffuse through all the layers of cells that make up the heart tissue. And the blood in the right chambers of the heart would never supply enough oxygen. The flow of blood through the numerous vessels that pierce the myocardium is called the **coronary circulation.** It is a specialized part of the systemic circulation (Figure 17–6). The vessels that serve the myocardium include the *left coronary artery,* which originates as a branch of the aorta. This artery runs under the left atrium and divides into the anterior descending and circumflex branches. The *anterior descending branch* supplies oxygenated blood to the walls of both ventricles. The *circumflex branch* distributes oxygenated blood to the walls of the left ventricle and left atrium. The *right coronary artery* also originates as a branch of the aorta. It runs under the right atrium and divides into the posterior descending and marginal branches. The *posterior descending branch* supplies the

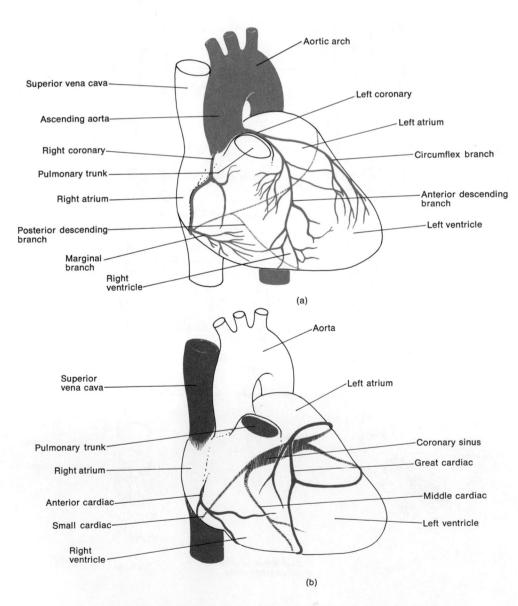

Figure 17–6. Coronary circulation. (a) Anterior view of arterial distribution. (b) Anterior view of venous drainage.

walls of the two ventricles with oxygenated blood. The *marginal branch* transports oxygenated blood to the myocardium of the right ventricle and right atrium. The left ventricle receives the most abundant blood supply because of the enormous work it must do.

As blood passes through the arterial system of the heart, it delivers oxygen and nutrients and collects carbon dioxide and wastes. Most of the deoxygenated blood, which carries the carbon dioxide and wastes, is collected by a large vein, the *coronary sinus*, which empties into the right atrium. The principal veins are the *great cardiac vein*, which drains the anterior aspect of the heart. It ascends and empties into the coronary sinus and the *middle cardiac vein*. This vein drains the posterior aspect of the heart and empties into the coronary sinus.

The majority of "heart problems" result from some foul-up in the coronary circulation. If a reduced oxygen supply weakens the cells, but does not actually kill them, the condition is called *ischemia*. *Angina pectoris* is ischemia of the myocardium. The name comes from the area in which the pain is felt. Remember that pain impulses originating from most visceral muscles are referred to an area on the surface of the body. Angina pectoris occurs when coronary circulation is somewhat reduced for some reason. Stress, which produces constriction of vessel walls, is a common cause. Equally common is strenuous exercise after a heavy meal. When any quantity of food is dumped into the stomach, the body increases blood flow to the digestive tract. The digestive glands can then receive enough oxygen for their increased activities, and the digested food can be quickly absorbed into the bloodstream. As a consequence, some blood is diverted away from other organs, including the heart. Exercise, however, increases heart muscle activity and thus increases its need for oxygen. Thus, doing heavy work while food is in the stomach can lead to oxygen deficiency in the myocardium. Angina pectoris weakens the heart muscle, but it does not produce a full-scale heart attack. The simple remedy of taking nitroglycerin, a drug that dilates vessels and thereby increases the area of blood flow, brings coronary circulation back to normal and stops the pain of angina. Because repeated attacks of angina can weaken the heart and lead to serious heart trouble, angina patients are told to avoid activities and stresses that bring on the attacks.

A much more serious problem is *myocardial infarction*, commonly called a "coronary" or "heart attack." *Infarction* means death of an area of tissue because of a drastically reduced or completely interrupted blood supply. Myocardial infarction results from a thrombus or embolus in one of the coronary arteries. The tissue on the far side of the obstruction dies, and the heart muscle loses at least some of its strength. The aftereffects depend partly on the size and location of the infarcted, or dead, area.

Portal circulation

Blood coming into the liver is derived from two sources. The hepatic artery delivers oxygenated blood from the systemic circulation, and the portal vein delivers deoxygenated blood from the digestive organs. The term **portal circulation** refers to this flow of blood from the digestive organs to the liver before returning to the heart (Figure 17–7). Portal blood is rich with substances that have been absorbed from the digestive tract. One of the roles of the liver is to monitor these substances before they are passed into the general circulation. For example, the liver stores nutrients such as glucose. It modifies other digested substances so that they may be more easily used by cells. And it detoxifies harmful substances that have been absorbed by the digestive tract.

The portal system of veins includes veins that drain blood from the pancreas, spleen, stomach, intestines, and gallbladder, and transport it to the portal vein of the liver. The *portal vein* is formed by the union of the superior mesenteric and splenic veins. The *superior mesenteric vein* drains blood from the small intestines and portions of the large intestine and stomach. The *splenic vein* drains the spleen and receives tributaries from the stomach, pancreas, and colon. The tributaries coming from the stomach are the *coronary*, *pyloric*, and *gastro-epiploic veins*. The *pancreatic veins* come from the pancreas, and the *inferior mesenteric veins* come from the colon. Before the portal vein enters the liver, it receives the *cystic vein* from the gallbladder. Ultimately, the blood leaves the liver through the *hepatic veins*, which enter the inferior vena cava.

Pulmonary circulation

The flow of deoxygenated blood from the right ventricle to the lungs and the return of oxygenated blood from the lungs to the left atrium is called the **pulmonary circulation** (Figure 17–8). The *pulmonary artery* or *pulmonary trunk* emerges

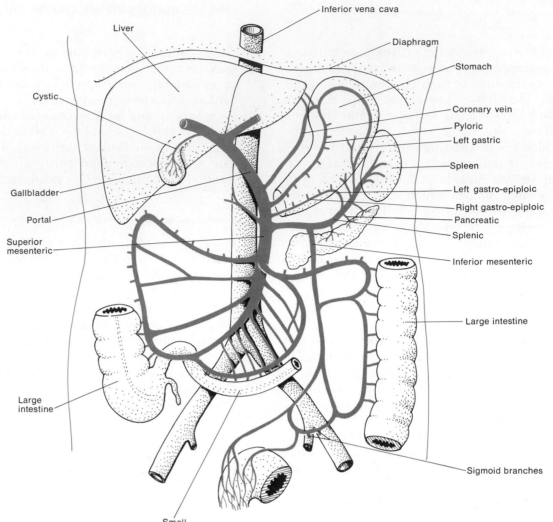

Figure 17–7. Portal circulation.

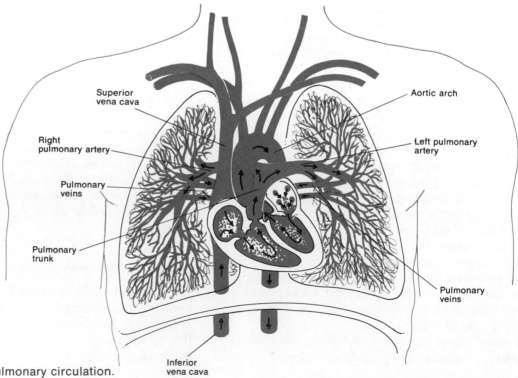

Figure 17–8. Pulmonary circulation.

from the right ventricle, passes upward, backward, and to the left. It then divides into two branches. These are the right pulmonary artery, which runs to the right lung, and the left pulmonary artery, which goes to the left lung. Upon entering the lungs, the branches divide and subdivide. They grow smaller in size and ultimately form capillaries around the air sacs in the lungs. Carbon dioxide is passed from the blood into the air sacs to be breathed out of the lungs. Oxygen breathed in by the lungs is passed from the air sacs into the blood. The capillaries then unite. They grow larger in size and become veins. Eventually, two *pulmonary veins* exit from each lung and transport the oxygenated blood to the left atrium. The pulmonary veins are the only postnatal veins that carry oxygenated blood. Contraction of the left ventricle then sends the blood into the systemic circulation.

Fetal circulation

The circulatory system of a fetus is necessarily different from that of an adult because the lungs, kidneys, and digestive tract of a fetus are nonfunctional. Instead, the fetus derives its oxygen and nutrients and eliminates its carbon dioxide and wastes through the maternal blood. This exchange of materials between fetus and mother constitutes the **fetal circulation** (Figure 17–9).

The exchange of materials between fetal and maternal circulation occurs through a structure called the *placenta*. The placenta is a large sac that encloses the fetus. It is attached to the navel of the fetus by the umbilical cord, and it communicates with the mother through countless small blood vessels that emerge from the walls of the uterus. The umbilical cord contains blood vessels that branch into capillaries within the placenta. Wastes from the fetal blood diffuse out of the capillaries, into the interstitial fluid of the placenta, and finally into capillaries belonging to the mother. Nutrients travel the opposite route, going from the maternal capillaries to the interstitial fluid of the placenta to the fetal capillaries. There is no mixing of maternal and fetal blood since all exchanges occur by diffusion through capillary walls.

Blood leaves the fetus by two *umbilical arteries*. These arteries are branches of the internal iliac arteries. The umbilical arteries pass through the umbilical cord. At the placenta, the blood picks up oxygen and nutrients and eliminates carbon dioxide and wastes. The oxygenated blood enters the fetus by way of the *umbilical vein*. This vein ascends to the liver of the fetus where it divides into two branches. Some of the blood flows through the branch that joins the portal vein and enters the liver. The fetal liver does not function, however, and it needs only enough blood for its own cells. Most of the blood, therefore, flows into the second branch, called the *ductus venosus*. The ductus venosus connects with the inferior vena cava.

Circulation through the lower portion of the fetus is not unlike the postnatal circulation. Deoxygenated blood returning from the lower regions is mingled with oxygenated blood from the ductus venosus in the inferior vena cava. This mixed blood then enters the right atrium. The circulation of blood through the upper portion of the fetus is also very similar to the postnatal flow. Deoxygenated blood returning from the upper regions of the fetus is collected by the superior vena cava, and it also passes into the right atrium.

However, unlike postnatal circulation, most of the blood does not pass into the right ventricle and go to the lungs. Because the fetal lungs do not operate, an extensive pulmonary circulation would be wasted effort. In the fetus, an opening called the *foramen ovale* exists in the septum between the right and left atrium. A valve in the inferior vena cava directs most of the blood through the foramen ovale so that it may be sent directly into the systemic circulation. The blood that does manage to descend into the right ventricle is pumped into the pulmonary artery, but only a small amount of this blood actually reaches the lungs. Most of the blood in the pulmonary artery is sent through the *ductus arteriosus*. This is a small vessel connecting the pulmonary artery with the aorta, and it enables another portion of the blood to bypass the fetal lungs. The blood in the aorta is carried to all parts of the fetus through its systemic branches. When the common iliac arteries branch into the external and internal iliacs, part of the blood flows into the internal iliacs. It then goes to the umbilical arteries and back to the placenta for another exchange of materials. Note that the only vessel which carries fully oxygenated blood is the umbilical vein.

At birth, when lung, digestive, and liver functions are established, the special structures of fetal circulation are no longer needed. Their postnatal fates are

1. The umbilical arteries atrophy to become the lateral umbilical ligament.
2. The umbilical vein becomes the round ligament of the liver.

3. The placenta is shed by the mother as the "after-birth."
4. The ductus venosus becomes a fibrous cord in the liver.
5. The foramen ovale normally closes shortly after birth.
6. The ductus arteriosus closes, atrophies, and becomes the ligamentum arteriosum.

Usually the ductus arteriosus closes shortly after birth. When it fails to close or closes imperfectly, some of the blood shuttles uselessly back and forth between the heart and lungs. This condition is called *patent ductus arteriosus.* A surgical procedure easily remedies this problem.

In addition to the heart and the veins, arteries,

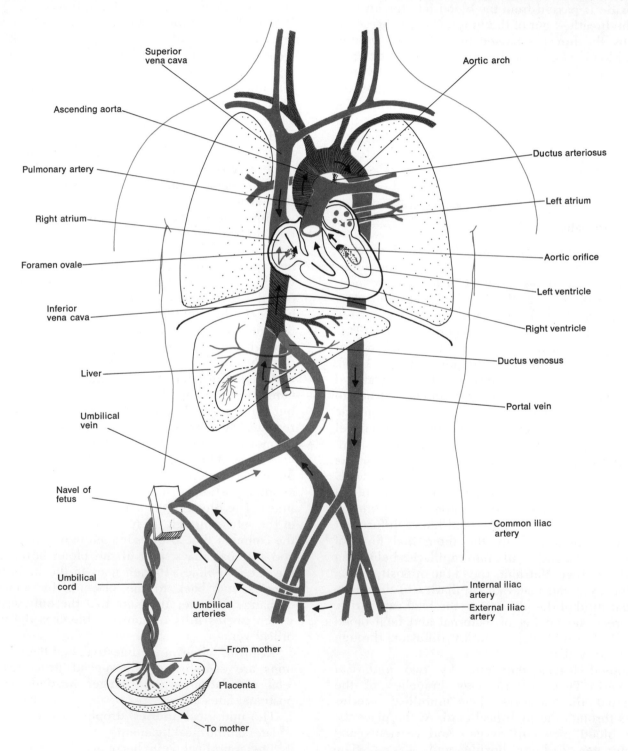

Figure 17–9. Fetal circulation.

and capillaries that route blood through the body, the circulatory system includes the structures of the lymph vascular system. Next, we shall examine the vessels, glands, and three organs of the lymph vascular system.

THE LYMPH VASCULAR SYSTEM

Lymph vessels, a series of small glands called nodes, and three organs—the tonsils, the thymus, and the spleen—make up the **lymph vascular system.** The functions of the lymph vascular system are to clean and return cellular wastes to the blood, to produce agranular leucocytes, and to develop immunities.

Lymphatic vessels

The architectural plan of the lymphatic vessels is similar to that of the blood vessels. Lymphatic vessels originate as tubes that end blindly in spaces between cells. The tubes are called **lymph capillaries.** (See Figure 16–4.) Lymph capillaries are distributed to most parts of the body. They are, however, slightly larger and more permeable than blood capillaries. Just as blood capillaries converge to form venules and veins, lymph capillaries unite to form larger and larger lymph vessels called **lymphatics** (Figure 17–10a). Lymphatics resemble veins in structure except that they have thinner walls, more valves, and contain lymph nodes at various intervals. Ultimately, lymphatics converge into two main channels— the thoracic duct and the right lymphatic duct.

The *thoracic duct*, or left lymphatic duct, begins as a dilation located in front of the second lumbar vertebra. This dilation is called the *cisterna chyli.* The thoracic duct receives lymph from the left side of the head, neck, and chest, the left arm, and the entire body below the ribs. It then empties the lymph into the left subclavian vein. A pair of valves at this juncture prevents the passage of venous blood into the thoracic duct. The *right lymphatic duct* drains lymph from the upper right side of the body and empties it into the right subclavian vein.

Lymphangiography is a procedure by which lymphatic vessels and lymph organs are filled with an opaque substance in order to be filmed. Such a film is called a *lymphangiogram.* Lymphangiograms are useful in detecting edema and carcinomas and in localizing lymph nodes for either surgical or radiotherapeutic treatment. A normal lymphangiogram of lymphatic vessels and a few nodes in the upper thighs and pelvis is shown in Figure 17–11.

Lymph nodes

The oval- or bean-shaped structures located along the lymphatics are the **lymph nodes,** or **lymph glands** (Figure 17–10b). They range from 1 millimeter to as long as 25 millimeters in length. Structurally, a lymph node contains a slight depression on one side called the *hilus.* An artery and vein enter and leave through the hilus. Each node is covered by a capsule of fibrous connective tissue that extends into the node. The capsular extensions, called trabeculae, divide the node into spaces referred to as *lymph sinuses.* The sinuses are filled with masses of lymphoid tissue called *lymph follicles.* Lymph moves into the node through several afferent vessels. It circulates slowly through the sinuses and usually leaves through a single efferent vessel.

As the lymph passes through the nodes, it is processed by the recticuloendothelial cells that line the sinuses. Reticuloendothelial cells are multifunctional. They can phagocytose microorganisms and cellular debris. They may develop into monocytes and lymphocytes. And they may develop directly into plasma cells and release antibodies. Thus, the lymph is filtered, and it picks up agranular leucocytes and antibodies. At times, however, the number of entering microbes is so great that the node itself may become infected. It then becomes enlarged and tender.

The majority of lymph nodes appear in groups or chains in certain areas of the body (Figure 17–10a). Among the larger and clinically important groups are the following: (1) The *cervical lymph nodes,* which are located by the sternocleidomastoid muscles, drain lymph from the head and neck. (2) The *submaxillary lymph nodes,* located in the floor of the mouth, drain the nose, lips, and teeth. (3) The *axillary lymph nodes* in the underarm and chest region drain the skin and muscles of the chest, including the breasts. (4) The *inguinal nodes* in the groin area drain the external genitals and the legs.

Tonsils

Tonsils are basically masses of lymphoid tissue embedded in mucous membrane. (See Figure 20–4.) Three sets are present in the body. The *pharyngeal tonsil* is located on the posterior wall of the passageway between the nose and throat. The *palatine tonsil* is situated on either side of the mouth cavity near the soft palate. This is the one commonly removed by a tonsillectomy. The *lingual tonsil* is located at the base of the tongue and may also have to be removed by a tonsillectomy. The tonsils are supplied with reticulo-

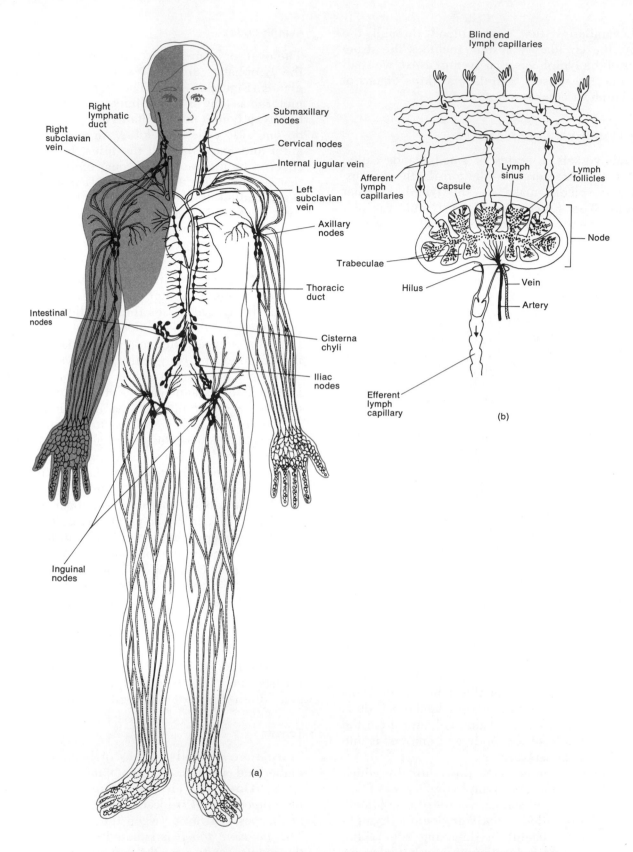

Figure 17–10. The lymph vascular system. (a) Location of the principal lymphatics and lymph nodes. The colored area indicates those portions of the body drained by the right lymphatic duct. All other areas of the body are drained by the thoracic duct. (b) Enlargement of a lymph node.

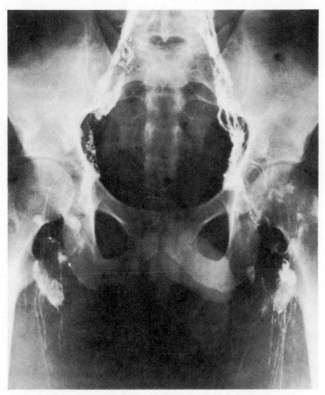

Figure 17–11. Normal lymphangiogram of the upper thighs and pelvis. Can you identify the lymphatics and lymph nodes? *(From Lester W. Paul and John H. Juhl, The Essentials of Roentgen Interpretation, 3d ed., Harper & Row, Publishers, Inc., New York, 1972.)*

endothelial cells that clean the lymph and produce lymphocytes.

Spleen

The oval-shaped *spleen* is the largest mass of lymphatic tissue in the body. It is situated in the left hypochondriac region below the diaphragm, and it is posterolateral to the stomach. The spleen is covered by a serous membrane that overlies a connective tissue capsule. Like the lymph nodes, it also contains a hilus. The splenic artery and vein and the efferent lymphatics pass through the hilus. The reticuloendothelial cells of the spleen phagocytose bacteria and worn-out red blood cells and platelets. They also produce lymphocytes, monocytes, and plasma cells. In addition, the spleen stores and releases blood in case of hemorrhage. The release seems to be purely sympathetic. The sympathetic impulses cause the spleen to contract.

Thymus gland

The thymus gland is a bilobed mass of lymphatic tissue in the upper thoracic cavity. It is found along the trachea behind the sternum. (See Figure 15–1.) The thymus is relatively large in children. It reaches its maximum size at puberty and then undergoes involution. Eventually, it is replaced by fat and connective tissue. The thymus, like the pancreas, is an organ that specializes in servicing two body systems. The pancreas serves both the endocrine and digestive systems. The thymus produces hormones that belong to the endocrine system, but it also performs a vital function as part of the lymphatic system. This function is related to immunity, the next topic of discussion.

IMMUNITY

The human body continually attempts to maintain homeostasis by counteracting various harmful stimuli in the environment. Frequently, these stimuli are disease-producing organisms, called *pathogens,* or their toxins. In general, your defenses against disease may be grouped into two broad areas: nonspecific and specific. These defenses provide you with your **immunity**—that is, your ability to overcome the disease-producing effects of certain organisms. Nonspecific defenses represent a wide variety of body reactions against a wide range of pathogens. Specific defense is the production of a specific antibody against a specific pathogen or its toxin.

Nonspecific defenses

A body reaction that deals with any of a variety of pathogens is a *nonspecific defense.* Conducting nonspecific defenses is a function of the skin, mucous membranes, phagocytic leucocytes, and reticuloendothelial cells. For example, the secretions of sweat and sebaceous glands are toxic to many microbes. The lacrimal glands of the eyes and the glands in the mucous membranes of the nose and mouth produce an enzyme called lysozyme. This enzyme is also capable of destroying many microbes. Many of the microbes that are swallowed with food are killed by the acid secretions of the stomach lining. Finally, throughout the respiratory tract, dust particles that carry microbes are trapped in the sticky mucus. Cilia that line the tract move the mucus up to the mouth where it may be swallowed or spit out. If microbes penetrate the defenses of the skin and mucous membranes, the role of the leucocytes and reticuloendothelial cells is to kill the microbes by phagocytosis.

Specific defense

Although the nonspecific defenses of the body are generally effective against microbes, they cannot fight the battle alone. Nor can they combat the toxins that are produced by pathogens. The body's second line of defense, then, is a *specific defense* that involves antibody production. As you recall, antibodies are proteins that inactivate materials called antigens. The antigens are usually proteins themselves. Pathogens and their toxins are examples of antigens. Tissues from another person, such as unmatched blood or a transplanted organ, can also be antigens. The antigen-antibody response is a very specific defense because only a particular antibody can combat a particular type of antigen.

Antibody formation

The role of the thymus gland is to teach cells how to produce antibodies. Long ago it was noticed that children who are born without a thymus cannot effectively fight disease. These children usually die from a serious infection before they are very old. It was also noticed that, if the thymus is removed from a mouse at the time of birth, the animal is unable to produce antigen-antibody responses. If the thymus is not removed until the mouse is several weeks old, the antigen-antibody responses are normal.

During the period ranging from fetal life to several years after birth, the thymus seems to set up the body's antibody system. Recall that certain wandering epithelial cells become entrapped in lymph nodes and other organs early in life. These are the cells that become agranular leucocytes or reticuloendothelial cells. Some investigators suspect that these wandering cells remain in the thymus for awhile before they settle down elsewhere in the body. During this stay, the thymus changes the DNA of the cells so that they are programmed to make antibodies. Once the thymus has programmed the wandering cells, its role is finished, and it starts to degenerate.

When an antigen enters the body sometime later in life, it stimulates the antibody-producing cells. The cells respond by producing the antibodies they have been programmed to make. The antigen, therefore, brings about its own destruction.

Active and passive immunity

Whether or not a microbe makes you sick often depends on whether you have been exposed to it before. The first time a cell is exposed to an antigen, its antibody response is a little slow to start. This is because the cell needs time to make

the proper adjustments in its protein manufacturing assembly lines. During this time, the microbes are free to multiply and produce toxins and other symptoms of the disease. If the same kind of microbe invades your body again in the future, your cells may still be geared to producing the antibody. In this case, the antigen-antibody response may occur before the microbes have a chance to bring on the symptoms of the disease. Such protection against future sickness is called *active immunity.* Active immunity may also be acquired through vaccinations with dead pathogens or with very low doses of their toxins. The proteins in the dead pathogens are capable of stimulating antibody production, but the dead microbes cannot hurt you. Toxins are given in doses just high enough to stimulate antibody-producing cells. The doses are too low, however, to cause disease.

Another form of immunity is called *passive immunity.* This happens when antibodies are injected from an animal or from another person who has previously been exposed to a disease. Examples are the globulin antibodies that are effective against hepatitis and measles. Passive immunity gives you only temporary protection. It is generally used when a person has been exposed to a disease to which he is not immune and when he does not have enough time to manufacture his own antibodies.

Many cells of the body produce antibodies. Both the lymphocytes and the reticuloendothelial cells are capable of forming plasma cells, which are good antibody producers. These cells and their antibodies circulate freely through the blood and lymph and can get at an antigen quickly. However, the life span of these cells is too short to provide immunity against future attacks. It is suspected that long-term immunity is provided by long-lived cells, such as muscle cells, which are also capable of producing antibodies.

The antigen-antibody response is essential to survival. However, under certain circumstances, it may create problems. Two such problems are related to allergies and the rejection of tissues that have been transplanted.

Allergy

When an individual becomes overly reactive to an antigen, he is said to be *allergic* or *hypersensitive.* With an allergy, the person reacts to antigens differently than he did at some earlier time or differently than do most other people. In addition, whenever an allergic reaction occurs, there is tissue injury.

The antigens that induce an allergic reaction are called *allergens*. Examples of allergens include certain foods; many antibiotics, such as penicillin; cosmetics; chemicals in plants, such as poison ivy; pollens; and even microbes. In the first stage of an allergic reaction, the cells of the body become sensitized by antibodies. That is, they become subject to injury if the allergen enters the body again at a later date. Sensitivity occurs when a person receives an initial dose of the allergen. When the allergen enters the body again in the second stage of an allergic reaction, tissue damage results.

In the allergic reaction, antibodies formed against the first dose of allergens remain attached to the cells that produced them. At this point, the cells become sensitized. When the antibodies react again with allergens introduced a second time, the antigen-antibody reaction destroys the cells as well as the allergens. Destruction of the cells triggers off several physiologic responses. Among these are the following:

1. Injured cells release histamine. Large quantities of histamine cause tissue inflammation and contraction of smooth muscle fibers. This occurs especially in the breathing tubes and blood vessels, causing them to constrict. Histamine also increases the permeability of blood vessels so that fluid moves from the vessels into the interstitial spaces, causing edema.
2. In a severe allergic reaction, *anaphylactic shock* may result. This is caused by the prolonged effects of histamine. The respiratory tubes are continuously constricted, and the occurrence of edema is accelerated. This lowers the blood volume even more. If anaphylactic shock is not counteracted, death may result. The effects of anaphylaxis may be reversed by administering epinephrine or antihistamines, drugs that inactivate histamine.

Tissue rejection

Another problem caused by the antigen-antibody response is the destruction of transplanted tissue. **Transplantation** involves the replacement of an injured or diseased tissue or organ. Usually, the body considers the proteins in the transplanted tissue or organ to be foreign. The body therefore produces antibodies against the transplanted tissues. This phenomenon is known as *tissue rejection.* Rejection of a transplant can often be avoided or reduced, if the genetic materials of the donor and recipient are similar. Rejection can also be somewhat reduced by administering drugs that inhibit the body's ability to form antibodies.

Types of transplants

The term *isograft* refers to a transplant in which the donor and recipient have identical genetic backgrounds. Isografts include transplants between identical twins. They also include the transplantation of tissues from one part of the body to another. Since the genetic makeup of the transplants is identical, or very closely related, there is no rejection. This type of transplant has been most successful.

An *allograft* refers to a transplant between individuals of the same species—but with different genetic backgrounds. The closer the relationship of the donor and recipient, the more successful the transplant. The success of this type of transplant has been moderate. Frequently, it is used as a temporary measure until the damaged or diseased tissue is able to repair itself. Skin transplants from other individuals and blood transfusions might properly be considered allografts.

The term *xenograph* refers to a transplantation between animals of different species. This type of transplantation is used primarily as a physiologic dressing over severe burns. Xenografts are primarily restricted to laboratory animals at present.

Since most people do not have an identical twin, most transplants are allografts. Allograft transplantation of vital organs is done only if the recipient's own organ is certain to fail within a short time. In most cases, allografts are eventually rejected by the recipient. The one organ allograft that has been quite successful, however, is the thymus. Children who are born without a thymus can now receive the gland from an aborted fetus. Since the thymus-deficient child cannot produce antibodies, he does not reject the transplant. Should his body reject the transplant later on, the rejection indicates that the child is now manufacturing antibodies and no longer needs the organ.

Immunosuppressive therapy

Scientists are looking for *immunosuppressive drugs* that can stop antibody destruction of the transplant but do not seriously interfere with antibody responses that protect us against microbes. To date, no single drug has been totally successful. There is, however, an exciting new development. It has been found that horses produce antibodies that react with lymphocytes of other species but that do not react with other kinds of foreign tissue. When this antibody was extracted and used in conjunction with other drugs, it increased the first-year survival rate of kidney transplant patients to 90 percent.

Implantation

The replacement of a tissue or organ with an artificial device is known as *implantation*. Artificial replacements are constructed of materials such as plastic, which do not simulate antibody production. In many cases, the artificial device functions quite well. Plastics are widely used to replace blood vessels, valves, and bones. The major problem has been to develop devices that can duplicate complex physiological activities of organs and yet remain small enough to be implanted. Artificial pacemakers, which control the beat of the heart, have been implanted for years. However, the artificial kidney, which duplicates the activities of a natural one, is much too large to be placed inside the body. It is hoped that someday the machine will be miniaturized to a point where it too can be implanted.

Medical terminology associated with the circulatory organs

Adenitis (*adeno* = gland) Enlarged, tender, and inflamed lymph nodes resulting from an infection.

Angiocardiography (*angio* = vessel; *cardio* = heart; *graph* = write, record) Recording of an image of the chambers of the heart by x-ray revealed by the direct injection of radiopaque dyes.

Arteriography Recording of an image of arteries by x-ray revealed by the direct injection of dyes.

Cardiomegaly (*megalo* = great, large) Heart enlargement.

Elephantiasis Great enlargement of a limb (especially lower limbs) and scrotum, resulting from obstruction of lymph glands or vessels; caused by a tiny parasitic worm.

Epistaxis A nosebleed.

Hematoma (*hemangioma*) (*haemo, hemato* = blood; *oma* = tumor. Leakage of blood from a vessel, which clots to form a solid mass or swelling in any tissue.

Hypersplenism Abnormal splenic activity involving highly increased blood cell destruction.

Lymphadenectomy (*ectomy* = cutting out, removal of) Removal of a lymph node.

Lymphadenopathy (*patho* = disease, suffering) Enlarged, sometimes tender lymph glands.

Lymphangitis Inflammation of the lymphatic vessels.

Lymphedema Accumulation of lymph fluid producing subcutaneous tissue swelling.

Lymphoma Any tumor composed of lymph tissue. Malignancy of reticuloendothelial cells of lymph nodes is called Hodgkin's disease.

Lymphangioma A benign tumor of the lymph vessels.

Lymphostasis (*stasis* = halt) A lymph flow stoppage.

Occlusion The closure or obstruction of the lumen of a structure, such as the lumen of a blood vessel.

Phlebitis Inflammation of a vein.

Thrombophlebitis (*thrombo* = clot) Inflammation of a vein with clot formation.

Chapter summary in outline

HEART

1. The pericardium, consisting of an outer fibrous layer and an inner serous layer, encloses the heart.

2. The walls of the heart have three layers. The chambers include two upper atria and two lower ventricles.

3. All valves of the heart prevent the back flow of blood. The blood flows through the heart from the superior and inferior vena cavae, to the right atrium, through the tricuspid valve to the right ventricle, through the pulmonary artery to the lungs, through the pulmonary veins to the left atrium, through the bicuspid valve to left ventricle, and out through the aorta.

BLOOD VESSELS

1. Arteries carry blood away from the heart. They are stronger and thicker than veins, consisting of a tunica interna, tunica media (which maintains elasticity and contractility), and tunica externa.

2. Many arteries anastomose, which means that the distal ends of two or more vessels unite. An alternate blood route from an anastomosis is called collateral circulation.

3. Capillaries are microscopic blood vessels through which materials are exchanged between blood and interstitial fluid. They unite to form venules, which in turn form veins to carry blood back to the heart.

4. Veins have less elastic tissue and smooth muscle than arteries, and they contain valves to prevent back flow of blood.

5. Weak valves can lead to varicose veins or hemorrhoids.

CIRCULATORY ROUTES

1. The systemic circulation takes oxygenated blood from the left ventricle through the aorta to all parts of the body except the lungs. It includes the coronary and portal circulations.

2. The coronary circulation takes oxygenated blood through the arterial system of the myocardium. Deoxygenated blood returns to the right atrium via the coronary sinus. Complications of this system are angina pectoris and myocardial infarction.

3. The portal circulation takes blood from the veins of the pancreas, spleen, stomach, intestines, and gallbladder to the portal vein of the liver. It enables the liver to utilize nutrients and detoxify harmful substances in the blood.

4. The pulmonary circulation takes deoxygenated blood from the right ventricle to the lungs and returns oxygenated blood from the lungs to the left atrium. It allows blood to be oxygenated for systemic circulation.

5. The fetal circulation involves the exchange of materials between the fetus and mother through the placenta.

LYMPH VASCULAR SYSTEM

1. This system consists of lymph vessels, lymph, lymph nodes, and lymph organs.

2. Lymphatic vessels are similar in structure to veins. All lymphatics deliver lymph to either the thoracic duct or right lymphatic duct.

3. Lymph nodes are oval-shaped structures located along lymphatics. Lymph passing through the nodes is filtered, and it picks up antibodies and agranular leucocytes.

4. Lymph organs that filter lymph and add white blood cells and antibodies are the tonsils, spleen, and thymus gland.

IMMUNITY

1. The ability to overcome the harmful effects of disease-producing organisms is provided by nonspecific and specific defenses.

2. Nonspecific defenses include the structure of the skin and mucous membranes and their secretions; the inflammatory response and phagocytosis; and the reticuloendothelial system.

3. Specific defenses are the production of antibodies against a specific pathogen or its toxin. Antibody-producing cells are programmed by the thymus.

4. An allergy is an overreaction to an antigen called an allergen. In allergic reactions, histamines are released by damaged tissues and bring about physiological responses that could cause anaphylactic shock.

5. Tissue rejection is the inactivation of foreign antigens (transplants) by antibodies. It is counteracted by the use of immunosuppressive drugs.

6. Types of transplants include isografts, allografts, and xenographs.

Review questions and problems

1. Describe the location of the heart in the mediastinum. Distinguish the subdivisions of the pericardium. What is the purpose of this structure?

2. Compare the three portions of the wall of the heart. Define atria and ventricles. What vessels enter or exit the atria and ventricles?

3. Discuss the principal kinds of valves in the heart and how they operate.

4. Describe the structural and functional differences among: arteries, capillaries, and veins.

5. Discuss the importance of the elasticity and contractility of arteries. What is an anastomosis?

6. Define varicose veins and hemorrhoids.

7. What is meant by a circulatory route? Define systemic circulation.

8. By means of a diagram, indicate the major divisions of the aorta, their principal arterial branches, and the regions supplied.

9. Trace a drop of blood from the aortic arch through its systemic circulatory route and back to the heart again. Remember that the major branches of the arch are the brachiocephalic artery, left common carotid artery, and left subclavian artery. In your answer, be sure to indicate which veins return the blood to the heart.

10. What is the circle of Willis? Why is it important?

11. What are visceral branches of an artery? Parietal branches?

12. What major organs are supplied by branches of the thoracic aorta? How is blood returned from these organs to the heart?

13. What organs are supplied by the celiac artery, superior mesenteric, renal, inferior mesenteric, inferior phrenic, and middle sacral? How is blood returned from these organs to the heart?

14. Trace a drop of blood from the common iliac arteries through their branches to the respective organs and back to the heart again.

15. What is a deep vein? A superficial vein? Define a venous sinus in relation to blood vessels. What are the three major groups of systemic veins?

16. Describe the route of blood in the coronary circulation. Distinguish between angina pectoris and myocardial infarction.

17. What is portal circulation? Describe the route by means of a diagram. Why is this route important?

18. Define pulmonary circulation. Prepare a diagram to indicate the route. What is the purpose of the route?

19. Discuss in detail the anatomy and physiology of fetal circulation. Be sure to indicate the function of the umbilical arteries, umbilical vein, ductus venosus, foramen ovale, and ductus arteriosus.

20. What is the fate of the special structures involved in fetal circulation once postnatal circulation is established?

21. Describe the cause and treatment of patent ductus arteriosus.

22. Identify the components of the lymph vascular system.

23. How do lymphatic vessels originate? Compare veins and lymphatics with regard to structure.

24. Construct a diagram to indicate the role of the thoracic duct and right lymphatic duct in draining lymph from different regions of the body.

25. What is a lymphangiogram? What is its diagnostic value?

26. Describe the structure of a lymph node. What functions do lymph nodes serve?

27. Identify four groups of clinically important lymph nodes.

28. Compare the functions of the tonsils, spleen, and thymus gland as lymphatic organs.

29. Define immunity. Distinguish between nonspecific and specific defenses that provide immunity.

30. Contrast the functions of the skin and mucous membranes, blood, and reticuloendothelial system in providing nonspecific defenses against pathogens.

31. Distinguish between active and passive immunity.

32. Describe the role of the thymus gland in antibody production.

33. Describe in detail the allergic reaction.

34. Define the various kinds of transplants.

35. Explain the term rejection as it applies to transplants. What immunosuppressive techniques are used to overcome rejection?

36. Refer to the glossary of medical terminology associated with the circulatory organs. Be sure you can define each term.

CHAPTER 18

THE CIRCULATORY SYSTEM: CARDIOVASCULAR PHYSIOLOGY

STUDENT OBJECTIVES

After you have read this chapter, you should be able to:

1. Define systole and diastole as the two principal events of the cardiac cycle

2. Describe the pressure changes associated with blood flow through the heart

3. Describe the events of the cardiac cycle as a function of time

4. Describe the sounds of the heart and their clinical significance

5. Describe the initiation and conduction of nerve impulse through the nodal system of the heart

6. Label and explain the deflection waves of a normal electrocardiogram

7. Define cardiac output and identify those factors that determine it

8. Define Starling's "law of the heart"

9. Contrast the effects of sympathetic and parasympathetic stimulation of the heart

10. Define the role of pressoreceptors and chemoreceptors in the control of the rate of heartbeat

11. Describe why blood flows through vessels

12. List the factors that resist the flow of blood through vessels

13. List the factors that assist the return of venous blood to the heart

14. List the forces responsible for the circulation of lymph

15. Define pulse and identify those arteries where pulse may be felt

16. Compare the several kinds of abnormal pulse rates

17. Define blood pressure

18. Describe one clinical method for recording systolic and diastolic pressure

19. Contrast the clinical significance of systolic, diastolic, and pulse pressures

20. List the causes and symptoms of aneurysms, arteriosclerosis, and hypertension

21. Describe the diagnosis of arteriosclerosis by the use of angiography

22. Define inadequate blood supply, faulty architecture, and malfunctions of conduction as primary reasons for heart trouble

23. Describe patent ductus arteriosus, septal defects, and valvular stenosis as congenital heart defects

24. List the four abnormalities of the heart present in tetralogy of Fallot

25. Define atrioventricular block, atrial flutter, atrial fibrillation, and ventricular fibrillation as abnormalities of the conduction system of the heart

26. Define circulatory shock

27. Describe the homeostatic mechanisms that compensate for circulatory shock

28. Draw a shock cycle to illustrate the effects of severe shock on the circulatory organs

29. Describe the use of hypothermia, the heart-lung bypass, artificial parts, and cardiac catheterization in cardiovascular surgery

30. Discuss the present status of heart transplants

31. Define medical terminology associated with the cardiovascular system

After discussing some of the pertinent structural features of the circulatory system, we can now look at the system in action. We shall attempt to answer questions such as: What is a heartbeat? How is a heartbeat recorded? What factors regulate heartbeat? Why does blood flow through vessels? What are meanings of pulse and blood pressure? We shall also examine a group of cardiovascular disorders that account for more than half of all deaths. Finally, we shall take a look at some of the challenging and exciting techniques currently used by cardiovascular surgeons.

CARDIAC PHYSIOLOGY

The first aspect of cardiovascular physiology we shall study is the action of the heart—the action that sustains life every second of every day. Let us see why the heartbeat is so vital to survival.

The cardiac cycle

In a normal heartbeat, the two atria contract simultaneously while the two ventricles relax. Then, when the two ventricles contract, the two atria relax. The term **systole** refers to the phase of contraction, and the term **diastole** refers to the phase of relaxation. A **cardiac cycle,** or complete heartbeat, consists of the systole and diastole of both atria, plus the systole and diastole of both ventricles followed by a short pause.

For purposes of discussion, we shall take the *atrial systole* as the starting point in the cardiac cycle (Figure 18–1a). During this period, the atria contract and force blood into the ventricles. Deoxygenated blood from the right atrium passes into the right ventricle through the open tricuspid valve. And oxygenated blood passes from the left atrium into the left ventricle through the open mitral valve. While the atria are contracting, the ventricles are in diastole. During *ventricular diastole,* the ventricles are filling with blood, and the semilunar valves in the aorta and pulmonary artery are closed.

When atrial systole and ventricular diastole are completed, the events are reversed. That is,

the atria go into diastole, and the ventricles go into systole. During *atrial diastole*, deoxygenated blood from the various parts of the body enters the right atrium through the superior vena cava, inferior vena cava, and coronary sinus. Simultaneously, oxygenated blood from the lungs enters the left atrium through the pulmonary veins. During the first part of atrial diastole, the atrioventricular valves are closed since the ventricles are in systole. In *ventricular systole,* the ventricles contract and force blood into their respective vessels. The right ventricle pumps deoxygenated blood to the lungs through the semilunar valve of the pulmonary artery. The left ventricle pumps oxygenated blood through the open semilunar valve of the aorta. At the end of the ventricular systole, the semilunar valves close, and both the atria and the ventricles relax.

Notice that two phenomena control the movement of blood through the heart. These are the opening and closing of the valves and the contraction and relaxation of the myocardium. Both these activities occur without any direct stimulation from the nervous system. The valves are controlled by pressure changes that occur within each heart chamber. The contraction of the cardiac muscle is stimulated by nervelike tissue that lies in the walls of the heart. Impulses from the nervous system influence only the rate of the heartbeat.

Pressure changes in the cycle

The pressure developed in a heart chamber is related primarily to the size of the chamber and the volume of blood it contains. The greater the volume of blood, the higher the pressure. As we discuss the pressure changes during the cardiac cycle, refer to Figure 18–1b. Here, pressure changes associated with the left side of the heart are indicated. Although the pressures in the right side of the heart are somewhat lower, the same results are achieved. The general principle to keep in mind during this discussion is that blood flows from an area of higher pressure to an area of lower pressure.

The pressure in the atria is referred to as *intraatrial pressure*. When the atria are in diastole, the pressure within them steadily increases. This happens because blood flows into the atria continuously from their respective vessels. As soon as the ventricles go into diastole, the atrioventricular valves open. Some of the blood drains into the ventricles, which are empty and consequently have a lower pressure. When atrial pressure builds up to point *a* in Figure 18–1b, the atria contract and send the remaining blood rushing into the ventricles.

As the ventricles fill with blood, their pressure, called *intraventricular pressure*, increases. As this pressure increases, it becomes greater than intraatrial pressure. At this point—point *b* in Figure 18–1b—a backflow of blood closes the atrioventricular valves. The ventricles start to contract, but the semilunar valves remain closed from the previous cardiac cycle. Now, no blood moves out into the aorta or pulmonary artery. The ventricles are closed chambers at this time.

Intraventricular pressure continues to rise until it is greater than the pressure in the aorta and pulmonary artery. This is the intraarterial pressure. At this point, the pressure forces the semilunar valves open, and blood is ejected from the ventricles into the great vessels.

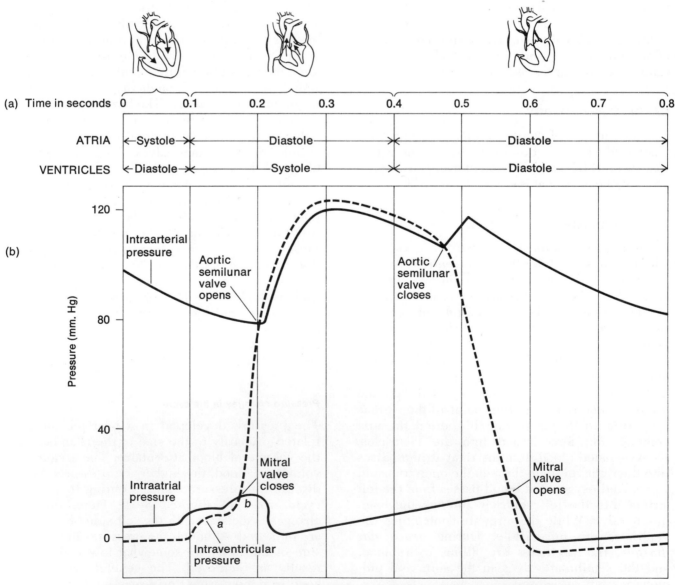

Figure 18–1. The cardiac cycle. (a) Systole and diastole of the atria and ventricles. (b) Intraatrial, intraventricular, and intraarterial pressure changes during the cardiac cycle. Note the relationship of the events of the cardiac cycle to time.

Once the ventricles eject their blood, intraventricular pressure decreases and falls below that in the great vessels. As a result, the semilunar valves are pushed closed by blood attempting to return from the vessels into the ventricles. Once again, the ventricles are closed chambers. As the ventricles relax, intraventricular pressure decreases until it becomes less than intraatrial pressure. At that point, intraatrial pressure forces the tricuspid and mitral valves open, blood fills the ventricles, and another cycle begins.

Timing of the cycle

We can now relate the events of the cardiac cycle to time. If we assume that the average heart beats 72 times per minute, then each beat with its short pause requires about 0.8 second (Figure 18–1a). During the first 0.1 second, the atria are contracting, and the ventricles are relaxing. The atrioventricular valves are open, and the semilunar valves are closed. For the next 0.3 second, the atria are relaxing and the ventricles contracting. During the first part of this period, all valves are closed, and during the second part the semilunars are open. The last 0.4 second of the cycle is the relaxation, or quiescent, period. All chambers are in diastole. And for the first part of the quiescent period, all valves are closed. During the latter part of the relaxation period, the atrioventricular valves open, and blood starts draining into the ventricles. When the heart beats at a faster rate than normal, the quiescent period is shortened accordingly.

Sounds of the heart

If you place the bell of a stethoscope on the surface of the skin about an inch below and a little to the median side of the left nipple, you will hear two distinct sounds. The first sound, which can be described as a *lubb* ($\overline{oo}$) sound, is a comparatively long, booming sound. The lubb is the sound of the atrioventricular valves closing soon after ventricular systole begins. The second sound, which is heard as a short, sharp sound, can be described as a *dup* ($\breve{u}$) sound. Dup is the sound of the semilunar valves closing toward the end of ventricular systole. A pause about two times longer comes between the second sound and the first sound of the next cycle. Thus, the cardiac cycle can be heard as a lubb, dup, pause; lubb, dup, pause; lubb, dup, pause. This is the sound of the heartbeat. But note that it comes from the closure of the valves and not from the contraction of the heart muscle.

Heart sounds provide valuable information about the valves of the heart. If the sounds are peculiar, they are referred to as *murmurs*. Murmurs are frequently the noise made by a little blood bubbling back up into an atrium because of the failure of one of the atrioventricular valves to close properly. However, murmurs do not always indicate a valve problem, and many have no clinical significance.

The conduction system

Skeletal and smooth muscle must receive impulses from the nervous system to initiate their contraction. Cardiac muscle is different. The heart is hooked up to the autonomic nervous system, but the autonomic neurons only increase or decrease the time it takes to complete a cardiac cycle. The walls of the chambers can go on contracting and relaxing, contracting and relaxing, without any direct stimulus from the nervous system. This is because the heart has a type of built-in private nervous system called the **conduction system**. The conduction system is composed of specialized tissues that generate the electrical impulses which stimulate the cardiac muscle fibers to contract. These tissues are the sinoatrial node, the atrioventricular node, the atrioventricular bundle, and the Purkinje fibers. The cells of the conduction system develop during embryological life from certain cardiac muscle cells. These cells lose their ability to contract and become specialists in impulse transmission.

A *node* is a compact mass of conducting cells. The *sinoatrial node*, known as the *SA node* or *pacemaker*, is located in the right atrium beneath the opening of the superior vena cava (Figure 18–2a). The SA node initiates each cardiac cycle and thereby sets the basic pace for the heart rate. This is why it is commonly called the pacemaker. However, the rate set by the SA node may be altered by nervous impulses from the autonomic nervous system or by certain hormones such as thyroid hormone or epinephrine. Once an electrical impulse is initiated by the SA node, the impulse spreads out over both atria and causes them to contract. From here, the impulse passes to the *atrioventricular (AV) node*, located toward the bottom of the interatrial septum. From the AV node, a tract of conducting fibers called a *bundle* runs to the top of the interventricular septum and then down both sides of the septum. This is called the *atrioventricular bundle*, or *bundle of His*. The bundle of His distributes the charge over the medial surfaces of the ventricles. Actual contraction of the ventricles is stimulated

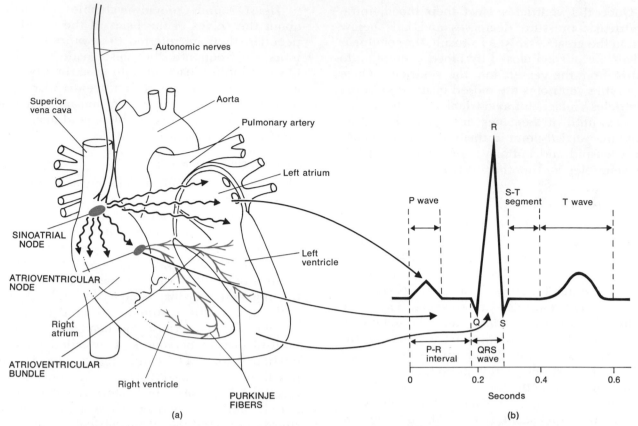

Figure 18–2. Conduction system of the heart. (a) Location of the nodes and bundles of the conduction system. (b) Recordings of a normal electrocardiogram. The significance of the recordings is as follows: P wave: passage of impulse from sinoatrial node through atria. P-R interval: time required for impulse to pass through atria, atrioventricular node, atrioventricular bundle, and Purkinje fibers. QRS wave: passage of the impulse through ventricles. S-T segment: time between end of spread of impulse and relaxation of ventricles. T wave: ventricular relaxation.

by the *Purkinje fibers.* The Purkinje fibers are individual conducting cells that emerge from the bundle of His and pass into the cells of the myocardium.

The electrocardiogram

Impulse transmission through the conduction system generates electrical currents that may be detected on the surface of the body. A recording of the electrical changes that accompany the cardiac cycle is called an **electrocardiogram (ECG).** The instrument used to record the changes is an *electrocardiograph.*

Each portion of the cardiac cycle produces a different electrical impulse. These are transmitted from the electrodes to a recording needle that graphs the impulses as a series of up-and-down waves called *deflection waves.* In a typical record (Figure 18–2b), three clearly recognizable waves accompany each cardiac cycle. The first wave,

called the *P wave,* is a small upward wave. It indicates the spread of an impulse from the SA node over the surface of the two atria. A fraction of a second after the P wave begins, the atria contract. Following this, there is a complex called the *QRS wave.* It begins as a downward deflection, continues as a large, upright, triangular wave, and ends as a downward wave at its base. This deflection represents the spread of the electrical impulse through the ventricles. The third recognizable deflection is a dome-shaped wave called the *T wave.* This wave indicates ventricular repolarization (relaxation).

In reading an electrocardiogram, it is exceedingly important to note time relationships between various waves. For example, refer to Figure 18–2b, and note the P–R interval. This interval, measured from the beginning of the P wave to the beginning of the Q wave, represents the conduction time from the beginning of atrial excitation to the beginning of ventricular excitation. The

P–R interval is the time required for an impulse to travel through the atria and atrioventricular node, to the atrioventricular bundle, and Purkinje fibers. The lengthening of this interval indicates partial blockage of conduction at the atrioventricular bundle. Other intervals and their significance are also indicated in the figure. The ECG is invaluable in diagnosing abnormal cardiac rhythms and conduction patterns, detecting the presence of fetal life, and determining multiple pregnancies.

Cardiac output

The heart is capable of living a rather independent life. But its output is nevertheless regulated by events occurring in the rest of the body. Cells must receive a certain amount of oxygenated blood each minute in order to maintain health and life. When they are very active, as during exercise, they need even more blood. When you are asleep, cellular need is reduced, and the heart cuts back on its output.

The amount of blood ejected from the left ventricle into the aorta per minute is called the **cardiac output,** or minute volume. Cardiac output is determined by two factors: (1) the amount of blood that is pumped by the left ventricle during each beat, and (2) the number of heartbeats per minute. The amount of blood ejected by a ventricle during each systole is called the **stroke volume.** In a resting adult, stroke volume averages 70 milliliters, and heart rate is about 72 beats per minute. The average cardiac volume, then, in a resting adult is

$$\text{Cardiac output} = \text{stroke volume} \times \text{ventricular systole/minute}$$
$$= 70 \text{ ml.} \times 72/\text{min.}$$
$$= 5,040 \text{ ml./min.}$$

In general, any factor that increases the heart rate or increases its stroke volume tends to increase cardiac output. Factors that decrease the heart rate or its stroke volume tend to decrease cardiac output. If stroke volume falls dangerously low, the body can compensate to some extent by increasing the heartbeat and vice versa.

Stroke volume

Stroke volume is determined by the force of the ventricular contraction. The more strongly the cardiac fibers contract, the more blood they eject. The strength of contraction is directly related to the amount of venous blood that is returned to the heart.

You may recall from Chapter 10 that, within limits, the contraction is more forceful when skeletal muscle fibers are stretched. The same principle applies to cardiac muscle. During exercise, for example, a large amount of blood enters the heart, and the increased diastolic filling stretches the fibers of the right ventricle. This increased length of the cardiac muscle fibers intensifies the force of the ventricular contraction—that is, the force of the beat. The result is that the increased incoming volume of blood is handled by an increased output through a more forceful ventricular contraction. As the increased amount of blood returns from the lungs to the left side of the heart, left ventricular stroke volume also increases. Thus, during exercise, cardiac output is increased. This phenomenon, by which the length of the cardiac muscle fiber determines the force of contraction, is referred to as **Starling's "law of the heart."**

In normal situations, the amount of blood returning to the heart regulates the stroke volume and, thereby, affects cardiac output. However, during certain pathological conditions, stroke volume may fall to a dangerously low level. For instance, if the ventricular myocardium is weak, or if it is damaged by an infarction, it cannot contract strongly. Or, blood volume may be reduced by excessive bleeding. Stroke volume then falls because the cardiac fibers are not sufficiently stretched. In these cases, the body attempts to maintain a safe cardiac output by increasing the rate of contraction. The rate of the heartbeat is controlled by the autonomic and endocrine systems.

Regulation of heart rate

Left to its own devices, the pacemaker sets a steady heart rate that never varies, regardless of the needs of the body. Consequently, a number of reflexes exist to quicken the heartbeat when tissues need more oxygen and to slow down the heart during periods of relative inactivity. The reflex arcs start in receptors located in blood vessels, pass to cardiac centers in the brain, and finally over autonomic nerves to the heart.

AUTONOMIC CONTROL. The pacemaker receives nerves from both the parasympathetic and sympathetic divisions of the autonomic nervous system. The parasympathetic neurons originate in the **cardioinhibitory center** of the medulla and travel with the vagus nerve to the heart. Stimulation of the pacemaker by the parasympathetic fibers slows down the cardiac cycle. The sympathetic pathway originates in the **cardioaccelera-tory center** of the medulla. It travels in a tract down the spinal cord and then passes over sympathetic nerves to the heart. Sympathetic stimulation counteracts parasympathetic stimulation and quickens the heartbeat. When neither of the cardiac centers is stimulated by sensory neurons, the cardioacceleratory center tends to dominate. The result is that sympathetic fibers have a free reign to speed up heart rate until receptors intervene to stimulate the cardioinhibitory center.

Pressoreceptors. Nerve cells that are capable of responding to changes in pressure are called **pressoreceptors.** We shall consider how three of these pressoreceptors can control the rate of heartbeat. The first of these belongs to the *carotid sinus reflex,* which is particularly concerned with maintaining normal blood pressure in the brain. The *carotid sinus* is a small widening of the internal carotid artery just above the point where it branches off from the common carotid. In the walls of the carotid sinus lie pressoreceptors. Any increase in blood pressure stretches the walls of the sinus, and the stretching stimulates the pressoreceptors. The impulses then travel from the pressoreceptors to the cardiac centers in the medulla. There, the impulses stimulate the cardioinhibitory center and inhibit the cardioacceleratory center. Consequently, more impulses pass from the cardioinhibitory center to the heart, and fewer impulses pass from the cardioacceleratory center to the heart. The result is a slowing of the heartbeat. There is a subsequent decrease in cardiac output, a decrease in arterial blood volume, and the restoration of blood pressure to normal.

But if blood pressure falls, reflex acceleration of the heart takes place. In this situation, the pressoreceptors in the carotid sinus do not stimulate the cardioinhibitory center, and the cardiac acceleratory center is free to dominate. The heart then beats faster to restore normal blood pressure.

The two other reflexes that we shall consider are the aortic reflex and the right heart reflex. The *aortic reflex* is concerned with general systemic blood pressure. It is initiated by pressoreceptors that lie in the walls of the aortic arch, and it operates as does the carotid sinus reflex. The *right heart (atrial) reflex* controls venous blood pressure. It is initiated by pressoreceptors located in the superior and inferior vena cavae and in the right atrium. It too operates in the same way as the carotid sinus reflex.

Chemoreceptors. Structures that are sensitive to chemicals are called **chemoreceptors.** For example, certain chemoreceptors in the blood vessels are sensitive to oxygen. When blood oxygen is low, they become stimulated. Impulses are conveyed to the cardiac center, and the heart rate is accelerated. This action increases cardiac output, and more blood is pumped to the lungs and to the cells of the body. Other chemoreceptors are sensitive to carbon dioxide concentrations. Any increase in the CO_2 concentration in the blood brings about inhibition of the cardioinhibitory center. The heart rate consequently increases.

OTHER INFLUENCES ON HEART RATE. It should be noted also that other factors may influence heart rate. Strong emotions, such as fear, anger, and anxiety, along with a multitude of physiological stressors, increase heart rate through the general stress response. For some reason, mental states, such as depression, tend to stimulate the cardioinhibitory center and decrease heart rate.

Certain chemicals produced by the body may act directly on the heart. Those that increase heart rate are also involved in the general stress response. For instance, thyroxin, a hormone produced by the thyroid gland, increases overall body metabolism, including that of the cells of the heart. Epinephrine (adrenalin) produced by the adrenal glands imitates sympathetic effects and increases the force and frequency of the heartbeat. Norepinephrine, the chemical transmitter of postsynaptic sympathetic fibers, also stimulates the pacemaker to accelerate. As you recall, norepinephrine is another adrenal cortex hormone. By contrast, acetylcholine, the chemical transmitter of parasympathetic neurons, decreases heart rate.

Still other factors affect the heartbeat. Sex is one factor, the heartbeat being somewhat faster in females. Age is another factor. The heartbeat is fastest at birth, moderately fast in youth, average in adulthood, and below average in old age. Muscular exercise is one more factor, with heartbeat increasing in proportion to the work done.

WHAT MAKES BLOOD FLOW?

Blood flows through its system of closed vessels because of differences in pressure in different parts of the system. It always flows *from regions of higher pressure to regions of lower pressure.* The mean (average) pressure in the aorta is about 100 millimeters Hg. This pressure continually decreases from the aorta through the arteries and arterioles to about 45 millimeters Hg (Figure

18–3). Capillary blood pressure averages from 10 to 30 millimeters Hg, which is greater than the pressure in venules but less than the pressure in arterioles. Thus, blood flows from the arterioles into the capillaries. In the venous system, the pressure falls very slowly from about 10 millimeters Hg in venules, to less than 10 millimeters Hg in veins, to 0 millimeters Hg in the right atrium. Thus, because of a continuous drop in pressure, blood flows from arteries to arterioles to capillaries to venules to veins to the heart. Other mechanisms also aid the flow of blood. For instance, when blood leaves the capillaries, it enters the venules and veins, which are larger in diameter and thereby offer less resistance to flow. Contraction of the skeletal muscles that surround the veins also helps to milk the blood back toward the heart.

How pressure develops

Before we explain how a pressure gradient is established throughout the circulatory system, you first need to learn how blood pressure occurs. The pressure of the blood is determined by two factors: the volume of blood delivered to a vessel in a given amount of time and the amount of re-

sistance exerted against the flow. These two factors and their relationship to pressure can be summarized as follows:

$$\text{Pressure} = \text{flow (vol/min.)} \times \text{resistance}$$

Note that the pressure is increased if either the flow or the resistance is increased. Although you may not have been aware of this formula, you have made use of it countless times in your everyday life. For instance, when you narrow the diameter of the nozzle on the garden hose, the water comes out in a narrower stream under higher pressure. By decreasing the diameter of the opening, you have increased the resistance to the flow of the water. City children use the same principle when they put their hands over a fire hydrant to change a large flowing stream of water into a spray under higher pressure. The hand against the moving water is the resistance.

In the circulatory system, the pressure is produced by the amount of blood pumped out of the heart in a minute and by the resistance offered by the walls of the blood vessels. Either an increase in cardiac output or a narrowing of the walls of the vessels will increase the pressure.

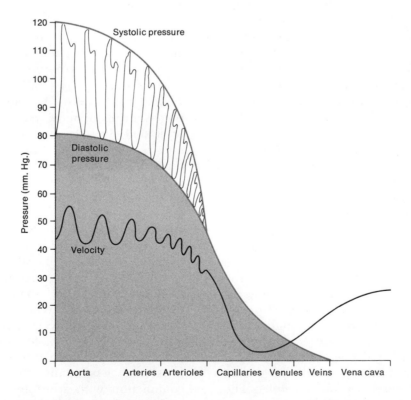

Figure 18–3. Relationship between blood pressure, blood vessel size, and blood velocity.

Heart action and blood volume

Cardiac output is the principal determinant of blood pressure. Without a delivery of blood to the vessels, there would be no pressure and no blood flow other than a slow oozing produced by the force of gravity. Cardiac output is dependent on the strength of the ventricular contraction and the rate of the heartbeat and on the amount of blood available to be pumped. The normal volume of blood in a human body is about 5 quarts. Any decrease in this volume, such as occurs from hemorrhage, decreases the amount of blood that is circulated through the arteries in each minute. As a result, blood pressure drops. Conversely, anything that increases blood volume, such as high salt intake and therefore water retention, increases blood pressure.

Resistance

The force applied by the vessel walls against the flowing blood is referred to as *resistance*. Resistance can be exerted in a number of different ways. And the body can also increase or decrease this part of the formula in order to adjust blood pressure.

EXTENSIBILITY AND ELASTICITY OF ARTERIAL WALLS. During ventricular systole, blood is forced from the ventricles into already filled arteries. To receive this extra supply of blood, the arterial walls, because of their extensibility, stretch outward. As soon as the force of systole is removed, the arterial walls, because of their elasticity, push back inward to resume their normal shape. The pushing inward puts pressure (resistance) on the blood and forces it into the capillaries.

The resistance offered by the inward push of the arterial walls does not increase blood pressure because it is cancelled out by the lack of resistance during distention. The purpose of the extensibility and elasticity of the arterial walls is to keep the blood pressure from fluctuating with every systole and diastole. As a spurt of blood comes out of the ventricle, it is captured in the distended arterial walls. During diastole, the inward push of the walls keeps the blood flowing into the capillaries. Blood is thereby funneled into the capillaries at an even rate throughout the cardiac cycle. When arteries lose their extensibility and elasticity and become rigid, large amounts of blood spurt into the capillaries during systole, and flow ceases during diastole. The spurting of blood into the capillaries can easily damage their thin, fragile walls.

VISCOSITY. Another form of resistance is the *viscosity*, or thickness, of the blood itself. In a thick fluid, molecules tend to be attracted to each other and resist flowing. As you know from experience, molasses pours more slowly than water even when the same force of gravity is applied. Any condition that increases the thickness of blood also increases blood pressure. One such condition is an unusually high number of red blood cells. A depletion of the plasma protein albumin, which is also responsible for the thickness of blood, produces a decrease in blood pressure.

PERIPHERAL RESISTANCE. The walls of the blood vessels themselves offer the most important force of resistance to blood flow. As blood is pushed into a vessel, it tries to spread out. The firm walls of the vessel offer resistance to spreading and channel the blood into a narrow stream. This is called *peripheral resistance*. The smaller the diameter of a vessel, the more resistance it offers to the blood. A very important function of arterioles is to control peripheral resistance and, therefore, blood pressure by changing their diameters. The center for this regulation is called the *vasomotor center*, located in the medulla. The term *vas* relates to vessel, whereas *motor* means mover or movement. Functionally, the vasomotor center consists of a vasoconstrictor and a vasodilator center.

Vasoconstrictor nerves conduct impulses from the medulla to the spinal cord and through sympathetic nerves to the arterioles. The impulses bring about a reduction in the diameter of blood vessels by stimulating the smooth muscle to contract. Normally, the nerves continually send impulses to the smooth muscle of arterioles. As a result, the arterioles are kept in a state of tonic contraction. However, if there is an increase in blood pressure, the vasoconstrictor nerves are inhibited. The arteriole diameter increases and blood pressure falls due to the reduction in peripheral resistance.

Vasodilator nerves, on the other hand, carry impulses that bring about an increase in the diameter of vessels. When there is an increase in blood pressure, the vasodilator nerves are stimulated, arteriole diameter increases, and blood pressure falls due to the decrease of peripheral resistance. Essentially, the vasomotor center, in conjunction with arterioles, adjusts the pressure with which blood flows in response to varying needs of the body.

Production of the pressure gradient

We said earlier that blood always flows from an area of greater pressure to an area of lesser pressure. What causes this continual decrease in pressure that directs the flow of the blood? Although resistance increases pressure at the spot where the resistance occurs, the fact is that the overall effect is a slow loss of pressure. The resistance of the walls of the arteries against the blood produces friction. And any time friction occurs some of the kinetic energy of the flowing fluid is converted to heat energy and escapes. Thus, as the blood passes through the arteries to the capillaries and on to the veins, it continually loses pressure as it rubs against the walls of the vessels.

Factors that aid circulation

The establishment of a pressure gradient is the primary reason why blood flows. But a number of other factors help the blood to return through the veins. These include an increased rate of flow, contractions of the skeletal muscles, and valves.

Rate of flow

If we rewrite the formula for pressure, we shall see that the less resistance offered the blood, the faster it flows. Thus,

$$\text{Flow} = \frac{\text{pressure}}{\text{resistance}}$$

A larger vessel offers less peripheral resistance than a smaller one. Thus, blood flows fastest through the aorta, which has the largest diameter and therefore the least peripheral resistance of any vessel in the body. (See Figure 18–3.) The blood flows most slowly through the capillaries, which have minute diameters and offer a great deal of resistance. The slow flow through the capillaries allows time for the exchange of materials and is another example of how body structure is admirably suited to function. But as the blood reaches the larger venules and finally the very large veins, it picks up more and more speed even though the pressure is decreasing. This is because the diameter of the veins increases as the blood nears the heart. The larger diameters of the veins offer much less resistance, which more than compensates for the progressive loss in blood pressure.

If we watch the blood circulate through the body, we see that blood pressure drops con-tinually until it reaches zero in the right atrium. The rate of flow, on the other hand, drops continuously until it reaches the venules. Then it rises until it reaches the atrium.

Muscles and valves

The contraction of skeletal muscles is very important in returning blood to the heart. As the muscles contract, they tighten around the walls of the veins running through them. This provides a pressure that squirts the blood forward from one part of a vein to another so that blood is moved upward toward the heart. This action is called milking. Individuals who are immobilized through injury or disease cannot take advantage of these skeletal muscular contractions. As a result, the return of venous blood to the heart is slower, and the heart has to work harder. Another very important factor in maintaining venous circulation is breathing. During each inspiration, pressure in the thoracic cavity decreases, and this draws blood toward the chest. Each expiration would drive blood back away from the chest, but the veins have valves, which further aid venous circulation. Veins in many places of the body, especially the extremities, contain valves that offer little resistance to blood flowing toward the heart. Any pressure on the valves from blood moving backward closes the valves and stops the backflow. These aids to circulation are illustrated in Figure 18–4.

Lymph Circulation

The flow of lymph from tissue spaces to the subclavian veins also depends upon a series of pressure gradients. The interstitial fluid has a higher pressure than the lymph in the lymph capillaries. Thus, the direction of flow is from interstitial fluid into lymph capillaries. The pressure in larger lymphatics is less than in smaller vessels so that lymph flows from lymphatic capillaries into larger lymph vessels. Similarly, the larger lymphatics have a higher pressure than the subclavian veins. Another factor that determines the flow of lymph is muscular movements. Skeletal muscle contractions compress lymph vessels and force lymph toward the subclavian veins. Smooth muscle movements of the digestive tract keep lymph flowing. A third factor that maintains lymph flow is respiratory movements. Finally, like veins, lymph vessels contain valves to prevent backflow.

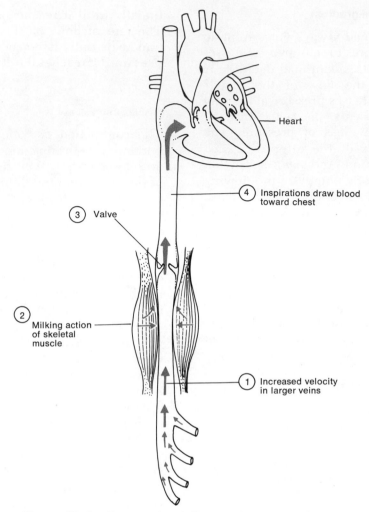

Figure 18–4. Factors that influence the return of venous blood to the heart through the interior vena cava.

Under normal circumstances, the amount of interstitial fluid remains fairly constant. The excessive accumulation of lymph in tissue spaces is referred to as *edema*. Several conditions may bring about edema. One cause is an obstruction to the flow of lymph from the tissue spaces to the veins. Infected lymph nodes may bring about this condition. Other causes are excessive lymph formation and increased permeability of capillary walls. A rise in capillary blood pressure, in which lymph is formed at a faster rate than it is passed into lymphatics, also may result in edema.

Now that you have learned about the dynamics of the circulatory system, let us apply some of this knowledge to the clinical setting. We shall describe two ways by which the health of the circulatory system can be routinely measured: the pulse and the blood pressure. Then we shall look at some circulatory disorders.

CHECKING CIRCULATION

The pulse tells us about the rate of the heartbeat. The blood pressure measures the pressure that the blood is under. First, let us discuss some aspects of the pulse.

Pulse

The alternate expansion and elastic recoil of an artery with each systole of the left ventricle is called the **pulse**. Pulse is strongest in the arteries closest to the heart. It becomes weaker as it passes over the arterial system, and it disappears altogether in the capillaries. The pulse may be felt in any artery that lies near the surface of the body and over a bone or other firm tissue. The radial artery at the wrist is most commonly used for this purpose. Other arteries that may be used for determining pulse are (1) the temporal artery, which is above and toward the outside of the eye; (2) the facial artery, which is at the lower jawbone on a line with the corners of the mouth; (3) the common carotid artery, which is on the one side of the neck; (4) the brachial artery along the inner side of the biceps; (5) the femoral artery near the pelvic bone; (6) the popliteal artery behind the knee; (7) the posterior tibial artery behind the medial malleolus; and (8) the dorsalis pedis artery over the instep of the foot.

The pulse rate is the same as the heart rate and averages between 70 and 80 per minute in the resting state. The term *tachycardia* is applied to a very rapid heart rate or pulse rate. *Tachy* means fast. The term *bradycardia, brady* meaning slow, indicates a very slow heart rate or pulse rate. In addition to recording the rate of the pulse, other factors should be noted. For example, the intervals between beats should be equal in

length. If a pulse is missed at regular or irregular intervals, the pulse is said to be irregular. Also, each pulse beat should be of equal strength. Irregularities in strength may indicate a lack of muscle tone in the heart or arteries.

Blood pressure

Although the term **blood pressure** may be defined as the pressure exerted by the blood on the walls of any blood vessel, in clinical settings it refers to the pressure only in the large arteries. Blood pressure is usually taken in the left brachial artery, and it is measured by a *sphygmomanometer*. The term *sphygmo* means pulse. A commonly used kind of sphygmomanometer (Figure 18–5a) consists of a rubber cuff attached by a rubber tube to a compressible hand pump or bulb. Another tube attaches to the cuff and to a column of mercury that is marked off in millimeters. This column measures the pressure. The cuff is wrapped around

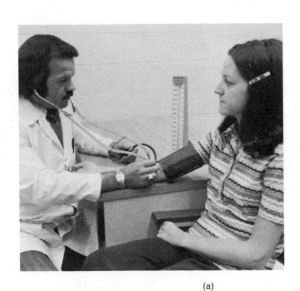

(a)

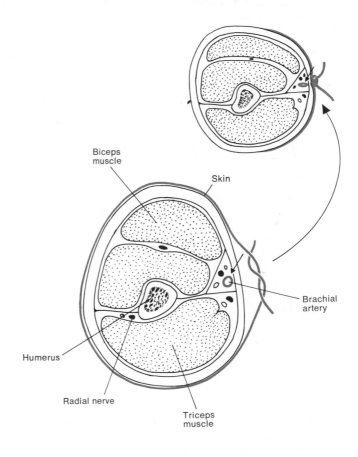

(b)

Figure 18–5. Measurement of blood pressure. (a) Use of a sphygmomanometer. (b) Pressure changes associated with the brachial artery. [*(a) Courtesy of Lenny Patti.*]

the arm over the brachial artery and inflated by squeezing the bulb. This causes a pressure on the outside of the artery (Figure 18–5b). The bulb is squeezed until the pressure in the cuff exceeds the pressure in the artery. At this point, the walls of the brachial artery are compressed tightly against each other, and no blood can flow through. Compression of the artery may be evidenced in two ways. First, if a stethoscope is placed below the cuff over the artery, no pulse can be heard. Secondly, no pulse can be felt by placing the fingers over the radial artery at the wrist.

Next, the cuff is deflated gradually until the pressure in the cuff equals the maximal pressure in the brachial artery. At this point, the artery opens, a spurt of blood passes through, and the pulse may be heard through the stethoscope. As the cuff pressure is further reduced, the sound suddenly becomes more faint or muffled. Finally, the sound becomes muffled and disappears altogether. When the first sound is heard, a reading on the mercury column is made. This sound corresponds to **systolic blood pressure.** This pressure is the force with which blood is pushing against arterial walls during ventricular contraction. The pressure recorded on the mercury column when the sounds suddenly become faint and muffled is called **diastolic blood pressure.** It measures the force of blood in arteries during ventricular relaxation. Whereas systolic pressure typically indicates the force of the left ventricular contraction, the diastolic pressure typically provides information about the resistance of blood vessels.

The average blood pressure of a young adult male is about 120 millimeters Hg systolic and about 80 millimeters Hg diastolic. For convenience and brevity, these pressures are indicated as 120/80. In young adult females, the pressures are 8 to 10 millimeters Hg less. The difference between the systolic and diastolic pressure is called **pulse pressure.** This pressure, which averages 40 millimeters Hg, provides information about the condition of the arteries. The higher the systolic pressure and the lower the diastolic pressure, the greater is the pulse pressure. The normal ratio of systolic pressure to diastolic pressure to pulse pressure is about 3:2:1.

It should be obvious to you that the cardiovascular system is exceedingly complex and is distributed to nearly every part of the body. Therefore, it is not surprising that disorders of the cardiovascular system also affect many other parts of the body. Let us now examine some of these disorders.

CARDIOVASCULAR DISORDERS

Diseases of the heart and blood vessels are the biggest single killers in the developed world. These diseases account for approximately 53 percent of all deaths. A recent comparison indicates that cardiovascular disease kills more people than cancer, accidents, pneumonia, influenza, and diabetes combined. Some of the cardiovascular problems involve aneurysms, arteriosclerosis, hypertension, and various heart disorders.

Aneurysm

A blood-filled sac formed by an outpouching in an arterial or venous wall is called an **aneurysm.** Aneurysms may occur in any major blood vessel of the body and include the following types:

1. Berry, which is a small aneurysm of a cerebral artery. If it ruptures, it may cause a hemorrhage below the dura mater (Figure 18–6 top). Hemorrhaging is one cause of stroke.
2. Ventricular, which is a focal dilatation of a ventricle of the heart (Figure 18–6 middle).
3. Aortic, which is a focal dilatation of the aorta (Figure 18–6 bottom).

Arteriosclerosis

A hardening of the arteries is described by the term **arteriosclerosis.** *Arterio* means artery, whereas *scler* means hard. One type of arteriosclerosis is responsible for the most important and prevalent of all clinical complications. In this type, the inner layer of the artery becomes thickened with soft fatty deposits, called *plaques.* The plaque looks like a pearly gray or yellow mound of tissue on the inside of the blood vessel wall. It usually consists of a core of lipid (mainly cholesterol) covered by a cap of fibrous (scar) tissue. As the plaques increase in size, they not only calcify, but they may also impede or cut off blood flow in affected arteries. This causes damage to the tissues supplied by these arteries. An additional danger is that the lipid core of the plaques may be washed into the bloodstream. There, it could become an embolus and obstruct small arteries and capillaries quite a distance away from the original site of formation. A third possibility is that the plaque will provide a roughened surface for clot formation.

Arteriosclerosis is generally a slow, progressive disease. It may start in childhood, and its development may produce absolutely no symptoms for

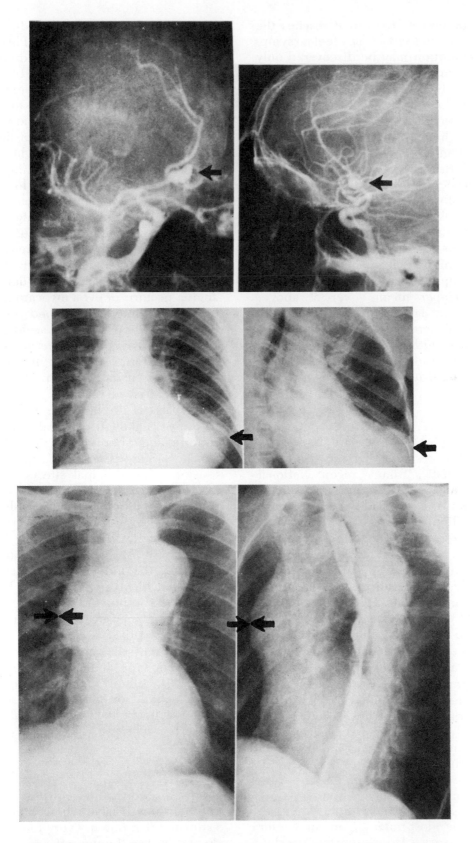

Figure 18–6.   Aneurysms. (Top) Aneurysm of the anterior cerebral artery. (Middle) Ventricular aneurysm. (Bottom) Aortic aneurysm. *(From Lester W. Paul and John H. Juhl, The Essentials of Roentgen Interpretation, 3d ed., Harper & Row, Publishers, Inc., New York, 1972.)*

20 to 40 years or longer. Even if it reaches the advanced stages, the individual may feel no symptoms, and the condition may be discovered only at postmortem examination. Diagnosis during life is made possible by injecting radiopaque substances into the blood and then taking x-rays of the arteries. This technique is called *angiography* or *arteriography*. The film is called an *arteriogram* (Figure 18–7a). Figure 18–7b indicates the most common sites of arteriosclerotic plaques.

Animal experiments have given us considerable scientific information about the plaques. They begin as yellowish fatty streaks of lipids that appear under the tunica intima. It is possible to produce the streaks in many animals by feeding them a diet that is high in fat and cholesterol. This raises the blood lipid levels—a condition called *hyperlipidemia*. *Hyper* means over or above, whereas *lipo* means fat. Hyperlipidemia is an important factor in increasing the risk of arteriosclerosis. Patients with high blood levels of cholesterol should be identified and treated with appropriate diet and drug therapy.

Hypertension

Hypertension, or high blood pressure, is the commonest of the diseases affecting the heart and blood vessels. Statistics from a recent National Health Survey indicate that hypertension afflicts at least 17 million American adults and perhaps as many as 22 million.

Primary hypertension, or essential hypertension, is a persistently elevated blood pressure that cannot be attributed to any particular organic cause. Specifically, the diastolic pressure continually exceeds 95 millimeters Hg. Approximately 85 percent of all hypertension cases fit this definition. The other 15 percent is called *secondary hypertension.* Secondary hypertension is caused by disorders such as arteriosclerosis, kidney disease, and adrenal hypersecretion. Arteriosclerosis increases blood pressure by reducing the elasticity of the arterial walls and by narrowing the space through which the blood can flow. Both kidney diseases and obstruction of blood flow to the kidney may cause the kidney to release renin into the blood. As you recall, this enzyme catalyzes the formation of angiotensin from a plasma protein. Angiotensin is a powerful blood-vessel constrictor. It is the most potent agent known for raising blood pressure. Aldosteronism, the hypersecretion of aldosterone, may also cause an increase in blood pressure. Aldosterone is the adrenal cortex hormone that promotes the retention of salt and water by the kidneys. It thus tends to increase plasma volume. Pheochromocytoma is a benign tumor of the adrenal medulla. It produces and releases into the blood large quantities of norepinephrine and epinephrine. These hormones also raise blood pressure by stimulating the heart and constricting blood vessels.

High blood pressure is of considerable concern because of the harm it can do to certain body organs such as the heart, brain, and kidneys if it remains uncontrolled for long periods. The heart is most commonly affected by high blood pressure. When pressure is high, the heart uses more energy in pumping. Because of the increased effort, the heart muscle thickens, and the heart becomes enlarged. The heart also needs more oxygen. If it cannot meet the demands put on it, angina pectoris or even myocardial infarction may occur. Continued high blood pressure may produce a cerebral vascular accident, or "stroke." In this case, severe strain has been imposed on the cerebral arteries that supply the brain. These arteries are usually less protected by the surrounding tissues than are the major arteries in other parts of the body. As a result, one or more of these weakened cerebral arteries may finally rupture, and a brain hemorrhage follows.

The kidney is another prime target of hypertension. The principal site of damage is in the arterioles that supply this vital organ. The continual high blood pressure pushing against the walls of the arterioles causes them to thicken, thus narrowing the lumen. The blood supply to the kidney is, thereby, gradually reduced. In response, the kidney may secrete renin, which raises the blood pressure even higher and complicates the problem. The reduced blood flow to the kidneys may eventually lead to the death of the kidney cells.

At present, the causes of primary hypertension are unknown. Medical science cannot cure it. However, almost all cases of hypertension, whether mild or very severe, can be controlled by a variety of effective drugs that reduce elevated blood pressure.

Heart disorders

It is estimated that one in every five persons who reaches 60 will have a **heart attack.** And it is also estimated that one in every four persons between 30 and 60 has the potential to be stricken. Heart disease is epidemic in this country, despite the fact that some of the causes can be foreseen and prevented.

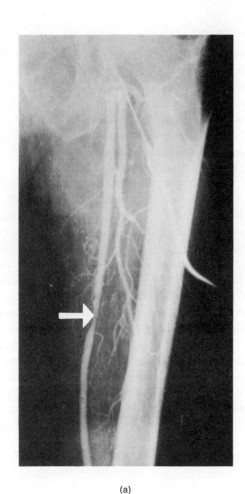

(a)

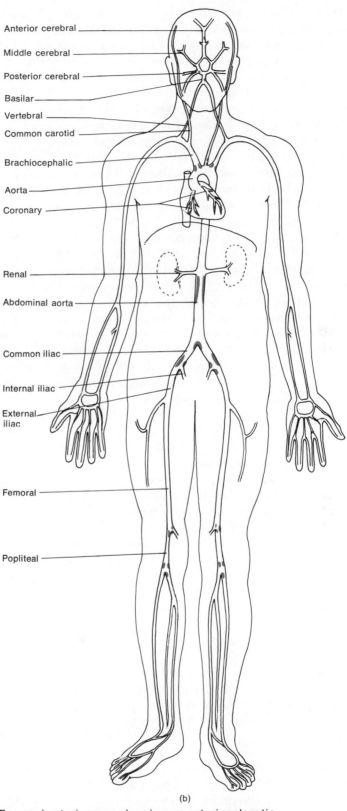

Anterior cerebral

Middle cerebral

Posterior cerebral

Basilar

Vertebral

Common carotid

Brachiocephalic

Aorta

Coronary

Renal

Abdominal aorta

Common iliac

Internal iliac

External iliac

Femoral

Popliteal

(b)

Figure 18–7. Arteriosclerosis. (a) Femoral arteriogram showing an arteriosclerotic plaque (arrow) in the middle third of the thigh. (b) Common sites of arteriosclerotic plaques as seen in a very schematic diagram of the arterial system. The plaques are shown in color. *(Arteriogram courtesy of Lester W. Paul and John H. Juhl, The Essentials of Roentgen Interpretation, 3d ed., Harper & Row, Publishers, Inc., New York, 1972.)*

Risk factors

The Framingham, Massachusetts Heart Study, which began in 1950 and is still going on, is the longest and most famous study ever made of the susceptibility of a community to heart disease. Approximately 13,000 people in the town have participated in the investigation by receiving examinations every 2 years since the study began. The results of this research indicate that people who develop combinations of certain risk factors eventually have heart attacks. These factors are high cholesterol blood level, high blood pressure, cigarette smoking, overweight, lack of exercise, and diabetes.

The first five risk factors—high cholesterol blood level that leads to arteriosclerosis, high blood pressure, cigarette smoking, overweight, and lack of exercise—all contribute to increasing the work load of the heart. The effects of high cholesterol blood level and hypertension have already been discussed. Cigarette smoking increases the work load through the effects of nicotine, which stimulates the adrenal gland to over-secrete aldosterone, epinephrine, and norepinephrine—powerful vasoconstrictors. Overweight people develop miles of extra capillaries to nourish their unwanted fat tissue. This means that the heart has to work harder to pump the blood through more vessels. Lack of exercise means that venous return gets less help from contracting skeletal muscles. In addition, regular exercise strengthens the smooth muscle of blood vessels and enables them to assist general circulation more efficiently. Exercise also increases cardiac efficiency and output. You may recall from Chapter 15 that in diabetes, fat metabolism dominates glucose metabolism. As a result of accelerated fat metabolism, cholesterol levels get progressively higher and result in plaque formation, a situation that may lead to hypertension.

People with three, four, or more risk factors form an especially high-risk group. The incidence of serious heart attacks in this high-risk group is far greater than in groups that have no risk factor or only one. The people who are most apt to develop arteriosclerosis and who, consequently, run the highest risk of all have three risk factors: high cholesterol, hypertension, and cigarette smoking. Other researchers list emotional stress as an important risk factor. But there is still controversy among medical people as to the relative importance of this factor.

The risk factors make a person more susceptible to heart trouble because they strain the heart or increase the likelihood that its oxygen supply will be shut off at some time. Generally, the immediate cause of the heart trouble is one of the following: (1) failure of the heart's blood supply, (2) faulty heart architecture, or (3) failure of the heart's conductivity. Of these three reasons, the first two are far more common than the third.

Failure of blood supply

Angina pectoris and myocardial infarction result from insufficient oxygen supply to the myocardium. Coronary artery disease takes about one in twelve of all Americans who die between the ages of 25 and 34. It claims almost one in four of all those who die between 35 and 44. It has been reported that 50 to 65 percent of all sudden deaths are due to coronary heart disease.

At least half of the deaths from myocardial infarction occur before the patient reaches the hospital. These early deaths could result from an irregular heart rhythm, which is called an *arrhythmia*. Sometimes this progresses to the stage called *cardiac arrest* or ventricular fibrillation, in which the heart stops functioning. An arrhythmia is an abnormal, irregular rhythm change of the heart, caused by disturbances in the conduction system. This abnormal rhythm of the heartbeat could result in cardiac arrest because in this condition the heart is not capable of supplying the oxygen demands of the body. Serious arrhythmias can be controlled, and the normal heart rhythm can be reestablished, if they are detected and treated early enough. Coronary care units have reduced hospital mortality rates from acute myocardial infarctions by about 30 to 20 percent or less by preventing or controlling serious arrhythmias.

Faulty architecture

Less than 1 percent of all new babies have a **congenital,** or **inborn, heart defect.** Even so, the total number in this country each year is estimated to be 30,000 to 40,000. Some of these infants may be able to live quite healthy and long lives without any need for repairing their hearts. But sometimes an inborn heart defect is so severe that an infant lives only a few hours. One of the more common of these defects is patent ductus arteriosus, which was described in Chapter 17. The seriousness of this condition is that the connection between the aorta and the pulmonary artery remains open instead of closing completely after birth. This results in aortic blood flowing into the lower-pressure pulmonary artery, thus increasing the pulmonary artery blood pressure. This increases considerably the work of both ventricles and overworks the heart.

Another common group of congenital problems are the septal defects. A **septal defect** is an opening in the septum that separates the interior of the heart into a left and right side. *Atrial septal defect* is a hole caused by the failure of the fetal foramen ovale to close off the two atria from one another. Because pressure in the right atrium is low, atrial septal defect generally allows a good deal of blood to flow from the left atrium to the right. This results in an overload of the pulmonary circulation, producing fatigability, increased respiratory infections, and growth failure, if it occurs early in life because the systemic circulation may be deprived of a considerable portion of the blood destined for the organ and tissues of the body. *Ventricular septal defect* is caused by an abnormal development of the interventricular septum. Pressure is normally somewhat lower in the right ventricle than in the left, so the blood initially pours from the left ventricle to the right. Deoxygenated blood subsequently gets mixed with the oxygenated blood that is pumped into the systemic circulation. Consequently, the victim suffers *cyanosis*, a blue or dark purple discoloration of the skin. Cyanosis results from insufficient oxygen in the blood. It occurs whenever deoxygenated blood reaches the cells because of

heart defect, lung defect, or suffocation. Septal openings can now be sewn shut or covered with synthetic patches.

A third defect is **valvular stenosis.** It is a narrowing, or *stenosis,* of one of the valves regulating blood flow inside the heart. Narrowing may occur in the valve itself, most commonly in the mitral valve, from rheumatic heart disease or the aortic valve from sclerosis or rheumatic fever. Or it may occur in an area near a valve. The seriousness of all types of stenoses stems from the fact that they all place a severe work load on the heart by making it work harder to push the blood through the abnormally narrow valve openings. As a result of mitral stenosis, blood pressure is increased and angina pectoris and heart failure may accompany the progress of this disorder. The majority of stenosed valves are totally replaced with artificial valves developed in recent years.

The last congenital defect that we shall discuss is tetralogy of Fallot. **Tetralogy of Fallot** is a combination of four defects causing a "blue baby." These are: a ventricular septal opening, an aorta that emerges from both ventricles instead of solely from the left ventricle, a stenosed pulmonary semilunar valve, and an enlarged right ventricle (Figure 18–8). Because of the ventricular septal defect, both oxygenated and unoxygenated blood are mixed in the ventricles. However, the tissues

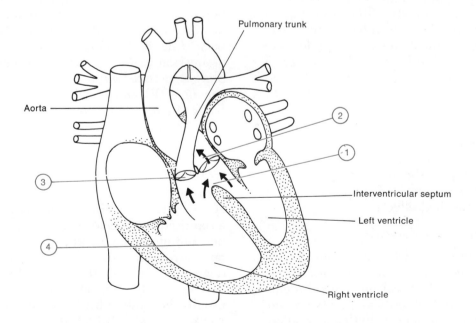

Figure 18–8. Tetralogy of Fallot. The four abnormalities associated with this condition are indicated by numbers. (1) Opening in the ventricular septum. (2) Origin of the aorta in both ventricles. (3) Stenosed pulmonary semilunar valve. (4) Enlarged right ventricle.

of the body are much more starved for oxygen than are those of a child with simple ventricular septal defect. Because the aorta also emerges from the right ventricle and the pulmonary artery is stenosed, very little blood ever gets to the lungs and pulmonary circulation is bypassed almost completely. Today it is possible to completely cure cases of tetralogy of Fallot when the patient is of proper age and condition. Open-heart operations are performed in which the narrowed pulmonary valve is cut open and the septal defect is sealed with a Dacron patch.

Another disorder that arises from some fault in the structure of the heart is rheumatic fever. The symptoms of rheumatic fever mimic many other diseases, but the most common and most serious effects are on the heart. **Rheumatic fever** is basically an acute inflammatory complication of a streptococcal infection. The infection can affect one or more of five major sites: the joints (arthritis), the brain (chorea), the heart (carditis), the subcutaneous tissues, and the skin.

Rheumatic fever occurs most frequently during school age, with the majority of the attacks occurring between the ages of 4 and 18 years. Factors such as overcrowding and malnutrition have been implicated. And statistics indicate that this disease is the most common cardiac abnormality of school children.

The most common serious effect of rheumatic fever is an inflammation of the heart that leaves permanent structural defects in the heart valves and chordae tendineae. This inflammation may cause edema, thickening, fusion, or other destruction of the valves. This leads to stenosis or to failure of the valves to close properly. The pericardial sac can also be adversely affected. In addition, there is a chance that an embolism could develop if a piece of scar tissue became dislodged in the circulating blood.

The long-term prognosis depends directly on the cardiac severity of the initial attack. Except for carditis, all symptoms of rheumatic fever subside without residual effects.

Problems arising from faulty conduction

As noted earlier, the term arrhythmia refers to any variation in the rate, rhythm, or synchrony of the heart. It arises when electrical impulses through the heart are blocked at critical points in the conduction system. One such arrhythmia is called a **heart block.** Perhaps the most common blockage is in the atrioventricular node, which conducts impulses from the atria to the ventricles. This disturbance is called *atrioventricular (AV)*

block. And it usually indicates a myocardial infarction, arteriosclerosis, rheumatic heart disease, diphtheria, or syphilis. In a first-degree AV block, which can be detected only by the use of an electrocardiograph, the transmission of impulses from the atria to the ventricles is delayed. Here, the P–R interval is greater than it should be. In a second-degree AV block, every second impulse fails to reach the ventricles so that the ventricular rate is about one-half that of the atrial rate. When ventricular contraction does not occur (dropped beat), oxygenated blood is not pumped efficiently to all parts of the body. The patient may feel faint and dizzy. Or he may collapse if there are many dropped ventricular beats. In a third-degree or complete AV block, impulses reach the ventricle at irregular intervals, and some never reach it at all. The result is that atrial and ventricle rates are out of synchronization (Figure 18–9a). The ventricles may go into systole at any time. This could occur when the atria are in systole or just before the atria go into systole. Or the ventricles may take a rest for a few cardiac cycles.

With complete AV block, many patients may have vertigo, unconsciousness, or convulsions. These symptoms result from a decreased cardiac output with consequent diminished cerebral blood flow and cerebral hypoxia or lack of sufficient oxygen. Among the causes of AV block are excessive stimulation by the vagus nerves that depresses conductivity of the junctional fibers, destruction of the AV bundle as a result of coronary infarct, arteriosclerosis, myocarditis, or depression caused by various drugs. Other heart blocks include *intraatrial (IA) block, interventricular (IV) block,* and *bundle branch (BBB) block.* In the latter condition, the ventricles do not contract together because of the delay in the impulse in the blocked branch.

Rhythms indicating heart trouble

Two rhythms that indicate heart trouble are atrial flutter and fibrillation. In **atrial flutter** the atrial rhythm averages between 240 and 360 beats per minute. The condition is essentially very rapid atrial contractions accompanied by a second-degree AV block. It is typically indicative of severe damage to heart muscle. Atrial flutter usually becomes fibrillation after a few days or weeks. **Atrial fibrillation** is an asynchronous contraction of the atrial muscles that causes the atria to contract irregularly and still faster. An electrocardiogram of atrial fibrillation is shown in Figure 18–9b. Atrial flutter and fibrillation occur

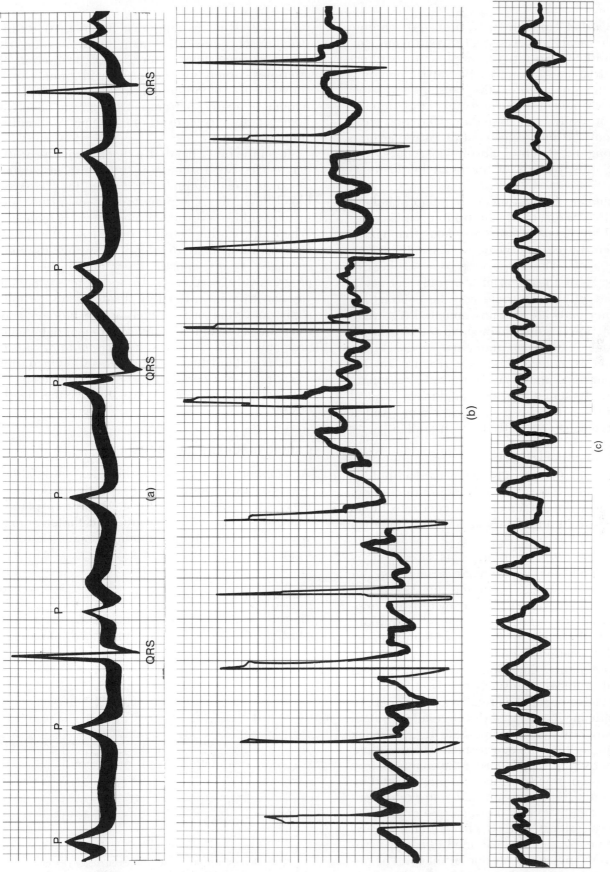

Figure 18–9. Abnormal electrocardiograms. (a) Complete heart block. There is no fixed ratio between atrial contractions (P waves) and ventricular contractions (QRS waves). (b) Atrial fibrillation. There is no regular atrial contraction and, therefore, no P wave. Since the ventricles contract irregularly and independently, the QRS wave appears at irregular intervals. (c) Ventricular fibrillation. In general, there is no rhythm of any kind.

in myocardial infarction, acute and chronic rheumatic heart disease, and hyperthyroidism. Atrial fibrillation results in complete uncoordination of atrial contraction so that atrial pumping ceases altogether. When the muscle fibrillates, the muscle fibers of the atrium quiver individually instead of contracting together. The quivering cancels out the pumping function of the atrium. In a strong heart, atrial fibrillation reduces the pumping effectiveness of the heart by 25 to 30 percent.

Ventricular fibrillation is another kind of rhythm that indicates heart trouble. It is characterized by asynchronous irregular, haphazard ventricular muscle contractions. The rate may be rapid or slow. The impulse travels to the different parts of the ventricles at different rates. Thus, part of the ventricle may be contracting, while other parts are still unstimulated. Ventricular contraction becomes ineffective and circulatory failure and death occur immediately unless the arrhythmia is reversed quickly (Figure 18–9c). Ventricular fibrillation may be caused by coronary occlusion. It sometimes occurs during surgical procedures on the heart or pericardium. And it may be the cause of death in electrocution.

SHOCK AND HOMEOSTASIS

The condition that results when cardiac output or blood volume are reduced to the point where body tissues do not receive an adequate blood supply is called **circulatory shock.** The principal cause of circulatory shock is loss of blood volume or decreased cardiac output. This is caused by loss of blood volume through hemorrhage or through the release of histamine due to damage to body tissues (trauma). The characteristic symptoms of circulatory shock are a pale, clammy skin; cyanosis of the ears and fingers; a feeble, though rapid pulse; shallow and rapid breathing; lowered body temperature; and some degree of mental confusion or unconsciousness.

If the cause of shock is relatively mild, certain homeostatic mechanisms of the circulatory system become operative and compensate for the shock so that no serious damage results. This is called the compensatory stage. During the compensatory stage, lowered blood pressure is compensated by constriction of blood vessels and water retention. This is accomplished by the secretion of renin by the kidneys, aldosterone by the adrenal cortex, epinephrine by the adrenal medulla, and vasopressin-ADH by the posterior pituitary. The result is that, even though some blood is lost from circulation, blood return to the heart is normal and cardiac output remains essentially unchanged.

It is interesting to note that although the constriction of veins and many arterioles occurs during compensation, there is no constriction of arterioles supplying the heart and brain. As a consequence, blood flow to the heart and brain is normal, or nearly so. Compensation is an effective homeostatic mechanism until about 900 milliliters of blood are lost.

If, on the other hand, the shock is severe, death may occur. For instance, if the return of venous blood is greatly diminished due to excessive blood loss, the compensatory mechanisms are insufficient. As a result, cardiac output is reduced. When the cardiac output decreases, the heart fails to pump enough blood to supply its own coronary vessels, and the heart muscle becomes progressively weakened. In addition, prolonged vasoconstriction ultimately leads to tissue hypoxia, and vital organs such as the kidneys and liver are damaged. Essentially, the initial shock promotes more shock, and a *circulatory shock cycle* is established (Figure 18–10). Once the shock reaches a certain level of severity, damage to the circulatory organs is so extensive that death ensues.

CARDIOVASCULAR SURGERY

A number of surgical techniques now allow surgeons to perform new diagnostic procedures, correct heart defects, and make heart transplants. First we shall look at cardiac catheterization, which has become an important diagnostic technique.

Cardiac catheterization

In this diagnostic procedure, the tip of a long plastic *catheter*, or tube, is introduced into a vein

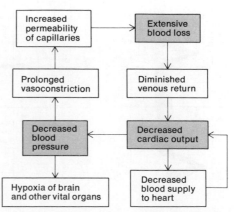

Figure 18–10. The circulatory shock cycle. Why does the cycle perpetuate itself until death results?

in the arm or leg. The catheter is radiopaque so that it can be seen with a fluoroscope. With the help of the fluoroscope, it is then threaded through the vena cava and into the right atrium, right ventricle, or pulmonary artery. The catheter can also be inserted into an artery of the arm or leg and worked up through the aorta to the left atrium and ventricle. When the catheter tip is in place, pressures may be recorded. This can tell the diagnostician about the functioning of the valves. An atrial or ventricular septal defect can be identified either by passing the tip of the catheter through the defect or by testing the oxygen content of the blood near the defect.

Open-heart Surgery

Before surgeons can correct even the simplest heart defect, they have to be able to open up the walls of the heart and expose the chambers. This is what is meant by open-heart surgery. But before any open-heart surgery could be done, techniques had to be developed to capture the blood spurting out of the open chamber and pump it back into the vessels. One such life-support technique is extracorporeal circulation. Coupled with hypothermia, it has now made possible both heart surgery and heart transplants.

Hypothermia

Hypothermia, or body cooling, slows metabolism and reduces the oxygen needs of the tissues. This allows the heart and brain to withstand short periods of interrupted or reduced blood flow. Lost blood is then replaced by transfusion during and after the operation.

Extracorporeal circulation

To sustain the patient for longer periods, surgeons turned to *extracorporeal circulation.* In this technique, blood bypasses the heart and lungs completely (Figures 18–11 and 18–12). It is pumped and oxygenated by a heart-lung machine located outside the body. Modern heart-lung machines may also chill the blood to produce hypothermia as well.

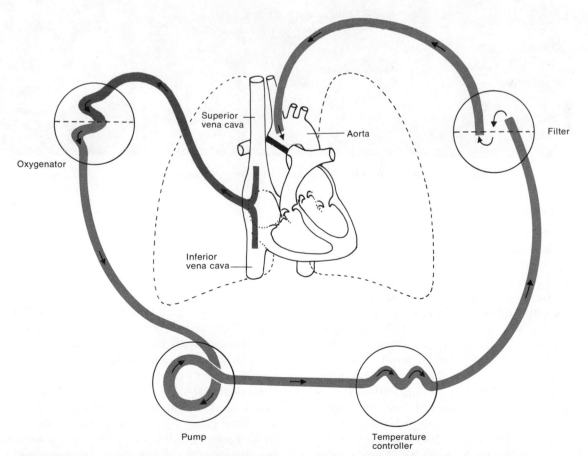

Figure 18–11. Diagram of the principle of the heart-lung bypass. Blood drawn from the vena cavae is oxygenated, rewarmed to body temperature or cooled, filtered to remove air and emboli, and returned to the aorta and coronary arteries at the proper pressure. If necessary, drugs, anesthetics, and transfusions may be added to the circuit.

Figure 18–12. Use of the heart-lung bypass in an operating room.

Artificial parts for the heart and blood vessels

The development of artificial blood vessels, heart valves, patches, and plugs made from synthetic materials has made it possible to repair many defects of the circulatory system. Replacing or bypassing diseased and damaged arteries with synthetic textile tubes has saved thousands of lives and limbs. The most important new development in treating coronary artery disease is the replacement of partially obstructed coronary vessels with synthetic tubes. This has saved many people from a critically ill, invalid life and allowed them to lead a normal one. Severely damaged aortic, mitral, or tricuspid valves have been replaced by artificial valves installed at the site of the natural valve in patients. When one or more major elements of the conduction system are disrupted, heart block of varying degrees

may result. Complete heart block results most commonly from a disturbance in AV conduction. The ventricles fail to receive any atrial impulses, causing the ventricles and atria to beat independently of each other. In patients with heart block, normal heart rate can be restored and maintained with an *artificial pacemaker* (Figure 18–13). Most pacemakers in current clinical use are compact devices, powered by long-lived batteries. Most of these devices are fixed-rate pacemakers. That is, they are set to pace the heart at a fixed rate — usually 80 beats per minute. Some more recent models make provisions for changing the firing rate of the artificial pacemaker. This allows for increased circulatory needs resulting from exertion or other factors.

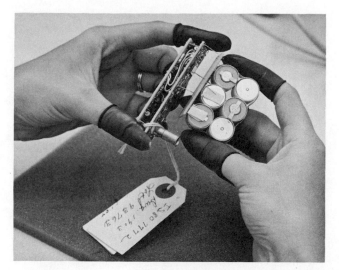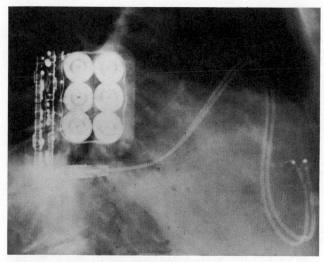

Figure 18–13. The cardiac pacemaker. (Left) Pacemaker with a six-cell powerpack prior to implantation. (Right) Roentgenogram of an implanted pacemaker. *(Photographs courtesy of General Electric.)*

Heart transplants

The first human heart transplant was performed in December 1967. Intense worldwide interest was aroused by this event. During the following 12 months, 101 such operations were performed in several countries. But in 1969 and 1970, the number of transplants declined sharply. The reason for the decline was the death rate. Of approximately 171 patients who have had this operation, 145 have died, and only 26 have survived. Only 12 patients have lived for more than 2 years with transplanted hearts. Seven have lived for more than 2.5 years. These statistics reflect the situation as of the middle of 1971. Sadly, the mortality rate for heart transplants is about the same as that for most other organ transplants. Two exceptions are kidney and thymus transplants. Transplanted kidneys are much more successful. The thymus, as you recall, shows a great deal of success. Rejection and the complications of immunosuppressive therapy presently constitute the most formidable problems following heart transplantation. Until the problems of rejection crises can be solved more easily, surgeons seem to be performing fewer heart transplants. Instead, they are trying to improve surgical bypass procedures for providing the heart muscle with an alternative supply of blood.

Medical terminology associated with the cardiovascular system

Bradycardia   (*brady* = slow) Slow heartbeat or pulse.

Cardiac arrest   Complete stoppage of the heartbeat.

Cyanosis   Slightly bluish, dark purple skin coloration due to oxygen deficiency in systemic blood.

Defibrillator   A mechanical device for applying electrical shock to the heart to terminate abnormal cardiac rhythms.

Murmurs   A heart sound produced by blood passing through a valve or an opening in the heart (septal defect). One of the primary means of clinically diagnosing heart disease.

Tachycardia   (*tachy* = rapid, fast) Rapid heart rate.

Chapter summary in outline

CARDIAC PHYSIOLOGY
Cardiac cycle

1. This cycle consists of the systole (contraction) and diastole (relaxation) of both atria plus the systole and diastole of both ventricles followed by a short pause.

2. Blood flows through heart from areas of higher to lower pressure.

3. With an average heartbeat of 72/minute, a complete cardiac cycle requires 0.8 second.

4. The first heart sound (lubb) represents the closing of the atrioventricular valves. The second sound (dup) represents the closing of semilunar valves.

5. The conduction system consists of tissue specialized for impulse conduction. Components are the sinoatrial node (pacemaker), atrioventricular node, bundle of His, and Purkinje fibers.

6. The record of electrical changes during cardiac cycle is referred to as an electrocardiogram (ECG). Normal ECG consists of a P wave (spread of impulse from SA node over atria), QRS wave (spread of impulse through ventricles), and T wave (ventricular repolarization).

Cardiac output

1. Cardiac output is the amount of blood ejected by the left ventricle into the aorta per minute. It is calculated as follows: cardiac output = stroke volume × number of beats/minute.

2. Stroke volume is the amount of blood ejected by a ventricle per beat. According to Starling's "law of the heart," the stretch of cardiac muscle fibers determines the force of contraction.

3. The pacemaker may be accelerated by sympathetic stimulation and slowed down by parasympathetic stimulation.

WHAT MAKES BLOOD FLOW?

1. Blood flows from regions of higher to lower pressure.

2. The established pressure gradient is aorta (100 millimeters Hg) to arterioles (45 millimeters Hg) to capillaries (10–30 millimeters Hg) to venules (10 millimeters Hg) to veins (less than 10 millimeters Hg) to right atrium (0 millimeter Hg).

3. Pressure is determined by volume of blood, resistance against flow, cardiac output, and extensibility and elasticity of arteries.

4. Rate of flow is determined by the diameter of a blood vessel. The greater the diameter, the faster the flow.

5. Return of venous blood to the heart is assisted by increasing diameters of veins as they approach the heart, skeletal muscle contractions, respirations, and valves.

CHECKING CIRCULATION

1. Pulse is the alternate expansion and elastic recoil of an artery. It may be felt in any artery that lies near the surface or over a hard tissue.

2. A normal rate is between 70 and 80 per minute.

3. Blood pressure is the pressure exerted by blood on the walls of an artery. It is measured by the use of a sphygmomanometer.

4. Systolic blood pressure is the force of blood recorded during ventricular contraction. Diastolic blood pressure is the force of blood recorded during ventricular relaxation. The average blood pressure is 120/80.

5. Pulse pressure is the difference between systolic and diastolic pressures. It averages 40 and provides information about the condition of arteries.

CARDIOVASCULAR DISORDERS

1. An aneurysm is a sac formed by an outpocketing of a portion of an arterial or venous wall.

2. Arteriosclerosis is the hardening of the arteries caused by formation of plaques.

3. Hypertension is high blood pressure. Primary hypertension cannot be linked to a specific organic cause. Secondary hypertension may be caused by arteriosclerosis, kidney disorders, excessive aldosterone secretion, and tumors.

4. Heart disorders related to inadequate blood supply are angina pectoris and myocardial infarction.

5. Congenital heart defects include patent ductus arteriosus, septal defects, valvular stenosis, and tetralogy of Fallot.

6. Rheumatic fever is an acute inflammation that may affect the heart valves and chordae tendineae.

7. Heart conditions relative to conduction problems include heart blocks, atrial flutter and fibrillation, and ventricular fibrillation.

SHOCK AND HOMEOSTASIS

1. Shock results when cardiac output is reduced or blood volume decreases to the point where body tissues become hypoxic.

2. Mild shock is compensated by vasoconstriction.

3. In severe shock, venous return is diminished; cardiac output decreases; the heart becomes hypoxic; prolonged vasoconstriction leads to hypoxia of other organs; shock cycle is intensified.

CARDIOVASCULAR SURGERY

1. Cardiac catheterization is the insertion of a catheter into a vein or artery to monitor pressure inside the heart, condition of valves, and septal defects.

2. Hypothermia (body cooling) and extracorporeal circulation (heart-lung bypass) allow open-heart surgery to be done.

3. Synthetic vessels, valves, and pacemakers are new life-saving devices.

4. Heart transplants, like most other transplants, are not successful because of rejection and the complications of immunosuppressive therapy.

Review questions and problems

1. Define systole and diastole and their relationship to the cardiac cycle.

2. Distinguish the principal events that occur during atrial systole, ventricular diastole, atrial diastole, and ventricular systole.

3. Describe the pressure changes associated with the movement of blood through the heart.

4. By means of a diagram, relate the events of the cardiac cycle to time. What is the quiescent period?

5. Describe the first and second heart sounds and indicate their clinical importance. Define a murmur.

6. What is cardiac output? How is it calculated? What factors alter cardiac output?

7. Define Starling's "law of the heart." Why is it important?

8. Describe the path of a nerve impulse through the conducting system of the heart.

9. Define and label the deflection waves of a normal electrocardiogram. Why is the ECG an important diagnostic tool?

10. Compare the effects of sympathetic and parasympathetic stimulation of the heart.

11. What is a pressoreceptor? Outline the operation of the carotid sinus reflex, the aortic reflex, and the right heart reflex.

12. Define a chemoreceptor. How do chemoreceptors operate? Cite two specific examples.

13. Explain why blood flows from arteries into capillaries into veins. Be sure to indicate pressures in the vessels in your explanation.

14. Relate the pressure of blood to volume and resistance. Can you state the relationship in an equation?

15. Relate cardiac output to blood pressure.

16. Define several factors that offer resistance to the flow of blood. Be sure to emphasize peripheral resistance in your response.

17. How is the vasomotor center in the medulla related to peripheral resistance?

18. Why does blood flow faster in arteries and veins than in capillaries?

19. Identify the factors that assist the return of venous blood to the heart.

20. What forces are responsible for the circulation of lymph?

21. Define pulse. Where may pulse be felt?

22. Contrast the following: tachycardia, bradycardia, and irregular pulse.

23. What is blood pressure? Describe how systolic and diastolic blood pressure may be recorded by means of a sphygmomanometer.

24. Compare the clinical significance of systolic and diastolic pressure. How are these pressures written?

25. Define pulse pressure. What does this pressure indicate?

26. What is an aneurysm? Distinguish three types on the basis of location.

27. Discuss the causes, symptoms, and diagnosis of arteriosclerosis.

28. Compare primary and secondary hypertension with regard to cause.

29. Describe the risk factors involved in heart disease.

30. Distinguish among the following congenital heart disorders: patent ductus arteriosus, septal defects, valvular stenosis, and tetralogy of Fallot.

31. What are the symptoms of rheumatic fever? How is the disease related to the heart?

32. Define an arrhythmia. What is a heart block? Distinguish the various kinds of heart block.

33. Describe atrial flutter and atrial fibrillation. What is ventricular fibrillation?

34. Define circulatory shock. What symptoms characterize circulatory shock?

35. How is the homeostasis of the body restored during mild shock?

36. Describe the effects of a severe shock on circulatory organs by drawing a shock cycle.

37. Explain the use of hypothermia and extracorporeal circulation in cardiovascular surgery.

38. Describe some artificial parts used in the treatment of cardiovascular disorders and explain why they might be needed.

39. What is cardiac catheterization? Why is it an important diagnostic tool?

40. Discuss the present status of heart transplants.

41. Refer to the glossary of medical terminology associated with the cardiovascular system, and be sure that you can define each term listed.

CHAPTER 19

THE RESPIRATORY SYSTEM: STRUCTURE, PHYSIOLOGY, AND DISORDERS

STUDENT OBJECTIVES

After you have read this chapter, you should be able to:

1. Identify the organs of the respiratory system

2. Compare the structure of the external and internal nose

3. Contrast the functions of the external and internal nose in filtering, warming, and moistening air

4. Differentiate between the three regions of the pharynx and describe their roles in respiration

5. Identify the anatomical features of the larynx related to respiration and voice production

6. Describe the tubes that form the bronchial tree with regard to structure and location

7. Contrast tracheotomy and intubation as alternate methods for clearing air passageways

8. Identify the coverings of the lungs and the gross anatomical features of the lungs

9. Describe the structure of a lobule of the lung

10. Describe the role of alveoli in the diffusion of respiratory gases

11. List the sequence of pressure changes involved in inspiration and expiration

12. Compare the volumes and capacities of air exchanged in respiration

13. Define the partial pressure of a gas

14. Describe the mechanisms of external and internal respiration based on differences in partial pressure of O_2 and CO_2

15. Describe how O_2 and CO_2 are carried by the blood

16. Describe the parts of the nervous system that control respiration

17. Compare the roles of the Hering-Breuer reflex and the pneumotaxic center in controlling respiration

18. Describe the effects of chemical stimuli and pressure in determining the rate of respiration

19. Define coughing, sneezing, sighing, yawning, crying, laughing, and hiccoughing as modified respiratory movements.

20. List the basic steps involved in heart-lung resuscitation

21. Define hay fever, bronchial asthma, emphysema, pneumonia, tuberculosis, hyaline membrane disease, and neoplasm as disorders of the respiratory system

22. Describe the effects of pollutants on the epithelium of the respiratory system

23. Describe the administration of medication by nebulization

24. Define medical terminology associated with the respiratory system

ells need a continuous supply of oxygen to carry out the activities that are vital to their survival. Many of these activities release quantities of carbon dioxide. Since an excessive amount of carbon dioxide is poisonous to cells, the gas must be eliminated quickly and efficiently. The two systems that are designed to supply oxygen and eliminate carbon dioxide are the circulatory system and the respiratory system. The **respiratory system** consists of organs that exchange gases between the atmosphere and blood. These organs are the nose, pharynx, larynx, trachea, bronchi, and lungs (Figure 19–1). In turn, the blood transports gases between the lungs and the cells. The overall exchange of gases between the atmosphere and the cells is called *respiration.* Both the respiratory and circulatory systems participate equally in respiration. Failure of either system has the same effect on the body: rapid

death of cells from oxygen starvation and disruption of homeostasis.

THE RESPIRATORY ORGANS

Each of the respiratory organs will now be described in some detail.

Nose

The **nose** has an external portion jutting out from the face and an internal portion lying hidden inside the skull. Externally, the nose consists of a supporting framework of bone and cartilage covered with skin and lined with mucous membrane. The bridge of the nose is formed by the nasal bones, which hold it in a fixed position. Because it has a framework of pliable cartilage, the rest of the external nose is quite flexible. On

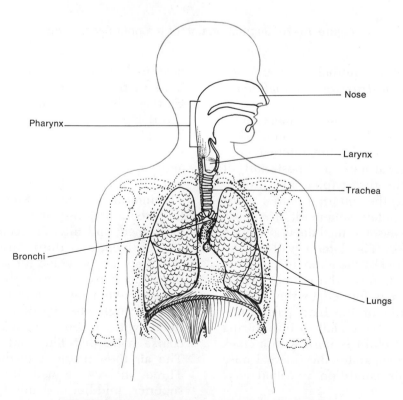

Figure 19–1. Organs of the respiratory system.

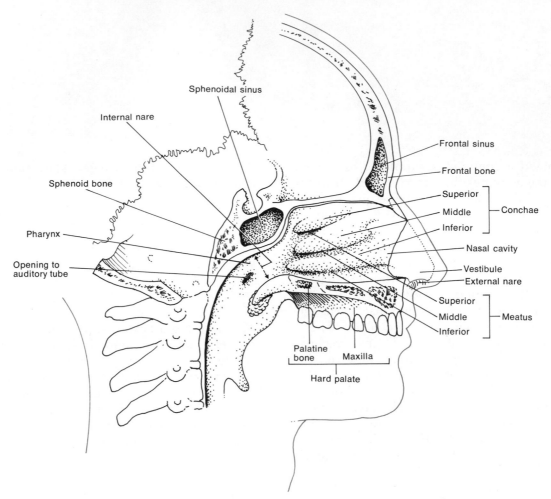

Figure 19–2. Sagittal section of the right nasal cavity.

the undersurface of the external nose are two openings called the *nostrils,* or *external nares* (Figure 19–2).

The internal region of the nose is a large cavity within the skull that lies below the cranium and above the mouth. Anteriorly, the internal nose merges with the external nose, and posteriorly it communicates with the throat (pharynx) through two openings called the *internal nares.* Four paranasal sinuses (frontal, sphenoid, maxillary, and ethmoid) and the nasolacrimal ducts also open into the internal nose. The lateral walls of the internal nose are formed by the ethmoid, maxillae, and inferior conchae bones. The ethmoid forms the roof, and the floor is formed by the palatine bones and the maxilla of the hard palate. Occasionally the palatine bones fail to fuse during embryonic life, and a child is born with a crack in the bony wall that separates the internal nose from the mouth. This condition is called *cleft palate.*

The inside of both the external and internal

nose is divided into right and left *nasal cavities* by a vertical partition called the *nasal septum.* Cartilage is the primary material making up the anterior portion of the septum. The remainder is formed by the vomer and the perpendicular plate of the ethmoid (See Exhibit 8–13). The anterior portions of the nasal cavities, which are just inside the nostrils, are called the *vestibules.* The interior structures of the nose are specialized for three functions. First, incoming air is warmed, moistened, and filtered. Second, olfactory stimuli are received, and third, large hollow resonating chambers are provided for speech sounds. These three functions are accomplished in the following manner. When air enters the nostrils, it passes first through the vestibule. The vestibule is lined with mucous membrane covered with coarse hairs that filter out large dust particles. The air then passes into the rest of the cavity. Three shelves formed by projections of the superior, middle, and inferior conchae or turbinates extend out of the lateral wall of the cavity.

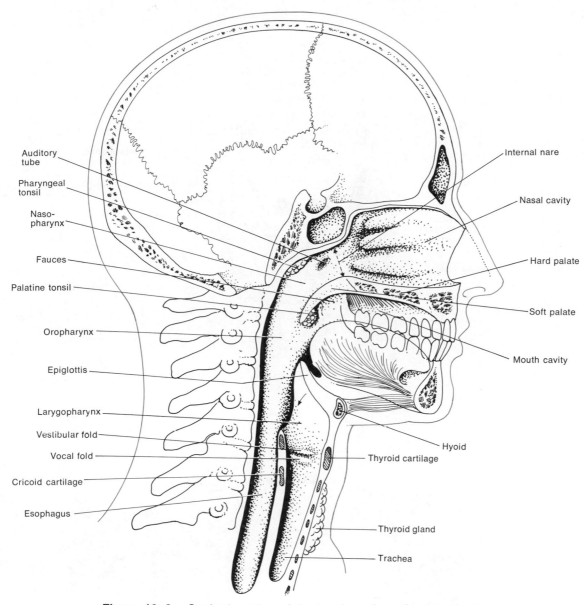

Auditory tube

Pharyngeal tonsil

Naso-pharynx

Fauces

Palatine tonsil

Oropharynx

Epiglottis

Larygopharynx

Vestibular fold

Vocal fold

Cricoid cartilage

Esophagus

Internal nare

Nasal cavity

Hard palate

Soft palate

Mouth cavity

Hyoid

Thyroid cartilage

Thyroid gland

Trachea

Figure 19–3. Sagittal section of the head, neck, and upper chest.

Underneath each shelf is a three-sided passageway. These passageways are called the *superior, middle,* and *inferior meati.* Mucous membrane lines the cavity and its shelves. The olfactory receptors lie in the membrane lining the upper portion of the cavity. Below the olfactory region, the membrane contains many goblet cells, cilia, and capillaries. As the air whirls around the turbinates and meati it is warmed by the capillaries. Mucus secreted by the goblet cells moistens the air and traps dust particles. The cilia move the resulting mucus-dust packages along to the throat so that they can be eliminated from the body. As the air passes through the top of the cavity, chemicals in the air may stimulate the olfactory receptors.

Pharynx

The **pharynx,** or throat, is a tube about 5 inches long that starts at the internal nares and runs partway down the neck (Figure 19–3). It lies just in back of the nasal cavity and mouth and just in front of the cervical vertebrae. Its walls are composed of skeletal muscles. The interior of the walls is lined with mucous membrane. As you might expect from such a structurally simple tube, the functions of the pharynx are limited to serving as a passageway for air and food and providing a resonating chamber for speech sounds.

The uppermost portion of the pharynx is called the *nasopharynx.* This part lies behind the nose

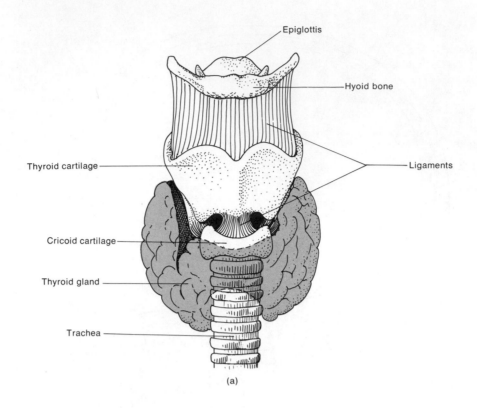

(a)

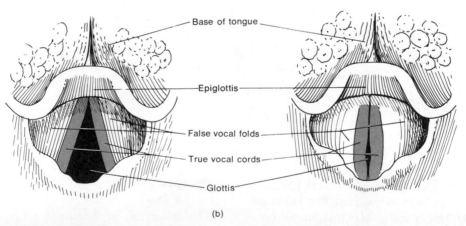

(b)

Figure 19–4. The larynx. (a) Anterior view. (b) Viewed from above. In the figure on the left the true vocal cords are relaxed; in the figure on the right the true vocal cords are pulled taut.

and extends down to the soft palate. There are four openings in its walls: two internal nares plus two openings that lead into the auditory tubes. The posterior wall of the nasopharynx also contains the pharyngeal tonsils, or adenoids. Through the internal nares the nasopharynx exchanges air with the nasal cavities and receives the packages of dust-laden mucus. Cilia in the walls of the nasopharynx move the mucus down toward the

mouth. The nasopharynx also exchanges small amounts of air with the auditory canal so that the pressure inside the middle ear equals the pressure of the atmospheric air flowing through the nose and pharynx.

The second portion of the pharynx, the *oropharynx*, lies behind the mouth and extends from the soft palate down to the hyoid bone. It receives only one opening, the *fauces*, or opening from the mouth. This portion of the pharynx is both respiratory and digestive in function since it is a common passageway for both air and food. Two pairs of tonsils, the palatine tonsils and the lingual tonsils, are found in the oropharynx. The lingual tonsils lie at the base of the tongue and are illustrated in Figure 20–4.

The lowest portion of the pharynx is called the *laryngopharynx.* The laryngopharynx extends downward from the hyoid bone and empties into the esophagus (food tube) posteriorly and into the larynx (voice box) anteriorly. Like the oropharynx, the laryngopharynx is both respiratory and digestive in function.

Larynx

The **larynx,** or voice box, is a short passageway that connects the pharynx with the trachea. It lies in the midline of the neck. The walls of the larynx are supported by pieces of cartilage. The three most prominent pieces are the large thyroid cartilage and the smaller epiglottis and cricoid cartilage (See Figure 19–3). The *thyroid cartilage,* or Adam's apple, consists of two fused plates that form the anterior wall of the larynx and give it its triangular shape (Figure 19–4a). In males the thyroid cartilage is bigger than it is in females.

The *epiglottis* is a large, leaf-shaped piece of cartilage lying on top of the larynx. The "stem" of the epiglottis is attached to the thyroid cartilage, but the "leaf" portion is unattached and free to move up and down like a door on a hinge. In fact, the epiglottis is sometimes called the trap door. As the larynx moves upward and forward during swallowing, the free edge of the epiglottis moves downward and forms a lid over the larynx (see Figure 19–3). In this way, the larynx is closed off and liquids and foods are routed into the esophagus and kept out of the trachea. If anything but air passes into the larynx, a cough reflex attempts to expel the material.

The *cricoid cartilage* is a ring of cartilage forming the lower walls of the larynx. It is attached to the first ring of cartilage of the trachea.

Like the other respiratory passageways, the larynx is lined with a ciliated mucous membrane. Dust not removed in the upper passages can be trapped by the mucus and moved back up toward the throat, where the mucus can be swallowed or spit out.

The mucous membrane of the larynx is arranged into two pairs of folds, an upper pair called the *false vocal folds,* and a lower pair called simply the *vocal folds* or *true vocal cords* (Figure 19–4b). The air passageway between the folds is called the *glottis.* Underneath the mucous membrane of the true vocal cords lie bands of elastic cartilage that are stretched between pieces of rigid cartilage like the strings on a guitar. Skeletal muscles are attached to the pieces of rigid cartilage and to the vocal folds themselves. When the muscles contract, they pull the strings of elastic cartilage tight and stretch the cords out into the air passageways so that the glottis is narrowed (Figure 19–4b). If air is directed against the vocal folds, they vibrate and set up sound waves in the column of air in the pharynx, nose, and mouth. The greater the pressure of air, the louder the sound.

Pitch is controlled by the tension on the true vocal cords. If the cords are pulled taut by the muscles, they vibrate more rapidly and a higher pitch results. Lower sounds are produced by decreasing the muscular tension on the cords. Vocal cords are usually thicker and longer in males than they are in females, and they vibrate more slowly. This is why men have a lower range of pitch than women.

Sound originates from the vibration of the true vocal cords. But other structures are necessary for converting the sound into recognizable speech. For instance, the pharynx, mouth, nasal cavities, and paranasal sinuses all act as resonating chambers that give the voice its human and individual quality. By constricting and relaxing the muscles in the walls of the pharynx we produce the vowel sounds. Muscles of the face, tongue, and lips help us to enunciate words.

Laryngitis is an inflammation of the larynx that is most often caused by a respiratory infection or by irritants, such as cigarette smoke. Inflammation of the vocal folds themselves causes hoarseness or loss of voice by interfering with the contraction of the cords or by causing them to swell to the point where they cannot vibrate freely. Many long-term smokers acquire a permanent hoarseness from the damage done by chronic inflammation.

Trachea

The **trachea,** or windpipe, is a tubular passageway for air, about $4\frac{1}{2}$ inches in length and 1 inch in diameter. It is located in front of the esophagus, and it extends from the larynx to the fifth thoracic vertebra (see Figure 19–1), where it divides into right and left primary bronchi.

The trachea is lined with ciliated mucous membrane providing the same protection against dust as the membrane lining the larynx. The

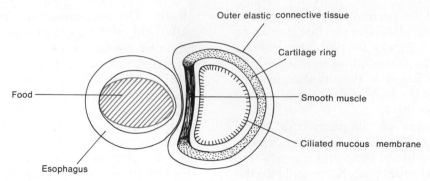

Figure 19–5. Cross section through the trachea and esophagus showing expansion of the esophagus when a large mass of food is swallowed.

walls of the trachea are composed of smooth muscle and elastic connective tissue. They are encircled by a series of horizontal rings of cartilage that look like a series of letter C's stacked one on top of the other. The open parts of the C's face the esophagus and permit the esophagus to expand into the trachea during swallowing. The solid parts of the C's provide a rigid support so that the tracheal walls do not collapse inward and obstruct the air passageway (Figure 19–5).

Occasionally the respiratory passageways are unable to protect themselves from obstruction. For instance, the rings of cartilage may be accidentally crushed, or the mucous membrane may become inflamed and swell so much that it closes off the air space. Inflamed membranes also secrete a great deal of mucus that may clog the lower respiratory passageways. Or a large object may be breathed in (aspirated) while the epiglottis is open. In any case, the passageways must be cleared quickly. If the obstruction is above the level of the chest, a *tracheotomy* may be performed. The first step in a tracheotomy is to make an incision through the neck and into the part of the trachea below the obstructed area. A metal tube is then inserted through the incision, and the patient breathes through the tube. Another method that may be employed is *intubation*. A tube is inserted into the mouth and passed down through the larynx and trachea. The firm walls of the tube push back any flexible obstruction, and the inside of the tube provides a passageway for air. If mucus is clogging the airways, it can be suctioned up through the tube.

Bronchi

The trachea terminates in the chest by dividing into a **right primary bronchus,** which goes to the right lung, and a **left primary bronchus,** which goes to the left lung (Figure 19–6 top). The right primary bronchus is more vertical, shorter, and wider than the left. As a result, foreign objects that enter the air passageways frequently lodge in it. Like the trachea, the primary bronchi contain incomplete rings of cartilage and are lined by a ciliated columnar epithelium.

Upon entering the lungs, the primary bronchi divide to form smaller bronchi, the *secondary bronchi,* one for each lobe of the lung. (The right lung has three lobes, the left lung has two.) The secondary bronchi continue to branch, forming still smaller tubes, the *bronchioles.* Bronchioles, in turn, branch into even smaller tubes, called the *terminal bronchioles.* The continuous branching of the trachea into primary bronchi, secondary bronchi, bronchioles, and terminal bronchioles resembles a tree trunk with its branches and is commonly referred to as the *bronchial tree.* As the branching becomes more extensive in the bronchial tree, several structural changes may be noted. First, rings of cartilage are replaced by plates of cartilage that finally disappear in the bronchioles. Second, as the cartilage decreases, the amount of smooth muscle increases. In addition, the epithelium changes from ciliated columnar to simple cuboidal in the terminal bronchioles. The fact that the walls of the bronchioles contain a great deal of muscle but no cartilage is clinically significant. During an asthma attack the muscles spasm. Because there is no supporting cartilage, the spasms tend to close off the air passageways.

Bronchography is a technique for examining the bronchial tree. The patient breathes in air that contains a safe doseage of a radioactive element. The element gives off rays that penetrate the chest walls and expose a film. The developed film, a *bronchogram,* provides a picture of the tree (see Figure 19–6 bottom).

Lungs

The **lungs** are paired, cone-shaped organs lying in the thoracic cavity (Figure 19–7a). They are

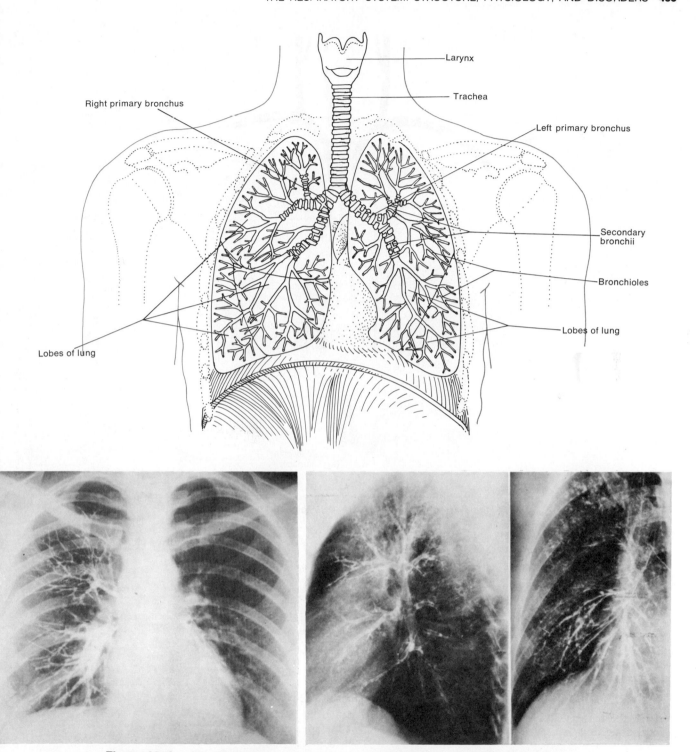

Figure 19–6. Air passageways to the lungs. (Top) Diagram of the bronchial tree in relationship to the lungs. (Bottom) Bronchograms of the lungs. *(Bronchograms from Lester W. Paul and John H. Juhl, The Essentials of Roentgen Interpretation, 3d ed., Harper & Row, Publishers, Inc., New York, 1972.)*

separated from each other by the heart and other structures in the mediastinum. Two layers of serous membrane, collectively called the *pleural membrane,* enclose and protect each lung. The outer layer is attached to the walls of the pleural cavity and is called the *parietal pleura.* The inner layer, the *visceral pleura,* covers the lungs themselves. Between the visceral and parietal pleura is a small space, the *pleural cavity,* which contains a lubricating fluid secreted by the membranes. This fluid prevents friction between the membranes and allows them to move easily on

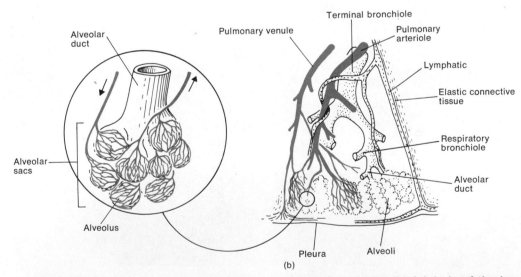

Apex

Superior lobe

Superior lobe

Thymus

Right lung

Left lung

Costal surface

Visceral pleura

Parietal pleura

Middle lobe

Cardiac notch

Heart

Inferior lobe

Inferior lobe

Base

Diaphragm

Mediastinal surfaces

(a)

Terminal bronchiole

Alveolar duct

Pulmonary venule

Pulmonary arteriole

Lymphatic

Elastic connective tissue

Alveolar sacs

Respiratory bronchiole

Alveolar duct

Alveolus

Pleura

Alveoli

(b)

Figure 19-7. The lungs. (a) Coverings and external anatomy. (b) A lobule of the lung.

one another during breathing. Inflammation of the pleural membranes, or *pleurisy,* causes friction during breathing that can be quite painful when the swollen membranes rub against each other.

The lungs extend from the diaphragm to a point just above the clavicles and lie against the ribs in front and back. The broad inferior portion of the lung, the *base,* is concave and fits over the convex area of the diaphragm. The narrow superior portion of the lung is referred to as the *apex.* The surface of the lung lying against the ribs, the *costal surface,* is rounded to match the curvature of the ribs. The *mediastinal (medial) surface* of each lung contains a vertical slit, the *hilum,* through which bronchi, pulmonary vessels, and nerves enter and exit. The blood vessels, bronchi, and nerves are held together by the pleura and connective tissue, and they constitute the *root* of the lung. Medially, the left lung also contains a concavity, the *cardiac notch,* in which the heart lies.

Divided by two fissures into three *lobes,* the superior, middle, and inferior lobes, the right lung is thicker and broader than the left. It is also somewhat shorter than the left because the diaphragm is higher on the right side to accommodate the liver that lies below it. The left lung— thinner, narrower, and longer than the right— is divided into a superior and an inferior lobe.

Each lobe of the lungs is broken up into many small compartments called *lobules* (Figure 19–7b). Every lobule is wrapped in elastic connective tissue and contains a lymphatic, an arteriole, a venule, and a branch from a terminal bronchiole. Terminal bronchioles subdivide into microscopic branches called *respiratory bronchioles.* These, in turn, subdivide into several *alveolar ducts* that terminate in a cluster of *alveolar sacs.* The walls of the alveolar sacs, called **alveoli,** are composed of a single layer of squamous epithelium supported by an extremely thin elastic basement membrane. Over the alveoli, the arteriole and venule disperse into a network of capillaries. The exchange of gases between the lungs and blood takes place by diffusion across the alveoli and the walls of the capillaries. It has been estimated that each lung contains 150 million alveolar sacs— a situation that provides an immense surface area for the exchange of gases.

PHYSIOLOGY OF RESPIRATION

The principal purpose of respiration is to supply the cells of the body with oxygen and to remove the carbon dioxide produced by cellular activities. To trace the path of oxygen from the atmosphere to the cells and that of carbon dioxide from the cells to the atmosphere, we shall consider three basic processes. The first process is ventilation, or breathing, which is the movement of air between the atmosphere and the lungs. The second and third processes involve the exchange of gases within the body. These processes are external respiration, which is the exchange of gases between the lungs and blood, and internal respiration, which is the exchange of gases between the blood and the cells of the body.

Ventilation

Ventilation or breathing is the process by which atmospheric gases are drawn down into the lungs and waste gases that have diffused into the lungs are expelled back up through the respiratory passageways. Air flows between the atmosphere and lungs for the same reason that blood flows through the body; that is, a pressure gradient exists. We breathe in when the pressure inside the lungs is less than the pressure in the atmosphere, and we breathe back out when the pressure inside the lungs is greater than the pressure in the atmosphere.

Inspiration

Breathing in is called **inspiration** or inhalation. Just before each inspiration the air pressure inside the lungs equals the pressure of the atmosphere, which is about 760 millimeters Hg. For air to flow into the lungs, the pressure inside the lungs must become lower than the pressure in the atmosphere. This is achieved by increasing the volume of the lungs. As you have observed, perhaps unknowingly, the pressure of a gas is inversely proportional to the volume of its container. If the size of a closed container is increased, the pressure of the air inside the container decreases. If the size of the container is decreased, then the pressure inside the container increases.

The first step toward increasing lung volume involves the contraction of the respiratory muscles, that is, the diaphragm and intercostal muscles (Figure 19–8a). The diaphragm is the sheet of skeletal muscle that forms the floor of the thoracic cavity. As it contracts it moves downward, thereby increasing the depth of the thoracic cavity. At the same time, the intercostal muscles contract, pulling the ribs upward and turning them slightly so the sternum is pushed forward. In this way the circumference of the thoracic cavity also is increased.

The overall increase in the size of the thoracic cavity causes its pressure, called *intrathoracic pressure,* to fall way below the pressure of the air inside the lungs (Figure 19–8b). Consequently, the walls of the lungs are sucked outward by the partial vacuum. Expansion of the lungs is aided by the pleural membranes. The parietal pleura lining the chest cavity tends to stick to the visceral pleura around the lungs and to pull the visceral pleura with it.

When the volume of the lungs increases, the

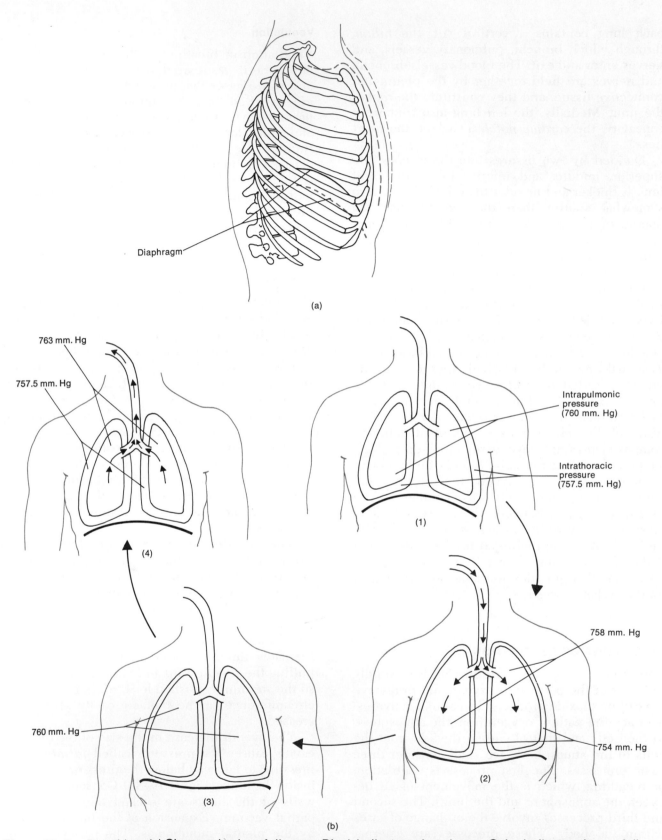

Diaphragm

(a)

763 mm. Hg

757.5 mm. Hg

(4)

Intrapulmonic pressure (760 mm. Hg)

Intrathoracic pressure (757.5 mm. Hg)

(1)

758 mm. Hg

754 mm. Hg

(2)

760 mm. Hg

(3)

(b)

Figure 19–8. Breathing. (a) Changes in size of rib cage. Black indicates relaxed cage. Color indicates shape of rib cage during inspiration. (b) Pressure changes. (1) Lungs and pleural cavity just before inspiration. (2) Chest expanded, and intrathoracic pressure decreased; lungs pulled outward and intrapulmonic pressure decreased. Air moves into lungs until intrapulmonic pressure equals atmospheric (3). (4) Chest relaxes, intrathoracic pressure rises, and lungs snap inward. Intrapulmonic pressure raised, forcing air out until intrapulmonic pressure equals atmospheric (1).

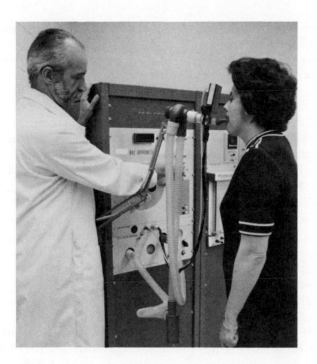

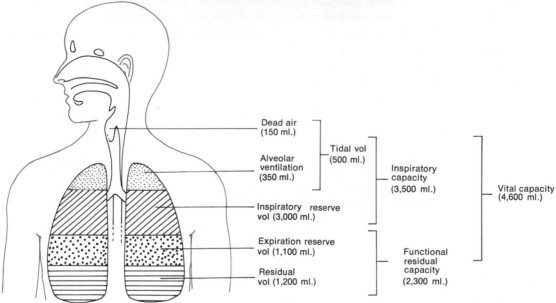

Figure 19–9. Air volumes exchanged in respiration. (Top) Determination of lung volumes using a spirometer. (Bottom) Respiratory air volumes. *(Photograph courtesy of Lenny Patti.)*

pressure inside the lungs, called the *intrapulmonic pressure*, drops from 760 to 758 millimeters Hg. A pressure gradient is thus established between the atmosphere and the alveolar sacs. Air rushes from the atmosphere into the lungs, and an inspiration takes place. Inspiration is frequently referred to as an active process because it is initiated by muscle contraction.

Expiration

Breathing out, called **expiration** or exhalation, is also achieved by a pressure gradient. But this time the gradient is reversed so that the pressure in the lungs is greater than the pressure of the atmosphere. Expiration starts when the respiratory muscles relax and the size of the chest cavity decreases in depth and circumference (Figure

19–8a). As the intrathoracic pressure returns to its preinspiration level, the walls of the lungs are no longer sucked out. The highly elastic basement membranes of the alveoli snaps back into their relaxed shape, and lung volume decreases (Figure 19–8b). Intrapulmonic pressure increases, and air moves from the area of higher pressure (the alveolar sacs) to the area of lower pressure (the atmosphere). Expiration is a passive process since no muscular contraction is required.

If you look again at Figure 19–8b, you will notice that the intrathoracic pressure is always a little less than the pressure inside the lungs or in the atmosphere. The pleural cavities are sealed off from the outside environment and cannot equalize their pressure with that of the atmosphere. Nor can the diaphragm and rib cage move inward enough to bring the intrathoracic pressure up to atmospheric pressure. Actually, maintenance of a lower intrathoracic pressure is vital to the functioning of the lungs. The alveoli are so elastic that at the end of an expiration they attempt to snap inward and collapse on themselves like the walls of a deflated balloon. Such a collapse, which would obstruct the movement of air, is prevented by the slightly lower pressure in the pleural cavities that keeps the alveoli slightly inflated.

Air volumes exchanged in respiration

In clinical practice the word respiration is used to mean one inspiration plus one expiration. The average healthy adult has 14–18 respirations a minute (that is, the individual inspires 14–18 times and expires 14–18 times). During each respiration the lungs exchange given volumes of air with the atmosphere. A lower than normal exchange volume is usually a sign of pulmonary malfunction. The apparatus commonly used to measure the amount of air exchanged during breathing is referred to as a *spirometer* (Figure 19–9 top). During normal quiet breathing, about 500 milliliters of air move into the respiratory passageways with each inspiration, and the same amount moves out with each expiration. This volume of air inspired (or expired) is called *tidal volume* (Figure 19–9 bottom). Actually, only about 350 milliliters of the tidal volume reach the alveolar sacs. The other 150 milliliters remain in the nose, pharynx, larynx, trachea, and bronchi dead air space and are known as *dead air.*

By taking a very deep breath, we can suck in a good deal more than 500 milliliters. This excess inhaled air, called the *inspiratory reserve volume,* averages 3,000 milliliters above the 500 milliliters

of tidal volume. Thus, the respiratory system can pull in as much as 3,500 milliliters of air. If we inhale normally and then exhale as forcibly as possible, we should be able to push out 1,100 milliliters of air in addition to the 500 milliliters tidal volume. This extra 1,100 milliliters is called the *expiratory reserve volume.* Even after the expiratory reserve volume is expelled, a good deal of air still remains in the lungs because the lower intrathoracic pressure keeps the alveolar sacs slightly inflated. This air, the *residual volume,* amounts to about 1,200 milliliters. Opening the thoracic cavity allows the intrathoracic pressure to equal the atmospheric pressure, forcing out the residual volume. The air still remaining is called the *minimal volume.*

The presence of minimal volume can be demonstrated by placing a piece of lung in water and watching it float. Minimal volume provides a medical and legal tool for determining whether a baby was born dead or died after birth. Fetal lungs contain no air. If a baby is born dead, no minimal volume will be observed, but if the child died after he took his first breath, a minimal volume will be detected.

Lung capacity can be calculated by combining various lung volumes. *Inspiratory capacity,* the total inspiratory ability of the lungs, is the sum of tidal volume plus inspiratory reserve volume. *Functional residual capacity* is the sum of residual volume plus expiratory reserve volume. *Vital capacity* is the sum of inspiratory reserve volume, tidal volume, and expiratory reserve volume. Finally, *total lung capacity* is the sum of all volumes.

The total air taken in during 1 minute is called the *minute volume of respiration.* It is calculated by multiplying the tidal volume by the normal breathing rate per minute:

$$\text{Minute volume of respiration} = \underset{\substack{\text{(tidal} \\ \text{volume)}}}{500} \times \underset{\substack{\text{(average} \\ \text{rate per} \\ \text{minute)}}}{16}$$

$$= 8,000 \text{ ml./min.}$$

The measurement of respiratory volumes and capacities is an essential tool for determining how well the lungs are functioning. For instance, during the early stages of the disorder called *emphysema,* many of the alveoli lose their elasticity. During expiration they fail to snap inward, and consequently they fail to force out a normal amount of air. Thus, the residual volume is increased at the expense of the expiratory reserve volume. Pulmonary infections can cause inflammation and an accumulation of fluid in the air spaces of the lungs. The fluid reduces the amount of space available for air and consequently decreases the vital capacity.

Gas exchange

As soon as the lungs fill with air, oxygen moves from the alveolar sacs to the blood, through the interstitial fluid, and finally to the cells. Carbon dioxide moves in just the opposite direction—from the cells, through interstitial fluid to the blood, and to the alveolar sacs. To understand why oxygen and carbon dioxide are able to move in reverse directions, you need to learn a little about the behavior of gases.

Partial pressure of a gas

Atmospheric air is a mixture of several gases—oxygen, carbon dioxide, nitrogen, water vapor, and a number of other gases—which appear in such small quantities that we shall ignore them. Each gas in the atmosphere exerts its own pressure regardless of whether or not the other gases are around. Atmospheric pressure is actually the sum of the pressures of all the gases that make up the air:

Atmospheric pressure (760 mm. Hg)
  $= O_2$ pressure $+ CO_2$ pressure
   $+ N_2$ pressure $+ H_2O$ vapor pressure

This relationship is true for any mixture of gases. For instance, the mixture of gases dissolved in the blood exerts a pressure equal to the sum of the pressures of each individual dissolved gas. In any mixture of gases, the pressure exerted by an individual gas is called the gas' *partial pressure.* For convenience, partial pressure is symbolized by the lowercase letter p followed by the formula for the gas. Thus, oxygen's partial pressure is written as pO_2, and carbon dioxide's partial pressure is written as pCO_2.

The partial pressure of a gas is found by multiplying the percent of the mixture that the particular gas constitutes by the total pressure of the mixture. For example, to find the partial pressure of oxygen in the atmosphere we simply multiply the percent of atmospheric air composed of oxygen (21 percent) by the total atmospheric pressure (760 millimeters Hg):

Atmospheric $pO_2 = 21$ percent $\times$ 760 mm. Hg
         $= 159.60$ or 160 mm. Hg

Now suppose that we want to calculate atmospheric pCO_2. Since the percent of atmospheric carbon dioxide is about 0.03, we multiply this figure by the total atmospheric pressure:

Atmospheric $pCO_2 = 0.03$ percent
           $\times$ 760 mm. Hg
         $= .228$ or .30 mm. Hg

When a mixture of gases diffuses across a semipermeable membrane, each gas diffuses from the area where its partial pressure is greater to the area where its partial pressure is less. Every gas is on its own and behaves as if the gases in the mixture did not exist.

External respiration

External respiration is the exchange of oxygen and carbon dioxide between the alveolar sacs and the blood (Figure 19–10). During inspiration, atmospheric air containing oxygen is brought down into the alveolar sacs. Meanwhile, venous blood, which is low in oxygen and high in carbon dioxide, is pumped through the pulmonary artery into the capillaries overlying the alveoli. If you look at Exhibit 19–1, you will see that the oxygen in the alveolar air has a partial pressure of 103 millimeters Hg, and the oxygen in venous (deoxygenated) blood has a partial pressure of only 40–45 millimeters Hg. Oxygen moves down its partial pressure gradient from the alveolar sacs to the blood until the blood's pO_2 reaches 100 millimeters Hg, the pO_2 of arterial blood. While the blood is being oxygenated, carbon dioxide is also moving down its partial pressure gradient. Thus, carbon dioxide diffuses from the venous blood, where its partial pressure is 45 millimeters Hg, to the alveolar sacs, where its partial pressure is 40 millimeters Hg.

External respiration is aided by several anatomic adaptations. The total thickness of the alveolar-capillary membranes is only 0.004 millimeters. Thicker membranes would inhibit diffusion. The blood and air are also given maximum surface exposure to each other. The total surface area of the alveoli is about 540 square feet, many more times the total surface area of the skin. Lying over the alveoli are countless capillaries—so many that 900 milliliters of blood are able to participate in gas exchange at any given time. Finally, the capillaries are so narrow that the red blood cells must flow through them in single file. This feature gives each red blood cell maximum exposure to the available oxygen.

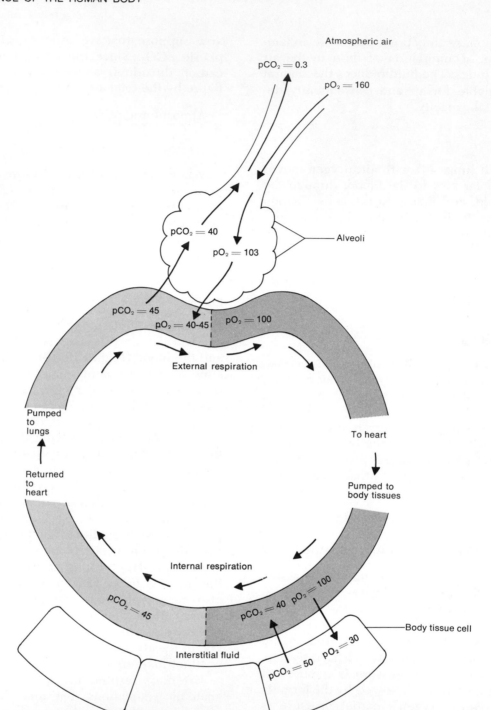

Figure 19–10. Partial pressures involved in external and internal respiration. All pressures are in millimeters Hg.

Exhibit 19–1. PARTIAL PRESSURES OF OXYGEN AND CARBON DIOXIDE*

	ATMOSPHERIC AIR	ALVEOLAR AIR	VENOUS BLOOD	ARTERIAL BLOOD	BODY TISSUES
pO_2	160	103	40–45	100	30
pCO_2	0.3	40	45	40	50

* All pressures are approximate under normal conditions.

The efficiency of external respiration depends on several factors. One of the most important is the alveolar pO_2 relative to the blood pO_2. As long as alveolar pO_2 is higher than venous blood pO_2, oxygen diffuses from the alveolar sacs into the blood. However, as a person ascends in altitude, the atmospheric pO_2 decreases. The alveolar pO_2 correspondingly decreases, and less oxygen diffuses into the blood. The common symptoms of altitude sickness—shortness of breath, nausea, and dizziness—are attributable to the lower concentrations of oxygen in the blood. Another factor that effects external respiration is the total surface area available for O_2-CO_2 exchange. Any pulmonary disorder that decreases the functional surface area formed by the alveolar-capillary membranes decreases the efficiency of external respiration. A third factor that influences external respiration is the minute volume of respiration (tidal volume times rate of respiration per minute). Certain drugs, such as morphine, slow down the respiration rate, thereby decreasing the amounts of oxygen and carbon dioxide that can be exchanged.

Internal respiration

As soon as the task of external respiration is completed, the blood moves through the pulmonary veins to the heart, where it is pumped out to the body tissues. In the capillaries of the body tissues a second exchange, called **internal respiration**, takes place. This is the exchange of O_2 and CO_2 between the blood and body tissues (see Figure 19–10). If you look again at Exhibit 19–1, you will see that the pO_2 of body tissues (30 millimeters Hg) is much lower than the pO_2 of the arterial blood (100 millimeters Hg). On the other hand, the pCO_2 of body tissues (50 millimeters Hg) is higher than the pCO_2 of the arterial blood (40 millimeters Hg). As you might expect, the two gases move down their concentration gradients. Oxygen diffuses from the blood through the interstitial fluid into the body tissues, and carbon dioxide diffuses from the body tissues through the interstitial fluid into the blood until the blood attains partial pressures that are typical of venous blood. The blood now returns to the lungs before it can exchange more gas with body tissues.

Carriage of respiratory gases

When oxygen and carbon dioxide enter the blood, certain physical and chemical changes occur that aid gas exchange. These changes work together, as you will see from the following discussion.

OXYGEN. When oxygen enters the blood, it initially dissolves in the plasma (Figure 19–11). When 0.5 milliliter of oxygen has dissolved in 100 milliliters of blood, the pO_2 of the blood approximately equals the pO_2 inside the alveolar sac. Because the partial pressures are equal, oxygen normally could not continue to move into the blood. However, the cells of the body need much more oxygen than this to survive. In fact, they need 20 milliliters of oxygen/100 milliliters of blood. To get around this problem, most of the oxygen quickly leaves the plasma and combines with the hemoglobin of red blood cells. Oxygen that has become attached to hemoglobin is no longer a free gas. It cannot behave as a gas and consequently cannot affect partial pressure. In this way, the pO_2 of the blood is lowered again, and more oxygen from the alveolar sacs can diffuse into the plasma. Most of these molecules are captured by hemoglobin until all the hemoglobin molecules are bound to oxygen. At this point, the pO_2 of the alveolar sacs and the plasma is equalized, and the blood leaves the lungs.

The chemical union between oxygen and hemoglobin is symbolized as follows:

$$\text{Hb} + \text{O}_2 \underset{\substack{\text{low } pO_2 \\ \text{high } H^+ \\ \text{high temp.}}}{\overset{\substack{\text{high } pO_2 \\ \text{low } pCO_2}}{\rightleftharpoons}} \text{HbO}_2$$

Reduced hemoglobin Oxygen (uncombined hemoglobin) — Oxyhemoglobin (combined hemoglobin)

The reduced hemoglobin represents hemoglobin that has not combined yet with oxygen. Oxyhemoglobin is the compound formed by the union of oxygen and hemoglobin. As you can see, this is a reversible reaction. When the pO_2 is high, as it is in the lungs, there are more oxygen molecules available to make contact with the hemoglobin, and thus more molecules of oxyhemoglobin can be formed.

When the blood reaches the tissue cells of the body, three factors conspire to split the oxyhemoglobin apart. The first factor is simply the low pO_2 in the cells. When an oxygen molecule separates from the oxyhemoglobin molecule, it moves into the tissue cells before it has a chance to recombine with the hemoglobin. All the free oxygen gas also moves into the tissue, so no other oxygen molecules are available to take its place.

The second factor is acidity. Oxyhemoglobin quickly splits apart when it is placed in an acid solution. As you will see, a large quantity of carbon dioxide is moving from tissue cells through the interstitial fluid into the blood. As the carbon dioxide is taken up by the blood, much of it is

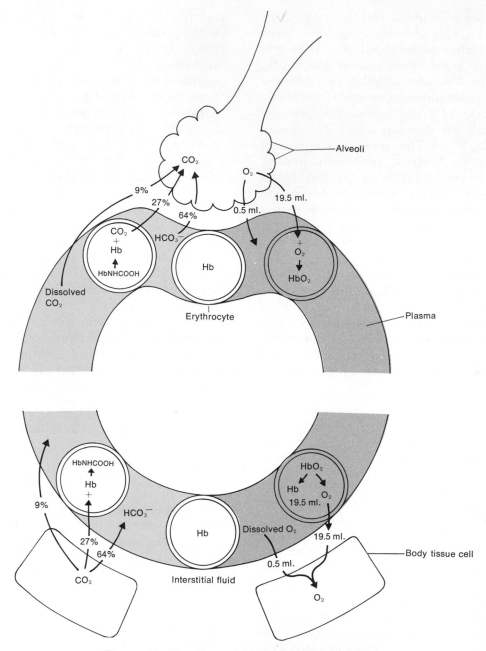

Figure 19–11. The carriage of respiratory gases.

temporarily converted into carbonic acid. Thus a high pCO_2 encourages the oxyhemoglobin to release its oxygen. Lactic acid, the reaction product of contracting muscles, also increases the blood acidity. Can you see now why active muscles are able to obtain oxygen more quickly than resting ones?

The third factor that encourages the blood to give up its oxygen is an increase in temperature. Heat energy is a by-product of the metabolic reactions of all cells, and contracting muscle cells release an especially large amount of heat.

Splitting of the oxyhemoglobin molecule is another example of how homeostatic mechanisms adjust body activities to cellular needs. Active cells require more oxygen, and active cells liberate more acid and heat. The acid and heat, in turn, stimulate the oxyhemoglobin to release its oxygen.

CARBON DIOXIDE. Venous blood contains about 56 milliliters of carbon dioxide/100 milliliters. Like oxygen, carbon dioxide is carried by the blood in several forms (see Figure 19–11). The smallest percentage, about 9 percent, is dissolved in plasma. A somewhat higher percentage, about 27 percent, combines with the protein part of the hemoglobin to form carbaminohemoglobin. This reaction may be represented as follows:

$$Hb \quad + \quad CO_2 \quad \rightleftharpoons \quad HbNHCOOH$$
Hemoglobin   Carbon dioxide   Carbaminohemoglobin

However, the greatest percentage of carbon dioxide, about 64 percent, is converted into the bicarbonate ion (HCO_3^-) in the following way:

$$CO_2 + H_2O \rightleftharpoons H_2CO_3 \rightleftharpoons H^+ + HCO_3^-$$
Carbon   Water   Carbonic   Hydrogen   Bicarbonate
dioxide   acid   ion   ion

As the carbon dioxide diffuses from tissues into blood plasma and then into red blood cells, an enzyme called carbonic anhydrase stimulates the major portion of the gas to combine with water to form *carbonic acid.* Inside red blood cells, carbonic acid dissociates into H^+ and HCO_3^- ions. The H^+ ions provide the acid stimulus for the release of oxygen from oxyhemoglobin. Some of the HCO_3^- ions remain within the cell and combine with potassium, the chief positive ion of intracellular fluid, to form potassium bicarbonate ($KHCO_3$). The majority of the bicarbonate ions, however, diffuse out into the plasma and combine with sodium, the principal positive ion of extracellular fluid to form sodium bicarbonate ($NaHCO_3$).

When the blood reaches the lungs, the above events reverse. The high pO_2 causes oxygen to replace the carbon dioxide on the hemoglobin molecule. Bicarbonate breaks apart and releases CO_2 Finally, the CO_2 that has been traveling dissolved in the plasma and the CO_2 that has been released from the reversed reactions diffuse down the partial pressure gradient and into the alveolar sacs.

Control of respiration: homeostatic mechanisms

Respiration is controlled by several mechanisms that help the body maintain homeostasis.

Nervous control

As noted earlier, the size of the thorax is affected by the action of the respiratory muscles. These muscles contract and relax in turn as a result of nerve impulses transmitted to them from centers in the brain. The area from which nerve impulses are sent to respiratory muscles is located in the medulla oblongata of the brain and is referred to as the *respiratory center.* This center is functionally divided into two regions: the *inspiratory center,* which causes the inspiratory muscles to contract and thus brings on inspiration, and the *expiratory center,* which inhibits the inspiratory center and thereby allows the inspiratory muscles to relax. The respiratory center, along with its "pacemaker" centers in the pons, regulates the rhythm of respiration in the following manner.

An area of the pons, called the *apneustic center,* continually sends stimulatory impulses to the inspiratory center. The stimulated inspiratory center, in turn, sends impulses over motor neurons that stimulate the respiratory muscles to contract, and inspiration occurs. As soon as the lungs are filled, the inspiratory impulses are shut off by two controls.

The first set of controls involves the stretch receptors in the lung tissue. When the lungs expand to a critical point, the stretch receptors are stimulated and impulses are sent along the vagus nerves to the expiratory center. The expiratory center then sends out inhibitory impulses to the inspiratory center. This causes relaxation of the inspiratory muscles, and expiration follows. As air leaves the lungs during expiration, the lungs are deflated and the stretch receptors are no longer stimulated. Thus the inspiratory center is no longer inhibited, and a new respiration begins. These events are called the *Hering-Breuer reflex.* The Hering-Breuer reflex controls the depth and rhythm of respiration. It is also a protective reflex because it prevents the lungs from inflating to the point of bursting.

Should the Hering-Breuer reflex fail to operate, breathing will still continue. If the vagus nerves are severed, inspirations are longer and deeper than normal, but expiration is eventually initiated by a back-up control. This second control is provided by the *pneumotaxic center* of the pons. When the inspiratory center is stimulated, it sends impulses to the pneumotaxic center as well as to the respiratory muscles. After a delay, the pneumotaxic center sends impulses to the expiratory center, which inhibits the inspiratory center.

The term applied to normal quiet breathing is *eupnea.* Eupnea is a function of the Hering-Breuer reflex, and it involves shallow, deep, or combined shallow and deep breathing. Shallow, or chest breathing, is called *costal breathing.* It consists of an upward and outward movement of the chest as a result of contraction of the intercostal muscles. Deep, or abdominal breathing, is called *diaphragmatic breathing.* It consists of the outward movement of the abdomen as a result of the contraction and descent of the diaphragm.

The respiratory center has connections with the cerebral cortex, which means that we can voluntarily alter our pattern of breathing. Or we can refuse to breathe at all for a short time. Voluntary control is protective because it enables us to prevent water or irritating gases from entering the lungs. However, the ability to willfully stop breathing is limited by the buildup of CO_2 in the blood, as we shall see shortly. When the pCO_2 increases to a certain critical level, the inspiratory center is stimulated, impulses are sent to inspiratory muscles, and breathing resumes whether or not the person wishes. Since the breath can be held for only a short period of time before the involuntary control centers take over, it is impossible for anyone to kill himself by holding his breath.

Chemical and pressure stimuli

Chemical stimuli, particularly CO_2, O_2, and H^+, determine how fast we breathe. All three stimuli may act directly on the respiratory center or on chemoreceptors located in the aortic and carotid bodies (see Chapter 18). Probably, the most important chemical stimulus that alters respirations is CO_2. Under normal circumstances, arterial blood pCO_2 is 40 millimeters Hg. If there is even a slight increase in pCO_2, a condition called *hypercapnia*, chemoreceptors in the medulla and in the carotid and aortic bodies are stimulated. Stimulation of the chemoreceptors causes the inspiratory center to become highly active, and the rate of respiration increases. This increased rate is called *hyperventilation*. Hyperventilation allows the body to expel more CO_2 until the pCO_2 is lowered to normal. Now let us consider the opposite situation. If arterial pCO_2 is lower than 40 millimeters Hg, the receptors are not stimulated and stimulatory impulses are not sent to the inspiratory center. Consequently, the center sets its own moderate pace until CO_2 accumulates and the pCO_2 rises above 40 millimeters Hg. A slow rate of respiration is called *hypoventilation*.

The oxygen receptors are sensitive only to large drops in the pO_2. If arterial pO_2 falls from a normal of 100 millimeters Hg to 70 millimeters Hg, the oxygen receptors become stimulated and send impulses to the inspiratory center. But if the pO_2 falls much below 70 millimeters Hg, the cells of the respiratory center suffer oxygen starvation and do not respond well to any of the chemical receptors. They send fewer impulses to the respiratory muscles, and the respiration rate decreases or breathing ceases altogether.

Within limits, any decrease in arterial blood pH stimulates chemoreceptors in the carotid and aortic bodies and the medulla. This results in an increase in the rate of respiration. The carotid and aortic bodies also contain pressoreceptors that are stimulated by a rise in blood pressure decreases the rate of respiration, and a concerned mainly with the control of circulation, they assume some function in the control of respiration. For example, a sudden rise in blood pressure decreases the rate of respiration, and a drop in blood pressure brings about an increase in the respiratory rate.

Some of the other factors that control respiration are

1. A sudden cold stimulus such as plunging into cold water causes a temporary cessation of breathing, called *apnea*.
2. A sudden, severe pain brings about apnea, but a prolonged pain triggers off the general stress syndrome and increases respiration rate.
3. Stretching of the anal sphincter muscle increases the respiratory rate. Medical personnel sometimes employ this technique to stimulate respiration during emergencies.
4. Irritation of the pharynx or larynx by touch or by chemicals brings about an immediate cessation of breathing followed by coughing.

MODIFIED RESPIRATORY MOVEMENTS

Respirations provide human beings with methods for expressing emotions such as laughing, yawning, sighing, and sobbing. Moreover, respiratory air can be used to expel foreign matter from the upper air passages through actions such as sneezing and coughing. Some of the modified respiratory movements that express emotion or clear the air passageways are listed in Exhibit 19-2. All these movements are reflexes, but some of them also can be initiated voluntarily.

HEART AND LUNG RESUSCITATION

A serious decrease in respiration or heart rate presents an urgent crisis because the body's cells cannot survive long if they are starved of oxygenated blood. In fact, if oxygen is withheld from the cells of the brain for 4–6 minutes, brain damage or death will result. Heart-lung resuscitation is the artificial reestablishment of normal or near

Exhibit 19–2. MODIFIED RESPIRATORY MOVEMENTS

MOVEMENT	COMMENT
Coughing	Preceded by a long-drawn and deep inspiration that is followed by a complete closure of the glottis — resulting strong expiration suddenly pushes glottis open and sends a blast of air through the upper respiratory passages. Stimulus for this reflex act could be a foreign body lodged in the larynx, trachea, or epiglottis.
Sneezing	Spasmodic contraction of muscles of expiration forcefully expels air through the nose and mouth. Stimulus may be an irritation of the nasal mucosa.
Sighing	A deep and long-drawn inspiration immediately followed by a shorter but forceful expiration.
Yawning	A deep inspiration through the widely opened mouth producing an exaggerated depression of the lower jaw. May be stimulated by drowsiness or fatigue, but precise stimulus-receptor cause is unknown.
Sobbing	Starts with a series of convulsive inspirations. Glottis closes earlier than normal after each inspiration so only a little air enters the lungs with each inspiration. Immediately followed by a single prolonged expiration.
Crying	An inspiration followed by many short convulsive expirations. Glottis remains open during the entire time, and the vocal cords vibrate. Accompanied by characteristic facial expressions.
Laughing	Involves the same basic movements as crying, but the rhythm of the movements and the facial expressions usually differ from those of crying. Laughing and crying are sometimes indistinguishable.
Hiccough	Spasmodic contraction of the diaphragm followed by a spasmodic closure of the glottis. Produces a sharp inspiratory sound. Stimulus is usually irritation of the sensory nerve endings of the digestive tract.

normal respiration and circulation. The two simplest techniques for heart-lung resuscitation are exhaled air ventilation and external cardiac compression. Both techniques can be administered by a layman at the site of the emergency, and both are highly successful. They can be used for any sort of heart or respiratory failure, whether the cause be drowning, strangulation, carbon monoxide or insecticide poisoning, overdose of a drug or anesthesia, electrocution, or myocardial infarction. However, the success of heart-lung resuscitation is directly related to the speed and efficiency with which it is applied. Delay may be fatal.

Exhaled air ventilation

A technique for reestablishing respiration is *exhaled air ventilation*. The first and most important step is immediate opening of the airway. This is accomplished easily and quickly by tilting the victim's head backward as far as it will go without being forced. The tilted position opens the upper air passageways to their maximum amount (Figure 19–12a). If the patient does not resume spontaneous breathing after his head has been tilted backward, immediately begin artificial ventilation by either the mouth-to-mouth or the mouth-to-nose method. In the more usual mouth-to-mouth method the nostrils are pinched together with the thumb and index finger of the hand. The rescuer then opens his mouth widely, takes a deep breath, makes a tight seal with his mouth around the patient's mouth, and blows in about twice the amount the patient normally breathes. He then removes his mouth and allows the patient to exhale passively. This cycle is repeated approximately 12 times per minute for adults. Atmospheric air contains about 21 percent O_2 and a trace of CO_2. Exhaled air still contains about 16 percent O_2 and 5 percent CO_2. This is more than adequate to maintain a victim's blood pO_2 and pCO_2 at normal levels if air is given at the prescribed rate and amount.

If the rescuer observes the following three signs, he knows that adequate ventilation is occurring:

1. The chest rises and falls with every breath.
2. He feels the resistance of the lungs as they expand.
3. He hears the air escape during exhalation.

The three most common errors in mouth-to-mouth resuscitation are *inadequate extension* of the victim's head, *inadequate opening* of the rescuer's mouth, and an *inadequate seal* around the patient's mouth or nose. If the rescuer is sure he has not made these errors and he is still unable to inflate the lungs, a foreign object is probably lodged in the respiratory passages. The rescuer's fingers should be swept through the patient's mouth to remove such material. An adult victim with this problem should next be rolled quickly onto his side. Firm blows should be delivered over his spine between the shoulder blades in an attempt to dislodge the obstruction. Then exhaled-air ventilation should be resumed quickly. A

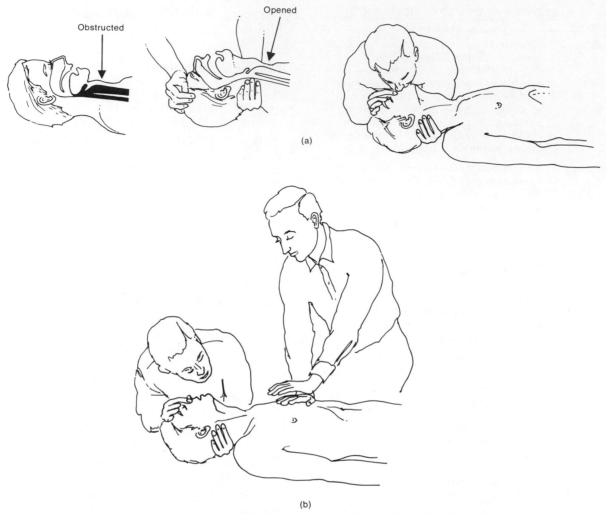

Figure 19–12. Heart-lung resuscitation. (a) Exhaled air ventilation. Shown on the left is the procedure for immediate opening of the airway. Shown on the right is the procedure for mouth-to-mouth respiration. (b) External cardiac compression technique in conjunction with exhaled air ventilation.

small child with an obstructive foreign object should be picked up quickly and inverted over the rescuer's forearm while firm blows are delivered over the spine between the shoulder blades. Then ventilation can be resumed.

If the patient is an infant or small child, the rescuer should make the following adjustments in his technique. The neck of an infant is so pliable that forceful backward tilting of the head may obstruct breathing passages, so the tilted position should not be exaggerated. The rescuer should also remember that the lungs of a small child do not have a large capacity. To avoid overinflating the child's lungs, he should cover both the victim's mouth and nose with his mouth and blow gently, using less volume, at a rate of 20 to 30 times a minute.

External cardiac compression

External cardiac compression, or closed-chest cardiac compression (CCCC), consists of the application of rhythmic pressure over the sternum (Figure 19–12b). The rescuer places the heels of his hands on the lower half of the sternum and presses down firmly and smoothly at least 60 times a minute. This action compresses the heart and produces an artificial circulation because the heart lies almost in the middle of the chest between the lower portion of the sternum and the spine. When properly done, external cardiac compression can produce systolic blood pressure peaks of over 100 millimeters Hg. It can also bring carotid arterial blood flow up to 35 percent of normal.

Complications that can occur from the use of cardiac compression include fracture of the ribs and sternum, laceration of the liver, and the formation of fat emboli. They can be minimized by adhering to the following precautions:

1. Never compress over the xiphoid process at the tip of the sternum. This bony prominence extends down over the abdomen, and pressure on it may cause laceration of the liver, which can be fatal.
2. Never let your fingers touch the patient's ribs when you compress. Keep your fingers off the patient, and place the heel of your hand in the middle of the patient's chest over the lower half of his sternum.
3. Never compress the abdomen and chest simultaneously since this action traps the liver and may rupture it.
4. Never use sudden or jerking movements to compress the chest. Compression should be smooth, regular, and uninterrupted, with 50 percent of the cycle compression and 50 percent relaxation.

Compression of the sternum produces some artificial ventilation but not enough for adequate oxygenation of the blood. Therefore exhaled air ventilation must always be used with it. This combination constitutes *heart-lung resuscitation.* When there are two rescuers, the most physiologically sound and practical technique is to have one rescuer apply cardiac compression at a rate of at least 60 compressions per minute. The other rescuer should exhale into the patient's mouth between every fifth and sixth compression. The sequential steps in emergency heart-lung resuscitation must be continued uniformly and without interruption until the patient recovers or is pronounced dead.

RESPIRATORY DISORDERS

Hay fever

An allergic reaction to the proteins contained in foreign substances such as plant pollens, dust, and certain foods is called **hay fever.** Allergic reactions are a special type of antigen-antibody response that initiate either a localized or a systemic inflammatory response. In hay fever the response is localized in the respiratory membranes. The membranes become inflamed, and a watery fluid drains from the eyes and nose.

Bronchial Asthma

Another disorder is **bronchial asthma.** This usually allergic reaction is characterized by attacks of wheezing and difficult breathing. Attacks are brought on by spasms of the smooth muscles that lie in the walls of the smaller bronchi and bronchioles, causing the passageways to close partially. The patient has trouble exhaling, and the alveoli may remain somewhat inflated during expiration. Usually the mucous membranes that line the respiratory passageways become irritated and secrete excessive amounts of mucus that may clog the bronchi and bronchioles and worsen the attack. About three out of four asthma victims are allergic to something they eat or to substances they breathe in, such as pollens, animal dander, house dust, or smog. Others are usually sensitive to the proteins of relatively harmless bacteria that inhabit the sinuses, nose, and throat.

Emphysema

One lung disease that starts with the deterioration of some of the alveoli is **emphysema.** The alveolar walls lose their elasticity and remain filled with air during expiration. The name of the disease means "blown up" or "full of air." Reduced forced expiratory volume is the first symptom. Later, alveoli in other areas of the lungs are damaged. The lungs become permanently inflated. To adjust to the increased lung size, the size of the chest cage increases. The patient has to work to exhale. Oxygen diffusion does not occur as easily across the damaged alveoli, blood pO_2 is somewhat lowered, and any mild exercise that raises the oxygen requirements of the cells leaves the patient breathless. Carbon dioxide diffuses much more easily across the alveoli than does oxygen, so the pCO_2 is not affected initially. But as the disease progresses, the alveoli degenerate and are replaced with thick fibrous connective tissue. Even carbon dioxide does not diffuse easily through this fibrous tissue. If the blood cannot buffer all the carbonic acid that accumulates, the blood pH drops. Or, unusually high amounts of carbon dioxide may dissolve in the plasma. High carbon dioxide levels are toxic to the brain cells. Consequently, the inspiratory center becomes less active and the respiration rate slows down, further aggravating the problem. The capillaries that lie around the deteriorating alveoli are compressed and damaged and may no longer be able to receive blood. As a result, pressure increases in the pulmonary artery, and the right atrium overworks as it attempts to force blood through the remaining capillaries.

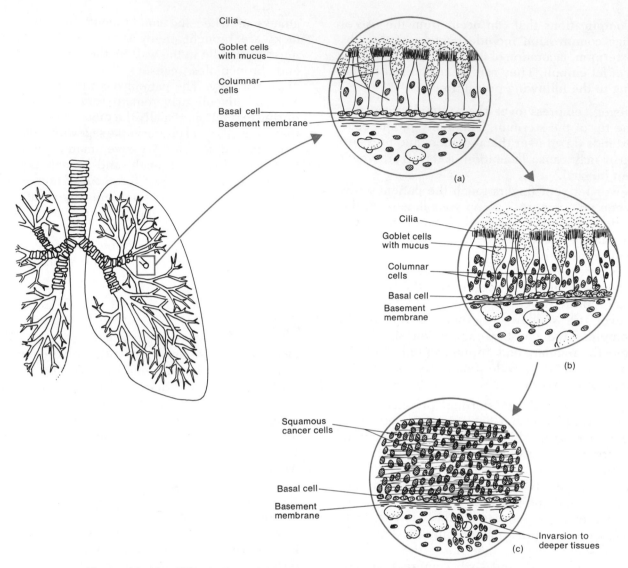

Figure 19–13. Effects of smoking on the respiratory epithelium. (a) Microscopic view of the normal epithelium of a bronchial tube. (b) Initial response of the bronchial epithelium to irritation by pollutants. (c) Advanced response of the bronchial epithelium.

Emphysema is generally caused by any of a number of long-term irritations. Air pollution, occupational exposure to industrial dusts, and cigarette smoke are the most common irritants. Chronic bronchial asthma also may produce alveolar damage. Cases of emphysema are becoming more and more frequent in the United States. The irony is that the disease can be prevented and the progressive deterioration can be stopped by eliminating the harmful stimuli.

Pneumonia

The term **pneumonia** means an acute infection or inflammation of the alveoli. In this disease the alveolar sacs fill up with fluid and dead white blood cells, reducing the amount of air space in the lungs. (Remember that one of the cardinal signs of inflammation is edema.) Oxygen has difficulty diffusing through the inflammed alveoli, and the blood pO_2 may be drastically reduced. Blood pCO_2 usually remains normal because carbon dioxide always diffuses through the alveoli more easily than oxygen does. If all the alveoli of a lobe are inflamed, the pneumonia is called *lobar pneumonia*. If only parts of the lobe are involved, it is called *lobular*, or *segmental*, *pneumonia*. If both the alveoli and the bronchial tubes are included, it is called *bronchopneumonia*.

The most common cause of pneumonia is the pneumococcus bacteria, but other bacteria or a fungus may be a source of the trouble. Viral pneumonia is caused by any of several viruses, including the influenza virus.

Tuberculosis

The bacterium called *Mycobacterium tuberculosis* produces an inflammation called **tuberculosis**. This disease is still one of the most serious of present-day illnesses and ranks as the number-one killer in the communicable disease category. Tuberculosis most often affects the lungs and the pleura. The bacteria destroy parts of the lung tissue, and the tissue is replaced by fibrous connective tissue. Because the connective tissue is inelastic and relatively thick, the affected areas of the lungs do not snap back during expiration, and larger amounts of air are retained. Gases no longer diffuse easily through the fibrous tissue.

Tuberculosis bacteria are spread by inhalation. Although they can withstand exposure to many disinfectants, they die quickly in sunlight. This is why tuberculosis is sometimes associated with crowded, poorly lit housing conditions. Many drugs are successful in treating tuberculosis. Rest, sunlight, and good diet are vital parts of treatment.

Hyaline Membrane Disease (HMD)

Sometimes called glassy-lung disease, **hyaline membrane disease** is responsible for approximately 20,000 newborn infant deaths per year. Autopsies reveal glassy membranes lining the alveoli and alveolar ducts. The membranes consist largely of fibrin deposits that prevent oxygen-carbon dioxide exchange and prevent the opening of the alveoli. Asphyxiation occurs for most infants within 72 hours of birth.

A new treatment currently being developed called PEEP—positive end expiratory pressure—could reverse the mortality rate from 90 percent deaths to 90 percent survival. This treatment consists of passing a tube through the air passage to the top of the lungs to provide needed oxygen-rich air at continuous pressures of up to 14 millimeters Hg. Continuous pressure keeps the baby's lungs open and available for gas exchange.

SMOKING AND THE RESPIRATORY SYSTEM

As part of ordinary breathing, many irritating substances are inhaled. Almost all pollutants, including inhaled smoke, have an irritating effect on the bronchial tubes and lungs and may be regarded as stresses.

Close examination of the epithelium of a bronchial tube reveals that it consists of three kinds of cells (Figure 19-13a). The uppermost cells are columnar cells that contain the cilia on their surfaces. At intervals between the ciliated columnar cells are the mucus-secreting goblet cells. The bottom of the epithelium normally contains two rows of basal cells above the basement membrane. The bronchial epithelium is important clinically because researchers have learned that one of the most common types of lung cancer, **bronchogenic carcinoma**, starts in the walls of the bronchi.

The stress of constant irritation by inhaled smoke and pollutants causes an enlargement of the goblet cells of the bronchial epithelium (Figure 19-13b). They respond by secreting excessive amounts of mucus. The basal cells respond to the stress by undergoing cell division so fast that the basal cells push into the area occupied by the goblet and columnar cells. As many as 20 rows of basal cells may be produced. Many researchers believe that if the stress is removed at this point, the epithelium can return to normal.

If the stress persists, more and more mucus is secreted and the cilia become less effective. As a result, mucus is not carried toward the throat; instead, it remains trapped in the bronchial tubes. The individual then develops a "smoker's cough." Moreover, the constant irritation from the pollutant slowly destroys the alveoli, which are replaced with thick, inelastic connective tissue. Mucus that has accumulated becomes trapped in the air sacs. Millions of the sacs rupture. This results in a loss of diffusion surface for the exchange of oxygen and carbon dioxide. The individual has now developed emphysema. If the stress is removed at this point, there is little chance for improvement. Any alveolar tissue that has been destroyed cannot be repaired. But removal of the stress can stop further destruction of lung tissue.

Assuming that the stress continues, the emphysema gets progressively worse, and the basal cells of the bronchial tubes continue to divide and break through the basement membrane. At this point the stage is set for bronchogenic carcinoma. Columnar and goblet cells disappear and may be replaced with squamous cancer cells (Figure 19-13c). If this happens, the malignant growth spreads throughout the lung and may block a bronchial tube. If the obstruction occurs in a large bronchial tube, very little oxygen enters the lung, and disease-producing bacteria thrive on the mucoid secretions. In the end, the patient may develop emphysema, carcinoma, and a host

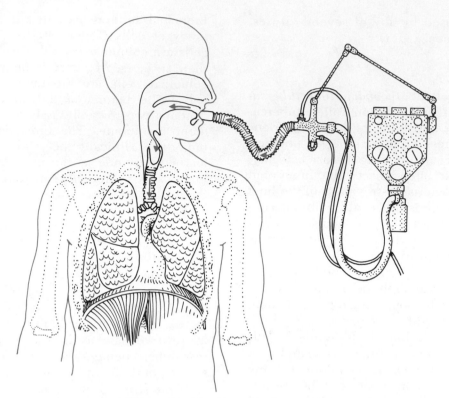

Figure 19–14. Use of a nebulizer.

of infectious diseases. Treatment involves surgical removal of the diseased lung. However, metastasis of the growth through the lymphatic or blood system may result in new growths in other parts of the body such as the brain and liver.

The processes that we have just described have been observed in some heavy smokers. One should realize, though, that there are other causes of emphysema, such as chronic asthma. There are also other factors that may be associated with lung cancer. For instance, breast, stomach, and prostate malignancies can metastasize to the lungs. People who apparently have not been exposed to pollutants do occasionally develop bronchogenic carcinoma. However, the occurrence of bronchogenic carcinoma is probably over 20 times higher in heavy cigarette smokers than it is in nonsmokers.

THERAPY BY NEBULIZATION

Many of the previously mentioned respiratory disorders are treated by means of a comparatively new method of treatment called *nebulization.* This procedure is the administering of medication, in the form of droplets that are suspended in air,

to selected areas of the respiratory tract. The patient inhales the medication as a fine mist (Figure 19–14). Droplet size is directly related to the number of droplets suspended in the mist. Smaller droplets (approximately 2 μ in diameter) can be suspended in greater numbers than can large droplets and will reach the alveolar ducts and sacs. The larger droplets (approximately 7–16 μ in diameter) will be deposited mostly in the bronchi and bronchioles. Droplets of 40 μ and larger will be deposited in the upper respiratory tract—the mouth, pharynx, trachea, and main bronchi. Nebulization therapy can be used with many different types of drugs, such as chemicals that relax the smooth muscle of the respiratory passageways, chemicals which reduce the thickness of mucus, and antibiotics.

Medical terminology associated with the respiratory system

Apnea   Absence of respirations.

Asphyxia   Oxygen starvation due to low atmospheric pO_2 or interference with ventilation, external respiration, or internal respiration.

Atelectasis   (*asis* = state or condition) A collapsed lung or portion of a lung.

Bronchitis (*bronch* = bronchus, trachea) Inflammation of the bronchi and bronchioles.

Cheyne-Stokes respiration Irregular breathing beginning with shallow breaths that increase in depth and rapidity, then decrease and cease altogether for 15–20 seconds. The cycle repeats itself again and again. Cheyne-Stokes is normal in infants. It is also often seen just before death from pulmonary, cerebral, cardiac, and kidney disease.

Diphtheria An acute bacterial infection that causes the mucous membranes of the pharynx, nasopharynx, and larynx to enlarge and become leathery. Enlarged membranes may obstruct airways and cause death from asphyxiation.

Dyspnea (*dys* = painful, difficult) Labored or difficult breathing. (Short-winded)

Eupnea (*eu* = good, normal) Normal quiet breathing.

Hypoxia Reduction in oxygen supply to cells.

Influenza Viral infection that causes inflammation of respiratory mucous membranes as well as fever.

Orthopnea Inability to breathe in a horizontal position.

Pneumothorax (*pneumo* = lung) Air in pleural space causing collapse of the lung. Most common cause is surgical opening of chest during heart surgery, making intrathoracic pressure equal atmospheric pressure.

Pulmonary edema Excess amounts of interstitial fluid in the lungs producing cough and dyspnea. Common in failure of left side of the heart.

Pulmonary embolism Presence of a blood clot or other foreign substance in a pulmonary arterial vessel stopping circulation to a part of the lungs.

Rales Sounds sometimes heard in the lungs that resemble bubbling or rattling. May be caused by air or an abnormal secretion in the lungs.

Respirator Metal chamber that entombs chest; also called "iron lung." Used to produce inspiration and expiration in patient with paralyzed respiratory muscles. Pressure inside chamber is rhythmically alternated to suck out and push in chest walls.

Chapter summary in outline

RESPIRATORY ORGANS
1. Respiratory organs include the nose, pharynx, larynx, trachea, bronchi, and lungs.
2. They act with the circulatory system to supply oxygen and remove carbon dioxide.

Nose
1. The external portion is made of cartilage and skin and lined with mucous membrane; openings to exterior are external nares.
2. The internal portion communicates with pharynx through internal nares and communicates with paranasal sinuses.
3. The nose is divided into cavities by a septum. Anterior portions of the cavities are called the vestibules.
4. The nose is adapted for the warming, moistening, and filtering of air; olfaction; and it assists in speech.

Pharynx
1. The pharynx, or throat, is a muscular tube lined by mucous membrane.
2. Anatomic regions are nasopharynx, oropharynx, and laryngopharynx.
3. The nasopharynx functions in respiration. The oropharynx and laryngopharynx function in digestion and respiration.

Larynx
1 The larynx is a passageway that connects the pharynx with the trachea.

2. Prominent cartilages are the thyroid, or Adam's apple, the epiglottis, which prevents food from entering the larynx, and the cricoid, which connects the larynx and trachea.
3. The larynx contains true vocal cords that produce sound. Taut cords produce high pitches, and relaxed cords produce low pitches.

Trachea
1. The trachea extends from the larynx to the primary bronchi.
2. It is composed of smooth muscle and C-shaped rings of cartilage and is lined with ciliated mucous membrane.

Bronchi
1. The bronchial tree consists of primary bronchi, secondary bronchi, bronchioles, and terminal bronchioles. Walls of bronchi contain rings of cartilage; walls of bronchioles do not.
2. A developed picture of the tree is called a bronchogram.

Lungs
1. Lungs are paired organs in the thoracic cavity. They are enclosed by the pleural membrane (parietal pleura is outer layer; visceral pleura is inner layer).
2. The right lung has three lobes; the left lung has two lobes and a depression, the cardiac notch. Each lobe consists of lobules, which contain lymphatics, arterioles, venules, terminal bronchioles, respiratory bronchioles, alveolar ducts, alveolar sacs, and alveoli.
3. Gas exchange occurs across alveoli-capillary membranes.

PHYSIOLOGY

Ventilation
1. Inspiration occurs when intrapulmonic pressure falls below atmospheric pressure. Contraction of the diaphragm and intercostals increases the size of the thorax and decreases the intrathoracic pressure. Decreased intrathoracic pressure causes a decreased intrapulmonic pressure.
2. Expiration occurs when intrapulmonic pressure is higher than atmospheric pressure. Relaxation of diaphragm and intercostals increases intrathoracic pressure, which causes an increased intrapulmonic pressure.
3. Among the air volumes exchanged in ventilation are tidal, inspiratory reserve, expiratory reserve, residual, and minimal volumes.
4. The minute volume of respiration is the total air taken in during 1 minute (tidal volume times 16 respirations per minute).

Exchange of gases
1. The partial pressure of a gas is the pressure exerted by that gas in a mixture of gases. It is symbolized by p.
2. In internal and external expiration O_2 and CO_2 move from areas of their higher partial pressure to areas of their lower partial pressure.
3. External respiration is aided by a very thin alveolar-capillary membrane, a large alveolar surface area (about 540 square feet), and a rich blood supply.

Carriage of gases
1. In each 100 milliliters of oxygenated blood, there are 20 milliliters of O_2, 0.5 milliliters are dissolved in plasma, and 19.5 milliliters are carried with hemoglobin as oxyhemoglobin (HbO_2).
2. In each 100 milliliters of deoxygenated blood, there are 56 milliliters of CO_2. About 9 percent of CO_2 is dissolved in plasma, about 27 percent combines with hemoglobin as carbaminohemoglobin ($HbNHCOOH$), and about 64 percent is converted to the bicarbonate ion (HCO_3^-).

Control of respiration
1. Nervous control is regulated by the respiratory centers in medulla and pons, which control the rhythm of respiration. The

Hering-Breuer reflex controls the depth and rhythm of respiration.

2. Chemical control is regulated by chemical stimuli (CO_2, O_2, and H^+ ions) in the blood.

3. Pressoreceptors in the carotid and aortic bodies also control rate of respiration.

MODIFIED RESPIRATIONS

1. Coughing, sneezing, sighing, yawning, sobbing, crying, laughing, and hiccoughing involve modified respiratory movements.

HEART-LUNG RESUSCITATION

1. Exhaled air ventilation is used to reestablish respiration. External cardiac compression is used to reestablish circulation.

RESPIRATORY DISORDERS

1. Hay fever is an allergic reaction of respiratory membranes.

2. Bronchial asthma occurs when spasms of smooth muscle in bronchial tubes result in partial closure of air passageways, inflammation, inflated alveoli, and excess mucous production.

3. Emphysema is characterized by deterioration of alveoli leading to loss of their elasticity. Symptoms are reduced expiratory volume, inflated lungs, and enlarged chest.

4. Pneumonia is an acute inflammation or infection of alveoli.

5. Tuberculosis is an inflammation of pleura and lungs produced by a specific bacterium.

6. Hyaline membrane disease is an infant disorder in which fibrin deposits occur in the alveolar ducts and alveoli.

7. Smoking
   (a) Pollutants, including smoke, act as stresses on the epithelium of the bronchi and lungs. Constant irritation results in excessive secretion of mucus and rapid division of bronchial basal cells.
   (b) Additional irritation may cause retention of mucus in bronchioles, loss of elasticity of alveoli, and less surface area for gaseous exchange.
   (c) In the final stages, bronchial epithelial cells may be replaced by cancer cells. The growth may block a bronchial tube and spread throughout the lung and other body tissues.

8. Many respiratory disorders are treated by nebulization—the administration of medication to the respiratory tract in the form of droplets suspended in air.

Review questions and problems

1. What organs comprise the respiratory system? What function do the respiratory and circulatory systems have in common?

2. Describe the structures of the external and internal nose and describe their functions in filtering, warming, and moistening air.

3. What is the pharynx? Differentiate the three regions of the pharynx, and indicate their roles in respiration.

4. Describe the structures of the larynx, and explain how these structures function in respiration and voice production.

5. Describe the location and structure of the trachea. What is a tracheotomy? Intubation?

6. What is the bronchial tree? Describe its structure. What is a bronchogram?

7. Where are the lungs located? Distinguish the parietal pleura from the visceral pleura. What is pleurisy?

8. Define each of the following parts of a lung: base, apex, costal surface, medial surface, hilum, root, cardiac notch, and lobe.

9. What is a lobule of the lung? Describe its composition and function in respiration.

10. Indicate several ways in which you think that the respiratory organs are structurally adapted to carry on their respiratory functions?

11. What are the basic differences among ventilation, external respiration, and internal respiration?

12. Discuss the basic steps involved in inspiration and expiration. Be sure to include values for all pressures involved.

13. What is a spirometer? Define the various lung volumes and capacities. How is the minute volume of respiration calculated?

14. Define the partial pressure of a gas. How is it calculated? What are the partial pressures of oxygen and carbon dioxide in the atmosphere, alveolar air, arterial blood, body tissues, and venous blood?

15. Construct a diagram to illustrate how and why the respiratory gases move during external and internal respiration.

16. How are oxygen and carbon dioxide carried by the blood?

17. Discuss how the inspiratory and expiratory centers are related to the Hering-Breuer reflex. What is the role of the apneustic center and the pneumotaxic center in controlling respiration?

18. How do the following chemical stimuli affect respiration: pCO_2, pO_2, and H^+ ions?

19. How do pressoreceptors affect the control of respiration?

20. How does the control of respiration demonstrate the principal of homeostasis?

21. Define the various kinds of modified respiratory movements?

22. What is the objective of heart and lung resuscitation? What cautions must be taken in exhaled air ventilation and external cardiac compression? Why?

23. For each of the following, list the outstanding clinical symptoms: hay fever, bronchial asthma, emphysema, pneumonia, tuberculosis, hyaline membrane disease, and bronchogenic carcinoma.

24. Discuss the stages in the destruction of respiratory epithelium as a result of continued irritation by pollutants.

25. What is nebulization? When is it used? Why is it effective?

26. Refer to the glossary of medical terminology associated with respiratory system. Be sure that you can define each term.

CHAPTER 20

THE DIGESTIVE SYSTEM: STRUCTURE, PHYSIOLOGY, AND DISORDERS

STUDENT OBJECTIVES

After you have read this chapter, you should be able to:

1. Define digestion as a chemical and mechanical process

2. Identify the organs of the alimentary canal and the accessory organs of digestion

3. Describe the structure of the wall of the alimentary canal

4. Define the mesentery, lesser omentum, and greater omentum as extensions of the peritoneum

5. Describe the role of the mouth in mechanical digestion

6. Identify the location of the salivary glands

7. Define the function of saliva in digestion

8. Define the action of salivary amylase

9. Describe the mechanisms that regulate the secretion of saliva

10. Identify the parts of a typical tooth

11. Compare deciduous and permanent dentitions

12. Discuss the sequence of events involved in swallowing

13. Describe the structural features of the stomach and the relationship between these features and digestion

14. Compare mechanical and chemical digestion in the stomach

15. Describe the factors that control the secretion of gastric juice

16. Describe the relationship of the pancreas to digestion

17. Define the role of the liver in digestion

18. Describe those structural features of the small intestine that adapt it for digestion and absorption

19. Describe the mechanisms involved in the hormonal control of digestion in the stomach and small intestine

20. Describe those digestive activities of the small intestine by which carbohydrates, proteins, and fats are reduced to their final products

21. Describe the mechanical movements of the small intestine

22. Define absorption

23. Compare the fates of absorbed nutrients

24. Describe those structural features of the large intestine that adapt it for absorption and feces formation and elimination

25. Describe the mechanical movements of the large intestine

26. Describe the processes involved in feces formation

27. Discuss the mechanisms involved in defecation

28. List the causes and symptoms of dental caries and periodontal disease

29. Contrast between the location and effects of gastric and duodenal ulcers

30. Describe the causes and dangers of peritonitis

31. Compare pancreatitis and cirrhosis as disorders of the accessory organs of digestion

32. Describe the location of tumors of the gastrointestinal tract

33. Define medical terminology associated with the digestive system

We all know that food is vital to life. Food is required for the chemical reactions that occur in every cell—both those that synthesize new enzymes, cell structures, bone, and all the other components of the body and those that release the energy needed for the building processes. However, the vast majority of foods we eat are simply too large to pass through the plasma membranes of the cells. Therefore, chemical and mechanical **digestion** must occur first.

Chemical digestion is a series of hydrolytic reactions that break down the large carbohydrate, lipid, and protein molecules which we eat into monosaccharides, glycerol and fatty acids, and

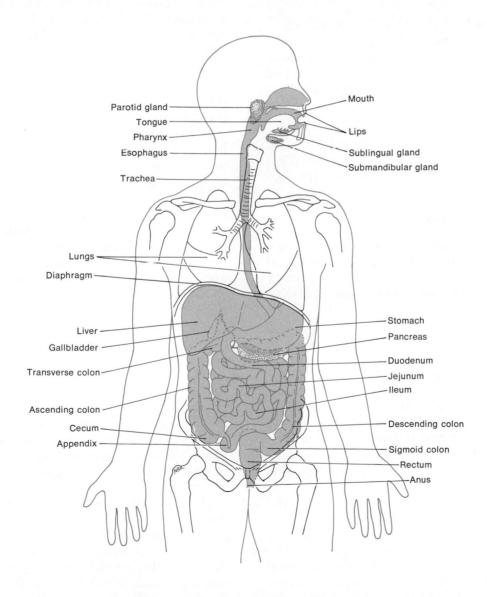

Figure 20–1. The organs of the digestive system.

amino acids, respectively. These products of digestion are small enough to pass through the walls of the digestive organs, into the blood and lymph capillaries, and eventually into the cells of the body. *Mechanical digestion* consists of various movements that aid chemical digestion. Food must be pulverized by the teeth before it is small and flexible enough to be swallowed. After it has been swallowed, the smooth muscles of the stomach and small intestine churn the food so it is thoroughly mixed with the enzymes that catalyze the hydrolysis reactions.

The function of the digestive system, then, is to prepare food for consumption by the cells. It does this through five basic activities:

1. Ingestion, or eating, which is taking food into the body
2. Movement of food along the digestive tract
3. Mechanical and chemical digestion
4. Absorption, the passage of digested food from the digestive tract into the circulatory and lympatic systems for distribution to cells
5. Defecation, the elimination of undigestible substances from the body

We shall discuss each of these activities as we describe the organs that are involved in them. But first let us orient ourselves by looking at the general plan of the digestive system.

GENERAL ORGANIZATION

The organs of digestion are traditionally divided into two main groups. First is the **alimentary canal** or **gastrointestinal tract,** a continuous tube running through the ventral body cavity and extending from the mouth to the anus (Figure 20-1). The length of a tract taken from a cadaver is about 9 meters (30 feet). In a living person it is somewhat shorter because the muscles lying in its walls are in a state of tonic contraction. Organs comprising the alimentary canal include the mouth, pharynx, esophagus, stomach, small intestine, and large intestine. The alimentary canal contains the food while it is being eaten, digested, and eliminated. Muscular contractions in the walls of the alimentary canal break down the food physically by churning it. Secretions produced by cells in the alimentary canal break down the food chemically.

The second group of organs comprising the digestive system are the *accessory organs*—the teeth, tongue, salivary glands, liver, gallbladder, and pancreas. Teeth are cemented to bone, protrude into the alimentary canal, and aid in the physical breakdown of food. The other accessory organs lie totally outside the canal and produce or store secretions—which aid in the chemical breakdown of the food—that are dumped into the canal through ducts.

Since the organs comprising the alimentary canal are structurally quite similar, let us first look at a generalized cross section of the canal. Later, as we follow the movement of food from the mouth to the anus, we shall describe the organs of the canal and their modifications, as well as the accessory organs, in more detail.

Histology of the alimentary canal

The walls of the alimentary canal, especially from the esophagus to the anal canal, have the same basic arrangement of tissues. The four coats of the alimentary canal from the inside out are the mucosa, submucosa, muscularis, and serosa or adventitia (Figure 20-2).

The **mucosa,** or inner lining of the alimentary canal, is a mucous membrane attached to a very thin layer of muscle. Two layers comprise the membrane: a lining epithelium, which is in direct contact with the food, and an underlying layer of connective tissue called the *lamina propria.* Under the lamina propria is a thin layer of muscle called the *muscularis mucosa.*

The epithelial layer is composed of non-keratinized cells that are stratified in the mouth and esophagus but are simple throughout the rest of the tract. The functions of the stratified epithelium are protection and secretion, and the functions of the simple epithelium are secretion and absorption. However, the lack of keratin allows some absorption to occur in all parts of the tract.

The lamina propria is made of connective tissue proper containing many blood and lymph vessels and scattered lymph nodules. This layer supports the epithelium, binds it to the muscularis mucosa, and provides it with a blood and lymph supply. The blood and lymph vessels are the avenues by which nutrients in the alimentary canal reach the other tissues of the body. The lymph tissue also gives protection against disease. Remember that the alimentary canal is in contact with the outside environment and that it contains food which often carries harmful bacteria. Unlike the skin, the mucous membrane of the tract is not protected from bacterial entry by keratin.

The muscularis mucosa contains muscle fibers that throw the mucous membrane of the intestine into small folds which increase the digestive and absorptive area. It also contains glandular epithelium that secretes products necessary for

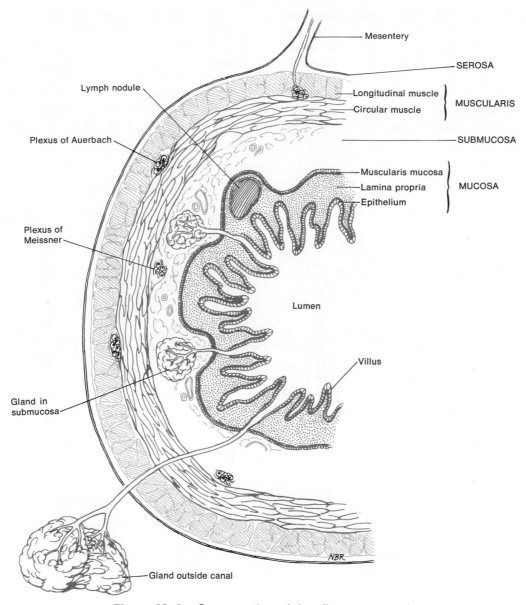

Figure 20–2. Cross section of the alimentary canal.

chemical digestion. With one exception, which will be described later, the other three coats of the intestine contain no glandular epithelium.

The **submucosa** consists of dense connective tissue binding the mucosa to the muscularis. It is highly vascular and contains a portion of the *plexus of Meissner,* which is a part of the autonomic nerve supply to the muscularis mucosa.

The **muscularis** of the mouth, pharynx, and esophagus consists in part of skeletal muscle that produces voluntary swallowing. Throughout the rest of the tract, the muscularis consists of smooth muscle that is generally found in two sheets: an inner ring of circular fibers and an outer sheet of longitudinal fibers. Contractions of the smooth muscles help to break down food physically, mix

it with digestive secretions, and propel it through the tract. The muscularis also contains the major nerve supply to the alimentary tract. This is called the *plexus of Auerbach,* and it consists of fibers from both autonomic divisions.

The **serosa,** the outermost layer of the canal, is a serous membrane composed of connective tissue and epithelium. This covering, also called the peritoneum, is worth discussing in more detail.

The peritoneum

The **peritoneum** is the largest serous membrane of the body. It consists of a layer of simple squamous epithelium and an underlying supporting layer of connective tissue. The *visceral peri-*

toneum covers some of the organs and constitutes their serosa; the *parietal peritoneum* lines the walls of the abdominopelvic cavity. The space between the parietal and visceral peritonea is called the *peritoneal cavity.*

Unlike the two other serous membranes of the body, the pericardium and pleura, the peritoneum contains large folds that weave in between the viscera. The folds bind the organs to each other and to the walls of the cavity and contain the blood and lymph vessels and the nerves that supply the abdominal organs. One extension of the peritoneum is called the *mesentery* and is an outward fold of the serous coat of the intestines (Figure 20–3a). Attached to the posterior abdominal wall is the tip of the fold. The mesentery binds the

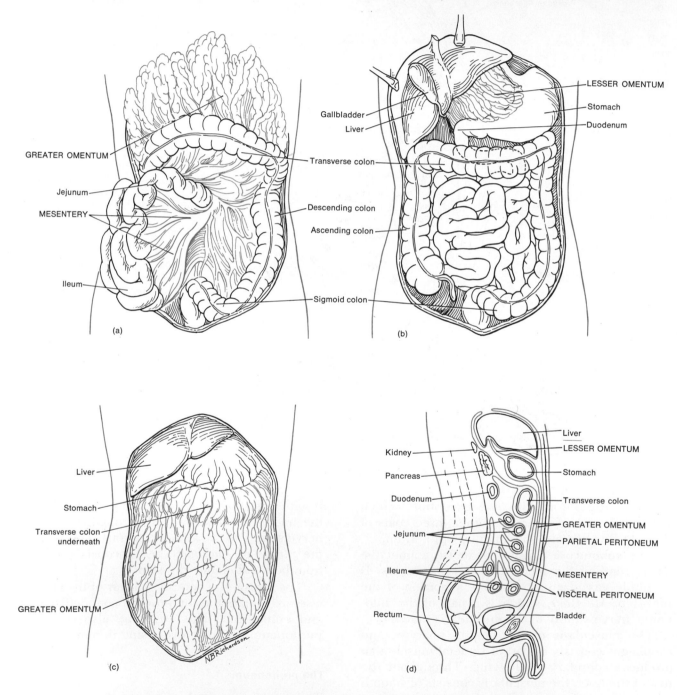

Figure 20–3. Extensions of the peritoneum. (a) Mesentery. The greater omentum has been lifted. (b) Lesser omentum. The liver and gallbladder have been lifted. (c) Greater omentum. (d) Sagittal section through the abdomen and pelvis indicating the relationship of the peritoneal extensions to each other.

small and most of the large intestines to the wall. It also carries blood vessels and lymphatics to the intestines.

Two other important peritoneal folds are the lesser omentum and the greater omentum. The *lesser omentum* arises as two folds in the serosa of the stomach and duodenum (Figure 20–3b). They extend anteriorly and connect with the visceral peritoneum of the liver. The lesser omentum suspends the stomach and duodenum from the liver (Figure 20–3d). An extension of the visceral peritoneum of the liver ties it, in turn, to the diaphragm and the upper abdominal wall. The *greater omentum* is a large fold in the serosa of the stomach that hangs down like an apron over the front of the intestines (Figure 20–3c). It then passes up to a part of the large intestine (the transverse colon), wraps itself around it, and finally attaches to the parietal peritoneum of the posterior wall of the abdominal cavity. Because the greater omentum contains large quantities of adipose tissue, it commonly is called the "fatty

apron." The greater omentum contains numerous lymph nodes. If an infection occurs in the intestine, plasma cells formed in the nodes combat the infection and help to prevent it from spreading to the peritoneum. Inflammation of the peritoneum *(peritonitis)* is a serious condition because the peritoneal membranes are continuous with each other, enabling the infection to spread to all the organs in the cavity (Figure 20–3d).

STRUCTURE AND PHYSIOLOGY OF THE DIGESTIVE ORGANS

The first portion of the alimentary canal we shall consider is the mouth—the first structure in the canal that is concerned with the physical and chemical digestion of food.

The mouth

The **mouth,** also referred to as the *oral* or *buccal cavity,* is formed by the cheeks, hard and soft

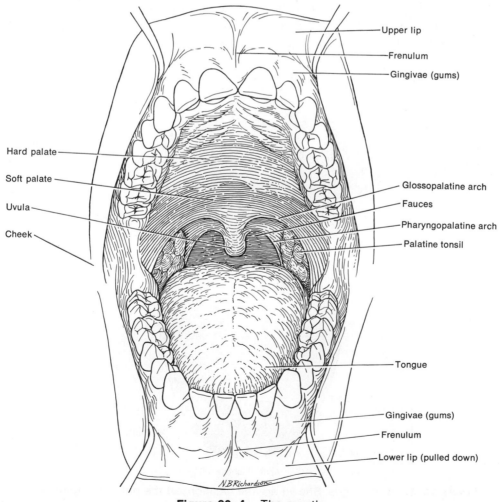

Figure 20–4. The mouth.

palates, and tongue (Figure 20–4). Forming the lateral walls of the oral cavity are the cheeks, which are muscular structures lined by stratified squamous epithelium. The anterior portions of the cheeks terminate in the lips, which surround the orifice (opening) of the mouth. The lips consist of striated muscle and connective tissue, covered with a mucosa of stratified squamous epithelium. During chewing the cheeks and lips help to keep food between the upper and lower teeth. They also assist in speech.

The **hard palate,** which constitutes the anterior portion of the roof of the mouth, is formed by the maxillae and palatine bones and is lined by mucous membrane (see Exhibit 8–13). The **soft palate** forms the posterior portion of the roof of the mouth. It is an arch-shaped muscular partition between the mouth and nasopharynx and is lined by mucous membrane. Hanging from the middle of the lower border of the soft palate is a finger-like muscular process called the *uvula.* On either side of the base of the uvula are two muscular folds that run down the lateral side of the soft palate. Anteriorly, the *glossopalatine arch* runs downward, laterally, and forward to the side of the base of the tongue. Posteriorly, the *pharyngo-*

palatine arch projects downward, laterally, and backward to the side of the pharynx. The palatine tonsils are situated between the arches, and the lingual tonsils are situated at the base of the tongue. At the posterior border of the soft palate, the mouth opens into the oropharynx through an opening called the *fauces.*

The **tongue,** together with its associated muscles, forms the floor of the oral cavity. It is composed of skeletal muscle covered with mucous membrane (Figure 20–5). The extrinsic muscles of the tongue originate outside the tongue and insert into it. They move the tongue from side to side and in and out and maneuver food for chewing and swallowing. The intrinsic muscles originate and insert within the tongue, and they alter the shape of the tongue for speech and swallowing. A fold of mucous membrane in the midline of the undersurface of the tongue, the *frenulum,* attaches the tongue to the floor of the oral cavity (see Figure 20–4). In individuals whose frenulum is too short, tongue movements are restricted, speech is faulty, and the person is said to be "tongue-tied." These functional problems can be corrected very easily by cutting the frenulum surgically.

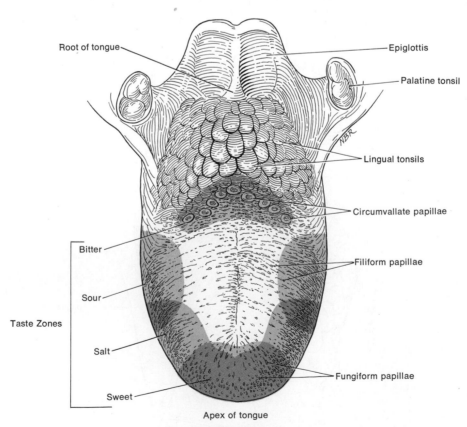

Figure 20–5. The tongue. Shown are the locations of the papillae and the four taste zones.

The upper surface and sides of the tongue contain projections of the lamina propria covered with epithelium called *papillae*. Taste buds are located within some papillae. *Filiform papillae* are conical projections distributed in parallel rows over the anterior two-thirds of the tongue and contain no taste buds. *Fungiform papillae* are mushroomlike elevations distributed among the filiform papillae and are more numerous near the tip of the tongue. They appear as red dots on the surface of the tongue, and most of them contain taste buds. *Circumvallate papillae* are arranged in the form of an inverted V on the posterior surface of the tongue, and all of them contain taste buds. Note the relative positions of the taste zones of the tongue in Figure 20–5.

Salivary glands

Saliva is a fluid that is continuously secreted by glands lying in or near the mouth. Ordinarily, just enough saliva is secreted to keep the mucous membranes of the mouth moist. But when food enters the mouth, secretion increases so the saliva can lubricate, dissolve, and chemically break down the food. The mucous membrane lining the mouth contains many small glands, the *buccal glands*, that secrete small amounts of saliva. However, the major portion of saliva is secreted by the **salivary glands**, which lie outside the mouth and pour their contents into ducts that empty into the oral cavity. There are three pairs of salivary glands: the parotid, submandibular, and sublingual glands (Figure 20–6). The *parotid glands* are located under and in front of the ears. Each secretes into the oral cavity via a duct that opens into the inside of the cheek opposite the upper second molar tooth. The *submandibular glands* are found beneath the base of the tongue in the posterior part of the floor of the mouth, and their ducts are situated on either side of the frenulum. The *sublingual glands* are anterior to the submandibular glands, and their ducts open into the floor of the mouth.

Saliva

The fluids secreted by the buccal glands and the three pairs of salivary glands constitute *saliva*. Amounts of saliva secreted daily vary considerably but range from 1,000 to 1,500 milliliters. Chemically, saliva is 99.5 percent water and 0.5

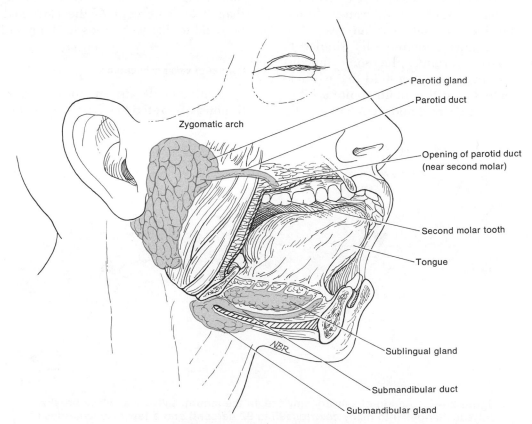

Parotid gland

Parotid duct

Zygomatic arch

Opening of parotid duct (near second molar)

Second molar tooth

Tongue

Sublingual gland

Submandibular duct

Submandibular gland

Figure 20–6. The salivary glands.

percent solutes. Among the solutes are salts, such as chlorides, bicarbonates, and phosphates of sodium and potassium. Some dissolved gases and various organic substances including urea and uric acid, serum albumin and globulin, mucin, the bacteriolytic enzyme lysozyme, and the digestive enzyme amylase are also present.

The water in saliva provides a medium for dissolving foods so they can be tasted and in which digestive reactions can take place. The chlorides in the saliva activate the amylase. The bicarbonates and phosphates buffer chemicals that enter the mouth and keep the saliva at a slightly acidic pH of 6.35 to 6.85. Urea and uric acid are found in saliva because the saliva-producing glands help the body to get rid of wastes like the sweat glands of the skin. Mucin is a protein that forms mucus when it is dissolved in water. Mucus lubricates the food so it can be easily turned in the mouth and swallowed. The enzyme lysozyme destroys bacteria, thereby protecting the mucous membrane from infection.

Saliva and digestion

Depending on the kinds of cells the gland contains, each saliva-producing gland supplies different ingredients to saliva. The parotids contain cells that secrete a thin watery liquid containing the enzyme salivary amylase. The submandibular glands contain cells similar to those found in the parotids plus some mucous cells. Therefore, they secrete a fluid that is thickened with mucus but still contains quite a bit of enzyme. The sublingual glands contain mostly mucous cells, so they secrete a much thicker fluid that contributes only a small amount of enzyme to the saliva.

The enzyme salivary amylase initiates the breakdown of carbohydrates, which is the only chemical digestion that occurs in the mouth. Recall from Chapter 2 that carbohydrates are starches and sugars and that they are classified as either monosaccharides, disaccharides, or polysaccharides. Monosaccharides are small molecules containing several carbon, hydrogen, and oxygen atoms. An example of a monosaccharide is glucose. Disaccharides consist of two monosaccharides linked together; polysaccharides are chains of three or more monosaccharides. The vast majority of carbohydrates that we eat are polysaccharides. Since only monosaccharides can be absorbed into the bloodstream, ingested disaccharides and polysaccharides must be broken down. The function of *salivary amylase* is to break the chemical bonds between some of the monosaccharides that make up the polysaccharides. In this way, the enzyme breaks the long-chain polysaccharides into shorter polysaccharides called *dextrins* (Figure 20–7). Given sufficient time, salivary amylase also can break down the dextrins into disaccharides. However, food usually is swallowed too quickly for more than 3 to 5 percent of the carbohydrates to be reduced to dissaccharides in the mouth.

Control of salivary secretion

Normally, moderate amounts of saliva are continuously secreted to keep the mucous membranes

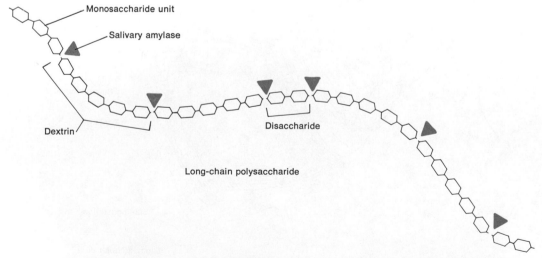

Figure 20–7. Action of salivary amylase. In the mouth, salivary amylase breaks polysaccharides into many dextrins (95 to 97 percent) and a few disaccharides (3 to 5 percent).

moist and to lubricate the movements of the tongue and lips during speech. The saliva is then swallowed and reabsorbed to prevent fluid loss. Dehydration, however, causes the salivary and buccal glands to cease secreting saliva to conserve water. The subsequent feeling of dryness in the mouth promotes sensations of thirst. Food stimulates the glands to secrete heavily. When food is taken into the mouth, chemicals in the food stimulate the taste receptors. Rolling a dry, indigestible object over the tongue produces friction, which also may stimulate the receptors. Impulses are conveyed from the receptors to two salivary centers in the brain stem. Returning autonomic impulses from one of the centers activate the secretion of saliva from the parotid glands, while returning autonomic impulses from the other center activate the submandibular and sublingual glands.

The smell or sight of food also serves as a stimulus for increased saliva secretion. This phenomenon is learned behavior and does not involve the taste buds. Memories stored in the cerebral cortex that associate the stimuli with food are stimulated. The cortex sends impulses to the brain stem, and the salivary glands are thereby activated. Psychological activation of the glands has some benefit to the body because it allows the mouth to start chemical digestion as soon as the food is ingested.

Saliva continues to be secreted heavily some time after food is swallowed. This is beneficial because the continued heavy flow of saliva washes out the mouth and dilutes and buffers chemical remnants of any irritating substances that might have been in the food.

Teeth

The **teeth,** or **dentes,** are located in sockets of the alveolar processes of the mandible and maxillae. The alveolar processes are covered by the *gingivae* (gums), which extend slightly into each socket (Figure 20–8). Periosteum, the fibrous membrane that covers all bone, lines the sockets. The periosteum lining the sockets is called the *periodontal membrane.*

A typical tooth consists of three principal

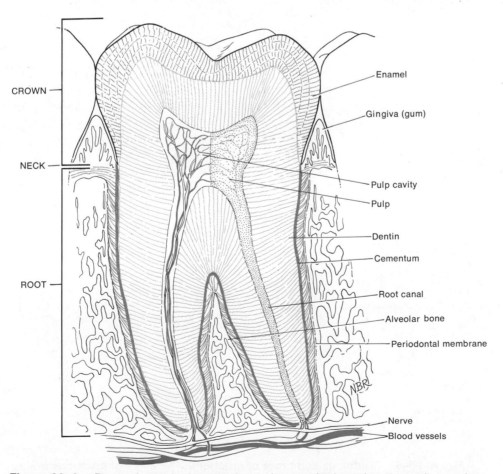

CROWN

NECK

ROOT

Enamel

Gingiva (gum)

Pulp cavity

Pulp

Dentin

Cementum

Root canal

Alveolar bone

Periodontal membrane

Nerve

Blood vessels

Figure 20–8. Parts of a typical tooth shown in a sagittal section through a molar.

portions (Figure 20–8). The *crown* is the portion above the level of the gums, the *root* consists of one to three projections embedded in the socket, and the *neck* is the constricted region between the crown and the root.

Teeth are composed primarily of *dentin,* a bonelike substance that gives the tooth its basic shape and rigidity. The dentin encloses a cavity. The enlarged part of the cavity, the *pulp cavity,* lies within the crown and neck and is filled with *pulp,* a connective tissue containing blood vessels, nerves, and lymphatics. Narrow extensions of the pulp cavity lying within the roots of the tooth are called the *root canals.* Each root canal has an opening at its base. Blood and lymph vessels and nerves from the peridontal membrane run through the opening in the root and travel up the root canal to the pulp cavity, where they provide the tooth with nourishment. The dentin of the crown is covered by *enamel* that consists primarily of calcium phosphate. Enamel is the hardest substance in the body and protects the tooth from the wear and tear of chewing. It is also an effective barrier against acids that easily dissolve the dentin. The dentin of the root

is covered by *cementum,* another bonelike substance. Cementum is bound to the underlying bone by the peridontal membrane. *Pyorrhea* is an inflammation of the periodontal membrane and adjacent gums. A prolonged, severe case of pyorrhea can weaken the periodontal membrane, erode the alveolar bone, and thereby cause the tooth to become loose.

Dentitions

Each individual has two *dentitions,* or sets of teeth. The first of these, the *deciduous* or *milk teeth,* begin to erupt at about 6 months of age, and one appears at about each month thereafter until all 20 are present. Figure 20–9a illustrates the deciduous teeth. The *incisors,* which are closest to the midline, are chisel-shaped and are adapted for cutting and biting food. Next to the incisors, moving posteriorly, are the conical-shaped *canines* or *cuspids,* which have a flat grinding surface called the cusp. Canines are used to tear and shred food. The incisors and canines have only one root apiece. Behind them lie the first and second *molars,* or *tricuspids,*

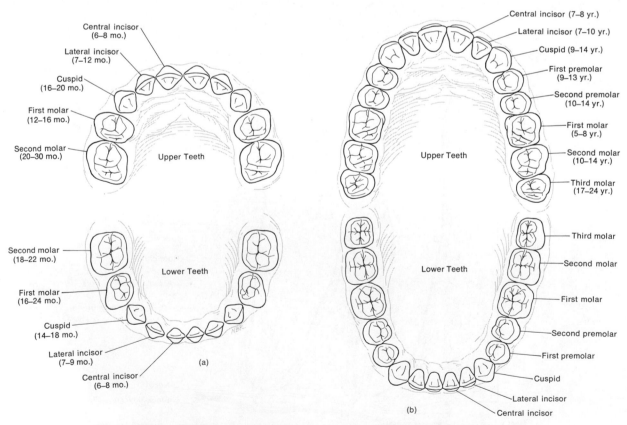

Figure 20–9. Dentitions and times of eruptions. The times of eruptions are indicated in parentheses. (a) Deciduous dentition. (b) Permanent dentition.

which have three or four cusps and two roots. The molars crush and grind food.

All the deciduous teeth are lost—generally between 6 and 13 years of age—and are replaced by the *permanent dentition* (Figure 20–9b). The permanent dentition contains 32 teeth that appear between the age of 6 and adulthood. It resembles the deciduous dentition with the following exceptions: The deciduous first molars are replaced with *premolars* or *bicuspids*, which have two cusps and two roots and are used for crushing and grinding. Behind the premolars lie three sets of molars. The *third molars*, or *wisdom teeth*, do not appear before the age of 17. In fact, in some people they do not come through at all.

Teeth and digestion

Through chewing, or *mastication*, the teeth pulverize food and mix it with saliva. As a result, the food is reduced to a soft, flexible mass, called a *bolus*, that is easily swallowed. Exhibit 20–1 summarizes the digestion that occurs in the mouth. As we continue our discussion of digestion, similar Exhibits will be presented, so you will have a complete summary of the digestive activities that take place from the mouth through the anus.

Exhibit 20–1. SUMMARY OF DIGESTION IN THE MOUTH

STRUCTURE	ACTIVITY	RESULT	STRUCTURE	ACTIVITY	RESULT
Cheeks	Keep food between teeth during mastication	Foods are uniformly chewed.	Buccal glands	Secrete saliva	Lining of mouth and pharynx moistened and lubricated.
Lips	Keep food between teeth during mastication	Foods are uniformly chewed.	Salivary glands	Secrete saliva	Same as above. Saliva softens and moistens food, coats food with mucin, cleanses mouth and teeth. Salivary amylase reduces carbohydrates to dextrins and disaccharides.
Tongue Extrinsic muscles	Move tongue from side to side and in and out	Maneuver food for mastication and deglutition.			
Intrinsic muscles	Alter shape of tongue	Deglutition.			
Taste buds	Serve as receptors for food stimulus	Nerve impulses from taste buds to brain to salivary glands stimulate the secretion of saliva.	Teeth	Pulverize food	Solid foods are reduced to smaller particles for chemical digestion and swallowing.

Deglutition: Mouth, pharynx, and esophagus

Swallowing, or *deglutition*, moves food from the mouth to the stomach. Swallowing starts with the bolus on the upper side of the tongue. Then the tip of the tongue rises and presses against the palate (Figure 20–10). The bolus slides by the force of gravity to the back of the mouth and is pulled through the fauces by muscles that lie in the pharynx. During this period the respiratory passageways close, and breathing is temporarily interrupted. The soft palate and uvula move upward to close off the nasopharynx, and the larynx is pulled forward and upward under the tongue. As the larynx rises, it meets the epiglottis, which seals off the glottis. The movement of the larynx also pulls the vocal cords together, further sealing off the respiratory tract, and widens the opening between the pharynx and esophagus. The bolus passes through the pharynx and enters the esophagus in 1 second. The respiratory passageways then reopen, and breathing resumes.

The **esophagus**, the third organ involved in deglutition, is a muscular, collapsible tube that lies behind the trachea (see Figure 20–1). It is about 23 to 25 centimeters (10 inches) long and begins at the end of the laryngopharynx, passes through the mediastinum in front of the vertebral column, pierces the diaphragm, and terminates in the upper portion of the stomach.

Food is pushed through the esophagus by muscular movements called **peristalsis** (Figure

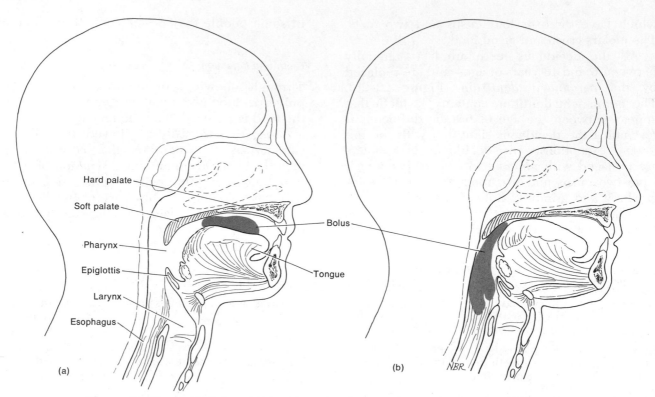

(a)

(b)

NBR

Figure 20–10. Deglutition. (a) Position of structures prior to swallowing. (b) During swallowing, the tongue rises against the palate, the nose is closed off, the larynx rises, the epiglottis seals off the larynx, and the bolus is passed into the esophagus.

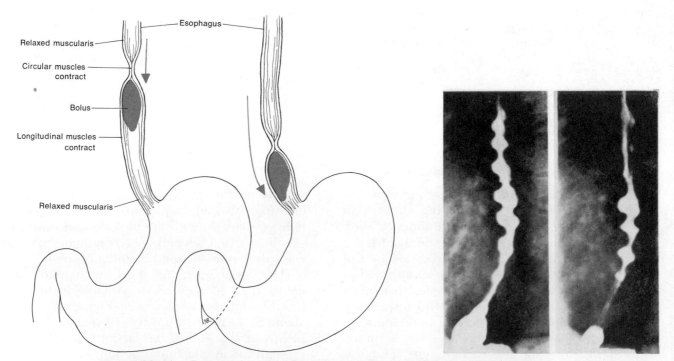

Figure 20–11. Peristalsis. (Left) Diagrammatic representation. (Right) Roentgenograms of peristalsis made during fluoroscopic examination while a patient was swallowing a barium "meal." [*(Right) from Lester W. Paul and John H. Juhl, The Essentials of Roentgen Interpretation, 3d ed., Harper & Row, Publishers, Inc., New York, 1972.*]

20-11). Peristalsis is a function of the muscularis, and it occurs as follows: In the section of the esophagus lying just above and around the top of the bolus, the circular muscle fibers contract. The contraction constricts the esophageal wall and squeezes the bolus downward. Meanwhile, longitudinal fibers lying around the bottom of the bolus and just below it also contract. Contraction of the longitudinal fibers shortens this lower section, pushing its walls outward so it can receive the bolus. The contractions are repeated in a wave that moves down the esophagus, pushing the food toward the stomach. Passage of the bolus is further facilitated by glands secreting mucus and by gravity. The passage of solid or semisolid food from the mouth to the stomach takes about 4 to 8 seconds. Very soft foods and liquids pass through in about 1 second.

A valve at the base of the esophagus, the *cardiac sphincter,* regulates the passage of the bolus from the esophagus into the stomach. A **sphincter** is an opening that has a thick circle of muscles around it. When the muscles are contracted, they constrict the opening so nothing can pass through. When the muscles relax, the passageway opens. The cardiac sphincter opens just long enough to permit passage of the bolus into the stomach. During the rest of the time, the sphincter is closed to prevent the contents of the stomach from being regurgitated into the esophagus. Exhibit 20-2 contains a summary of the digestion-related activities of the pharynx and esophagus.

Exhibit 20-2. DIGESTIVE ACTIVITIES OF PHARYNX AND ESOPHAGUS

STRUCTURE	ACTIVITY	RESULT
Pharynx	Deglutition	Food is passed from the pharynx into the esophagus. Air passageways are closed off and opening to the esophagus is widened.
Esophagus	Peristalsis	Bolus is forced down the esophagus into the stomach.
	Secretion of mucus	Bolus passes smoothly down esophagus.

Stomach

The **stomach** is a J-shaped enlargement of the alimentary canal directly under the diaphragm in the epigastric, umbilical, and left hypochondriac regions of the abdomen (see Figure 20-1). The superior portion of the stomach is a continuation of the esophagus, and the inferior portion empties into the duodenum, the first part of the small intestine. Within each individual, the position and size of the stomach vary continually. For instance, the diaphragm pushes the stomach downward with each inspiration and pulls it upward with each expiration. When the stomach is empty it is about the size of a large sausage, but when food enters, the stomach can stretch itself immeasureably to accommodate its contents.

Anatomy

The stomach is divided into four areas: the cardia, fundus, body, and pylorus (Figure 20-12 left). The *cardia* surrounds the cardiac sphincter, and the rounded portion above and to the left of the cardia is the *fundus.* Below the fundus, the large central portion of the stomach is called the *body,* and the more narrow, inferior region is the *pylorus.* The concave medial border of the stomach is called the *lesser curvature,* and the convex lateral border is referred to as the *greater curvature.* The pylorus communicates with the duodenum via a sphincter called the *pyloric valve.*

Two abnormalities of the pyloric valve sometimes are found in infants. One of these abnormalities, *pylorospasm,* is characterized by failure of the muscle fibers encircling the opening to relax normally. As a result, ingested food does not pass easily from the stomach to the small intestine. The stomach becomes overly full, and the infant vomits frequently to relieve the pressure. Pylorospasm is treated by drugs that relax the muscle fibers of the valve. The other abnormality, called *pyloric stenosis,* is a narrowing of the pyloric valve caused by a tumorlike mass that apparently is formed by enlargement of the circular muscle fibers. The mass obstructs the passage of food and must be surgically corrected.

The wall of the stomach is composed of the same four basic layers as the rest of the alimentary canal, with certain modifications. When the stomach is empty, the mucosa lies in large folds that can be seen with the naked eye. These folds are called *rugae* (Figure 20-12 left). As the stomach fills and distends, the rugae gradually smooth out and disappear. Microscopic inspection of the mucosa reveals a layer of simple columnar epithelium containing many narrow openings that extend down into the lamina propria. These pits

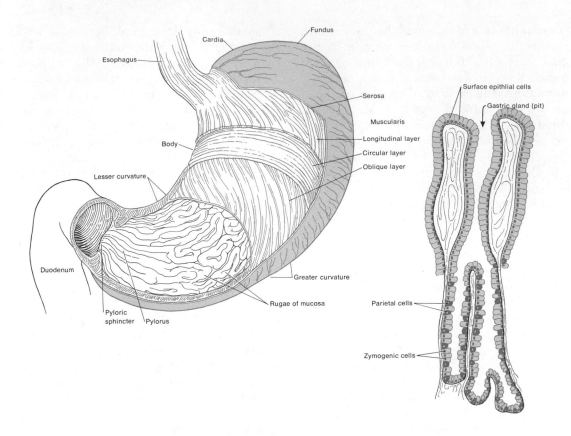

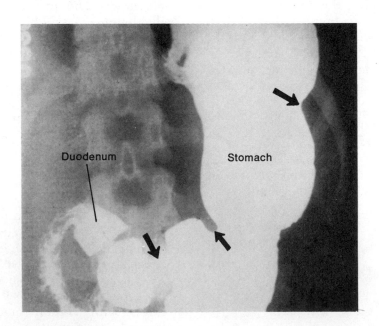

Figure 20–12. The stomach. (Left) External and internal anatomy. (Right) Microscopic view of gastric glands from the fundus region. (Bottom) Roentgenogram of a normal stomach. Note the peristaltic waves. [*(Bottom) from Lester W. Paul and John H. Juhl, The Essentials of Roentgen Interpretation, 3d. ed., Harper & Row, Publishers, Inc., New York, 1972.*]

are called *gastric glands,* and they are lined with three kinds of secreting cells: zymogenic, parietal, and mucous (Figure 20–12 right). The *zymogenic,* or *chief cells,* secrete the digestive enzymes of the stomach. Hydrochloric acid, which stimulates one of the digestive enzymes, is produced by the *parietal cells.* The *mucous cells* secrete mucus and the intrinsic factor, a substance involved in the absorption of vitamin B_{12}. Secretions of the gastric glands are collectively called *gastric juice.*

The submucosa of the stomach is composed of loose areolar connective tissue, and it connects the mucosa to the muscularis. The muscularis, unlike that in other areas of the alimentary canal, has three layers of smooth muscle: an outer longitudinal layer, a middle circular layer. and an inner oblique layer (Figure 20–12 left). This arrangement of the fibers allows the stomach to contract in a variety of ways to churn food, break it into small particles, mix it with gastric juice, and pass it to the duodenum. The serosa covering the stomach is part of the peritoneum. At the lesser curvature the two layers of the peritoneum come together and extend upward to the liver as the lesser omentum. At the greater curvature, the peritoneum continues downward as the greater omentum hanging over the intestines.

Digestion in the stomach

Several minutes after food enters the stomach, gentle, rippling, peristaltic movements called *mixing waves* pass over the stomach. These waves occur about every 15 to 25 seconds and serve to macerate food, mix it with the secretions of the digestive glands, and reduce it to a thin liquid called *chyme.* Relatively few mixing waves are observed in the fundus, which is primarily a storage area. Foods may remain in the fundus for an hour or more without becoming mixed with gastric juice. During this time, salivary digestion continues.

The principal chemical activity of the stomach is to begin the digestion of proteins. See Fig. 20–18a. In the adult, this is achieved primarily through the enzyme *pepsin.* Pepsin breaks the peptide bonds between the amino acids making up proteins. Thus a protein chain of many amino acids is broken down into fragments containing 4–12 amino acids – longer fragments are called *proteoses,* and the shorter ones are called *peptones.* Pepsin is most effective in the very acidic environment of the stomach, which has a pH of 1. It becomes inactive in an alkaline environment.

As you know, living cells are composed in part of proteins. What keeps pepsin from digesting the cells of the stomach along with the food? First of all, pepsin is secreted in an inactive form called *pepsinogen,* so it cannot digest the proteins in the zymogenic cells that produce it. When pepsinogen comes in contact with the hydrochloric acid secreted by the parietal cells, it is converted to active pepsin. Once pepsin has been activated, the cells of the stomach are protected by mucus. The mucus coats the mucosa and forms a barrier between the gastric juice and the cells. Sometimes the mucus fails to do its job, and the pepsin and hydrochloric acid eat a hole known as a *gastric ulcer* in the stomach wall.

Another enzyme found in gastric juice is *renin,* a milk-curdling agent. Renin breaks some of the peptide bonds between the amino acids in casein, the protein found in milk. The partially digested product then reacts with calcium to form insoluble clots called curds, and the remaining liquid portion of milk is called whey – words memorialized in the nursery rhyme "Little Miss Muffet." In the human stomach pepsin then acts on the curd to break the protein into peptones and proteoses. Renin plays an important role in the digestive processes of infants but probably is not used in adult digestive processes because it works best at a pH of 5 to 6, which is typical of an infant stomach. In the much more acidic adult stomach, renin becomes relatively inactive, and hydrochloric acid takes over the job of milk curdling. Renin taken from the stomachs of young animals is used to solidify milk in the making of certain dairy products.

The third enzyme of the stomach is *gastric lipase.* Gastric lipase splits the butterfat molecules found in milk. Like renin, this enzyme operates best at a pH of 5 to 6 and has a limited role in the adult stomach. Adults rely exclusively on an enzyme found in the small intestine to digest fats.

As digestion proceeds in the stomach, more vigorous peristaltic waves begin at about the middle of the stomach, pass downward, reach the pyloric valve, and sometimes go into the duodenum. The actual movement of chyme from the stomach into the duodenum depends on a pressure gradient between the two organs. When the pressure in the stomach (intragastric pressure) is greater than that in the duodenum (intraduodenal pressure), chyme is forced into the duodenum. Peristaltic waves are largely responsible for increased intragastric pressure. It is estimated that 2–5 milliliters of chyme are passed into the duodenum with each peristaltic wave. When intraduodenal pressure exceeds intragastric pressure, the pyloric valve closes and prevents the regurgitation of chyme from the duodenum to the stomach. The stomach empties all its contents into the duodenum about 2–6 hours after ingestion. Food rich in carbohydrate leaves the stomach in a few hours. Protein foods are somewhat slower, and emptying is slowest after a meal containing large amounts of fat. The stomach wall is impermeable to the passage of most materials into the blood, so most substances are not absorbed until they reach the small intestine. However, the stomach does participate in the absorption of some water and salts, certain drugs, and alcohol.

Regulation of gastric secretion

The secretion of gastric juice is regulated by both nervous and hormonal mechanisms. Seeing, smelling, tasting, or thinking of food stimulates the cerebral cortex to send impulses to the medulla. The medulla relays impulses over the parasympathetic fibers in the vagus nerve to stimulate the gastric glands to secrete. Psychic stimulation is important because it prepares the stomach for digestion.

Once the food reaches the stomach, both nervous and hormonal mechanisms ensure that gastric secretion continues. Food of any kind stimulates receptors lying in the walls of the stomach. These receptors send impulses over a reflex arc to the medulla and back to the gastric glands, and they may send messages directly to the glands as well. Emotions such as anger, fear, and anxiety may slow down digestion in the stomach because they stimulate the sympathetic nervous system that, in turn, inhibits the impulses of the parasympathetic fibers. Protein foods stimulate the pyloric mucosa to secrete a hormone called *gastrin*. Gastrin is absorbed into the bloodstream and carried to the gastric glands, where it stimulates secretion of large amounts of digestive enzymes and hydrochloric acid.

Some investigators believe that when proteoses and peptones leave the stomach and enter the duodenum, they stimulate the intestinal mucosa to release a gastrinlike hormone which stimulates the gastric glands to continue their secretion. However, this mechanism produces relatively small amounts of gastric juice. Exhibit 20–3 summarizes the chief activities of the stomach.

The next step in the breakdown of food is digestion in the small intestine. However, chemical digestion in the small intestine is dependent not only on its own secretions but on those from three organs that lie outside the alimentary canal. These organs—the pancreas, liver, and gallbladder—will be discussed before we continue with the alimentary canal.

Pancreas

The **pancreas** is a soft, oblong-shaped organ about 12.5 centimeters (6 inches) long and 2.5 centimeters (1 inch) thick. It lies along the greater curvature of the stomach and is connected by a duct to the duodenum (Figure 20–13a). The

Exhibit 20–3. SUMMARY OF GASTRIC DIGESTION

STRUCTURE	ACTIVITY	RESULT
Mucosa		
1. Rugae	Provide a large surface area for stretching of stomach	Allow for distention of stomach
2. Mucous cells	Secrete mucus Secrete intrinsic factor	Prevents digestion of stomach wall Required for erythrocyte formation
3. Zymogenic cells	Secrete pepsinogen	Its active form (pepsin) digests proteins into proteoses and peptones
4. Parietal cells	Secrete hydrochloric acid	In its presence, pepsinogen is converted into pepsin
Muscularis	Mixing waves	Macerate food, mix it with gastric juice, and reduce food to chyme
	Peristaltic waves	Force chyme through the pyloric valve into the duodenum
Cardiac valve	Regulates passage of bolus from the esophagus into the stomach	Prevents backflow of food from stomach to esophagus
Pyloric valve	Opens to permit passage of chyme into duodenum	Prevents backflow of food from duodenum to stomach

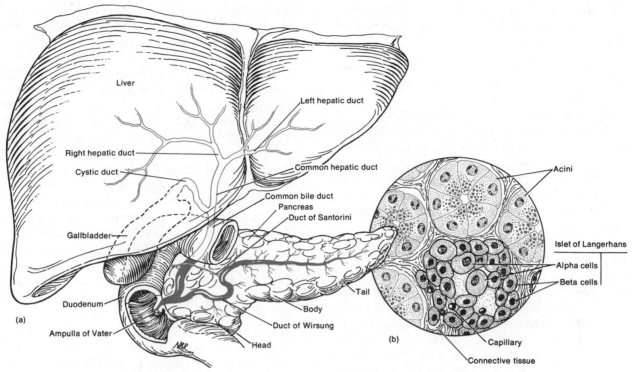

Figure 20–13. The pancreas. (a) The pancreas, liver, and gallbladder in relation to the duodenum. (b) Microscopic view of pancreatic cells.

pancreas is made up of small clusters of glandular epithelial cells (Figure 20–13b). Some of the clusters, called islets of Langerhans, form the endocrine portions of the pancreas and consist of alpha and beta cells that secrete glucagon and insulin. The other masses of cells, called *acini*, are the exocrine portions of the organ. Secreting cells of the acini release a mixture of digestive enzymes called *pancreatic juice*, which is dumped into small ducts attached to the acini. Pancreatic juice eventually leaves the pancreas through a single large tube called the *pancreatic duct*, or *duct of Wirsung*. In many people the pancreatic duct often unites with the common bile duct from the liver and gallbladder and enters the duodenum in a small, raised area called the *ampulla of Vater*. An accessory duct, the *duct of Santorini*, also leads from the pancreas and empties into the duodenum about an inch above the ampulla of Vater.

The functions of the pancreas are twofold. The acini secrete enzymes that digest food in the small intestine, and the alpha and beta cells secrete glucagon and insulin, which control the fate of digested and absorbed carbohydrates.

Liver

The **liver** performs so many vital functions that we cannot live long without it. The most important

of these are listed below. You already know some of these functions. The rest will be more meaningful to you after you have read the chapters on nutrition and excretion.

1. The liver manufactures the anticoagulant heparin and most of the other plasma proteins.
2. The reticuloendothelial cells of the liver phagocytoze worn-out red blood cells and some bacteria.
3. Liver cells contain enzymes that either break down poisons or transform them into less harmful compounds. For example, when amino acids are burned for energy, they leave behind toxic nitrogenous wastes that are converted to urea by the liver cells. Moderate amounts of urea are harmless to the body and are easily excreted by the kidneys and sweat glands.
4. Newly absorbed nutrients are collected in the liver. It can change any excess monosaccharides into glycogen or fat, both of which can be stored. In addition it can transform glycogen, fat, and protein into glucose and vice versa, depending on the body's needs.
5. The liver stores glycogen, copper, iron, and vitamins A, D, E, and K. It also stores some poisons that cannot be broken down and excreted. (This is why high levels of DDT are found in the livers of animals, including man, who eat sprayed fruits and vegetables.)

6. Finally, the liver manufactures bile, which is used in the small intestine for the digestion and absorption of fats.

The liver is the largest single organ in the body, weighing about 1.4 kilograms (4 pounds) in the average adult. It is located under the diaphragm and occupies most of the right hypochondrium and part of the epigastrium of the abdomen. The liver is covered largely by peritoneum and completely by a dense connective tissue layer that lies beneath the peritoneum. Anatomically, the liver is divided into two principal lobes—the *right lobe* and the *left lobe*—separated by the *falciform ligament* (Figure 20–14a).

The lobes of the liver are made up of numerous functional units called *lobules*, which may be seen under a microscope (Figure 20–14b). A lobule consists of cords of *hepatic* (liver) *cells* arranged in a radial pattern around a *central vein*. Between the cords are endothelial-lined spaces called *sinusoids* through which blood passes. The sinusoids are also partly lined with phagocytic cells, termed *Kupffer cells*, that destroy worn-out white and red blood cells.

The liver receives a double supply of blood. From the hepatic artery it obtains oxygenated blood, and from the portal vein it receives deoxygenated blood containing newly digested nutrients (see Figure 17–7). Branches of both the hepatic artery and the portal vein carry the blood into the sinusoids of the lobules, where oxygen, most of the nutrients, and certain poisons are extracted by the hepatic cells. Nutrients are stored or used to make new materials, and the poisons are stored or detoxified. Products manufactured by the hepatic cells and nutrients needed by other cells are secreted back into the blood. The blood then drains into the central vein and eventually passes into the hepatic vein. Unlike the other products of the liver, bile normally is not secreted into the bloodstream.

Bile is manufactured by the hepatic cells and secreted into *bile capillaries* that empty into small ducts. These small ducts eventually merge to form the larger *right* and *left hepatic ducts*, which unite to leave the liver as the *common hepatic duct*. Further on, the common hepatic duct joins the *cystic duct* from the gallbladder, and the two tubes become the *common bile duct*, which empties into the duodenum (see Fig. 20–13a). The *sphincter of Oddi* is a valve in the common bile duct. When the small intestine is empty the sphincter closes, and the bile is forced up the cystic duct to the gallbladder, where it is stored.

Gallbladder

The **gallbladder** is a sac attached to the underside of the liver (see Figure 20–14a). Its inner walls

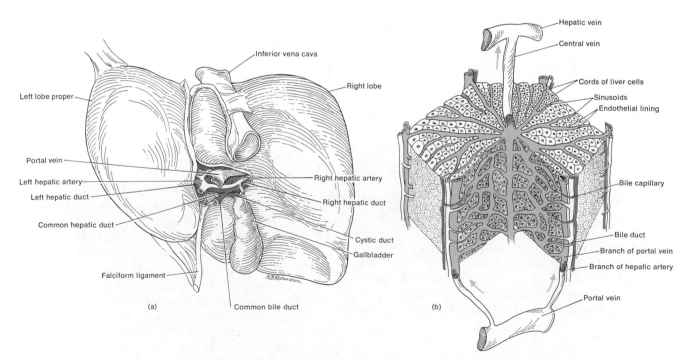

Figure 20–14. The liver. (a) External anatomy in posteroinferior view. (b) Microscopic appearance of a lobule.

consist of a mucous membrane arranged in rugae resembling those of the stomach. When the gallbladder fills with bile, the rugae allow it to expand to the size and shape of a pear. The middle, muscular coat of the wall consists of smooth muscle fibers. Contraction of these fibers ejects the bile into the cystic duct. The outer coat is the peritoneum.

Small intestine

The major portions of digestion and absorption occur within a long tube called the **small intestine.** The small intestine begins at the pyloric valve of the stomach, coils through the central and lower part of the abdominal cavity, and eventually merges with the large intestine (See Figure 20–1). In a living human it averages about 2.5 centimeters (1 inch) in diameter and 3 meters (10 to 12 feet) in length.

Anatomy

The small intestine is divided into three segments: duodenum, jejunum, and ileum. The *duodenum*, the broadest part of the small intestine, originates at the pyloric valve of the stomach and extends about 25 centimeters (10 to 12 inches) until it merges with the jejunum. The *jejunum* is about 1 meter (3 to 4 feet) long and extends to the ileum. The final portion of the small intestine, the *ileum*, measures about 2 meters (6 to 7 feet) and joins the large intestine at the *ileocecal valve*. A roentgenogram of the normal small intestine is shown in Figure 20–15 top.

The walls of the intestine are composed of the same four coats that make up most of the alimentary canal. However, both the mucosa and submucosa are modified to allow the small intestine to complete the processes of digestion and absorption. The mucosa contains many pits lined with glandular epithelium. These pits—the *intestinal glands*, or *crypts of Lieberkühn*—secrete the intestinal digestive enzymes (Fig. 20–15 bottom). The submucosa contains *Brunner's glands*, which secrete mucus to protect the walls of the small intestine from the action of the enzymes. *Succus entericus* is the name for the composite of all the intestinal secretions.

Since almost all the absorption of nutrients occurs in the small intestine, its walls need to be specially equipped to do this job. The epithelium covering and lining the mucosa consists of simple columnar epithelium. Some of the epithelial cells have been transformed to goblet cells, which

secrete additional mucus. The rest contain microvilli—fingerlike projections of the plasma membrane (see Figure 3–7a). Digested nutrients diffuse more quickly into the intestinal wall because the microvilli increase the surface area of the plasma membrane.

The mucosa lies in a series of 2.5 centimeters (1 inch) high projections called *villi*, giving the intestinal mucosa its velvety appearance (Figure 20–15 bottom). The enormous number of villi (4 to 5 million) vastly increases the surface area of the epithelium available for the epithelial cells specializing in absorption. Each villus is lined with the lamina propria, the connective tissue layer of the mucosa. Embedded in this connective tissue are an artery, a venule, a capillary, and a *lacteal* (lymphatic vessel). Nutrients that diffuse through the adjacent epithelial cells are able to pass through the walls of the capillary and lacteal and enter the blood.

A third set of projections called *plicae circulares* further increase the surface area for absorption. The plicae are deep folds in the mucosa and submucosa (Figure 20–15 left). Some of the folds extend all the way around the circumference of the intestine, and others extend only part way around.

The muscularis of the small intestine consists of two layers of smooth muscle. The outer, thinner layer contains longitudinally arranged fibers, and the inner, thicker layer contains circularly arranged fibers. Except for a major portion of the duodenum, the serosa, or visceral peritoneum, completely covers the small intestine.

There is an abundance of lymphatic tissue in the walls of the small intestine. Single lymph nodules, called *solitary lymph nodules*, are most numerous in the lower part of the ileum. Aggregated lymph nodules, referred to as *Peyer's patches*, are also most numerous in the ileum.

Chemical digestion

The digestion of carbohydrates, proteins, and lipids in the small intestine requires the combined actions of the secretions from the pancreas, liver, and intestinal glands.

SECRETIONS. Each day the liver secretes about 800 to 1,000 milliliters of the yellow, brownish, or olive-green liquid called *bile*. Bile consists of water, bile salts, bile acids, a number of lipids, and two pigments called biliverdin and bilirubin. Bile is partially an excretory product and partially a digestive secretion. When red blood cells

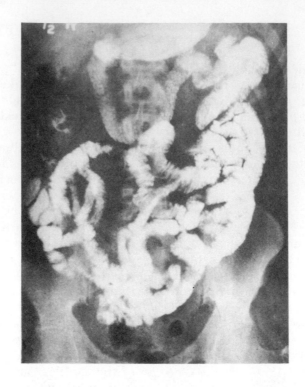

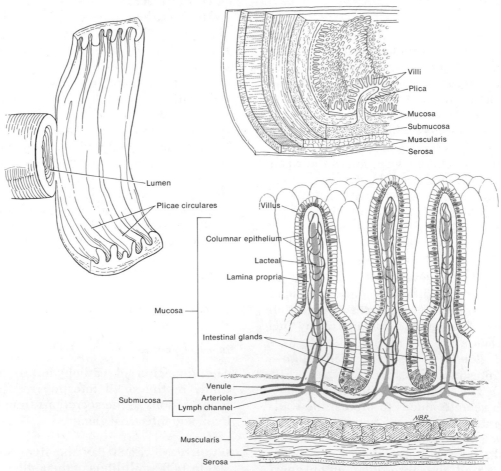

Figure 20–15. The small intestine. (Top) Roentgenogram of the normal small intestine one half hour after taking a barium "meal." (Left) Section of small intestine cut open to expose plicae circulares. (Middle) Villi in relation to the coats of the small intestine. (Bottom) Enlarged aspect of several villi. [*(Top) from Lester W. Paul and John H. Juhl, The Essentials of Roentgen Interpretation, 3d ed., Harper & Row, Publishers, Inc., New York, 1972.*]

are broken down, iron, globin, and bilirubin are released. The iron and globin are recycled, but the bilirubin is excreted into the bile ducts. Bilirubin eventually is broken down in the intestines, and its breakdown products give feces their color. If the liver is unable to remove bilirubin from the blood or if the bile ducts are obstructed, large amounts of bilirubin circulate through the bloodstream and collect in other tissues, giving the skin and eyes a yellow color called *jaundice*. Other substances found in bile aid in the digestion of fats and are required for their absorption.

Each day the pancreas produces 1,200 to 1,500 milliliters of a clear, colorless liquid called *pancreatic juice*. Pancreatic juice consists mostly of water, some salts, sodium bicarbonate, and enzymes. The sodium bicarbonate gives pancreatic juice a slightly alkaline pH (7.1 to 8.2) that stops the action of pepsin and creates the proper environment for the enzymes in the small intestine. The enzymes in pancreatic juice include a carbohydrate-digesting enzyme, several protein-digesting enzymes, and the only active fat-digesting enzyme in the adult body.

The intestinal juice, or *succus entericus,* is a clear yellow fluid secreted in amounts of about 2 to 3 liters/day. It has a pH of 7.6, which is slightly alkaline, and contains water, mucus, and enzymes that complete the digestion of carbohydrates and proteins.

CONTROL OF SECRETION. The small intestine starts to prepare itself for digestion as soon as food enters the mouth. Taste buds send impulses to the brain when they are stimulated. One way the brain responds is by sending impulses over the vagus nerve that stimulate the pancreas to secrete its juice. Thus pancreatic enzymes enter the small intestine within 1 to 2 minutes after food is ingested.

When the chyme leaves the stomach and enters the duodenum, chemicals in the chyme stimulate endocrine cells in the duodenal mucosa to secrete any of a number of hormones. Most of the hormones stimulate secretion of digestive juices. One of them slows down stomach contractions. The hormonal control of digestion is summarized in Exhibit 20–4 and illustrated in Figure 20–16.

THE DIGESTIVE PROCESS. When chyme reaches the small intestine, the carbohydrates and proteins have been only partially digested and are not ready for absorption. Lipid digestion has not even begun. Digestion in the small intestine continues as follows:

1. *Carbohydrates.* In the mouth, the carbohydrates are broken down into dextrins containing several monosaccharide units (Figure 20–17). Even though the action of salivary amylase may continue in the stomach, very few of the carbohydrates are reduced to disaccharides by the time chyme leaves the stomach. *Pancreatic amylase,* an enzyme in pancreatic juice, breaks dextrins into the disaccharides maltose, sucrose, and lactose.

Exhibit 20–4. SUMMARY OF THE HORMONAL CONTROL OF DIGESTION IN THE STOMACH AND SMALL INTESTINE

HORMONE	WHERE PRODUCED	STIMULANT	ACTION
Gastrin	Pyloric mucosa	Partially digested proteins	Causes gastric glands to secrete gastric juice
Gastrinlike hormone	Intestinal mucosa	Partially digested proteins	Same as above
Secretin	Duodenal mucosa	Acidity of chyme	Stimulates secretion of pancreatic juice rich in carbonate and promotes production of bile by the liver
Pancreozymin	Duodenal mucosa	Acidity of chyme	Stimulates secretion of pancreatic juice rich in enzymes
Enterocrinin	Duodenal mucosa	Acidity of chyme	Stimulates the secretion of succus entericus
Cholecystokinin	Intestinal mucosa	Combination of acid and fat	Causes ejection of bile from the gallbladder and opening of the sphincter of Oddi
Enterogastrone	Duodenal mucosa	Fats	Inhibits the secretion of gastric juice and decreases gastric motility

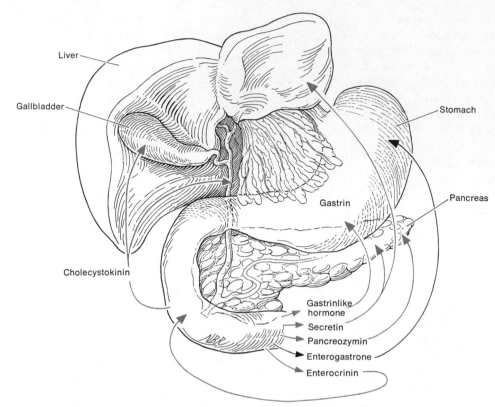

Figure 20–16. Sources and actions of gastrointestinal hormones. The arrows extend from the sources of the hormones to the target organs affected. Colored arrows indicate stimulation of the target organ. The black arrow indicates inhibition of the target organ.

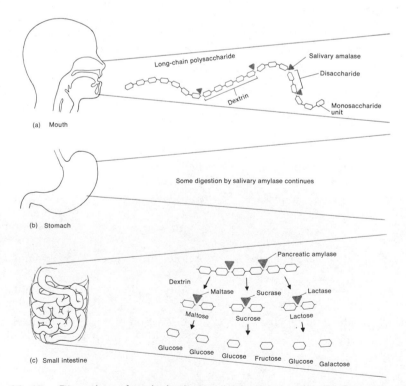

Figure 20–17. Digestion of carbohydrates. (a) In the mouth. (b) Some breakdown continues in the stomach. (c) In the small intestine.

Next, three enzymes in the intestinal juice digest the disaccharides into monosaccharides. *Maltase* splits maltose into two molecules of glucose, *sucrase* breaks sucrose into a molecule of glucose and a molecule of fructose, and *lactase* digests lactose into a molecule of glucose and a molecule of galactose. This completes the digestion of carbohydrates.

2. *Proteins.* Protein digestion starts in the stomach, where most of the proteins are fragmented into short chains of amino acids called peptones and proteoses (Figure 20–18). Three enzymes found in pancreatic juice continue the digestion. *Trypsin* digests any intact proteins into peptones and proteoses, breaks the peptones and proteoses into dipeptides (containing only two amino acids), and breaks some of the dipeptides into single amino acids. *Chymotrypsin* duplicates trypsin's activities. *Carboxypeptidase,* the third enzyme, reduces whole or partially digested proteins to amino acids. To prevent these enzymes from digesting the proteins in the cells of the pancreas, they are secreted in inactive forms—trypsin as *trypsinogen,* activated by an intestinal enzyme called *enterokinase,* chymotrypsin as *chymotrypsinogen,* activated by trypsin, and carboxypeptidase as *procarboxypeptidase,* also activated by trypsin. Protein digestion is completed by an intestinal enzyme called *erepsin,* which converts all the remaining dipeptides into single amino acids. Single amino acids can be absorbed.

3. *Lipids.* In an adult, almost all lipid digestion occurs in the small intestine. The first step in the process is the *emulsification* of fats, which is a function of bile (Figure 20–19). Bile salts break the globules of fat into tiny droplets so the fat-splitting enzyme can get at the lipid molecules more easily. In the second step, *pancreatic lipase,* an enzyme found in pancreatic juice, hydrolyzes each fat molecule into glycerol and fatty acids, the end products of fat digestion.

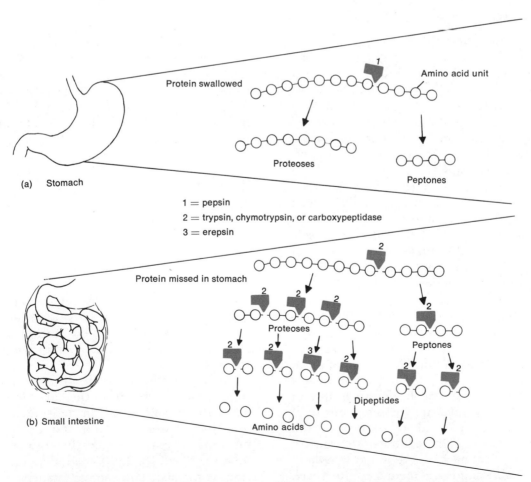

Figure 20–18. Digestion of proteins. (a) In the stomach. (b) In the small intestine.

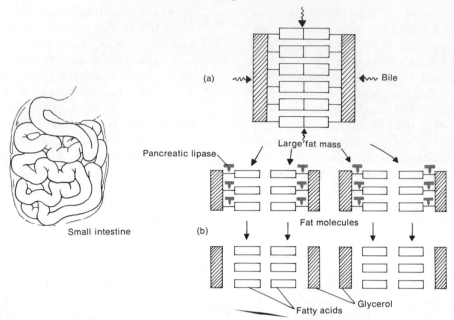

Figure 20-19. Digestion of fats in the small intestine. (a) Bile emulsifies large masses of fat into smaller fragments so that pancreatic amylase can break down the fat molecules. (b) The enzyme pancreatic lipase breaks fats into fatty acids and glycerol.

Mechanical digestion

In the small intestine, three distinct types of movement occur as a result of contractions of the longitudinal and circular muscles. These movements are *rhythmic segmentation, pendular movements,* and peristalsis. Rhythmic segmentation and pendular movements are strictly localized contractions occurring in areas containing food. The two movements act to mix the chyme with the digestive juices and bring every particle of food into contact with the mucosa for absorption. They do not push the intestinal contents along the tract. Rhythmic segmentation starts with the contractions of some of the circular muscle fibers in a portion of the intestine, an action that constricts the intestine into segments. Next, muscle fibers that encircle the middle of each segment also contract, dividing each segment into two smaller segments. Finally, the fibers which contracted first relax, and each small segment unites with an adjoining small segment so that large segments are reformed. This sequence of events is repeated at a rate of 12 to 16 times a minute, sloshing the chyme back and forth and back and forth. Pendular movements consist of alternating contractions and relaxations of the longitudinal muscles. The contractions cause a portion of the intestine to shorten and lengthen, an action that also sends the chyme spilling back and forth.

The third kind of movement, peristalsis, propels the chyme onward through the intestinal tract. Peristaltic movement in the intestine is similar to that in the esophagus. In the intestine, these waves may be as slow as 5 centimeters (2 inches)/minute or as fast as 50 centimeters (20 inches)/second.

Absorption

All the chemical and mechanical phases of digestion occurring from the mouth down through the small intestine are directed toward changing foods into forms that can diffuse through the epithelial cells lining the mucosa into the underlying blood and lymph vessels. The diffusible forms are monosaccharides (glucose, fructose, and galactose), amino acids, fatty acids, and glycerol. Passage of these digested nutrients from the alimentary canal into the blood or lymph is called **absorption.**

About 90 percent of all absorption takes place throughout the length of the small intestine. The other 10 percent occurs in the stomach and large intestine. Absorption of materials in the small intestine occurs specifically through the villi (Figure 20-15 bottom). and depends on diffusion, osmosis, and active transport (refer to Chapter 3). Monosaccharides and amino acids are absorbed into the blood capillaries of the villi (Figure 20–20) and are transported in the bloodstream to the liver via the portal system. Fatty acids and glycerol do not enter the bloodstream immediately. Glycerol is a water-soluble compound that passes rather easily into the lacteals of the villi. Fatty acids are not soluble and must combine with bile salts before they become soluble and pass into the lacteals. Once inside the lacteals, the fatty acids break apart from the bile salts and recombine with glycerol to form fats. The fat is carried as tiny droplets called *chylomicrons* to the thoracic duct, into the subclavian vein, and eventually into the systemic

circulation. It is finally delivered to the liver through the hepatic artery. Most of the products of carbohydrate, protein, and lipid digestion are processed by the liver before they are delivered to the other cells of the body. We shall describe this process in the next chapter. Large amounts of water, electrolytes, mineral salts, and some vitamins also are absorbed in the small intestine.

In summary, then, the principal chemical activity of the small intestine is to digest all foods into forms that are usable by body cells (see Exhibit 20–5). Any undigested materials that are left behind are processed in the large intestine.

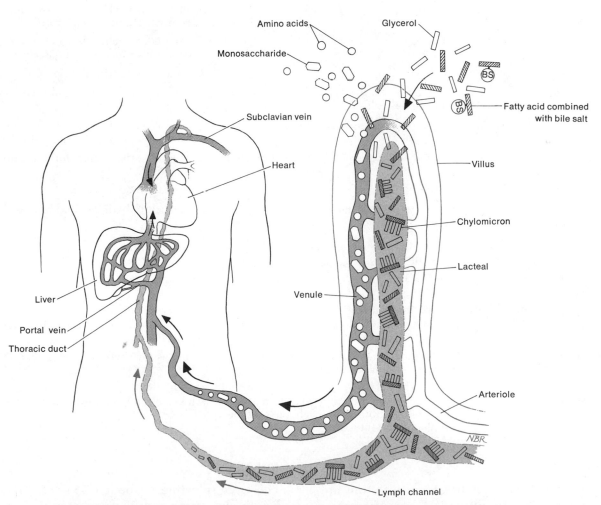

Figure 20–20. Absorption. The end products of carbohydrate and protein digestion pass into the capillary of a villus; the end products of fat digestion pass into the lacteal.

Exhibit 20–5. SUMMARY OF DIGESTION AND ABSORPTION IN THE SMALL INTESTINE

STRUCTURE	DESCRIPTION	FUNCTION
Pancreas (pancreatic juice)		
Trypsin	Protein-digesting enzyme activated by the intestinal enzyme enterokinase	Digests intact proteins into proteoses and peptones. Digests partially digested proteins into dipeptides plus some amino acids.
Chymotrypsin	Protein-digesting enzyme activated by trypsin	Same as above.
Carboxypeptidase	Protein-digesting enzyme activated by trypsin	Reduces proteins to amino acids.
Pancreatic amylase	Enzyme that digests carbohydrates	Converts dextrins into the disaccharides maltose, sucrose, and lactose.
Pancreatic lipase	Fat-splitting enzyme	Converts fats into fatty acids and glycerol.
Liver (bile)	Bile salts	Emulsifies fats in preparation for digestion by pancreatic lipase. Bile salts also allow fatty acids to be absorbed.
Small Intestine		
Mucosa and submucosa		
Intestinal glands	Secrete succus entericus	
Maltase		Converts disaccharide maltose into monosaccharide glucose.
Sucrase		Converts disaccharide sucrose into monosaccharides glucose and frutose.
Lactase		Converts disaccharide lactose into monosaccharides glucose and galactose.
Erepsin		Changes dipeptides into amino acids.
Microvilli	Projections of the plasma membranes of epithelial cells	Increase surface area for absorption.
Villi	Fingerlike projections of the mucous membrane	Serve as sites for the absorption of digested foods and increase absorptive area.
Plicae circulares	Circular folds of mucosa and submucosa	Increase surface area for digestion and absorption.
Goblet cells	Secrete mucus	Lubricate foods.
Intestinal glands	Secrete intestinal digestive juices	Digest carbohydrates and proteins.
Brunner's glands	Secrete mucus	Lubricate foods.
Solitary lymph nodules	Lymphatic tissue associated with the small intestine	Filter lymph.
Peyer's patches	Same as above	Same as above.
Muscularis		
Rhythmic segmentation	Alternating contractions of circular fibers produce segmentation and resegmentation of portions of small intestine	Mixes chyme with digestive juices and brings food into contact with mucosa for absorption.
Pendular movement	Contractions of longitudinal muscle pull portions of intestine forward and backward	Same as for rhythmic segmentation.
Peristalsis	Waves of contraction and relaxation of circular and longitudinal muscle passing the length of small intestine	Moves chyme forward.

Large intestine

The overall functions of the large intestine are the completion of absorption, the manufacture of some vitamins, the formation of feces, and the expulsion of feces from the body.

Anatomy

The **large intestine** is about 1.5 meters (5 feet) in length and averages 6.5 centimeters (2.5 inches) in diameter. It extends from the ileum to the anus and is attached to the posterior abdominal wall by its mesentery. Structurally, the large intestine is divided into four principal regions: the cecum, colon, rectum, and anal canal (Figure 20–21). Let us now look at the parts of the large intestine in the order in which food passes through them.

The opening from the ileum into the large intestine is guarded by a fold of mucous membrane called the *ileocecal valve*. This structure allows materials from the small intestine to pass into the large intestine but prevents them from moving

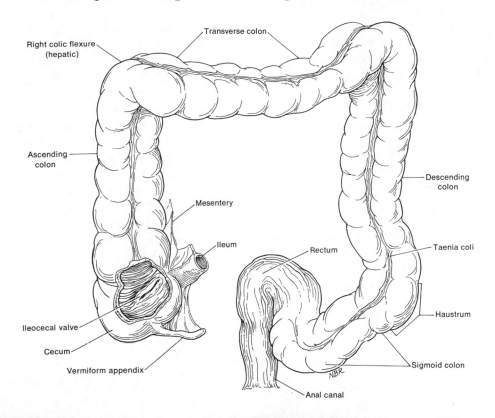

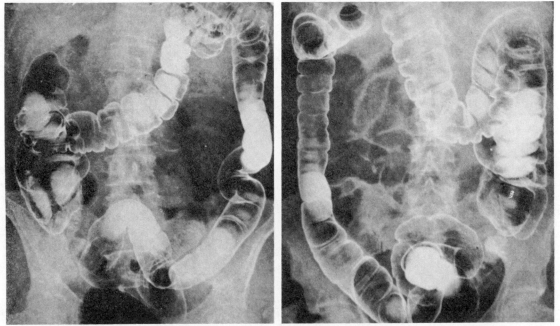

Figure 20–21.   The large intestine. (Top) Anatomy of the large intestine. (Bottom) Roentgenogram of the large intestine in which several haustra are clearly visible. [*(Bottom) from Lester W. Paul and John H. Juhl, The Essentials of Roentgen Interpretation, 3d ed., Harper & Row, Publishers, Inc., New York, 1972.*]

in the opposite direction. Hanging below the ileocecal valve is the *cecum*, a blind pouch about 6 centimeters (2 to 3 inches) long. Attached to the cecum is a twisted, coiled tube, called the *vermiform appendix* (*vermis* = worm). Inflammation of the appendix is called *appendicitis*.

The open end of the cecum merges with a long tube called the *colon*. Based on location, the colon is divided into ascending, transverse, descending, and sigmoid portions. The *ascending colon* ascends on the right side of the abdomen, reaches the undersurface of the liver, and turns abruptly to the left. The colon continues across the abdomen to the left side as the *transverse colon*. It curves beneath the lower end of the spleen on the left side and passes downward to the level of the iliac crest as the *descending colon*. The *sigmoid colon* is the S-shaped portion that begins at the iliac crest, projects inward to the midline, and terminates as the *rectum* at about the level of the third sacral vertebra.

The rectum, the last 20 centimeters (7 to 8 inches) of gastrointestinal tract, lies anterior to the sacrum and coccyx. The terminal 2 to 3 centime-

ters of the rectum is referred to as the *anal canal* (Fig. 20–22). Internally, the mucous membrane of the anal canal is arranged in longitudinal folds called *anal columns* that contain a network of arteries and veins. Inflammation and enlargement of the anal veins is known as *hemorrhoids* or *piles*. The opening of the anal canal to the exterior is called the *anus*. It is guarded by an internal sphincter of smooth muscle and an external sphincter of skeletal muscle. Normally the anus is closed except during the elimination of the wastes of digestion.

The wall of the large intestine differs from that of the small intestine in several respects. No villi or circular folds are found in the mucosa, which does, however, contain simple columnar epithelium with numerous goblet cells. These cells secrete mucus that lubricates the colonic contents as they pass through the colon. Solitary lymph nodes also are found in the mucosa. The submucosa of the large intestine is similar to that found in the rest of the alimentary canal. The muscularis consists of an external layer of longitudinal muscles and an internal layer of

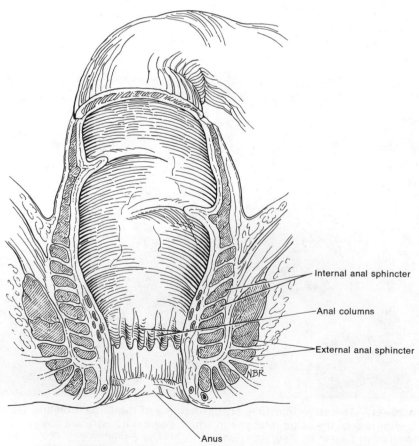

Figure 20–22. Longitudinal section through the anal canal.

circular muscles. Unlike other parts of the digestive tract, however, the longitudinal muscles do not form a continuous sheet around the wall but are broken up into three flat bands called *taenia coli.* Each band runs the length of the large intestine. Tonic contractions of the bands gather the colon into a series of pouches called *haustra,* which give the colon its puckered appearance. The serosa of the large intestine is part of the visceral peritoneum.

Activities of the large intestine

The principal activities of the large intestine are concerned with mechanical movements, absorption, and the formation and elimination of feces.

MOVEMENTS. Movements of the colon begin when substances enter through the ileocecal valve. Since chyme moves through the small intestine at a fairly constant rate, the time required for a meal to pass into the colon is determined by gastric evacuation time. As food passes through the ileocecal valve, it fills the cecum and accumulates in the ascending colon.

One movement characteristic of the large intestine is *haustral churning.* In this process the haustra remain relaxed and distended while they fill up. When the distention reaches a certain point, the walls contract and squeeze the contents into the next haustrum. Peristalsis also occurs, although at a slower rate than in other portions of the tract (3 to 12 contractions per minute). A final type of movement is *mass peristalsis,* a very strong peristaltic wave that drives the colonic contents into the sigmoid colon and into the rectum if the sigmoid colon is full. Food in the stomach initiates this reflex action in the colon. Thus, mass peristalsis usually takes place three or four times a day, during or immediately after a meal is eaten.

ABSORPTION AND FORMATION OF FECES. By the time the intestinal contents arrive at the large intestine, digestion and absorption are almost complete.

By the time the chyme has remained in the large intestine for about 3 to 10 hours, it has become solid or semisolid as a result of absorption and is now known as *feces.* Chemically, feces consist of water, inorganic salts, mucus, epithelial cells from the mucosa of the alimentary canal, bacteria, products of bacterial decomposition, and undigested parts of food not attacked by bacteria.

Mucus is secreted by the glands of the large intestine, but no enzymes are secreted. The mucus serves as a lubricant to aid the movement of the colonic materials and acts as a protective covering for the mucosa.

Chyme is prepared for elimination in the large intestine by the action of bacteria. These bacteria ferment any remaining carbohydrates and release hydrogen, carbon dioxide, and methane gas. They also convert remaining proteins to amino acids and break down the amino acids into simpler substances, such as indole, skatole, hydrogen sulfide, and fatty acids. Some of the indole and skatole is carried off in the feces and contributes to its smell, and the rest is absorbed. Bacteria also decompose bilirubin, the breakdown product of red blood cells that is excreted in bile, to simpler pigments which give feces its color. Intestinal bacteria also aid in the synthesis of several vitamins needed for normal metabolism, including some B vitamins (riboflavin, nicotinic acid, biotin, and folic acid) and vitamin K.

Although most water absorption occurs in the small intestine, the large intestine absorbs enough to make it an important organ in maintaining the water balance of the body. Intestinal water absorption is greatest in the cecum and ascending colon. The large intestine also absorbs inorganic solutes plus some products of bacterial action, including vitamins and large amounts of indole and skatole. The indole and skatole are transported to the liver, where they are converted to less toxic compounds that are excreted in the urine.

DEFECATION. Under normal circumstances, the rectum remains empty until *defecation,* the emptying of the rectum. Eventually, however, mass peristaltic movement pushes fecal material from the sigmoid colon into the rectum. The resulting distention of the rectal walls stimulates pressure-sensitive receptors, initiating a reflex and bringing on further peristaltic movements in the sigmoid colon plus contractions of the longitudinal muscles in the rectum. Contraction of the longitudinal rectal muscles shortens the rectum, and the peristaltic movements force more feces into the rectum, thereby increasing the pressure inside it. The pressure forces the sphincters open, and the feces are expelled through the anus. Voluntary contractions of the diaphragm and abdominal muscles aid defecation by increasing the pressure inside the abdomen, which pushes the walls of the sigmoid colon and rectum inward. Activities of the large intestine are summarized in Exhibit 20–6.

Exhibit 20–6. SUMMARY OF DIGESTIVE ACTIVITIES OF THE LARGE INTESTINE

STRUCTURE	ACTION	FUNCTION
Mucosa	Secretes mucus	Lubricates the colon and protects the mucosa.
	Absorbs water and other soluble compounds	Water balance is maintained and feces become solidified. Vitamins and minerals are obtained, and toxic substances are sent to the liver to be detoxified.
	Bacterial action	Undigested carbohydrates, proteins, and amino acids are broken down so they can be expelled in feces or absorbed and detoxified by liver. Certain vitamins are synthesized.
Muscularis	Haustral churning	Haustra fill and contract, moving contents from haustra to haustra.
	Peristalsis	Contractions of circular and longitudinal muscles continually move contents along length of colon.
	Mass peristalsis	Strong peristaltic wave forces contents into sigmoid colon and rectum.
	Defecation	Contractions in sigmoid colon and rectum rid the body of feces.

DISORDERS OF THE DIGESTIVE SYSTEM

Now that we have discussed the structure and physiology of the digestive system, we shall look at some disorders that are related to it.

Dental caries

Dental caries, or tooth decay, involve a gradual disintegration of the enamel and dentin (Figure 20–23 top). If this condition remains untreated, various microorganisms may invade the pulp cavity, causing infection and inflammation of the living tissue. If the pulp is destroyed, the tooth is pronounced "dead."

No individual microbe is responsible for dental caries, but oral bacteria that create a pH of 5.5 or lower start the process. Acids can come directly from foods, such as the ascorbic acid of citrus fruits, or they may be breakdown products of carbohydrates. Microbes that digest carbohydrates include two bacteria, *Lactobacillus acidophilus* and streptococci, as well as some yeasts. Research suggests that the streptococci break down carbohydrates into *dental plaque*, a polysaccharide which adheres to the tooth surface. When other bacteria digest the plaque, acid is produced. Saliva cannot reach the tooth surface to buffer the acid because the plaque covers the teeth.

Certain measures can be taken to prevent dental caries. First, the diet of the mother during pregnancy is very important in forestalling tooth decay of the newborn. Simple, balanced meals are the best diet during pregnancy. Supplementation with multivitamins, with emphasis on vitamin D and the minerals calcium and phosphorus

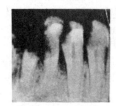

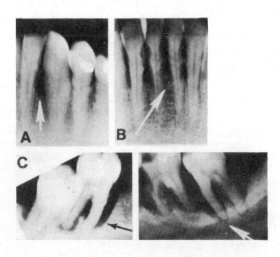

Figure 20–23. Diseases of the teeth. (Top) Dental caries. (Middle and bottom) Periodontitis in which the alveolar bone (arrow) has been destroyed. *(From Lester W. Paul and John H. Juhl, The Essentials of Roentgen Interpretation, 3d ed., Harper & Row, Publishers, Inc., New York, 1972.)*

is customary because they are responsible for normal bone and teeth development.

Other preventive measures have centered around fluoride treatment because teeth are less susceptible to acids when they are permeated with fluoride. Fluoride may be incorporated in the drinking water or applied topically to erupted teeth. Maximum benefit often occurs when fluoride is used in drinking water during the period when teeth are being calcified. Excessive fluoride may cause a light brown to brownish-black discoloration of the enamel of the permanent teeth called mottling.

Brushing the teeth immediately after eating removes the plaque from flat surfaces before the bacteria have a chance to go to work. Dentists also suggest that the plaque between the teeth be removed every 24 hours with dental floss.

Periodontal diseases

Periodontal disease is a collective term for a variety of conditions characterized by inflammation and/or degeneration of the gingivae, alveolar bone, periodontal membrane, and cementum. The initial symptoms are enlargement and inflammation of the soft tissue. Without treatment, the soft tissue may deteriorate and the alveolar bone may be resorbed, causing loosening of the teeth and receding of the gums (Figure 20–23 middle and bottom).

Periodontal diseases are frequently caused by local irritants, such as bacteria, impacted food, cigarette smoke, or by a poor "bite." The latter may put a strain on the tissues supporting the teeth. Methods of prevention and treatment include good mouth care to remove plaque and other sources of irritation. Periodontal diseases may also be caused by allergies, vitamin deficiences, and a number of systemic disorders, especially those that affect bone, connective tissue, or circulation. In these cases, the systemic disorder must be treated as well.

Peptic ulcers

An **ulcer** is a craterlike lesion in a membrane. Ulcers that develop in areas of the alimentary canal exposed to acid gastric juice are called *peptic ulcers*. Peptic ulcers occasionally develop in the lower end of the esophagus. However, most of them occur on the lesser curvature of the stomach, in which case they are called gastric ulcers, or in the first part of the duodenum, where they are called *duodenal ulcers* (Figure 20–24).

The cause of ulcers is obscure. However, hypersecretion of acid gastric juice seems to be the immediate cause in the production of duodenal ulcers and in the reactivation of healed ulcers. Hypersecretion of acid gastric juice is not implicated as much in gastric ulcer patients because the stomach walls are highly adapted to resist gastric juice through their secretion of mucus. A possible cause of gastric ulcers is hyposecretion of mucus. Hypersecretion of pepsin also may contribute to ulcer formation.

Among the factors believed to stimulate an increase in acid secretion are certain foods or medications, such as alcohol, coffee or aspirin, and overstimulation of the vagus nerve. Normally, the mucous membrane lining the stomach and duodenal walls resists the secretions of hydrochloric acid and pepsin, and no ulcer develops. In some people, however, this resistance breaks down, and an ulcer develops.

The danger inherent in ulcers is the erosion of the muscular portion of the wall of the stomach or duodenum. This could damage blood vessels and possibly produce fatal hemorrhage. If an ulcer erodes all the way through the wall, the condition is called *perforation*. Perforation allows bacteria and partially digested food to pass into the peritoneal cavity, producing peritonitis.

Peritonitis

Peritonitis is an acute inflammation of the serous membrane lining the abdominal cavity and covering the abdominal viscera. One possible cause is contamination of the peritoneum by pathologic bacteria from the external environment. This contamination could result from accidental or surgical wounds in the abdominal wall or from perforation of organs exposed to the outside environment. Another possible cause is perforation of the walls of organs that contain bacteria or chemicals which are normally beneficial to the organ but are toxic to the peritoneum. For example, the large intestine contains colonies of bacteria that live on undigested nutrients and break them down so they can be eliminated more easily. But if the bacteria enter the peritoneal cavity, they attack the cells of the peritoneum for food and produce acute infection. As another example, the normal bacteria of the female reproductive tract protect the tract by giving off acid wastes that produce an acid environment unfavorable to many yeasts, protozoa, and bacteria which might otherwise attack the tract. However, these acid-producing bacteria are harmful to the peritoneum. A third cause may be chemical irritation. The peritoneum does not have any natural barriers that keep it from being irritated

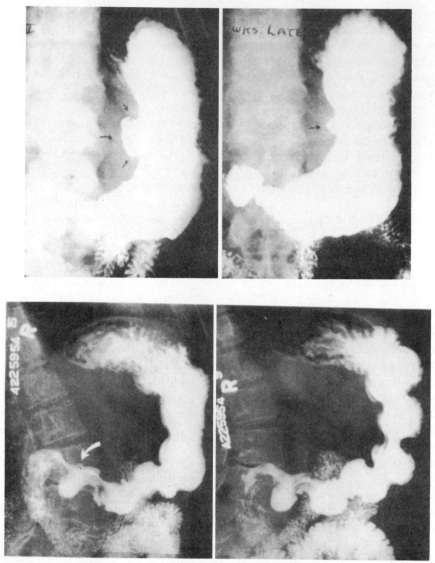

Figure 20-24. Peptic ulcers. (Top) Gastric ulcers (arrows). Left, at the time of diagnosis. Right, three weeks after treatment. (Bottom) Duodenal ulcer (arrow). Left, prior to treatment. Right, after treatment. *(Courtesy of Lester W. Paul and John H. Juhl, The Essentials of Roentgen Interpretation, 3d ed., Harper & Row, Publishers, Inc., New York, 1972.)*

or digested by chemical substances such as bile and digestive enzymes. However, it does contain a great deal of lymphatic tissue and can fight infection fantastically well. The danger stems from the fact that the peritoneum is in contact with most of the abdominal organs. If the infection gets out of hand, it may destroy vital organs and bring on death. For these reasons, perforation of the alimentary canal from an ulcer or perforation of the uterus from an incompetent abortion are considered serious. If a surgeon plans to do extensive surgery on the colon, he may give the patient high doses of antibiotics for several days preceding surgery to kill intestinal bacteria and reduce the risk of peritoneal contamination.

Cirrhosis

Cirrhosis is a chronic disease of the liver in which the parenchymal liver cells are replaced by fibrous connective tissue, a process called *stromal repair*. Often there is a lot of replacement by adipose connective tissue as well. The liver has a high ability for parenchymal regeneration, so stromal repair occurs whenever any parenchymal cell is killed or when damage to the cells occurs continuously over a long time. These conditions could be caused by *hepatitis* (inflammation of the liver), certain chemicals that may destroy liver cells, parasites that sometimes infect the liver, and alcoholism. Malnutrition, particularly deficiencies of essential amino acids, is common among alcoholics. It is not known whether degeneration of the cells is caused by alcohol, malnutrition, or both.

Tumors

Both benign and malignant **tumors** occur in all parts of the gastrointestinal tract. The benign growths are much more common, but the malignant tumors are responsible for 30 percent of all deaths from cancer in the United States. To achieve relatively early diagnosis, complete periodic routine examinations are necessary. Cancers of the mouth usually are detected through routine dental checkups.

A regular physical checkup should include rectal examination. Fifty percent of all rectal carcinomas are within reach of the finger, and 75 percent of all colonic carcinomas can be seen with the sigmoidoscope (Figure 20–25). Both the

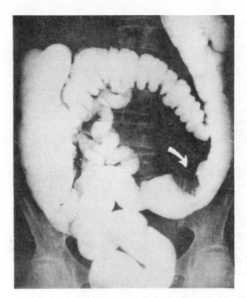

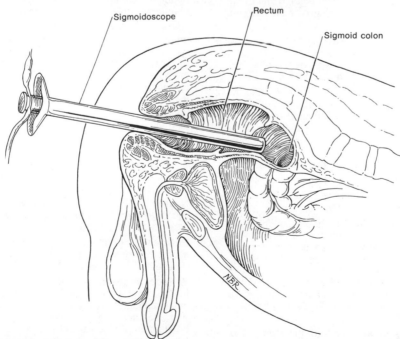

Figure 20–25. Tumors of the alimentary canal. (Top) Carcinoma of the sigmoid colon is indicated at the arrow. (Bottom) Detection of carcinomas by use of the sigmoidoscope. [*(Top) from Lester W. Paul and John H. Juhl, The Essentials of Roentgen Interpretation, 3d ed., Harper & Row, Publishers, Inc., New York, 1972.*]

fiberoptic sigmoidoscope and the more recent fiberoptic endoscope are flexible tubular instruments composed of a light and many tiny glass fibers. They allow visualization, magnification, and even photography of almost the entire length of the gastrointestinal tract and have been invaluable in the correct diagnosis of a wide range of gastrointestinal disorders without surgery.

Another test in a routine examination for intestinal disorders is the filling of the gastrointestinal tract with barium, which is either swallowed or given in an enema. Barium, a mineral, shows up on x-rays the same way that calcium appears in bones. Tumors as well as ulcers can be diagnosed this way. The only definitive treatment of gastrointestinal carcinomas is surgery.

Medical terminology associated with the digestive system

Calculus (*calc* = stone) A stone in an organ. Enteroliths (*entero* = intestine; *lith* = stone, calcification) are intestinal stones or calculi. Cholelithiases (*chole* = bile, gall) are calculi composed of bile salts, lecithin, and cholesterol that are formed in the gallbladder or its ducts and are also called *gallstones*.

Cholecystitis Inflammation of the gallbladder that often leads to infection. Some cases are caused by obstruction of the cystic duct with bile stones. Stagnating bile salts irritate the mucosa. Dead mucosal cells provide medium for bacteria.

Colitis An inflammation of the colon and rectum. Inflammation of the mucosa reduces absorption of water and salts, producing watery, bloody feces, and—in severe cases—dehydration and salt depletion. Irritated muscularis spasms produce cramps.

Colostomy The cutting of the colon in half and bringing the upper, lower, or both halves through the abdominal wall to the exterior surface of the body. Feces are eliminated through the upper end. A temporary colostomy may be done to allow a badly inflamed colon to rest and heal. Later the two halves are rejoined, and the abdominal opening is closed. If the rectum is removed for malignancy, the colostomy provides a permanent outlet for feces.

Constipation Infrequent or difficult defecation.

Diarrhea Frequent defecation of liquid feces.

Diverticulosis The presence of diverticula, or abnormal sacs or outpockets of the intestinal mucosa into the muscularis. They occur when the intestinal wall is weakened or when there is hypertrophy (increase in size of an organ or structure not involving tumor formation) of segments of the circular muscle of the colon. Diverticulitis is an inflammation of a diverticulum occurring when the diverticular neck becomes obstructed by edema or feces, producing a static condition favoring bacterial reproduction and infection.

Flatus Excessive amounts of air (gas) in the stomach or intestine, usually expelled through the anus. If the gas is expelled through the mouth, it is called *belching* (burping). Flatus may result from gas released during the breakdown of foods in the stomach or from swallowing air or gas-containing substances such as carbonated drinks.

Heartburn A burning sensation in the region of the esophagus and stomach. It may result from regurgitation of gastric contents into the lower end of the esophagus or from distention stemming from causes such as the retention of regurgitated food and gastric contents in the lower esophagus.

Hepatitis (*hepato* = liver) A liver inflammation. It may be caused by organisms such as viruses, bacteria, and protozoa or by the absorption of materials, such as carbon tetrachloride and certain anesthetics and drugs, that are toxic to liver cells.

Hernia Protrusion of an organ or part of an organ through a membrane or through the wall of a cavity, usually the abdominal cavity. *Diaphragmatic hernia* is the protrusion of the lower esophagus, stomach, or intestine into the thoracic cavity through the hole in the diaphragm that allows passage of the esophagus. *Umbilical hernia* is the protrusion of abdominal organs through the naval area of the abdominal wall. *Inguinal* hernia is the protrusion of the hernial sac containing the intestine into the inguinal opening. It may extend into the scrotal compartment, causing strangulation of the herniated part.

Mumps Viral disease causing painful inflammation and enlargement of the salivary glands particularly the parotids. In adults, the sex glands and pancreas may be involved. Inflammation of testes may cause male sterility.

Nausea Discomfort preceding vomiting. Possibly, it is caused by distention or irritation of the gastrointestinal tract, most commonly the stomach.

Pancreatitis Inflammation of the pancreas. The pancreas secretes active trypsin instead of trypsinogen, and the trypsin digests the pancreatic cells and blood vessels.

Periodontal disease Diseases of the tissues surrounding the teeth. The following diseases are included: *Gingivitis* is an inflammation of the gingivae characterized by swelling, redness, and bleeding. *Periodontitis* is a progression of gingivitis to destruction of alveolar bone. *Gingivostomatitis,* better known as trench mouth, is an ulceration of the gingivae, the oral or pharyngeal mucosa, or the tonsils.

Vomiting Expulsion of stomach (and sometimes duodenal) contents through the mouth by reverse peristalsis. The abdominal muscle walls forcibly empty the stomach.

Chapter summary in outline

DIGESTION

1. Digestion is a series of chemical and mechanical processes by which foods are reduced to a form that the body can use.

2. Digestion occurs in the organs of the alimentary canal and depends on the functioning of accessory organs as well.

3. The basic arrangement of tissues in the alimentary canal from the inside outward is mucosa, submucosa, muscularis, and serosa.

4. Extensions of the peritoneum include the mesentery, lesser omentum, and greater omentum.

MOUTH AND SALIVARY GLANDS

1. The mouth is formed by cheeks, palates, and tongue which aid mechanical digestion.

2. The teeth project into the mouth and are adapted for mechanical digestion.

3. The salivary glands produce saliva that lubricates foods and starts the chemical digestion of carbohydrates.

PHARYNX AND ESOPHAGUS

1. Both organs assume a role in deglutition, or swallowing.

2. When a bolus is swallowed, the respiratory tract is sealed off and the bolus moves into the esophagus.

3. Peristaltic movements of the esophagus pass the bolus into the stomach.

STOMACH

1. The stomach begins at the bottom of the esophagus and ends at the pyloric valve.

2. Adaptations of the stomach for digestion include rugae that permit distention; glands which produce mucus, hydrochloric acid, and enzymes; and a three-layered muscularis for efficient mechanical movements.

3. Nervous and hormonal mechanisms initiate the secretion of gastric juice.

4. Proteins are chemically digested into peptones and proteoses through the action of pepsin in the stomach.

5. The stomach also stores food, produces the intrinsic factor, and carries on some absorption.

PANCREAS, LIVER, AND GALLBLADDER

1. Pancreatic acini produce enzymes that enter the duodenum via the pancreatic duct. Pancreatic enzymes digest proteins, carbohydrates, and fats.

2. Cells of the liver produce bile, which is needed to emulsify fats. Bile is stored in the gallbladder and passed into the duodenum via the common bile duct.

SMALL INTESTINE

1. This organ extends from the pyloric valve to the ileocecal valve.

2. It is very highly adapted for digestion and absorption. Its glands produce enzymes and mucus, and its wall contains microvilli, villi, and plicae circulares.

3. The enzymes of the small intestine digest carbohydrates, proteins, and fats into the end products of digestion: monosaccharides, amino acids, fatty acids, and glycerol.

4. The entrance of chyme into the small intestine stimulates the secretion of several hormones that coordinate the secretion and release of bile, pancreatic juice, and intestinal juice and inhibit gastric activity.

5. Mechanical digestion in the small intestine involves rhythmic segmentation, pendular movements, and peristalsis.

6. Absorption is the passage of the end products of digestion from the alimentary canal into the blood or lymph.

7. Absorption in the small intestine occurs through the villi. Monosaccharides and amino acids pass into the blood capillary, and fatty acids and glycerol pass into the lacteal.

LARGE INTESTINE

1. This organ extends from the ileocecal valve to the anus.

2. Mechanical movements of the large intestine include haustral churning, mass peristalsis, and peristalsis.

3. The large intestine functions in the synthesis of several vitamins and in water absorption from chyme, leading to feces formation.

4. The elimination of feces from the large intestine is called defecation. Defecation is a reflex action aided by voluntary contractions of the diaphragm and abdominal muscles.

DISORDERS

1. Dental caries are started by acid-producing bacteria.

2. Periodontal diseases are characterized by inflammation and/or degeneration of gingivae, alveolar bone, peridontal membrane, and cementum.

3. Peptic ulcers are craterlike lesions that develop in the mucous membrane of the alimentary canal in areas exposed to gastric juice.

4. Peritonitis is inflammation of the peritoneum.

5. In cirrhosis, parenchymal cells of the liver are replaced by fibrous connective tissue.

6. Tumors may be detected by sigmoidoscope and barium x-rays.

Review questions and problems

1. Define digestion. Distinguish between chemical and mechanical digestion.

2. In what respect is digestion an important component of your homeostatic mechanism?

3. Identify the organs of the alimentary canal in sequence. How does the alimentary canal differ from the accessory organs of digestion?

4. Describe the structure of each of the four coats of the alimentary canal.

5. What is the peritoneum? Describe the location and function of the mesentery, lesser omentum, and greater omentum.

6. What structures form the oral cavity? How do each of the structures contribute to digestion?

7. Make a simple diagram of the tongue. Indicate the location of the papillae and the four taste zones.

8. Describe the location of the salivary glands and their ducts. What are buccal glands?

9. Briefly explain the mechanisms involved in the control of saliva secretion.

10. Describe the composition of saliva and the role of each of its components in digestion. What is the pH of saliva?

11. By means of a labeled diagram, outline the action of salivary amylase in the mouth.

12. What are the principal portions of a typical tooth? What are the functions of each of the parts?

13. Compare deciduous and permanent dentitions with regard to numbers of teeth and times of eruption.

14. Contrast the functions of incisors, cuspids, premolars, and molars. What is pyorrhea?

15. What is a bolus? How is it formed?

16. Define deglutition. List the sequence of events involved in passing a bolus from the mouth to the stomach.

17. Describe the location of the stomach. List and briefly explain the anatomic features of the stomach.

18. Distinguish between pyloric stenosis and pylorospasm.

19. What is the importance of rugae, zymogenic cells, parietal cells, and mucous cells in the stomach?

20. What is chyme? Why are protein-digesting enzymes secreted in an inactive form?

21. By means of a labeled diagram, outline the action of pepsin in the stomach.

22. Describe the actions of renin and gastric lipase in the infant stomach.

23. What is a gastric ulcer? How is it formed?

24. What forces operate to move chyme through the pyloric valve into the duodenum?

25. What factors control the secretion of gastric juice?

26. Where is the pancreas located? Describe the duct system by which the pancreas is connected to the duodenum.

27. What are pancreatic acini? Contrast their functions with those of the islets of Langerhans.

28. Where is the liver located? What are the principal functions of the liver?

29. Draw a labeled diagram of a liver lobule.

30. How is blood supplied to and drained from the liver?

31. Once bile has been formed by the liver, how is it collected and transported to the gallbladder for storage?

32. Where is the gallbladder located? How is the gallbladder connected to the duodenum?

33. What are the subdivisions of the small intestine? How are the coats of the small intestine adapted for digestion and absorption?

34. Describe the movements that occur in the small intestine.

35. By means of a labeled diagram, outline the chemical digestion that occurs in the small intestine.

36. List the hormones, and their actions, that control digestion in the stomach and small intestine.

37. Why is the small intestine considered the most important area of the digestive tract?

38. Define absorption. How are the end products of carbohydrate and protein digestion absorbed? How are the end products of fat digestion absorbed?

39. Pretend that you have just eaten a roast beef sandwich with butter. Describe or diagram the chemical changes that occur in the sandwich as it passes through the mouth, stomach, and small intestine. Give the names of the enzymes involved and the names of the glands that secrete the enzymes. Also include the role of bile. In formulating your response remember that roast beef is a protein, the bread is a carbohydrate, and the butter is a fat.

40. What routes are taken by absorbed nutrients to reach the liver?

41. What are the principal subdivisions of the large intestine? How does the muscularis of the large intestine differ from that of the rest of the digestive tract?

42. Describe the mechanical movements that occur in the large intestine?

43. Explain the activities of the large intestine that change chyme into feces.

44. Define defecation. How does defecation occur?

45. Define dental caries. How are they started? What is dental plaque?

46. What are three preventive measures that can be taken against dental caries?

47. Define periodontal disease, and describe the best method of prevention.

48. What is a peptic ulcer? Distinguish between gastric and duodenal ulcers.

49. Describe some of the suspected causes of ulcers. What is perforation?

50. Define peritonitis. Explain some possible causes. Why is peritonitis a potentially dangerous condition?

51. Define cirrhosis.

52. How are tumors of the alimentary canal detected?

53. Refer to the glossary of medical terminology associated with the digestive system and be sure that you can define each term.

CHAPTER 21

NUTRITION : THE UTILIZATION OF FOODS

STUDENT OBJECTIVES

After you have read this chapter, you should be able to:

1. Define a nutrient and list the functions of the six classes of nutrients

2. Define metabolism and contrast between the physiological effects of catabolism and anabolism

3. Describe the fate of glucose as it is catabolized via glycolysis and the Krebs cycle

4. Define glycogenesis as an example of glucose anabolism into glycogen

5. Define glycogenolysis as an example of glycogen catabolism into glucose

6. Define gluconeogenesis as a conversion of lipids and proteins into glucose

7. Describe fat storage in adipose tissue

8. Discuss the mode by which glycerol, a product of fat digestion, may be converted to carbohydrate

9. Describe the catabolism of fatty acids via beta oxidation and ketogenesis

10. Define ketosis and list its effects on the body

11. Define lipogenesis as the synthesis of lipids from glucose and amino acids

12. Describe the mechanism involved in protein synthesis

13. Discuss the catabolism of amino acids, noting especially their conversion into acetyl coenzyme A, pyruvic acid, and acids of the Krebs cycle

14. Describe the hormonal control of metabolism by contrasting the roles of insulin, glucagon, STH, ACTH, TSH, epinephrine, and sex hormones

15. Compare the sources, functions, and importance of minerals in metabolism

16. Define a vitamin and differentiate between fat-soluble and water-soluble vitamins

17. Compare the sources, functions, deficiency symptoms, and disorders of the principal vitamins

18. Define phenylketonuria (PKU), cystic fibrosis, and celiac disease as disorders related to faulty metabolism

19. Define the causes and treatment of obesity

In the last chapter you learned how food is digested and absorbed. Now you will learn what happens to the food after it reaches the cells of the body. You also will learn what nutrients are needed for survival and why they are needed.

Nutrients are chemical substances in food that provide energy, act as building blocks in forming new body components, or assist body processes. There are six major classes of nutrients: carbohydrates, lipids, proteins, minerals, vitamins, and water. Carbohydrates, proteins, and lipids are the raw materials for reactions occurring inside cells. The cells either break them down to release energy or use them to build new structures and new regulatory substances, such as hormones and enzymes. Some minerals and many vitamins are used by enzyme systems that catalyze the reactions undergone by carbohydrates, proteins, and lipids. Many minerals have other functions that we shall describe later on. Water has four major functions. It acts as a reactant in hydrolysis reactions, as a solvent and suspending medium, as a lubricant, and as a coolant.

METABOLISM WITHIN CELLS

In its broadest sense, the word **metabolism** refers to all the chemical activities of the body. Since chemical reactions either release or require energy, the body's metabolism may be thought of as an energy-balancing act. Accordingly, metabolism has two phases, catabolism and anabolism.

Catabolism

Catabolism is the term for processes that provide energy. Digestion is a catabolic process because the breaking of bonds releases energy. However, digestion occurs within a cavity lying outside the cells. Thus, the energy is not directly available to the cells, and it is mostly dissipated as heat. In this chapter we shall be concerned only with the catabolic processes occurring within cells. Catabolism within cells consists of three steps.

The first, *oxidation* or *cellular respiration,* is the breakdown of absorbed nutrients resulting in a release of energy. Therefore, oxidation reactions are decomposition reactions. Glucose is the body's favorite nutrient for oxidation, but fats and proteins are also oxidized. The second step in catabolism is the manufacture of ATP from ADP. This synthetic reaction utilizes the energy obtained from oxidation and provides a method for storing it. The final step in catabolism is the decomposition of ATP, releasing great quantities of energy.

Anabolism

Anabolism is just the opposite of catabolism. Anabolism consists of a series of synthetic reactions whereby small molecules are built up into larger ones that form the body's structural and functional components. Anabolic reactions are building processes that occur within cells. They require energy, and the energy is supplied by the body's catabolism. One example of an anabolic process is the formation of peptide bonds between amino acids—thereby building up the amino acids into the protein portions of cytoplasm, enzymes, and antibodies. Fats also participate in the body's anabolism. For instance, fats can be built up into the lipids that form the middle layer of the plasma membrane. They are also a part of the steroid hormones.

Almost every metabolic reaction requires the proper enzyme to proceed. In this chapter we shall look at the minerals and vitamins that work with these enzymes, and we shall describe some disorders that occur when an enzyme is missing. But first we shall describe the catabolic and anabolic processes that carbohydrates, fats, and proteins undergo within the cells.

CARBOHYDRATE METABOLISM

During the process of digestion, carbohydrates are hydrolyzed to become the simple sugars—glucose, fructose, and galactose—that are then absorbed into the capillaries of the villi of the small intestine and carried through the portal vein to the liver, where fructose and galactose are converted to glucose. Thus, the story of carbohydrate metabolism is really the story of glucose metabolism.

Since glucose is the body's most direct source of energy, the fate of absorbed glucose depends on the body cells' energy needs. If the cells require immediate energy, the liver releases some of the glucose back into the bloodstream so it can be oxidized by the cells. The glucose not needed for immediate use is handled in several ways. First, the liver can convert excess glucose to glycogen that can be stored in the liver and skeletal muscle cells. Second, if the glycogen storage areas are filled up, the liver cells can transform the glucose to fat that can be stored in adipose tissue. Later, when the cells need more energy, the glycogen and fat can be converted back to glucose and oxidized. Third, excess glucose can be excreted in the urine. Normally, this happens only when a meal containing mostly carbohydrates and no fats is eaten. Without the inhibiting effect of fats, the stomach empties its contents very quickly, and the carbohydrates arc all digested at the same time. As a result, large numbers of monosaccharides suddenly flood into the bloodstream. Unable to process all of them simultaneously, the liver excretes them.

Glucose catabolism

The oxidation of glucose is also known as cellular respiration. It occurs in every cell in the body and provides the cell's source of energy. The complete oxidation of glucose occurs in two successive phases: glycolysis and the Krebs cycle.

Glycolysis

The term *glycolysis* refers to a series of chemical reactions that convert glucose into *pyruvic acid*. *Glyco* refers to sugar, and *lysis* means breakdown. The overall reaction for glycolysis can be written like this:

$$\boxed{\text{C-C-C-C-C-C}} \rightarrow 2\;\boxed{\text{C-C-C}} + 2\;\text{ATP}$$

Glucose   Pyruvic
                       acid

Many details of the intermediate reactions for glycolysis are shown in Figure 21–1. We wish to emphasize strongly that even though details of intermediate reactions have been included in the figure, we do not expect you to learn them. They have been included only to illustrate the complexity of the process. Notice in Figure 21–1a that the glucose molecule containing six carbon atoms is eventually broken down into two molecules of pyruvic acid, both of which contain three carbon atoms. Since the reactions leading to pyruvic acid formation are decomposition reactions, energy is released. Most of this energy goes into making two molecules of ATP that are used subsequently by the cell. The rest of the energy is expended as heat energy, some of which helps to maintain body temperature. Glycolysis occurs in the cytoplasm of the cell, and it is an *anaerobic* process; that is, it does not require oxygen.

The remaining steps in the oxidation of glucose are *aerobic;* that is, they do require oxygen. The immediate fate of pyruvic acid, then, depends on the availability of oxygen. When the body is exercising strenuously, glycolysis occurs so rapidly that the lungs and blood cannot supply enough oxygen to break down all the pyruvic acid. The excess is converted to lactic acid and stored (see Figure 21–1a). Moderate amounts of lactic acid are buffered by the body so that the pH is not significantly disturbed. In addition, several protective mechanisms prevent excessive amounts of the acid from building up. The first of these involves the liver, which changes lactic acid and pyruvic acid back to glucose. The second mechanism stems from the fact that rapid glycolysis increases the blood pCO_2, thus increasing the respiratory rate. Eventually, the person becomes so short of breath that he is forced to stop exercising. Finally, lactic acid itself contributes to muscle fatigue and makes the person want to rest. After the exercise has stopped, the person breathes heavily until the blood pCO_2 returns to normal and the cells receive enough oxygen to break down the pyruvic acid. This is the phenomenon that we referred to earlier as the oxygen debt. When the debt has been repaid, the lactic acid is changed back to pyruvic acid. The pyruvic acid then goes through a transitional process involving a special substance called coenzyme A and enters the Krebs cycle.

Coenzyme A

As you know, an enzyme is a protein. However, many enzymes are attached to nonprotein compounds. If a nonprotein group detaches from the enzyme and acts as a carrier molecule, it is called

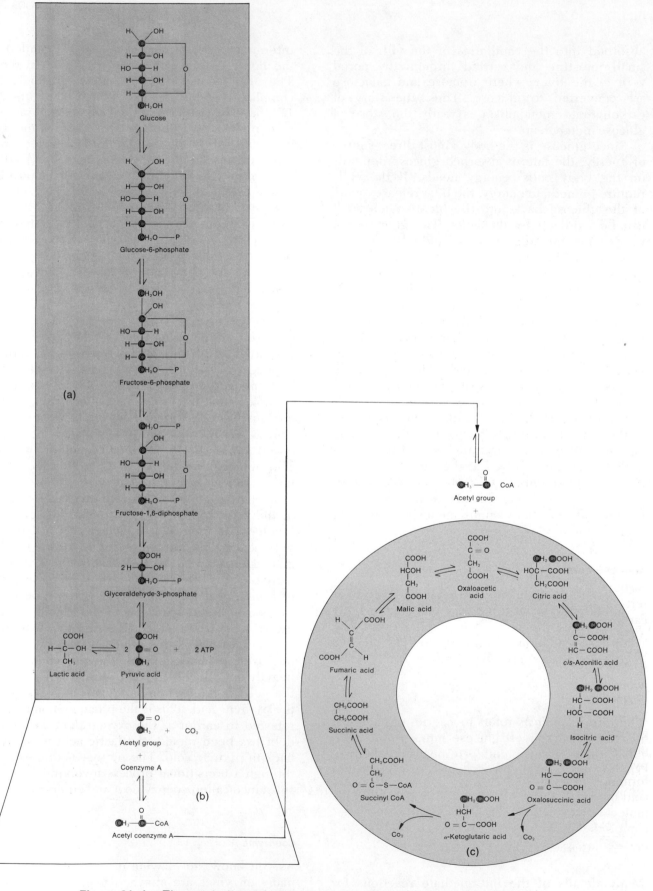

Figure 21-1. The catabolism of glucose via glycolysis and the Krebs cycle. (a) Glycolysis. (b) Intermediate step. (c) Krebs cycle. The carbon atoms have been colored so you can trace the entrance of glucose into the Krebs cycle. Refer to Figure 21-3 to see how glycolysis and the Krebs cycle are related to the metabolism of other nutrients.

a *coenzyme*. Enzymes work together with their coenzymes. The enzyme catalyzes the reaction, and the coenzyme attaches to the end product of the reaction and carries it to the next reaction. Each step in the oxidation of glucose requires a different enzyme and often a coenzyme as well. However, we are interested in only one coenzyme, a substance called *coenzyme A,* which is abbreviated as *CoA*.

During the transitional step between glycolysis and the Krebs cycle, pyruvic acid is prepared for entrance into the cycle. Essentially, pyruvic acid is converted to a two-carbon compound by the loss of carbon dioxide. This two-carbon fragment, called an acetyl group, attaches itself to coenzyme A, and the whole complex is called *acetyl coenzyme A*. It is in this form that pyruvic acid enters the Krebs cycle (Figure 21–1b). The lost carbon atom combines with oxygen supplied by the lungs and blood and becomes carbon dioxide, which is removed by the blood and exhaled.

The Krebs Cycle

The *Krebs cycle,* or *citric acid cycle,* is a cyclic series of reactions that occur on the mitochondria of the cells (Figure 21–1c). Coenzyme A carries an acetyl group to a spot on one of the mitochondria, detaches itself, and goes back into the cytoplasm to pick up another fragment. The acetyl group combines with a substance called oxaloacetic acid to form citric acid. As the compound proceeds through the cycle, the acetyl carbons are lost one by one until there is nothing left of the original glucose molecule. As each carbon is split off, more and more molecules of ATP are formed. The carbons combine with oxygen to form carbon dioxide. Oxaloacetic acid is left intact, ready to combine with another acetyl group.

The decomposition of the acetyl group in the Krebs cycle produces far more energy than does glycolysis. During glycolysis, one molecule of glucose is decomposed into two molecules of pyruvic acid, and two molecules of ATP are formed. But during the Krebs cycle, each pyruvic acid molecules releases enough energy to form 18 molecules of ATP—a total of 36 molecules. The complete oxidation of glucose, then, can be summarized as follows:

$$C_6H_{12}O_6 +\ 6O_2\ \rightarrow 6CO_2 + 6H_2O + 38ATP$$

Glucose   Oxygen   Carbon   Water
                            dioxide

Glycolysis and the Krebs cycle provide all the energy for cellular activities. The Krebs cycle provides most of the energy, and because it is an aerobic process, the cells cannot carry on their activities for very long without sufficient oxygen. Cells die quickly when they are deprived of oxygen because without it they cannot produce enough energy to continue their vital activities.

Glucose anabolism

Eventually, most of the glucose in the body is catabolized to supply energy. However, glucose participates in a number of anabolic reactions. One is the synthesis of one large molecule of *glycogen* from many glucose molecules. Another anabolic process is the manufacture of glucose molecules from the breakdown products of lipids and proteins.

Glucose storage

When glucose is absorbed, it is not always needed immediately for energy. If it is not needed, it is combined with many other molecules of glucose to form a long-chain molecule called glycogen. This process is called *glycogenesis* (Figure 21–2). The term glycogenesis is formed from the terms *glyco,* meaning sugar, and *genesis,* meaning origin. The body can store about 400 grams (1 pound) of glycogen in the liver and skeletal muscle cells. When the body needs energy, the stored glycogen is broken down into glucose to be catabolized. The process of converting glycogen back to glucose is called *glycogenolysis* (Figure 21–2). Here, the term *lysis* means breakdown. Glycogenolysis usually occurs between meals.

Conversion of lipids and proteins to glucose

When your body runs out of glycogen, it is time to eat another meal. If you do not eat, your body starts catabolizing fats and proteins. Actually, the body normally catabolizes some of its fats and a few of its proteins. But large-scale fat and protein catabolism does not happen unless you are starving, eating meals that contain very few carbohydrates, or suffering from one of the endocrine disorders discussed later.

In order for fats and proteins to be catabolized, they first must be converted to glucose, a process called *gluconeogenesis*. Figure 21–3 shows how gluconeogenesis is related to other metabolic reactions. Notice that fats and proteins can be transformed into compounds that also are intermediate products in the catabolism of glucose. For example, glycerol, one of the products of

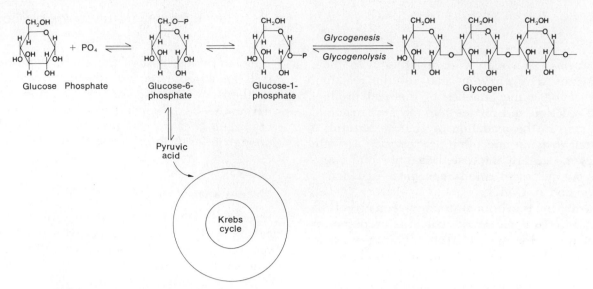

Figure 21-2. Glycogenesis and glycogenolysis. When glucose is needed, skeletal muscle cells can change glycogen into glucose-6-phosphate. This substance can be converted to pyruvic acid and enter the Krebs cycle in any body cell for oxidation to carbon dioxide and water. Liver cells also have the necessary enzymes to change glycogen all the way back to glucose, which can enter the blood for distribution to body cells.

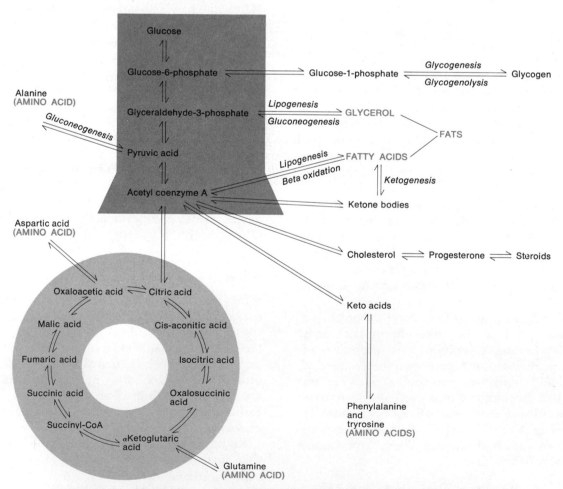

Figure 21-3. Metabolism of carbohydrates, lipids, and proteins. Note especially that glucose, fatty acids and glycerol, and amino acids must be converted into glucose breakdown products before they are catabolized. Mechanisms by which one type of nutrient is converted into another are indicated above and below the directional arrows. The catabolism of carbohydrates, lipids, and proteins are considered separately in Figures 21-1, 21-4, and 21-5, respectively.

fat digestion, may be converted into glyceraldehyde-3-phosphate, and the amino acid alanine may be converted into pyruvic acid. Also notice that most of the reactions in the figure are reversible. This has several important consequences. Most important, it means that the intermediate products of glucose catabolism also can be resynthesized into glucose. This is exactly what happens during gluconeogenesis, which is therefore an anabolic process. Only the liver has the proper enzymes for initiating gluconeogenesis, although in some cases enzymes existing in other cells are needed to complete the process.

The fact that many of the reactions in Figure 21–3 are reversible also means that many conversions of carbohydrates, fats, and proteins may occur. We shall describe how this is accomplished as we look at fat and protein metabolism.

LIPID METABOLISM

The chief function of carbohydrates is to provide energy. Although lipids are also an important source of energy, they are only second best. Lipids are used more frequently as building blocks to form essential structures.

Fat storage

When fats are eaten, they are digested into glycerol and fatty acids. As soon as the glycerol and fatty acids are absorbed by the lacteals of the villi, they recombine to form fat molecules. If the body has no immediate need for the fat, it is stored in adipose tissue. About 50 percent of your stored fat is stored in subcutaneous tissue, approximately 12 percent around the kidneys, about 10–15 percent in the omenta, approximately 20 percent in genital areas, and about 5–8 percent between muscles. Fat is also stored behind the eyes, in the furrows of the heart, and in the folds of the large intestine. Fat does not remain stationary until it is needed for energy. It is continually reabsorbed into the lymph, transported through the blood, and redeposited in other adipose tissue cells. Some researchers estimate that as much as one half of the total body-fat reserve changes position daily.

Lipid catabolism

Fats stored in fat depots constitutes the largest reserve of energy. The body can store much more fat than it can glycogen. Extremely obese people have been known to store a couple hundred pounds of fat. In addition, fats represent a more concentrated energy source than does glucose. For example, the oxidation of 1 gram of carbohydrate yields 4.1 kilocalories, whereas the oxidation of 1 gram of fat yields 9.4 kilocalories. A **kilocalorie,** or **Calorie,** is the amount of heat necessary to raise the temperature of one kilogram (1,000 grams) of water from 14.5° to 15.5°C. The kilocalorie, or Calorie with a capital C, is the same unit used to describe calories in foods. Thus, the energy yield of fats is more than twice that of carbohydrates. Despite these facts, fats are only the body's second-favorite source of energy because they are more difficult to catabolize than carbohydrates.

Glycerol

When fat molecules are metabolized, first they are separated into glycerol and fatty acids. The glycerol and fatty acids are then catabolized separately (Figure 21–4). Glycerol is converted easily by the liver cells to a compound called glyceraldehyde-3-phosphate, one of the compounds also formed during the catabolism of glucose. The liver cells then transform glyceraldehyde-3-phosphate into glucose and release it so it can be catabolized by other cells. This is one example of the process we called gluconeogenesis. Glyceraldehyde-3-phosphate is an intermediate product in the conversion of glycerol to glucose and is also a link in the conversion of glucose to fats.

Fatty acids

Fatty acids, however, are catabolized differently. The first step in fatty-acid catabolism involves a series of reactions called *beta oxidation* (Figure 21–4). During beta oxidation, enzymes in the liver cells remove pairs of carbon atoms from the long chain of carbon atoms comprising a fatty acid molecule. The liver cells convert some of the two-carbon fragments into acetyl CoA, which can be converted to glucose and catabolized by any cell. Most of the fragments, however, are converted into either *acetone* or into substances classified as *keto acids. Ketone bodies* is a collective term for keto acids and acetone; the formation of ketone bodies is called *ketogenesis.* The liver does not have the proper enzymes for converting the ketone bodies into an intermediate glucose product, so it releases them into the bloodstream. The ketone bodies are then catabolized, via the Krebs cycle, by the other body cells that have an enzyme which converts

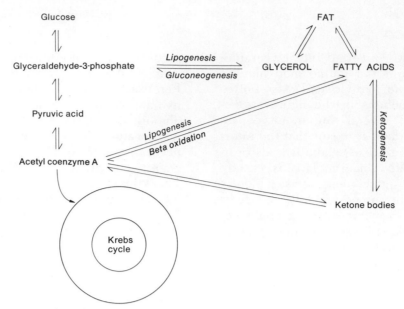

Figure 21-4. Metabolism of lipids. Glycerol may be converted to glyceraldehyde-3-phosphate and enter the Krebs cycle. Fatty acids undergo beta oxidation and ketogenesis and enter the Krebs cycle via acetyl coenzyme A.

ketone bodies into acetyl coenzyme A. Whether the process is direct or indirect, acetyl CoA links fatty acid with glucose metabolism.

Ketosis

Ketone bodies are normal intermediate products of fatty acid metabolism. But because the body prefers glucose as a source of energy, they are generally produced in very small quantities (1.5 to 2.0 milligrams/100 milliliters of blood). When the number of ketone bodies in the blood rises above normal, a condition called *ketosis*, the keto acids must be buffered by the body. If too many accumulate, they use up the body's buffers and the blood pH falls. Thus, extreme or prolonged ketosis can lead to acidosis, or abnormally low blood pH. The causes of ketosis include starvation, low carbohydrate diets, and metabolic abnormalities. One of the most common abnormalities is diabetes mellitus, or insufficient insulin. As you remember, insulin is required for more than minimal glucose consumption by the cells, and it discourages fat catabolism. When a diabetic fails to take sufficient insulin and becomes seriously insulin-deficient, one of the telltale signs is a sweet smell of acetone on his breath.

Lipid anabolism

Liver cells can synthesize lipids from glucose and amino acids through a process called *lipogenesis*. The steps in the conversion of glucose to lipids are just the reverse of the steps that transform fats into glucose. The links are glyceraldehyde-3-phosphate, which can be converted to glycerol, and acetyl CoA, which can be converted to fatty acids (see Figure 21-3). Amino acids are transformed to lipids in the same way, but first they must be converted to glucose or to a glucose-breakdown product. The resulting glycerol and fatty acids can undergo anabolic reactions to become fat that can be stored or go through a series of anabolic reactions which produce other types of lipids. Such lipids may become the middle layer of the plasma membrane, a part of the steroid hormones, or some other component of the body. Exhibit 2-4 lists some of the many ways in which lipids are used.

PROTEIN METABOLISM

Proteins are primarily body builders. Generally, the body uses very little protein for energy, as long as it ingests sufficient quantities of carbohydrates and fats, or as long as it has a supply of stored fat.

Protein catabolism

A certain amount of protein catabolism occurs in the body each day, although much of this is only partial catabolism. For instance, proteins are extracted from worn-out cells and broken down into free amino acids. Some of the amino acids are converted into other kinds of amino acids, peptide bonds are reformed, and new proteins are made.

However, if other energy sources are used up, the body can catabolize large amounts of protein to carbon dioxide and water. The first step in the

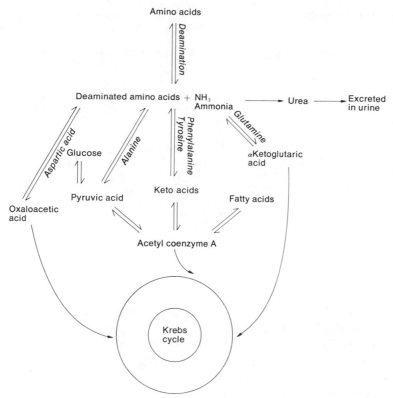

Figure 21–5. Metabolism of proteins. The deaminated amino acid may follow several pathways (see Figure 21–3).

catabolism of amino acids consists of removing the amino group (NH_2) from the amino acid (Figure 21–5), a process called *deamination* that occurs in the liver. The liver cells then convert the NH_2 to ammonia (NH_3) and finally to urea that is excreted in the urine. The fate of the remaining part of the amino acid depends on what kind of amino acid it is. If you look at Figure 21–3, you will see that some amino acids are converted to keto acids and then to acetyl CoA, other amino acids are changed to pyruvic acid, and still others are converted to acids of the Krebs cycle.

Protein anabolism

Protein anabolism involves the formation of peptide bonds between amino acids to produce new proteins. Protein anabolism, or synthesis, is carried out on the ribosomes of almost every cell in the body, directed by the cells' DNA and RNA. (The details of this process were described in Chapter 3.) The synthesized proteins are the primary constituents of cell structures, enzymes, antibodies, and many glandular secretions. Exhibit 2–5 shows the many ways in which the body uses proteins. Because proteins are a primary ingredient of most cell structures, high-protein diets are essential during the growth years and when tissue has been damaged by disease or injury.

CONTROL OF METABOLISM

The metabolism of carbohydrates, lipids, and proteins is regulated by a number of hormones. Two hormones that are particularly important, insulin and glucagon, are secreted in response to blood glucose level. Under normal circumstances, the blood sugar level ranges from 80–120 milligrams/100 milliliters of blood. After a meal, the level may rise to 120–130 milligrams/100 milliliters. The increase stimulates the pancreas to secrete insulin, which pushes the blood glucose level down to normal within a few hours. Insulin lowers blood sugar by stimulating the liver cells to convert glucose to glycogen and fat, stimulating all other cells to catabolize glucose, and inhibiting the catabolism of fats and proteins. When the body fails to produce enough insulin, the blood sugar is somewhat lowered as the glucose is excreted in the urine. This is why sugar in the urine is a sign of untreated diabetes mellitus. The body also excretes glucose if a large amount of carbohydrates are absorbed all at once. In this case, the condition is called *alimentary glycosuria*. Alimentary glycosuria is temporary and is not considered pathological.

After a period of fasting, the blood glucose level falls below normal. The pancreas then releases glucagon that stimulates the liver to change glycogen to glucose and to catabolize fats.

Exhibit 21–1. LISTING OF SELECTED MINERALS, COMMENTS, AND IMPORTANCE IN THE BODY

MINERAL	COMMENTS	IMPORTANCE
Calcium	Most abundant cation in body. Appears in combination with phosphorus in ratio of 2:1.5. About 99 percent is stored in bone and teeth. Remainder stored in muscle, other soft tissues, and blood plasma. Blood calcium level controlled by thyrocalcitonin and parathyroid hormone. Most is excreted in feces and small amount in urine. Recommended daily intake for children is 1.2–1.4 gm.° and 800 mg.° for adults. Good sources are milk, egg yolk, shellfish, and green leafy vegetables.	Formation of bones and teeth, blood clotting, and normal muscle and nerve activity.

The anterior pituitary gland also assumes a role in the regulation of metabolism through the actions of somatotropin, ACTH, and thyroid-stimulating hormone. Somatotropin (STH), or growth hormone, encourages new tissue to be laid down by stimulating protein anabolism. Growth hormone also stimulates the body cells to catabolize fats instead of carbohydrates. This action causes blood sugar levels to increase. Adrenocorticotropic hormone (ACTH) stimulates the adrenal cortex to increase its secretion of glucocorticoid hormones. The chief effect of the glucocorticoids is to increase the amount of energy available to cells. As you recall, large quantities of these hormones are released during extreme stress—a time when the body needs energy to combat the stressor. Glucocorticoids stimulate the liver to convert glycogen to glucose, encourage fat to be removed from its depots and to be converted to glucose by the liver, and accelerate the breakdown of proteins into amino acids. Since the resulting glucose is released into the bloodstream, glucocorticoids are hyperglycemic. The third anterior pituitary hormone, thyroid-stimulating hormone (TSH), stimulates the thyroid to release thyroxin, a hormone that encourages glucose catabolism and is therefore hypoglycemic.

The adrenal medulla, the ovaries, and the testes also produce hormones that affect metabolism. Epinephrine, the adrenal medulla hormone, accelerates the conversion of glycogen to glucose and is therefore hyperglycemic. Testosterone, the male hormone, and progesterone, the female hormone, encourage protein anabolism. Progesterone is secreted in particularly large amounts during pregnancy, when uterine and mammary tissues are growing.

Hormones are the primary regulators of metabolism. However, hormonal control is ineffective without the proper minerals and vitamins. Some minerals and many vitamins are components of the enzyme systems that catalyze the metabolic reactions.

MINERALS

Minerals are inorganic elements. They may appear in combination with each other or in combination with organic compounds. Minerals constitute about 4 percent of the total body weight, and they are concentrated most heavily in the skeleton. Minerals known to perform functions essential to life include calcium, phosphorus, sodium, chlorine, potassium, magnesium, iron, sulfur, iodine, manganese, cobalt, copper, and zinc. Other minerals, such as aluminum, silicon, arsenic, and nickel are also present in the body, but their functions have not yet been determined.

Calcium and phosphorus form part of the structure of bone. But since minerals do not form long-chain compounds, they are otherwise poor building materials, Their chief role is to help regulate body processes. Calcium, iron, magnesium, and manganese are constituents of some coenzymes. Magnesium also serves as a catalyst for the conversion of ADP to ATP. Without these minerals, metabolism would come to a screeching halt, and the organism would die. Minerals such as sodium and phosphorus work in buffer systems. Sodium regulates the osmosis of water and, along with other ions, is involved in the generation of nerve impulses. Exhibit 21–1 describes the functions of some minerals vital to the body. Note that the body generally uses the ions of the minerals rather than the un-ionized form. Some minerals, such as chlorine, are toxic or even fatal to the body if they are ingested in the un-ionized form.

MINERAL	COMMENTS	IMPORTANCE
Phosphorus	About 80 percent found in bones and teeth, remainder distributed in muscle, brain cells, and blood. More functions than any other mineral. Most excreted in urine, small amount eliminated in feces. Recommended daily intake is 1.2–1.4 gm. for children and 1,200 mg. for adults. Good sources are dairy products, meat, fish, poultry, and nuts.	Formation of bones and teeth. Constitutes important buffer system of blood. Plays important role in muscle contraction and nerve activity. Component of many enzymes. Involved in transfer and storage of energy (ATP). Component of DNA and RNA.
Iron	About 66 percent is found in hemoglobin of blood, remainder distributed in skeletal muscles, liver, spleen, and enzymes. Normal loses of iron occur by shedding of hair, epithelial cells, and mucosae cells and in sweat, urine, feces, and bile. Recommended daily intake for children is 7–12 mg. and for adults 10–15 mg. Sources are meat, liver, shellfish, egg yolk, beans, legumes, dried fruits, nuts, and cereals.	As component of hemoglobin, carries O_2 to body cells. Component of coenzymes involved in formation of ATP from catabolism.
Iodine	Essential component of thyroxin. Excreted in urine. Estimated daily requirement: 0.15–0.30 mg. Adequate sources are iodized salt, seafoods, cod-liver oil, and vegetables grown in iodine-rich soils.	Required by thyroid gland to synthesize thyroxin, the hormone that regulates metabolic rate.
Copper	Some stored in liver and spleen. Most excreted in feces. Daily requirement about 2 mg. Good sources include eggs, whole-wheat flour, beans, beets, liver, fish, spinach, and asparagus.	Required with iron for synthesis of hemoglobin. Component of enzyme necessary for melanin pigment formation.
Sodium	Most found in extracellular fluids, some in bones. Excreted in urine and perspiration. Recommended daily intake of NaCl (table salt) is 5 gm.	As most abundant cation in extracellular fluid, strongly affects distribution of water through osmosis. Part of bicarbonate buffer system.
Potassium	Principal cation in intracellular fluid. Most is excreted in urine. Recommended daily intake not known. Normal food intake supplies required amounts.	Functions in transmission of nerve impulses and muscle contraction.
Chlorine	Found in extracellular and intracellular fluids, principal anion of extracellular fluid. Most excreted in urine. Normal intake of NaCl supplies required amounts.	Assumes role in acid–base balance of blood, water balance, and formation of HCl in the stomach.
Magnesium	Component of soft tissues and bone. Excreted in urine and feces. Suggested minimum daily intake: 250–350 mg. Widespread in various foods.	Required for normal functioning of muscle and nervous tissue. Participates in bone formation. Constituent of many coenzymes.
Sulfur	Constituent of many proteins (such as insulin) and some vitamins (thiamine and biotin). Excreted in urine. Good sources include beef, liver, lamb, fish, poultry, eggs, cheese, and beans.	As component of hormones and vitamins, regulates various body activities.
Zinc	Important component of certain enzymes. Widespread in many foods.	Necessary for normal growth.
Fluorine	Component of bones, teeth, and other tissues.	Appears to improve tooth structure.
Manganese	Distribution throughout body similar to that of copper. Human daily requirement about 2.5 mg.	Activates several enzymes. Needed for hemoglobin synthesis. Required for growth, reproduction, and lactation.
Cobalt	Constituent of vitamin B_{12}.	As part of B_{12}, required for stimulation of erythropoeisis.

* 1 gram (gm.) = 1,000 milligrams (mg.)
  1,000 gm. = 2.2 pounds
  1 ounce = 28.25 gm.

Exhibit 21-2. SUMMARY OF VITAMINS

VITAMIN	COMMENT AND SOURCE	FUNCTION	RELATED SYMPTOMS AND DEFICIENCY DISORDERS
I. Fat-soluble			
A	Formed from provitamin carotene (and other provitamins) in intestinal tract. Requires bile salts and fat for absorption. Stored in liver. Sources of carotene and other provitamins include yellow and green vegetables; sources of vitamin A include fish, liver oils, milk, and butter.	1. Maintains general health and vigor of epithelial cells.	Deficiency results in atrophy and keratinization of epithelium, leading to dry skin and hair, increased incidence of ear, sinus, respiratory, urinary, and digestive infections, inability to gain weight, drying of cornea with ulceration (xerophthalmia), nervous disorders, and skin sores.
		2. Essential for formation of rhodopsin, light-sensitive chemical in rods of the retina.	Night blindness or decreased ability for dark adaptation.
		3. Growth of bones and teeth by apparently helping to regulate activity of osteoblasts and osteoclasts.	Slow and/or faulty development of bones and teeth.

VITAMINS

Organic nutrients required in minute amounts to maintain growth and normal metabolism are called *vitamins*. Unlike carbohydrates, fats, or proteins, vitamins do not provide energy or serve as building materials. The essential function of vitamins is the regulation of physiologic processes. Accordingly, some vitamins act as enzymes, others are precursors (forerunners) of enzymes, and still others serve as coenzymes.

Vitamins cannot be synthesized by the body from its own resources and must be obtained from a variety of sources. One source of vitamins is ingested foods—for example, vitamin C in citrus fruits. Another source is vitamin pills that contain several or all vitamins. Other vitamins, such as vitamin K, are produced by bacteria in the gastrointestinal tract. The body can assemble some vitamins if the raw materials are provided. Such raw materials are called *provitamins*. For example, vitamin A is produced by the body from the provitamin carotene, a chemical present in spinach, carrots, liver, and milk.

You will soon discover that vitamins are found in varying quantities in different foods and that no single food contains all required vitamins.

This is probably one of the best reasons for eating a balanced diet. The term *avitaminosis* refers to a condition in which there is a deficiency of any vitamin in the diet.

On the basis of solubility, vitamins are divided into two principal groups: fat-soluble and water-soluble. *Fat-soluble* vitamins are absorbed along with digested dietary fats by the lacteals of the villi of the small intestine. In fact, they cannot be absorbed unless they are ingested with some fat. Fat-soluble vitamins are generally stored in cells, particularly the cells of the liver, so reserves can be built up. Examples of fat-soluble vitamins are vitamins A, D, E, and K. *Water-soluble* vitamins, by contrast, are absorbed along with water in the gastrointestinal tract and dissolve in the body fluids. As the blood is filtered by the kidneys, excess quantities of the vitamins are excreted in the urine. Thus, the body does not store water-soluble vitamins well. However, it has the dubious advantage that very few cases of water-soluble vitamin overdose have been observed. Examples of water-soluble vitamins are the B vitamins and vitamin C. Exhibit 21-2 lists the principal vitamins, their sources, functions, and related disorders.

VITAMIN	COMMENT AND SOURCE	FUNCTION	RELATED SYMPTOMS AND DEFICIENCY DISORDERS
D	In presence of sunlight, provitamin D_3 (derivative of cholesterol) converted to vitamin D. Requires moderate amounts of bile salts and fat for absorption. Stored in tissues to slight extent. Most excreted via bile. Sources include liver oils of bony fish, egg yolk, and fortified milk.	Essential for absorption and utilization of calcium and phosphorus from gastrointestinal tract. May work in conjunction with the parathyroid hormone that controls calcium metabolism.	Defective utilization of calcium by bones leads to rickets in children and osteomalacia in adults. Possible loss of muscle tone.
E (tocopherols)	Stored in liver, adipose tissue, and muscles. Requires bile salts and fat for absorption. Sources include fresh nuts and wheat germ, seed oils, and green leafy vegetables.	Believed to inhibit catabolism of certain fatty acids that help form cell structures, especially membranes. Involved in formation of DNA, RNA, and red blood cells. Believed to help protect liver from such toxic chemicals as carbon tetrachloride.	Catabolism of certain fatty acids on exposure to oxygen (for example, fatty acids in membranes of red blood cells possibly leading to hemolytic anemia). Deficiency also causes muscular dystrophy in monkeys and sterility in rats.
K	Produced in considerable quantities by intestinal bacteria. Requires bile salts and fat for absorption. Stored in liver and spleen. Other sources include spinach, cauliflower, cabbage, and liver.	Believed to serve as coenzyme essential for synthesis of prothrombin by liver and thus for normal blood clotting. Also known as antihemorrhagic vitamin.	Delayed clotting time results in excessive bleeding.

II. Water-soluble

B

B₁ (thiamine)	Rapidly destroyed by heat. Not stored in body. Excessive intake is eliminated in urine. Sources include whole-grain cereals, eggs, pork, nuts, liver and yeast.	Acts as coenzyme for 24 different enzymes involved in carbohydrate metabolism of pyruvic acid to CO_2 and H_2O. Essential for synthesis of acetylcholine.	Improper carbohydrate metabolism leads to build-up of pyruvic and lactic acids and insufficient energy for cells, especially muscle and nerve cells. Deficiency leads to two syndromes: (1) *Beriberi*—partial paralysis of smooth muscle of gastrointestinal tract causing digestive disturbances, skeletal muscle paralysis, atrophy of limbs. (2) *Polyneuritis*—reflexes related to kinesthesia are impaired, impairment of sense of touch, decreased intestinal motility, stunted growth in children, poor appetite.

Exhibit 21–2. SUMMARY OF VITAMINS *(cont'd.)*

VITAMIN	COMMENT AND SOURCE	FUNCTION	RELATED SYMPTOMS AND DEFICIENCY DISORDERS
B_2 (riboflavin)	Not stored in large amounts in tissues. Most is excreted in urine. Small amounts are supplied by bacteria of gastrointestinal tract. Other sources include yeast, liver, beef, veal, lamb, eggs, whole-wheat products, asparagus, peas, beets, and peanuts.	Component of certain coenzymes concerned with carbohydrate and protein metabolism, especially in cells of eye, integument, mucosa of intestine, and blood.	Deficiency may lead to improper utilization of oxygen resulting in blurred vision, cataracts, and corneal ulcerations. Also dermatitis and cracking of skin, lesions of intestinal mucosa, and development of one type of anemia.
Niacin (nicotinamide)	Derived from amino acid tryptophan. Sources include yeast, meats, liver, fish, whole-grain breads and cereals, peas, beans, and nuts.	Essential component of coenzyme concerned with energy-releasing reactions. In lipid metabolism, inhibits production of cholesterol and assists in fat breakdown.	Principal deficiency is *pellegra*, characterized by dermatitis, diarrhea, and psychological disturbances.
B_6 (pyridoxine)	Formed by bacteria of gastrointestinal tract. Stored in liver, muscle, and brain. Other sources include salmon, yeast, tomatoes, yellow corn, spinach, whole-grain cereals, liver, and yogurt.	May function as coenzyme in fat metabolism. Essential coenzyme for normal amino acid metabolism. Assumes role in production of circulating antibodies.	Most common deficiency symptom is dermatitis of eyes, nose, and mouth. Other symptoms are retarded growth and nausea.
B_{12} (cyanocobalamin)	Only B vitamin not found in vegetables; only vitamin containing cobalt. Absorption from gastrointestinal tract dependent on HCl and intrinsic factor secreted by gastric mucosa. Sources include liver, kidney, milk, eggs, cheese, and meat.	Coenzyme necessary for red blood cell formation, formation of amino acid methionine, entrance of some amino acids into Krebs cycle, and manufacture of choline, (chemical similar in function to acetylcholine).	Pernicious anemia and malfunction of nervous system due to degeneration of axons of spinal cord.
Pantothenic acid	Stored primarily in liver and kidneys. Some produced by bacteria of gastrointestinal tract. Other sources include kidney, liver, yeast, green vegetables, and cereal.	As constituent of coenzyme A, essential for transfer of pyruvic acid into Krebs cycle, conversion of lipids and amino acids into glucose, and synthesis of cholesterol and steroid hormones.	Experimental deficiency tests indicate fatigue, muscle spasms, neuromuscular degeneration, and insufficient production of adrenal steroid hormones.

VITAMIN	COMMENT AND SOURCE	FUNCTION	RELATED SYMPTOMS AND DEFICIENCY DISORDERS
Folic acid	Synthesized by bacteria of gastrointestinal tract. Other sources include green leafy vegetables and liver.	Component of enzyme systems synthesizing purines and pyrimidines built into DNA and RNA. Essential for normal production of red and white blood cells.	Production of abnormally large red blood cells— macrocytic anemia.
Biotin	Synthesized by bacteria of gastrointestinal tract. Other sources include yeast, liver, egg yolk, and kidneys.	Essential coenzyme for conversion of pyruvic acid to oxaloacetic acid, and synthesis of fatty acids and purines.	Mental depression, muscular pain, dermatitis, fatigue, and nausea.
C (ascorbic acid)	Very rapidly destroyed by heat. Some stored in glandular tissue and plasma. Sources include citrus fruits, tomatoes, and green vegetables.	Exact role in promoting many activities not understood. Promotes many metabolic reactions, particularly protein metabolism, including laying down of collagen in formation of connective tissue. Possibly as coenzyme, combines with poisons, rendering them harmless until excreted, and works with antibodies.	Scurvy: Many symptoms related to poor connective tissue growth and repair including tender swollen gums, loosening of teeth (alveolar processes also deteriorate), poor wound healing, bleeding (vessel walls fragile because of connective tissue degeneration), and retardation of growth. Anemia and low resistance to infection of scurvy.

NUTRITIONAL AND METABOLIC DISORDERS

Disorders involving minerals have many causes, including inadequate diet or absorption, endocrine imbalance, and kidney malfunction. Chapters 7 and 15 described some mineral disorders involving bone and the endocrine system; in Chapters 22 and 23 we shall look at how the body's major ions affect osmosis and pH. For the time being, let us look at some inherited disorders that have wide-ranging effects on the body's metabolism.

Phenylketonuria (PKU)

By definition, **phenylketonuria** is an inborn error of metabolism characterized by an elevation of the amino acid phenylalanine in the blood and is frequently associated with mental retardation. The DNA of people with phenylketonuria lacks the gene that normally programs the manufacture of the enzyme phenylalanine hydroxylase. This enzyme is necessary for the conversion of phenylalanine into the amino acid tyrosine, an amino acid that enters the Krebs cycle. As a result, phenylalanine cannot be metabolized, and what is not used in protein synthesis builds up in the blood. High levels of phenylalanine are toxic to the brain during the early years of life when the brain is developing, and mental retardation is produced. Mental retardation can be prevented, when the condition is detected early, by restricting the child to a diet that supplies only the amount of phenylalanine necessary for growth.

Cystic fibrosis

Cystic fibrosis is an inherited disease of the exocrine glands, affecting the pancreas, respiratory system, and salivary and sweat glands. It is characterized by the production of thick exocrine secretions that do not drain easily from the passageways. The buildup of the secretions leads to inflammation and replacement of injured cells with connective tissue that blocks the passageways. One of the prominent features is blockage of the pancreatic ducts so that the digestive enzymes cannot reach the intestine. Since pancreatic juice contains the only fat-digestion enzyme, the person fails to absorb fats or fat-soluble vitamins and thus suffers from vitamin A, D, and K deficiency diseases. Calcium also needs fat to be absorbed, so tetany also may result.

A child suffering from cystic fibrosis is given pancreatic extract and large doses of vitamins A, D, and K. The therapeutic diet is low, but not lacking, in fats and high in carbohydrates and proteins that can be used for energy and can also be converted by the liver into the lipids essential for life processes.

Celiac disease

Cystic fibrosis was once confused with an allergy to gluten—the protein in wheat, rye, barley, and oats—called **celiac disease.** The allergy causes changes in the mucosa of the small intestine that decrease the absorption of all nutrients. The condition is easily remedied by administering a diet that excludes all cereal grains except rice and corn.

OBESITY

In the United States and other affluent countries, a growing number of citizens live in highly mechanized urban areas, where the demands and opportunities for regular physical activities are diminishing increasingly. This same environment provides a variety of appetizing, calorically rich foods that are increasingly available and easily obtained. During the last half century, this trend has shown little sign of slowing down.

In this setting, *obesity* has become a serious problem, both for individuals and for the community. There is little doubt that even moderate obesity is hazardous to health. Medicine has preached against fat since Hippocrates, who is supposed to have advised that "persons who are naturally very fat are apt to die earlier than those who are slender." Statistics from modern insurance companies indicate that Hippocrates was correct. The now famous Framingham Study of the Public Health Service showed a high correlation between obesity and sudden death. Obesity seems solidly accepted as one of the "risk factors" in coronary artery disease.

Despite the fact that obesity is a disorder that usually can be diagnosed by inspection, it is not easily defined. There is no overall agreement as to the degree of overweight that divides obesity from nonobesity. One definition holds that an obese individual is one who is 20 percent or more over his so-called desirable weight when the extra weight is in the form of stored fat.

Because storage of fat is a normal function of adipose tissue, it is difficult to determine the point at which the quantity of stored fat becomes excessive. It is usually assumed that the "normal" or "best" weight for an individual is achieved between the ages of 18 and 25 years. Therefore, any significant positive deviation from this norm could be considered "obesity." Fatness, however, cannot always be predicated on the basis of weight. Height-weight relationships do not necessarily take into consideration an individual's build.

Much obesity, particularly when it is severe, has its roots in childhood. Children may become obese at any time, but this disorder is said to develop most commonly in three phases of childhood: in the latter part of infancy, at the time of starting school, and during adolescence. Obese children are likely to become obese adults. The more severe the obesity in childhood, the more likely it is to persist into adult life.

Causes

Causes of obesity can be classified under two broad headings: regulatory and metabolic. People with *regulatory obesity* have no apparent metabolic abnormality that can account for the obesity. They just seem to ingest more high-energy-releasing foods than their bodies need. Causes include neurotic overeating, cultural dietary habits in otherwise normal people, and inactivity. Occasionally, regulatory obesity is caused by a disorder in the hypothalamus that destroys or reduces sensations of satiety. *Metabolic obesity* results primarily from a disorder that reduces the catabolism of carbohydrates and/or fats. An example is hyposecretion of thyroxin. It is not yet certain to what degree these metabolic disorders may be caused by changes in diet or physical activity.

Regulatory obesity seems to be far more common than metabolic obesity. However, in the clinical setting, obesity seems more usefully categorized according to factors such as age of onset, degree of severity, presence of an associated pertinent disorder such as diabetes, hypertension, osteoarthritis, and hyperlipidemia.

Treatment

Reduction of body weight involves restricting Calorie intake to a level well below that of energy expenditure. The goals during weight decrease are

1. Loss of body fat with a minimal accompanying breakdown of lean tissue
2. Maintenance of physical and emotional fitness during the reducing period
3. Establishment of eating and exercise habits that will help the formerly obese individual maintain his weight at the recommended level

A number of factors must be considered in the formulation of a reducing diet. These include the individual's degree of overweight, age, state of physical fitness, normal level of physical activity, and the presence of related illness such as hypertension, coronary heart disease, diabetes mellitus, or gastrointestinal disorders.

The most drastic reducing diet is one that provides no Calories at all. We know now that diets without Calories must provide sufficient amounts of water, electrolytes (especially potassium), and vitamins to prevent dehydration, muscular weakness, mental confusion, heart function abnormalities, and other complications. Thus people treated with the "no-Calorie" diet should be hospitalized during the period of total Calorie starvation. The person remains at rest or limits himself to light activity during the fasting period. During the total Calorie restriction, fat is burned at an appreciable rate, but significant quantities of lean tissue also are broken down and lost.

In recent years, "unbalanced" diets, particularly diets that are very low in carbohydrate, have achieved both popularity and notoriety and have been praised by some as being remarkably effective and condemned by others for promoting side effects. Two of the latest unbalanced diets are the Dr. Stillman "quick weight loss" diet — high on water and low on food — and Dr. Robert Atkins' method — high protein, high fat, and extremely low carbohydrate. Both diets produce ketosis because neither contains enough carbohydrate. The Dr. Stillman diet can also produce hypercholesterolemia (an excess of cholesterol in the blood). This diet is essentially limited to fat and protein. However the fat is almost entirely of animal origin and thus is highly saturated. The average amount of cholesterol consumed per day in this diet is more than twice the amount in the average American diet.

The following points are essential for proper and effective dieting:

1. Have a complete physical examination to determine whether or not a special diet is needed.
2. Get a list of the nutrients that are required daily and foods containing them. One such booklet is "Nutritive Value of Foods," Home and Garden Bulletin No. 72, U.S. Dept. of Agriculture.
3. Use the Calorie chart contained in nutrition booklets and devise a high nutrition diet that has less Calories than you have been taking in.

Medical authorities agree that a sensible diet results in the loss of approximately 2 pounds a week and not much more.

Surgery may be the answer to gross obesity that does not respond to dieting. Surgical procedures are considered when a person weighs twice as much as he or she should and has maintained this weight for at least 5 years. The surgical procedures are all drastic measures and primarily involve removing portions of the stomach or intestinal tract.

Chapter summary in outline

METABOLISM
1. All chemical activities of the body involve anabolism or catabolism of nutrients.
2. Nutrients include carbohydrates, lipids, proteins, water, minerals, and vitamins.

Carbohydrate
1. Carbohydrate metabolism is primarily concerned with glucose.
2. Glucose is broken down via glycolysis (anaerobic) and the Krebs cycle (aerobic) to produce energy in the form of ATP. Lack of oxygen results in oxygen debt.
3. Glucose is stored as glycogen. The transformation of glucose to glycogen is called glycogenesis. Glycogen breakdown is called glycogenolysis. Conversion of lipids and proteins to glucose is referred to as gluconeogenesis.

Lipids
1. Fats are stored in adipose tissue, mostly in subcutaneous tissue.

2. In fat catabolism, glycerol is converted into glucose, and fatty acids undergo beta oxidation and transformation into ketone bodies. Ketone bodies are transformed by nonliver cells to acetyl CoA and enters the Krebs cycle. The presence of excess ketone bodies is called ketosis.

3. Synthesis of lipids from glucose and amino acids is called lipogenesis.

Proteins

1. Amino acids are built into proteins that serve as cell structures, enzymes, antibodies, and glandular secretions.

2. Protein catabolism involves the deamination of amino acids. Amino acids may then be converted to keto acids, pyruvic acid, and acids of the Krebs cycle.

Control of Metabolism

1. Metabolism is controlled by hormones such as insulin, glucagon, STH, ACTH, TSH, epinephrine, and sex hormones.

MINERALS

1. Minerals are inorganic elements that help to regulate body processes.

2. Principal minerals and their functions are: Calcium and phosphorus are necessary for growth of bones and teeth. Iron and copper are used in the synthesis of hemoglobin. Iodine is necessary for thyroxin synthesis. Sodium is used in water balance and in buffers. Potassium is necessary for nerve impulse transmission. Chlorine is required for acid-base balance. Magnesium is used for proper muscle and nerve functioning. Sulfur is a component of hormones.

VITAMINS

1. Vitamins are organic nutrients that regulate metabolism. Many function in enzyme systems.

2. Fat-soluble vitamins are absorbed with fats and include A, D, E, and K.

3. Water-soluble vitamins are absorbed with water and include the B vitamins and vitamin C.

4. Representative physiological functions are: vitamin A—healthy epithelium and vision; D—proper utilization of calcium; K—blood clotting; B_1, B_2, niacin, B_6—regulation of energy metabolism; B_{12} and folic acid—blood cell formation.

DISORDERS

1. Phenylketonuria is an elevation of the amino acid phenylalanine in the blood due to a defective gene. High levels of phenylalanine are toxic to the brain and lead to mental retardation.

2. Cystic fibrosis is characterized by the production of thick secretions leading to inflammation and tissue repair with fibrous tissue that blocks ducts. Blockage of pancreatic duct results in faulty absorption of fats, vitamins A, D, and K, and calcium.

3. Celiac disease is an intolerance to wheat and rye proteins, such as gluten, resulting in changes in the mucosa of the small intestine that decrease absorption of all nutrients.

OBESITY

1. Causes of obesity are regulatory (overeating) or metabolic disorders.

2. Treatment includes various diets or, in extreme cases, surgery.

Review questions and problems

1. Define a nutrient. List the six classes of nutrients and indicate the function of each.

2. What is metabolism? Distinguish between catabolism and anabolism and give examples of each.

3. Explain what happens to glucose during glycolysis and the Krebs cycle.

4. What is the importance of lactic acid in glucose catabolism? Relate your answer to oxygen debt.

5. Define glycogenesis and glycogenolysis. Under what circumstances does each occur?

6. Why is gluconeogenesis important? Give specific examples to substantiate your answer.

7. Indicate some areas where fat is stored in the body.

8. What is a kilocalorie? Relate your definition to fat and carbohydrate catabolism.

9. Explain how glycerol participates in gluconeogenesis.

10. Define beta oxidation. What are ketone bodies?

11. What is ketosis? What is the importance of the condition to the body?

12. Define lipogenesis. Why is the process important?

13. Briefly describe the mechanism involved in protein synthesis.

14. What is the importance of proteins to the body?

15. Define deamination. Explain the conversions that occur between amino acids and keto acids, pyruvic acid, and acids of the Krebs cycle.

16. Indicate the role of the following hormones in the control of metabolism: insulin, glucagon, STH, ACTH, TSH, epinephrine, and sex hormones.

17. What is a mineral? Briefly describe the functions of the following minerals: calcium, phosphorus, iron, iodine, copper, sodium, potassium, chloride, magnesium, sulfur, zinc, fluoride, manganese, and cobalt.

18. Define a vitamin. List and explain the ways in which we obtain vitamins. Distinguish between a fat-soluble and water-soluble vitamin.

19. What are the functions of vitamin A? Relate its functions to health of the epithelium, night blindness, and growth of bones and teeth.

20. How is sunlight related to vitamin D? What are the functions of vitamin D?

21. What is believed to be the principal physiological activity of vitamin E?

22. How does vitamin K function in blood clotting?

23. Relate the roles of vitamin B_1 to beriberi and polyneuritis.

24. How does vitamin B_2 function in the body?

25. Why does a niacin deficiency cause pellegra?

26. Relate the role of vitamin B_6 to dermatitis.

27. How does vitamin B_{12} function in red blood cell formation?

28. What are the principal physiological effects of pantothenic acid?

29. What relationship exists between folic acid and macrocytic anemia?

30. What are the functions of biotin in the body?

31. Relate vitamin C deficiency to the symptoms of scurvy, indicating the functions of vitamin C.

32. Define phenylketonuria. What causes the disorder? Why is it dangerous? How is the danger prevented?

33. What is cystic fibrosis? How does it affect vitamin absorption?

34. Define celiac disease. How is it corrected?

35. Formulate a definition for obesity. Distinguish between metabolic and regulatory obesity.

36. Discuss some disadvantages of unbalanced reducing diets.

CHAPTER 22

THE URINARY SYSTEM: STRUCTURE, PHYSIOLOGY, AND DISORDERS

STUDENT OBJECTIVES

After you have read this chapter, you should be able to:

1. Identify the gross anatomical features of the kidneys

2. Define the structural adaptations of a nephron for urine formation

3. Describe the blood and nerve supply to the kidneys

4. Describe the process of urine formation

5. Define glomerular filtration, tubular reabsorption, and tubular secretion

6. Compare the chemical composition of plasma, glomerular filtrate, and urine

7. Define the forces that support and oppose the filtration of blood in the kidneys

8. Discuss renal suppression as a disorder resulting from a decreased filtration pressure

9. Describe the physiological role of tubular reabsorption

10. Compare the obligatory and facultative reabsorption of water

11. Describe tubular excretion as a mechanism of elimination and a control of blood pH

12. Define kidney excretion of H^+ and NH_4^+ as a means of maintaining the pH of the body while conserving bicarbonate

13. Compare the lungs, integument, and alimentary canal as organs of excretion that help maintain body pH

14. Describe the effects of blood pressure, diet, temperature, and emotions on urine production

15. List the physical characteristics of urine

16. List the normal chemical constituents of urine

17. Define albuminuria, glycosuria, hematuria, pyuria, ketosis, casts, and calculi

18. Describe the structure and physiology of the ureters

19. Describe the physiology of micturition

20. Compare the causes of incontinence, retention, and suppression

21. Describe the structure and physiology of the urethra

22. Discuss the causes of ptosis, kidney stones, gout, glomerulonephritis, pyelitis, and cystitis

23. Discuss the operational principle of hemodialysis

24. Define medical terminology associated with the urinary system

The metabolism of nutrients results in the production of wastes by body cells, including carbon dioxide and excesses of water and heat. Protein catabolism produces toxic nitrogenous wastes, such as ammonia and urea. In addition, too many of the essential ions such as sodium, chloride, sulfate, phosphate, and hydrogen tend to be accumulated in the body. All the toxic materials and the excess essential materials must be eliminated by the body.

The primary function of the urinary system is to keep the body in homeostasis by controlling the concentration and volume of blood by removing and restoring selected amounts of water and solutes. It also excretes selected amounts of various wastes. Two kidneys, two ureters, one urinary bladder, and a single urethra comprise the system (Figure 22–1). The kidneys control the concentration and volume of the blood and remove wastes from the blood, manufacturing urine in the process. Urine drains out of each kidney through its ureter and is stored in the urinary bladder until it is expelled from the body through the urethra. Other systems that aid in waste elimination are the respiratory, integumentary, and digestive systems. When we describe the characteristics of urine, we shall look also at the excretory functions of these other systems.

THE KIDNEYS

The paired **kidneys** are reddish organs that resemble lima beans in shape (see Figure 22–1). They are found just above the waist, between the parietal peritoneum and the posterior wall of the abdomen. Since they are external to the peritoneal lining of the abdominal cavity, their placement is described as *retroperitoneal*. Relative to the vertebral column, the kidneys are located between the levels of the last thoracic and third lumbar vertebrae, with the right kidney slightly lower than the left because of the relatively large area occupied by the liver.

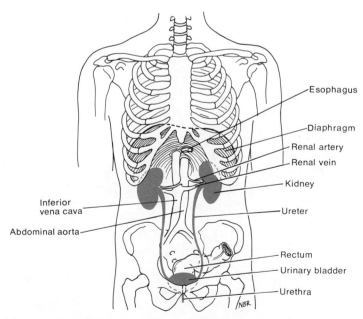

Esophagus

Diaphragm

Renal artery

Renal vein

Kidney

Inferior vena cava

Abdominal aorta

Ureter

Rectum

Urinary bladder

Urethra

Figure 22–1. Organs of the urinary system.

Gross anatomy

The gross anatomy of the kidneys may be studied both externally and internally. Let us examine their gross external anatomy first.

External

The average adult kidney measures about 11.25 centimeters (4 inches) long, 5.0–7.5 centimeters (2 to 3 inches) wide, and 2.5 centimeters (1 inch) thick. Its concave medial border faces the vertebral column. Near the center of the concave border is a notch called the *hilum*, through which the ureter leaves the kidney. Blood and lymph vessels and nerves also enter and exit the kidney through the hilum.

Three layers of tissue surround each kidney. The innermost layer, the *renal capsule*, is a smooth, transparent, fibrous membrane that adheres to the kidney and is continuous with the outer coat of the ureter at the hilum. It serves as a barrier against trauma and the spread of infection to the kidney. The second layer, the *adipose capsule*, is a mass of fatty tissue surrounding the renal capsule. It also protects the kidney from trauma and holds it firmly in place in the abdominal cavity. The outermost layer, the *renal fascia,* is a thin layer of fibrous connective tissue that anchors the kidneys to their surrounding structures and to the abdominal wall. Some individuals, especially thin ones in whom either the adipose capsule or renal fascia is deficient, may develop a condition called *ptosis* (dropping) of one or both kidneys. Ptosis is dangerous because it may cause kinking of the ureter with reflux of urine and back pressure. Ptosis of the kidneys below the rib cage also makes the individuals susceptible to blows and penetrating injuries.

Internal

If you make a longitudinal section through a kidney, you will see an outer, reddish area called the *cortex* and an inner, reddish brown region called the *medulla* (Figure 22–2). Within the medulla are 8 to 18 striated, triangular-shaped structures termed *renal*, or *medullary, pyramids.* The bases of the pyramids face the cortical area,

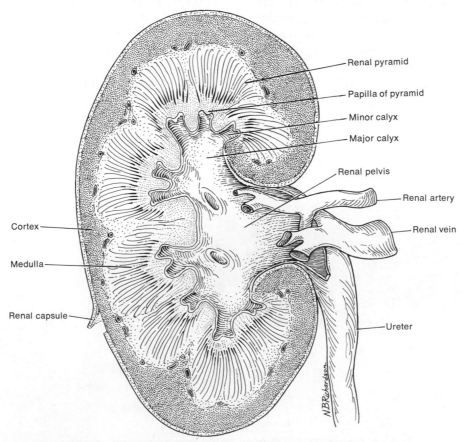

Figure 22–2. Longitudinal section through the right kidney illustrating gross internal anatomy.

and their apices, called *renal papillae,* are directed toward the center of the kidney. The cortex is the smooth-textured area extending from the renal capsule to the bases of the pyramids and into the spaces between the pyramids. Together, the cortex and renal pyramids constitute the parenchyma of the kidney. Structurally, the parenchyma of each kidney consists of approximately 1 million microscopic units called nephrons, collecting ducts, and their associated vascular supply. Nephrons are the functional units of the kidney. They form the urine and regulate the blood composition.

In the center of the kidney is a large cavity called the *renal pelvis* of the kidney. The edge of the pelvis is divided into cuplike extensions called the *major* and *minor calyces.* Each minor calyx collects urine from collecting ducts. From the calyx, the urine drains into the body of the pelvis and out through the ureter.

The nephron

The physiological unit of the kidney is referred to as a **nephron** (Figure 22–3). Essentially, each nephron is a *renal tubule* plus its associated blood supply. The parts of a nephron are as follows: glomerular capsule, proximal convoluted tubule, descending limb of Henle, loop of Henle, ascending limb of Henle, and distal convoluted tubule. Let us examine each of these parts of a nephron in detail. It begins as a double-walled globe called the *glomerular,* or *Bowman's capsule,* lying in the cortex of the kidney. The inner wall of the capsule consists of simple squamous epithelium surrounding a capillary network called the *glomerulus.* A space separates the inner wall from the outer one, which is also composed of simple squamous epithelium. Collectively, the Bowman's capsule and the enclosed glomerulus are called a *renal corpuscle.*

You will discover that the different kinds of epithelium found in the nephron are adapted to perform specialized functions. Simple squamous epithelium provides a semipermeable membrane that offers minimal resistance to the passage of molecules. Water and solutes in the blood are filtered easily through the inner wall of the Bowman's capsule and pass into the space between the inner and outer walls. From here the fluid drains into the renal tubule, which is subdivided into a number of sections.

The first section of the renal tubule, the *proximal convoluted tubule,* also lies in the cortex. Convoluted means that the tubule is highly coiled rather than straight, and the word "proximal" refers to the fact that the tubule is nearest its point of origin at the Bowman's capsule. The wall of the proximal convoluted tubule consists of cuboidal epithelium with microvilli. These cytoplasmic extensions, like those of the small intestine, increase the surface area for reabsorption and secretion. (As you will learn soon, much of the fluid extracted by the renal corpuscle is reabsorbed as it passes through the tubule.)

The second section of the renal tubule, the *descending limb of Henle,* dips into the medulla. It consists of squamous epithelium. The tubule then bends into a C-shaped structure called the *loop of Henle.* As the tubule straightens out, it increases in diameter and ascends toward the cortex as the *ascending limb of Henle,* which consists of cuboidal and columnar epithelium. In the cortex, the tubule again becomes convoluted. Because of its distance from the point of origin at the Bowman's capsule, this section is referred to as the *distal convoluted tubule.* Like those of the proximal tubule, the cells of the distal tubule are cuboidal with microvilli. The distal tubule terminates by merging with a straight *collecting duct.* In the medulla, the collecting ducts receive the distal tubules of several nephrons, pass through the renal pyramids, and open into the calyces of the pelvis through a series of *papillary ducts.*

Blood and nerve supply

Because the nephrons are responsible for removing wastes from the blood and regulating its fluid and electrolyte content, it should not seem surprising that they are abundantly supplied with blood vessels. The two *renal arteries* transport about one-fourth the total cardiac output to the kidneys (see Figure 22–1). Thus, approximately 1,200 milliliters of blood pass through the kidneys each minute. Before or immediately after entering through the hilum, the renal artery divides into several branches that enter the parenchyma and pass between the renal pyramids. Further divisions of the branches produce a series of *interlobular arteries* (see Figure 22–3a). The interlobular arteries enter the cortex and divide into *afferent arterioles.* One afferent arteriole is distributed to each glomerular capsule, where the arteriole breaks up into the capillary network termed the *glomerulus.* The glomerular capillaries then reunite to form an *efferent arteriole,* leading away from the capsule, that is smaller in diameter than the afferent arteriole. This situation is unique

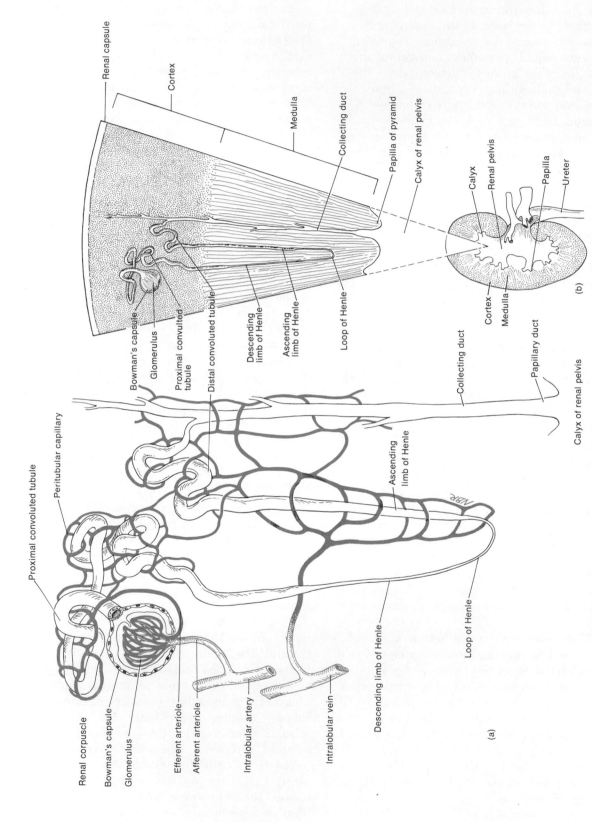

Renal capsule

Cortex

Medulla

Collecting duct

Papilla of pyramid

Calyx of renal pelvis

Bowman's capsule

Glomerulus

Proximal convoluted tubule

Distal convoluted tubule

Descending limb of Henle

Ascending limb of Henle

Loop of Henle

Calyx

Renal pelvis

Papilla

Ureter

Cortex

Medulla

Papillary duct

Collecting duct

Calyx of renal pelvis

(b)

Proximal convoluted tubule

Peritubular capillary

Ascending limb of Henle

Renal corpuscle

Bowman's capsule

Glomerulus

Efferent arteriole

Afferent arteriole

Intralobular artery

Intralobular vein

Descending limb of Henle

Loop of Henle

(a)

Figure 22-3. The nephron. (a) Microscopic appearance of an isolated nephron. (b) Position of a nephron in relation to the cortex and medulla.

516

because blood usually flows out of capillaries into venules and not into other arterioles. Each efferent arteriole divides to form a second network of capillaries, called the *peritubular capillaries.* The peritubular capillaries supply the renal tubule and then eventually reunite to form *intralobular veins.* The blood then drains through veins running between the pyramids and leaves the kidney through a single *renal vein* that exits at the hilum.

The nerve supply to the kidneys is derived from the *renal plexus* of the autonomic system. Nerves from the plexus accompany the renal arteries and their branches and are distributed to the vessels. Because the nerves are vasomotor, they regulate the circulation of blood in the kidney by regulating the diameters of the small blood vessels.

Physiology

The major work of the urinary system is done by the nephrons, while the other parts of the system are primarily passageways and storage areas. Nephrons carry out three important functions. They control the concentration and volume of the blood by removing selected amounts of water and solutes, help to regulate blood pH, and remove

some types of toxic wastes from the blood. As the nephrons go about these activities, they remove many materials from the blood, return the ones that the body requires, and eliminate the remainder. The eliminated materials are collectively called urine. A good way to learn how the nephrons perform their regulatory functions is to follow the formation of urine from the renal corpuscle to the collecting ducts. Although the collecting ducts are not part of the nephron, they are involved in urine formation. Urine formation requires three principal processes: glomerular filtration, tubular reabsorption, and tubular secretion.

Glomerular Filtration

The first step in the production of urine is *filtration.* You will recall from Chapter 2 that filtration is the forcing of fluids and dissolved substances through a membrane by an outside pressure. Filtration occurs in the renal corpuscle of the kidneys. When blood enters the glomerulus, the blood pressure forces water and dissolved blood components through the walls of the capillaries and on through the adjoining inner wall of the Bowman's capsule (Figure 22–4a). The resulting fluid is called the *filtrate.* In a healthy person, the

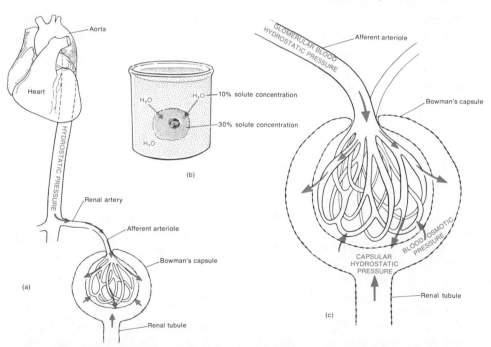

Figure 22–4. Glomerular filtration. (a) Heart action and resistance of walls of blood vessels provide the hydrostatic pressure for filtration. The hydrostatic pressure is counteracted by the walls of Bowman's capsule and the already present filtrate. (b) The development of osmotic pressure. When water moves into the cell, it swells as its osmotic pressure increases. (c) Application of hydrostatic pressure and osmotic pressure to glomerular filtration.

filtrate consists of all the materials present in the blood except for the formed elements and proteins, which are too large to pass through the capillary walls. Exhibit 22–1 compares the constituents of plasma, glomerular filtrate, and urine.

Exhibit 22–1. ANALYSIS OF CHEMICAL DIFFERENCES AMONG PLASMA, FILTRATE, AND URINE*

CHEMICAL	PLASMA	FILTRATE	URINE
Sodium	0.3	0.3	0.35
Potassium	0.02	0.02	0.15
Chloride	0.37	0.37	0.60
Sulfate	0.002	0.002	0.18
Glucose	0.01	0.01	0
Ammonia	0.001	0.001	0.04
Urea	0.03	0.03	2.0
Creatinine	0.001	0.001	0.075
Protein	7,000–9,000	10–20	0

* All values are in mg. per 100 ml.

Renal corpuscles are specially structured for filtering the blood. First, each capsule contains a tremendous length of highly coiled glomerular capillaries, presenting a very large surface area over which filtration can occur. Second, the efferent arteriole is smaller in diameter than the afferent arteriole, so a large amount of resistance to the outflow of blood from the glomerulus is established. Consequently, blood pressure is higher in the glomerular capillaries than it is in other capillaries of the body. It is estimated that glomerular blood pressure averages about 75 millimeters Hg, whereas the blood pressure of other capillaries averages only about 30 millimeters Hg. Third, the walls separating the blood from the space in the Bowman's capsule are very thin (0.1 μ). Finally, the glomerular capillaries have a great many pores. These pores aid the filtration of large molecules, but they are not quite large enough to allow the blood proteins or blood cells to pass through.

The filtering of the blood depends on a number of opposing pressures. The chief one is the *glomerular blood hydrostatic pressure.* **Hydrostatic** (from the word *hydro*, meaning water) **pressure** is the force that a fluid under pressure exerts against the walls of its container. Glomerular blood hydrostatic pressure, then, means the blood pressure in the glomerulus. This pressure tends to move fluid out of the glomeruli at a force averaging 75 millimeters Hg.

However, glomerular blood hydrostatic pressure is opposed by two other forces. The first of these, *capsular hydrostatic pressure,* develops in the following way. When the filtrate is forced into

the space between the walls of the Bowman's capsule, it meets up with two forms of resistance that limit its free flow: the walls of the capsule and the fluid that already has partially filled the renal tubule. As a result, some of the filtrate is pushed back into the capillary. The amount of "push" is the capsular hydrostatic pressure. It usually measures about 20 millimeters Hg (Figure 22–4c).

The second force opposing filtration into the Bowman's capsule is the *blood osmotic pressure. Osmotic pressure* is the pressure that develops in a contained solution because of water movement by osmosis into the solution. For example, let us look at a cell placed in a hypotonic solution. As water moves from the outside solution into the cell, the volume inside the cell increases, forcing the cell membrane outward (Figure 22–4b). Osmotic pressure always develops in the solution with the higher concentration of solutes. Hydrostatic pressure develops because of a force lying outside a solution, and osmotic pressure develops because of the concentration of the solution itself. Since the blood contains a much higher concentration of proteins than the filtrate does, water tends to move out of the filtrate and back into the blood vessel. This blood osmotic pressure is normally about 30 millimeters Hg (Figure 22–4c).

To determine how much filtration finally occurs, we have to subtract the forces that oppose filtration from the glomerular blood hydrostatic pressure. The net result is called the *effective filtration pressure,* which is abbreviated as P_{eff}:

$$P_{eff} = \begin{pmatrix} \text{Glomerular} \\ \text{blood} \\ \text{hydrostatic} \\ \text{pressure} \end{pmatrix} - \begin{pmatrix} \text{Capsular} & \text{Blood} \\ \text{hydrostatic} + \text{osmotic} \\ \text{pressure} & \text{pressure} \end{pmatrix}$$

By substituting the values just discussed, a normal P_{eff} may be calculated as follows:

$P_{eff} = (75 \text{ mm. Hg}) - (20 \text{ mm. Hg} + 30 \text{ mm. Hg})$
$P_{eff} = (75 \text{ mm. Hg}) - (50 \text{ mm. Hg})$
$P_{eff} = 25 \text{ mm. Hg}$

This means that a pressure of 25 millimeters Hg causes a normal amount of plasma to filter from the glomerulus into the Bowman's capsule. This is about 125 milliliters of filtrate per minute.

Certain conditions may alter one or more of these pressures and thus decrease or increase the P_{eff}. For example, in some forms of kidney disease the glomerular capillaries become so permeable that the plasma proteins are able to pass from the blood into the filtrate. As a result, the capsular filtrate exerts an osmotic pressure that draws water out of the blood. Thus, if a capsular osmotic pressure develops, the P_{eff} will increase. At the same time, blood osmotic pressure decreases, further increasing the P_{eff}.

The P_{eff} also is affected by changes in the general arterial blood pressure. For example, severe hemorrhaging produces a drop in general blood pressure that also decreases the glomerular blood hydrostatic pressure. If the blood pressure falls to the point where the hydrostatic pressure in the glomeruli reaches 50 millimeters Hg, no filtration occurs because the glomerular blood hydrostatic pressure equals the opposing forces. Such a condition is called *renal suppression*.

A final factor that may affect the P_{eff} is the regulation of the size of the afferent and efferent arterioles. In this case, glomerular blood hydrostatic pressure is regulated separately from the general blood pressure. Sympathetic impulses and minute doses of epinephrine cause constriction of both the afferent and efferent arterioles. However, intense sympathetic impulses and large doses of epinephrine cause greater constriction of afferent than efferent arterioles. This intense stimulation results in a decrease in glomerular hydrostatic pressure even though blood pressure in other parts of the body may be normal or even higher than normal. Intense sympathetic stimulation is most likely to occur during the alarm reaction of the general stress syndrome. Blood may also be shunted away from the kidneys during hemorrhage.

Tubular reabsorption

The amount of filtrate that flows out of all the renal corpuscles of both kidneys in each minute is called the *glomerular filtration rate* (GFR). In the normal adult, the rate is about 125 milliliters/minute. But as the filtrate passes through the renal tubules, it is reabsorbed into the blood at a rate of about 124 milliliters/minute. Thus, only about 1 to 3 percent of the filtrate actually leaves the body. The movement of the filtrate back into the blood is called *tubular reabsorption*. Tubular reabsorption is carried out by epithelial cells throughout the renal tubule. It is a very discriminating process. Only specific amounts of specific substances are reabsorbed, depending on the needs of the body at a particular moment. The maximum amount of a substance that can be reabsorbed under any condition is called the substance's *tubular maximum (Tm)*. Materials that are reabsorbed include water, glucose, amino acids, any proteins that managed to get into the filtrate, and ions such as Na^+, K^+, Ca^{++}, Cl^-, and HCO_3^-. Tubular reabsorption is an important process because it allows the body to retain most of its nutrients. Wastes such as urea are only partially reabsorbed.

Reabsorption is carried out through both passive and active mechanisms of transport. It is believed that glucose is reabsorbed by an active process involving a carrier system. The carrier, probably an enzyme, exists in the membranes of the tubular epithelial cells in a fixed and limited amount. Normally, all the glucose filtered by the glomeruli (125 milligrams/125 milliliters of filtrate per minute) is reabsorbed by the tubules. However, the capacity of the carrier system is limited. If the plasma concentration of glucose is significantly above normal, the glucose transport mechanism cannot reabsorb it all, and the excess remains in the urine. If there is a malfunction in the tubular carrier mechanism, glucose appears in the urine even though the blood sugar level may be normal. This condition is called glycosuria.

Sodium ions are actively transported from all parts of the tubule. However, sodium reabsorption in the distal convoluted tubules varies with the concentration of the extracellular fluid. When the Na^+ concentration of the blood is low, the adrenal cortex is stimulated to increase its secretion of aldosterone, a hormone that stimulates the epithelial cells of the distal convoluted tubule to transport greater quantities of Na^+ from the tubules into the peritubular capillaries. Conversely, an excess of Na^+ in the blood inhibits the secretion of aldosterone. As a result, the distal tubule cells are no longer stimulated to transport sodium ions, and large quantities are lost in the urine (Figure 22–5).

The reabsorption of chloride and other anions is controlled by the active transport of Na^+. When Na^+ moves out of the tubules into the peritubular blood, the blood momentarily becomes more electropositive than the filtrate. Chloride, a negatively charged ion, follows the positive sodium ion out of the tubule by electrostatic attraction. Thus, the movement of Na^+ ions influences the movement of Cl^- and other anions into the peritubular blood (see Figure 22–5).

Water reabsorption is controlled both by sodium transport and by water carrier molecules. As Na^+ ions are transported into the blood, the osmotic pressure of the peritubular blood becomes higher than that of the filtrate. Water, according to the law of osmosis, follows the Na^+ ions into

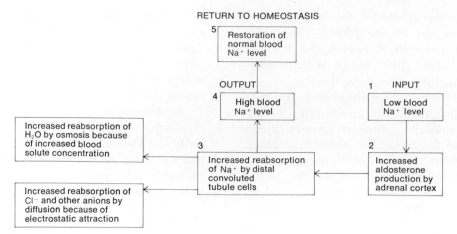

Figure 22–5. Control of sodium reabsorption and the effects of sodium reabsorption on water and chloride ions.

the blood in order to reestablish the osmotic equilibrium between the filtrate and the blood. About 80 percent of the water is reabsorbed by this method. This portion of reabsorbed water occurs as a function of the law of osmosis, so it is referred to as *obligatory* ("have to") *reabsorption.* The renal tubule has no control over osmosis.

However, passage of most of the remaining water in the filtrate can be regulated. The cells of the distal and collecting tubules contain carrier molecules controlled by vasopressin-ADH. When blood-water concentration is low, the posterior pituitary releases vasopressin-ADH, a hormone that increases the permeability of the plasma membranes of the distal tubule and collecting duct cells. As the membranes become more permeable, more carrier molecules can pass in and out of the cell, carrying water molecules with them. This kind of reabsorption, called *facultative*

("optional") *reabsorption,* is responsible for the fate of about 18 percent of the water in the filtrate. It is one of the major mechanisms for controlling water content of the blood (Figure 22–6).

Tubular secretion

The third process involved in urine formation is *tubular secretion,* or *tubular excretion.* Whereas tubular reabsorption removes substances from the filtrate, tubular secretion adds materials to the filtrate. In man, these secreted substances include K^+, H^+, ammonia, creatine, and the drugs penicillin and para-aminohippuric acid, among others. Tubular secretion has two principal effects: it rids the body of certain materials, and it controls the blood pH.

In general, normal dietary conditions provide more acid-producing foods and fewer alkali-

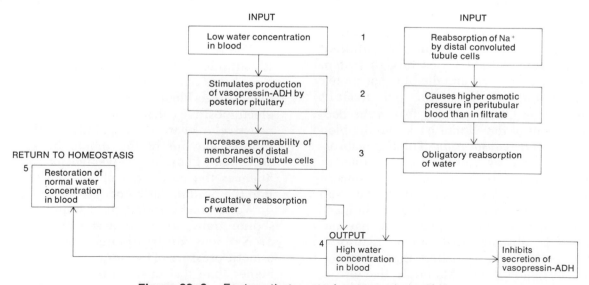

Figure 22–6. Factors that control water reabsorption.

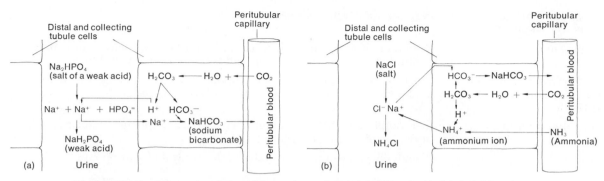

Figure 22–7. The role of the kidneys in maintaining blood pH. (a) Acidification of urine and conservation of sodium bicarbonate by the elimination of H^+ ions. (b) Acidification of urine and conservation of sodium bicarbonate by the elimination of NH_4^+.

producing foods. Therefore, the body is faced with the problem of maintaining the normal pH of blood (7.35–7.45) while certain mechanisms that tend to lower it are in operation. To raise blood pH, the renal tubules secrete hydrogen and ammonium ions into the filtrate. Both these substances make the urine acidic.

Secretion of H^+ occurs through the formation of carbonic acid (Figure 22–7a). A certain amount of carbon dioxide normally diffuses from the peritubular blood into the cells of the distal tubules and collecting ducts. Once inside the epithelial cells, the CO_2 combines with water to form carbonic acid (H_2CO_3), which then dissociates into H^+ ions and bicarbonate ions (HCO_3^-). A low blood pH stimulates the cells to secrete the hydrogen ion into the urine. As the H^+ enters the urine, it displaces another positive ion, usually Na^+, and forms a weak acid or a salt of the acid that is eliminated in urine. The displaced Na^+ or other positive ion diffuses from the urine into the tubule cell, where it combines with the bicarbonate ion to form sodium bicarbonate ($NaHCO_3$), which then is absorbed into the blood. As a result of this exchange mechanism, not only is H^+ eliminated by the body and Na^+ conserved, but Na^+ is conserved in the form of sodium bicarbonate, which can buffer other H^+ ions in the blood. Review Chapter 2 if you have forgotten how the sodium bicarbonate buffer system works.

A second important mechanism for raising blood pH is secretion of the ammonium ion (Figure 22–7b). Ammonia, in certain concentrations, is a poisonous waste product derived from the deamination of amino acids. The liver converts much of the ammonia to a less toxic compound called urea. Urea and ammonia both become part of the glomerular filtrate and are subsequently expelled from the body. Any ammonia still re-

maining in the bloodstream is secreted into the renal tubules in the following manner. When ammonia (NH_3) diffuses into the distal and collecting tubule cells, it combines with H^+ to form the ammonium ion (NH^{4+}). (The H^+ may come from the dissociation of H_2CO_3 that we just discussed.) The cells secrete NH_4^+ into the filtrate, where it takes the place of a positive ion, usually Na^+, in a salt and is eliminated. The displaced Na^+ diffuses into the renal cells and combines with HCO_3^- to form sodium bicarbonate. When the blood pH is low, the kidney cells also can deaminate an amino acid called glutamine, thus increasing NH_4^+ secretion.

As a result of H^+ and NH_4^+ secretion, urine normally has an acidic pH of 6. The relationship of renal tubule excretion of H^+ and NH_3 to blood pH level is summarized in Figure 22–8. Exhibit 22–2 summarizes filtration, reabsorption, and secretion in the nephrons.

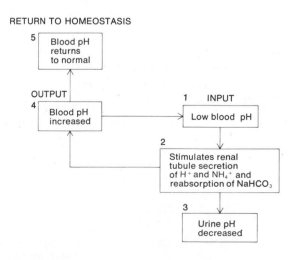

Figure 22–8. Summary of kidney mechanisms that maintain the homeostasis of blood pH.

Exhibit 22–2. SUMMARY OF FILTRATION, REABSORPTION, AND SECRETION

REGION OF NEPHRON	ACTIVITY
Renal corpuscle	Filtration of glomerular blood under hydrostatic pressure results in formation of a filtrate free of plasma proteins and cellular elements of blood.
Proximal convoluted tubule and descending and ascending limbs of Henle	Reabsorption of physiologically important solutes such as Na^+, K^+, Cl^-, HCO_3^-, and glucose. Obligatory reabsorption of water by osmosis.
Distal convoluted tubule	Reabsorption of Na^+. Secretion of H^+, NH_3, K^+, creatine, and certain drugs. Conservation of sodium bicarbonate.
Collecting duct	Facultative reabsorption of water under control of vasopressin-ADH.

HOMEOSTASIS OF EXCRETION

Excretion is one of the primary ways in which the volume, pH, and chemistry of the body fluids are kept in homeostasis. The kidneys assume a good deal of the burden for excretion, but they also share this responsibility with several other organ systems.

Other excretory organs

The lungs, integument, and alimentary canal all perform specialized excretory functions (Exhibit 22–3). Primary responsibility for regulating body temperature through the excretion of water is assumed by the skin. The lungs maintain blood–gas homeostasis through the controlled excretion of CO_2. One way in which the kidneys maintain homeostasis is by coordinating their activities with the activities of the other excretory organs. For example, when the integument increases its excretion of water, the renal tubules increase their reabsorption of water, and blood volume is maintained. Or, if the lungs fail to eliminate enough carbon dioxide, the kidneys attempt to compensate. They change some of the CO_2 into hydrogen ions, which are excreted, and into sodium bicarbonate, which becomes a part of the blood buffer systems.

Exhibit 22–3. EXCRETORY ORGANS OF THE BODY AND PRODUCTS ELIMINATED

EXCRETORY ORGANS	PRODUCTS ELIMINATED PRIMARY	SECONDARY
Kidneys	Water, soluble salts from protein catabolism, and inorganic salts	Heat and carbon dioxide
Lungs	Carbon dioxide	Heat and water
Skin	Heat	Carbon dioxide, water, and salts
Alimentary tract	Solid wastes and secretions	Carbon dioxide, water, salts, and heat

Urine

The kidneys perform their homeostatic functions through the manufacture of **urine.** Urine is a fluid that contains a very high concentration of solutes. In a healthy person its volume, pH, and solute concentration vary with the needs of the internal environment. During certain pathological conditions, the characteristics of urine may change drastically. An analysis of the volume and physical and chemical properties of urine tells us much about the state of the body.

Urine volume

The volume of urine eliminated per day in the normal adult varies between 1,000–1,800 milliliters, or 2–3½ pints. Urine volume is influenced by a number of factors, including blood pressure, blood solute concentration, diet, environmental temperature, diuretics, mental state, and general health.

BLOOD PRESSURE. In the walls of each afferent arteriole lie a group of modified smooth muscle cells called the *juxtaglomerular apparatus*. These cells are particularly sensitive to changes in the blood pressure. It is believed that when renal blood pressure falls below normal, the juxtaglomerular apparatus secretes renin (Figure 22–9). Renin converts the plasma protein angiotensinogen into angiotensin I, which subsequently is transformed to angiotensin II. Angiotensin II raises blood pressure in two ways. It causes constriction of arteries throughout the body, and it stimulates the adrenal cortex to secrete aldosterone. Aldosterone, as you have learned, stimulates the epithelial cells of the distal convoluted tubules to transport Na^+ back into the blood. As a result, obligatory reabsorption of water increases, blood volume increases, and urine volume decreases. By raising the blood pressure, the juxtaglomerular apparatus ensures that the

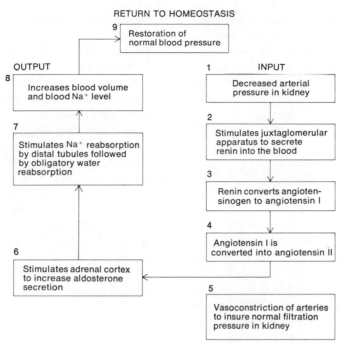

Figure 22–9. The role of the juxtaglomerular apparatus in maintaining normal blood volume.

kidney cells receive enough oxygen and that glomerular hydrostatic pressure is high enough to maintain a normal P_{eff}. The juxtaglomerular apparatus also serves as an important regulator of blood pressure throughout the body.

BLOOD CONCENTRATION. The concentrations of water and solutes in the blood also affect urine volume. If you have gone without water all day and the water concentration of your blood becomes low, osmotic receptors in the hypothalamus become active. They stimulate the posterior pituitary to release vasopressin-ADH. This hormone stimulates renal tubule cells to transport water out of the filtrate and into the blood. Thus urine volume decreases, and water is conserved.

If you have just drunk an excessive amount of liquid, urine volume may be increased through two mechanisms. First, the blood-water concentration increases above normal. This means that the osmoreceptors in the hypothalamus are no longer stimulated to secrete vasopressin-ADH, and facultative water reabsorption stops. Second, the excess water causes the blood pressure to rise. In response, the renal vessels dilate, more blood is brought to the glomeruli, and the filtration rate increases.

The concentration of Na^+ in the blood also affects urine volume. As you have seen, sodium concentration determines aldosterone secretion,

which, in turn, affects both sodium reabsorption and the obligatory reabsorption of water.

TEMPERATURE. When the temperature of the internal or external environment rises above normal, the cutaneous vessels dilate and fluid diffuses from the capillaries to the surface of the skin. As water volume decreases, vasopressin-ADH is secreted, and facultative reabsorption increases. In addition, the increase in temperature stimulates the abdominal vessels to constrict, so the blood flow to the glomeruli and filtration decrease. Both mechanisms serve to decrease the volume of urine.

On the other hand, if the body is exposed to low temperatures, the cutaneous vessels constrict, and the abdominal vessels dilate. More blood is shunted to the glomeruli, glomerular blood hydrostatic pressure increases, and urine volume increases.

DIURETICS. Certain chemicals increase urine volume by inhibiting the facultative reabsorption of water. Such chemicals are called *diuretics*, and the abnormal increase in urine flow is called *diuresis*. Some diuretics act directly on the tubular epithelium as they are carried through the kidneys. Others act indirectly by inhibiting the secretion of vasopressin-ADH as they circulate through the brain. Examples of diuretics are coffee, tea, and alcoholic beverages.

EMOTIONS. The effects of some emotional states of the individual on urine volume are well known. In cases of extreme nervousness, for example, there is an enormous discharge of urine because impulses from the brain cause dilation of the renal vessels, resulting in an increased glomerular filtration rate.

Physical characteristics

Normal urine is usually a yellow- or amber-colored, transparent liquid with a characteristic and aromatic odor. The color is caused by the presence of urochrome, a pigment derived from the destruction of hemoglobin by reticuloendothelial cells. The color varies considerably with the ratio of solutes to water in the urine. For example, the less water there is, the darker the color of the urine. Fever decreases urine volume in the same way that high environmental temperatures do, sometimes making the urine quite concentrated. It is not uncommon for a feverish person to have dark yellow or brown urine. The color of urine may also be affected by diet, such as a reddish color from beets, and by the presence of abnormal constituents, such as certain drugs. A red or brown to black color may indicate the presence of red blood cells or hemoglobin from bleeding in the urinary system.

Freshly voided urine is usually transparent, but a turbid or (cloudy) urine does not necessarily indicate a pathological condition since turbidity may result from mucin secreted by the lining of the urinary tract. It should be noted, however, that the presence of mucin above a critical level usually denotes an abnormality. A persistent turbidity due to the presence of blood, pus, or bacteria is indicative of some pathology.

The aromatic odor of urine may be modified by specific substances excreted into it. For example, the digestion of asparagus adds a substance called methyl mercaptan to urine that gives the urine a characteristic odor. In cases of diabetes, urine has a "sweetish" odor because of the presence of acetone. If it is allowed to stand around, urine develops an ammonialike odor due to ammonium carbonate formation as a result of urea decomposition.

The reaction of normal urine is usually slightly acid, averaging about 6.0 in pH. It ranges between 5.0 and 7.8 and rarely becomes more acid than 4.5 or more alkaline than 8. Variations in urine pH are closely related to diet. Moreover, these variations are due to differences in the end products of metabolism that appear in the urine. Whereas a high-protein diet increases acidity, a diet composed largely of vegetables increases alkalinity. As indicated earlier, ammonium carbonate forms in standing urine. Since it can dissociate into NH_4^+ and form a relatively strong base, the presence of ammonium carbonate tends to make urine more alkaline.

Specific gravity is the ratio of the weight of a volume of a substance to the weight of an equal volume of distilled water. Water has a specific gravity of 1.000. The specific gravity of urine depends on the amount of solid materials in solution and ranges from 1.008 to 1.030 in normal urine. The greater the concentration of solutes, the higher the specific gravity. The lower the concentration, the lower the specific gravity. Exhibit 22-4 summarizes the physical characteristics of normal urine.

Exhibit 22-4. PHYSICAL CHARACTERISTICS OF NORMAL URINE

Volume	1,000–1,800 ml. in 24 hours, but varies considerably with many factors
Color	Yellow or amber colored, but varies with quantity voided and diet
Turbidity	Transparent when freshly voided, becomes turbid upon standing
Odor	Aromatic, becomes ammonialike upon standing
Reaction	Ranges in pH from 5.0–7.8, average 6.0, varies considerably with diet
Specific gravity	Ranges from 1.008–1.030

Chemical composition

Water accounts for about 95 percent of the total volume of urine. The remaining 5 percent consists of solutes derived from cellular metabolism and from outside sources such as drugs. The solutes are described in Exhibit 22-5.

Abnormal constituents

If some of the chemical processes of the body are not operating efficiently, traces of particular substances that are not normally present may appear in the urine. Or, normal constituents may appear in abnormal amounts. Analyzing the physical and chemical properties of a patient's urine often provides information that aids diagnosis. Such an analysis is called a *urinalysis*.

ALBUMIN. Protein albumin is one of the things a technician looks for when he does a urinalysis.

Exhibit 22-5. PRINCIPAL SOLUTES IN URINE OF AN ADULT MALE ON A MIXED DIET

CONSTITUENT IN GM./L.	COMMENTS
A. Organic Urea—30.000	Comprises 60–90 percent of all nitrogenous material. Derived primarily from de-amination of proteins (ammonia combines with CO_2 to form urea).
Creatinine—1.0	Normal alkaline constituent of blood. Primarily derived from creatine (nitrogenous substance in muscle tissue that stores energy).
Uric acid—0.7	Product of catabolism of nucleic acids derived from food or cellular destruction. Because of insolubility, tends to crystallize and is common component of kidney stones.
Hippuric acid—0.7	Form in which benzoic acid (toxic substance in fruits and vegetables) is believed to be eliminated from body. High vegetable diets increase quantity of hippuric acid excreted.
Indican—0.01	Potassium salt of indole. Indole results from putrefaction of protein in large intestine and is carried by blood to liver where it is probably changed to indican (less poisonous substance).
Acetone bodies—0.04	Also called ketone bodies. Normally found in urine in very small amounts. In cases of diabetes and acute starvation, ketone bodies appear in high concentrations.
Other substances—2.945	May be present in minute quantities depending on diet and general health. Include carbohydrates, pigments, fatty acids, mucin, enzymes, and hormones.
B. Inorganic NaCl—15.000	Principal inorganic salt. About 15 gm. excreted daily, but concentration varies with intake.
K^+—3.3 Mg^{++}—0.1 Ca^{++}—0.3	Appear as salts of chlorides, sulfates, and phosphates.
$SO_4^=$—2.5	Derived from amino acids.
$PO_4^=$—2.5	Occur in urine as sodium compounds (monosodium and disodium phosphate) that serve as buffers in blood.
NH_4^+—0.7	Occurs in urine as ammonium salts. Derived from protein catabolism and from glutamine in kidneys. Amount produced by kidney may vary with need of body for conserving Na^+ ions to offset acidity of blood and tissue fluids.

Albumin is a normal constituent of plasma, but it usually does not appear in urine because the particles are too large to pass through the pores in the capillary walls. The presence of albumin in the urine is called *albuminuria*. Albuminuria indicates an increase in the permeability of the glomerular membrane. Conditions that lead to albuminuria include injury to the glomerular membrane as a result of disease, increased blood pressure, and irritation of kidney cells by substances such as bacterial toxins, ether, or heavy metals. Other proteins, such as globulin and fibrinogen, may also appear in the urine under certain conditions.

GLUCOSE. The presence of sugar in the urine is termed *glycosuria*. Normal urine contains such small amounts of glucose that clinically it may be considered absent. The most common cause of glycosuria is a high blood sugar level. Remember that glucose is filtered into the Bowman's capsule. Later, in the renal tubules, the tubule cells actively transport the glucose back into the blood.

However, the number of glucose carrier molecules is limited. If a person ingests more carbohydrates than his body can convert to glycogen or fat, more sugar is filtered into the Bowman's capsule than can be removed by the carriers. This condition, called *temporary* or *alimentary glycosuria*, is not considered pathological. Other nonpathological causes include emotional upsets, pregnancy, and lactation. Emotional upsets can cause excessive amounts of epinephrine to be secreted. Epinephrine stimulates the breakdown of glycogen and the liberation of glucose from the liver. During pregnancy and lactation, the breasts produce lactose, some of which may be absorbed into the blood and raise the blood sugar above normal limits. One pathological kind of glycosuria is that resulting from diabetes mellitus. In this case, there is a frequent or continuous elimination of glucose because the pancreas fails to produce sufficient insulin. Overactivity of the posterior lobe of the pituitary gland is also a pathological cause of glycosuria. When glycosuria occurs with normal blood sugar levels, the problem lies in failure of the kidney tubular cells to reabsorb glucose.

ERYTHROCYTES. The appearance of red blood cells in the urine is called *hematuria*. Hematuria generally indicates a pathological condition. One possible cause is acute inflammation of the urinary organs as a result of disease or of irritation from stones in the organs. Whenever blood is found in the urine, additional tests are performed to ascertain the part of the urinary tract that is bleeding. One should also make sure that the sample was not contaminated with blood from the vagina.

LEUCOCYTES. The presence of leucocytes and other components of pus in the urine, referred to as *pyuria*, is evidence that there is some kind of infection in the kidney or other urinary organs. Again, the source of the pus must be located, and care should be taken to make sure the urine was not contaminated.

KETONE BODIES. Ketone, or acetone, bodies appear in normal urine in very small amounts. However, their appearance in urine in unusually high quantities, a condition called *ketosis*, or *acetonuria*, may indicate a number of abnormalities. For example, it may be caused by diabetes mellitus, starvation, or simply too little carbohydrate in the diet. Whatever the cause, excessive quantities of fatty acids are oxidized in the liver, and the ketone bodies are filtered from the plasma into the Bowman's capsule.

CASTS. Microscopic examination of urine may reveal the presence of *casts*—tiny masses of material that have hardened within and assumed the shape of the lumens of the tubules, later flushed out of the tubules by a buildup of filtrate behind them. Casts are named on the basis of either the substances that compose them or of their appearance. Accordingly, there are white-blood-cell casts, red-blood-cell casts, epithelial casts that contain cells from the walls of the tubes, granular casts which contain decomposed cells that form granules, and fatty casts from cells which have become fatty.

CALCULI. Occasionally, the salts found in urine may solidify into insoluble stones called *calculi*. They may be formed in any portion of the urinary tract from the kidney tubules to the external opening. Conditions leading to calculi formation include the ingestion of excessive amounts of mineral salts, a decrease in the amount of water, and abnormally alkaline or acid urine.

THE URETERS

Once urine is formed by the nephrons, it drains through the collecting ducts into the calyces surrounding the renal papillae. The minor calyces join with the major calyces that unite to become the renal pelvis (see Figure 22–2). From the pelvis, the urine drains into the ureters and is carried down to the urinary bladder. From the bladder, the urine is discharged from the body through the single urethra.

Structure

The body has two **ureters**—one for each kidney. Each ureter is an extension of the pelvis of the kidney and runs 25–30 centimeters (10 to 12 inches) to the bladder (see Figure 22–1). As the ureters descend, their thickened walls progressively increase in diameter, but at their widest point they measure less than 1.7 centimeters ($\frac{1}{2}$ inch) in diameter. Like the kidneys, the ureters are retroperitoneal in placement. Three coats of tissue form the walls of the ureters. A lining of mucous membrane, the *mucosa*, and transitional epithelium is the inner coat. The solute concentration and pH of urine differs drastically from the internal environment of the cells that form the walls of the ureters. Mucus secreted by the mucosa prevents the cells from coming in contact with the urine. Throughout most of the length of the ureters, the second or middle coat, the *muscularis*, is composed of inner longitudinal and outer circular layers of smooth muscle. The muscularis of the proximal one-third of the ureters also contains a layer of outer longitudinal muscle. Peristalsis is the major function of the muscularis. The third, or external, coat of the ureters is a *fibrous coat*. Extensions of the fibrous coat anchor the ureters in place.

Physiology

The principal function of the ureters is to carry urine from the renal pelvis into the urinary bladder. Urine is carried through the ureters primarily by peristaltic contractions of the muscular walls of the ureters, but hydrostatic pressure and gravity also contribute. Peristaltic waves pass from the kidney to the urinary bladder, varying in rate from one to five per minute, depending on the amount of urine formation.

THE URINARY BLADDER

The **urinary bladder** (Figure 22–10) is a hollow muscular organ situated in the pelvic cavity posterior to the symphysis pubis. In the male

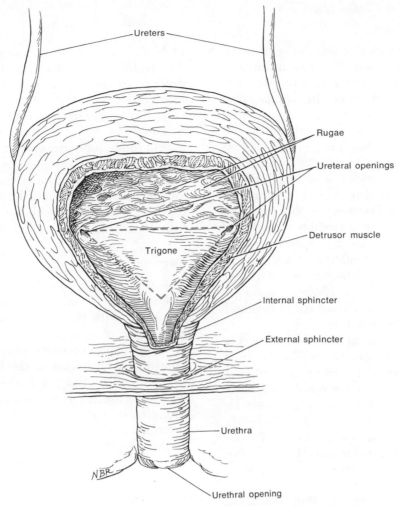

Figure 22–10. The urinary bladder and urethra.

Structure

it is directly in front of the rectum, whereas in the female it is also in front and under the uterus and in front of the vagina. It is a freely movable organ, but it is held in position by folds of the peritoneum. The shape of the urinary bladder depends on the volume of urine it contains. When empty it looks like a deflated balloon. It becomes more spherical when slightly distended, and as urine volume increases in the bladder, it becomes more pear-shaped and rises in the pelvic cavity.

Structure

At the base of the bladder is a small triangular area with its apex pointing anteriorly. The opening to the urethra is found in the apex of the triangle. At the two points that form the base of the triangle, the ureters drain into the bladder. This triangular area is called the *trigone*.

Four coats comprise the walls of the bladder. The mucosa, the innermost coat, is a mucous membrane containing transitional epithelium. Recall from Chapter 3 that transitional epithelium

is able to stretch—a marked advantage for an organ that must continually inflate and deflate. Stretchability is further enhanced by the rugae, or folds in the mucosa that appear when the bladder is empty. The second coat, the submucosa, is a layer of dense connective tissue that connects the mucous and muscular coats. The third coat—a muscular one called the *detrusor muscle*—consists of three layers: inner longitudinal, middle circular, and outer longitudinal muscles. In the area around the opening to the urethra, the circular fibers form an *internal sphincter* muscle. Below the internal sphincter is the external sphincter that is composed of skeletal muscle. The outermost coat, the serous, is formed by the peritoneum and covers only the superior surface of the organ.

Physiology

Urine is expelled from the bladder by an act called *micturition*. This response is brought about by a combination of involuntary and voluntary nervous impulses. The average capacity of the bladder is

700–800 milliliters. When the amount of urine in the bladder exceeds about 200–400 milliliters, stretch receptors in the bladder wall transmit impulses to the lower portion of the spinal cord. These impulses initiate a conscious desire to expel urine and an unconscious reflex referred to as the *micturition reflex*. In the micturition reflex, parasympathetic impulses transmitted from the spinal cord reach the bladder wall and internal urethral sphincter, bringing about contraction of the bladder and relaxation of the internal sphincter. Then the conscious portion of the brain sends impulses to the external sphincter, the sphincter relaxes, and urination takes place. Although emptying of the bladder is controlled by reflex, it may be initiated voluntarily and may be started or stopped at will because of cerebral control of the external sphincter.

A lack of voluntary control over micturition is referred to as *incontinence*. In infants about 2 years and under, incontinence of urine is normal because they have not developed voluntary control over the external sphincter muscle. Infants void whenever the bladder is sufficiently distended to arouse a reflex stimulus. Proper training overcomes incontinence if the latter is not caused by emotional stress or irritation of the bladder.

Involuntary micturition in the adult may occur as a result of unconsciousness, injury to the spinal nerves controlling the bladder, irritation due to abnormal constituents in urine, disease of the bladder, and emotional stress due to failure of the detrusor muscle to relax.

Retention is a term used to describe a condition in which there is a failure to void urine. Retention may be due to an obstruction in the urethra or neck of the bladder, nervous contraction of the urethra, or lack of sensation to urinate.

A condition far more serious than retention is *suppression*, or *anuria* — the failure of the kidneys to secrete urine. It usually occurs when blood plasma is prevented from reaching the glomerulus as a result of inflammation of the glomeruli. Anuria also may be caused by a low filtration pressure.

THE URETHRA

The **urethra** is a small tube leading from the floor of the bladder to the exterior of the body (see Figure 22–10). In females, it lies directly behind the symphysis pubis and is embedded in the anterior wall of the vagina. Its undilated diameter is about 6 millimeters ($\frac{1}{4}$ inch), and its length is approximately 3.8 centimeters ($1\frac{1}{2}$

inches). The female urethra is directed obliquely downward and forward, and the opening of the urethra to the exterior, the *urinary meatus*, is located between the clitoris and vaginal opening.

In males, the urethra is about 20 centimeters (8 inches) long, and it follows a different course from that of the female urethra. Immediately below the bladder it runs vertically through the prostate gland. It then penetrates the penis and takes a curved course through its body. Unlike the female urethra, the male urethra serves as a common tube for urinary and reproductive systems.

Structure

The walls of the female urethra consist of three coats: an inner mucous coat that is continuous internally with that of the vulva, an intermediate thin layer of spongy tissue containing a plexus of veins, and an outer muscular coat which is continuous with that of the bladder and consists of circularly arranged fibers of smooth muscle. The male urethra is composed of an inner mucous membrane that is continuous with the mucous membrane of the bladder, and an outer submucous tissue which connects the urethra with the structures through which it passes.

Physiology

Since the urethra is the terminal portion of the urinary system, it serves as the passageway for discharging urine from the body. The male urethra also serves as the duct through which reproductive fluid (semen) is discharged from the body.

DISORDERS OF THE URINARY SYSTEM

Disorders of the urinary system may be categorized as physical, chemical, or infectious. Let us now examine some of the more common disorders.

Physical

Floating kidney (ptosis) occurs when the kidney no longer is held in place securely by the adjacent organs or by its covering of fat and drifts or slips from its normal position. Pain occurs if the ureter is twisted or bent. Such an abnormal orientation also may obstruct the flow of urine.

Chemical

If urine becomes too concentrated, some of the chemicals that are normally dissolved in it may crystallize out, forming **kidney stones (renal calculi)**. Some common constituents of stones are uric acid, calcium oxalate, and calcium phosphate. The stones usually form in the pelvis of the kidney, where they cause pain, hematuria, and pyuria. Severe pain occurs when a stone passes through a ureter and stretches its walls. Ureteral stones are seldom completely obstructive because they are usually needle-shaped and urine can flow around them.

Gout, as you may recall, is a hereditary condition associated with a high level of uric acid in the blood. When the purine-type nucleic acids are catabolized, a certain amount of uric acid is produced as a waste. Some people, however, seem to produce excessive amounts of uric acid, and others seem to have trouble excreting normal amounts. In either case, uric acid accumulates in the body and tends to solidify into crystals that are deposited in the joints and kidney tissue. Gout is further aggravated by excessive use of diuretics, dehydration, and starvation.

Nephrons can be poisoned by various chemicals, such as mercury, gold, uranium, and other heavy metals, and blood transfusions can produce an agglutination reaction that blocks the nephrons. These common causes of toxicity of the kidneys have produced complete kidney shutdown. The body fluids become extremely acidic, and severe edema, coma, and death in 8 to 14 days follow.

Infectious

Glomerulonephritis is an inflammation of the kidney that involves the glomeruli. One of the most common causes of glomerulonephritis is an allergic reaction to the toxins given off by streptococci bacteria that have recently infected another part of the body, especially the throat. The glomeruli become so inflamed, swollen, and engorged with blood that the glomerular membranes become highly permeable and allow blood cells and proteins to enter the filtrate. Thus the urine contains many erythrocytes and much protein. The glomeruli may be permanently changed, leading to chronic renal disease and renal failure.

Pyelitis is an inflammation of the kidney pelvis and its calyces, and **pyelonephritis** is the interstitial inflammation of one or both kidneys. The latter usually involves both the parenchyma and the renal pelvis, due to bacterial invasion from the middle and lower urinary tracts or the bloodstream.

Cystitis is an inflammation of the urinary bladder involving principally the mucosa and submucosa. It may be caused by bacterial infection, chemicals, or mechanical injury.

HEMODIALYSIS

If the kidneys are impaired so severely by disease or injury that they are unable to excrete nitrogenous wastes and regulate pH and electrolyte concentration of the plasma, then the blood must be filtered by an artificial device. Such filtering of the blood is called *hemodialysis.* As you remember from Chapter 2, *dialysis* means using a semipermeable membrane to separate large particles from smaller ones. One of the most well-known devices for accomplishing dialysis is the kidney machine (Figure 22–11). When the machine is in operation, a tube connects it with a much smaller tube implanted in the patient's radial artery. The blood is pumped from the artery and through the tubes to one side of a semiporous cellophane membrane. The other side of the membrane is continually washed with an artificial solution called the dialyzing solution.

All substances (including wastes) in the blood except protein molecules and erythrocytes can diffuse back and forth across the semipermeable membrane. The electrolyte level of the blood is controlled by keeping the dialyzing solution electrolytes at the same concentration as that found in normal plasma. Any excess blood electrolytes move down the concentration gradient and into the dialyzing solution. If the blood electrolyte level is normal, it is in equilibrium with the dialyzing solution, and no electrolytes are gained or lost. Since the dialyzing solution contains no wastes, substances such as urea move down the concentration gradient and into the dialyzing solution. Thus, wastes are removed, and normal electrolyte balance is maintained.

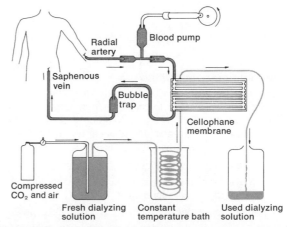

Figure 22–11. Diagrammatic representation of the operation of an artificial kidney. The blood route is indicated in color. The route of the dialyzing solution is indicated in gray.

One additional advantage of the kidney machine is that an individual's nutritional status can be bolstered by placing large quantities of glucose in the dialyzing solution. While the blood gives up its wastes, the glucose diffuses into the blood. Thus, the kidney machine beautifully accomplishes the principal function of the fundamental unit of the kidney—the nephron.

Obvious drawbacks to this artificial kidney system include the fact that the blood must be anticoagulated while dialysis is occurring, and a very large amount of blood must flow through this apparatus to make it work. To date, an artificial kidney that is capable of becoming a permanent implant has not yet been devised.

Use of the artificial kidney replaced *peritoneal dialysis*. In this process, the dense capillary network of the peritoneal cavity serves as the dialyzing membrane. By means of a tube, a sterile dialyzing solution containing the proper electrolyte mixture is placed through a small opening into the abdominal cavity. This mixture usually is allowed to remain in the peritoneal cavity for 30 to 45 minutes and is then pumped back out through the tube. The procedure is repeated until the normal plasma concentrations of the various substances have been achieved. Peritoneal dialysis must be done in a hospital and is generally used during short-term kidney shutdown. A kidney machine can be used at home.

In the next chapter we shall examine the distribution of fluids and electrolytes in parts of the body other than the urinary system and its immediate vicinity.

Medical terminology associated with the urinary system

Cystoscope (*cyst* = bladder; *scope* = to view) Instrument used for examination of the urinary bladder.

Dysuria (*dys* = painful) Painful urination.

Nephrosis (*neph* = kidney) Any disease of the kidney, but usually one that is degenerative.

Oliguria (*olig* = scanty) Scanty urine.

Polyuria (*poly* = many, much) Excessive amounts of urine.

Stricture Narrowing of the lumen of a canal or hollow organ, as the ureter or urethra.

Uremia (*emia* = condition of blood) Toxic levels of urea in the blood resulting from severe malfunction of the kidneys.

Chapter summary in outline

KIDNEYS
1. The primary homeostatic function of the urinary system is to regulate the concentration and volume of blood by removing and restoring selected amounts of water and solutes. It also excretes wastes to maintain homeostasis.
2. The functional unit of the kidneys is the nephron.
3. The kidneys form urine by glomerular filtration, tubular reabsorption, and tubular secretion.
4. The primary force behind filtration is hydrostatic pressure.
5. If the hydrostatic pressure falls to 50 millimeters, renal suppression occurs.
6. Most substances in plasma are filtered by Bowman's capsule. Normally, blood cells and plasma proteins are not filtered.
7. Tubular reabsorption retains substances needed by the body, including water, glucose, and ions.
8. Chemicals not needed by the body are discharged into urine by tubular secretion.
9. The kidneys help to maintain pH by excreting H^+ and NH_4^+ ions. In exchange, the kidneys conserve sodium bicarbonate.

HOMEOSTASIS OF EXCRETION
1. Besides the kidneys, the lungs, integument, and alimentary canal assumes excretory functions.
2. Urine volume is influenced by blood pressure, diet, diuretics, temperature, and emotions.
3. The physical characteristics of urine evaluated in a urinalysis are color, odor, turbidity, pH, and specific gravity.
4. Chemically, normal urine contains about 95 percent water and 5 percent solutes. The solutes include urea, creatinine, uric acid, indican, acetone bodies, salt, and ions.
5. Abnormal conditions diagnosed through urinalysis include albuminuria, glycosuria, hematuria, pyuria, ketosis, casts, and calculi.

URETERS, URINARY BLADDER, AND URETHRA
1. The paired ureters convey urine from the kidneys to the urinary bladder, mostly by peristaltic contractions.
2. The urinary bladder stores urine. The expulsion of urine from the bladder is called micturition.
3. A lack of control over micturition is called incontinence, failure to void urine is referred to as retention, and inability of the kidneys to produce urine is called suppression.
4. The urethra extends from the floor of the bladder to the exterior and discharges urine from the body.

DISORDERS

1. Ptosis or floating kidney occurs when the kidney slips from its normal position.
2. Renal calculi are kidney stones.
3. Gout is a high level of uric acid in the blood.
4. Glomerulonephritis is an inflammation of the glomeruli of the kidney.
5. Pyelitis is an inflammation of the kidney pelvis and calyces.
6. Cystitis is an inflammation of the urinary bladder.

HEMODIALYSIS

1. Filtering blood through an artificial device is called hemodialysis.
2. The kidney machine filters the blood of wastes and adds nutrients.
3. In peritoneal dialysis, the dense capillary network of the peritoneal cavity serves as a dialyzing membrane.

Review questions and problems

1. What are the functions of the urinary system? What organs comprise the system?
2. Describe the location of the kidneys. Why are they said to be retroperitoneal?
3. Prepare a labeled diagram that illustrates the principal external and internal gross features of the kidney.
4. What is a nephron? List and describe, in order, the parts of a nephron from the Bowman's capsule to the collecting duct.
5. How are nephrons supplied with blood?
6. What is glomerular filtration? Define the filtrate.
7. Set up an equation to indicate how effective filtration pressure is calculated. What is the cause of renal suppression?
8. What are the major chemical differences among plasma, filtrate, and urine?
9. Define tubular reabsorption. Why is the process physiologically important?
10. What chemical substances are normally reabsorbed by the kidneys?
11. Describe how glucose and sodium are reabsorbed by the kidneys. Where does the process occur?
12. How is chloride reabsorption related to sodium reabsorption?
13. Distinguish between obligatory and facultative reabsorption of water. How is facultative reabsorption controlled?
14. Define tubular secretion. Why is the process important?
15. Explain the mechanisms by which the kidneys help to control body pH.
16. Contrast the functions of the lungs, integument, and alimentary tract as excretory organs.
17. What is urine? Describe the effects of blood pressure, blood concentration, temperature, diuretics, and emotions on the volume of urine formed.
18. Describe the following physical characteristics of normal urine: color, turbidity, pH, and specific gravity. Give numeric values where possible.
19. Describe the chemical composition of normal urine.
20. Define each of the following: albuminuria, glycosuria, hematuria, pyuria, ketosis, casts, and calculi.
21. Describe the structure and function of the ureters.
22. How is the urinary bladder adapted to its storage function? What is micturition?
23. Contrast the causes of incontinence, retention, and suppression.
24. Compare the position of the urethra in the male and female. What is the function of the urethra?
25. Define each of the following: ptosis, kidney stones, gout, glomerulonephritis, pyelitis, and cystitis.
26. What is hemodialysis? Briefly describe the operation of an artificial kidney.
27. What is peritoneal hemodialysis?
28. Refer to the glossary of medical terminology associated with the urinary system. Be sure that you can define each term.

CHAPTER 23

FLUID AND ELECTROLYTE DYNAMICS

STUDENT OBJECTIVES

After you have read this chapter, you should be able to:

1. Define a body fluid

2. Distinguish between intracellular fluid and extracellular fluid

3. Define the processes available for fluid intake and fluid output

4. Compare the mechanisms involved in regulating fluid intake and fluid output

5. Compare the effects of nonelectrolytes and electrolytes on body fluids

6. Calculate the concentration of ions in a body fluid

7. Contrast the electrolytic composition of the three major fluid compartments

8. Define the factors involved in the movement of fluid between plasma and interstitial fluid and between interstitial fluid and intracellular fluid

9. Define the relationship between electrolyte imbalance and fluid imbalance

10. Describe the role of buffers, respirations, and kidney excretion in maintaining body pH

11. Define the kinds of acid-base imbalances and their effects on the body

12. Describe appropriate treatments for acidosis and alkalosis

13. Define medical terminology associated with fluid and electrolyte dynamics

The term **body fluid** refers to the body water and its dissolved substances. Fluid composes from 45 to 75 percent of the body weight. About two-thirds of the fluid is located within cells and is termed *intracellular fluid* or *ICF*. The other third, called *extracellular fluid*, or *ECF*, encompasses all the rest of the body fluids (see Figure 1-2)—interstitial fluid, plasma and lymph, cerebrospinal fluid, synovial fluid, the fluids of the eyes and ears, and the glomerular filtrate. The different kinds of body fluids are separated into distinct compartments, whose walls are the semipermeable membranes provided by the plasma membranes of cells. A compartment may be as small as the interior of a single cell or as large as the combined interiors of the heart and vessels.

Water comprises the bulk of all the body fluids. When we say that the body is in *fluid balance*, we mean that the body contains the required amount of water and that the water is distributed to the various compartments according to their needs. You are already well aware of some of the consequences of a fluid imbalance. Consider what happens when the blood volume becomes abnormally low or when too much water moves into or out of a cell.

Osmosis is the primary way in which water moves in and out of body compartments. The concentration of solutes in the fluids is therefore an important determinant of fluid balance. Most solutes in body fluids are electrolytes, that is, compounds that dissociate into ions. Fluid balance, then, means water balance, but it also implies electrolyte balance. The two are inseparable.

WATER

Water is by far the largest single constituent of the body, comprising from 45 to 75 percent of the total body weight. The percentage varies from person to person and depends primarily on the amount of fat present and age. Since fat is basically waterfree, leaner individuals have a greater proportion of water to total body weight. Water proportion also decreases with age. An infant has the highest amount of water per body weight. In an average adult male, water constitutes about 65 percent of the body weight, and in an average adult female, it constitutes about 55 percent. Females have more subcutaneous fat than males do. As the person ages, the amount of fluid decreases.

Fluid intake and output

The primary source of body fluid is water derived from ingested liquids and foods that has been absorbed from the alimentary canal. This water amounts to about 2,300 milliliters daily. Another source of fluid is water produced through catabolism. This amounts to about 200 milliliters daily. Thus, total fluid input averages about 2,500 milliliters/day (Figure 23-1).

There are several avenues of fluid output. The kidneys lose about 1,500 milliliters/day, the skin about 500 milliliters/day, the lungs about 300 milliliters/day and the gastrointestinal tract about 200 milliliters/day. Fluid output thus totals 2,500 milliliters/day. Under normal circumstances, fluid intake equals fluid output so that the body maintains a constant volume of fluid.

Regulation of intake

Fluid intake is regulated by the presence or absence of thirst. When water loss is greater than water intake, the condition is called *dehydration*. Dehydration stimulates thirst through both local and general responses. Locally, it leads to a decrease in the flow of saliva that produces a dryness of the mucosa of the mouth and pharynx (Figure 23-2). Dryness is interpreted by the brain as a sensation of thirst. In addition, dehydration raises blood osmotic pressure. It is believed that receptors in the thirst center of the hypothalamus are stimulated by the increase in osmotic pressure and initiate impulses that also are interpreted as a sensation of thirst. The

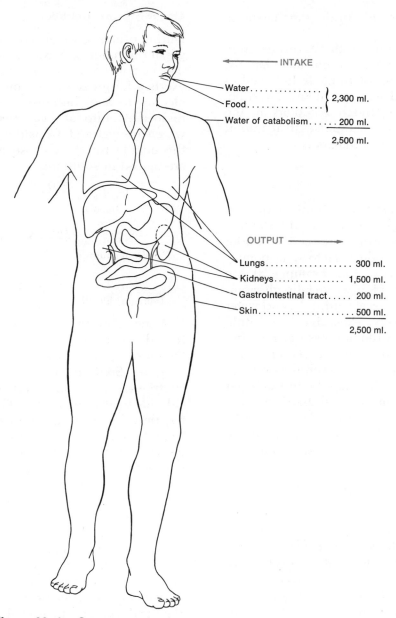

Figure 23–1. Sources and volumes involved in fluid intake and output.

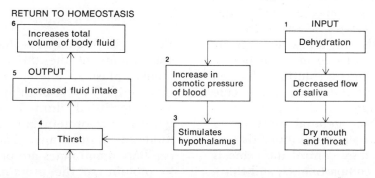

Figure 23–2. Regulation of fluid volume by the adjustment of intake to output.

response, a desire to drink fluids, thus balances the fluid loss.

The initial quenching of thirst results from wetting the mucosa of the mouth and pharynx, but the major inhibition of thirst is believed to occur as a result of distention of the intestine. Apparently, stimulated stretch receptors in the walls of the intestine send impulses that inhibit the thirst center in the hypothalamus.

Regulation of output

Under normal circumstances, fluid output is adjusted by vasopressin-ADH and aldosterone, both of which regulate urine production. Under abnormal conditions, other factors also may influence output heavily. For example, if the body is dehydrated, blood pressure falls, glomerular filtration rate decreases accordingly, and water is conserved. Conversely, excessive blood fluid results in an increase of blood pressure, glomerular filtration rate, and fluid output. Hypertension produces the same effect. Hyperventilation leads to increased fluid output through the loss of water vapor by the lungs. Vomiting and diarrhea result in fluid loss from the gastrointestinal tract. Finally, fever and destruction of extensive areas of the skin bring about excessive water loss through the skin.

ELECTROLYTES

The body fluids contain a variety of dissolved chemicals. Some of these chemicals are compounds with covalent bonds — that is, the elements that compose the molecule share electrons and do not form ions. Such compounds are called nonelectrolytes. Nonelectrolytes include most organic compounds, such as glucose, urea, and creatine. Other compounds, called **electrolytes**, have at least one ionic bond. When they dissolve in the body fluid, they dissociate into positive and negative ions. The positive ions are called cations, and the negative ones anions. Acids, bases, and salts are electrolytes. Most electrolytes are inorganic compounds, but a few are organic. For example, some proteins form ionic bonds. When the protein is put in solution, the ion detaches and the rest of the protein molecule carries the opposite charge.

Electrolytes serve three general functions in the body. First, many are essential minerals (see Exhibit 21–1). Second, they control the osmosis of water between body compartments, and third, they help to maintain the acid–base balance required for normal cellular activities.

Electrolyte concentration

During osmosis, water moves to the area with the greater number of particles in solution. A particle may be a whole molecule or an ion. An electrolyte exerts a far greater effect on osmosis than a nonelectrolyte does because an electrolyte molecule dissociates into at least two particles, both of which are charged. Consider what happens when the nonelectrolyte glucose and two electrolytes are placed in solution:

$$C_6H_{12}O_6 \xrightarrow{H_2O} C_6H_{12}O_6$$
Glucose

$$NaCl \xrightarrow{H_2O} Na^+ + Cl^-$$
Sodium chloride

$$CaCl_2 \xrightarrow{H_2O} Ca^{++} + Cl^- + Cl^-$$
Calcium chloride

Notice that glucose does not break apart when it is dissolved in water. A molecule of glucose, therefore, contributes only one particle to the solution. Sodium chloride, on the other hand, contributes two ions, or particles, and calcium chloride contributes three. Thus, calcium chloride has three times as great an effect on solute concentration as glucose. Just as important, once the electrolyte dissociates, its ions can attract other ions of the opposite charge. For example, if equal amounts of Ca^{++} and Na^+ are placed in solution, the calcium ion will attract twice as many chloride ions to its area as the sodium ion.

To determine how much effect an electrolyte has on concentration, we must look at the concentrations of its individual ions. The concentration of an ion is commonly expressed in *milliequivalents per liter* or mEq./L. — the number of electrical charges in each liter of solution. The mEq./L. equals the number of charges the ion carries times the number of ions in solution. Ion concentration can be calculated as shown in Exhibit 23–1.

Distribution of electrolytes

Figure 23–3 compares the principal chemical constituents of the three major fluid compartments: plasma, interstitial fluid, and intracellular fluid. The chief difference between plasma and interstitial fluid is that plasma contains quite a few protein anions, whereas interstitial fluid has hardly any. This is because normal capillary membranes are practically impermeable to protein, so the protein stays in the plasma and does not move out of the blood into the interstitial fluid. Plasma also contains more Na^+

Exhibit 23-1. CALCULATING THE CONCENTRATION OF AN ION IN SOLUTION

The number of milliequivalents of an ion in each liter of solution is expressed by the following equation:

$$\text{mEq./L.} = \frac{\text{milligrams of ion per liter of solution} \times \text{number of charges on one ion}}{\text{atomic weight of ion}}$$

The atomic weight of an element indicates how heavy it is compared with the element carbon. Dividing the total weight of the solute by its atomic weight tells us how many ions we have. For instance, the atomic weight of calcium is 40, whereas that of sodium is 23. Calcium is therefore a heavier element, and 100 grams of calcium contains fewer atoms than do 100 grams of sodium. The atomic weights of the elements have been calculated by scientists and can be found in a periodic table.

Using the preceding formula, we can calculate the mEq./L. for calcium. In 1 liter (L.) of plasma there are normally 100 milligrams (mg.) of calcium. Thus, by substituting this value in the formula we arrive at:

$$\text{mEq./L.} = \frac{100 \times \text{number of charges}}{\text{atomic weight}}$$

The atomic weight of calcium is 40, and its number of charges is 2. By substituting these values we arrive at:

$$\text{mEq./L.} = \frac{100 \times 2}{40}$$

$$= 5 \text{ mEq./L. for calcium}$$

Let us now find the mEq./L. for sodium:

Mg. Na^+/liter of plasma = 3,300 mg./L.
Number of charges = 1
Atomic weight = 23
$$\text{mEq./L.} = \frac{3,300 \times 1}{23}$$
$$= 143.0$$

Comparing the mEq./L. of sodium with that of calcium, we can see that even though calcium has a greater number of charges than sodium, the body retains many more sodium ions than it does calcium ions. Therefore, the millequivalent for sodium in plasma is higher.

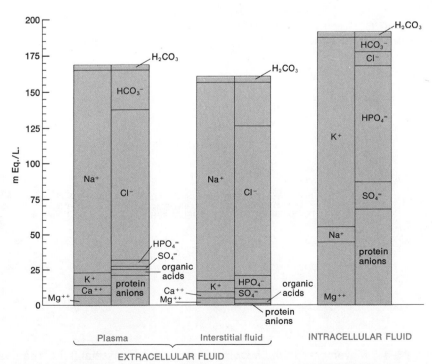

Figure 23-3. Comparison of electrolyte concentrations in plasma, interstitial fluid, and intracellular fluid. The height of each column represents the total electrolyte concentration.

ions but fewer Cl⁻ ions than does the interstitial fluid. In most other respects the two fluids are similar.

However, intracellular fluid varies considerably from extracellular fluid. In extracellular fluid, the most abundant cation is Na^+, and the most abundant anion is Cl^-. In intracellular fluid, the most abundant cation is K^+, and the most abundant anion is HPO_4^-. Also, there are more protein anions in intracellular fluid than there are in extracellular fluid.

MOVEMENT OF BODY FLUIDS

Movement between plasma and interstitial compartments

The movement of water between plasma and interstitial fluid occurs across capillary membranes (Figure 23–4) and is dependent on four principal pressures: (1) blood hydrostatic pressure, (2) interstitial fluid hydrostatic pressure, (3) blood osmotic pressure, and (4) interstitial fluid osmotic pressure.

Blood hydrostatic pressure is synonymous with the blood pressure in a capillary. You are already familiar with blood hydrostatic pressure in the glomeruli. As you know, it tends to force fluid out of the plasma compartment. In most capillaries, blood hydrostatic pressure is about 35 millimeters Hg at the arterial end of a capillary and about 15 millimeters Hg at the venous end. *Interstitial fluid hydrostatic pressure* is the pressure of the interstitial fluid against the cells of the tissue. It pushes the fluid out of the interstitial compartment under a pressure of 2 millimeters Hg at the arterial end of a capillary and at 1 millimeter Hg at the venous end.

Blood osmotic pressure pulls water into the plasma. It averages 25 millimeters Hg at both ends of the capillary. *Interstitial fluid osmotic pressure* forces water into the interstitial compartment. It is 0 at the arterial end of a capillary and 3 millimeters Hg at the venous end. The protein anions in plasma are chiefly responsible for the higher osmotic pressure of blood.

The difference between the two forces that move fluid out of the blood and the two forces that push it into the blood is called the *effective filtration pressure*, or P_{eff}. Effective filtration pressure determines the direction of fluid movement and is represented by the following equation:

$$P_{eff} = \begin{pmatrix} \text{blood} & \text{interstitial} \\ \text{hydrostatic} + \text{fluid osmotic} \\ \text{pressure} & \text{pressure} \end{pmatrix}$$
$$- \begin{pmatrix} \text{interstitial} & \text{blood} \\ \text{fluid hydrostatic} + \text{osmotic} \\ \text{pressure} & \text{pressure} \end{pmatrix}$$

By substituting the values already given, we can calculate the effective filtration pressure at the arterial end of a capillary as follows:

$$P_{eff} = (35 + 0) - (2 + 25)$$
$$= (35) - (27)$$
$$= 8 \text{ mm. Hg}$$

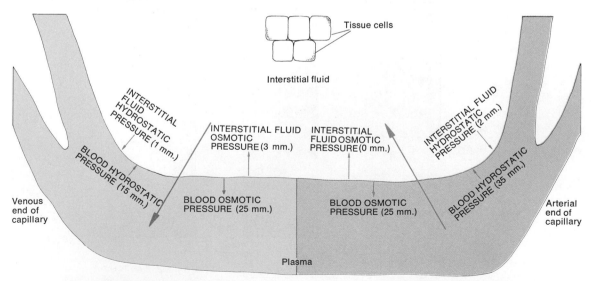

Figure 23–4. Movement of fluid between plasma and interstitial fluid. The heavy arrows indicate the directions of movement at the arterial and venous ends of the capillary.

This means that at the arterial end of a capillary, fluid moves from the plasma into the interstitial fluid because the sum of the blood hydrostatic pressure and interstitial fluid osmotic pressure forces is greater than the sum of the interstitial fluid hydrostatic pressure and blood osmotic pressure forces.

Now let us calculate the effective filtration pressure at the venous end of a capillary:

$$P_{eff} = (15 + 3) - (1 + 25)$$
$$= (18) - (26)$$
$$= -8 \text{ mm. Hg}$$

Note here that fluid moves from the interstitial fluid back into the plasma because the sum of the interstitial fluid hydrostatic pressure and blood osmotic pressure forces is greater than the sum of the blood hydrostatic pressure and interstitial fluid osmotic pressure forces.

Note that the amount of fluid leaving the plasma at the arterial end of a capillary equals the amount returned to plasma at the venous end of a capillary. This fluid balance between plasma and interstitial fluid exists under normal conditions and is referred to as *Starling's law of the capillaries.*

Occasionally, the balance between interstitial fluid and plasma is disrupted. *Edema*, the abnormal increase in interstitial fluid resulting in tissue swelling, is an example of this type of fluid imbalance (Figure 23–5). One cause of edema, hypertension, raises the blood hydrostatic pressure. Another cause is inflammation. As part of the inflammatory response, capillaries become more permeable and allow proteins to leave the plasma and enter the interstitial fluid. Consequently, the blood osmotic pressure falls, and the interstitial osmotic pressure rises.

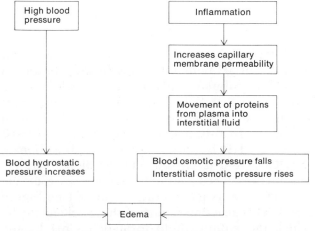

Figure 23–5. Conditions that produce edema.

Movement between interstitial and intracellular compartments

Water movement between the interstitial fluid and the intracellular fluid results from the same types of pressures that exist between the plasma and the interstitial fluid. If you look at Figure 23–3, you will see that the intracellular fluid has a higher osmotic pressure than does the interstitial fluid. In addition, the principal cation inside the cell is K^+, whereas the principal cation outside the cell is Na^+. Normally, the higher intracellular osmotic pressure is balanced by forces that move water out of the cell, so the amount of water inside the cell does not change. When a fluid imbalance between these two compartments occurs, it is usually caused by a change in the Na^+ or K^+ concentration. Let us examine how a sodium electrolyte imbalance leads to a fluid imbalance.

Sodium balance in the body normally is controlled by vasopressin-ADH and by aldosterone. Vasopressin-ADH regulates extracellular fluid electrolyte concentration by regulating the amount of water reabsorbed into the blood by the kidney tubules. Aldosterone regulates extracellular fluid volume by regulating the amount of sodium reabsorbed by the blood from the kidney tubules. Certain conditions, however, may result in a decrease in the sodium concentration in interstitial fluid. For instance, during sweating the skin excretes sodium as well as water. Sodium also may be lost through vomiting and diarrhea. Coupled with little or no sodium intake, these conditions can quickly produce a sodium deficit (Figure 23–6). The decrease in sodium concentration in the interstitial fluid lowers the interstitial fluid osmotic pressure and establishes an effective filtration pressure gradient between the interstitial fluid and the intracellular fluid. Water moves from the interstitial fluid into the cells, producing two results that can be quite serious.

The first result, an increase in intracellular water concentration called *overhydration*, is particularly disruptive to nerve cell function. In fact, severe overhydration, or *water intoxication*, produces neurological symptoms ranging from disoriented behavior to convulsions, coma, and even death. The second result of the fluid shift is a loss of interstitial fluid volume that leads to a decrease in the interstitial hydrostatic pressure. As the interstitial hydrostatic pressure drops, water moves out of the plasma, resulting in a loss of blood volume that may lead to shock.

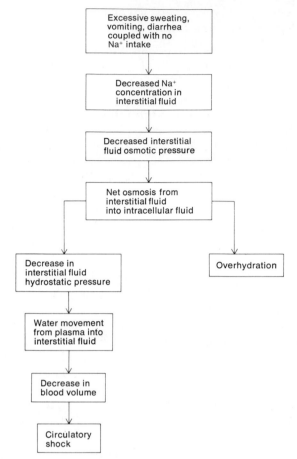

Figure 23–6. Interrelations between fluid imbalance and electrolyte imbalance.

ACID-BASE BALANCE

In addition to controlling water movement, electrolytes also help to regulate the acid–base balance of the body. The overall acid–base balance of the body is maintained by controlling the H^+ concentration of body fluids, particularly of extracellular fluid. In a healthy individual, the pH of the extracellular fluid is stabilized between 7.35 and 7.45. Homeostasis of this very narrow range is essential to survival and is dependent upon three major mechanisms: (1) buffer systems, (2) respirations, and (3) kidney excretion.

Buffer systems

A **buffer system** consists of a weak acid and a salt of that acid, and it functions to prevent drastic changes in the pH of a body fluid by changing strong acids and bases into weak acids and bases and salts. Recall that a strong acid dissociates into H^+ ions more easily than does a weak acid. Strong acids, therefore, lower pH more than weak ones do because strong acids contribute more H^+ ions. Similarly, strong bases raise pH more than weak ones because strong bases dissociate more easily into OH^- ions. The principal buffer systems of the body fluids are the carbonic acid-bicarbonate system, the phosphate system, the hemoglobin-oxyhemoglobin system, and the protein system.

The *carbonic acid-bicarbonate buffer system* is an important regulator of blood pH. Recall from Chapter 2 that this system is based on the weak acid H_2CO_3 and on salts of the acid, primarily sodium bicarbonate ($NaHCO_3$). The following equations should refresh your memory:

$$\underset{\text{Weak acid}}{H_2CO_3} + \underset{\text{Strong base}}{NaOH} \rightleftharpoons H_2O + \underset{\text{Salt of weak acid}}{NaHCO_3}$$

$$\underset{\text{Salt of weak acid}}{NaHCO_3} + \underset{\text{Strong acid}}{HCl} \rightleftharpoons \underset{\text{Salt}}{NaCl} + \underset{\text{Weak acid}}{H_2CO_3}$$

Because normal body processes tend to acidify the blood, the body needs more bicarbonate salt than it needs carbonic acid. In fact, when extracellular pH is at its midrange of normal (7.4), bicarbonate salt outnumbers carbonic acid by 20 to 1.

The *phosphate buffer system* acts in basically the same manner as the bicarbonate buffer system. Its two components are sodium dihydrogen phosphate (NaH_2PO_4) and sodium monohydrogen phosphate (Na_2HPO_4). The dihydrogen phosphate ion acts as the weak acid and is capable of buffering strong bases:

$$\underset{\text{Weak acid}}{NaH_2PO_4} + \underset{\text{Strong base}}{NaOH} \rightleftharpoons \underset{\text{Water}}{H_2O} + \underset{\text{Salt of weak acid}}{NaHPO_4}$$

The monohydrogen phosphate ion acts as the salt of a weak base and is capable of buffering strong acids:

$$\underset{\text{Salt of weak base}}{Na_2HPO_4} + \underset{\text{Strong acid}}{HCl} \rightleftharpoons \underset{\text{Salt}}{NaCl} + \underset{\text{Weak acid}}{NaH_2PO_4}$$

The phosphate buffer system is an important regulator of pH both within red blood cells and in the kidney tubular fluids. You may recall from the last chapter that NaH_2PO_4 is formed when excess H^+ ions in the kidney tubules combine with Na_2HPO_4. In this reaction, the sodium released from Na_2HPO_4 forms sodium bicarbonate ($NaHCO_3$) and is passed into the blood. The H^+ ion that replaces sodium becomes part of the NaH_2PO_4 that is passed into the urine (see Figure 22–7). This reaction is one of the mechanisms by which the kidneys help to maintain pH by the acidification of urine.

The *hemoglobin-oxyhemoglobin buffer system* is an effective method for buffering carbonic acid in the blood. When blood moves from the arterial end of a capillary to the venous end, the CO_2 given up by body cells enters the erythrocytes and combines with water to form carbonic acid (Figure 23–7). Simultaneously, oxyhemoglobin gives up its oxygen to the body cells, becomes reduced hemoglobin, and carries a negative charge. The hemoglobin anion attracts the H^+ from the carbonic acid and becomes an acid that is even weaker than carbonic acid. When the hemoglobin-oxyhemoglobin system is active, the exchange reaction that occurs shows why the erythrocyte tends to give up its oxygen when pCO_2 is high.

The *protein buffer system* is the most abundant buffer in body cells and plasma. Proteins are composed of amino acids. An amino acid is an organic compound that contains at least one carboxyl group (COOH) and at least one amino group (NH_2). The carboxyl group acts like an acid and can dissociate in this way:

$$NH_2 - \boxed{\text{amino acid}} - COO^- H^+$$

The hydrogen ion is then able to react with any excess OH^- in the solution to form water. On the other hand, the amino group has a tendency to act as a base:

$$COOH - \boxed{\text{amino acid}} - NH_3^+ OH^-$$

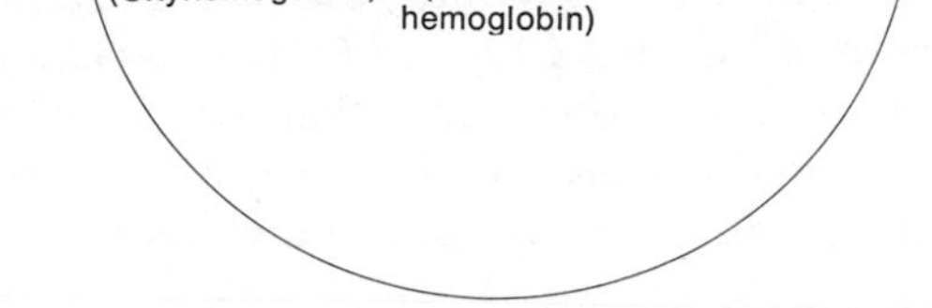

The OH^- can dissociate, react with any excess H^+ ions, and also form water. Thus proteins act as both acidic and basic buffers.

Respirations

Respirations also assume a role in maintaining the pH of the body. An increase in the CO_2 concentration in body fluids lowers the pH. This is illustrated by the following equation:

$$CO_2 + H_2O \rightleftharpoons H_2CO_3 \rightleftharpoons H^+ + HCO_3^-$$

Conversely, a decrease in the CO_2 concentration of body fluids raises the pH. It is on the basis of these effects that respirations help to control pH.

The pH of body fluids may be adjusted by a change in the rate of breathing. If the rate of breathing is increased, more CO_2 is exhaled, and the blood pH subsequently rises. Slowing down the respiration rate means less CO_2 is exhaled, and the blood pH falls. Doubling the breathing rate increases the pH by about 0.23, and reducing the breathing rate to one-quarter its normal rate lowers the pH by 0.4. If you consider that breathing rate can be altered from about zero to eight times the normal rate, it should become obvious that alterations in the pH of body fluids may be greatly influenced by respiration rate.

The pH of body fluids, in turn, affects the rate of breathing (Figure 23–8). If, for example, the blood becomes more acidic, the increase in H^+ ions stimulates the respiratory center in the medulla, and respirations increase. The same effect is achieved if the blood concentration of CO_2 increases. On the other hand, if the pH of the blood increases, the respiratory center is inhibited and respirations decrease, and a decrease in the

Figure 23–7. Hemoglobin-oxyhemoglobin buffer system. Oxyhemoglobin gives up its oxygen in an acid medium and buffers the acid. The HCO_3^- may remain in the cell and combine with K^+, or it may move out of the cell and combine with Na^+. In this way, much of the CO_2 is carried back to the lungs in the form of potassium bicarbonate or sodium bicarbonate.

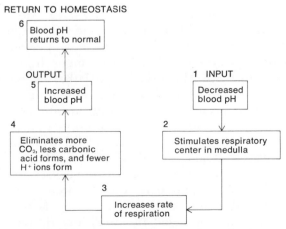

Figure 23–8. Relationship between pH and respirations.

CO_2 concentration of blood has the same effect. The respiratory mechanism normally can eliminate one to two times more acid or base than can all the buffers combined.

Kidney excretion

Inasmuch as the role of the kidneys in maintaining pH has already been considered in Chapter 22, we shall simply refer you to that section. Review Figure 22–7 very carefully.

A summary of the mechanisms that maintain body pH is presented in Exhibit 23–2.

Exhibit 23–2. SUMMARY OF MECHANISMS THAT MAINTAIN BODY pH

MECHANISM	COMMENTS
Buffer systems	Consist of a weak acid and a salt of a weak acid. Prevent drastic changes in the pH of a body fluid.
Carbonic acid-bicarbonate buffer system	Important regulator of blood pH.
Phosphate buffer system	Important buffer in red blood cells and kidney tubule cells.
Hemoglobin-oxyhemoglobin buffer system	Buffers carbonic acid in the blood.
Protein buffer system	Most abundant buffer in body cells and plasma.
Respirations	Regulate CO_2 level of body fluids.
Increased	Raises pH.
Decreased	Lowers pH.
Kidneys	Excrete H^+ and NH_4^+ and conserve bicarbonate.

Acid-base imbalances

Normal blood pH range is 7.35–7.45. Any considerable deviation from this value falls under the category of acidosis or alkalosis. A person has **acidosis** when his blood pH ranges from 7.35 down to 6.80. **Alkalosis** is a pH that ranges between 7.45 to 8.00. There are a number of causes of acidosis and alkalosis. *Respiratory acidosis* occurs as a result of any condition that decreases the movement of CO_2 from the blood to the alveoli of the lungs and therefore causes a buildup of CO_2, H_2CO_3, and H^+ ions. Possible causes include emphysema, pulmonary edema, injury to the respiratory center of the medulla, or disorders of the muscles involved in breathing. *Respiratory alkalosis* is caused by an increased minute volume of respiration in excess of that required to handle

increased metabolic demands of the body. It therefore results in a decrease of CO_2. Conditions leading to respiratory alkalosis include oxygen deficiency due to high altitude and aspirin overdose—factors that stimulate the respiratory center.

Metabolic acidosis results from an abnormal increase in acidic metabolic products other than CO_2 and/or from the loss of bicarbonate ion from the body. Ketosis is a good example of metabolic acidosis brought on by an increase in the production of acidic metabolic products. Acidosis due to loss of bicarbonate may occur with diarrhea and with renal tubular dysfunction. *Metabolic alkalosis* is caused by a nonrespiratory loss of acid by the body or by the excessive intake of alkaline drugs. Excessive vomiting of gastric contents results in a substantial loss of HCl and is probably the most frequent cause of metabolic acidosis.

The principal physiological effect of acidosis is depression of the central nervous system. If the pH of the blood falls below 7, depression of the nervous system is so acute that the individual becomes disoriented and comatose. In fact, patients with severe acidosis usually die in a state of coma. On the other hand, the major physiological effect of alkalosis is overexcitability of the nervous system. The overexcitability occurs both in the central nervous system and in peripheral nerves. Because of the overexcitability, nerves conduct impulses repetitively even when not stimulated by normal stimuli, resulting in extreme nervousness, muscle spasms, and even convulsions.

The primary treatments for respiratory acidosis are aimed at increasing the exhalation of CO_2. Excessive secretions may be suctioned out of the respiratory tract, and artificial respiration may be given. Treatment of metabolic acidosis consists of intravenous solutions of sodium bicarbonate and correcting the cause of the acidosis.

The treatment of respiratory alkalosis is aimed at increasing the level of CO_2 in the body. One corrective measure is to have the patient rebreathe a mixture of his own CO_2 and O_2 from a paper bag. Treatment for metabolic alkalosis consists of giving a medication containing the chloride ion and correcting the cause of the alkalosis.

Medical terminology associated with fluid and electrolyte dynamics

Acidosis Abnormal decrease in pH ranging from 7.35 down to 6.80.

Alkalosis Abnormal increase in pH ranging from 7.45 up to 8.00.

Buffer system A weak acid and the salt of that acid which prevents drastic changes in the pH of a body fluid.

Edema (*ede* = swelling) A larger than normal volume of interstitial fluid producing swelling of the tissue.

Electrolyte A chemical substance that dissociates into ions. Example: NaCl

Extracellular fluid (ECF) (*extra* = outside) That portion of body fluid found outside body cells; ECF includes plasma interstitial fluid, cerebrospinal fluid, and synovial fluid.

Intracellular fluid (ICF) (*intra* = within) That portion of body fluid found inside cells.

Metabolic acidosis Abnormal decrease in pH due to a build-up of metabolic acids in the blood and/or loss of bicarbonate.

Metabolic alkalosis Abnormal increase in pH due to a loss of acid by the body or the excessive intake of alkaline substances.

Respiratory acidosis Abnormal decrease in pH due to a decrease in the rate of respiration or, in rare cases, the inability of CO_2 to diffuse into the alveoli. Results in a buildup of CO_2, H_2CO_3, and H^+.

Respiratory alkalosis Abnormal increase in pH due to an increase in the minute volume of respiration; results in a decrease of CO_2.

Chapter summary in outline

WATER

1. Water, together with substances dissolved in it, constitutes body fluid.

2. Primary sources of fluid intake are ingested liquids and foods and water produced by catabolism.

3. Avenues of fluid output are the kidneys, skin, lungs, and gastrointestinal tract.

4. The stimulus for fluid intake is dehydration resulting in thirst sensations. The stimulus for fluid output is secretion of aldosterone and vasopressin-ADH.

ELECTROLYTES

1. Electrolytes are compounds that dissolve in body fluids and produce either cations (positive ions) or anions (negative ions).

2. Electrolyte concentration is expressed in milliequivalents per liter.

3. Plasma, interstitial fluid, and intracellular fluid contain varying kinds and amounts of electrolytes.

4. Electrolytes are needed for normal metabolism, proper fluid movement between compartments, and regulation of pH.

MOVEMENT OF BODY FLUIDS

1. At the arterial end of a capillary, fluid moves from plasma into interstitial fluid. At the venous end of a capillary, fluid moves in the opposite direction. Fluid balance between plasma and interstitial fluid is called Starling's law of the capillaries.

2. Fluid movement between interstitial and intracellular compartments depends on the movement of sodium and the secretion of aldosterone and vasopressin-ADH.

3. Fluid imbalance may lead to edema and overhydration.

ACID-BASE BALANCE

1. The overall acid-base balance of the body is maintained by controlling the H^+ concentration of body fluids, especially that of extracellular fluid.

2. The normal pH of extracellular fluid is 7.35–7.45.

3. Homeostasis of pH is maintained by buffers, respirations, and kidney excretion.

4. Acid-base imbalances are acidosis and alkalosis.

Review questions and problems

1. Define body fluid. What are the two principal fluid compartments of the body? How are the compartments separated?

2. What is meant by fluid balance? How are fluid balance and electrolyte balance related?

3. Describe the avenues of fluid intake and fluid output. Be sure to indicate volumes in each case.

4. Discuss the role of thirst in the regulation of fluid intake.

5. Explain how aldosterone and vasopressin-ADH adjust fluid output.

6. Define a nonelectrolyte and an electrolyte. Give specific examples of each.

7. Distinguish between a cation and an anion. Give several examples of each.

8. How is the ionic concentration of a fluid expressed? Calculate the ionic concentration of sodium in a body fluid.

9. List and describe the functions of electrolytes in the body.

10. Describe some of the major differences in the electrolytic concentrations of the three fluid compartments of the body.

11. Explain the forces involved in moving fluid between plasma and interstitial fluid. Summarize these forces by setting up an equation to express effective filtration pressure.

12. What is Starling's law of the capillaries?

13. Explain the factors involved in fluid movement between the interstitial fluid and the intracellular fluid.

14. Define edema. What are some of its causes?

15. Describe an example of how electrolyte balance is related to fluid balance.

16. Explain how the following buffer systems help to maintain the pH of body fluids: carbonic acid-bicarbonate, phosphate, hemoglobin-oxyhemoglobin, and protein.

17. Describe how respirations are related to the maintenance of pH.

18. Briefly discuss the role of the kidneys in maintaining pH.

19. Define acidosis and alkalosis. Distinguish between respiratory and metabolic acidosis and alkalosis.

20. What are the principal physiological effects of acidosis and alkalosis?

21. How are acidosis and alkalosis treated?

22. Refer to the glossary of medical terminology associated with fluid and electrolyte dynamics. Be sure that you can define each term.

Bibliography

Unit 4

Armstrong, W. M. and A. S. Nunn: *Intestinal Transport of Electrolytes, Amino Acids and Sugar,* Charles C Thomas, Publisher, Springfield, Ill., 1971.

"Arteriosclerosis," National Institutes of Health, DHEW Publication 72–137, 1971.

Atherton, J. C.: "Renal Physiology," *Br. J. Anaesth.,* **42**:236, 1972.

Audrey, Burgess: *The Nurse's Guide to Fluid and Electrolyte Balance,* McGraw-Hill Book Company, New York, 1970.

Avery, M. E. et al.: "The Lung of the Newborn Infant," *Scientific American,* **228**:75, April 1973.

"Battling the Hyaline Membrane," *Medical World News,* Nov. 13, 1970.

Bennett, B. et al.: "The Normal Coagulation Mechanism," *Med. Clin. North Amer.,* **59**:95, 1972.

Berne, R. M. and M. N. Levy: *Cardiovascular Physiology,* The C. V. Mosby Company, St. Louis, 1972.

Beutler, E.: "Sickle Cell Anemia: How to Detect and Combat It," *Consultant,* March 1972.

Blumenthal, S. and M. J. Jesse: "Prevention of Atherosclerosis: A Pediatric Problem," *Hospital Practice,* April 1973.

Bortoff, A.: "Digestion," *Ann. Rev. Physiol.,* **34:**261, 1972.

Burke, S. R.: *The Composition and Function of Body Fluids,* The C. V. Mosby Company, St. Louis, 1972.

Chapman, W. H.: *Urinary System: An Integrated Approach,* W. B. Saunders Company, Philadelphia, 1972.

Cherniack, R.: *Respiration in Health and Disease,* W. B. Saunders Company, Philadelphia, 1972.

Chidsey, C. A. III: "Calcium Metabolism in the Normal and Failing Heart," *Hospital Practice,* August 1972.

Child, J. et al.: "Blood Transfusions," *Amer. J. Nurs.,* **72:**1602, September 1972.

"Clinical Notes on Respiratory Diseases," *The American Thoracic Society,* **11:**2, New York, 1972.

Coodley, E. L.: "Anatomy of Circulation," part 1, *Consultant,* May 1972.

————: "Anatomy of Circulation," part 2, The Failing Heart, *Consultant,* July 1972.

————: "Anatomy of Circulation," part 3, Diagnosis and Management of Heart Failure, *Consultant,* September 1972.

Cooper, M. D. and A. R. Lawton, III: "The Development of the Immune System," *Scientific American,* **231:**5, November 1974.

Cunningham, D. J.: *Textbook of Anatomy,* G. J. Romanes (ed.), Oxford University Press, Oxford, 1971.

Davenport, H. W.: *Physiology of the Digestive Tract: An Introductory Text,* 3d ed., Year Book Medical Publishers, Inc., Chicago, 1971.

————: "Why the Stomach Does Not Digest Itself," *Scientific American,* **226:**87, January 1972.

Deaton, J. G.: "How Blood Clots," *Consultant,* May 1973.

Del Greco, F. and F. A. Krumlovsky: "Malignant Hypertension," *Hospital Medicine,* **9:**7, July 1973.

Fajans, S. S. and J. C. Floyd, Jr.: "Hypoglycemia: How to Manage a Complex Disease," *Modern Medicine,* Oct. 15, 1973.

Fine, R. N. and B. M. Korsch: "Renal Transplantation in Children," *Hospital Practice,* March 1974.

French, A. B.: "Nutritional Management of Gastrointestinal Disease," *Modern Medicine,* Oct. 2, 1972.

Gallstones: "Current Concepts in Therapy," *Medical World News,* Mar. 29, 1974.

Genest, J.: "Basic Mechanisms in Benign Essential Hypertension," *Hospital Practice,* August 1974.

Grundy, S. M.: "Cholesterol-Bile Acid Interactions in Gallstone Pathogenesis, *Hospital Practice,* December 1973.

Grushkin, C. M. et al.: "Hemodialysis in Small Children," *J. Amer. Med. Assoc.,* **221:**869, Aug. 21, 1972.

Hall, R. L.: "Food Additives," *Nutrition Today,* July/August 1973.

Hurst, J. W. (editor-in-chief): *The Heart: Arteries and Veins,* 3d ed., McGraw-Hill Book Company, New York, 1974.

Iber, F. L.: "In Alcoholism, the Liver Sets the Pace," *Nutrition Today,* January/February 1971.

Kaye, D. (ed.): *Urinary Tract Infection and Its Management,* The C. V. Mosby Company, St. Louis, 1972.

Kretchmer, N.: "Lactose and Lactase," *Scientific American,* October 1972.

Laragh, J. H.: "An Approach to the Classification of Hypertensive States," *Hospital Practice,* January 1974.

Lehman, J., Sr.: "Auscultation of Heart Sounds," *Amer. J. Nurs.,* **72:**1242, July 1972.

Longmore, D.: *The Heart,* McGraw-Hill Book Company, New York, 1971.

Maxwell, M. H. and C. R. Kleeman: *Clinical Disorders of Fluid and Electrolyte Metabolism,* 2d ed., McGraw-Hill Book Company, New York, 1972.

Meares, E. M.: "New Concepts in Treating Prostatic Diseases," *Modern Medicine,* Mar. 18, 1974.

"Mild and Severe Renal Failure," *Medical World News,* Nov. 8, 1974.

Moore, C.: "Cigarette Smoking and Cancer of the Mouth, Pharynx, and Larynx," *J. Amer. Med. Assoc.,* **218:**553, Oct. 25, 1971.

Morey, L. W.: "Causative Factors in Atherosclerosis," *J. Am. Osteopath. Assoc.,* **71:**506, 1972.

Moser, R. H. (ed.): "Standards for Cardiopulmonary Resuscitation (CPR) and Emergency Cardiac Care (ECC)," *J. Amer. Med. Assoc.,* **227:**833 (supplement), Feb. 18, 1974.

Motley, H. L.: "Pulmonary Emphysema," *Hospital Medicine,* **9:**11, November 1973.

"Poietic Puzzles," editorial in *MD,* May 1974.

Ratnoff, O. D.: "New Insight on the 'Royal Disease,'" *Modern Medicine,* July 9, 1973.

"Respiratory Infection," *Medical Tribune,* Feb. 20, 1974.

Rosett, D. (special ed.): "Urinary Infection," *Medical Tribune,* Apr. 24, 1974.

"Roughage in the Diet," *Medical World News,* Sept. 6, 1974.

Safer, P.: *Respiratory Therapy: Resuscitation and Intensive Care,* F. A. Davis Company, Philadelphia, 1972.

Seltzer, C. C.: "The Effect of Cigarette Smoking on Coronary Heart Disease," *Arch. Environ. Health,* 20, March 1970.

Sergis, E. and M. W. Hilgartner: "Hemophilia," *Amer. J. Nurs.,* **72:**11, November 1972.

Share, L. et al.: "Regulation of Body Fluids," *Ann. Rev. Physiol.,* **34:**235, 1972.

Shwachman, H.: "Changing Concepts of Cystic Fibrosis," *Hospital Practice,* January 1974.

Simone, J. V.: "Childhood Leukemia: The Changing Prognosis," *Hospital Practice,* July 1974.

Spain, D. M. et al.: "Emphysema in Apparently Healthy Adults," *J. Amer. Med. Assoc.,* **224:**322, Apr. 16, 1973.

Tobian, L. J. Jr.: "Experimental Models for the Study of Hypertension," *Hospital Practice,* February 1974.

Tumen, H. J.: "Alcoholic Liver Disease," *Hospital Medicine,* September 1974.

Weiss, W. et al.: "Risk of Lung Cancer According to Histologic Type and Cigarette Dosage," *J. Amer. Med. Assoc.,* **222:**799, 1972.

Weldy, N. J.: *Body Fluids and Electrolytes,* The C. V. Mosby Company, St. Louis, 1972.

Winick, M.: "Childhood Obesity," *Nutrition Today,* May/June 1974.

Young, V. R. and N. S. Scrimshaw: "The Physiology of Starvation," *Scientific American,* **225:**14, October 1971.

CONTINUITY

CHAPTER 24

REPRODUCTION OF THE HUMAN ORGANISM

STUDENT OBJECTIVES

After you have read this chapter, you should be able to:

1. Define reproduction

2. List the organs that comprise the male and female systems of reproduction

3. Describe the role of the scrotum in protecting the testes

4. Describe the testes as glands that produce sperm and the male hormone testosterone

5. Describe the physiological effects of testosterone

6. Trace the course of sperm cells through the system of ducts that lead from the testes to the exterior

7. Contrast the functions of the seminal vesicles, prostate gland, and Cowper's glands in secreting constituents of seminal fluid

8. Describe the chemical composition of seminal fluid

9. Describe the penis as the organ of copulation

10. Describe the ovaries as glands that produce ova and female sex hormones

11. Describe the function of the uterine tubes in transporting a fertilized ovum to the uterus

12. Identify the anatomical portions of the uterus and the ligaments that maintain its normal position

13. Discuss the physiological effects of estrogens and progesterone

14. Describe the principal events of the menstrual and ovarian cycles

15. Correlate the activities of both menstrual and ovarian cycles

16. Discuss the hormonal interactions that control the menstrual and ovarian cycles

17. Describe the role of the vagina in the menstrual flow and copulation

18. Identify the components of the vulva

19. Define the anatomical boundaries of the perineum

20. Describe the structure and development of the mammary glands

21. Describe such symptoms and causes of disorders of the reproductive systems as venereal diseases, prostate disorders, impotence, infertility, menstrual disorders, and cancer of the breasts and cervix

22. Define medical terminology associated with the reproductive systems

In its broadest sense, **reproduction** may be viewed as the self-perpetuation of genetic molecules, that is, molecules which determine the characteristics of all living forms. Reproduction is the mechanism by which the thread of life is sustained. It is the process by which a single cell duplicates its genetic material, allowing an organism to grow and to repair itself, and in this sense, reproduction enables the individual organism to maintain its own life. But reproduction is also the process by which genetic material is passed from generation to generation. In this regard, reproduction maintains the life of the species. In this last unit we shall be concerned with the continuity of life from generation to generation. In this chapter we shall describe the male and female reproductive systems, and in the following chapter we shall look at the formation of the ovum(egg) and sperm, conception and fetal development, and inheritance.

THE MALE REPRODUCTIVE SYSTEM

The organs of the male reproductive system (Figure 24–1) are the testes, or male gonads, which produce sperm, a number of ducts which either store or transport sperm to the exterior, accessory glands that add secretions comprising the semen, and several supporting structures.

Scrotum

The **scrotum** is a pouching of the abdominal wall. It is the supporting structure for the primary male reproductive organs. Externally, it looks like a single pouch of skin (see Figure 24–1). Internally, it is divided by a septum into two sacs, each of which contains a single testis. The testes are the organs that produce sperm. Sperm require a temperature that is lower than body temperature. Because the scrotum is isolated from the body cavities, it is not as warm as the rest of the abdominal cavity. Thus, it supplies an environment about 3°F below body temperature. The subcutaneous tissue of the scrotum contains several small bundles of smooth muscle fibers. Exposure to cold causes contraction of the smooth muscle fibers, moving the testes closer to the abdomen where they can absorb body heat. This causes the skin of the scrotum to appear more wrinkled. Exposure to warmth reverses the process.

The testes

The **testes** are paired oval-shaped glands measuring about 5 centimeters (2 inches) in length and 2.5 centimeters (1 inch) in diameter (Figure 24–2a). During most of fetal life they lie in the abdominal cavity, but about 2 months prior to birth they descend into the scrotum. When the testes do not descend, the condition is referred to as *cryptorchidism.* Cryptorchidism results in sterility because the sperm cells are destroyed by the higher body temperature of the abdominal cavity. Undescended testes can be placed in the scrotum by administering hormones or by surgical means.

The testes are covered by a dense layer of white fibrous tissue that extends inward and divides each testis into a series of internal compartments called *lobules.* Each lobule contains one to three tightly coiled tubules, the *seminiferous tubules,* that produce the sperm—a process called *spermatogenesis.* A cross section through a seminiferous tubule reveals that the tubule is packed with sperm cells in various stages of development (Figure 24–2b). The most immature cells, the *spermatogonia,* are located against the basement membrane. Moving toward the center of the tube, one can see layers of progressively more mature cells. By the time a sperm cell, or **spermatozoan,** has reached full maturity, it is in the lumen of the tubule and begins to move through a series of ducts. Embedded between the developing sperm cells in the tubules are *Sertoli cells* that produce secretions for the supplying of nutrients to the spermatozoa. Between the seminiferous tubules are clusters of *interstitial cells of Leydig.* These cells secrete the male hormone testosterone.

The functions of the testes are to produce sperm and produce testosterone.

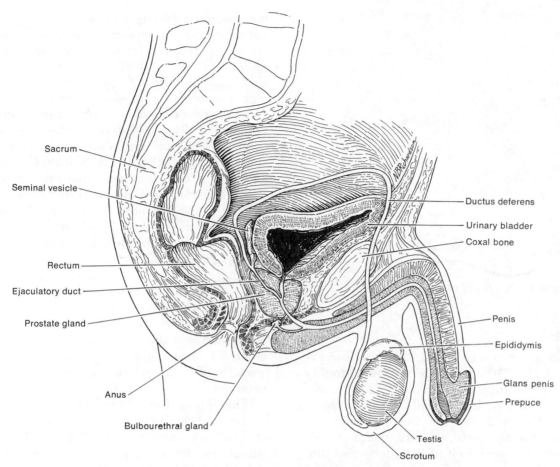

Figure 24–1. Male organs of reproduction seen in sagittal view.

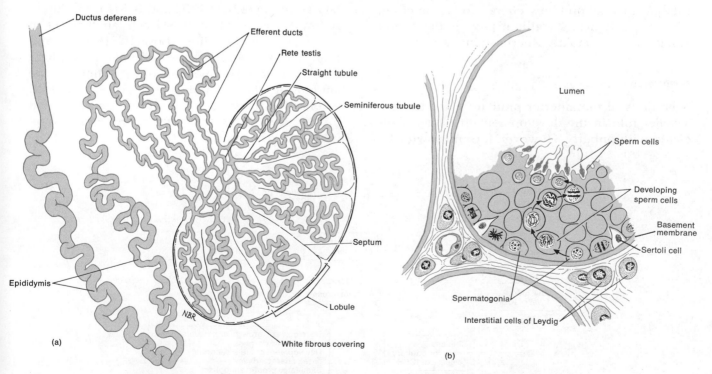

Figure 24–2. The testes. (a) Sectional view of a testis showing its system of tubes. (b) Cross section of a seminiferous tubule showing the stages of spermatogenesis.

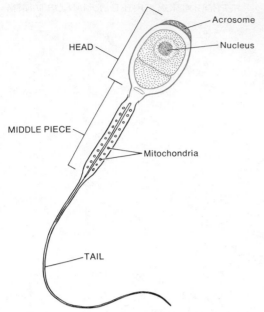

Figure 24–3. Parts of a spermatozoan.

Sperm

Spermatozoa, once ejaculated, have a life expectancy of 24–72 hours. A spermatozoan is highly adapted for reaching and penetrating a female ovum. It is composed of a head, a middle piece, and a tail (Figure 24–3). The head contains the nucleus and the acrosome, which contains chemicals that effect penetration of the sperm cell into the ovum. Numerous mitochondria are contained in the middle piece. Mitochondria carry on the catabolism that provides energy for locomotion. The tail, a typical flagellum, propels the sperm along its way. Actually, sperm swim very well.

Testosterone

Secretions of the anterior pituitary gland assume a major role in the developmental changes associated with puberty. However, it is unknown why the anterior pituitary start to secrete and release gonadotrophic hormones at the onset of puberty. Once secreted, the gonadotrophic hormones have profound effects on male reproductive organs. We shall examine the effects of gonadotrophic hormones on female reproductive organs later.

At the onset of puberty, the follicle stimulating hormone (FSH) is released by the anterior pituitary. It acts on the seminiferous tubules to initiate spermatogenesis (Figure 24–4). However, spermatogenesis is not completed under the influence of FSH alone. Interstitial cell stimulating hormone (ICSH) is also secreted by the anterior pituitary at the onset of puberty. This hormone also acts on the seminiferous tubules and further assists the tubules to develop mature sperm, but the chief function of ICSH is to stimulate the interstitial cells of Leydig to secrete testosterone.

Testosterone has a number of effects on the body. It controls the development, growth, and maintenance of the male sex organs; the development of male secondary sex characteristics; bone growth; protein anabolism; and normal sexual behavior.

The interaction of the anterior pituitary hormones and testosterone illustrates the operation of feedback mechanisms (Figure 24–4). ICSH stimulates the production of testosterone, but once the testosterone concentration in the blood is increased to a certain level, the anterior pituitary secretion of both FSH and ICSH is inhibited. A low blood level of ICSH sets another set of reactions into operation. Low ICSH levels inhibit the secretion of testosterone. Low blood levels of testosterone, in turn, stimulate the anterior pituitary secretion of ICSH. As the blood level of ICSH increases, the testes are stimulated to produce testosterone. Thus, the stimulatory-inhibitory cycle is complete.

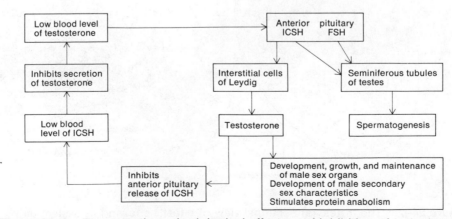

Figure 24–4. The secretion, physiological effects, and inhibition of testosterone.

Ducts

When the sperm mature, they move through the seminiferous tubules to the **straight tubules** (see Figure 24–2a). The straight tubules form a network of ducts in the center of the testis called the **rete testis.** Some of the cells lining the rete testis possess a flagellum that probably helps to push the sperm along. The sperm are next transported out of the testis through a series of coiled **efferent ducts** that empty into a single tube called the epididymis.

The epididymis

The two **epididymides** are highly coiled tubes, measuring about 6 meters (20 feet) in length, that lie tightly packed within the scrotum. Each epididymis attaches to its testis, where it receives the efferent ducts, and then it descends along the posterior side of the testis, makes a loop, and ascends. The epididymis is lined with pseudostratified columnar epithelium, and its wall contains smooth muscle. Functionally, the epididymis stores spermatozoa in anticipation of ejaculation and propels spermatozoa toward the urethra during ejaculation. Propulsion of the sperm is accomplished by peristaltic contractions of the smooth muscle.

The ductus deferens

The terminal portion of the epididymis is less coiled and considerably thicker. At this point it is referred to as the **ductus (vas) deferens** (see Figure 24–2a). The ductus deferens ascends along the posterior border of the testis, penetrates the inguinal canal, and enters the abdomen, where it extends over the top and down the posterior surface of the bladder (see Figure 24–1). Histologically, the ductus deferens is lined with pseudostratified epithelium and contains a heavy coat of three layers of muscle. Peristaltic contractions of the muscular coat propel the spermatozoa toward the urethra.

Traveling with the ductus deferens are the testicular artery, autonomic nerves, veins that drain the testes, lymphatics, and a small, circular band of skeletal muscle called the *cremaster muscle.* These structures together constitute the **spermatic cord,** a supporting structure of the male reproductive system. The cremaster muscle elevates the testes ever so slightly during sexual stimulation and exposure to cold.

The ejaculatory duct

Behind the urinary bladder, each ductus deferens joins its **ejaculatory duct** (Figure 24–5). Both ejaculatory ducts eject spermatozoa into the urethra. The urethra is the terminal duct of the system, serving as a common passageway for both spermatozoa and urine.

Accessory glands

Whereas the ducts of the male reproductive system store and transport sperm cells, the *accessory glands* secrete the liquid portion of semen. The first of the accessory glands we shall consider are the paired **seminal vesicles** (Figure 24–5). These glands are convoluted pouchlike structures lying posterior to and at the base of the urinary bladder, in front of the rectum. They secrete the viscous component of semen and pass it into the ejaculatory duct.

The **prostate gland** is a single, doughnut-shaped gland about the size of a chestnut (see Figure 24–5). It is inferior to the urinary bladder and surrounds the upper portion of the urethra. The prostate secretes an alkaline fluid that constitutes the largest fraction of semen into the urethra. In older men, the prostate sometimes enlarges to the point where it compresses the urethra and obstructs urine flow. At this stage, surgical removal of part of or the entire gland (prostatectomy) usually is indicated. The prostate gland is also a common tumor site in older males.

The paired **bulbourethral,** or **Cowper's glands** are about the size and shape of peas. They are located beneath the prostate on either side of the urethra (Figure 24–5). Like the prostate, the Cowper's glands secrete an alkaline fluid into the urethra.

Semen

In our discussion of the male reproductive organs we have mentioned a fluid called semen. **Semen,** or **seminal fluid,** is a mixture of sperm and the secretions of the seminal vesicles, the prostate gland, and the bulbourethral glands. The average volume of semen for each ejaculation is 3–4 milliliters, and the average number of spermatozoa ejaculated is about 400 million. When the number of spermatozoa falls below approximately 100 million, the male is likely to be physiologically sterile. Even though only a single spermatozoan fertilizes an ovum, it is hypothesized that fertilization requires the combined action of a tremendous

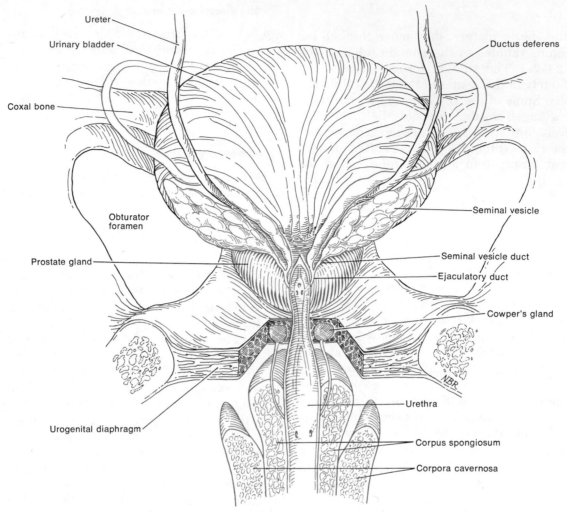

Figure 24–5. Posterior view of the urinary bladder showing relations of some male reproductive organs.

number of spermatozoa. The ovum is enclosed by cells that form a barrier between it and the sperm. An enzyme called hyaluronidase is secreted by the acrosomes of sperm. Hyaluronidase is believed to dissolve the cell covering of the ovum, giving the sperm a passageway into its interior. Apparently, vast numbers of sperm are required to secrete an effective amount of the enzyme.

Semen has a pH range of 7.35 to 7.50; that is, it is slightly alkaline. The prostatic secretion gives semen a milky appearance, and fluids from the seminal vesicles and bulbourethral glands give it a mucoid consistency. Semen provides spermatozoa with a transportation medium and acts as a buffer to neutralize the acid environment of the female reproductive system.

Penis

The **penis** is used to introduce spermatozoa into the female vagina (Figure 24–6). The distal end

of the penis is a slightly enlarged region called the *glans*. Covering the glans is the loosely fitting skin, called the *prepuce* or *foreskin*, that is removed during circumcision. Internally, the penis is composed of three cylindrical masses of tissue bound together by fibrous tissue. The two dorsally located masses are called the *corpora cavernosa penis*, and the smaller ventral mass, the *corpus spongiosum penis*, contains the urethra. All three masses of tissue are spongelike and contain venous sinuses. Under the influence of sexual excitation, the arteries supplying the penis dilate, and large quantities of blood enter the venous sinuses. Expansion of these spaces compresses the veins draining the penis so most entering blood is retained. These vascular changes result in an *erection*. The penis returns to its flaccid state when the arteries constrict and pressure on the veins is relieved. Erection prevents urination by the male during sexual excitation and ejaculation.

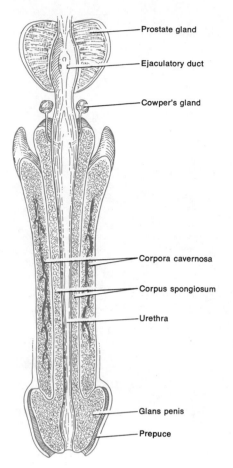

Figure 24-6. Internal structure of the penis.

Labels on figure: Prostate gland; Ejaculatory duct; Cowper's gland; Corpora cavernosa; Corpus spongiosum; Urethra; Glans penis; Prepuce

THE FEMALE REPRODUCTIVE SYSTEM

The female organs of reproduction (Figure 24–7) include the ovaries, which produce ova; the uterine tubes, which transport the ova to the uterus (or womb); the vagina; and some external organs that constitute the vulva. The mammary glands, or breasts, also are considered part of the female reproductive system.

Ovaries

The **ovaries,** or female gonads, are paired glands resembling almonds in size and shape. They are positioned in the upper pelvic cavity, one on each side of the uterus (see Figure 24–7). The ovaries are maintained in position by a series of ligaments (Figure 24–8). They are suspended by a part of the broad ligament of the uterus, a fold of peritoneum called the *mesovarium;* anchored to the uterus by the *ovarian ligament;* and attached to the pelvic wall by the *suspensory ligament.* Each ovary also contains a *hilus,* the point of entrance for blood vessels and nerves.

In microscopic view it can be seen that each ovary has an outer layer of simple epithelium called the *germinal epithelium* (Figure 24–9). The interior of the ovary is filled with connective tissue in which the ovarian follicles are embedded. *Ovarian follicle* is a general term for an

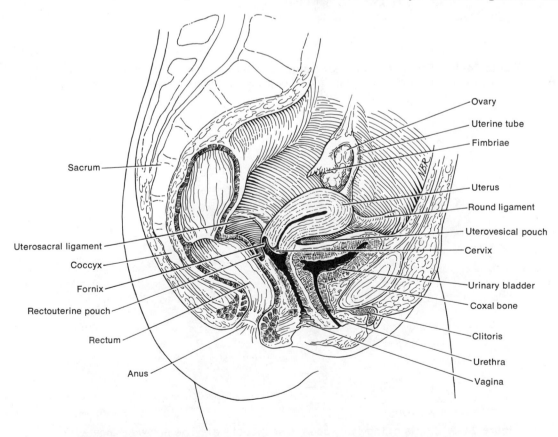

Labels on figure: Sacrum; Uterosacral ligament; Coccyx; Fornix; Rectouterine pouch; Rectum; Anus; Ovary; Uterine tube; Fimbriae; Uterus; Round ligament; Uterovesical pouch; Cervix; Urinary bladder; Coxal bone; Clitoris; Urethra; Vagina

Figure 24-7. Female organs of reproduction seen in sagittal section.

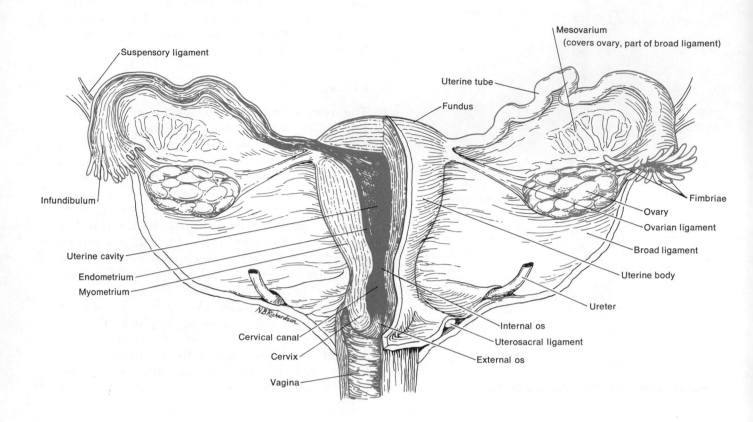

Figure 24–8. The uterus and associated female reproductive structures. The left side of the figure has been sectioned to show internal structures.

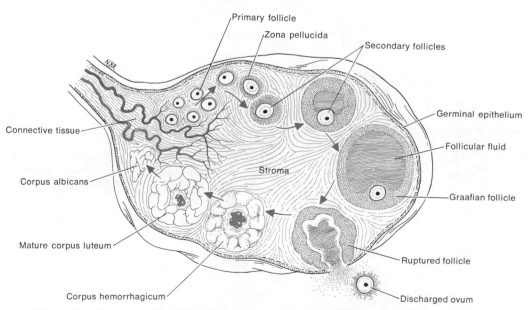

Figure 24–9. Parts of the ovary seen in sectional view. The arrows indicate the developmental stages that occur as part of the ovarian cycle.

ovum in any stage of development, along with its surrounding group of epithelial cells.

The ovaries produce mature ova and secrete the female sexual hormones. In this respect, they are analogous to the testes of the male.

Uterine tubes

The female body contains two **uterine,** or **fallopian, tubes,** which transport the ova from the ovaries to the uterus (see Figure 24–8). Measuring about 4 inches long, the tubes are positioned between the folds of the broad ligaments of the uterus. The open end of each tube, called the *infundibulum,* lies very close to the ovary but is not attached to it and is surrounded by a fringe of fingerlike projections called *fimbriae.* From the infundibulum the tube extends inward and downward and attaches to the upper side of the uterus.

Histologically, the uterine tubes are composed of three layers. The internal *mucosa* contains ciliated columnar cells that create a current to draw the ova into the uterine tubes, and secretory cells, which are believed to aid the nutrition of the ovum. The middle layer, the *muscularis,* is composed of a circular region of smooth muscle. Peristaltic contractions of the muscle serve to move the ovum down into the uterus. The outer layer of the uterine tubes is a *serous membrane.*

About once a month a mature ovum is released into the area of the abdominal cavity near the infundibulum of the uterine tube, a process called **ovulation.** The ovum then is drawn into the uterine tube because of the current created by the beating fimbriae. Under normal circumstances, the ovum is fertilized in the uterine tubes. This may occur at any time up to 24 hours following ovulation. The ovum, fertilized or unfertilized, descends into the uterus within about 72 hours. Sometimes a fertilized ovum is not drawn into the uterine tubes and becomes implanted in the pelvic cavity. Pelvic implantations usually fail because the developing ovum does not make vascular connections with the maternal blood supply. On occasion, fertilized ova fail to descend to the uterus and begin development in the uterine tubes. In this instance, the pregnancy must be terminated surgically before the tube ruptures. Both pelvic and tubular implantations are referred to as *ectopic pregnancies.*

Uterus

The organ of the female reproductive system that assumes a role in menstruation, implantation of a fertilized ovum, development of the fetus during pregnancy, and labor is the **uterus.** Situated between the bladder and the rectum, the uterus is an inverted, pear-shaped organ (Figure 24–7). Before the first pregnancy, the adult uterus measures approximately 7.5 centimeters (3 inches) long, 5 centimeters (2 inches) wide, and 1.75 centimeter (1 inch) thick. Anatomical subdivisions of the uterus include the dome-shaped portion above the uterine tubes called the *fundus,* the major tapering central portion called the *body,* and the inferior narrow portion opening into the vagina called the *cervix.* Between the body and the cervix is the constricted *isthmus* (see Figure 24–8). The interior of the body of the uterus is called the *uterine cavity,* and the interior of the narrow cervix is called the *cervical canal.* The junction of the uterine cavity with the cervical canal is the *internal os;* the *external os* is the place where the cervix opens into the vagina.

Normally the uterus is flexed between the uterine body and the cervix (see Figure 24–7). In this position, the body of the uterus projects forward and slightly upward over the urinary bladder, and the cervix projects downward and backward, joining the vagina at nearly a right angle. Several structures that are either extensions of the parietal peritoneum or fibromuscular cords, referred to as ligaments, maintain the position of the uterus (see Figures 24–7 and 24–8). The paired *broad ligaments* are each a double fold of parietal peritoneum attaching the uterus to either side of the pelvic cavity. Uterine blood vessels and nerves pass through the broad ligaments. The paired *uterosacral ligaments,* also peritoneal extensions, connect the sacrum to the uterus, one on either side of the rectum. The *cardinal (lateral cervical) ligament* extends below the base of the broad ligament between the pelvic wall and the cervix and vagina. This ligament contains smooth muscle, uterine blood vessels, and nerves and is the chief ligament supporting the position of the uterus and keeping the uterus from dropping down into the vagina. The *round ligaments* are bands of fibrous connective tissue between the layers of the broad ligament. They extend from a point on the uterus just below the uterine tubes to a portion of the external genitalia. Although the ligaments normally maintain the position of the uterus, they also afford the uterine body with some movement. As a result, the uterus may become malpositioned. For example, a backward tilting of the uterus called *retroflexion* or a forward tilting called *anteflexion* may occur.

Histologically, the uterus consists of three layers of tissue. The outer layer, derived from the peritoneum, is referred to as the *serous layer* and covers all the uterus except the cervix. Laterally, the serosa becomes the broad ligament. Anteriorly, it is reflected over the urinary bladder and forms a shallow pouch, the *uterovesicle pouch.* Posteriorly, it is reflected on to the rectum and forms a deep pouch, the *rectouterine pouch,* or *pouch of Douglas,* which is the lowest point in the pelvic cavity (see Figure 24–7).

The middle layer of the uterus, the *myometrium,* forms the bulk of the uterine wall (see Figure 24–8). This layer consists of smooth muscle fibers and is thickest in the fundus and thinnest in the cervix. During childbirth, coordinated contractions of the muscles dilate the cervix and help to expel the fetus from the body of the uterus.

The inner layer of the uterus, the *endometrium,* is a mucous membrane composed of two principal layers. The *functionalis,* the layer closer to the uterine cavity, is the layer shed during menstruation. The other layer, the *basalis,* is maintained during menstruation and produces a new functionalis following menstruation.

To understand how menstruation occurs and how the fetus receives nourishment from its mother, you need to know a little about the blood supply to the uterus. Branches of the uterine artery, called arcuate arteries, are arranged in a circular fashion underneath the serosa and give off branches that penetrate the myometrium (Figure 24–10). Just before these branches enter the endometrium, they divide into two kinds of arterioles. One branch terminates in the basalis and supplies it with the materials necessary to regenerate the functionalis. The other branch penetrates the functionalis and changes markedly during the menstrual cycle. These changes will be described shortly.

Endocrine relations

Before continuing our discussion of the female reproductive organs, we shall correlate the principal events of the menstrual cycle with those of the ovarian cycle and changes in the endometrium. These are all hormonally controlled events.

The term **menstrual cycle** refers to a series of changes that occur in the endometrium of a nonpregnant female. Each month the endometrium is prepared to receive a fertilized ovum. An implanted ovum eventually develops into a fetus and normally remains in the uterus until delivery. If no fertilization occurs, a portion of the endometrium is shed. The **ovarian cycle** is a monthly series of events associated with the maturation of an ovum.

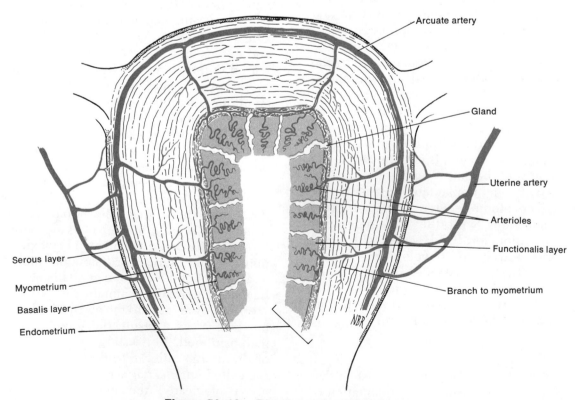

Figure 24–10. Blood supply of the uterus.

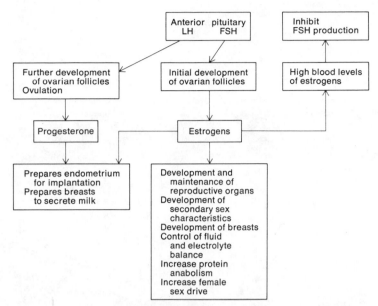

Figure 24–11. Physiological effects of estrogens and progesterone.

Gonadotrophic hormones of the anterior pituitary gland initiate the menstrual cycle, ovarian cycle, and other changes associated with puberty in the female (Figure 24–11). FSH stimulates the initial development of the ovarian follicles and the secretion of estrogens by the follicles. Another anterior pituitary hormone, the leuteinizing hormone (LH), stimulates the further development of ovarian follicles, brings about ovulation, and stimulates progesterone production by ovarian cells. The female sex hormones, estrogens and progesterone, effect the body in different ways. Estrogens have four main functions. First is the development and maintenance of female reproductive organs, especially the endometrium, secondary sex characteristics, and the breasts. Second, they control fluid and electrolyte balance. Third, they increase protein anabolism. Fourth, they cause an increase in the female sex drive. High levels of estrogens in the blood inhibit the secretion of FSH by the anterior pituitary gland. This inhibition provides the basis for the action of one kind of contraceptive pill. Progesterone, the other female sex hormone, works with estrogens to prepare the endometrium for implantation and to prepare the breasts for milk secretion. Let us now examine some of the details of the menstrual and ovarian cycles.

The duration of the menstrual cycle is variable among different females, normally ranging from 24 to 35 days. For purposes of our discussion, it will be assumed that the average duration of the cycle is 28 days. Events occurring during the menstrual cycle may be divided into three phases:

(1) the menstrual phase, (2) the preovulatory phase, and (3) the postovulatory phase (Figure 24–12).

The *menstrual phase*, also called *menstruation* or the *menses*, is the periodic discharge of blood (25–65) milliliters), tissue fluid, mucus, and epithelial cells. It lasts approximately for the first 5 days of the cycle. The discharge is associated with endometrial charges in which the functionalis layer degenerates and patchy areas of bleeding develop. Small areas of the functionalis detach one at a time (total detachment would result in hemorrhage), the uterine glands discharge their contents and collapse, and tissue fluid is discharged. The menstrual flow passes from the uterine cavity to the cervix, through the vagina, and ultimately to the exterior. Generally the flow terminates by the fifth day of the cycle. At this time the entire functionalis is shed, and the endometrium is very thin because only the basalis remains.

During the menstrual phase, the ovarian cycle is also in operation. Ovarian follicles, called *primary follicles*, begin their development (see Figure 24–9). At the time of birth each ovary contains about 200,000 such follicles, each consisting of an ovum surrounded by a layer of cells. During the early part of the menstrual phase, a primary follicle starts to produce very low levels of estrogens. A clear membrane, the zona pellucida, also develops around the ovum. Later in the menstrual phase (4–5 days) the primary follicle develops into a *secondary follicle* as the cells of the surrounding layer increase in number and

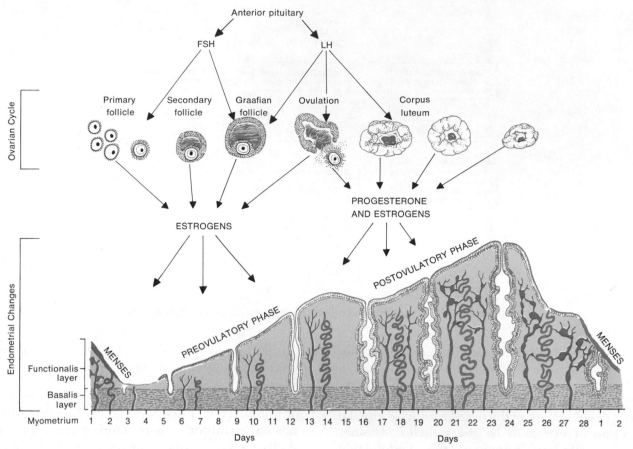

Figure 24-12. The menstrual and ovarian cycles.

secrete a fluid called the follicular fluid. This fluid forces the ovum to the edge of the follicle. The production of estrogens by the secondary follicle elevates the estrogen level of the blood slightly. Ovarian follicle development is the result of FSH production by the anterior pituitary, and during this part of the ovarian cycle, FSH secretion is maximal. Although a number of follicles begin development each cycle, only one will attain maturity.

The *preovulatory phase*, the second phase of the menstrual cycle, is the period of time between the end of menstruation and ovulation. This phase of the menstrual cycle is more variable in length than are the other phases. It lasts from days 6 to 13 in a 28-day cycle. During the preovulatory phase, a secondary follicle in the ovary matures into a *graafian follicle*, a follicle ready for ovulation. During the maturation process, the follicle increases its estrogen production. Early in the preovulatory phase, FSH is the dominant hormone, but, close to the time of ovulation, LH is secreted in increasing quantities. Moreover, small amounts of progesterone may be produced by the graafian follicle a day or two before ovulation.

FSH and LH stimulate the ovarian follicles to produce more estrogens, and this increase in estrogens stimulates the repair of the endometrium. During this process of repair, basilar cells undergo mitosis and produce a new functionalis. As the endometrium thickens, the endometrial glands are short and straight, and the arterioles become coiled and increase in length as they penetrate the functionalis. Because the proliferation of endometrial cells occurs during the preovulatory phase, the phase is also referred to as the *proliferative phase*. Still another name for this phase is the *follicular phase* because of increasing estrogen secretion by the developing follicle. Functionally, estrogen is the dominant hormone during this phase of the menstrual cycle.

Ovulation, the rupture of the graafian follicle with release of the ovum into the pelvic cavity, occurs on day 14 in a 28-day cycle. Just prior to ovulation, the high estrogen level that developed during the preovulatory phase inhibits FSH secretion by the anterior pituitary. Concurrently, LH secretion by the anterior pituitary is greatly increased. As FSH secretion is inhibited and LH and estrogen secretion increases, ovulation occurs. Following ovulation, the graafian follicle collapses, and blood within it forms a clot called the corpus hemorrhagicum. The clot is eventually absorbed by the remaining follicular cells. In time, the follicular cells enlarge, change character, and form the *corpus luteum.*

The *postovulatory phase* of the menstrual cycle is fairly constant in duration and lasts from days 15 to 28 in a 28-day cycle. It represents the period of time between ovulation and the onset of the next menses. Following ovulation, the level of estrogen in the blood drops slightly, and LH secretion stimulates the development of the corpus luteum. The corpus luteum then secretes increasing quantities of estrogens and progesterone, the latter being responsible for the preparation of the endometrium to receive a fertilized ovum. Preparatory activities include the filling of the endometrial glands with secretions that cause the glands to appear tortuously coiled, vascularization of the superficial endometrium, thickening of the endometrium, and an increase in the amount of tissue fluid. These preparatory changes are maximal about 1 week after ovulation, and they correspond to the anticipated arrival of the fertilized ovum. During the preovulatory phase, FSH secretion gradually increases and LH secretion decreases. The functionally dominant hormone during this phase is progesterone.

If fertilization and implantation do not occur, the rising levels of progesterone and estrogens inhibit LH secretion, and as a result the corpus luteum degenerates and becomes the *corpus albicans.* The decreased secretion of progesterone and estrogens by the degenerating follicle then initiates another menstrual period. In addition, the decreased progesterone and estrogen levels in the blood bring about a new output of the anterior pituitary hormones, especially FSH, and a new ovarian cycle is initiated. A summary of these hormonal interactions is presented in Figure 24–13.

If, however, fertilization and implantation do occur, the corpus luteum is maintained for nearly 6 months, and for most of this time it continues to secrete progesterone. Maintenance of the corpus luteum is accomplished by *chorionic gonadotrophin,* a hormone produced by the developing fetus, until the placenta can secrete estrogen to support pregnancy and progesterone to support pregnancy and breast development for lactation.

The menstrual cycle normally occurs one each month from *menarche,* the first menstrual cycle, to *menopause,* the complete cessation of menstruation. Menopause typically occurs between 45 and 50 years of age and results from failure of the ovaries to respond to the stimulation of gonadotrophic hormones from the anterior pituitary. The onset of menopause may be characterized by "hot flashes," copious sweating, headache, muscular pains, and emotional instability. Ultimately, menopause results in some degree of atrophy of the ovaries, uterine tubes, uterus, vagina, external genitalia, and breasts.

Vagina

We now shall continue our discussion of the female reproductive organs, with consideration of the vagina. The **vagina** serves as a passageway for the menstrual flow, the receptacle for the penis during copulation, and as the lower portion of the birth canal. It is a muscular, tubular organ lined with mucous membrane, measures about 10 centimeters (4 inches) in length (Figures 24–7 and 24–8), and is situated between the bladder and the rectum. It is directed upward and backward where it attaches to the uterus. The mucosa of the vagina lies in a series of transverse folds, the *rugae,* and is capable of a good deal of extension. The muscularis is composed of smooth muscle that can stretch considerably. This distention is important because the vagina receives the penis during intercourse and serves as the lower portion

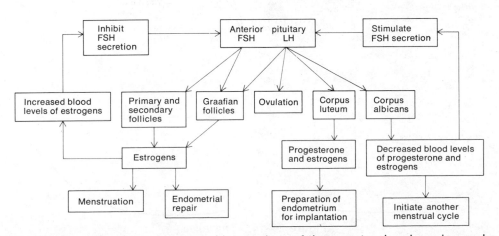

Figure 24–13. Summary of hormonal interactions of the menstrual and ovarian cycles.

of the birth canal. At the lower end of the vaginal opening (*vaginal orifice*) is a thin fold of vascularized mucous membrane called the *hymen*, which forms a border around the orifice, partially closing it (Figure 24–14). Sometimes the hymen completely covers the orifice, a condition called *imperforate hymen*, and surgery is required to open the orifice to permit the discharge of the menstrual flow. The mucosa of the vagina contains large amounts of glycogen that, upon decomposition, produce organic acids. These acids create a low pH environment in the vagina, a situation that retards microbial growth. However, the acidity is also injurious to sperm cells. For this reason, the buffering action of semen is important. Semen neutralizes the acidity of the vagina to ensure survival of the sperm.

Vulva

The term **vulva** or **pudendum** is a collective designation for the external genitalia of the female (Figure 24–14).

The *mons pubis (veneris)* is an elevation of adipose tissue covered by coarse pubic hair situated over the symphysis pubis. It lies in front of the vaginal and urethral openings. From the mons pubis, two longitudinal folds of skin, the *labia majora*, extend downward and backward. The labia majora contain an abundance of adipose tissue and sebaceous and sweat glands; they are covered by hair on their upper outer surfaces. Medial to the labia majora are two folds of skin called the *labia minora*. Unlike the labia majora, the labia minora are devoid of hair and have relatively few sweat glands. They do, however, contain numerous sebaceous glands.

The *clitoris* is a small, cylindrical mass of erectile tissue, blood vessels, and nerves. It is located just behind the junction of the labia minora. A layer of skin called the *prepuce* is formed at the point where the labia minora unite and covers the body of the clitoris. The exposed portion of the clitoris is referred to as the *glans*. The clitoris is homologous to the penis of the male in that it is capable of enlargement upon tactile stimulation and assumes a role in sexual excitement of the female.

The cleft between the labia is called the *vestibule*. Within the vestibule are the hymen, vaginal orifice, urethral orifice, and the openings of several ducts. The vaginal orifice occupies the greater portion of the vestibule and is bordered by the hymen. In front of the vaginal orifice and behind the clitoris is the *urethral orifice*. Behind and to either side of the urethral orifice are the

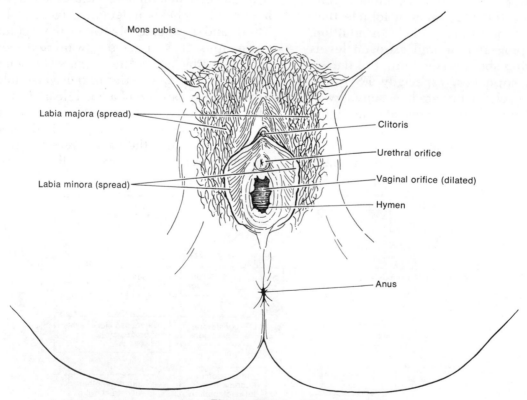

Figure 24–14. The vulva.

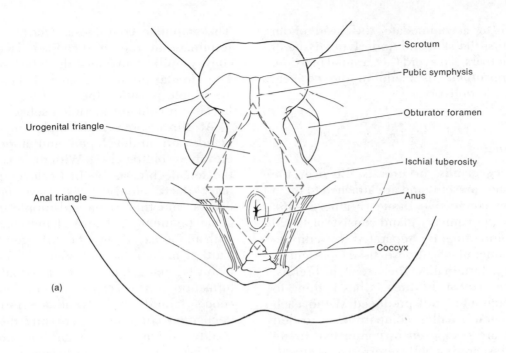

Scrotum

Pubic symphysis

Obturator foramen

Urogenital triangle

Ischial tuberosity

Anal triangle

Anus

Coccyx

(a)

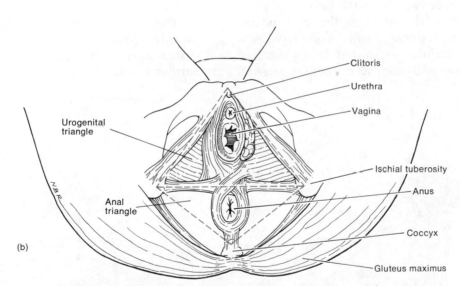

Clitoris

Urethra

Vagina

Urogenital triangle

Ischial tuberosity

Anus

Anal triangle

Coccyx

(b)

Gluteus maximus

Figure 24–15. The perineum. (a) Borders as seen in the male. (b) Dissected view of some regions of the female perineum.

openings of the ducts of the *lesser vestibular glands*. These glands secrete mucus. On either side of the vaginal orifice itself are two small glands, the *greater vestibular glands*. These glands open by a duct into the space between the hymen and labia minora and produce a mucoid secretion that serves as a lubricant during sexual intercourse. Whereas the lesser vestibular glands are homologous to the male prostate, the greater vestibular glands are homologous to the male Cowper's glands.

Perineum

The **perineum** is the diamond-shaped area at the lower end of the trunk between the thighs and buttocks of both males and females (Figure 24–15). It is surrounded anteriorly by the symphysis pubis, laterally by the ischial tuberosities, and posteriorly by the coccyx. A transverse line drawn between the ischial tuberosities divides the perineum into an anterior *urogenital triangle* that contains the external genitalia and a posterior *anal triangle* that contains the anus. If the vagina is too small to accommodate the head of an emerging fetus, the skin and underlying tissue of the perineum tears. To avoid this, a small incision, called an *episiotomy,* is made in the perineal skin just prior to delivery.

Mammary glands

The **mammary glands, or breasts,** lie over the pectoralis major muscles and are attached to them by a layer of connective tissue (Figure 24–16). Internally, each mammary gland consists of 15–20 *lobes,* or compartments, separated by adipose tissue. The amount of adipose tissue present is the principal determinant of the size of the breasts. However, the size of the breasts has nothing to do with the amount of milk produced. Within each lobe are several smaller compartments, called *lobules,* that are composed of connective tissue in which milk-secreting cells referred to as *alveoli* are embedded. Alveoli are arranged in grapelike clusters. They convey the milk into a series of *secondary tubules.* From here, the milk passes into the *mammary ducts.* As the mammary ducts approach the nipple, expanded sinuses called *ampullae,* where milk may be stored, are present.

The ampullae continue as *lactiferous ducts* that terminate in the *nipple.* Each lactiferous duct conveys milk from one of the lobes to the exterior. The circular pigmented area of skin surrounding the nipple is called the *areola.* It appears rough because it contains modified sebaceous glands.

At birth, both male and female mammary glands are undeveloped and appear as slight elevations on the chest. With the onset of puberty, the female breasts begin to develop—the mammary ducts elongate, extensive fat deposition occurs, and the areola and nipple grow and become pigmented. These changes are correlated with an increased output of estrogen by the ovary. Further mammary development occurs at sexual maturity, with the onset of ovulation and the formation of the corpus luteum. During adolescence, lobules and alveoli are formed and fat deposition continues, increasing the size of the glands. Although these changes are associated with estrogen and progesterone secretion by the ovaries, ovarian secretion is ultimately controlled by FSH.

The essential function of the mammary glands is milk secretion. Because this function is associated only with pregnancy, we shall discuss it further in Chapter 25.

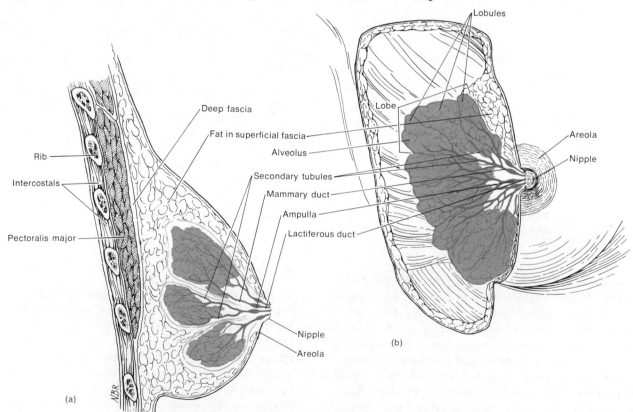

Figure 24–16. The mammary glands seen in (a) sagittal section and (b) front view partially sectioned.

DISORDERS OF THE REPRODUCTIVE SYSTEMS

In discussing the disorders that occur in the reproductive system, we shall consider venereal diseases first. Then we shall look at some common diseases involving the male and female reproductive tracts, respectively.

Venereal diseases

The term venereal comes from Venus, the goddess of love. **Venereal diseases** represent a group of infectious diseases that are spread primarily through sexual intercourse. With the exception of the common cold, venereal diseases are ranked as the number 1 communicable diseases in the United States. Gonorrhea and syphilis are the two most common venereal diseases.

Gonorrhea, more commonly known as "clap," is an infectious disease that primarily affects the mucous membrane of the urogenital tract, the rectum, and occasionally the eye. The disease is caused by the bacterium *Neisseria gonorrhoeae.* Males usually suffer inflammation of the urethra, with pus and painful urination. Fibrosis sometimes occurs in an advanced stage of gonorrhea, causing stricture of the urethra. There also may be involvement of the epididymis and prostate gland. In females, infection may occur in the urethra, vagina, and cervix, and there may be a discharge of pus. If the uterine tubes become involved, sterility and pelvic inflammation may result. Females commonly can harbor the disease and be asymptomatic.

Discharges from the involved mucous membranes are the source of infection, and the bacteria are transmitted by direct contact, usually sexual. The bacteria may be transmitted to the eyes of a newborn when the baby passes through the birth canal. Initially, the infant develops conjunctivitis. But later, other structures of the eye may become involved, and blindness may result. Administration of a 1 percent silver nitrate solution or penicillin in the eyes of the infant is very effective in preventing their infection. Penicillin is also the drug of choice for the treatment of gonorrhea in adults.

Syphilis is an infectious disease caused by the bacterium *Treponema pallidum.* It also is acquired through sexual contact. The early stages of the disease primarily affect the organs that are most likely to have made sexual contact—the genital organs, the mouth, and the rectum. The point where the bacteria enter the body is marked by a lesion called a *chancre.* In males, it usually occurs on the penis, and in females, it usually occurs in the vagina or cervix. The chancre heals without scarring. Following the initial infection, the bacteria enter the bloodstream and are spread throughout the body. In some individuals, an active secondary stage of the disease occurs, characterized by lesions of the skin and mucous membranes and often fever. The signs of the secondary stage also go away without medical treatment. During the next several years, the disease progresses without symptoms and is said to be in a latent phase. When symptoms again appear, anywhere from 5 to 40 years after the initial infection, the person is said to be in the tertiary stage of the disease. Tertiary syphilis may involve the circulatory system, skin, bones, viscera, and the nervous system. The effects of the syphilis bacterium on the nervous system have already been discussed in Chapter 13.

Male disorders

Among the common male reproductive disorders are those involving the prostate gland and sexual functions.

Prostate disorders

The prostate gland is susceptible to infection, enlargement, and benign and malignant tumors. Because the prostate surrounds the urethra, any of these disorders can cause obstruction to the flow of urine. Prolonged obstruction also may result in serious changes in the bladder, ureters, and kidneys.

Acute and chronic infections of the prostate gland are common in postpubescent males, many times in association with inflammation of the urethra. In **acute prostatitis** the prostate gland becomes swollen and very tender. Appropriate antibiotic therapy, bed rest, and above-normal fluid intake are effective in treatment.

Chronic prostatitis is one of the most common chronic infections in men of the middle and later years. On examination, the prostate gland feels enlarged, soft, and extremely tender. The surface outline is irregular and may be hard. This disease frequently produces no symptoms, but the prostate is believed to harbor infectious microorganisms responsible for some allergic conditions, arthritis, and inflammations of nerves (neuritis), muscles (myositis), and the iris (iritis).

An **enlarged prostate** gland occurs in approximately one-third of all males over 60 years of age. The enlarged gland is from two to four times larger than normal. The cause is unknown, and the enlarged condition usually can be detected by rectal examination.

Tumors of the male reproductive system primarily involve the prostate gland. Both benign and malignant growths are common in elderly men. Both types of tumors put pressure on the urethra, making urination painful and difficult. At times, the excessive back pressure destroys kidney tissue and gives rise to an increased susceptibility to infection. Therefore, even if the tumor is benign, surgery is indicated to remove the prostate or parts of it if the tumor is obstructive and perpetuates urinary tract infections.

Male sexual function abnormalities

Included here are those disorders that prevent satisfactory performance of the sex act or interfere with male fertility. **Impotence** is the inability of an adult male to attain or hold an erection long enough for normal intercourse. Impotence could be the result of physical abnormalities of the penis; systemic disorders such as syphilis; vascular disturbances; neurological disorders; or psychic factors such as fear of causing pregnancy, fear of venereal disease, religious inhibitions, and emotional immaturity. Impotence does not mean infertility.

Infertility, or **sterility,** is an inability to fertilize the ovum and does not imply impotence. Male fertility requires viable spermatozoa, adequate production of spermatozoa by the testes, unobstructed transportation of sperm through the seminal tract, and satisfactory deposition within the vagina. The tubules of the testes are sensitive to many factors—x-rays, infections, toxins, malnutrition, and others—that may cause degenerative changes and produce male sterility.

If inadequate spermatozoa production is suspected, a sperm analysis should be performed. Analysis includes measuring the volume of semen, counting the number of sperm per milliter, evaluating sperm motility at 4 hours after ejaculation, and determining the percentage of abnormal sperm forms (not to exceed 20 percent).

Female disorders

Common disorders of the female reproductive system include menstrual abnormalities, ovarian cysts, leukorrhea, infertility, breast tumors, and cervical cancer.

Abnormalities of menstruation

Disorders of the female reproductive system frequently include menstrual disorders. This is hardly surprising because proper menstruation reflects not only the health of the uterus but the health of the glands that control it, that is, the ovaries and the pituitary gland.

Amenorrhea is the absence of menstruation in a woman. If the woman has never menstruated, the condition is called *primary amenorrhea.* Primary amenorrhea can be caused by endocrine disorders, most often in the pituitary gland and hypothalamus, or by genetically caused abnormal development of the ovaries or uterus. *Secondary amenorrhea* is cessation of uterine bleeding in women who have previously menstruated. The first cause considered is pregnancy. If that is ruled out, various endocrine disturbances are considered.

Dysmenorrhea is painful menstruation caused by contractions of the uterine muscles. A primary cause is believed to be low levels of progesterone. Recall that progesterone prevents uterine contractions. It can also be caused by pelvic inflammatory disease, uterine tumors, cystic ovaries, or congenital defects.

Abnormal uterine bleeding includes menstruation of excessive duration and/or excessive amount, too-frequent menstruation, intermenstrual bleeding, and postmenopausal bleeding. These abnormalities may be caused by disordered hormonal regulation, emotional factors, and systemic diseases.

Ovarian cysts

Ovarian cysts are tumors of the ovary that contain fluid. Follicular cysts may occur in the ovaries of elderly people, in ovaries that have inflammatory diseases, and in menstruating females. They have thin walls and contain a serous albuminous material. Cysts may also arise from the corpus luteum or the endometrium. *Endometriosis* is a painful disorder characterized by endometrial tissue or cysts in abnormal locations, such as in the uterine tubes, ovaries, vagina, peritoneum, or any other place in the body outside the uterus.

Leukorrhea

Leukorrhea is a nonbloody vaginal discharge that may occur at any age and affects most women at some time. It is not a disease; it is a symptom of infection or congestion of some portion of the reproductive tract. It may be a normal discharge in some women. If it is evidence of an infection, it may be caused by a protozoan microorganism called *Trichomonas vaginalis*, a yeast, a virus, or a bacterium.

Female infertility

Female infertility, or the inability to conceive, occurs in about 10 percent of married females in the United States. Once it is established that ovulation occurs regularly, the reproductive tract is examined for functional and anatomical disorders to determine the possibility of union of the sperm and the ovum in the oviduct.

Diseases of the breasts

The breasts are highly susceptible to cysts and tumors. Men are also susceptible to breast tumor, but certain breast cancers are 100 times more common in women than in men. Usually these growths can be detected early by the woman who inspects and palpates her breasts regularly. To **palpate** means to feel or examine by touch. Unfortunately, so few women practice periodic self-examination that many growths are discovered by accident and often too late for proper treatment.

In the female, the benign *fibroadenoma* is the third most common tumor of the breast. It occurs most frequently in young people. Fibroadenomas have a firm rubbery consistency and are easily moved about within the mammary tissue. The usual treatment is excision of the growth. The breast itself is not removed.

Breast cancer has the highest fatality rate of all cancers affecting women, but it is rare in men. In the female, breast cancer is rarely seen before age 30, and its occurrence rises rapidly after menopause.

Breast cancer is generally not painful until it becomes quite advanced, so often it is not discovered early, or if it is noted, it is ignored. Any lump, be it ever so small, should be reported to a doctor at once. If there is no evidence of *metastasis* (the spread of cancer cells from one part of the body to another or from one organ to another), the treatment of choice is a *modified* or *radical mastectomy*. A radical mastectomy involves removal of the affected breast, along with the underlying pectoral muscles and the axillary lymph nodes. Metastasis of cancerous cells is usually through the lymphatics or blood. Radiation treatments may follow the surgery to ensure the destruction of any remaining stray cancer cells.

Cervical cancer

Another common disorder of the female reproductive system is cancer of the uterine cervix. It ranks third in frequency after breast and skin cancers. **Cervical cancer** starts with a change in the shape of the cervical cells called *cervical dysplasis*. Cervical dysplasis is not a cancer in itself, but the abnormal cells tend to become malignant.

Early diagnosis of cancer of the uterus is accomplished by the *Papanicolaou test,* or "Pap" smear. In this generally painless procedure, a few cells from the vaginal fornix (that part of the vagina surrounding the cervix) and the cervix are removed with a swab and examined microscopically. Malignant cells have a characteristic appearance and are indicative of an early stage of cancer, even before any symptoms occur. Estimates indicate that the "Pap" smear is more than 90 percent reliable in detecting cancer of the cervix. Treatment of cervical cancer may involve complete or partial removal of the uterus, called a *hysterectomy,* or radiation treatments.

Medical terminology associated with the reproductive systems

Castration Excision of the testes or ovaries.

Copulation Sexual intercourse. Coitus refers to sexual intercourse among human beings.

Hermaphroditism Presence of both male and female sex organs in one individual.

Hysterectomy Removal of the uterus.

Impotence Inability for sexual intercourse in the man.

Menarche Beginning of menstruation.

Menopause Cessation of menstruation.

Salpingitis Inflammation or infection of the uterine tube.

Vaginitis Inflammation of the vagina.

Chapter summary in outline

MALE REPRODUCTIVE SYSTEM

1. The scrotum provides an appropriate temperature for the testes.
2. The major functions of the testes are sperm production and the secretion of testosterone.
3. FSH and ICSH maintain the growth and development of the male reproductive organs.
4. Sperm cells are conveyed from the testes to the exterior through the seminiferous tubules, epididymis, ductus deferens, ejaculatory duct, and urethra.
5. The seminal vesicles, prostate, and Cowper's glands secrete the liquid portion of semen.
6. Semen is a mixture of sperm and secreted liquids.
7. The penis serves as the organ of copulation.

FEMALE REPRODUCTIVE SYSTEM

1. The ovaries produce ova and secrete estrogens and progesterone.
2. The uterine tubes convey ova from the ovaries to the uterus and are the sites of fertilization.
3. The normal position of the uterus is maintained by a series of ligaments.
4. The uterus is associated with menstruation, implantation of a fertilized ovum, development of the fetus, and labor.
5. The function of the menstrual cycle is to prepare the endometrium each month for the reception of an ovum.
6. The ovarian cycle produces a mature ovum each month.
7. FSH and LH control the ovarian cycle. Estrogens and progesterone control the menstrual cycle.
8. The vagina serves as a passageway for the menstrual flow, as the lower portion of the birth canal, and as the receptacle for the penis.
9. The vulva is a collective designation for the external genitalia of the female.
10. The perineum is a diamond-shaped area at the lower end of the trunk between the thighs and buttocks.
11. The mammary glands function in the secretion of milk.

DISORDERS

1. Venereal diseases are a group of infectious diseases spread primarily through sexual intercourse.
2. Conditions that affect the prostate are prostatitis, enlarged prostate, and tumors.
3. Impotence is the inability of the male to attain or hold an erection long enough for intercourse.
4. Infertility is the inability of a male's sperm to fertilize an ovum.
5. Menstrual disorders include amenorrhea, dysmenorrhea, and abnormal bleeding.
6. Ovarian cysts are tumors that contain fluid.
7. Leukorrhea is a nonbloddy vaginal discharge that may be caused by an infection.
8. The mammary glands are susceptible to benign fibroadenomas and malignant tumors. The removal of a malignant breast, pectoral muscles, and lymph nodes is called a radical mastectomy.
9. Cervical cancer can be diagnosed by a "Pap" test. Complete or partial removal of the uterus is called a hysterectomy.

Review questions and problems

1. Define reproduction. List the male and female organs of reproduction.
2. Describe the function of the scrotum in protecting the testes from temperature fluctuations. What is cryptorchidism?
3. Describe the internal structure of a testis. Where are the sperm cells made?
4. Identify the principal parts of a spermatozoan. List the function of each.
5. Explain the effects of FSH and ICSH on the male reproductive system.
6. Describe the physiological effects of testosterone. How is the testosterone level in the blood controlled?
7. Trace the course of a sperm cell through the male system of ducts from the seminiferous tubules to the urethra.
8. What is the spermatic cord?
9. Briefly explain the functions of the seminal vesicles, prostate gland, and Cowper's glands.
10. What is seminal fluid? What is its function?
11. How is the penis structurally adapted as an organ of copulation?
12. How are the ovaries held in position in the pelvic cavity? What is ovulation?
13. What is the function of the uterine tubes? Define an ectopic pregnancy.
14. Diagram the principal parts of the uterus.
15. Describe the arrangement of ligaments that hold the uterus in its normal position. Explain the two major malpositions of the uterus.
16. Discuss the blood supply to the uterus. Why is an abundant blood supply important?
17. Define menstrual cycle and ovarian cycle. What is the function of each?
18. Briefly outline the major events of the menstrual cycle and correlate them with the events of the ovarian cycle.
19. Explain, by means of a labeled diagram, the principal hormonal interactions involved in the menstrual and ovarian cycles.
20. What is the function of the vagina?
21. List the parts of the vulva.
22. What is the perineum? Define episiotomy.
23. Describe the passage of milk from the areolar cells of the mammary gland to the nipple.
24. Define venereal disease. Distinguish between gonorrhea and syphilis.
25. Describe several disorders that affect the prostate gland.
26. Distinguish between impotence and infertility.
27. What are some of the causes of amenorrhea, dysmenorrhea, and abnormal uterine bleeding?
28. What are ovarian cysts? Define endometriosis.
29. What are some possible causes of leukorrhea?
30. Distinguish between a fibroadenoma and a malignant tumor of the breast.
31. What is a radical mastectomy?
32. What is a "Pap" smear? What is a hysterectomy?
33. Refer to the glossary of medical terminology at the end of the chapter. Be sure that you can define each term.

CHAPTER 25

DEVELOPMENT AND INHERITANCE

STUDENT OBJECTIVES

After you have read this chapter, you should be able to:

1. Define meiosis

2. Contrast the events of spermatogenesis and oogenesis

3. Describe and discuss the role of the male and female in sexual intercourse

4. Describe the activities associated with fertilization and implantation

5. Discuss the formation of the primary germ layers, embryonic membranes, and placenta

6. List the body structures produced by the primary germ layers

7. Describe the function of the embryonic membranes

8. Describe the roles of the placenta and umbilicus during embryonic and fetal growth

9. Discuss the principal body changes associated with the growth of the fetus

10. Compare the sources and functions of the hormones secreted during pregnancy

11. Describe the three stages of labor

12. Describe the physiology of lactation

13. Define inheritance and describe the inheritance of PKU, sex, and color blindness

14. Describe the cause and symptoms of Down's syndrome

15. Contrast the various kinds of birth control and their relative effectiveness

16. Define medical terminology associated with development and inheritance

Now that we have studied the organs of reproduction, we shall look at developmental processes. The term *developmental processes* refers to a sequence of events starting with the fertilization of an ovum or egg and ending with the formation of a complete organism. As we look at this sequence, we shall consider how reproductive cells are produced, the role of the male and female prior to fertilization, and events associated with pregnancy, birth, and lactation. Finally, we shall say a few words about inheritance.

MEIOSIS: THE FORMATION OF GAMETES

Each human being develops from the union of an ovum and a sperm. Ova and sperm are collectively called *gametes*, and they differ radically from all the other cells in the body in that they have only half the normal number of chromosomes in their nuclei. *Chromosome number* is the number of chromosomes contained in each nucleated cell that is not a gamete. Chromosome numbers vary from species to species. The human chromosome number is 46, which means that each brain cell, stomach cell, heart cell, and every other cell contains 46 chromosomes in its nucleus. In other words, there are 23 pairs of chromosomes in each cell other than a gamete. The ovum or sperm has only one-half of each pair. Of these 46 chromosomes, 23 contain the genes that are necessary for programming all the activities of the body. In a sense, the other 23 are a duplicate set. Another word for chromosome number is *diploid number* (the prefix *di-* means two), symbolized as $2n$. That duplicate set of chromosomes has an important genetic significance, but now we shall look only at how it relates to the development of the gametes.

Suppose that a sperm containing 46 chromosomes fertilizes an egg that also contains 46 chromosomes. The offspring then would have 96 chromosomes, the grandchildren 192 chromosomes, and so on. In reality, the chromosome number does not double with each generation because of a special kind of nuclear division called **meiosis**. Meiosis occurs in sex cells before they become mature. It causes a developing sperm or ovum to relinquish its duplicate set of chromosomes so that the mature gamete has only 23. This is called the *haploid number*, meaning "one-half," and is symbolized as n.

In the testes the formation of haploid spermatozoa by meiosis is called *spermatogenesis*. The seminiferous tubules are lined with immature cells called *spermatogonia* (Figure 25–1a). Spermatogonia contain the diploid chromosome number and are the precursor cells for all the sperm that the man will produce. During childhood these cells are relatively inactive. When the male reaches puberty, the spermatogonia embark on a lifetime of active division. Some of the spermatogonia undergo developmental changes and become known as *primary spermatocytes*. Each primary spermatocyte then undergoes a cellular division before replicating its DNA. As a result, the two daughter cells, called *secondary spermatocytes*, contain only half the chromosomes that the parent cell contained. This is the unique nuclear division called meiosis. Each secondary spermatocyte then duplicates its DNA and goes through a mitotic division to produce two cells called *spermatids*. The spermatids contain the same number of chromosomes as the secondary spermatocytes—thus, they are also haploid. Without further cell division each spermatid develops a head with an acrosome and a flagellum. It is now a mature sperm or spermatozoan.

In the ovary, the formation of a haploid ovum by meiosis is referred to as *oogenesis* (Figure 25–1b). The precursor cell in this sequence is a diploid cell called the *oogonium*. Starting with puberty, one oogonium undergoes developmental changes each month and becomes a *primary oocyte*. The primary oocyte undergoes meiosis and forms two cells of unequal size, both of which are haploid. The larger cell is called a *secondary oocyte*, and the smaller is referred to as the *first polar body*. After replicating its DNA, the secondary oocyte undergoes a cell division and forms

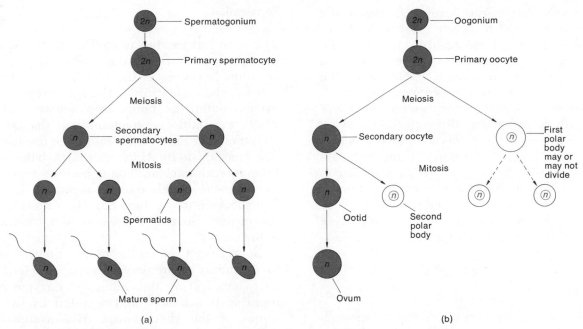

Figure 25–1. Meiosis. (a) Production of mature sperm by spermatogenesis in the testes. (b) Production of a mature ovum by oogenesis in the ovaries.

two more cells of unequal size. The larger haploid cell is an *ootid*, and the smaller is the *second polar body*. In time, the ootid develops into a mature ovum and the polar bodies disintegrate. Thus, in the female one oogonium produces a single ovum, whereas in the male each spermatogonium produces four sperm.

SEXUAL INTERCOURSE

Fertilization of an ovum is accomplished through *sexual intercourse*, in which spermatozoa are deposited in the vagina. The male role in the sexual act starts with erection, the enlargement and stiffening of the penis. An erection may be initiated in the cerebrum by stimuli such as anticipation, memory, and visual sensation, or it may be a reflex brought on by stimulation of the touch receptors in the penis, especially in the glans. In either case, parasympathetic impulses that pass from the sacral portion of the spinal cord to the penis cause dilation of the arteries of the penis, allowing blood to fill the cavernous spaces. These impulses also cause the Cowper's glands to secrete mucus that affords lubrication for intercourse. However, the major portion of lubricating fluid is produced by the female.

Tactile stimulation of the penis brings about emission and ejaculation. When sexual stimulation becomes extremely intense, rhythmic sympathetic impulses leave the spinal cord at the levels of the first and second lumbar vertebrae and pass to the genital organs. These impulses cause peristaltic contractions of the ducts in the testes, the epididymis, and the vas deferens that propel spermatozoa into the urethra—a process called *emission*. Simultaneously, peristaltic contractions of the seminal vesicles and prostate expel semen and prostatic fluid along with the spermatozoa. All these mix with the mucus of the Cowper's glands, resulting in the fluid called semen. Other rhythmic impulses sent from the spinal cord at the levels of the first and second sacral vertebrae reach the skeletal muscles at the base of the penis, and the penis expels the semen from the urethra to the exterior. The propulsion of semen from the urethra to the exterior constitutes an *ejaculation*. A number of sensory and motor activities accompany ejaculation, including a rapid heart rate, an increase in blood pressure, an increase in respiration, and pleasurable sensations. These activities, together with the muscular events involved in ejaculation, are referred to as an *orgasm*.

The female role in the sex act also involves erection, lubrication, and orgasm. Stimulation of the female, as in the male, depends on both psychic and tactile responses. Under appropriate conditions, stimulation of the female genitalia, especially the clitoris, results in *erection* and widespread sexual arousal. This response is controlled by parasympathetic impulses sent from the spinal cord to the external genitalia. These impulses also pass to the Bartholin's glands and vaginal mucosa, which secrete most of the *lubrication* during sexual intercourse. When tactile stimulation of the genitalia reaches maximum intensity, reflexes are initiated that cause the female *orgasm* or *climax*. Female orgasm is analogous to male ejaculation, except that there is no expulsion of semen, although there may be an increased secretion of cervical mucus. As part of the female orgasm, the perineal muscles contract rhythmically from spinal reflexes similar to those that occur in the male ejaculation. There is speculation that these same impulses also cause peristaltic movements of the uterus and uterine tubes, thus helping to transport the spermatozoa toward the ovum.

PREGNANCY

Once spermatozoa and ova are developed through meiosis and the spermatozoa are deposited in the vagina, pregnancy can occur. **Pregnancy** is a sequence of events including fertilization, implantation, embryonic growth, and, normally, fetal growth that terminates in birth.

Fertilization

The term **fertilization** is applied to the union of the sperm nucleus and the nucleus of the ovum. It normally occurs in the uterine tube when the ovum is about one-third of the way down the tube, usually within 24 hours after ovulation (Figure 25–2). Peristaltic contractions and the action of cilia transport the ovum through the uterine tube. The mechanism by which sperm reach the uterine tube is still unclear. Some believe that sperm swim up the female tract by means of the whiplike movements of their flagella; others believe sperm are transported by muscular contractions of the uterus.

Sperm must remain in the female genital tract for 4–6 hours before they are capable of fertilizing an ovum. During this time, the enzyme hyaluronidase is activated and secreted by the acrosomes of the spermatozoa. Hyaluronidase apparently dissolves parts of the membrane covering the ovum (Figure 25–2a). Normally, only one spermatozoan fertilizes an ovum because once penetration is achieved, the ovum develops a fertilization membrane that is impermeable to the

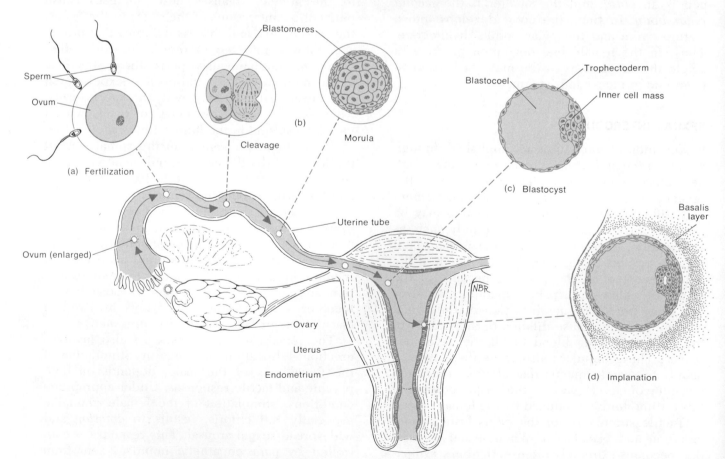

Figure 25–2. Fertilization, cleavage, and implantation of an ovum.

entrance of other spermatozoa. When the spermatozoan has entered the ovum, the tail is shed and the nucleus in the head develops into a structure called the *male pronucleus*. The nucleus of the ovum also develops into a *female pronucleus*. After the pronuclei are formed, they fuse to produce a *segmentation nucleus*—a process termed fertilization. The segmentation nucleus contains 23 chromosomes from the male pronucleus and 23 chromosomes from the female pronucleus. Thus, the fusion of the haploid pronuclei restores the diploid number. The fertilized ovum, consisting of a segmentation nucleus, cytoplasm, and enveloping membrane, is referred to as a *zygote*.

Immediately after fertilization, rapid cell division of the zygote takes place (Figure 25–2b). This early division of the zygote is called *cleavage*. The progressively smaller cells produced are called *blastomeres*. Successive cleavages produce a solid mass of cells, the *morula*, which is only slightly larger than the original zygote.

Implantation

As the morula descends through the uterine tube, it continues to divide and eventually forms a hollow ball of cells. At this stage of development the mass is referred to as a *blastocyst* (Figure 25–2c). The blastocyst is differentiated into an outer covering of cells called the *trophectoderm* and an *inner cell mass*, and the internal cavity is referred to as the *blastocoel*. Whereas the trophectoderm ultimately will form the membranes composing the fetal portion of the placenta, the inner cell mass will develop eventually into the embryo. About the fifth day after fertilization, the blastocyst enters the uterine cavity.

The attachment of the blastocyst to the endometrium occurs 7 to 8 days following fertilization and is called **implantation** (Figure 25–2d). At this time, the endometrium is in its postovulatory phase. During implantation, the cells of the trophectoderm secrete an enzyme that enables the blastocyst to literally "eat a hole" in the uterine lining and become buried in the endometrium, usually on the posterior wall of the fundus of the uterus. The portion of the endometrium to which the blastocyst adheres and in which it becomes implanted is the basalis layer. Implantation enables the blastocyst to absorb nutrients from the glands and blood vessels of the endometrium for its subsequent growth and development.

Embryonic period

The first 2 months of development are considered the **embryonic period**. During this period the developing human is called an *embryo*. After the second month it will be called a *fetus*. By the end of the embryonic period the rudiments of all the principal adult organs are present, the embryonic membranes are developed, and the placenta is functioning. Let us now examine these events in more detail.

Beginnings of organ systems

Following implantation, the inner cell mass of the blastocyst begins to differentiate into the three primary germ layers: the ectoderm, endoderm, and mesoderm. The **primary germ layers** are the embryonic tissues from which all tissues and organs of the body will develop. The fetal membranes, structures that lie outside the embryo and protect and nourish it, also develop from these three germ layers.

In the human being, the formation of the germ layers happens so quickly that it is difficult to determine the exact sequence of events. Before implantation, a layer of *ectoderm* (the trophectoderm) already has formed around the blastocoel (Figure 25–3a). The trophectoderm will become part of the chorion—one of the fetal membranes. Within 8 days after implantation, the inner cell mass moves downward so a space called the amnionic cavity lies between the inner cell mass and the trophectoderm (Figure 25–3b). The bottom layer of the inner cell mass develops into an *endodermal* germ layer.

About the twelfth day after fertilization the striking changes shown in Figure 25–3c appear. A layer of cells from the inner cell mass has grown around the top of the amnionic cavity. These cells will become the amnion, another fetal membrane. The cells below the cavity are called the *embryonic disc;* these cells will form the embryo. The embryonic disc contains scattered ectodermal, mesodermal, and endodermal cells in addition to the endodermal layer observed in Figure 25–3b. Notice in Figure 25–3c that the cells of the endodermal layer have been dividing rapidly, so groups of them now extend downward in a circle. This circle is the yolk sac, another fetal membrane. The *mesodermal* cells also have been dividing, and many have left the area of the embryonic disc and can be seen around the structures that are becoming fetal membranes.

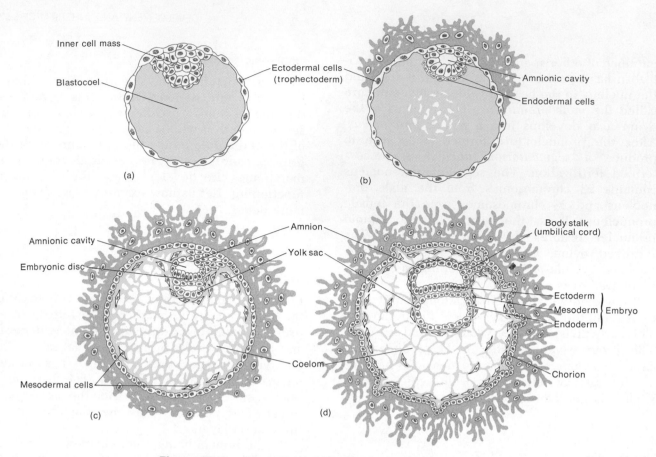

Figure 25–3. Formation of the three primary germ layers.

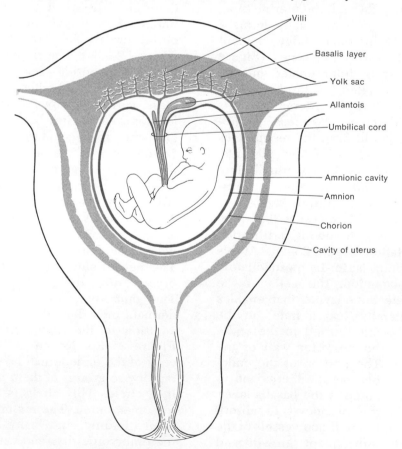

Figure 25–4. Embryonic membranes.

About the fourteenth day, the scattered cells in the embryonic disc separate into three distinct layers: the upper ectoderm, the middle mesoderm, and the lower endoderm (Figure 25–3d). At this time the two ends of the embryonic disc draw together, squeezing off the yolk sac. The resulting cavity inside the disc is the endoderm-lined *primitive gut*. The mesoderm within the disc soon splits into two layers, and the space between the layers becomes the coelom, or body cavity.

As the embryo develops, the endoderm becomes the epithelium lining the digestive tract and a number of other organs. The mesoderm forms the peritoneum, muscle, bone, and other connective tissue, and the ectoderm develops into the skin and nervous system. Exhibit 25–1 provides more details about the fates of these primary germ layers.

Embryonic membranes

During the embryonic period, the *embryonic membranes* form. These membranes lie outside the embryo and will protect and nourish the fetus. The membranes are the yolk sac, the amnion, the chorion, and the allantois (Figure 25–4).

The human *yolk sac* is an endoderm-lined membrane that encloses the yolk. In many species the yolk provides the primary or exclusive nutrient for the embryo, and consequently, the ova of these animals contain a great deal of yolk. However, the human embryo receives its nourishment from the endometrium. The human yolk sac is small, and during an early stage of development it becomes a nonfunctional part of the umbilical cord.

The *amnion* is a thin, protective membrane that initially overlies the embryonic disc. As the embryo grows, the amnion entirely surrounds the embryo and becomes filled with a fluid called *amniotic fluid*. Amniotic fluid serves as a shock absorber for the fetus. The amnion usually ruptures just before birth and it and its fluid constitute the so-called "bag of waters."

The *chorion* derives from the trophectoderm of the blastocyst and its associated mesoderm. It surrounds the embryo and, later, the fetus. Eventually the chorion becomes the principal part of the placenta, the structure through which materials are exchanged between the mother and fetus. The amnion also surrounds the fetus and eventually fuses to the inner layer of the chorion.

The *allantois* is a small vascularized membrane. Later its blood vessels serve as connections in the placenta between the mother and fetus.

The umbilicus and the placenta

Development of a functioning placenta is the third major event of the embryonic period. This is accomplished by the third month of pregnancy. The **placenta** is formed by the chorion of the embryo and the basalis layer of the endometrium of the mother (Figure 25–5). It provides an exchange of nutrients and wastes between the fetus and mother and secretes the hormones necessary to maintain pregnancy.

During embryonic life, fingerlike projections of the chorion, called *chorionic villi*, grow into the basalis layer of the endometrium. These villi contain fetal blood vessels, and they con-

Exhibit 25–1. STRUCTURES PRODUCED BY THE THREE PRIMARY GERM LAYERS

ENDODERM	MESODERM	ECTODERM
Epithelium of digestive tract and its glands	Skeletal, smooth, and cardiac muscle	Epidermis of skin
Epithelium of urinary bladder and gall bladder	Cartilage, bone, and other connective tissues	Hair, nails, and skin glands
Epithelium of pharynx, auditory tube, tonsils, larynx, trachea, bronchi, and lungs	Blood, bone marrow, and lymphoid tissue	Lens of eye
	Epithelium of blood vessels and lymphatics	Receptor cells of sense organs
Epithelium of thyroid, parathyroid, and thymus glands	Epithelium of coelomic and joint cavities	Epithelium of mouth, nostrils, sinuses, oral glands, and anal canal
Epithelium of vagina, vestibule, urethra, and associated glands	Epithelium of kidneys and ureters	Enamel of teeth
Adenohypophysis	Epithelium of gonads and associated ducts	Entire nervous tissue, except adenohypophysis
	Epithelium of adrenal cortex	

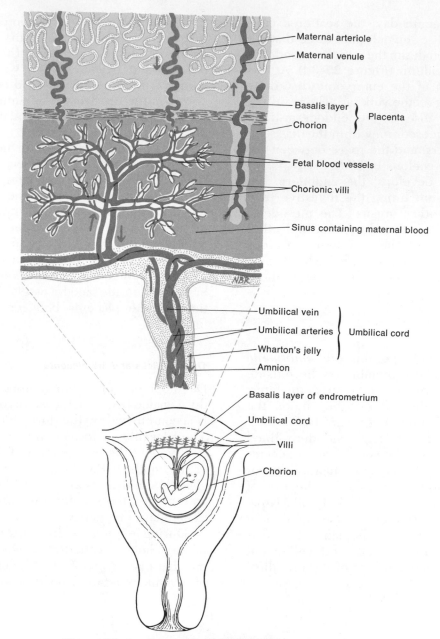

Maternal arteriole
Maternal venule

Basalis layer } Placenta
Chorion

Fetal blood vessels

Chorionic villi

Sinus containing maternal blood

Umbilical vein
Umbilical arteries } Umbilical cord
Wharton's jelly
Amnion

Basalis layer of endrometrium

Umbilical cord

Villi

Chorion

Figure 25–5. Structure of the placenta and umbilicus.

tinue growing until they are bathed in the maternal blood in the sinuses of the basalis. Thus, maternal and fetal blood vessels are brought into close proximity. It should be noted, however, that maternal and fetal blood do not mix. Oxygen and nutrients from the mother's blood diffuse across the walls and into the capillaries of the villi. From the capillaries the nutrients circulate into the umbilical vein. Wastes leave the fetus through the umbilical arteries, pass into the capillaries of the villi, and diffuse into the maternal blood. The **umbilical cord** consists of an outer layer of amnion containing the umbilical arteries and umbilical vein, supported internally by mucous connective tissue called Wharton's jelly. At delivery, the placenta becomes detached from the uterus and is referred to as the "afterbirth." At this time, the umbilicus is severed, leaving the baby on its own.

Fetal growth

During the **fetal period,** rapid growth of organs established by the primary germ layers occurs. No new structures are formed, but there is accelerated growth and maturation of structures that were laid down during the embryonic period. At the beginning of the fetal period, the organism takes on a human appearance. A summary of changes associated with the fetal period is presented in Exhibit 25-2.

Hormones of pregnancy

Following fertilization, the corpus luteum is maintained until the fourth month of pregnancy. For most of this time it continues to secrete estrogens and progesterone. Both these hormones maintain the lining of the uterus during pregnancy and prepare the mammary glands to secrete milk. The amounts of estrogens and progesterone secreted by the corpus luteum, however, are only slightly higher than those produced after ovulation in a normal menstrual cycle. As you will see shortly, the high levels of estrogens and progesterone needed to maintain pregnancy and initiate lactation are provided by the placenta.

During pregnancy, the chorion of the placenta secretes a hormone called the *chorionic gonadotrophic hormone*, or *CG*. This hormone is excreted in the urine of pregnant women from about the middle of the first month of pregnancy, reaching its peak of excretion during the third month. The CG level decreases sharply during the fourth and fifth months and then gradually levels off until

birth. Excretion of CG in the urine serves as the basis for some pregnancy tests. The primary role of CG seems to be to maintain the activity of the corpus luteum, especially with regard to continuous progesterone secretion—an activity necessary for the continued attachment of the fetus to the lining of the uterus (Figure 25-6).

As noted earlier, the placenta provides the high levels of estrogens and progesterone needed for

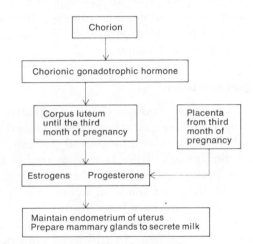

Figure 25-6. Hormones of pregnancy.

Exhibit 25-2. SUMMARY OF REPRESENTATIVE CHANGES ASSOCIATED WITH FETAL GROWTH

END OF MONTH	APPROXIMATE SIZE AND WEIGHT	REPRESENTATIVE CHANGES
1	$^3/_{16}$ inch	Eyes, nose, and ears not yet visible. Backbone and vertebral canal form. Small buds that will develop into arms and legs form. Heart forms and starts beating. Body systems begin to form.
2	$1^1/_4$ inches $^1/_{30}$ ounce	Eyes far apart, eyelids fused, nose flat. Ossification begins. Limbs become distinct as arms and legs and digits are well formed. Major blood vessels form. Many internal organs continue to develop.
3	3 inches 1 ounce	Eyes almost fully developed but eyelids still fused, nose develops a bridge, and external ears are present. Ossification continues. Appendages are fully formed and nails develop. Heartbeat can be detected. Body systems continue to develop.
4	$6^1/_2$ to 7 inches 4 ounces	Head large in proportion to rest of body. Face takes on human features and hair appears on head. Skin bright pink. Many bones ossified, and joints begin to form. Continued development of body systems.
5	10 to 12 inches $^1/_2$ to 1 pound	Head less disproportionate to rest of body. Fine hair (laguno hair) covers body. Skin still bright pink. Very rapid development of body systems.
6	11 to 14 inches $1^1/_4$ to $1^1/_2$ pounds	Head becomes less disproportionate to rest of body. Eyelids separate and eyelashes form. Skin wrinkled and pink.
7	13 to 17 inches $2^1/_2$ to 3 pounds	Head and body become more proportionate. Skin wrinkled and pink. Seven-month fetus (premature baby) is capable of survival.
8	$16^1/_2$ to 18 inches $4^1/_2$ to 5 pounds	Subcutaneous fat deposited. Skin less wrinkled. Testes descend into scrotum. Bones of head are soft. Chances of survival are much greater at end of eighth month.
9	20 inches 7 to $7^1/_2$ pounds	Additional subcutaneous fat accumulates. Laguno hair shed. Nails extend to tips of fingers and maybe even beyond.

the maintenance of pregnancy. The placenta begins to secrete these hormones no later than the sixtieth day of pregnancy. They are secreted in increasing quantities until they reach their maximum levels at the time of birth. Once the placenta is established, the secretion of CG is cut back drastically during the fourth and fifth months. This decrease results in disintegration of the corpus luteum, which is no longer needed because the placenta supplies the levels of estrogens and progesterone needed to maintain the pregnancy. Following delivery, levels of estrogens and progesterone in the blood decrease to their nonpregnant values.

Parturition and labor

The time that the embryo or fetus is carried in the uterus is called *gestation*. It is assumed that the total human gestation period is 280 days from the beginning of the last menstrual period. The term **parturition** refers to birth. Parturition is preceded by a sequence of events commonly called **labor**. The onset of labor stems from a complex interaction of many factors, especially hormones. Because we already have considered hormonal changes during pregnancy, it will be necessary to review only a few significant changes at this point. Just prior to birth, the muscles of the uterus contract rhythmically and forcefully. Both placental and ovarian hormones play a dominant role in these contractions. You may recall from the last chapter that estrogen stimulates uterine contractions, whereas progesterone inhibits them. Until the effects of progesterone are effectively diminished, labor cannot take place. At the end of gestation, however, there is barely sufficient estrogen in the mother's blood to overcome the inhibiting effects of progesterone, and labor commences. Coupled with this, oxytocin from the posterior pituitary gland stimulates uterine contractions. This mechanism may be reviewed by referring to Figure 15–5.

Uterine contractions occur in waves, quite similar to peristaltic waves, that start at the top of the uterus and move downward. These waves serve to expel the fetus. *True labor* begins when pains occur at regular intervals. The pains correspond to uterine contractions, and they first are felt every 30 minutes and last for about 1 minute. Then the intervals shorten to about every minute or two, and the contraction intensifies. Another sign of true labor is localization of pain in the back, which is intensified by walking. The final indication of true labor is the "show" and dilata-

tion of the cervix. The "show" is a discharge of a blood-containing mucus that accumulates in the cervical canal during pregnancy. Cervical dilatation will be discussed shortly. In *false labor*, by contrast, pain is felt in the abdomen at long, irregular intervals. The pain does not intensify and is not altered significantly by walking. Also, in false labor there is no "show," and cervical dilatation does not occur.

The first stage of labor, called the *stage of dilatation*, is the period of time from the onset of labor to the complete dilatation of the cervix (Figure 25–7). During this stage there are regular contractions of the uterus, a rupturing of the amniotic sac, and complete dilatation (10 centimeters) of the cervix. The next stage of labor, the *stage of expulsion*, is the period of time from complete cervical dilatation to delivery. In the final stage, the *placental stage*, the placenta or "afterbirth" is expelled. A few minutes after delivery, powerful uterine contractions expel the placenta. These contractions also constrict blood vessels that were torn during delivery. In this way, the possibility of hemorrhage is reduced.

Lactation

The term **lactation** refers to the secretion of milk by the mammary glands. During pregnancy, the mammary glands are prepared for lactation by estrogens and progesterone. When the levels of estrogens and progesterone in the mother's blood decrease at birth, the hormone prolactin stimulates the secretory cells of the mammary glands to produce milk (see Figure 15–5). Lactation starts 3 to 4 days after delivery. Once initiated by prolactin, it is stimulated and maintained by the suckling action of the infant. Suckling initiates impulses to the posterior pituitary via the hypothalamus. In the posterior pituitary, the impulses stimulate the release of oxytocin, causing the ejection of milk. Prolactin formation is inhibited by progesterone. Lactation usually prevents the occurrence of the female sexual cycles for the first few months following delivery.

INHERITANCE

Inheritance, as you probably know, is the passage of hereditary traits from one generation to another. It is the process by which you acquired your characteristics from your parents and will transmit your characteristics to your children. The branch of biology that deals with inheritance is called *genetics*.

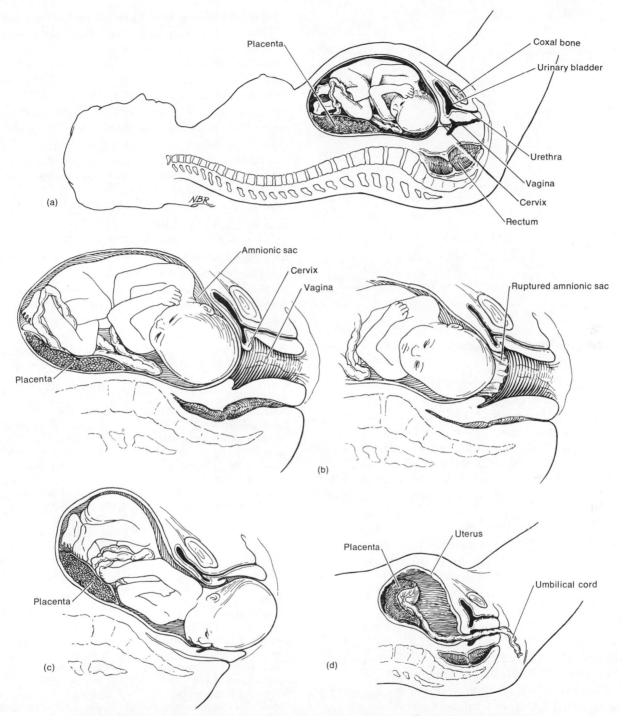

Figure 25–7. Parturition. (a) Fetal position prior to birth. (b) Dilatation. Protrusion of amnionic sac through partly dilatated cervix. Amnionic sac ruptured and complete dilatation of cervix. (c) Stage of expulsion. (d) Placental stage.

Genotype and phenotype

The nuclei of all human cells except gametes contain 23 pairs of chromosomes, that is, the diploid number. One chromosome from each pair comes from the mother, and the other comes from the father. The two chromosomes that belong to a pair are called *homologous chromosomes.* Homologs contain genes that control the same traits. For instance, if a chromosome contains a gene for height, its homolog, or mate, will also contain a gene for height.

To explain the relationship of genes to heredity, we shall look at the disorder called PKU

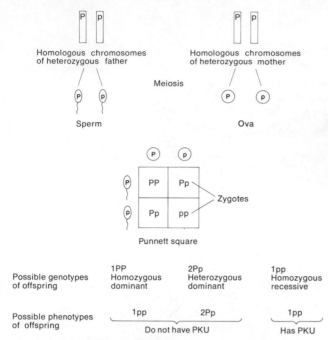

Possible genotypes of offspring

1PP Homozygous dominant 2Pp Heterozygous dominant 1pp Homozygous recessive

Possible phenotypes of offspring

1pp 2Pp 1pp

Do not have PKU Has PKU

Figure 25–8. The inheritance of PKU.

(Figure 25–8). People with PKU are unable to manufacture the enzyme phenylalanine hydroxylase. It is believed that PKU is brought about by an abnormal gene, which can be symbolized as p. The normal gene will be symbolized as P. The chromosome that is concerned with directions for phenylalanine hydroxylase production will have either p or P on it. It's homolog will also have p or P. Thus, every individual will have one of the following genetic makeups, or *genotypes:* PP, Pp, or pp. Although people with genotypes of Pp have the abnormal gene, only those with genotype pp suffer from the disorder. The reason is that the normal gene dominates over and inhibits the abnormal one. A gene that dominates is called the *dominant gene*, and the trait expressed is said to be a dominant trait. The gene that is inhibited is called the *recessive gene*, and the trait expressed is called a recessive trait.

By tradition, we symbolize the dominant gene with a capital letter and the recessive one with a lowercase letter. If an individual has the same genes on homologous chromosomes (for example, PP or pp), he is said to be *homozygous* for the trait. If however, the genes on homologous chromosomes are different (for example, Pp), he is said to be *heterozygous* for the trait. The word *phenotype* refers to how the genetic makeup is expressed in the body. A person with Pp has a different genotype than one with PP, but both have the same phenotype, which in this case is normal production of phenylalanine hydroxylase.

To determine how gametes containing haploid chromosomes unite to form diploid fertilized eggs, special charts called *Punnett squares* are used. Usually, the male gametes (sperm cells) are placed to the side of the chart, and the female gametes (ova) are placed to the top of the chart (Figure 25–8). The four spaces on the chart represent the possible combinations of male and female gametes that could form fertilized eggs. Possible combinations are determined simply by "dropping" the female gamete on the left into the two boxes below it and by "dropping" the female gamete on the right into the two spaces under it. The upper male gamete is then moved across to the two spaces in line with it, and the lower male gamete is moved across to the two spaces in line with it.

Several dominant and recessive traits that are inherited in human beings are listed in Exhibit 25–3.

Exhibit 25–3. HEREDITARY TRAITS IN HUMAN BEINGS

DOMINANT	RECESSIVE
Curly hair	Straight hair
Dark hair	Light hair
Nonred hair	Red hair
Coarse body hair	Normal body hair
Normal skin pigmentation	Albinism
Brown eyes	Blue or gray eyes
Near or farsightedness	Normal vision
Normal hearing	Deafness
Normal color vision	Color blindness
Normal blood clotting	Hemophilia
Broad lips	Thin lips
Large eyes	Small eyes
Short stature	Tall stature
Polydactylism (extra digits)	Normal digits
Brachydactylism (short digits)	Normal digits
Syndactylism (webbed digits)	Normal digits
Normal muscle tone	Muscular dystrophy
Hypertension	Normal blood pressure
Diabetes insipidus	Normal excretion
Huntington's chorea	Normal nervous system
Normal mentality	Schizophrenia
Nervous temperament	Calm temperament
Average intellect	Genius or idiocy
Migraine headaches	Normal
Normal resistance to disease	Susceptibility to disease
Enlarged spleen	Normal spleen
Enlarged colon	Normal colon
A or B blood factor	O blood factor
Rh blood factor	No Rh blood factor

Normal traits do not always dominate over abnormal ones, but genes for severe disorders are more frequently recessive than they are dominant. People who have severe disorders very often do not live long enough to pass the abnormal gene on to the next generation. In this way, expression of the gene tends to be weeded out of the population. Huntington's chorea is one example of a major disorder caused by a dominant gene. This disorder is characterized by degeneration of nervous tissue, usually leading to mental disturbance and death. It is interesting to note that the first signs of Huntington's chorea do not occur until adulthood, very often after the person has already produced offspring.

Inheritance of sex

Microscopic examination of the chromosomes in cells reveals that one pair differs in males and in females (Figure 25–9a). In females, the pair consists of two rod-shaped chromosomes designated as X chromosomes. One X chromosome also is present in males, but its mate is a hook-shaped structure called a Y chromosome. The XX pair in the female and the XY pair in the male are called the *sex chromosomes*. All other chromosomes are called *autosomes*.

As you have probably guessed, the sex chromosomes are responsible for the sex of the individual (Figure 25–9b). When a spermatocyte undergoes

meiosis to reduce its chromosome number, one of the daughter cells will contain the X chromosome and the other will contain the Y chromosome. Oocytes have no Y chromosomes and produce only X-containing ova. If the ovum is subsequently fertilized by an X-bearing sperm, the offspring will be female (XX). Fertilization by a Y sperm produces a male (XY).

Color blindness and sex-linked inheritance

The sex chromosomes also are responsible for the transmission of a number of nonsexual traits. Genes for these traits appear on X chromosomes, but many of these genes are absent from Y chromosomes. This produces a pattern of heredity that is different from the pattern we described earlier. Let us consider color blindness as an example. The gene for *color blindness* is a recessive one, and we shall designate it as c. Normal vision, designated as C, dominates. The C/c genes are located on the X chromosome. The Y chromosome, however, does not contain the segment of DNA that programs this aspect of vision. Thus the ability to see colors depends entirely on the

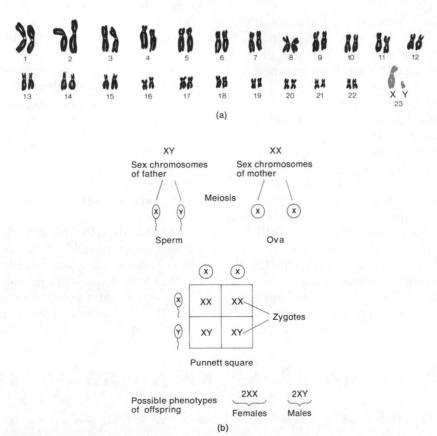

Figure 25–9. Inheritance of sex. (a) Human male chromosomes. Sex chromosomes are indicated in color. (b) Sex determination.

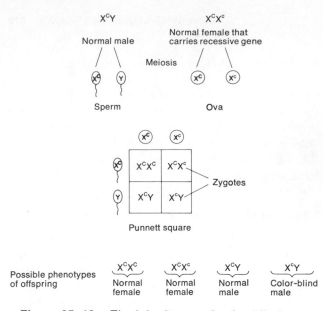

Figure 25-10. The inheritance of color blindness.

X chromosomes. The genetic possibilities are

$X^C X^C$	Normal female
$X^C X^c$	Normal female that carries the recessive gene
$X^c X^c$	Color-blind female
$X^C Y$	Normal male
$X^c Y$	Color-blind male

As you can see, only females who have two X^c chromosomes are color blind. In $X^C X^c$ females the trait is inhibited by the normal, dominant gene. Males, on the other hand, do not have a second X chromosome that would inhibit the trait. Therefore, all males with an X^c chromosome will be color blind. The inheritance of color blindness is illustrated in Figure 25-10.

Traits that are inherited in the manner we have just described are called *sex-linked traits.* Another example of a sex-linked trait is *hemophilia*, a condition in which the blood fails to clot or clots very slowly after a surface or internal injury. It is a much more serious defect than color blindness because people with severe hemophilia can bleed to death from even a small cut. Like the trait for color blindness, hemophilia is caused by a recessive gene. If H represents normal clotting, and h represents abnormal clotting, then

$X^h X^h$ females will have the disorder. Males with $X^H Y$ will be normal, and males with $X^h Y$ will be hemophiliac. Actually, clotting time varies somewhat among hemophiliacs, so the condition may be affected by other genes as well.

A few other sex-linked traits in human beings are nonfunctional sweat glands, certain forms of diabetes, some types of deafness, uncontrollable rolling of the eyeballs, absence of central incisors night blindness, one form of cataract, white forelocks, juvenile glaucoma, and juvenile muscular dystrophy.

Chromosome disorders

The amniotic fluid that bathes the fetus has many living cells floating in it. These cells are derived mostly from the skin and respiratory tract of the fetus. *Amniocentesis* is a technique of withdrawing some of this amniotic fluid by hypodermic needle puncture of the uterus, usually 16-20 weeks after conception. A small amount of amniotic fluid is removed, and its cells are examined for biochemical defects and/or for abnormalities in chromosome number or structure. All disorders of chromosome number usually can be detected by this procedure early in pregnancy, and approximately 40 different diseases involving enzyme deficiencies usually can be detected as well.

One example of a chromosome disorder that may be diagnosed through amniocentesis is *Down's syndrome*, or mongolism (Figure 25-11). This disorder is characterized by mental retardation; retarded physical development (short stature and stubby fingers); distinctive facial structures (large tongue, broad skull, slanting eyes, and round head); and malformation of the heart, ears, and feet. Sexual maturity is rarely attained.

Individuals with the disorder usually have 47 chromosomes instead of the normal 46, and the extra chromosome is responsible for the syndrome. All the chromosomes of a person with Down's syndrome are in pairs except for the twenty-first pair. Here the chromosomes are present in triplicate.

Figure 25-11. Down's syndrome. Appearance of chromosomes of an individual with Down's syndrome.

BIRTH CONTROL

Methods of **birth control** include removal of the gonads and uterus, sterilization, contraception, and abstinence. *Castration,* or removal of the testes in the male, *hysterectomy,* or removal of the uterus, and *oophorectomy,* or removal of the ovaries in the female are all absolute preventive methods. Once performed, these operations cannot be reversed, and it is impossible for the individuals to produce offspring. However, removal of the testes or ovaries has adverse effects on individuals because of the importance of these organs in the endocrine system, and these operations generally are performed only if the organs are diseased. Castration before

puberty prevents the development of secondary sex characteristics.

One means of *sterilization* of males is called *vasectomy,* a simple operation in which a portion of each ductus deferens is removed. In this procedure an incision is made in the scrotum, the tubes are located, and each is tied in two places. Then the portion between the ties is cut out. Sperm production can continue in the testes, but the sperm cannot reach the exterior.

Sterilization in females generally is achieved by performing a similar operation on the uterine tubes. The tubes are squeezed, and a small loop called a knuckle is made. A suture is tied very

Exhibit 25–4. SUMMARY OF CONTRACEPTIVE METHODS

METHOD	COMMENTS
Removal of gonads and uterus	Results in irreversible sterility. Generally performed if organs are diseased rather than as contraceptive methods because of importance of hormones produced by gonads.
Sterilization	Procedure involving severing ductus deferens in males and uterine tubes in females.
Natural contraception	Abstinence from intercourse during time of month woman is fertile. Under ideal circumstances, effectiveness in some women with regular menstrual cycles may approach that of mechanical and chemical contraceptives. Extremely difficult to determine fertile period. Effectiveness can be greatly increased by measuring and recording body temperature each morning before getting up. Small rise in temperature will show that ovulation has occurred 1 or 2 days earlier and ovum can no longer be fertilized.
Mechanical contraception Condom	Thin, strong sheath or cover, made of rubber or similar material, worn by male to prevent sperm from entering vagina. Failures caused by sheath tearing or slipping off after climax. Rates in effectiveness with diaphragm if used correctly and consistently.
Diaphragm	Flexible hemispherical rubber dome used in combination with spermacidal cream or jelly. Inserted into vagina to cover cervix, providing barrier to sperm. Must be left in place at least 6 hours after intercourse, and may be left in place as long as 24 hours. Must be fitted by physician and refitted every 2 years and after each pregnancy. Offers high level of protection; rate of two to three pregnancies per 100 women per year is estimated for consistent users. If motivation is weak, much higher pregnancy rates must be expected. Occasional failures caused by improper insertion or displacement during sexual intercourse.
Intrauterine devices	Small objects (loops, spirals, rings) made of plastic or stainless steel and inserted into uterus by physician. May be left in place indefinitely. Do not require continued attention by user. Some women cannot use them satisfactorily because of expulsion, bleeding, or discomfort. Not recommended for women who have not had child because uterus is too small and cervical canal too narrow. Infrequently, insertion may lead to mild or severe inflammation of pelvic organs.
Chemical contraception Foams, creams, jellies, suppositories, vaginal douches	Sperm-killing chemicals inserted into vagina to coat vaginal surfaces and cervical opening. Provide protection for about 1 hour. Significantly effective when used alone, but believed to be more effective when used in combination with diaphragm or condom.
Oral contraceptives	Except for total abstinence or surgical sterilization, most effective contraceptive known to man. Side effects include nausea, occasional light bleeding between periods, breast tenderness or enlargement, fluid retention, and weight gain. Should not be used by women who might be thrombosis-prone.

tightly at the base of the knuckle and the knuckle is then cut. After 4 or 5 days the suture is digested by body fluids and the two severed ends of the tubes separate. The ovum thus is prevented from passing to the uterus, and the sperm cannot reach the ovum. Sterilization normally does not affect sexual performance or enjoyment.

Contraceptives include all methods of preventing fertilization without destroying fertility— by "natural," mechanical, and chemical means. The natural methods are complete or periodic abstinence. An example of periodic abstinence is the rhythm method, which takes advantage of the fact that a fertilizable ova is available only during a period of 3–5 days in each menstrual cycle. During this time the couple refrains from intercourse. One of the difficulties in this method is that few women have absolutely regular cycles. Another problem with the rhythm method is that some women occasionally ovulate during the "safe" times of the month, such as during menstruation.

Mechanical means of contraception include the condom used by the male and the diaphragm by the female. The *condom* is a nonporous, elastic covering placed over the penis that prevents deposition of sperm in the female reproductive tract. The *diaphragm* is a dome-shaped structure that fits over the cervix and is generally used in conjunction with some kind of sperm-killing chemical. The diaphragm stops the sperm from passing into the cervix, and the chemical kills the sperm cells.

Another mechanical method of contraception is an *intrauterine device (IUD)*. These are small objects such as loops, spirals, and rings made of plastic, copper, or stainless steel that are inserted into the cavity of the uterus. It is not clear how IUDs operate. Some investigators believe that they cause changes in the lining of the uterus that, in turn, produces a substance which destroys either the sperm or the fertilized ovum.

Chemical means of birth control include the use of various foams, creams, jellies, suppositories, and douches that make the vagina and cervix unfavorable for sperm survival. Of the newer chemical means, *oral contraception* (the pill) has found rapid and widespread use. Although several types of pills are available, the most commonly used one is the combination pill. This pill contains a high concentration of progesterone and a low concentration of estrogens. These two hormones act on the anterior pituitary to decrease the secretion of FSH and LH. The low levels of FSH and LH are not adequate to initiate follicle maturation or ovulation. Consequently, pregnancy cannot occur in the absence of a mature ovum.

A summary of contraceptive methods is presented in Exhibit 25–4.

Medical terminology associated with development and inheritance

Abortion   Premature expulsion from the uterus of the products of conception—embryo or nonviable fetus.

Autosome   Any chromosome that is not a sex chromosome (not an X or Y chromosome). Man has 22 pairs of autosomes.

Cesarean section   The removal of the baby and the placenta through an abdominal incision in the uterine wall.

Endometrectomy   The removal of the uterine mucous lining; also called curettage. A variation of this procedure is dilation and curettage (D and C).

Lethal gene   A gene that, when expressed, results in death either in the embryonic state or shortly after birth.

Mutation   A permanent heritable change in a gene that causes the gene to have a different effect.

Chapter summary in outline

MEIOSIS

1. Development is a sequence of events starting with fertilization and terminating with the formation of an organism.
2. Gametes are formed by meiosis, in which the chromosome number is reduced from diploid to haploid.
3. Meiosis in the testes is called spermatogenesis. Meiosis in the ovaries is called oogenesis.

SEXUAL INTERCOURSE

1. The role of the male in the sex act involves erection, emission, and ejaculation.
2. The female role in the sex act involves erection, lubrication, and climax.

PREGNANCY

1. Pregnancy is a sequence of events that includes fertilization, implantation, embryonic growth, fetal growth, and birth.
2. Fertilization, the union of the ovum and sperm, normally occurs in the uterine tubes.
3. The attachment of a fertilized ovum to the endometrium of the uterus is called implantation.
4. During embryonic growth, the primary germ layers and embryonic membranes are formed and the placenta is functioning.
5. The primary germ layers—ectoderm, mesoderm, and endoderm—form all tissues of the developing organism.
6. Embryonic membranes include the yolk sac, amnion, chorion, and allantois.
7. Fetal and maternal materials are exchanged through the placenta.
8. Pregnancy is maintained by the chorionic gonadotrophic hormone, estrogens, and progesterone secreted by the placenta.
9. The birth of a baby involves dilatation of the cervix, expulsion of the fetus, and delivery of the placenta.
10. Lactation, or the secretion of milk by the mammary glands, is influenced by estrogens and progesterone, prolactin, and oxytocin.

INHERITANCE

1. Inheritance is the passage of hereditary traits from one generation to another.
2. The genetic makeup of an organism is called its genotype, and the traits expressed are called its phenotype.
3. Dominant genes control a particular trait; expression of recessive genes is inhibited by dominant genes.
4. Sex is determined by the Y chromosome of the male.
5. Color blindness and hemophilia primarily affect males because there is no counterbalancing dominant genes on the Y chromosomes.
6. Down's syndrome is a chromosomal disorder in which the individual has 47 instead of 46 chromosomes.

BIRTH CONTROL

1. Methods include removal of gonads; sterilization; rhythm; use of condom, diaphragm, and intrauterine devices; chemicals that kill sperm; and the pill.
2. Contraceptive pills of the combination type contain estrogens and progesterone in concentrations that decrease the secretion of FSH and LH and thereby inhibit ovulation.

Review questions and problems

1. Define development. What is the importance of meiosis to development?
2. Compare the events associated with spermatogenesis and oogenesis.
3. Explain the role of erection, emission, and ejaculation in the male sex act. How do erection, lubrication, and orgasm contribute to the female sex act?
4. Define fertilization. Where does it normally occur? How is a morula formed?
5. What is implantation? How does the fertilized ovum implant itself?
6. Define the embryonic period. Describe some of the body structures formed by the ectoderm, mesoderm, and endoderm.
7. What is an embryonic membrane? Describe the functions of the four embryonic membranes.
8. Explain the importance of the placenta and umbilicus to fetal growth.
9. Briefly outline some of the major developmental changes that occur during fetal growth.
10. Compare the sources and functions of estrogens and progesterone, and the chrionic gonadotrophic hormone. How does the placenta serve as an endocrine organ during pregnancy?
11. Define gestation. Distinguish between false and true labor.
12. Describe what happens during the stage of dilatation, the stage of expulsion, and the placental stage of delivery.
13. What is lactation? How is the female prepared for lactation?
14. Define inheritance. What is genetics?
15. Define the following terms: genotype, phenotype, dominant, recessive, homozygous, and heterozygous.
16. What is a Punnett square?
17. List several dominant and recessive traits that are inherited in human beings.
18. Set up Punnett squares to show the inheritance of the following traits: sex, color blindness, and hemophilia.
19. What is amniocentesis? Why is the procedure important?
20. Describe the cause and symptoms of Down's syndrome.
21. Briefly describe the following methods of birth control: removal of the gonads, sterilization, and rhythm.
22. Distinguish between a condom, diaphragm, and IUD as methods of mechanical contraception.
23. List several examples and functions of chemical contraceptives.
24. Explain the operation of the combination contraceptive pill.
25. Refer to the glossary of medical terminology associated with development and inheritance. Be sure that you can define each term.

Bibliography

Unit 5

Arey, L. B.: *Developmental Anatomy*, 7th ed., W. B. Saunders Company, Philadelphia, 1965.

Beck, M. B. et al.: "Abortion: A National Public and Mental Health Problem—Past, Present, and Proposed Research," *Amer. J. Pub. Health*, **59**:2131, December 1969.

Caspersson, T. and L. Zech: "Chromosome Identification by Fluorescence," *Hospital Practice*, September 1972.

Colman, L. and A. D. Colman: *Pregnancy: The Psychological Experience*, Herder and Herder, New York, 1971.

Elstein, M.: "Sterilization and Family Planning," *The Practitioner*, **205**:30, July 1970.

Fraser, F. C.: "Genetic Counseling," *Hospital Practice*, January 1971.

Friedmann, T.. "Prenatal Diagnosis of Genetic Disease," *Scientific American,* November 1971.

Gardner, E. J.: *Principles of Genetics,* 4th ed., John Wiley & Sons, Inc., New York, 1972.

Hawkins, D. F.: "Intra-Uterine Contraception," *The Practitioner,* **205:**20, July 1970.

Hill, H.: "Oral Contraception," *The Practitioner,* **205:**5, July 1970.

Kaback, M. M. and J. S. O'Brien: "Tay-Sachs: Prototype for Prevention of Genetic Disease," *Hospital Practice,* March 1973.

Leis, H. P. Jr., et al.: "Diagnosis of Breast Cancer," *Hospital Medicine,* November 1974.

Maas, J. M.: "The Oral Contraceptives: Their Composition, Properties, Selection, and Performance," *Clinical Medicine,* March 1970.

MacLeod, J.: "The Parameters of Male Fertility," *Hospital Practice,* December 1973.

Mastroianni, L., Jr., and C. Noriega: "Observations on Human Ova and the Fertilization Process," *Amer. J. Obstet. Gynec.,* **107:**682, July 1, 1970.

McKusick, V. A.: "The Nosology of Genetic Disease," *Hospital Practice,* July 1971.

Mickal, A. and J. H. Bellina: "Pelvic Inflammatory Disease," *Hospital Medicine,* August 1973.

Moody, P. A.: *Genetics in Man,* W. W. Norton & Company, Inc., New York, 1967.

Nadler, H. L.: "Prenatal Detection of Genetic Disorders," *Modern Medicine,* June 24, 1974.

Page, E. W. et al.: *Human Reproduction: The Core Content of Obstetrics, Gynecology and Prenatal Medicine,* W. B. Saunders Company, Philadelphia, 1972.

Patten, B. M.: *Human Embryology,* 2d ed., McGraw-Hill Book Company, New York, 1953.

Pregnancy, Birth and the Newborn Baby, Boston Children's Medical Center, Delacorte Press, Dell Publishing Co., Inc., New York, 1972.

"Prenatal Diagnosis: Problems and Outlook," *J. Amer. Med. Assoc.,* **222:**132, Oct. 9, 1972.

"Research on Immunity to Syphilis, *Infectious Diseases,* **4:**8, August 1974.

Rugh, R. and L. B. Shettles: *From Conception to Birth,* Harper & Row, Publishers, Inc., New York, 1971.

Savel, L. E.: "The Present Status of Contraception," *J. Med. Soc. N.J.,* **68:**635, August 1971.

Segal, S. J.: "The Physiology of Human Reproduction," *Scientific American,* **231:**53, September 1974.

Scanlon, J.: "Human Fetal Hazards from Environmental Pollution with Certain Non-Essential Trace Elements," *Clin. Pediat.,* **11:**135, March 1972.

Scheinfeld, A.: *Heredity in Humans,* J. P. Lippincott, Philadelphia, 1972.

Shearman, R. P. (ed.): *Human Reproductive Physiology,* Blackwell Scientific Publications, Ltd., Oxford, 1972.

Stewart, E.: "Mechanical Methods of Contraception," *The Practitioner,* **205:**13, July 1970.

Tietze, C. and S. Lewit: "Abortion," *Scientific American,* **220:**3, January 1969.

"Treating Menopausal Women and Climacteric Men," *Medical World News,* June 28, 1974.

Tyler, E. T.: "Fertility Control with Drugs," *Postgraduate Medicine,* **47:**82, January 1970.

Van Vleck, D. B.: *The Crucial Generation—Your Challenges and Your Choices,* Optimum Population Inc., Charlotte, Vt., April 1973.

Wils, C.: "Genetic Load," *Scientific American,* March 1970.

GLOSSARY OF PREFIXES, SUFFIXES, AND COMBINING FORMS

It is possible to build many hundreds of medical words if you learn a few basic parts that can be combined in a variety of ways. A long complicated medical word will seem less difficult to analyze after you learn the meaning of these fundamental parts. Look at the word carefully, and say it out loud several times, so that each new word will become familiar to you.

A "prefix" is a part of a word that precedes the word root and changes its meaning. A "suffix" or word ending is a part that follows the word root and changes its meaning, or adds to it. A "combining form" is the basic word root such as *abdomin-*, referring to the belly region; and *aden-*, pertaining to a gland. Word roots are often followed by a vowel to facilitate pronunciation, as in *abdomino-* and *adeno-*, in which case they are then called combining forms.

We have also included here one of the methods used to learn the proper pronunciation of the medical terms that we consider essential for the proper understanding of *Human Anatomy and Physiology*. These pronunciations are not the only ones used because there is and probably always will be some conflict between doctors and dictionaries regarding pronunciation of medical terms.

Pronunciation appears in the parentheses immediately following the words. The strongest accented syllable appears in capital letters, for example, bī-LAT-er-al, TIB-ē-al. If the words are long, and if there is a secondary accent, it is noted by a double quote mark ("), for example: re"trō-per-i-ton-Ē-al.

Any additional secondary accents are noted by a double quote mark ("), for example: stur"nō-klī"dō-MAS-toyd.

If the vowels are pronounced with a long sound, this is indicated by a *line above the vowel*. For example, these vowels would be pronounced as follows:

ā as in *māke, ē* as in *bē, ī* as in *īvy,*
ō as in *pōle, ū* as in *pūre*

All other vowels not so marked will be pronounced with the short sound as follows:

e as in *met* and *bet, i* as in *bit* and *sip,*
o as in *not, u* as *bud*

Other phonetics are as follows: *a* as in *father* is written *ah,* and the diphthong *oi* is written *oy.*

Many medical terms are "compound" words; that is, they are made up of more than one root or combining form. Examples of such compound words are leucocyte (white blood cell) and chemotherapy (treatment of disease by administering chemicals).

The following list includes some of the most commonly used word roots, prefixes, suffixes and combining forms used in making medical terms.

Word roots and combining forms

Acou-, Acu- hearing: acoustics (ah-KOOS-tiks), the science of sounds or hearing.

Acr-, Acro- extremity: acromegaly (ak-rō-MEG-ah-lē), hyperplasia of the nose, jaws, fingers, and toes.

Aden-, Adeno- gland: adenoma (ad-i-NŌ-mah), a tumor with a glandlike structure.

Alg-, Algia- pain: neuralgia (nū-RAL-jē-ah), pain along the course of a nerve.

Angi- vessel: angiocardiography (an-jēō-kar-dē-OG-rah-fē), roentgenography of the great blood vessels and heart after intravenous injection of radiopaque fluid.

Arthr-, Arthro- joint: arthropathy (ar-THRO-pah-thē), disease of a joint.

Aut-, Auto- self: autolysis (aw-TOL-li-sis), destruction of cells of the body by their own enzymes, especially after death.

Bio- life, living: biopsy (BĪ-op-se), examination of tissue removed from a living body.

Blast- germ, bud: blastocyte (BLAS-tō-sīt), an embroynic or undifferentiated cell.

Blephar- eyelid: blepharitis (blef-ah-RĪ-tis), inflammation of the eyelids.

Bronch- trachea, windpipe: bronchoscopy (brong-KOS-kō-pē), direct visual examination of the bronchi.

Bucc- cheek: buccocervical (bū-kō-SER-vē-kal), pertaining to the cheek and neck.

Capit- head: decapitate (dē-KAP-i-tāte), to remove the head.

Carcin- cancer: carcinogenic (kar-sin-ō-JEN-ic), causing ing cancer.

Cardi-, Cardia-, Cardio- heart: cardiogram (KAR-dē-ō-gram), a recording of the force and form of the heart's movements.

Cephal- head: hydrocephalus (hī-drō-SEF-ah-lus), enlargement of the head due to an abnormal accumulation of fluid.

Cerebro- brain: cerebrospinal fluid (ser"i-brō-SPĪ-nal), fluid contained within the cranium and spinal canal.

Cheil- lip: cheilosis (kī-LŌ-sis), dry scaling of the lips

Chole- bile, gall: cholecystogram, (kō-le-SIS-tō-gram), roentgenogram of the gallbladder.

Chondr-, Chondri-, Chondrio- cartilage: chondrocyte (KON-drō-sıt), a cartilage cell.

Chrom-, Chromat-, Chromato- color: hyperchromic (hī-per-KRŌ-mik), highly colored.

Crani- skull: craniotomy (krā-nē-OT-ō-mē), surgical opening of the skull.

Cry-, Cryo- cold: cryosurgery (kr-IŌ-ser-jerē), surgical procedure using a very cold liquid nitrogen probe.

Cut- skin: subcutaneous (sub-kyoo-TĀ-nē-us), under the skin.

Cysti-, Cysto- sac or bladder: cystoscope (SIS-tō-skōp), instrument for interior examination of the bladder.

Cyt-, Cyto-, Cyte- cell: cytology (sī-TOL-ō-jē), the study of cells.

Dactyl-, Dactylo- digits (usually fingers, but sometimes toes): polydactylia (pol-ē-DAK-til-ē-ah), above normal number of fingers or toes.

Derma-, Dermato- skin: dermatosis (der-mah-TŌ-sis), any skin disease.

Entero- intestine: enteritis (en-ter-Ī-tis), inflammation of the intestine.

Gastr- stomach: gastrointestinal (gas"tro-in-TES-ti-nal), pertaining to the stomach and intestine.

Gloss-, Glosso- tongue: hypoglossal (hī-pō-GLOS-al), located under the tongue.

Hem-, Hemat- blood: hematoma (he-mah-TŌ-mah), blood beneath the skin.

Hepar-, Hepato- liver: hepatitis (hep-ah-TĪ-tis), inflammation of the liver.

Hist-, Histio- tissue: histology (his-TOL-ō-jē), the study of tissues.

Homeo-, Homo- unchanging, the same, steady: homeostasis (hō-mē-ō-STĀ-sis), achievement of a steady state.

Hyster- uterus: hysterectomy (his-ter-EK-tō-mē), surgical removal of the uterus.

Ileo-, ileum (part of intestine): ileocecal valve (il"e-ō-SĒ-kal), folds at the opening between ileum and cecum.

Ilio- ilium (flaring portion of hip bone): iliosacral (il"e-ō-SĀK-ral), pertaining to ilium and sacrum.

Lachry-, Lacri- tears: nasolacrimal (nā-zō-LAK-rē-mal), pertaining to the nose and lacrimal apparatus.

Leuco-, Leuko- white: leucocyte (LŪ-kō-sīt), white blood cell.

Lip-, Lipo- fat: lipoma (lī-PŌ-mah), a fatty tumor.

Mamm- breast: mammography (mam-OG-rah-fē), roentgenography of the mammary gland.

Mast- breast: mastitis (mas-TĪ-tis), inflammation of the mammary gland.

Meningo- membrane: meningitis (men-in-JĪ-tis), inflammation of the membranes of spinal cord and brain.

Metro- uterus: endometrium (en-dō-ME-trē-um), lining of the uterus.

Morpho- form, shape: morphology (mōr-FOL-ō-jē), study of form and structure of things.

Myelo- marrow, spinal cord: poliomyelitis (pō"lē-ō-mī-el-I-tis), inflammation of the gray matter of the spinal cord.

Myo- muscle: myocardium (mī-ō-KAR-dē-um), heart muscle.

Nephro- kidney: nephrosis (ne-FRŌ-sis), degeneration of kidney tissue.

Neuro- nerve: neuroblastoma (nū"-rō-BLAS-tō-mah), a malignant tumor of the nervous system composed of embryonic nerve cells.

Oculo- eye: binocular (bī-NOK-ū-lar), pertaining to the two eyes.

Odont- tooth: orthodontic (ōr-thō-DON-tik), pertaining to the proper positioning and relationship of the teeth.

Ophthalm- eye: ophthalmology (of-thal-MO-lō-jē), science of the eye and its diseases.

Oss-, Osseo-, Osteo- bone: osteoma (os-tē-Ō-mah), a bone tumor.

Oto- ear: otosclerosis (ō"tō-skle-RŌ-sis), formation of bone in the labyrinth of the ear.

Patho- disease: pathogenic (path-ō-JEN-ik), giving origin to disease.

Phag-, Phago- to eat: phagocytosis (fa"gō-sī-TŌ-sis), a cell that eats microorganisms, other cells, and foreign particles.

Philic-, Philo- to like, have an affinity for: hemophilic (hē-mō-FI-lik), have an affinity for bleeding.

Phleb- vein: phlebitis (fleb-Ī-tis), inflammation of the veins.

Pneumo- lung: pneumothorax (nū-mō-THŌ-raks), air in the thoracic cavity.

Pod- foot: podiatry (pō-DĪ-a-trē), the diagnosis and treatment of foot disorders.

Procto- anus, rectum: proctoscopy (prok-TO-skō-pē), instrumental examination of the rectum.

Pseud-, Pseudo- false: pseudoangina (SŪ-dō-an-jī"nah), false angina.

Psycho- soul or mind: psychiatry (sī-KĪ-a-trē), treatment of mental disorders.

Pyo- pus: pyuria (pī-ū-RĒ-ah), pus in the urine.

Scler-, Sclero- hard: arteriosclerosis (ar-tē"rē-ō-skle-RŌ-sis), hardening of the arteries.

Sep-, Septic- poison or toxic substance: septicemia (sep-ti-SĒ-mē-ah), presence of bacterial toxins in the blood (blood poisoning).

Soma-, Somato- body: somatotropic (sō-ma-tō-TRŌ-pik), having a stimulating effect on body growth.

Stasis-, Stat- stand still: homeostasis (hō-mē-ō-STĀ-sis), achievement of a steady state.

Therm- heat: thermometer (ther-MO-me-ter), instrument used to measure and record heat.

Viscer- organ: visceral (VIS-er-al), pertaining to the abdominal organs.

Prefixes

A-, An- without, lack of, deficient: aseptic (ā-SEP-tik), without infection; anesthesia (an-es-THĒ-zhē-ah), without sensation.

Ab- away from, from: abnormal (ab-NOR-mal), away from normal.

Ad- to, near, toward: adduction (ad-DUK-shun), movement of an extremity toward the axis of the body.

Ambi- both sides: ambidextrous (am-be-DEKS-trus), able to use either hand.

Ante- before: antepartum (an-te-PAR-tum), before delivery of a baby.

Anti- against: anticoagulant (an"te-kō-AG-ū-lant) a substance that prevents coagulation of blood.

Bi- two, double, both: biceps (BĪ-seps), a muscle with two heads of origin.

Brachi- arm: brachialis (brā-kē-AL-is), muscle for flexing forearm.

Brachy- short: brachyesophagus (brā-kē-e-SOF-ah-gus), short esophagus.

Brady- slow: bradycardia (brā-dē-KAR-dē-ah), abnormal slowness of the heartbeat.

Cata- down, lower, under, against: catabolism (ka-TAB-ō-lizm), metabolic breakdown into simpler substances.

Circum- around: circumrenal (ser-kum-RĒ-nal), around the kidney.

Con- with, together: congenital (kon-JEN-i-tal), born with a defect.

Contra- against, opposite: contraception (kon-tra-SEP-shun), the prevention of conception.

Crypt- hidden concealed: cryptorchidism (krip-TOR-ki-dizm), undescended or hidden testes.

Di-, diplo- two: diploid (DI-ployd) having double the haploid number of chromosomes.

Dis- separation, apart, away from: disarticulate (dis-ar-TI-kū-lāt), to separate at joint.

Dys- painful, difficult: dyspnea (disp-NĒ-ah), difficult breathing.

E-, Ec-, Ex- out from, out of: eccentric (eks-EN-trik), not located at the center.

Ecto-, Exo- outside: ectopic pregnancy (ek-TOP-ik), gestation outside the uterine cavity.

Em-, En- in or on: empyema (em-pī-E-mah), pus in a body cavity.

End-, Endo- inside: endocardium (en-dō-KAR-dē-um), membrane lining the inner surface of the heart.

Epi- upon, on above: epidermis (ep-i-DER-mis), outermost layer of skin.

Erythro- red: erythrocyte (e-RITH-rō-sīt), red blood cell.

Eu- well: eupnea (Ū-pnē-ah), normal breathing.

Ex-, Exo- out, away from: exocrine (EKS-ō-krin), excreting outwardly or away from.

Extra- outside, beyond, in addition to: extracellular (eks"tra- SEL-ū-lar), outside of the cell.

Galacto- milk: galactose (gal-AK-tōs), a milk sugar.

Glyco- sugar: glycosuria (glī-kō-SUR-ē-ah), sugar in the urine.

Gyn-, Gyne-, Gynec- female sex (women): gynecology (gī"ne-KOL-o-jē), the study of the diseases of the female.

Hemi- half: hemiplegia (hem-ē-PLĒ-jē-ah), paralysis of only one-half of the body.

Heter-, Hetero- other, different: heterogenous (he-ter-ō-JĒN-ē-us), composed of different substances.

Hydr- water: hydrocele (HĪ-drō-sēl), accumulation of fluid in a saclike cavity.

Hyper- beyond, excessive: hyperglycemia (hī"per-glī-SĒ-mē-ah), excessive amount of sugar in the blood.

Hypo- under, below, deficient: hypodermic (hī-pō-DER-mik), below the skin or dermis.

Idio- self, one's own, separate: idiopathic (id-ē-ō-PATH-ik), a disease without recognizable cause.

Inter- among, between: intercostal (in-ter-KOS-tal), between the ribs.

Intra- within, inside: intracellular (in"trah-SEL-ū-lar), inside the cell.

Iso- equal, like: isogenic (īso-JEN-ik), alike in morphological development.

Macro- large, great: macrophage (MAK-rō-faj), large phagocytic cell.

Mal- bad, abnormal: malnutrition (mal-nū-TRISH-un), lack of necessary food substances.

Mega-, Megalo- great, large: magakaryocyte (meg"ah-KAR-ē-ō-sīt), giant cell of bone marrow.

Meta- after, beyond: metacarpus (met-ah-KAR-pus), the part of the hand between the wrist and fingers.

Micro- small: microtome (MĪ-krō-tōm), instrument for preparing very thin slices of tissue for microscopic examination.

Necro- corpse, dead: necrosis (nē-KRŌ-sis), death of areas of tissue or bone surrounded by healthy parts.

Neo- new: neonatal (nē-ō-NĀT-al), pertaining to the first 4 weeks after birth.

Oo- egg: oocyte (Ō-ō-sīt), original egg cell.

Oligo- small, deficient: oliguria (ō-lig-UR-ē-ah), abnormally small amount of urine.

Ortho- straight, normal: orthopnea (or-THOP-nē-ah), inability to breathe in any position except when straight or erect.

Para- near, beyond, apart from, beside: paranasal (par-ah-NĀ-zal), near the nose.

Ped- children: pediatrician (pē-dē-at-TRISH-ē-an), a medical specialist in treatment of children's diseases.

Per- through: percutaneous (per-kyōō-TĀ-nē-us), through the skin.

Peri- around: pericardium (per-ē-KAR-dē-um), membrane or sac around the heart.

Poly- much, many: polycythemia (pol″e-si-THĒ-me-ah), an excess of red blood cells.

Post- after, behind: postnatal (pōst-NĀ-tal), after birth.

Pre-, Pro- before, in front of: prenatal (prē-NĀ-tal), before birth.

Retro- backward, located behind: retroperitoneal (re″tro-per-i-ton-E-al), located behind the peritoneum.

Rhin- nose: rhinitis (rī-NĪ-tis), inflammation of the nasal mucosa.

Semi- half: semicircular canals (semi-SER-kyōō-lar), canals in the shape of a half circle.

Sten- narrow: stenosis (ste-NŌ-sis), narrowing of a duct or a canal.

Sub- under, beneath, below: submucosa (sub-myōō-KŌ-sah), tissue layer under a mucous membrane.

Super- above, beyond: superficial (su-per-FISH-ē-al), confined to the surface, not through.

Supra- above, over: suprarenal (su-prah-RE-nal), adrenal gland above the kidney.

Sym-, Syn- with, together, joined: syndrome (SIN-drom), all the symptoms of a disease considered as a whole.

Tachy- quick, rapid: tachycardia (tak-ē-KAR-dē-ah), rapid heart action.

Tox-, Toxic- poison: toxemia (toks-Ē-mē-ah), poisonous condition of the blood.

Trans- across, through beyond: transudation (trans-ū-DĀ-shun), oozing of a fluid through pores.

Tri- three: trigone (TRĪ-gōn), a triangular space, as at the base of the bladder.

Trich- hair: trichosis (tri-KŌ-sis), disease of the hair.

Zoo- animal: zoology (zō-OL-ō-jē), the study of animals.

Suffixes

-able capable, of, having ability to: viable (VĪ-ah-bel), capable of living.

-ac, -al pertaining to: cardiac (KAR-dē-ak) (cardial), pertaining to the heart.

-ary connected with: ciliary (SIL-ē-ār-ē), resembling any hairlike structure.

-asis, -asia, -esis, -osis condition or state of: hemostasis (hē-mō-STĀ-sis), stopping of bleeding or circulation.

-cel, -cele swelling, an enlarged space or cavity: meningocele (men-in-JŌ-sēl), enlargement of the meninges.

-cid, -cide, -cis cut or kill, destroy: germicide (GERM-i-cīd), a substance that kills germs.

-ectasia, -ectasis stretching, dilatation: bronchiectasis (brong-kē-EK-tah-sis), dilatation of a bronchus or bronchi.

-ectomize, -ectomy cutting out, excision of, removal of: thyroidectomy (thī-royd-EK-tō-mē), surgical removal of thyroid gland.

-emia condition of blood: lipemia (lī-PĒ-mē-ah), abnormally high concentration of fat in the blood.

-ferent bear or carry: efferent (EF-er-ent), carrying away from a center.

-form shape: fusiform (FŪZ-i-form), spindle-shaped.

-gen an agent that produces or originates: pathogen (PATH-ō-jen), a microorganism or substance capable of producing a disease.

-genic produced from, producing: pyogenic (pī-ō-JEN-ik), produced from pus.

-gram a record, that which is recorded: electrocardiogram (ē-lek″trō-CAR-dē-ō-gram), a record of normal heart action.

-graph an instrument for recording: electroencephalograph (ē-lek″trō-en-SEF-ah-lō-graf), an instrument for recording electrical activity of the brain.

-ia a state or condition: hypermetropia (hī″per-me-TRŌ-pē-ah), condition of farsightedness.

-iatrics, -iatry medical practice specialties: pediatrics (pē-dē-AT-riks), medical science relating to care of children and treatment of diseases peculiar to them.

-ism condition or state: rheumatism (RŪ-mah-tizm), inflammation, especially of muscle and joints.

-itis inflammation: neuritis (nū-RI-tis), inflammation of a nerve or nerves.

-logy, -ology the study or science of: physiology (fiz-ē-OL-ō-jē), the study of function of body parts.

-lyso, -lysis solution, dissolve, loosening: hemolysis (hē-MOL-is-is), dissolution of the red blood cells.

-malacia softening: osteomalacia (os″tē-o-mal-Ā-she-ah), softening of bone.

-oma tumor: fibroma (fī-BRŌ-mah), a tumor composed mostly of fibrous tissue.

-ory pertaining to: sensory (SEN-sō-rē), pertaining to sensation.

-ose full of: adipose (AD-i-pōz), characterized by excessive amounts of fat.

-pathy disease: neuropathy (nū-RŌ-pah-thē), disease of the peripheral nervous system.

-penia deficiency: thrombocytopenia (throm″bo-sī-tō-PĒ-nē-ah), deficiency of thrombocytes in the blood.

-phobe, -phobia fear of, aversion for: hydrophobia (hī-drō-FŌ-bē-ah), fear of water.

-plasia, -plasty reconstruction of: rhinoplasty (RĪ-nō-plas-tē), surgical reconstruction of the nose.

-plegia, -plexy stroke, paralysis: apoplexy (AP-ō-plek-sē), sudden loss of consciousness, and paralysis.

-pnea to breathe: apnea (AP-nē-ah), temporary absence of respiration, following a period of overbreathing.

-poiesis formation of: hematopoiesis (hē-mat"ō-PĪ-es-is), formation and development of red blood cells.

-rrhea flow, discharge: diarrhea (dı-ah-RĒ-ah), abnormal frequency of bowel evacuation, the stools with a more or less fluid consistency.

-scope an instrument used to look into or examine a part: bronchoscope (BRONG-kō-skōp), instrument used to examine the interior of a bronchus.

-stomy creation of a mouth or artificial opening: tracheostomy (trā-kē-OST-ō-mē), creation of an opening in the trachea.

-tomy cutting into, incision into: appendectomy (apen-DEC-tō-mē), surgical removal of the appendix.

-trophy a state relating to nutrition: hypertrophy (hı-PER-trō-fē), a condition resulting from increase in size of cells.

-tropic turning toward, influencing, changing: gonadotropic (gō-na-dō-TRŌ-pik), influencing the gonads.

-uria urine: polyuria (pol-ē-Ū-rē-ah), excessive secretion of urine.

GLOSSARY OF TERMS

Abdomen (ab-DŌ-men) The area between the diaphragm and pelvis.

Abduct (ab-DUKT) To draw away from the axis or midline of the body or one of its parts.

Abortion (ab-OR-shun) The premature loss or removal of the embryo, or of the nonviable fetus; any failure in the normal process of developing or maturing.

Abscess (AB-ses) A localized collection of pus and liquefied tissue in a cavity.

Absorption (ab-SŌRP-shun) The taking up of liquids by solids, or of gases by solids or liquids.

Acapnia (a-KAP-ne-ah) Decrease in the normal amount of carbon dioxide in the blood.

Accommodation (ak-kom-mō-DĀ-shun) A change in the shape of the eye lens so that vision is more acute; an adjustment of the eye lens for various distances; focusing.

Acetabulum (as"i-TAB-ye-lem) The rounded cavity on the external surface of the innominate bone that receives head of femur.

Acetone (AS-e-tōne) bodies Organic compounds that may be found in excessive amounts in the urine and blood of diabetics. Found whenever too much fat in proportion to carbohydrate is being oxidized. Also called ketone bodies (KĒ-tōne).

Acetylcholine (a-sēt-il-KŌ-lēn) Substance found normally in many animal tissues, believed to be the substance released from nerves to activate muscles and some glands. Liberated at synapses in the central nervous system.

Achilles (ah-KIL-ez) reflex Flexion of foot and contraction of calf muscles following blow upon tendon of Achilles.

Achilles tendon The tendon of the soleus and gastrocnemius muscles, at the back of the heel.

Achlorhydria (ā-klōr-HĪ-drē-ah) Absence of hydrochloric acid in the gastric juice.

Acid (AS-id) A proton donor; excess hydrogen ions producing a pH less than 7.

Acidosis (as-i-DŌ-sis) A serious disorder in which the normal alkaline substances of the blood are reduced in amount.

Acromion (ah-KRŌ-mē-on) The lateral, triangular projection of spine of scapula, forming point of the shoulder, and articulating with the clavicle.

Actin (AK-tin) One of the proteins in muscle fiber, the other being myosin.

Action potential The momentary difference in electrical potential between active and resting parts of a nerve fiber.

Actomyosin (ak-tō-MĪ-ō-sin) The combination of actin and myosin in a muscle.

Acuity (ak-Ū-i-tē) Clearness, or sharpness, usually of vision.

Acute (ak-ŪT) Having rapid onset, severe symptoms, and a short course; not chronic.

Adam's apple The laryngeal prominence.

Adaptation (ad-ap-TĀ-shun) The adjustment of the pupil of the eye to light variations.

Addison's (AD-i-sonz) disease One due to deficiency in the secretion of adrenocortical hormones.

Adenohypophysis (ad"i-nō-hī-POF-i-sis) The anterior portion of the pituitary gland.

Adenoids (AD-i-noyds) The pharyngeal tonsils.

Adenosine triphosphate-ATP (àd-EN-ō-sēn, trī-FOS-fāte) Made up of sugar, adenine, nitrogen, and phosphorus. The breakdown of ATP provides the energy for muscle contraction and many other physiological processes.

Adhesion (ad-HĒ-zhun) Abnormal joining of parts to each other.

Ad libitum (ad-LIB-it-um) At pleasure; the amount desired.

Adolescence (ad-ō-LES-ens) The period from the beginning of puberty until adult life.

Adrenal (ad-RĒ-nal) glands Two glands, one on top of each kidney; also called the suprarenal glands.

Adrenalin (ad-REN-ah-lin) Proprietary name for epinephrine; the active secretion of the medulla of the adrenal gland.

Adrenergic (ad-ren-ER-jik) Applied to nerve fibers that when stimulated, release epinephrine (adrenalin) or an epinephrinelike substance at their terminations.

Adrenocorticotrophic (ad-rēn"ō-kōr-ti-ko-TRŌF-ik) hormone-ACTH Hormone produced by the anterior pituitary gland, which influences the adrenal glands.

Adsorption (ad-SORP-shun) Process whereby a gas or a dissolved substance becomes concentrated at the surface of a solid or at the interfaces of a colloid system.

Adventitia (ad-ven-TISH-yah)   The outermost covering of a structure or organ.

Aerobic (ayer-Ō-bik)   Living only in the presence of oxygen.

Afferent (AF-er-ent)   Carrying impulses toward a center; nerves that carry a message toward the central nervous center, or toward ganglia.

Agglutination (ag-glū-tin-Ā-shun)   Clumping of microorganisms or blood corpuscles; an immunity response; an antigen-antibody reaction.

Agglutinin (ag-GLŪ-tin-in)   A specific principle or antibody in blood serum of an animal affected with a microbic disease that is capable of causing the clumping of bacteria.

Agglutinogen (ag-GLŪ-tin-ō-jen)   A substance inherited that clumps only the blood of parent and child; it is present on the surface of red cells.

Agonist (AG-ō-nist)   The muscle directly engaged in contraction as distinguished from muscles, which have to relax at the same time.

Albinism (AL-bin-ism)   Abnormal, nonpathological absence of pigment in skin, hair and eyes, partial or total.

Albumin (al-BŪ-min)   A protein substance found in nearly every animal or plant tissue and fluid.

Albuminuria (al-bū"min-UR-ēa)   Presence of albumin in the urine.

Aldosterone (al-DOS-tē-rōne)   Powerful salt-retaining hormone of the adrenal cortex.

Alimentary (al-i-MENT-ar-ē)   Of or pertaining to nutrition.

Alkaline (AL-ka-lin)   Pertaining to an alkali, (metallic hydroxide), or having the reactions of one.

Alkalosis (al-kah-LŌ-sis)   Increased bicarbonate content of the blood, due to excess of alkalies or withdrawal of acid or chlorides from the blood.

Allantois (al-AN-tō-is)   A kind of elongated bladder, between the chorion and amnion of the fetus.

Allergic (ah-LER-jik)   Pertaining to or sensitive to an allergen.

Alveolar (al-VĒ-ō-lar)   A small depression.

Alveolus (al-VĒ-ō-lus)   A small hollow or cavity; an air cell.

Ameboid (am-Ē-boyd)   Having the appearance and characteristics of an ameba.

Amenorrhea (ā-men-ō-RĒ-ah)   Absence or suppression of menstruation.

Amino (am-Ē-no) acids   One of the compounds, of which about 22 different ones are known, derived from the fatty acids.

Amniocentesis (am"nē-ō-sen-TĒ-sis)   Removal of amniotic fluid by inserting a needle transabdominally into the amniotic cavity.

Amnion (AM-nē-on)   The inner of the fetal membranes, a thin transparent sac that holds the fetus suspended in amniotic fluid. Also called the "bag of waters."

Amorphous (ā-MŌR-fus)   Without definite shape or differentiation in structure; pertains to solids without crystalline structure.

Amphiarthrosis (am"fē-ar-THRŌ-sis)   A form of articulation midway between diarthrosis and synarthrosis, in which the articulating bony surfaces are separated by an elastic substance to which both are attached, so that the mobility is slight, but may be exerted in all directions.

Amphoteric (am-fō-TER-ik)   Affecting both red and blue litmus paper.

Ampulla (am-PŪ-lah)   A saclike dilatation of a canal.

Amyl nitrate (Ā-mil, NĪ-trāte)   An organic compound that produces dilatation of the blood vessels when inhaled. Used in attacks of angina pectoris.

Anabolism (ah-NA-bō-lizm)   The building up of the body substance.

Anaerobe (an-Ā-er-ōb)   A microorganism that thrives best or lives only without oxygen.

Analgesia (an-al-JĒ-zē-ah)   Absence of normal sense of pain.

Anaphylactic (an"a-fil-AC-tik)   Pertaining to increasing susceptibility to any foreign protein introduced into the body; decreasing immunity.

Anastomosis (ah-nas-tō-MŌ-sis)   A communication between either blood vessels, lymphatics or nerves. An end-to-end union or joining together.

Anatomy (ah-NAT-ō-mē)   The structure or study of structure of the body and the relationship of its parts to each other.

Androgens (AN-drō-jen)   Substance producing or stimulating male characteristics, such as the male hormone.

Anemia (ah-NĒM-ē-ah)   A decrease in certain elements of the blood, especially red cells and hemoglobin.

Aneurysm (AN-ū-rizm)   A saclike enlargement of a blood vessel caused by a weakening of the wall.

Angina pectoris (AN-ji-nah, Pek-tō-ris)   An agonizing pain in the chest that may or may not involve heart or artery disease.

Angiography (an-jē-OG-rah-fē)   The injection of a contrast medium into the common carotid or vertebral artery that demonstrates the cerebral blood vessels in the x-rays, and may detect brain tumors with specific vascular patterns.

Angiotensin (an-jē-ō-TEN-sin)   A blood substance that produces vasoconstriction.

Angstrom (Å) (ANG-strum)   One tenth of a millimicron or about one two hundred fifty millionth of an inch.

Anion (AN-ī-un)   An ion, carrying a negative charge.

Anisocytosis (an"eso-sī-TŌ-sis)   An abnormal condition in which there is a lack of uniformity in the size of red blood cells.

Ankylose (ANG-ke-lōs) Immobilization of a joint by pathological or surgical process.

Anomaly (ah-NOM-ah-lē) An abnormality that may be a developmental (congenital) defect; a variant from the usual standard.

Anorexia (an-ō-REK-sē-ah) Loss of appetite.

Anoxemia (an-oks-ĒM-ē-ah) Lack of oxygen in the blood.

Anoxia (an-OK-sē-ah) Deficiency of oxygen.

Anterior (an-TĒ-rē-ōr) In front of or the ventral surface.

Antibody (AN-ti-bodē) A specific substance produced by and in an animal or person that produces immunity to the presence of an antigen.

Antidiuretic (an-tī-dīur-ET-ik) Substance that inhibits urine formation.

Antigen (AN-ti-jen) Any substance that when introduced into the tissues or blood induces the formation of antibodies, or reacts with them.

Antimetabolite (an-tī-met-AB-ō-līt) Any substance resembling a normal metabolite but is foreign to the body, which compete with, replaces, or antagonizes the regular metabolite.

Antioxidants (an-tī-OKS-i-dants) Substances that inhibit oxidation.

Antrum (AN-trum) Any nearly closed cavity or chamber, especially one within a bone, such as a sinus.

Anuria (an-U-rē-ah) Absence of urine formation.

Anus (A-nus) The distal end and outlet of the rectum.

Aorta (ā-OR-tah) The main systemic trunk of the arterial system of the body, emerging from the left ventricle.

Aperture (AP-er-chur) An opening or orifice.

Apex (A-peks) The pointed end of a conical structure.

Aphasia (a-FA-sze-ah) Loss of ability to express oneself properly through speech or loss of verbal comprehension.

Apocrine (AP-ō-krin) Pertaining to cells that lose part of their cytoplasm while secreting.

Aponeurosis (ap″o-nū-RŌ-sis) A sheetlike layer of connective tissue connecting a muscle to the part that it moves, or functioning as a sheath enclosing a muscle.

Appendage (ah-PEN-dij) A part or thing attached.

Aqueduct (ĀK-we-duct) A canal or passage, especially for the conduction of a liquid.

Aqueous (AK-we-us) humor The watery fluid that fills the anterior and posterior chambers of the eye.

Arachnoid (ar-AK-noyd) The middle of the three coverings (meninges) of the brain.

Areola (ah-RĒ-ō-lah) Any tiny space in a tissue; the pigmented ring around the nipple of the breast.

Areolar (ah-RĒ-ō-lar) A type of connective tissue.

Arrhythmia (ah-RITH-mē-ah) Irregular heart action causing absence of rhythm.

Arthrosis (ar-THRŌ-sis) A joint or articulation.

Articulate (ar-TIK-ū-lāt) To join together as a joint, so as to permit motion between parts.

Arytenoid (ar-IT-en-oyd) Resembling a ladle.

Ascites (ah-SĪ-tēz) Serous fluid in the peritoneal cavity.

Asphyxia (as-FIX-ē-ah) Unconsciousness due to interference with the oxygen supply of the blood.

Aspirate (AS-pir-āte) To remove by suction.

Astereognosis (as-ter″ē-ō-GNŌ-sis) Inability to recognize objects or forms by touch.

Asthenia (as-THE-nē-ah) Lack or loss of strength; debility.

Astigmatism (ah-STIG-mah-tizm) An irregularity of the lens or cornea of the eye, causing the image to be out of focus, producing faulty vision.

Astrocyte (AS-trō-sīt) Star-shaped cell forming the neuroglia fibers.

Ataxia (a-TAKS-ē-ah) The lack of muscular coordination; lack of precision.

Atherosclerosis (ath″erō-skle-RŌ-sis) A disease involving mostly the lining of large arteries, in which yellow patches of fat are deposited, forming plaques that decrease the size of the opening. Also called arteriosclerosis.

Atresia (ah-TRĒ-zē-ah) The abnormal closure of a passage, or the absence of a normal body opening.

Atrium (Ā-trē-um) A cavity or sinus.

Atrophy (AT-rō-fē) A wasting away or decrease in size of a part, due to a failure or abnormality of nutrition.

Auricle (AW-re-kul) The flap or pinna of the ear.

Auscultation (aws-kul-TĀ-shun) Examination by listening to sounds within the body.

Axilla (ak-SIL-ah) The small hollow beneath the arm where it joins the body at the shoulder; the armpit.

Azygos (AZ-ī-gōs) An anatomical structure that is not paired; occurring singly.

Bactericide (bak-TĒ-ri-cīd) That which kills bacteria.

Bacteriophage (bac-TĒ-rē-ō-faj) A nonspecific agent capable of destroying bacteria.

Baroreceptor (bar-ō-rēs-EP-tōr) Receptor stimulated by pressure change.

Bartholin's (BAR-tō-linz) glands Two small mucous glands located one on each side of the vaginal opening.

Basal (BĀ-zel) metabolic (met-ah-BOL-ik) rate (BMR) The minimum amount of metabolism needed to maintain the vital processes in operation.

Base A nonacid, a proton acceptor; characterized by excess of OH ion and a pH greater than 7.

Basement membrane A thin layer of solid substance underlying the epithelium of mucous surfaces.

Basophil (BĀ-sō-fil) A white blood cell characterized by a pale nucleus and large, densely basophilic granules.

Bel   A measure of sound intensity.

Benign (bē-NĪN)   Not malignant.

Bicipital (bī-CIP-i-tal)   Having two heads, as in muscle.

Bifurcate (bī-FUR-kāt)   Having two branches or divisions; forked.

Bilateral (bī-LAT-er-al)   Pertaining to, affecting, or relating to two sides of the body.

Bile (BĪ-el)   A secretion of the liver.

Bilirubin (bil-i-ROO-bin)   The orange or yellowish pigment in bile.

Biliverdin (bil-i-VER-din)   A greenish pigment in bile formed in oxidation of bilirubin.

Blastocyst (BLAS-tō-sist)   A stage in the development of a mammalian embryo.

Blastomere (BLAS-tō-mēr)   One of the cells resulting from the cleavage or segmentation of a fertilized ovum.

Blastula (BLAS-tū-lah)   An early stage in the development of an ovum.

Blind spot   Special area in the retina in which there are no light receptor cells.

Blood-brain barrier   A special mechanism that prevents the passage of materials from the blood to the cerebrospinal fluid and brain.

Bolus (BŌ-lus)   A soft, rounded mass; usually food ready to swallow.

Bradycardia (brā-dē-KAR-dē-ah)   Slow heart action.

Bronchiole (BRONG-kē-ōl)   The smaller divisions of the bronchi.

Bronchogram (BRONG-kō-gram)   A roentgenogram of the lungs and bronchi.

Bronchus (BRONG-kus)   One of the two large branches of the trachea.

Brownian movement   Random movement of particles distinguished from self motility of living microorganisms.

Buccal (BŪK-al)   Pertaining to the cheek or mouth.

Buffer (BU-fer)   A substance that tends to preserve the original hydrogen-ion concentration, which otherwise would change by adding acids or bases.

Bunion (BUN-yun)   Inflammation and thickening of the bursa of the joint of the toes.

Bursa (BUR-sah)   A sac or pouch in connective tissue, especially about joints.

Cachexia (kah-KEK-sē-ah)   A state of ill health, malnutrition, and wasting.

Calcify (KAL-si-fī)   To harden by deposits of calcium salts.

Calcitonin (kal-si-TON-in)   Hormone secreted by the parathyroid glands.

Calculus (KAL-kū-lus)   A stone formed within the body, as in the gallbladder, kidney or urinary bladder.

Calorie (KAL-ō-rē)   A unit of heat. The small calorie is the standard unit and is the amount of heat necessary to raise 1 gram of water 1° Centigrade. The large calorie (kilocalorie) is used in metabolic and nutrition studies; it is the amount of heat necessary to raise 1 kilogram of water 1° Centigrade.

Calyx (KĀL-iks)   Any cuplike division of the kidney pelvis into the kidney tissue itself.

Canal (kan-al)   A narrow tube, channel or passageway.

Canaliculus (kan-al-IK-ū-lus)   A small channel or canal, as in bones, where they connect the lacunae.

Cancellous (KAN-sel-us)   Having a reticular or latticework structure, as in spongy tissue of bone.

Cancer (KAN-ser)   A malignant tumor of epithelial origin, tending to infiltrate and give rise to new growths or metastases; also called carcinoma (kar-si-NŌ-mah).

Capillary (KAP-i-lar-ē)   A tiny blood vessel.

Carbohydrate (car-bō-HĪD-rāt)   An organic compound containing carbon, hydrogen, and oxygen in a particular amount and arrangement.

Caries (KĀ-ri-ēz)   Decay and death of a tooth or bone associated with inflammation and the formation of abscesses in the periosteum and surrounding tissues.

Carotid (kah-ROT-id)   The main artery in the neck extending to the head.

Carpal (KAR-pul)   Pertaining to the carpus or wrist.

Casein (KĀS-ēin)   The main protein in milk, seen in milk curds.

Cast   A solid mold of a part; can originate from different areas of the body, and be composed of different material.

Castration (kas-TRĀ-shun)   The removal of the testes or ovaries.

Catalysis (kat-AL-is-is)   Decomposition produced chemically by a substance not affected by the reaction.

Cataract (KAT-ah-rakt)   Loss of transparency of the crystalline lens of the eye, or its capsule, or both.

Catheter (KATH-i-ter)   A tube that can be inserted into a body cavity through a canal to remove fluids such as urine or blood.

Cation (KAT-ī-on)   An ion, carrying a positive charge.

Caudal (KAW-dal)   Pertaining to any tail-like structure; inferior in position.

Cecum (SĒ-kum)   A blind pouch at the junction of the small intestines with the ascending colon, and to which the ileum is attached.

Celiac (SĒ-lī-ak)   Pertaining to the abdominal regions.

Cellulitis (sel-ū-LĪ-tis)   Inflammation of cellular or connective tissue; especially those just under the skin.

Cellulose (SEL-ū-lōs)   A fibrous form of carbohydrate constituting the main plant structure.

Centigrade (SEN-ti-grād)   A unit of measurement consisting of 100 gradations between the boiling and freezing point, which is 0 degree.

Centimeter (SEN-ti-mēt-er)   The hundredth part of a meter.

Centrifugation (sen-tri-fū-GĀ-shun) A machine used for separating heavier materials from lighter ones as blood cells from plasma.

Cephalic (se-FA-lik) Pertaining to the head; superior in position.

Cerumen (se-RU-men) Ear wax.

Cervix (SER-viks) Any neck or constricted portion of an organ; especially the lower cylindrical part of the uterus.

Chalazion (kah-LĀ-zē-on) A small tumor of the eyelid.

Chemotherapy (kēm″o-THER-a-pē) The treatment of disease by chemicals that affect the microorganisms causing the disease without harming the patient.

Chiasm (KĪ-azm) A crossing; especially the crossing of the optic nerve fibers.

Choana (ko-AN-a) Funnel-shaped, like the posterior openings of the nasal fossa.

Cholecystectomy (kō″le-sis-TEK-tō-mē) Surgical removal of the gallbladder.

Cholesterol (kō-LES-ter-ōl) An organic fatlike compound found in many parts of the body.

Cholinergic (Kōl-in-ER-jik) Nerve endings of the nervous system that liberate acetylcholine at a synapse.

Cholinesterase (kōl″in-ES-ter-ās) A substance that hydrolyzes acetylcholine.

Chorion (KŌR-ē-on) The outermost envelope of the fertilized ovum that serves a protective and nutritive function.

Chromatography (krō″mat-OG-rah-fē) Separating chemical substances and particles by differential movement through a two-phase system.

Chronic (KRO-nik) Long drawn out; applied to a disease that is not acute.

Chyle (kīl) The milky fluid found in the lacteals of the small intestine after digestion.

Chyme (kīm) The semifluid mixture of partly digested food and digestive secretions found in the stomach and small intestine during digestion of a meal.

Cicatrix (SIK-ah-triks) A scar left by a healed wound.

Ciliary (SIL-ē-ār-ē) Pertaining to any hairlike processes; an eyelid, an eyelash.

Circadian (ser-KĀD-ē-an) Daily; occurring on a 24-hour cycle.

Circle of Willis Union of the anterior and posterior cerebral arteries (branches of the carotid) forming an anastomosis at the base of the brain.

Circumcision (ser-kum-SIZH-un) Removal of the foreskin, the fold over the glans penis.

Cirrhosis (si-RŌ-sis) A liver disorder in which the cells are destroyed.

Cochlea (KŌK-lē-ah) A winding cone-shaped tube forming a portion of the inner ear.

Coitus (KŌ-i-tus) Sexual intercourse; also coition and copulation.

Colostomy (kō-LOS-tō-mē) The surgical creation of a new opening from the colon to the body surface.

Colostrum (kō-LOS-trum) The first milk secreted at the end of pregnancy.

Coma (KŌ-mah) Profound unconsciousness from which one cannot be roused.

Contralateral (kon″trah-LAT-er-al) Situated on, or pertaining to the opposite side.

Convoluted (KON-vō-lu-ted) Rolled together or coiled.

Corpus (KOR-pus) The principal part of any organ; any mass or body.

Costal (KOS-tal) Pertaining to a rib.

Cramp A spasmodic, especially a tonic contraction of one or many muscles, usually painful.

Crenation (krē-NĀ-shun) The conversion of normally round red blood corpuscles into shrunken, knobbed, starry forms, as when blood is mixed with a salt solution of 5 percent strength.

Cretinism (KRĒ-tin-izm) Severe congential thyroid deficiency leading to physical and mental retardation.

Curare (ku-RAH-rē) A very toxic extract that paralyzes muscle by acting on the motor end plates.

Cutaneous (kyōō-TĀ-nē-us) Pertaining to the skin.

Cyanosis (sī-an-O-sis) Slightly bluish or dark purple discoloration of the skin and the mucous membrane, due to an oxygen deficiency.

Cystitis (sis-TĪ-tis) Inflammation of the urinary bladder.

Cytopenia (sī-tō-PĒ-nī-ah) Reduction or lack of cellular elements in the circulating blood.

Debility (dē-BIL-i-tē) Weakness of tonicity in functions or organs of the body.

Deciduous (dē-SID-ū-us) Anything that is cast off at maturity; especially the first set of teeth.

Decubitus (dē-KYŌŌ-be-tus) The lying-down position; a decubitus ulcer is caused by pressure when a patient is confined to bed for a long period of time.

Defecation (def-e-KĀ-shun) The discharge of excreta (feces) from the rectum.

Deglutition (dē-glū-TISH-un) The act of swallowing.

Dehydration (dē-hīd-RĀ-shun) A condition due to excessive water loss from the body or its parts.

Dens (denz) Tooth.

Dentine (DEN-tēn) The osseous tissues of a tooth, enclosing the pulp cavity.

Dentition (den-TI-shun) The eruption of teeth; the number, shape, and arrangement of teeth. Also called teething.

Dermatome (DER-mah-tōm) An instrument for incising the skin, or for cutting thin transplants of skin.

Detritus (de-TRĪ-tus) Any broken down or degenerative tissue or carious matter.

Diagnosis (dī-ag-NŌ-sis) Recognition of disease states from symptoms, inspection, palpation, posture, reflexes, general appearance, abnormalities, and other means.

Dialysis (dī-AL-i-sis) The process of separating crystalloids (smaller particles) from colloids (larger particles) by the difference in their rates of diffusion through a semipermeable membrane.

Diapedesis (dī″ah-pe-DĒ-sis) The passage of blood cells through intact blood vessel walls.

Diaphragm (DĪ-a-fram) Any partition that separates one area from another, especially the dome-shaped musculomembranous partition between the thoracic and abdominal cavities.

Diarthrosis (dī-ar-THRO-sis) An articulation in which opposing bones move freely, as in a hinge joint.

Diastole (dī-AS-tō-lē) The relaxing dilatation period of the heart muscle, especially of the ventricles.

Differentiation (dif″er-en-shē-Ā-shun) Acquirement of functions different from those of the original type.

Dilate (DĪ-lāte) To expand or swell; dilatation (noun).

Diplopia (dip-LŌ-pē-ah) Double vision.

Dissect (DĪ-sekt) To separate tissues and parts of a cadaver (a corpse) for anatomical study.

Distal (DIS-tal) Farthest from the center, from the medial line, or from the trunk.

Diuretic (dī-ūr-ET-ik) Any agent that increases the secretion of urine.

Diurnal (dī-UR-nal) Daily.

Diverticulum (dī-ver-TIK-ū-lum) A sac or pouch in the walls of a canal or organ; especially the colon.

Dorsal (DOR-sal) Pertaining to the back.

Dropsy (DROP-sē) A condition rather than a disease where there is an abnormal accumulation of water in the tissues and cavities.

Dysfunction (dis-FUNK-shun) Absence of complete normal function.

Dysmenorrhea (dis″men-ō-RĒ-ah) Painful or difficult menstruation.

Dystrophia (dis-TRŌ-fē-ah) Progressive weakening of a muscle.

Ectopic (ek-TOP-ik) Out of the normal location.

Edema (ē-DĒ-mah) An abnormal accumulation of fluid in the body tissues.

Effusion (ef-Ū-zhun) The escape of fluid from the lymphatics or blood vessels into a cavity or into tissues.

Electrolyte (ē-LEK-trō-līt) A solution that conducts electricity by means of either positive or negatively charged ions.

Embolism (EM-bō-lizm) Obstruction, or closure of a vessel by a transported blood clot, a mass of bacteria, or other foreign material.

Embryo (EM-brē-ō) The young of any organism in an early stage of development; in human beings between the second and eighth weeks, inclusive.

Emesis (EM-e-sis) Vomiting.

Emphysema (em-fi-SĒ-mah) A swelling or inflation of air passages with resulting stagnation of air in parts of the lungs; loss of elasticity in the alveoli.

Emulsification (ē-mul″si-fi-CĀ-shun) The breaking down of large fat globules in the intestine to smaller uniformly distributed particles.

Endogenous (en-DO-jen-us) Growing from or beginning within the organism.

Endosteum (en-DOS-tē-um) The membrane that lines the medullary cavity of bones.

Enuresis (en-ūr-E-sis) Involuntary discharge of urine, complete or partial, after the age of 3 years.

Enzyme (EN-zīm) A substance that causes chemical changes; an organic catalyst, usually a protein.

Epiphysis (ē-PIF-i-sis). The end of a long bone, usually larger in diameter than the shaft (the diaphysis); epiphyses is plural.

Episiotomy (ē-piz-ē-OT-ō-mē) Incision of perineum at end of second stage of labor to avoid tearing of the perineum.

Epistaxis (ep-ē-STAKS-is) Hemorrhage from the nose; nosebleed.

Epithelium (ep-i-THĒ-lē-um) The tissue that forms the outer part of the skin, lines blood vessels, hollow organs, and passages that lead to the outside of the body; epithelial (adjective).

Erythropoiesis (ē-rith″rō-PĪ-e-sis) The production of red blood cells.

Estrogen (ES-trō-jen) Any substance that induces estrogenic activity or stimulates the development of secondary female characteristics; female hormones.

Etiology (ē-tē-OL-ō-jē) The study of the causes of disease, including theories of origin, and what organisms, if any, are involved.

Euthanasia (ū-than-Ā-zē-ah) The proposed practice of ending a life in case of incurable disease.

Eversion (e-VER-zhun) Turning outward.

Exogenous (ex-OJ-en-us) Originating outside of an organ or part.

Exophthalmos (ek-sof-THAL-mus) An abnormal protrusion or bulging of the eyeball.

Exteroceptor (eks-ter-ō-SEP-tor) A sense organ adapted for the reception of stimuli from outside the body.

Extravasation (eks-trah-va-SĀ-shun) The process of escaping from a vessel into the tissues; especially blood, lymph, or serum.

Exudate (EKS-ū-dāt) Escaping fluid or semifluid material that oozes out of a blood vessel, which may contain serum, pus and cellular debris.

Falciform (FAL-si-form) Sickle-shaped; as in falciform ligament.

Fascia (FASH-ē-ah) A fibrous membrane covering, supporting and separating muscles.

Febrile (FEB-ril) Feverish; pertaining to a fever.

Feces (FE-sēz) Material discharged from the bowel (stool), which is made up of bacteria, secretions, and food residue.

Fenestration (fen-is-TRA-shun) Surgical procedure where an artificial opening is made into the labrynth of the ear, for some conditions of deafness.

Fetus (FE-tus) The latter stages of the developing young of an animal. In human beings, the child in utero from the third month to birth.

Fibrillation (fī-bre-LA-shun) Irregular twitching movements of individual muscle cells (fibers) or small groups of muscle fibers, preventing effective action by an organ or a muscle.

Fibrinolysis (fī-brin-OL-i-sis) Action of a proteolytic enzyme that converts insoluble fibrin into soluble substance.

Fibroblast (FĪ-bro-blast) A flat, long connective tissue cell that forms the fibrous tissues of the body.

Fibrosis (fī-BRŌ-sis) Abnormal formation of fibrous tissue.

Filtration (fil-TRA-shun) The passage of a liquid through a filter, or a membrane that acts like a filter.

Fimbriae (FIM-brē-ē) Any fringelike structure; especially the lateral ends of the uterine tubes (oviducts).

Fissure (FISH-ūr) A groove, fold, or slit that may be normal or abnormal.

Fistula (FIS-tū-lah) An abnormal passage between two organs or between an organ cavity and the outside.

Flaccid (FLAK-sid) Relaxed, flabby or soft, lacking or having defective muscle tone.

Flagellum (fla-JEL-um) A hairlike, motile process on the extremity of a bacterium or protozoon. Plural is flagella.

Flatus (FLA-tus) Gas or air in the digestive tract; commonly used to denote passage of gas by rectum.

Follicle (FOL-i-kul) A small secretory sac or cavity.

Fontanel (fon-tah-NEL) A soft area in a baby's skull; a membrane covered spot where bone formation has not yet occurred.

Foramen (fōr-A-men) A passage or opening; a communication between two cavities of an organ, or a hole in a bone for passage of vessels or nerves.

Fossa (FOS-ah) A furrow or shallow depression.

Fovea (FŌV-ē-ah) A pit or cuplike depression.

Frenulum (FREN-ū-lum) A small fold of mucous membrane that connects two parts and limits movement.

Fundus (FUN-dus) The part of a hollow organ farthest from the opening.

Gamete (GAM-ēt) A male or female reproductive cell; the spermatozoan or ovum.

Gangrene (GANG-grēn) Death of tissue, accompanied by bacterial invasion and putrefaction; usually due to blood vessel obstruction.

Gene (jēn) One of the biological units of heredity; an ultramicroscopic, self-reproducing DNA particle located in a definite position on a particular chromosome.

Genitalia (jen-i-TA-lē-ah) Reproductive organs.

Genotype (JEN-ō-tīp) The basic hereditary combination of genes of an organism.

Gerontology (je-ron-TOL-ō-gē) The study of old age.

Gestation (jes-TA-shun) The period of intrauterine fetal development.

Gingivitis (jin-je-VI-tis) Inflammation of the gums.

Glaucoma (glaw-KŌ-mah) An eye disorder in which there is increased pressure due to an excess of fluid within the eye.

Glomerulus (glō-MER-ū-lus) A rounded mass of nerves or blood vessels, especially the microscopic tuft of capillaries that is surrounded by the expanded part of each kidney tubule.

Glucosuria (gloo-kō-SŪ-rē-ah) Abnormal amount of sugar in the urine.

Goiter (GOY-ter) An enlargement of the thyroid gland.

Gonad (GŌ-nad) A term referring to the female sex glands, or ovaries, and the male sex glands, or testes.

Gradient (GRA-dē-ent) A slope or a grade; in the body, gradients refer to the difference in concentration or electrical charges across a semipermeable membrane.

Groin (groyn) The depression between the thigh and the trunk; the inguinal region.

Gyrus (JĪ-rus) One of the tortuous elevations (convolutions) of the cerebral cortex region of the brain. Plural is gyri.

Half-life The time required for a radioactive substance to lose one-half its energy.

Haustra (HAWS-tra) The sacculated elevations of the colon.

Hematuria (hem-at-U-rē-ah) Blood in the urine.

Hemorrhoids (HEM-ō-royds) Dilated or varicosed blood vessels (usually veins) in the anal region; also called piles.

Hernia (HER-nē-ah) The protrusion or projection of an organ or a part of an organ through the wall of the cavity that normally contains it.

Hilus (HĪ-lus) An area, depression, or pit where blood vessels and nerves enter or leave the organ.

Homogeneous (hō-mō-JEN-ē-us) Having similar or the same consistency and composition throughout.

Homologous (hō-MOL-ō-gus) Similar in structure and origin, but not necessarily in function.

Hordeolum (hor-DE-ō-lum) Inflammation of a sebaceous gland of the eyelid; a sty.

Hyaluronidase (hī''al-ū-RON-i-dāse) An enzyme that breaks down hyaluronic acid, increasing the permeability of connective tissues by dissolving the substances which hold body cells together.

Hydrostatic (hī-drō-STAT-ik) Pertaining to the pressure of liquids in equilibrium and that exerted on liquids.

Hypercapnia (hī-per-KAP-nē-ah) Abnormal amount of carbon dioxide in the blood.

Hyperemia (hī-per-Ē-mē-ah) An excess of blood in an area or part of the body.

Hyperplasia (hī-per-PLĀ-zē-ah) An abnormal increase in the number of normal cells in a tissue or organ, increasing its size.

Hypoxia (hī-POKS-ē-ah) Lack of an adequate amount of oxygen; also anoxia.

Immunity (im-Ū-ni-tē) The state of being resistant, to injury, particularly by poisons, foreign proteins, and invading parasites.

Impotence (IM-pō-tens) Weakness. inability to copulate; failure of sexual power.

Incontinence (in-KON-tin-ens) Inability to retain urine, semen, or feces, through loss of sphincter control.

Insertion (in-SER-shun) The manner or place of attachment of a muscle to the bone that it moves.

In situ (in, SIT-ū) In position.

Integument (in-TEG-ū-ment) A covering, especially the skin.

Intercellular (in-ter-SEL-ū-lar) Between the cells of a structure.

Intracellular (in-tra-SEL-ū-lar) Within cells.

Intubation (in-tu-BĀ-shun) Insertion of a tube into the larynx through the glottis for entrance of air, or to dilate a stricture.

Intussusception (in"tus-sus-SEP-shun) The infolding (invagination) of one part of the intestine within another segment.

In utero (in, Ū-ter-ō) Within the uterus.

Invagination (in-vaj-in-Ā-shun) The pushing of the wall of a cavity into the cavity itself.

In vitro (in, VIT-rō) In a glass, as in a test tube.

In vivo (in, VIV-ō) In the living body.

Ipsilateral (ip-sē-LAT-er-al) On the same side, affecting the same side of the body.

Ischemia (is-KĒM-ē-ah) A lack of sufficient blood to a part, due to obstruction of the circulation to it.

Isotonic (iso-TON-ik) Having equal tension or tone; the existence of equality of osmotic pressure between two different solutions or between two elements in a solution.

Isotope (Ī-so-tōpe) A chemical element that has the same atomic number as another, but a different atomic weight. Radioactive isotopes change into other elements with the emission of certain radiations.

Jaundice (JAWN-dis) A condition characterized by yellowness of skin, white of eyes, mucous membranes, and body fluids.

Karyotype (KAR-ē-ō-tīp) Chromosome characteristics of an individual or of a group of cells.

Keratin (KER-a-tin) A special insoluble protein found in the hair, nails, and other horny tissues of the epidermis.

Ketosis (kē-TŌ-sis) Abnormal condition marked by excessive production of ketone bodies.

Kilogram (KIL-ō-gram) Equivalent to 1,000 grams; about 2.2 pounds avoirdupois.

Kinesthesia (kin-es-THĒ-szē-ah) Ability to perceive extent, direction, or weight of movement; muscle sense.

Krebs cycle The citric acid cycle; a series of energy-yielding steps in the catabolism of carbohydrates.

Kyphosis (kī-FŌ-sis) An increased curvature of the chest, giving a hunchback appearance.

Labia (LĀ-be-ah) A lip, liplike or a structure like one. Plural is labium.

Labyrinth (LAB-i-rinth) Intricate communicating passageways; especially the internal ear.

Lacrimal (LAK-rem-al) Pertaining to tears.

Lactation (lak-TĀ-shun) The period of suckling in mammals.

Lacteal (LAK-tē-al) Related to milk; one of many intestinal lymph vessels that take up fat from digested food.

Lacuna (la-KU-nah) A small hollow space, such as that found in bones, in which lie the osteoblasts. Plural is lacunae.

Lamina (LAM-in-ah) A thin, flat layer or membrane, as the flattened part of either side of the arch of a vertebra. Plural is laminae.

Laryngoscope (lar-INJ-ō-skōp) An instrument for examining the larynx.

Latent (LAY-tent) period The period elapsing between the application of a stimulus and the response.

Lateral (LAT-er-al) On the outer side.

Lesion (LĒ-zhun) Any diseased change in tissue formation locally.

Leukemia (lū-KĒ-mē-ah) A cancerlike disease of the blood forming organs characterized by a rapid and abnormal increase in the number of white blood cells, plus many immature cells in the circulating blood.

Leukocyte, leucocyte (LŪ-kō-sīt) A white blood cell.

Leukocytosis (lū-kō-sī-TŌ-sis) An increase in the number of white blood cells, characteristic of many infections and other disorders.

Leukopenia (lū-kō-PĒ-ne-ah) A decrease of the number of white blood cells, below 5,000 per cubic millimeter.

Libido (li-BĒ-dō) The sexual drive, conscious or unconscious.

Liter (LĒ-ter) Volume occupied by 1 kilogram of water at standard atmospheric pressure; equivalent to 1.057 quarts.

Lobe (lōbe) A curve or rounded projection.

Lordosis (lor-DŌ-sis) Abnormal anterior convexity of the spine.

Lumbar (LUM-bar)   Region of the back and side between the ribs and pelvis; loins.

Lumen (LŪ-men)   The space within an artery, vein, intestine or tube.

Lymphocyte (LIM-fō-sīt)   A type of white blood cell formed in lymph nodes instead of in bone marrow.

Macula (MAK-ū-lah)   A discolored spot or a colored area.

Malaise (ma-LĀYZ)   Discomfort, uneasiness, indisposition, often indicative of infection.

Malignant (mah-LIG-nant)   Referring to diseases that tend to become worse and cause death; especially the invasion and spreading of cancer.

Manometer (man-OM-e-ter)   A device used for determining liquid or gaseous pressure.

Mastication (mas"ti-KĀ-shun)   Chewing.

Meatus (mē-Ā-tus)   A passage or opening, especially the external portion of a canal.

Medial (MĒ-dē-al)   Relating to the middle or center position.

Median (MĒD-ē-an)   A vertical plane dividing the body into right and left halves.

Melanin (MEL-an-in)   The dark pigment found in some parts of the body, such as the skin.

Melanoma (mel-an-Ō-mah)   A tumor containing melanin, usually dark colored.

Menopause (MEN-o-pawz)   The termination of the menstrual cycles.

Metabolism (me-TAB-ō-lizm)   The physical and chemical changes or processes by which living substance is maintained, producing energy, for the use of the organism.

Metastasis (me-TA-sta-sis)   The transfer of disease from one organ or part of the body to another part that is not connected with it.

Microgram (MĪ-kro-gram)   One one-millionth of a gram, or $1/1,000$ of a milligram.

Micron (MĪ-kron)   One one-millionth of a meter or $1/1,000$ of a millimeter; $1/25,000$ of an inch.

Micturition (mik-tu-RI-shun)   The act of expelling urine from the bladder; urination.

Millimeter (MIL-i-mē-ter)   One one-thousandth of a meter; about $1/25$ inch.

Mittelschmerz (MIT-el-shmerz)   Abdominal pain that supposedly indicates the release of an egg from the ovary.

Milliliter (MIL-ē-lē-ter)   One-thousandth of a liter; equivalent to 1 cubic centimeter.

Morbid (MOR-bid)   Diseased; pertaining to disease.

Mucin (MŪ-sin)   A protein found in mucus and other parts of the body.

Mucus (MŪ-kus)   The thick fluid secretion of the mucous glands, and mucous membranes.

Multiparous (mul-TIP-ar-us)   Having borne more than one child.

Myology (mī-OL-ō-jē)   The science or study of the muscles and their parts.

Myopia (mī-Ō-pē-ah)   Defect in vision so that objects can only be seen distinctly when very close to the eyes; nearsightedness.

Narcosis (nar-KO-sis)   Unconscious state due to narcotics.

Nebulization (ne"būl-i-ZĀ-shun)   Treatment with spray method.

Nephritis (ne-FRĪ-tis)   Kidney inflammation.

Nulliparous (nul-LIP-ar-us)   Never having borne a child.

Nystagmus (nis-TAG-mus)   Constant, involuntary, rhythmic movement of the eyeballs; horizontal, rotary, or vertical.

Odontoid (ō-DON-toyd)   Toothlike.

Olfactory (ōl-FAK-tō-rē)   Pertaining to smell.

Oogenesis (ō-ō-JEN-e-sis)   Formation and development of the ovum.

Ophthalmic (of-THAL-mik)   Pertaining to the eye.

Orbit (OR-bit)   The bony pyramid-shaped cavity of the skull that holds the eyeball.

Organelle (or-gan-EL)   A tiny specific particle of living material present in most cells and serving a specific function.

Orgasm (OR-gazm)   A state of highly emotional excitement; especially that which occurs at the climax of sexual intercourse.

Orifice (OR-i-fis)   Any aperture or opening.

Ossicle (OS-i-kul)   Any small bone; as the three tiny bones of the ear.

Ossification (os"i-fi-KĀ-shun)   Formation of bone substance.

Osteomyelitis (os"tē-ō-mī-i-LĪ-tis)   Inflammation of bone marrow, or of the bone and marrow.

Osteoporosis (os"tē-ō-pō-RŌ-sis)   Increased porosity of bone.

Ostium (OS-tē-um)   Any small opening; especially entrance into a hollow organ or canal.

Otic (Ō-tik)   Pertaining to the ear.

Ovulation (ō-vū-LĀ-shun)   The discharge of a mature egg cell (ovum) from the follicle of the ovary.

Ovum (Ō-vum)   The female reproductive or germ cell.

Oxidation (ok-si-DĀ-shun)   The combining of oxygen and food in the tissues.

Pacchionian (pak-kē-Ō-nē-an) bodies   Small growths of the arachnoid tissue of the cerebrum.

Palate (PAL-at)   The horizontal structure separating the mouth and the nasal cavity; the roof of the mouth.

Palliative (PAL-ē-ah-tiv) Serving to relieve or alleviate without curing.

Palpate (PAL-pāt) To examine by touch; to feel.

Papilla (pah-PIL-ah) Any small projection or elevations; plural is papillae.

Parenchyma (par-EN-ki-mah) The essential parts of any organ concerned with its function.

Parenteral (par-EN-ter-al) Situated or occurring outside of the intestines; as by a subcutaneous method.

Paries (PĀ-rēs) The enveloping wall of any structure; especially hollow organs.

Parous (PA-rus) Having borne at least one child.

Paroxysm (PAR-oks-sizm) A sudden periodic attack or recurrence of symptoms of a disease.

Parturition (par-tu-RISH-un) Act of giving birth to young; childbirth, delivery.

Pectoral (PEK-tō-ral) Pertaining to the chest, or breast.

Perineum (per-i-NĒ-um) The pelvic floor; the space between the anus and the scrotum in the male, and the anus and the vulva in the female.

Periphery (pe-RIF-er-ē) Outer part or a surface of the body; part away from the center.

Peritonitis (per-i-tōn-Ī-tis) Inflammation of the peritoneum, the membranous coat lining the abdominal cavity and going into the viscera.

pH The symbol commonly used in expressing hydrogen ion concentration.

Phalanges (fah-LAN-jēz) Bones of a finger or toe.

Phlebotomy (fleb-OT-ō-mē) The cutting of a vein to allow the escape of blood.

Pilonidal (pī-lo-NĪ-dal) Containing hairs resembling a tuft inside a cyst or sinus.

Pinna (PIN-nah) The projecting part of the external ear; the auricle.

Pinocytosis (pi″nō-sī-TŌ-sis) The absorption of liquids by cells.

Pleura (PLOOR-ah) Serous membrane that enfolds the lungs and lines the walls of the chest and diaphragm.

Plexus (PLEK-sus) A network of nerves, veins, or lymphatic vessels.

Pons (ponz) A process of tissue connecting two or more parts.

Postpartum (pōst-PAR-tum) After parturition; occurring after the delivery of a baby.

Presbyopia (prez-bē-Ō-pē-ah) Defect of vision due to advancing age; loss of elasticity of the lens of the eye.

Pressor (PRES-or) Stimulating the activity of a function; especially vasomotor, usually accompanied by an increase in blood pressure.

Primordial (prī-MŌR-dē-al) Existing first; especially primordial egg cells in the ovary.

Prognosis (prog-NŌ-sis) A forecast of the probable results of a disorder; the outlook for recovery.

Prolapse (PRŌ-laps) A dropping or falling down of an organ; especially the uterus or rectum.

Proliferation (prō-lif″er-Ā-shun) Rapid and repeated reproduction of new parts, especially of cells.

Prosthesis (PROS-thē-sis) Replacement of a missing part by an artificial substitute.

Protuberance (prō-TŪ-ber-ans) A part that is prominent beyond a surface, like a knob.

Proximal (PROK-se-mal) Near the point of origin; referring to the nearest part.

Psychosomatic (sī-ko-sō-MA-tik) Pertaining to the interrelationship between the mind and body.

Puberty (PŪ-ber-tē) Period of life at which the reproductive organs become functionally operative.

Pulmonary (PUL-mōn-ary) Concerning or affected by the lungs.

Pulse (puls) Throbbing caused by the regular contraction and alternate expansion of an artery; the periodic thrust felt over arteries in time with the heartbeat.

Pus Liquid product of inflammation containing leukocytes, or their remains, and debris of dead cells.

Pyorrhea (pī-ō-RĒ-ah) A discharge or flow of pus; especially the tooth sockets and the tissues of the gums.

Pyrexia (pī-REK-sē-ah) A condition in which the temperature is above normal.

Radiography (rā-dē-OG-ra-fē) The making of x-ray pictures.

Ramus (RĀ-mus) A branch; especially a nerve or blood vessel. Plural is rami.

Receptor (rē-SEP-tor) A nerve ending that receives a stimulus.

Recumbent (rē-KUM-bent) One who is lying down.

Regurgitation (rē-gur-ji-TĀ-shun) Return of solids or fluids to the mouth from the stomach; flowing backward of blood through incompletely closed heart valves.

Renal (RĒ-nal) Pertaining to the kidney.

Resuscitation (rē″sus-i-TĀ-shun) Act of bringing a person back to full consciousness.

Reticulum (rē-TIK-ū-lum) A network of connective tissue cells and fibers.

Retraction (rē-TRAK-shun) A shortening. The act of drawing backward or state of being drawn back.

Retroversion (re-trō-VER-zhun) A turning backward of an entire organ; especially the uterus.

Rickets (RIK-ets) A disease of metabolism affecting children, characterized by ineffective nutrition, and often resulting in deformities.

Roentgen (RENT-gen) The international unit of radiation; a standard quantity of x or gamma radiation.

Salpingitis (sal-pin-JĪ-tis) Inflammation of the uterine (fallopian) tube or of the auditory (eustachian) tube.

Sarcoma (sar-KŌ-mah)   A connective tissue tumor, often highly malignant.

Sciatica (sī-AT-ik-ah)   Inflammation and pain along the sciatic nerve, felt at back of thigh running down the inside of the leg.

Sclerosis (skle-RŌ-sis)   A hardening with loss of elasticity of the tissues.

Scoliosis (skō-lē-Ō-sis)   An abnormal curvature sideways (laterally) from the normal vertical line of the spine.

Sebum (SĒ-bum)   A fatty secretion of the sebaceous glands of the skin.

Sella turcica (SEL-ah, TUR-si-kah)   A saddlelike depression on the middle upper surface of the sphenoid bone enclosing the pituitary gland.

Senescence (sen-ES-ens)   The process of growing old; or the period of old age.

Senility (se-NIL-i-tē)   The state of being old.

Septum (SEP-tum)   A wall dividing two cavities.

Sigmoid (SIG-moyd)   Shaped like the Greek letter sigma(σ).

Simmond's (SIM-mundz) disease   Condition in which atrophy of the pituitary gland causes premature senility and mental symptoms.

Sinus (SĪ-nus)   A hollow in a bone or other tissue; a channel for blood, any cavity having a narrow opening.

Sinusoid (SĪ-nus-oyd)   A blood space in certain organs, as the liver or spleen.

Spasm (spazm)   An involuntary, convulsive, muscular contraction.

Specific gravity   Weight of a substance compared with an equal volume of water. Water is represented by 1.000.

Spermatogenesis (sper″mah-tō-JEN-e-sis)   The formation and development of the spermatozoa.

Spermicidal (sper-mi-SĪ-dal)   Killing spermatozoa.

Sphincter (SFINK-ter)   A circular muscle constricting an orifice.

Sphygmomanometer (sfig″mō-man-OM-e-ter)   An instrument for measuring arterial blood pressure.

Spirometer (spī-ROM-et-er)   An apparatus used to measure air capacity of the lungs.

Sputum (SPŪ-tum)   Substance ejected from the mouth containing saliva and mucus.

Squamous (SKWĀ-mus)   Scalelike.

Stasis (STĀ-sis)   Stagnation or halt of normal flow of fluids, as blood, urine, or of the intestinal mechanism.

Sterility (ster-IL-it-ē)   Infertility; absence of reproductive power.

Stratum (STRA-tum)   A layer.

Stricture (STRIK-tur)   A local contraction of a tubular structure.

Stroma (STRO-mah)   The tissue that forms the ground substance, foundation, framework of an organ; as opposed to its functional parts.

Sulcus (SUL-kus)   A groove or depression between parts; especially a fissure between the convulutions of the brain. Plural is sulci.

Suppuration (sup-ū-RĀ-shun)   The process of pus formation.

Suture (SU-chur)   A type of joint, especially in the skull, where bone surfaces are closely united.

Symphysis (SIM-fi-sis)   A line of union; a cartilagenous joint such as that between the bodies of the pubic bones.

Syncytium (sin-SIT-ī-um)   A multinucleated mass of protoplasm produced by the merging of cells.

Systemic (sis-TEM-ik)   Affecting the whole body; generalized.

Systole (sis-TŌ-lē)   Heart muscle contraction; especially that of the ventricles.

Tactile (TAK-til)   Pertaining to the sense of touch.

Tetany (TET-an-ē)   A nervous condition characterized by intermittent tonic muscular contractions of the extremities.

Thoracic (thor-A-sik)   Pertaining to the chest.

Thrombocyte (THROM-bō-sīt)   A tiny particle found in the circulating blood; a blood platelet, believed to be part of the process of blood clotting.

Thrombophlebitis (throm″bō-fle-BĪ-tis)   A disorder in which inflammation of a vein wall is followed by the formation of a blood clot (thrombus).

Thymectomy (thy-MEK-tō-mē)   Surgical removal of the thyroid gland.

Tinnitus (tin-Ī-tus)   A ringing or tinkling sound in the ears.

Toxic (TOK-sik)   Pertaining to, resembling, or caused by poison; poisonous.

Trabecula (tra-BEK-ū-lah)   Fibrous cord of connective tissue, serving as supporting fiber by forming a septum extending into an organ from its wall or capsule.

Trace elements   Organic elements normally found in minute traces in blood and tissues; examples are fluorine, copper, and manganese.

Transplantation (trans-plan-TĀ-shun)   The transfer or implantation of body tissue from one part of a person's body to another, or from one individual to another.

Transverse (trans-VERS)   Lying across; crosswise.

Trauma (TRAW-mah)   An injury or a wound that may be produced by external force or by shock, as in psychic trauma.

Tumor (TŪ-mor)   A swelling or enlargement.

Ulcer (UL-ser)   An open lesion upon the skin or mucous membrane of the body, with loss of substance and necrosis of the tissue.

Umbilical (um-BIL-i-kal)   Pertaining to the umbilicus or navel.

Umbilicus (um-BIL-i-kus)   A small scar on the abdomen that marks the former attachment of the umbilical cord to the fetus, the navel.

Uremia (ū-RĒ-mē-ah)   Toxic condition from urea and other waste products in the blood.

Urticaria (ur-ti-KĀ-rē-ah)   A skin reaction to certain foods, drugs, or other substances to which a person may be allergic; hives.

Uvula (Ū-vū-lah)   A soft, fleshy mass, especially the V-shaped pendant part hanging down from the soft palate.

Varicocele (VAR-i-kō-sēl)   A twisted vein; especially veins of the spermatic cord.

Varicose (VAR-i-kōs)   Pertaining to an unnatural swelling, as in the case of a varicose vein.

Vas   A vessel or a duct.

Vascular (VAS-kū-lar)   Pertaining to or containing many vessels.

Ventral (VEN-tral)   Pertaining to the anterior or front side of the body; opposite of dorsal.

Vertigo (VUR-ti-go)   Sensation of dizziness, a whirling motion of oneself or of external objects.

Vesicle (VES-i-kal)   A small bladder or sac containing liquid.

Vestibule (VES-tib-ūl)   A small space or cavity at the beginning of a canal; especially the inner ear, larynx, mouth, nose, vagina.

Villus (VIL-lus)   One of the short vascular hairlike proccesses found on certain membranous surfaces. Plural is villi.

Viscosity (vis-KOS-i-tē)   The state of being sticky or thick.

Wheal (hwēl)   More or less round elevated lesion of the skin.

INDEX